Generalized Geologic and Tectonic Map of North America

SEDIMENTARY UNITS

Thick deposits in structurally negative areas

Paleozoic and Mesozoic active margin deposits

Synorogenic and postorogenic deposits

Paleozoic and Mesozoic passive margin deposits

Late Precambrian deposits
Of Middle and Upper Proterozoic ages

VOLCANIC AND PLUTONIC UNITS

Postorogenic volcanic cover

Ultramafic rocks

Granitic plutons
Ages are generally within the span of the tectonic cycle of the foldbelt in which they lie

SPECIAL UNITS

Former subduction complex rocks of the Pacific border

Exposed parts of Ouachita foldbelt

Probable western extension of Innuitian foldbelt
In cores of northern Alaska ranges

PLATFORM AREAS

Ice cap of Quaternary age
On Precambrian and Paleozoic basement

Platform deposits on Precambrian basement
In central craton

Platform deposits on Paleozoic basement
In Atlantic and Gulf coastal plains

Platform deposits within the Precambrian
Mainly in the Canadian Shield

PRECAMBRIAN

Basement igneous and metamorphic complexes mainly of Precambrian age

Grenville foldbelt
Deformed 880–1,000 m.y. ago

Kenoran foldbelts
Deformed 2,390–2,600 m.y. ago

Anorthosite bodies
Plutons composed almost entirely of plagioclase

STRUCTURAL SYMBOLS

Normal fault
Hachures on downthrown side

Subsea fault

Salt domes and salt diapirs
In Gulf coastal plain and Gulf of Mexico

Strike-slip fault
Arrows show relative lateral movement

▲
Volcano

Thrust fault
Barbs on upthrown side

✳
World's oldest rock

Axes of seafloor spreading

Contours on basement surfaces beneath platform areas
All contours are below sea level except where marked with plus symbols. Interval is 1,000 meters

Modified from the Generalized Tectonic Map of North America by P.B. King and Gertrude J. Edmonston, U.S. Geological Survey Map I-688

Eleventh Edition

PHYSICAL GEOLOGY

Charles C. Plummer
Emeritus of California State University at Sacramento

Diane H. Carlson
California State University at Sacramento

The Late David McGeary
Emeritus of California State University
at Sacramento

McGraw Hill **Higher Education**

Boston Burr Ridge, IL Dubuque, IA Madison, WI New York San Francisco St. Louis
Bangkok Bogotá Caracas Kuala Lumpur Lisbon London Madrid Mexico City
Milan Montreal New Delhi Santiago Seoul Singapore Sydney Taipei Toronto

The McGraw·Hill Companies

Mc Graw Hill Higher Education

PHYSICAL GEOLOGY, ELEVENTH EDITION

This book is printed on acid-free paper.

2 3 4 5 6 7 8 9 0 VNH/VNH 0 9 8 7 6

ISBN-13   978–0–07–282692–0
ISBN-10   0–07–282692–4

Publisher: *Margaret J. Kemp*
Senior Sponsoring Editor: *Daryl Bruflodt*
Managing Developmental Editor: *Shirley R. Oberbroeckling*
Developmental Editor: *Liz Recker*
Associate Marketing Manager: *Todd L. Turner*
Project Manager: *Jodi Rhomberg*
Senior Production Supervisor: *Sherry L. Kane*
Lead Media Project Manager: *Judi David*
Senior Media Technology Producer: *Jeffry Schmitt*
Senior Designer: *David W. Hash*
Cover/Interior Designer: *Jamie E. O'Neal*
Cover Photo: © *Getty Images/David Muench*
Lead Photo Research Coordinator: *Carrie K. Burger*
Photo Research: *Chris Hammond/PhotoFind, LLC*
Supplement Producer: *Melissa M. Leick*
Compositor: *Carlisle Communications, Ltd.*
Typeface: *10.5/12 Times Roman*
Printer: *Von Hoffmann Corporation*

Library of Congress Cataloging-in-Publication Data

Plummer, Charles C., 1937–
  Physical geology / Charles C. Plummer, Diane H. Carlson, David McGeary. — 11th ed.
    p. cm.
  Includes index.
  ISBN 978–0–07–282692–0 — ISBN 0–07–282692–4 (hard copy : alk. paper)
  1. Physical geology—Textbooks. I. Carlson, Diane H. II. McGeary, David. III. Title.

QE28.2.P58   2007
550—dc22
                                      2005024100
                                         CIP

www.mhhe.com

Dedicated to the memory of
David McGeary

BRIEF CONTENTS

Volcanism and Extrusive Rocks 84

Weathering and Soil 116

Sediment and Sedimentary Rocks 138

Metamorphism, Metamorphic Rocks, and Hydrothermal Rocks 170

Mass Wasting 222

Time and Geology 194

Streams and Floods 246

Ground Water 282

Glaciers and Glaciation 306

Deserts and Wind Action 338

Waves, Beaches, and Coasts 360

14

Geologic Structures 380

15

Earthquakes 404

16

Earth's Interior and Geophysical Properties 440

17

The Sea Floor 466
18

Plate Tectonics 490
19

Mountain Belts and the Continental Crust 526
20

Geologic Resources 550

The Earth's Companions 582

PREFACE

WHY USE THIS BOOK?

One excellent reason is that it's tried and true. Since the book was published in 1979, approximately 500,000 students have read this text as an introduction to physical geology. Proportionately, geology instructors have relied on this text for over 5,000 courses to explain, illustrate, and exemplify basic geologic concepts to both majors and non-majors. Today, the 11th edition continues to provide contemporary perspectives that reflect current research, recent natural disasters, unmatched illustrations, and unparalleled learning aids. We have worked closely with contributors, reviewers, and our editors to publish the most accurate and current text possible. The most exciting element of the new edition is the presentation of 300+ new illustrations, created by the artistic skill of Cindy Shaw. Ideas that shaped the development and articulation of new figures resulted from the numerous recommendations of a group of geology instructors.

APPROACH

Our purpose is to clearly present the various aspects of physical geology so that students can understand the logic of what scientists have discovered as well as the elegant way the parts are interrelated to explain how Earth, as a whole, works.

This approach is epitomized by our treatment of plate tectonics. Plate tectonics is central to understanding how the Earth works. Rather than providing a full-fledged presentation of plate tectonics at the beginning of the text and overwhelming students, *Physical Geology* presents the essentials of plate tectonics in the first chapter. Subsequent chapters then detail interrelationships between plate tectonics and major geologic topics. Chapter 19, typically covered late in the course, presents a full synthesis of plate tectonics. By this time, students have learned the many aspects of physical geology and can appreciate the elegance of plate tectonics as a unifying theory.

NEW TO THE ELEVENTH EDITION

Superior Art Program

Geology is a visually oriented science and one of the best ways a student can learn it is by studying illustrations and photographs. This new edition includes an updated art program that will not only aid to understanding, but also engage a student's interest.

In this new edition, 300 illustrations have been revised or created from scratch. An art focus group composed of geology professors originally met with the authors and illustrator to determine which pieces needed to be updated. Once the pieces were rendered, the members of the focus group and other geology professors provided feedback on how to make the illustrations as effective and accurate as possible.

The revised and new pieces of art were created by Cindy Shaw from Richland, Washington. Cindy used her expertise as a geological illustrator to provide realistic and beautiful illustrations.

This edition also includes over 130 new photos. This book has been enhanced by the photographs of Dr. Parvinder Sethi of the Geology Department, Radford University, Virginia.

"A Geologist's View" Features

Seventeen photos in the text are accompanied by an illustration depicting how a geologist would view the scene. Students gain experience understanding how the trained eye of a geologist views a landscape to comprehend the geologic events that have occurred.

New Animations

McGraw-Hill is proud to bring you an assortment of 44 outstanding animations like no others. These include 20 new animations and 23 animations retained from previous editions. These animations are located on ARIS and also on the Digital Content Manager. A special animation icon has been placed beside every figure in the text that has a corresponding animation. These animations offer students a fresh dynamic method of learning about geology concepts such as dynamics of groundwater movement, isostacy, plate tectonics, and more.

Three Page Fold Out

This has been added to the back of the text for students' reference. The front side of the foldout contains a geographic map of the world. This fold out is constructed so students can easily leave it folded out and refer to it while reading the text. By referencing this fold out students gain a better sense of the location of the places that are mentioned within the text. The North America Tapestry of Time and Terrain Map is located on the back of this fold out.

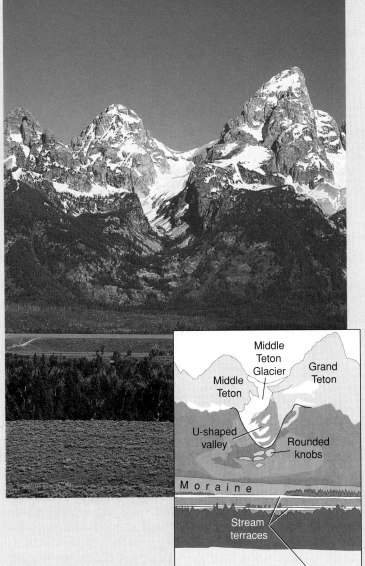

Updated Content

Significant content changes for the eleventh edition include:

Chapter 1—Begins with a discussion of the Indian Ocean tsunami of 2004. (The tsunami is covered in more depth in the earthquake chapter.) In chapter 1, it serves to relate human concerns to the importance of geology and to demonstrate Earth systems interrelationships.

Chapter 1—The introduction to plate tectonics has been expanded to provide the essentials of plate tectonics necessary for understanding how plate tectonics relates to subsequent topics in the book. Our expansion in chapter 1 includes: (1) More information on transform boundaries. (2) How divergent boundaries may begin in continental lithosphere as well as continuously create oceanic crust. (3) The distinctions between the three types of convergent boundaries-ocean-ocean convergence, ocean-continental convergence, and continental-continental convergence.

Chapter 2—Was largely rewritten to give minerals and their chemical constituents more appeal to introductory students. The nature of atoms and basics of chemistry section was redone to be consistent with today's introductory chemistry courses.

Chapter 3—We have added a figure with photos of the six most common igneous rocks placed around a copy of the previously introduced classification diagram. This should help the reader visually correlate classification of igneous rocks and their characteristics. An additional figure (figure 3.8) reproduces the granite/rhyolite field for igneous rock classification. In this, two compositions (one silica-rich and the other relatively deficient in silica) are shown as lines in the diagram. These relate to bar charts on the sides of the diagram showing the percentages of minerals for each of these compositions.

Chapter 4—We have expanded upon the great, caldera-forming eruptions and their production of prodigious amounts of pumice. We have added a section on submarine eruptions that produce large, oceanic plateaus which are comparable to basalt plateaus on the continents.

Chapter 5—Portions on weathering have been rewritten to emphasize Earth systems. A new section on the factors that affect weathering has been added. A new Earths Systems Box describes the relationship between weathering, the carbon cycle, and global climate.

Chapter 6—We have emphasized the importance of sediment and sedimentary rock in interpreting geologic environments. We also put more emphasis economic importance on sedimentary material used for building material and fossil fuels. The description of the different kinds of sediment and sedimentary rocks has been rewritten to capture the interest of the introductory student, and photomicrographs of sedimentary textures have been added. The box on sedimentary rocks on Mars has been completely rewritten to include the latest photos and discussion of the possibility of water-deposited rocks and evidence of extraterrestrial life on Mars. A new section on fossils has been added to expand our treatment of the preservation of organisms within the sedimentary record.

Chapter 7—We have added a section showing how experimentally determined mineral phase diagrams are used to infer the geothermal gradients during metamorphism. Another new section discusses pressure and temperature paths in time, explaining, using geothermometry and geobarometry, the growth of minerals while pressure and temperature changed. We can use the information to infer the timing of heating, cooling, burial and uplift during a mountain-building episode and relate it to a plate tectonic setting.

Chapter 8—We introduce the new, International Commission on Stratigraphy's recommendation that Tertiary be dropped from the Geologic Time Scale and replaced by Paleogene and Neogene. However, we have not thrown out the traditional time scale as it is too early to tell whether the new system will catch on among geologists.

Chapter 9—We changed the classification and nomenclature for materials and mass wasting processes to conform to that used by the U.S. Geologic Survey. New examples of mass wasting include the landslides of 2003–2005 in southern California caused by episodes of heavy rain and New Hampshire's loss of its beloved symbol, the Old Man of the Mountain, to rockfall.

Chapter 10—The art has undergone a major revision to increase the realism of figures illustrating the processes involved in fluvial erosion and deposition.

Chapter 11—Many figures have been redone to improve clarity and many new photos have been added. The box on Darcy's Law has been simplified while maintaining the details of ground water flow useful for students who continue their studies in this important specialty in geology. The discussion of ground water contamination has been updated. The geothermal energy section includes the latest innovative technology used at The Geysers in California to increase production by injecting waste water from nearby communities.

Chapter 12—We added an Earth Systems Box "Global Warming and Glaciers." While it includes some of the material from the previous edition's box on ice cores, the emphasis is on the global climate changes for the past 400,000 years as determined from ice core analysis. The box includes a graph showing the changes of Antarctic temperature and the relationship to greenhouse gases during that time period. We also discuss the effect of human contribution to greenhouse gases and current global warming.

Chapter 13—We have added a discussion and photos of the deadly flash flood that hit Death Valley National Park in August 2004 causing tremendous damage. The box on Wind Action on Mars has been updated to include the latest discoveries and photos from the 2004 Mars Rover Opportunity.

Chapter 14—Many of the figures have been redrawn to increase the clarity and realism of shoreline processes. Photos and a discussion of the damage done by Hurricane Katrina in October of 2005 and the series of hurricanes that struck Florida in 2004 are also included to highlight the sometimes dangerous interaction of Earth systems.

Chapter 15—The section on stress and strain and the development of geologic structures has been rewritten to improve clarity for the introductory student. The sections on folds, joints, and faults have also been rewritten and accompanying figures have been redrawn to help students visualize the three-dimensional architecture of the lithosphere. Line drawings illustrating what a geologist sees when looking at geologic structures have been added to photos of geologic structures.

Chapter 16—Our treatment of tsunamis was expanded to include incredible photos and accounts of the 2004 Indian Ocean tsunami. New figures incorporate the latest research on how tsunami waves are generated. Maps of the Pacific show the paths of previous tsunamis as well as the monitoring system in place that prevents the tremendous loss of life seen in communities surrounding the Indian Ocean. The tsunami section also discusses how some of the largest tsunamis have been generated by landslides and volcanic eruptions. The section on earthquakes and plate boundaries has been expanded to include intraplate earthquakes and the potential hazards in the eastern and central United States. A new figure summarizes the relationship of all the plate boundaries and more clearly shows the depth of earthquakes at each type of boundary.

Chapter 17—Most of the figures have been redone to make the study of geophysical investigations relating to Earth's interior more interesting for the reader. The discussion on isostatic adjustment of the lithosphere has been expanded and illustrated with revised figures that are more realistic.

Chapter 18—We have continued to update the Sea Floor Chapter and this edition includes additional new photos and also new maps of features on the sea floor.

Chapter 19—The figures have undergone a major overhaul to more accurately show the details of the different plate boundaries. There is a new box on indentation tectonics and 'mushy' plate boundaries with examples from the Himalayan collision zone and the San Andreas Fault zone. The relationship between plate tectonics and ore deposits is now in a box as is the discussion of back arc spreading.

Chapter 20—Has been overhauled to emphasize that a particular mountain belt is a product of the interaction of *orogenies, isostasy,* and *weathering and erosion.* We eliminated the use of "stages" in the evolution of mountain belts.

Chapter 21—Has been rewritten so that it is much more appealing to introductory students. Some basic concepts of thermodynamics are introduced at the beginning of the energy resources section to allow for a better understanding of the problems inherent in coal, petroleum and other energy resources.

Chapter 22—Has been modified so that there is less emphasis on astronomy and more on geology. We have added new or improved images of geologic features on Mars. New images from recent or ongoing spacecraft missions include pictures of Titan, Triton, Pluto, Charon, and a recently acquired high-resolution image of a comet nucleus.

KEY FEATURES

- Chapter Introductions—Each chapter begins with a "Purpose Statement," and an explanation of how the chapter relates to the Earth systems and how the material relates to the concepts in other chapters.

- Environmental Geology Boxes—Discuss topics that relate the chapter material to environmental issues, including impact on humans (e.g., *Radon—A Radioactive Health Hazard*).

- In Greater Depth Boxes—Discuss phenomena that are not necessarily covered in a geology course (e.g., *Precious Gems*) or present material in greater depth (e.g., *Calculating the Age of a Rock*).

- Earth Systems Boxes—Highlight the interrelationships between the geosphere, the atmosphere, and other Earth systems (e.g., Oxygen Isotopes and Climate Change).

- Planetary Geology Boxes—Compare features elsewhere in the solar system to their Earthly counterparts (e.g., *Stream Features on the Planet Mars*).

- Animations—Key concepts are further enhanced on the Online Learning Center. These are identified in the text by the icon.

- Integration of the World Wide Web—The Internet has revolutionized the way we obtain knowledge, and this book makes full use of its potential to help students learn. We have URLs for appropriate web-

sites throughout the book—within the main body of text, at the end of many boxes, and at the end of chapters. We have made the process student-friendly by having all websites that we mention in the book posted as links in this book's ARIS website. (We also include all URLs in the textbook for those who wish to go directly to a site.)

- Internet Exercises—These are located on the text's ARIS and allow students to investigate appropriate sites as well as raise interest for further, independent exploration on a topic. ARIS also includes additional readings and video resources. By placing these on the website, we can update them after the book has been published. We expect to add more sites and exercises to our website as we discover new ones after the book has gone to press. ARIS also features online quizzes, flashcards, animations, and other interactive items to help a student succeed in a geology course.

- Study Aids are found at the end of each chapter and include:
  - *Summaries* bring together and summarize the major concepts of the chapter.
  - *Terms to Remember* include all the boldfaced terms covered in the chapter so that students can verify their understanding of the concepts behind each term
  - *Testing Your Knowledge Quizzes* allow students to gauge their understanding of the chapter (The answers to the multiple choice portions are posted on the website.)
  - *Expanding Your Knowledge Questions* stimulate a student's critical thinking by asking questions with answers that are not found in the textbook.
  - *Exploring Web Resources* describe some of the best sites on the web that relate to the chapter.

SUPPLEMENTS

- NEW McGraw-Hill's ARIS—Assessment, Review, and Instruction System for *Physical Geology* (*http://www.mhhe.com/plummer11e/.*)

This is a complete, online tutorial, electronic homework, and course management system, designed for greater ease of use than any other system available. Free on adoption of *Physical Geology,* instructors can create and share course materials and assignments, quizzes, tutorials, animations, flash cards, and Internet activities directly tied to text-specific materials in *Physical Geology,* but instructors can also edit questions, import their own content, and create announcements and due dates for assignments. ARIS has automatic grading and reporting of easy-to-assign homework, quizzing, and testing. All student activity within McGraw-Hill's ARIS is automatically recorded and available to the instructor through a fully integrated grade book that can be downloaded to Excel.

- NEW Discovery Channel DVD—This exciting DVD offers short (3–5 minute) videos on topics ranging from conservation to volcanoes. Begin your class with a quick peek at science in action.

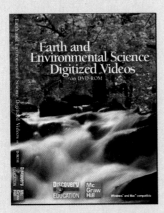

- Digital Content Manager—Available in both CD-ROM and DVD versions this contains every illustration, photograph, and table from the text, 44 animations, active art, lecture outlines, and additional photos. The software makes customizing your multimedia presentation easy. You can organize figures in any order you want; add labels, lines, and your own artwork; integrate material from other sources; edit and annotate lecture notes; and have the option of placing your multimedia lecture into another presentation program such as PowerPoint.

- Instructor's Testing and Resource CD-ROM—McGraw-Hill's EZ Test is a flexible and easy-to-use electronic testing program. The program allows instructors to create tests from book specific items. It accommodates a wide range of question types and instructors may add their own questions. Multiple versions of the test can be created and any test can be exported for use with course management systems such as WebCT, BlackBoard, or PageOut. EZ Test Online is a new service and gives you a place to easily administer your EZ Test created exams and quizzes online. The program is available for Windows and Macintosh environments.

- Instructor's Manual—The Instructor's Manual is found on the *Physical Geology* ARIS Site and on the Instructor's Testing and Resource CD, and can be accessed only by instructors.

- Classroom Performance System and Questions—McGraw-Hill has partnered with eInstruction to provide the revolutionary Classroom Performance System (CPS) and to bring interactivity into the classroom. CPS is a wireless response system that gives the instructor and students immediate feedback from the entire class. The wireless response pads are essentially remotes that are easy to use and engage students. CPS allows you to motivate student preparation, interactivity, and active learning so you can receive immediate feedback and know what students understand. A text-specific set of questions, formatted for both CPS and Powerpoint, is available via download from the Instructor area of the *Physical Geology* ARIS site.

- Transparencies—This collection contains two hundred and fifty illustrations from the text, all enlarged for excellent visibility in the classroom.

- Slides—This collection contains one hundred illustrations and photographs from the text.

Packaging Opportunities

McGraw-Hill offers packaging opportunities that not only provide students with valuable course-related material, but also a substantial cost savings. Ask your McGraw-Hill sales representative for information on discounts and special ISBNs for ordering a package that contains one of the following laboratory manuals:

- *Physical Geology Laboratory* Manual, Twelfth Edition, by Zumberge et al.

- *Laboratory Manual for Physical Geology,* Fifth Edition, by Jones/Jones

Custom Publishing

Did you know that you can design your own text or laboratory manual using any McGraw-Hill text and your personal materials to create a custom product that correlates specifically to your syllabus and course goals? Contact your McGraw-Hill sales representative to learn more about this option.

ACKNOWLEDGEMENTS

We have tried to write a book that will be useful to both students and instructors. We would be grateful for any comments by users, especially regarding mistakes within the text or sources of good geological photographs.

Tom Arny wrote the planetary geology chapter for the 10th edition. This chapter was revised and updated by Steve Kadel. Rick Hazlett revised chapters 2, 19, and 21. Scott Babcock helped revise the chemistry portion of chapter 2. Bret Bennington worked on the revision of chapters 5 and 6. Peggy Johnson helped revise the plate tectonic material in chapter 1. We greatly appreciate the publisher's "book team," whose names appear on the copyright page. Their guidance, support, and interest in the book were vital for the completion of this edition.

Thank you also to Cindy Shaw and Dr. Parvinder Sethi for their contributions to the superior art program of this edition. Cindy illustrated many of the new and revised pieces of art. Dr. Sethi provided many of the new photographs.

Diane Carlson would like to thank her husband, Reid Buell, for his support and technical assistance with several chapters. Charles Plummer thanks his wife, Beth Strasser, for assistance with photography in the field and for her anthropological perspective. We thank Susan Slaymaker for writing the planetary geology material originally in early editions.

We are also very grateful to the following reviewers of the eleventh edition for their careful evaluation and useful suggestions for improvement.

Scott Babcock *Western Washington University*
Kathryn Baldwin *Washington State University*
Laura Barnhart *Mississippi State University*
J Bret Bennington *Hofstra University*
Juk Bhattacharyya *University of Wisconsin—Whitewater*
Phyllis Camilleri *Austin Peay State University*
Mark J. Camp *University of Toledo*
Stan Celestian *Glendale Community College*
Chu-Yung Chen *University of Illinois @ Urbana-Champaign*
Renee M. Clary *University of Louisiana at Lafayette*
Jack Deibert *Austin Peay State University*
Grenville Draper *Florida International University*
William J. Frazier *Columbus State University*
Francisco Gomez *University of Missouri—Columbia*
N. Gary Hemming *Queens College*
Charles Herzig *El Camino College*
Roger D. Hoggan *Brigham Young University—Idaho*
Curtis L. Hollabaugh *State University of West Georgia*
Joseph Holliday *El Camino College*
Mary Hubbard *Kansas State University*
Paul F. Hudak *University of North Texas*
Robert B. Jorstad *Eastern Illinois University*
Jeffrey Karson *Duke University*
David T. King, Jr. *Auburn University*
J. Steven Kite *West Virginia University*

Emily M. Klein *Duke University*
Mark A. Kulp *University of New Orleans*
Nan Lindsley-Griffin *University of Nebraska—Lincoln*
Kelly Liu *Kansas State University*
David N. Lumsden *The University of Memphis*
Kathleen Marsaglia *California State University—Northridge*
Jamie Martin-Hayden *University of Toledo*
Kevin Mickus *Southwest Missouri State*
Dan Moore *Brigham Young University—Idaho*
Trent Morrell *Laramie County Community College*
Kirsten Nicolaysen *Kansas State University*
Charles G. Patterson *Red Rocks Community College*
Robert W. Pinker *Johnson City Community College*
Karen L. Savage *California State University–Northridge*
David R. Schwimmer *Columbus State University*
Craig R. Scott *California State University—Northridge*
Wm Jay Sims *University of Arkansas–Little Rock*
Jeffrey Snyder *Bowling Green State University*
Kenneth F. Steele *University of Arkansas*
Don Stierman *University of Toledo*
John P. Stimac *Eastern Illinois University*
Barry Weaver *University of Oklahoma*

Art Review Team

We extend a special thanks to the Art Review Team whose members initially met in Dubuque and continued to give constructive criticism to the art work as it was rendered and revised by our artist, Cindy Shaw. The team members included:

Scott Babcock *Western Washington University*
J Bret Bennington *Hofstra University*
Vincent S. Cronin *Baylor University*
Steve Kadel *Glendale Community College*
Ronald H. Konig *University of Arkansas*
Richard L. Orndorff *Eastern Washington University*
Karen L. Savage *California State University–Northridge*

Ancillary Contributors

Special thanks also to those reviewers who contributed to the production of the ancillaries that accompany this eleventh edition. These contributors include:

J Bret Bennington
Stephen Boss *University of Arkansas*
James Martin Hayden *University of Toledo*
Steve Kadel *Glendale Community College*
Julie Libarkin *The University of Ohio—Athens*
David N. Lumsden *The University of Memphis*

MEET THE AUTHORS

Charles Plummer at Thengboche, in the Himalayan Mountains of Nepal.

Diane Carlson along the bank of the American River, east of Sacramento, California. Note the rounded stream cobbles that were left from dredge mining the River in search of placer gold deposits.

CHARLES PLUMMER Professor Charles "Carlos" Plummer grew up in the shadows of volcanoes in Mexico City. There, he developed a love for mountains and mountaineering that eventually led him into geology. He received his B.A. degree from Dartmouth College. After graduation, he served in the U.S. Army as an artillery officer. He resumed his geological education at the University of Washington, where he received his M.S. and Ph.D. degrees. His geologic work has been in mountainous and polar regions, notably Antarctica (where a glacier is named in his honor). He taught at Olympic Community College in Washington and worked for the U.S. Geologic Survey before joining the faculty at California State University, Sacramento.

At CSUS, he taught optical mineralogy, metamorphic petrology, and field courses as well as introductory courses. He retired from teaching in 2003. He skis, has a private pilot license, and is certified for open-water SCUBA diving. (plummercc@csus.edu)

DIANE CARLSON Professor Diane Carlson grew up on the glaciated Precambrian shield of northern Wisconsin and received an A.A. degree at Nicolet College in Rhinelander and B.S. in geology at the University of Wisconsin at Eau Claire. She continued her studies at the University of Minnesota–Duluth, where she focused on the structural complexities of high-grade metamorphic rocks along the margin of the Idaho batholith for her master's thesis. The lure of the West and an opportunity to work with the U.S. Geological Survey to map the Colville batholith in northeastern Washington led her to Washington State University for her Ph.D. Dr. Carlson

accepted a position at California State University, Sacramento, after receiving her doctorate and teaches physical geology, structural geology, environmental geology, and field geology. Professor Carlson is a recipient of the Outstanding Teacher Award from the CSUS School of Arts and Sciences. She is also actively engaged in researching the structural and tectonic evolution of part of the Foothill Fault System in the northern Sierra Nevada of California. (carlsondh@csus.edu)

DAVID MCGEARY Dave McGeary died in December 2002. He was born in 1940 and grew up in the town of State College, Pennsylvania. He received his B.A. in geology from Williams College in 1962. He earned an M.S. degree from University of Illinois and a Ph.D. in marine geology at Scripps Institution in La Jolla, California. While at Scripps, he taught SCUBA diving. He began his college teaching career at Sacramento State College (later to become California State University, Sacramento) in 1969. Dave and Elly, his wife, had two sons born during his early years at Sacramento State College.

Dave was known as a demanding but brilliant teacher. He developed and taught a broad range of courses, most of them outside his specialty of marine geology. He loved teaching in the field. His weeklong field trips to classic geology locales (e.g., Yellowstone, Grand Canyon) were legendary. He organized and taught field method courses in the Mojave Desert that were known for their rigor. He retired from CSUS in 1992 and from coauthoring this book in 1995. After his retirement, he indulged his love of the theater. He played leading roles in various community productions and traveled with Elly to see performances in New York and London.

LIST OF FEATURES

Chapter 1

Reading Boxes

Environmental Geology 1.1: Delivering Alaskan Oil—The Environment VERSUS the Economy
Environmental Geology 1.2: The 1991 Eruption of Mount Pinatubo—Geologists Save Thousands of Lives
In Greater Depth 1.3: Geology as a Career
In Greater Depth 1.4: Plate Tectonics and the Scientific Method

Animation

Figure 1.9: Divergence of Plates at Mid-Ocean Ridge

Chapter 2

Reading Boxes

In Greater Depth 2.1: Atomic Number, Atomic Mass Number, Isotopes and Atomic Weight
Earth Systems 2.2: Oxygen Isotopes and Climate Change
In Greater Depth 2.3: Bonding
In Greater Depth 2.4: Elements in the Earth
Environmental Geology 2.5: Asbestos—How Hazardous?
Environmental Geology 2.6: Clay Minerals that Swell
In Greater Depth 2.7: Precious Gems
Web Box 2.8: On Time with Quartz
In Greater Depth 2.9: Water and Ice—Molecules and Crystals

Animations

Figures 2.7 and 2.8: Silicate Mineral Structures

Chapter 3

Reading Boxes

In Greater Depth 3.1: Pegmatite—A Rock Made of Giant Crystals
Environmental Geology 3.2: Harnessing Magmatic Energy

Animation

Figure 3.26: How Subduction Causes Volcanism

Chapter 4

Reading Boxes

Environmental Geology 4.1: Mount St. Helens Blows Up
In Greater Depth 4.2: Volcanoes and Flying
Planetary Geology 4.3: Extraterrestrial Volcanic Activity
Environmental Geology 4.4: Popocatepetl—Will It Erupt Big Time?
Environmental Geology 4.5: A Tale of Two Volcanoes—Lives Lost and Lives Saved in the Caribbean
Environmental Geology 4.6: Fighting a Volcano in Iceland—and Winning

Chapter 5

Reading Boxes

Environmental Geology 5.1: Acid Rain
Earth Systems 5.2: Weathering, the Carbon Cycle, and Global Change

Chapter 6

Reading Boxes

Environmental Geology 6.1: Valuable Sedimentary Rocks
Planetary Geology 6.2: Sedimentary Rocks: The Key to Mars' Past
Web Box 6.3: Transgression and Regression

Animations

Figure 6.28: Migration of Sand Grains to Form Ripples, Dunes, and Crossbeds
Figure 6.31: Formation of a Graded Bed

Chapter 7

Reading Boxes

Planetary Geology 7.1: Impact Craters and Shock Metamorphism
In Greater Depth 7.2: Index Minerals
Web Box 7.3: Metamorphic Facies and the Relationship to Plate Tectonics
Environmental Geology 7.4: The World's Largest Humanmade Hole—The Bingham Canyon Copper Mind

Animation

Figure 7.24: Hydrothermal Ore Vein Formation

Chapter 8

Reading Boxes

Earth Systems 8.1: Highlights of the Evolution of Life through Time
Earth Systems 8.2: Demise of the Dinosaurs—Was It Extraterrestrial?
Environmental Geology 8.3: Radon, A Radioactive Health Hazard
In Greater Depth 8.4: Calculating the Age of a Rock

Animation

Figure 8.25: The Geologic History of the Earth Scaled to a Single Year

Chapter 9

Reading Boxes

Environmental Geology 9.1: Disaster in the Andes
Environmental Geology 9.2: Los Angeles, A Mobile Society
Environmental Geology 9.3: Failure of the St. Francis Dam—A Tragic Consequence of Geology Ignored.

Animation

Figure 9.1: Types of Earth Movements

Chapter 10

Reading Boxes

Environmental Geology 10.1: A Controlled Flood in the Grand Canyon: A Bold Experiment to Restore Sediment Movement in the Colorado River
In Greater Depth 10.2: Estimating the Size and Frequency of Floods
Planetary Geology 10.3: Stream Features on the Planet Mars

Animations

Figure 10.13: Modes of Sediment Transport
Figure 10.20: River Meander Development

Chapter 11

Reading Boxes

In Greater Depth 11.1: Darcy's Law and Fluid Potential
Environmental Geology 11.2: Prospecting for Ground Water
Environmental Geology 11.3: Hard Water and Soapsuds

Animations

Figure 11.7: Basic Dynamics of Groundwater Movement
Figure 11.18a: Landfill and Cone Depression
Figure 11.18b, c, d: Cone of Depression and Saltwater Intrusion during Groundwater Pumping

Chapter 12

Reading Boxes

Environmental Geology 12.1: Glaciers as a Water Resource
Environmental Geology 12.2: Water Beneath Glaciers: Floods, Giant Lakes, and Galloping Glaciers
Earth Systems 12.3: Global Warming and Glaciers
Planetary Geology 12.4: Mars on a Glacier
Earth Systems 12.5: Causes of Glacial Ages
In Greater Depth 12.6: The Channeled Scablands

Animations

Figure 12.6: Dynamics of Glacial Advance and Retreat
Figure 12.9b: Crevasse Formation in Glaciers
Figure 12.28: Formation of Glacial Features by Deposition at a Wasting Ice Front
Figure 12.33: Glacial Maximum and Deglaciation

Chapter 13

Reading Boxes

Environmental Geology 13.1: Expanding Deserts
Earth Systems 13.2: Desert Pavement and Desert Varnish
Planetary Geology 13.3: Wind Action on Mars

Chapter 14

Reading Boxes

Environmental Geology 14.1: The Effects of Rising Sea Level

Animations

Figure 14.8: Seasonal Beach Cycle
Figure 14.9: Wave Refraction and Longshore Movement of Sand and Water

Chapter 15

Reading Boxes

In Greater Depth 15.1: Is There Oil Beneath My Property? First Check the Geologic Structure
In Greater Depth 15.2: California's Greatest Fault—The San Andreas

Animations

Figure 15.17: Styles of Folding
Figure 15.21: Styles of Faulting
Figure 15.23: Normal Faulting
Figure 15.25c: Reverse and Thrust Faults

Chapter 16

Reading Boxes

In Greater Depth 16.1: Earthquake Engineering
Environmental Geology 16.2: What to Do Before, During, and After an Earthquake
Environmental Geology 16.3: Waiting for the Big One in California

Animations

Figure 16.4: Earthquake Focus
Figure 16.5: Earthquake Waves
Figure 16.6: Seismometer
Figure 16.7: Seismometer
Figure 16.8, 16.9, 16.10: Locating Earthquake Epicenter

Chapter 17

Reading Boxes

In Greater Depth 17.1: Deep Drilling on Continents
In Greater Depth 17.2: A CAT Scan of the Mantle
Planetary Geology 17.3: Meteorites
In Greater Depth 17.4: Earth's Spinning Inner Core

Animations

Figures 17.8 and 17.9: P and S Wave Shadow Zones
Figure 17.11: Isostacy-Basic Principle
Figure 17.12: How Isostacy, Orogeny, and Metamorphism Are Interrelated
Figure 17.13: Isostatic Rebound after Deglaciation

Chapter 18

Reading Boxes

Earth Systems 18.1: Does the Earth Breathe?
Environmental Geology 18.2: Geologic Riches in the Sea

Chapter 19

Reading Boxes

Earth Systems 19.1: Plate Tectonics and Sea Level
Web Box 19.2: Backarc Spreading
Earth Systems 19.3: Indentation Tectonics and "Mushy" Plate Boundaries
Earth Systems 19.4: The Relationship Between Plate Tectonics and Ore Deposits

Animations

Chapter 20

Reading Boxes

Chapter 21

Reading Boxes

Chapter 22

Animations

Introducing Geology, the Essentials of Plate Tectonics, and Other Important Concepts

Geology uses the scientific method to explain natural aspects of the Earth—for example, how mountains form or why oil resources are concentrated in some rocks and not in others. This chapter briefly explains how and why Earth's surface and its interior are constantly changing. The chapter relates the changes to the major geological topics of interaction of the atmosphere, water and rock, the modern theory of plate tectonics, and geologic time. These concepts form a framework for the rest of the book. Understanding the "big picture" presented here will aid you in comprehending the chapters that follow.

Mount Robson, 3,954 meters (12,972 feet) above sea level, is the highest peak in the Canadian Rocky Mountains. *Photo © J. A. Kraulis/Masterfile*

Strategy for Using This Textbook

- As authors, we try to be thorough in our coverage of topics so the textbook can serve you as a resource. Your instructor may choose, however, to concentrate only on certain topics for *your* course. Find out which topics and chapters you should focus on in your studying and concentrate your energies there.

- Your instructor may present additional material that is not in the textbook. Take good notes in class.

- Do not get overwhelmed by terms. (Every discipline has its own language.) Don't just memorize each term and its definition. If you associate a term with a concept or mental picture, remembering the term comes naturally when you understand the concept. (You remember names of people you know because you associate personality and physical characteristics with a name.) You may find it helpful to learn the meanings of frequently used prefixes and suffixes for geological terms. These can be found in appendix G.

- **Boldfaced** terms are ones you are likely to need to understand because they are important to the entire course.

- *Italicized* terms are not as important but may be necessary to understand the material in a particular chapter.

- Pay particular attention to illustrations. Geology is a visually oriented science, and the photos and artwork are at least as important as the text. You should be able to sketch important concepts from memory.

- Find out to what extent your instructor expects you to learn the material in the boxes. They offer an interesting perspective on geology and how it is used, but much of the material might well be considered optional for an introductory course and not vital to your understanding of major topics. Many of the "In Greater Depth" boxes are meant to be challenging—do not be discouraged if you need your instructor's help in understanding them.

- Read through the appropriate chapter before going to class. Reread it after class, concentrating on the topics covered in the lecture or discussion. Especially concentrate on concepts that you do not fully understand. Return to previously covered chapters to refresh your memory on necessary background material.

- Use the end of chapter material for review. The Summary is just that, a summary. Don't expect to get through an exam by only reading the summary and not the rest of the chapter. Use the Terms to Remember to see if you can visually or verbally associate the appropriate concept with each term. Answer the Testing Your Knowledge questions in writing. Be honest with yourself. If you are fuzzy on an answer, return to that portion of the chapter and reread it. Remember that these are just a sampling of the kind of questions that might be on an exam.

- Geology, like most science, builds on previously acquired knowledge. You must retain what you learn from chapter to chapter. If you forget or did not learn significant concepts covered early in your course, you will find it frustrating later in the course. (To verify this, turn to chapter 20 and you will probably find it intimidating; but if you build on your knowledge as you progress through your course, the chapter material will fall nicely into place.)

- Get acquainted with the book's online Learning Center at www.mhhe.com/plummer11e. Go to the "Student Edition" and follow the instructions to log on (you must first register using the code that is packaged with your textbook). You will find the online quizzes useful for review and the web exercises interesting.

- Be curious. Geologists are motivated by a sense of discovery. We hope you will be too.

WHO NEEDS GEOLOGY?

Geology, the scientific study of Earth, benefits you and everyone else on this planet. The clothes you wear, the radio you listen to, the food you eat, the car you drive exist because of what geologists have discovered about Earth. Earth can also be a killer. You might have survived an earthquake, flood, or other natural disaster thanks to action taken based on what scientists have learned about these hazards. Before getting into important scientific concepts, we will look at some of the ways geology has and will continue to benefit you.

Supplying Things We Need

We depend on the Earth for energy resources and the raw materials we need for survival, comfort, and pleasure. Every manufactured object relies on Earth's resources—even a pencil (figure 1.1). The Earth, at work for billions of years, has localized material into concentrations that humans can mine or extract. By learning how the Earth works and how different kinds of substances are distributed and why, we can intelligently search for metals, sources of energy, and gems. Even maintaining a supply of sand and gravel for construction purposes depends on geology.

The economic systems of Western civilization currently depend on abundant and cheap energy sources. Nearly all our vehicles and machinery are powered by petroleum, coal, or nuclear power and depend on energy sources concentrated unevenly in the Earth. The U.S. economy in particular is geared to petroleum as a cheap source of energy. In a few decades, Americans have used up most of their country's known petroleum reserves, which took nature hundreds of

FIGURE 1.1

Earth's resources necessary to make a wooden pencil.

millions of years to store in the Earth. The United States, and most other industrialized nations, are now heavily dependent on imported oil. When fuel prices jump, people who are not aware that petroleum is a nonrenewable resource become upset and are quick to blame oil companies, politicians, and oil-producing countries. (The Gulf Wars of 1991 and 2003 were at least partially fought because of the industrialized nations' petroleum requirements.) To find more of this diminishing resource will require more money and increasingly sophisticated knowledge of geology. Although many people are not aware of it, we face similar problems with diminishing resources of other materials, notably metals such as iron, aluminum, copper, and tin, each of which has been concentrated in a particular environment by the action of the Earth's geologic forces.

Just how much of our resources do we use? According to the Mineral Information Institute, for every person living in the United States, 18,000 kilograms (40,000 pounds; for metric conversions, go to appendix E) of resources, not including energy resources are mined annually. The amount of each commodity mined is 4,400 kilograms stone, 3,500 kilograms sand and gravel, 325 kilograms limestone for cement, 160 kilograms clays, 165 kilograms salt, 760 kilograms other non-

metals, 545 kilograms iron, 19 kilograms aluminum, 9 kilograms copper, 5 kilograms each for lead and zinc, 3 kilograms manganese, and 11 kilograms other metals. Americans' yearly per capita consumption of energy resources is over 8,000 kilograms (17,000 pounds); of this, 3,500 kilograms is petroleum, 2,300 kilograms coal, 2,250 kilograms natural gas, and .02 kilograms uranium.

Protecting the Environment

Our demands for more energy and metals have, in the past, led us to extract them with little regard for effects on the balance of nature within the Earth and therefore on us, Earth's residents. Mining of coal, if done carelessly, for example, can release acids into water supplies. Understanding geology can help us lessen or prevent damage to the environment—just as it can be used to find the resources in the first place.

The environment is further threatened because these are nonrenewable resources. Petroleum and metal deposits do not grow back after being harvested. As demands for these commodities increase, so does the pressure to disregard the ecological damage caused by the extraction of the remaining deposits.

Problems involving petroleum illustrate this. Oil companies employ geologists to discover new oil fields, while the public and government depend on other geologists to assess the potential environmental impact of petroleum's removal from the ground, the transportation of petroleum (see box 1.1), and disposal of any toxic wastes from petroleum products.

Avoiding Geologic Hazards

Almost everyone is, to some extent, at risk to natural hazards, such as earthquakes or hurricanes. Earthquakes, volcanic eruptions, landslides, floods, and tsunamis are the most dangerous *geologic hazards*. Each is discussed in detail in appropriate chapters. Here, we will give some examples to illustrate the role that geology can play in mitigating geologic hazards.

Prior to December 26, 2004, "tsunami" may not have been part of your vocabulary. As of that date, the world became sadly aware of the enormous destructive power of *tsunamis* (huge ocean waves, usually caused by displacement of the sea floor). Earth's largest earthquake in forty years took place off the coast of northern Indonesia (figure 1.2). Its shaking caused widespread destruction in Banda Aceh province and would have been a major disaster in its own right. But the earthquake was overshadowed by the tsunamis that followed. A tsunami, caused by the earthquake, began forming when a large segment of sea floor was displaced along a fault. (Earthquakes and tsunamis are fully explained in chapter 16.) The energy transferred into ocean waves was enormous. Tsunamis radiated in all directions from the displaced sea floor. Huge waves crashed into the Indonesian coastline almost immediately, adding thousands to the death toll from the earthquake. Other waves traveled at the speed of a jetliner to the distant shores of the Indian Ocean rim countries and to the east coast of Africa. As

ENVIRONMENTAL GEOLOGY 1.1

Delivering Alaskan Oil—The Environment VERSUS the Economy

In the 1960s, geologists discovered oil beneath the shores of the Arctic Ocean on Alaska's North Slope. It is now the United States' largest oil field. Thanks to the Alaska pipeline, completed in 1977, Alaska has supplied as much as 20% of the United States' domestic oil.

In the late 1970s before Alaskan oil began to flow, the United States was importing almost half its petroleum, at a loss of billions of dollars per year to the national economy. (By 1997, the United States was importing more than half of the petroleum it uses, despite Alaskan oil in the market.) The drain on the country's economy and the increasing cost of energy can be major causes of inflation, lower industrial productivity, unemployment, and the erosion of standards of living. At its peak, over 2 million barrels of oil a day flowed from the Arctic oil fields, which meant that over $10 billion a year that would have been spent importing foreign oil was kept in the American economy.

Despite its important role in the American economy, some considered the Alaska pipeline and the use of oil tankers as unacceptable threats to the area's ecology.

Geologists with the U.S. Geological Survey conducted the official environmental impact investigation of the proposed pipeline route in 1972. After an exhaustive study, they recommended against its construction, partly because of the hazards to oil tankers and partly because of the geologic hazards of the pipeline route. Their report was overruled. The Congress and the president of the United States exempted the pipeline from laws that require a favorable environmental impact statement before a major project can begin.

The 1,250-kilometer-long pipeline crosses regions of ice-saturated, frozen ground and major earthquake-prone mountain ranges that geologists regard as serious hazards to the structure.

Building anything on frozen ground creates problems. The pipeline presented enormous engineering problems. If the pipeline were placed on the ground, the hot oil flowing through it could melt the frozen ground. On a slope, mud could easily slide and rupture the pipeline. Careful (and costly) engineering minimized these hazards. Much of the pipeline is elevated above the ground (box figure 1). Radiators conduct heat out of the structure. In some places, refrigeration equipment in the ground protects against melting.

Records indicate that a strong earthquake can be expected every few years in the earthquake belts crossed by the pipeline. An earthquake could rupture a pipeline—especially a conventional pipe as in the original design. When the Alaska pipeline was built, however, in several places sections were specially jointed to allow the pipe to shift as much as 6 meters without rupturing. In 2002, a major earthquake caused the pipeline to shift several meters, resulting in minor damage to the structure, but the pipe did not rupture.

The original estimated cost of the pipeline was $900 million, but the final cost was $7.7 billion, making it the costliest privately financed construction project in history. The redesigning and construction that minimized the potential for an environmental disaster were among the reasons for the increased cost. Some minor spills from the pipeline have occurred. For instance, in January 1981, 5,000 barrels of oil were lost when a valve ruptured. In

BOX 1.1 ■ FIGURE 1

The Alaska pipeline. *Photo by David Applegate*

2001, a man fired a rifle bullet into the pipeline, causing it to rupture and spill 7,000 barrels of oil into a forested area.

When the tanker *Exxon Valdez* ran aground in 1989, over 240,000 barrels of crude oil were spilled into the waters of Alaska's Prince William Sound. It was the worst-ever oil spill in U.S. waters. The spill, with its devastating effects on wildlife and the fishing industry, dramatically highlighted the conflicts between maintaining the energy demands of the American economy and conservation of the environment. The 1972 environmental impact statement had singled out marine oil spills as being the greatest threat to the environment. Based on statistical studies of tanker accidents worldwide, it gave the frequency with which large oil spills could be expected. The *Exxon Valdez* spill should not have been a surprise.

In the early 2000s, as the United States once again faces an energy crisis, one of the "fixes" being proposed is to allow exploitation of oil in the Arctic National Wildlife Refuge on Alaska's North Slope. The rhetoric in the early stages of the debate is more self-serving or emotional than scientific. At one extreme are those who feel that any significant, potential oil field should be developed without regard to environmental damage. At the other extreme are those who instinctively assume that any intrusion on an ecological environment is unacceptable. We can hope that the enormous amount of data from the Alaskan pipeline and the drilling of the North Slope oil field (which has been producing decreasing amounts of oil with ongoing pumping) will be used to help transcend the politics.

Additional Resources

The Alyeska pipeline company's site.
• www.alyeska-pipe.com/

U.S. Geological Survey fact sheet on the Arctic National Wildlife Refuge.
• http://pubs.usgs.gov/fs/fs-0028-01/fs-0028-01.htm

South Asia Earthquake and Tsunami
Potential land affected in tsunami inundation zone

Strongest Epicenter
Date: 26/12/2004
Time: 00:58:53 UTC
Location:3.26N 95.82E
Magnitude:8.9

Estimated land affected in tsunami inundation zones (sq km)*

	Forest	Agriculture	Urban area	Other	Total
Andaman Islands	545	15	7	471	1,038
Bangladesh	965	9,224	178	1,509	11,876
India	1,266	8,097	356	938	10,657
Indonesia	2,882	4,587	32	2,061	9,562
Malaysia	758	1,397	104	266	2,525
Maldives	0	0	0	160	165
Nicobar	58	2	0	122	182
Srilanka	135	1,724	107	1,134	3,100
Thailand	1,480	625	0	299	2,404
Total	8,089	25,676	784	6,960	41,509

* Stats derived from SRTM90 data 0 to 5km from coastline, and below 10m in height, crossed with South East Asia regional landcover, at 1km scale

Legend
- National Capital
- Main coastal cities
- Earthquake epicenters
- Affected coast by Tsunami
- Estimated tsunami inundation zone
- Areas under 20 meters elevation and within 5 km from the coastal line
- Provinces

Land Cover Legend
- Urban area
- Cropland and plantation
- Forest
- Shrub
- Grassland
- Bare soil
- Swamp
- Water body

Affected Countries in orange

2500 km from Main Epicenter

1 Cm equals 176 Km

EUROPEAN COMMISSION
DIRECTORATE-GENERAL
Joint Research Centre

Earthquake epicenters data from USGS as 03 Jan 2005 9:47:18 UTC
Population density data from landscan 2002
Sources: Regional land cover map generated by a partnership coordinated by JRC; part of Global Land Cover 2000 Project.

Disclaimer: this map does not reflect the official opinion of the European Communities or other European Community institutions. Neither the European Commission nor any person or company acting on the behalf of the European Commission is responsible for the use that may be made of the information contained in this map.

A

FIGURE 1.2

The earthquake and tsunami of December 26, 2004. (A) Map of the Indian Ocean region showing the epicenter of the quake and the countries and shorelines where people were killed.

B

C

FIGURE 1.2 (CONTINUED)

(B) Marina beach in Madras, India inundated by the tsunami. (C) Tsunami survivors carry items they saved from the rubble at a commercial area of Banda Aceh in northwest Indonesia. *Photo B © AFP/Getty Images; photo C © AP/Wide World Photos*

explained in chapter 16, in the deep ocean, a tsunami has a small wave height and travels rapidly—it is not noticed by people on boats. As it propagates into shallower water, it slows down and the wave heights get larger. When the tsunamis reached Thailand, India, Sri Lanka, and eight other countries, waves as high as 14 meters (40 feet) rapidly inundated coastal communities. When the seas returned to normal, over an estimated 220,000 people were dead and millions injured. The damage to homes and property was incalculable. For more on this and other tsunamis, go to the book's website (www.mhhe.com/plummer11e). Here you can see a computer animation of the spread of tsunamis, videos, and more photos.

The tsunami was among the worst natural disasters in recorded history. What made it truly exceptional was the death and destruction in so many countries over such a large segment of the Earth. Could the death toll have been reduced through knowledge of geology? Most definitely. At one beach resort in Thailand, a ten-year-old English schoolgirl on holiday with her family noticed that the sea began withdrawing. A few weeks earlier her geography class had learned about tsunamis. She knew that a drop in sea level often precedes the arrival of the first giant wave. She told her mother and they then spread the alarm throughout the resort. Everyone ran to higher ground. This was the only part of this segment of the Thai coastline where there were no casualties. The girl's knowledge of tsunamis saved around a hundred lives. Thousands of people died elsewhere because they had no idea what was going on when the water withdrew and then began rising. Many actually moved closer to the shoreline to see what was going on.

The Pacific Rim countries, where tsunamis are more common, have a sophisticated warning system that alerts all coastal regions after a submarine earthquake takes place and a tsunami is likely. For example, if an earthquake produces a tsunami in Alaska, or Chile, it will take hours to reach Hawaii. This gives plenty of time for threatened Hawaiian beaches to be evacuated. Undoubtedly, a similar early warning system will be put in place for the Indian Ocean. But even without a formal warning system in place, it is amazing that, in this age of instantaneous worldwide communication, the death toll was so high. While the Indonesian coast was being ravaged there was little, if any, communication to India, Sri Lanka, or other distant countries about a tsunami which would take hours for its transoceanic crossing.

Volcanic eruptions, like earthquakes and tsunamis, are products of Earth's sudden release of energy. They can be dangerous; however, their biggest dangers are not what most people think. Neither falling volcanic debris nor lava flows are as big a killer as pyroclastic flows or volcanic mudflows. As described in the volcano chapter, a *pyroclastic flow* is a hot, turbulent mixture of expanding gases and volcanic ash that flows rapidly down the side of a volcano. Pyroclastic flows often reach speeds of over 100 kilometers per hour and are extremely destructive. A *mudflow* is a slurry of water and rock debris that flows down a stream channel.

Mount Pinatubo's eruption in 1991 was the second largest volcanic eruption of the twentieth century (box 1.2). Geologists successfully predicted the climactic eruption (figure 1.3) in time for Philippine officials to evacuate people living near the mountain. Tens of thousands of lives were saved from pyroclastic flows and mudflows.

FIGURE 1.3

The major eruption of Mount Pinatubo on June 15, 1991, as seen from Clark Air Force Base, Philippines. *Photo by Robert Lapointe, U.S. Air Force*

By contrast, one of the worst volcanic disasters of the 1900s took place after a relatively small eruption of Nevado del Ruiz in Colombia in 1985. Hot volcanic debris blasted out of the volcano and caused part of the ice and snow capping the peak to melt. The water and loose debris turned into a mudflow. The mudflow overwhelmed the town of Armero at the base of the volcano, killing 23,000 people (figure 1.4). Colombian geologists had previously predicted such a mudflow could occur and published maps showing the location and extent of expected mudflows. The actual mudflow that wiped out the town matched that shown on the geologists' map almost exactly. Unfortunately, government officials had ignored the map and the geologists' report; otherwise, the tragedy could have been averted.

Understanding Our Surroundings

It is a uniquely human trait to want to understand the world around us. Most of us get satisfaction from understanding our cultural and family histories, how governments work or do not work. Music and art help link our feelings to that which we have discovered through our life. The natural sciences involve understanding the physical and biological universe in which we live. Most scientists get great satisfaction from their work because, besides gaining greater knowledge from what has been discovered by scientists before them, they can find new truths about the world around them. Even after a basic geology course, you can use what you learn to explain and be able to

FIGURE 1.4

Most of the town of Armero, Colombia and its residents are buried beneath up to 8 meters of mud from the 1985 mudflow. *Photo © Jacques Langevin/Corbis*

ENVIRONMENTAL GEOLOGY 1.2

The 1991 Eruption of Mount Pinatubo— Geologists Save Thousands of Lives

When minor steam eruptions began in April 1991, Mount Pinatubo was a vegetation-covered mountain that had last erupted 400 years earlier. As the eruptions intensified, Filipino geologists thought a major eruption might be developing. Geologic field work completed in earlier years indicated that prehistoric eruptions of the volcano tended to be large and violent. Under a previous arrangement for cooperation, American geologists joined their Philippine colleagues and deployed portable seismographs to detect and locate small earthquakes within the volcano and tiltmeters to measure the bulging of the volcano. These and other data were analyzed by state-of-the-art computer programs.

Fortunately, it took two months for the volcano to reach its climactic eruption, allowing time for the scientists to work with local officials and develop emergency evacuation plans. Geologists had to educate the officials about the principal hazards—mudflows and pyroclastic flows.

In June, explosions, ash eruptions, and minor pyroclastic flows indicated that magma (molten rock) was not far underground and a major eruption was imminent. Some 80,000 people were evacuated from the vicinity of the volcano. The U.S. military evacuated and later abandoned Clark Air Force Base, which was buried by ash. The climactic eruption occurred on June 15, when huge explosions blasted the top off the volcano and resulted in large pyroclastic flows (figure 1.3). Volcanic debris was propelled high into the atmosphere. A typhoon 50 kilometers away brought heavy rains, which mixed with the ash and resulted in numerous, large mudflows.

The estimated volume of magma that erupted from the climactic eruption was 5 cubic kilometers, making it the world's largest eruption since 1917. Its effects extended beyond the Philippines. Fine volcanic dust and gas blasted into the high atmosphere were carried around the world and would take years to settle out. For a while, we got more colorful sunsets worldwide. Because of the filtering effect for solar radiation, worldwide average temperature was estimated to drop by 0.5°C for two years, more than countering the long-term warming trend of the Earth's climate.

The death toll from the eruption was 374. Of these, 83 were killed in mudflows. Most of the rest died because roofs collapsed from the weight of ash. In addition, 358 people died from illness related to the eruptions. More than 108,000 homes were partly or totally destroyed. The death toll probably would have been in the tens of thousands had the prediction and warning system not been so successful. Although Mount Pinatubo is quiet now, lives and property are still being lost to mudflows, more than a decade after the big eruption.

Additional Resources

Volcano World

The site contains a wealth of information on volcanoes, including Mount Pinatubo.

- http://volcano.und.nodak.edu/

In the Path of a Killer Volcano is a first-rate videotape produced for the *Nova* television series. Available from Films for the Humanities and Science, Princeton, New Jersey.

---

appreciate what you see around you, especially when you travel. If, for instance, you were traveling through the Canadian Rockies, you might see the scene in this chapter's opening photo and wonder how the landscape came to be.

You might wonder: (1) why there are layers in the rock exposed in the cliffs; (2) why the peaks are so jagged; (3) why there is a glacier in a valley carved into the mountain; (4) why this is part of a mountain belt that extends northward and southward for thousands of kilometers; (5) why there are mountain ranges here and not in the central part of the continent. After completing a course in physical geology, you should be able to answer these questions as well as understand how other kinds of landscapes formed.

EARTH SYSTEMS

The awesome energy released by an earthquake or volcano is a product of forces within the Earth that move firm rock. Earth-quakes and volcanoes are only two consequences of the ongoing changing of Earth. Ocean basins open and close. Mountain ranges rise and are worn down to plains through slow, but very effective, processes. Studying how Earth works can be as exciting as watching a great theatrical performance. The purpose of this book is to help you understand how and why those changes take place. More precisely, we concentrate on *physical geology*, which is the division of geology concerned with Earth materials, changes in the surface and interior of the Earth, and the dynamic forces that cause those changes. Put another way, physical geology is about how Earth works.

But to understand geology, we must also understand how the solid Earth interacts with water, air, and living organisms. For this reason, it is useful to think of Earth as being part of a system. A *system* is an arbitrarily isolated portion of the universe that can be analyzed to see how its components interrelate. The *solar system* is a part of the much larger universe. The solar system includes the Sun, planets, the moons orbiting planets, and asteroids (see chapter 22).

The **Earth system** is a small part of the larger solar system, but it is, of course, very important to us. The Earth system has its components, which can be thought of as its subsystems. We refer to these as *Earth systems* (plural). These systems, or *"spheres,"* are the atmosphere, the hydrosphere, the biosphere, and the geosphere. You, of course, are familiar with the **atmosphere,** the gases that envelop Earth. The **hydrosphere** is the water on or near Earth's surface. The hydrosphere includes the oceans, rivers, lakes, and glaciers of the world. Earth is unique among the planets in that two-thirds of its surface is covered by oceans. The **biosphere** is all of the living or once-living material on Earth. The **geosphere,** or **solid Earth system,** is the rock and other inorganic Earth material that make up the bulk of the planet. This book concentrates on the geosphere; to understand geology, however, we must understand the interaction between the solid Earth and the other systems (spheres).

The Indian Ocean tsunami involved the interaction of the geosphere and the hydrosphere. The faulting of the sea floor and the earthquake took place in the geosphere. Energy was transferred into giant waves in the hydrosphere. The hydrosphere and geosphere again interacted when waves inundated distant shores.

All four of the Earth systems interact with each other to produce soil, such as we find in farms, gardens, and forests. The solid "dirt" is a mixture of decomposed and disintegrated rock and organic matter. The organic matter is from decayed plants—from the biosphere. The geosphere contributes the rock that has broken down while exposed to air (the atmosphere) and water (the hydrosphere). Air and water also occupy pore space between the solid particles.

AN OVERVIEW OF PHYSICAL GEOLOGY—IMPORTANT CONCEPTS

The remainder of this chapter is an overview of physical geology that should provide a framework for most of the material in this book. Although the concepts probably are totally new to you, it is important that you comprehend what follows. You may want to reread portions of this chapter while studying later chapters when you need to expand or reinforce your comprehension of this basic material. You will especially want to refresh your understanding of plate tectonics when you learn about the plate tectonic setting for the origin of rocks in chapters 3 through 7.

The Earth can be visualized as a giant machine driven by two engines, one internal and the other external. Both are *heat engines,* devices that convert heat energy into mechanical energy. Two simple heat engines are shown in figure 1.5. An automobile is powered by a heat engine. When gasoline is ignited in the cylinders, the resulting hot gases expand, driving pistons to the far end of cylinders. In this way, the heat energy of the expanding gas has been converted to the mechanical energy of the moving pistons, then transferred to the wheels, where the energy is put to work moving the car.

Earth's *internal* heat engine is driven by heat moving from the hot interior of the Earth toward the cooler exterior. Moving plates and earthquakes are products of this heat engine.

Earth's *external* heat engine is driven by solar power. Heat from the Sun provides the energy for circulating the atmosphere and oceans. Water, especially from the oceans, is

A

Expanding steam

Spinning pinwheel
(mechanical energy)

Teapot

Heat energy

B

FIGURE 1.5

Two examples of simple heat engines. (*A*) A "lava lamp." Blobs are heated from below and rise. Blobs cool off at the top of the lamp and sink. (*B*) A pinwheel held over steam. Heat energy is converted to mechanical energy. *Photo by C. C. Plummer*

IN GREATER DEPTH 1.3

Geology as a Career

If someone says that she or he is a geologist, that information tells you almost nothing about what he or she does. This is because geology encompasses a broad spectrum of disciplines. Perhaps what most geologists have in common is that they were attracted to the outdoors. Most of us enjoyed hiking, skiing, climbing, or other outdoor activities before getting interested in geology. We like having one of our laboratories being Earth itself.

Geology is a collection of disciplines. When someone decides to become a geologist, she or he is selecting one of those disciplines. The choice is very large. Some are financially lucrative; others may be less so but might be more satisfying. Following are a few of the areas in which geologists work.

Petroleum geologists work at trying to determine where existing oil fields might be expanded or where new oil fields might exist. A petroleum geologist can make over $90,000 a year working on wave-lashed drilling platforms in the North Sea off the coast of Norway. Mining geologists might be concerned with trying to determine where to extend an existing mine to get more ore or trying to find new concentrations of ore that are potentially commercially viable. Environmental geologists might work at mitigating pollution or preventing degradation of the environment. Marine geologists are concerned with understanding the sea floor. Some go down thousands of meters in submersibles to study geologic features on the sea floor. Hydrogeologists study surface and underground water and assist in either increasing our supply of clean water or isolating or cleaning up polluted water. Glaciologists work in Antarctica studying the dynamics of glacier movement or collecting ice cores through drilling to determine climate changes that have taken place over the past 100,000 years or more. Other geologists who work in Antarctica might be deciphering the history of a mountain range, working on skis and living in tents (box figure 1). Volcanologists sometimes get killed or injured while trying to collect gases or samples of lava from a volcano. Some sedimentologists scuba dive in places like the Bahamas, skewering lobsters for lunch while they collect sediment samples. One geologist was the only scientist to work on the moon. Geophysicists interpret earthquake waves or gravity measurements to determine the nature of Earth's interior. Seismologists are geophysicists who specialize in earthquakes.

Engineering geologists determine whether rock or soil upon which structures (dams, bridges, buildings) are built can safely support those structures. Paleontologists study fossils and learn about when extinct creatures lived and the environment in which they existed.

Teaching is an important field in which geologists work. Some teach at the college level and are usually involved in research as well. Demand is increasing for geologists to teach Earth science (which includes meteorology, oceanography, astronomy as well

BOX 1.3 ■ FIGURE 1

Geologists investigating the Latady Mountains, Antarctica. *Photo by C. C. Plummer*

as geology) in high schools. More and more secondary schools are adding Earth science to their curriculum and need qualified teachers.

Many geologists enjoy the challenge and adventure of field work, but some work comfortably behind computer screens or in laboratories with complex analytical equipment. Usually, a geologist engages in a combination of field work, lab work, and computer analysis.

Geologists tend to be happy with their jobs. In surveys of job satisfaction in a number of professions, geology rates near or at the top. A geologist is likely to be a generalist who solves problems by bringing in information from beyond his or her specialty. Chemistry, physics, and life sciences are often used to solve problems. Problems geologists work on tend to be ones in which there are few clues. So the geologist works like a detective, piecing together the available data to form a plausible solution. In fact, some geologists work at solving crimes—forensic geology is a branch of geology dedicated to criminal investigations.

Not all people who major in geology become professional geologists. Physicians, lawyers, and businesspeople who have majored in geology have felt that the training in how geologists solve problems has benefited their careers.

Additional Resource

For more information, go to the American Geological Institute's career site at

•   www.agiweb.org/careers.html

evaporated due to solar heating. When moist air cools, rain or snow falls.

Over long periods of time, moisture at the Earth's surface helps rock disintegrate. Water washing down hillsides and flowing in streams loosens and carries away the rock particles. In this way, mountains originally raised by Earth's internal forces are worn away by processes driven by the external heat engine.

We will look at how the Earth's heat engines work and show how some of the major topics of physical geology are related to the *internal* and *surficial* (on the Earth's surface) processes powered by the heat engines.

Internal Processes: How the Earth's Internal Heat Engine Works

The Earth's internal heat engine works because hot, buoyant material deep within the Earth moves slowly upward toward the cool surface and cold, denser material moves downward. Visualize a vat of hot wax, heated from below (figure 1.6). As the wax immediately above the fire gets hotter, it expands, becomes less dense (that is, a given volume of the material will weigh less), and rises. Wax at the top of the vat loses heat to the air, cools, contracts, becomes denser, and sinks. A similar process takes place in the Earth's interior. Rock that is deep within the Earth and is very hot rises slowly toward the surface, while rock that has cooled near the surface is denser and sinks downward. Instinctively, we don't want to believe that rock can flow like hot wax. However, experiments have shown that under the right conditions, rocks are capable of being molded (like wax or putty). Deeply buried rock that is hot and under high pressure can deform, like taffy or putty. But the deformation takes place very slowly. If we were somehow able to strike a rapid blow to the deeply buried rock with a hammer, it would fracture, just as rock at Earth's surface would.

Earth's Interior

As described in more detail in chapter 17, the **mantle** is the most voluminous of Earth's three major concentric zones (see

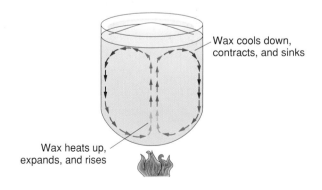

FIGURE 1.6
Movement of wax due to density differences caused by heating and cooling (shown schematically).

figure 1.7). Although the mantle is solid rock, parts of it flow slowly, generally upward or downward, depending on whether it is hotter or colder than adjacent mantle.

The other two zones are the **crust** and the **core.** The crust of the Earth is analogous to the skin on an apple. The thickness of the crust is insignificant compared to the whole Earth. We have direct access to only the crust, and not much of the crust at that. We are like microbes crawling on an apple, without the ability to penetrate its skin. Because it is our home and we depend on it for resources, we are concerned more with the crust than with the inaccessible mantle and core.

Two major types of crust are *oceanic crust* and *continental crust.* The crust under the oceans is much thinner. It is made of rock that is somewhat denser than the rock that underlies the continents.

The lower parts of the crust and the entire mantle are inaccessible to direct observation. No mine or oil well has penetrated through the crust, so our concept of the Earth's interior is based on indirect evidence.

The crust and the uppermost part of the mantle are relatively rigid. Collectively, they make up the **lithosphere.** (To help you remember terms, the meanings of commonly used prefixes and suffixes are given in appendix G. For example, *lith* means "rock" in Greek. You will find *lith* to be part of many geologic terms.) The uppermost mantle underlying the lithosphere, called the **asthenosphere,** is soft and therefore flows more readily than the underlying mantle. It provides a "lubricating" layer over which the lithosphere moves (*asthenos* means "weak" in Greek). Where hot mantle material wells upward, it will uplift the lithosphere. Where the lithosphere is coldest and densest, it will sink down through the asthenosphere and into the deeper mantle, just as the wax does in figure 1.6. The effect of this internal heat engine on the crust is of great significance to geology. The forces generated inside the Earth, called **tectonic forces,** cause deformation of rock as well as vertical and horizontal movement of portions of the Earth's crust. The existence of mountain ranges indicates that tectonic forces are stronger than gravitational forces. (Mount Everest, the world's highest peak, is made of rock that formed beneath an ancient sea.) Mountain ranges are built over extended periods, as portions of the Earth's crust are squeezed, stretched, and raised.

Most tectonic forces are mechanical forces. Some of the energy from these forces is put to work deforming rock, bending and breaking it, and raising mountain ranges. The mechanical energy may be stored (an earthquake is a sudden release of stored mechanical energy) or converted to heat energy (rock may melt, resulting in volcanic eruptions). The working of the machinery of the Earth is elegantly demonstrated by plate tectonics.

The Theory of Plate Tectonics

From time to time a theory emerges within a science that revolutionizes that field. (As explained in box 1.4, a *theory* in science is a concept that has been highly tested and in all

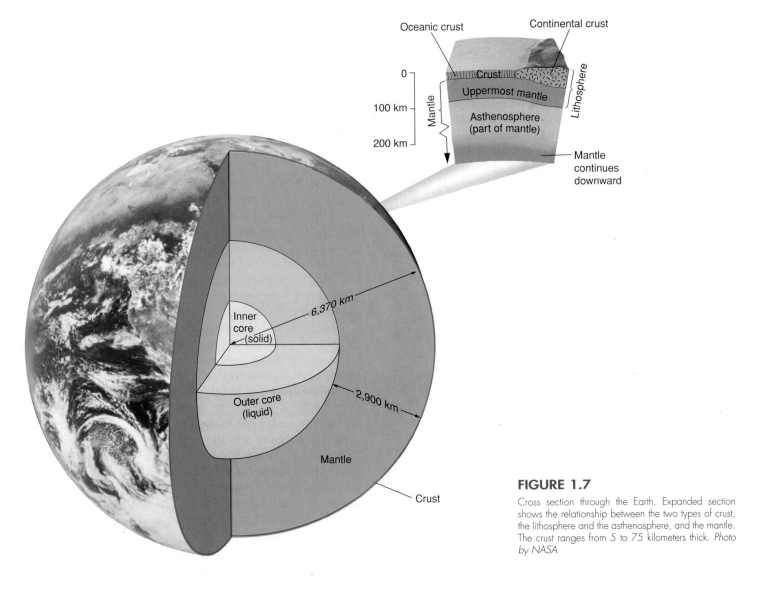

FIGURE 1.7

Cross section through the Earth. Expanded section shows the relationship between the two types of crust, the lithosphere and the asthenosphere, and the mantle. The crust ranges from 5 to 75 kilometers thick. *Photo by NASA*

likelihood is true. In common usage, the word *theory* is used for what scientists call a *hypothesis*—that is, a tentative answer to a question or solution to a problem.) The theory of plate tectonics is as important to geology as the theory of relativity is to physics, the atomic theory to chemistry, or evolution to biology. The plate tectonic theory, currently accepted by virtually all geologists, is a unifying theory that accounts for many seemingly unrelated geological phenomena. Some of the disparate phenomena that plate tectonics explains are where and why we get earthquakes, volcanoes, mountain belts, deep ocean trenches, and midoceanic ridges.

Plate tectonics was seriously proposed as a hypothesis in the early 1960s, though the idea was based on earlier work—notably, the hypothesis of *continental drift*. In the chapters on igneous, sedimentary, and metamorphic rocks, as in the chapter on earthquakes, we will expand on what you learn about the theory here to explain the origin of some rocks and why volcanoes and earthquakes occur. Chapter 19 is devoted to plate tectonics and will show that what you learned in many previous chapters is interrelated and explained by plate tectonic theory.

Plate tectonics regards the lithosphere as broken into *plates* that are in motion (see figure 1.8). The plates, which are much like segments of the cracked shell on a boiled egg, move relative to one another along *plate boundaries* that slide upon the underlying asthenosphere. Much of what we observe in the rock record can be explained by the type of motion that takes place along plate boundaries. Plate boundaries are classified into three types based on the type of motion occurring between the adjacent plates. These are summarized in table 1.1.

Divergent Boundaries

The first type of place boundary, a **divergent boundary,** involves two plates that are moving apart from each other. Most divergent boundaries coincide with the crests of submarine mountain ranges, called **mid-oceanic ridges** (figure 1.8). The mid-Atlantic ridge is a classic, well-developed example. Motion along a mid-oceanic ridge causes small to moderate earthquakes.

Although most divergent boundaries present today are located within oceanic plates, a divergent boundary typically

TABLE 1.1	Three Types of Plate Boundaries	
Boundary	**What Takes Place**	**Result**
Divergent	Plates move apart	Creation of new ocean floor with submarine volcanoes; mid-oceanic ridge; small to moderate earthquakes
Convergent	Plates move toward each other	Destruction of ocean floor; creation and growth of mountain range with volcanoes; subduction zone; Earth's greatest earthquakes and tsunamis
Transform	Plates move sideways past each other	No creation or destruction of crust; small to large earthquakes

initiates within a continent. It begins when a split, or *rift*, in the continent is caused either by extensional (stretching) forces within the continent or by the upwelling of hot asthenosphere from the mantle below (figure 1.9A). Either way, the continental plate pulls apart and thins. Initially, a narrow valley is formed. Fissures extend into a magma chamber. **Magma** (molten rock) flows into the fissures and may erupt onto the floor of the rift. With continued separation, the valley deepens, the crust beneath the valley sinks, and a narrow sea floor is formed (Figure 1.9B). The new ocean floor is created from the solidification of magma in the fissures and eruption on the ocean floor. Rock that forms when magma solidifies is **igneous rock.** The igneous rock that solidifies on the sea floor and in the fissures becomes *oceanic crust.* As the two sides of the split continent continue to move apart, new fissures develop, magma fills them, and more oceanic crust is formed. As the ocean basin widens, the central zone where new crust is created remains relatively high. This is the mid-oceanic ridge that will remain as the divergent boundary as the continents continue to move apart and the ocean basin widens (figure 1.9C).

A mid-oceanic ridge is higher than the deep ocean floor (figure 1.9C) because the rocks, being hotter at the ridge, are less dense. A *rift valley,* bounded by tensional cracks, runs along the crest of the ridge. The magma in the chamber below the ridge that squeezes into fissures comes from partial melting of the underlying asthenosphere. Continued pulling apart of the ridge crest develops new cracks, and the process of filling and cracking continues indefinitely. Thus, new oceanic crust is continuously created at a divergent boundary. All of the mantle material does *not* melt—a solid residue remains under the newly created crust. New crust and underlying solid mantle make up the lithosphere that moves away from the ridge crest, traveling like the top of a conveyor belt. The rate of motion is generally 1 to 18 centimeters per year (approximately the growth rate of a fingernail), slow in human terms but quite fast by geologic standards.

The top of a plate may be composed exclusively of oceanic crust or might include a continent or part of a continent. For example, if you live on the North American plate, you are riding westward relative to Europe because the plate's divergent boundary is along the mid-oceanic ridge in the North Atlantic Ocean (figure 1.8). The western half of the North Atlantic sea floor and North America are moving together in a westerly direction away from the mid-Atlantic ridge plate boundary.

Convergent Boundaries

The second type of boundary, one resulting in a wide range of geologic activities, is a **convergent boundary,** wherein plates move toward each other (figure 1.10). By accommodating the addition of new sea floor at divergent boundaries, the destruction of old sea floor at convergent boundaries ensures the Earth does not grow in size. Examples of convergent boundaries include the Andes mountain range, where the Nazca plate is subducting beneath the South American plate, and the Cascade Range of Washington, Oregon, and northern California, where the Juan de Fuca plate is subducting beneath the North American plate. Convergent boundaries, due to their geometry, are the sites of the largest earthquakes on Earth.

It is useful to describe convergent boundaries by the character of the plates that are involved: **ocean-continent, ocean-ocean,** and **continent-continent.** The difference in density of oceanic and continental rock explains the contrasting geological activities caused by their convergence.

Ocean-Continent Convergence

If one plate is capped by oceanic crust and the other by continental crust, the less-dense, more-buoyant continental plate will override the denser, oceanic plate (figure 1.10). The oceanic plate bends beneath the continental plate and sinks along what is known as a **subduction zone,** a zone where an oceanic plate descends into the mantle beneath an overriding plate. Deep *oceanic trenches* are found where oceanic lithosphere bends and begins its descent. These narrow, linear troughs are the deepest parts of the ocean floor.

In the region where the top of the subducting plate slides beneath the asthenosphere, melting takes place and magma is created. Magma is less dense than the overlying solid rock. Therefore, the magma created along the subduction zone works its way upward and either erupts at volcanoes on the Earth's surface to solidify as *extrusive* igneous rock, or solidifies within the crust to become *intrusive* igneous rock. Hot rock, under high pressure, near the subduction zone that does not melt may change in the solid state to a new rock—**metamorphic rock.**

Near the edge of the continent, above the rising magma from the subduction zone, a major mountain belt, such as the Andes or Cascades, forms. The mountain belt grows due to the volcanic activity at the surface, the emplacement of bodies of

45   90   135   180

60

Eurasian plate

BERING SEA

OKHOTSK SEA

Himalayas

Arabian plate

30

Philippine Sea plate

African plate

0

Pacific plate

Indian-Australian plate

30

Antarctic plate

60

45   90   135   180

Transform boundary

FIGURE 1.8

Plates of the world and the three types of plate boundaries. Arrows indicate direction of plate motion.

16

135　　　　　　　90　　　　　　　45　　　　　　　0

Eurasian
plate

60

Juan
de
Fuca
plate

San Andreas
fault

North American plate

Pacific
plate

Caribbean
plate

African
plate

30

Cocos plate

Mid - Atlantic

South
American
plate

0

Pacific
Rise

Nazca plate

Andes
Mountains

Peru-Chile
Trench

Ridge

30

East

Pacific
plate

Scotia plate

60

Antarctic plate

135　　　　　　　45

Convergent
boundary

Divergent
boundary

17

intrusive igneous rock at depth, and intense compression caused by plate convergence. Layered sedimentary rock that may have formed on an ocean floor especially shows the effect of intense squeezing (for instance, the "folded and faulted sedimentary rocks" shown on figure 1.10). In this manner, rock that may have been below sea level might be squeezed upward to become part of a mountain range.

Ocean-Ocean Convergence

If both converging plates are oceanic, the denser plate will subduct beneath the less-dense plate (figure 1.11). A portion of a plate becomes colder and denser as it travels farther from the

mid-oceanic ridge where it formed. After subduction begins, molten rock is produced just as it is in an ocean-continent subduction zone; however, in this case, the rising magma forms volcanoes that grow from an ocean floor rather than on a continent. The resulting mountain belt is called a *volcanic island arc*. Examples include the Aleutian Islands in Alaska and the islands of Indonesia and Sumatra, the site of the great earthquake that caused the devastating tsunami of 2004, described earlier.

Continent-Continent Convergence

If both converging plates are continental, a quite different geologic deformation process takes place at the plate boundary. Continental lithosphere is much less dense than the mantle below and, therefore, neither plate subducts. The buoyant nature of continental lithosphere causes the two colliding continental plates to buckle and deform with significant vertical uplift and thickening as well as lateral shortening. A spectacular example of continent-continent collision is the Himalayan mountain belt. The tallest peaks on Earth are located here and they continue to grow in height due to continued collision of the Indian continent with the continental Eurasian plate.

Continent-continent convergence is preceded by oceanic-continental convergence (figure 1.12). An ocean basin between two continents closes because oceanic lithosphere is subducted beneath one of the continents. When the continents collide, one becomes wedged beneath the other. India collided with Asia around 40 million years ago, yet the forces that propelled them together are still in effect. The rocks continue to be deformed and squeezed into higher mountains.

A–Continent undergoes extension. The crust is thinned and a rift valley forms.

B–Continent tears in two. Continent edges are faulted and uplifted. Basalt eruptions form oceanic crust.

C–Continental sediments blanket the subsiding margins to form continental shelves. The ocean widens and a mid-oceanic ridge develops, as in the Atlantic Ocean.

FIGURE 1.9

A divergent boundary begins as a continent is pulled apart. As separation of continental crust proceeds, oceanic crust develops and an initially narrow sea floor grows larger in time.

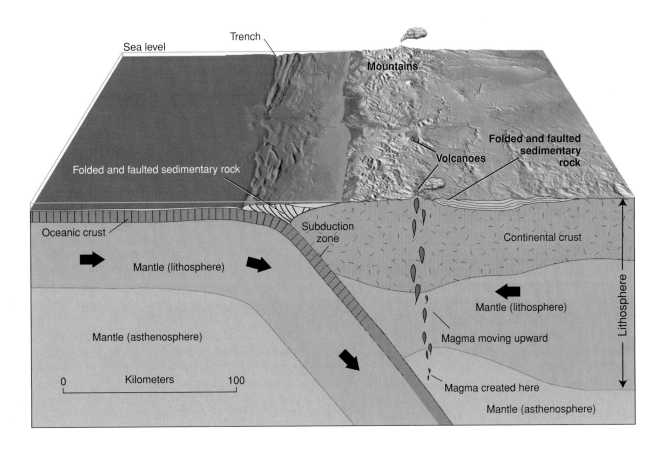

FIGURE 1.10

Block diagram of an ocean-continent convergent boundary. Oceanic lithosphere moves from left to right and is subducted beneath the overriding continental lithosphere. Magma is created by partial melting of the asthenosphere.

FIGURE 1.11

A volcanic island arc forms as a result of oceanic-oceanic plate convergence.

Transform Boundaries

The third type of boundary, a **transform boundary** (figure 1.13), occurs where two plates slide horizontally past each other, neither toward nor away from each other. The San Andreas fault in California and the Alpine fault of New Zealand are two examples of this type of boundary. Earthquakes resulting from motion along transform faults vary in size depending on whether the fault cuts through oceanic or continental crust and on the length of the fault. The San Andreas transform fault has generated large earthquakes, but the more numerous and much shorter transform faults within ocean basins generate much smaller earthquakes.

The significance of transform faults was first recognized in ocean basins. Here they occur as fractures perpendicular to offset mid-oceanic ridges (figure 1.8). As

Trench

Accretionary wedge (fine-grained sediments scraped off oceanic crust)

Ocean becomes narrower

Continental crust

Upper-mantle lithosphere

A Ocean-continent convergence

Young mountain belt (Himalaya)

Tibetan Plateau

Foreland basin

(India)

Mt. Everest

Suture zone

Indian Continental crust

Asian continental crust

Upper-mantle lithosphere

Asthenosphere

100 km

Thrust faults

B Continent-continent collision

(Surface vertical scale exaggerated 8x)

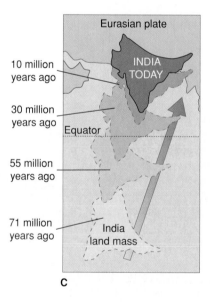

Eurasian plate

INDIA TODAY

10 million years ago

30 million years ago

Equator

55 million years ago

71 million years ago

India land mass

C

FIGURE 1.12

Continent-continent convergence is preceded by the closing of an ocean basin while ocean-continent convergence takes place. *C* shows the position of India relative to the Eurasian plate in time. The convergence of the two plates created the Himalaya. Some of the features shown accretionary wedge, foreland basin are described in chapters 6 and 19.

FIGURE 1.13

Transform faults (transform boundaries between plates) are the segments of the fractures between offset ridge crests. Oceanic crust is created at the ridge crests and moves away from the crest as indicated by the heavy arrows. The pairs of small arrows indicate motion on adjacent sides of fractures. Earthquakes take place along the transform fault because rocks are moving in opposite directions. The fractures extend beyond the ridges, but here the two segments of crust are moving in the same direction and rate and there are no earthquakes—these are not part of transform faults.

shown in figure 1.13, the motion on either side of a transform fault is a result of rock that is created at and moving away from each of the displaced oceanic ridges. Although most transform faults are found along mid-oceanic ridges, occasionally a transform fault cuts through a continental plate. Such is the case with the San Andreas fault, which is a boundary between the North American and the Pacific plates.

Box 1.4 outlines how plate tectonic theory was developed through the *scientific method.* If you do not have a thorough understanding of how the scientific method works, be sure to study the box.

Surficial Processes: The Earth's External Heat Engine

Tectonic forces can squeeze formerly low-lying continental crustal rock along a convergent boundary and raise the upper part well above sea level. Portions of the crust also can rise because of **isostatic adjustment,** vertical movement of sections of Earth's crust to achieve balance. That is to say, lighter rock will "float" higher than denser rock on the underlying mantle. Isostatic adjustment is why an empty ship is higher above water than an identical one that is full of cargo. Continental crust, which is less dense than oceanic crust, will tend to float higher over the underlying mantle than oceanic crust (which is why the oceanic crust is below sea level and the continents are above sea level). After a portion of the continental crust is pulled downward by tectonic forces, it is out of isostatic balance. It will then rise slowly due to isostatic adjustment when tectonic forces are relaxed.

When a portion of crust rises above sea level, rocks are exposed to the atmosphere. Earth's external heat engine, driven by solar power, comes into play. Circulation of the atmosphere and hydrosphere is mainly driven by solar power. Our weather is largely a product of the solar heat engine. For instance, hot air rises near the equator and sinks in cooler zones to the north and south. Solar heating of air creates wind; ocean waves are, in turn, produced by wind. When moist air cools, it rains or snows. Rainfall on hillsides flows down slopes and into streams. Streams flow to lakes or seas. Glaciers grow where there is abundant snowfall at colder, high elevations and flow downhill because of gravity.

Where moving water, ice, or wind loosens and removes material, **erosion** is taking place. Streams flowing toward oceans remove some of the land over which they run. Crashing waves carve back a coastline. Glaciers grind and carry away underlying rock as they move. In each case, rock originally brought up by the Earth's internal processes is worn down by surficial processes (figure 1.14). As material is removed through erosion, isostasy works to move the landmass upward, just as part of the submerged portion of an iceberg floats upward as ice melts. Or, going back to our ship analogy, as cargo is unloaded, the ship rises in the water.

Rocks formed at high temperature and under high pressure deep within the Earth and pushed upward by isostatic and tectonic forces are unstable in their new environment. Air and water tend to cause the once deep-seated rocks to break down and form new materials. The new materials, stable under conditions at the Earth's surface, are said to be in **equilibrium**—that is, adjusted to the physical and chemical conditions of their environment so that they do not change or alter with time. For example, much of an igneous rock (such as granite) that formed at a high temperature tends to break down chemically to clay. Clay is in equilibrium—that is to say it is stable—at the Earth's surface.

IN GREATER DEPTH 1.4

Plate Tectonics and the Scientific Method

Although the hypothesis was proposed only a few decades ago, plate tectonics has been so widely accepted and disseminated that most people have at least a rough idea of what it is about. Most nonscientists can understand the television and newspaper reports (and occasional comic strip, such as that in box figure 1) that include plate tectonics in reports on earthquakes and volcanoes. Our description of plate tectonics implies little doubt about the existence of the process. The theory of plate tectonics has been accepted as scientifically verified by geologists. Plate tectonic theory, like all knowledge gained by science, has evolved through the processes of the **scientific method.** We will illustrate the scientific method by showing how plate tectonics has evolved from a vague idea into a theory that is so likely to be true that it can be regarded as "fact."

The basis for the scientific method is the belief that the universe is orderly and that by *objectively* analyzing phenomena, we can discover their workings. Science is a deeply human endeavor that involves creativity. A scientist's mind searches for connections and thinks of solutions to problems that might not have been considered by others. At the same time, a scientist must be aware of what work has been done by others, so that science can build on those works. Here, the scientific method is presented as a series of steps. A scientist is aware that his or her work must satisfy the requirements of the steps but does not ordinarily go through a formal checklist.

1. A question is raised or a problem is presented.

2. Available information pertinent to the question or problem is analyzed. Facts, which scientists call **data,** are gathered.

3. After the data have been analyzed, tentative explanations or solutions that are consistent with the observed data, called **hypotheses,** are proposed.

4. One predicts what would occur in given situations if a hypothesis were correct.

5. Predictions are tested. Incorrect hypotheses are discarded.

6. A hypothesis that passes the testing becomes a **theory,** which is regarded as having an excellent chance of being true. In science, however, nothing is considered proven

absolutely. All scientifically derived knowledge is subject to being proven false. (Can you imagine what could prove that atoms and molecules don't exist?) A thoroughly and rigorously tested theory becomes, for all intents and purposes, a fact, even though scientists still call it a theory (e.g., atomic theory).

Like any human endeavor, the scientific method is not infallible. Objectivity is needed throughout. Someone can easily become attached to the hypothesis he or she has created and so tend subconsciously to find only supporting evidence. As in a court of law, every effort is made to have observers objectively examine the logic of both procedures and conclusions. Courts sometimes make wrong decisions; science, likewise, is not immune to error.

The following outline shows how the concept of plate tectonics evolved:

Step 1: A question asked or problem raised. Actually, a number of questions were being asked about seemingly unrelated geological phenomena.

What caused the submarine ridge that extends through most of the oceans of the world? Why are rocks in mountain belts intensely deformed? What sets off earthquakes? What causes rock to melt underground and erupt as volcanoes? Why are most of the active volcanoes of the world located in a ring around the Pacific Ocean?

Step 2: Gathering of data. Early in the twentieth century, the amount of data was limited. But through the decades, the information gathered increased enormously. New data, most notably information gained from exploration of the sea floor in the mid-1900s, forced scientists to discard old hypotheses and come up with new ideas.

Step 3: Hypotheses proposed. Most of the questions being asked were treated as separate problems wanting separate hypotheses. Some appeared interrelated. One hypothesis, **continental drift,** did address several questions. It was advocated by Alfred Wegener, a German scientist, in a book published in the early 1900s.

Wegener postulated that the continents were all once part of a single supercontinent called Pangaea. The hypothesis explained why the coastlines of Africa and South America look like separated parts of a jigsaw puzzle. Some 200 million years ago, this supercontinent broke up, and the various continents slowly drifted into their present positions. The hypothesis suggested that the rock within mountain belts becomes deformed as the leading edge of a continental crust moves against and over the stationary oceanic crust. Earthquakes were presumably caused by continuing movement of the continents.

Until the 1960s, continental drift was not widely accepted. It was scoffed at by many geologists who couldn't conceive of how a continent could be plowing over oceanic crust. During the 1960s, after new data on the nature of the sea floor

FRANK & ERNEST® by Bob Thaves

BOX 1.4 ■ FIGURE 1

Plate tectonics sometimes show up in comic strips. *FRANK & ERNEST reprinted by permission of Newspaper Enterprise Association, Inc.*

became available, the idea of continental drift was incorporated into the concept of plate tectonics. What was added in the plate tectonic hypothesis was the idea that oceanic crust, as well as continental crust, was shifting.

Step 4: Prediction. An obvious prediction, if plate tectonics is correct, is that if Europe and North America are moving away from each other, the distance measured between the two continents is greater from one year to the next. But we cannot stretch a tape measure across oceans, and, until recently, we have not had the technology to accurately measure distances between continents. So, in the 1960s, other testable predictions had to be made. Some of these predictions and results of their testing are described in the chapter on plate tectonics. One of these predictions was that the rocks of the oceanic crust will be progressively older the farther they are from the crest of a midoceanic ridge.

Step 5: Predictions are tested. Experiments were conducted in which holes were drilled in the deep-sea floor from a specially designed ship. Rocks and sediment were collected from these holes, and the ages of these materials were determined. As the hypothesis predicted, the youngest sea floor (generally less than a million years old) is near the mid-oceanic ridges, whereas the oldest sea floor (up to about 200 million years old) is farthest from the ridges (box figure 2).

This test was only one of a series. Various other tests, described in some detail later in this book, tended to confirm the hypothesis of plate tectonics. Some tests did not work out exactly as predicted. Because of this, and more detailed study of data, the original concept was, and continues to be, modified. The basic premise, however, is generally regarded as valid.

Step 6: The hypothesis becomes a theory. Most geologists in the world considered the results of this and other tests as positive, indicating that the concept is not reasonably disputable and very probably true. It then became the plate tectonic theory.

During the last few years, plate tectonic theory has been further confirmed by the results of very accurate satellite surveys that determine where points on separate continents are relative to one another. The results indicate that the continents are indeed moving relative to one another. Europe and North America *are* moving farther apart.

Although it is unlikely that plate tectonic theory will be replaced by something we haven't thought of yet, aspects that fall under plate tectonics' umbrella (for instance, exactly how does magma form at a convergent plate boundary?) continue to be analyzed and revised as new data become available.

Important Note

Words used by scientists do not always have the same meaning when used by the general public. A case in point is the word *theory*. To most people, a "theory" is what scientists regard as a "hypothesis." You may remember news reports about an airliner that exploded offshore from New York in 1996. A typical statement on television was: "One theory is that a bomb in the plane exploded; a second theory is that the plane was shot down by a missile fired from a ship at sea; a third theory is that a spark ignited in a fuel tank and the plane exploded." Clearly, each "theory" is a hypothesis in the scientific sense of the word. This has led to considerable confusion for nonscientists about science. You have probably heard the expression, "It's just a theory." Statements such as, "Evolution is just a theory," are used to imply that scientific support is weak. The reality is that theories such as evolution and plate tectonics have been so overwhelmingly verified that they come as close as possible to what scientists accept as being indisputable facts. They would, in laypersons' terms, be "proven."

BOX 1.4 ■ FIGURE 2
Ages of rocks from holes drilled into the oceanic crust. (Vertical scale of diagram is exaggerated).

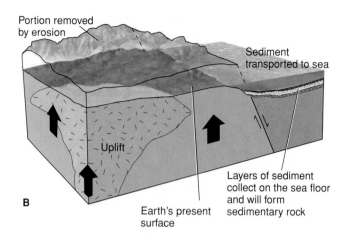

FIGURE 1.14

Erosion, deposition, and uplift. (A) Magma has solidified deep underground to become igneous rock. (B) As the surface erodes, sediment is transported to the sea to become sedimentary rock. Isostatic adjustment causes uplift of the continent. Erosion and uplift expose the igneous rock at the surface.

The product of the breakdown of rock is **sediment,** loose material. Sediment may be transported by an agent of erosion, such as running water in a stream. Sediment is deposited when the transporting agent loses its carrying power. For example, when a river slows down as it meets the sea, the sand being transported by the stream is deposited as a layer of sediment.

In time, a layer of sediment deposited on the sea floor becomes buried under another layer. This process may continue, burying our original layer progressively deeper. The pressure from overlying layers compresses the sediment, helping to consolidate the loose material. With the cementation of the loose particles, the sediment becomes *lithified* (cemented or otherwise consolidated) into a **sedimentary rock.** Sedimentary rock that becomes deeply buried in the Earth may later be transformed by heat and pressure into metamorphic rock.

GEOLOGIC TIME

We have mentioned the great amount of time required for geologic processes. As humans, we think in units of time related to personal experience—seconds, hours, years, a human lifetime. It stretches our imagination to contemplate ancient history that involves 1,000 or 2,000 years. Geology involves vastly greater amounts of time, often referred to as *deep time.*

To be sure, some geological processes occur quickly, such as a great landslide or a volcanic eruption. These events occur when stored energy (like the energy stored in a stretched rubber band) is suddenly released. Most geological processes, however, are slow but relentless, reflecting the pace at which the heat engines work. It is unlikely that a hill will visibly change in shape or height during your lifetime (unless through

human activity). However, in a geologic time frame, the hill probably is eroding away quite rapidly. "Rapidly" to a geologist may mean that within a few million years, the hill will be reduced nearly to a plain. Similarly, in the geologically "recent" past of several million years ago, a sea may have existed where the hill is now. Some processes are regarded by geologists as "fast" if they are begun and completed within a million years.

The rate of plate motion is relatively fast. If new magma erupts and solidifies along a mid-oceanic ridge, we can easily calculate how long it will take that igneous rock to move 1,000 kilometers away from the spreading center. At the rate of 1 centimeter per year, it will take 100 million years for the presently forming part of the crust to travel the 1,000 kilometers.

Although we will discuss geologic time in detail in chapter 8, table 1.2 shows some reference points to keep in mind. The Earth is estimated to be about 4.55 (usually rounded to 4.5 or 4.6) billion years old (4,550,000,000 years). Fossils in rocks indicate that complex forms of animal life have existed in abundance on Earth for about the past 544 million years. Reptiles became abundant about 230 million years ago. Dinosaurs evolved from reptiles and became extinct about 65 million years ago. Humans have been here only about the last 3 million years. The eras and periods shown in table 1.2 comprise a kind of calendar for geologists into which geologic events are placed (as explained in the chapter on geologic time).

Not only are the immense spans of geologic time difficult to comprehend, but very slow processes are impossible to duplicate. A geologist who wants to study a certain process cannot repeat in a few hours a chemical reaction that takes a million years to occur in nature. As Mark Twain wrote in *Life on the Mississippi,* "Nothing hurries geology."

TABLE 1.2	Some Important Ages in the Development of Life on Earth

Millions of Years before Present	Noteworthy Life	Eras	Periods
4	Earliest hominids	Cenozoic	Quaternary Tertiary
65	First important mammals Extinction of dinosaurs		
	First dinosaurs	Mesozoic	Cretaceous Jurassic Triassic
251			
300	First reptiles	Paleozoic	Permian Pennsylvanian Mississippian Devonian Silurian Ordovician Cambrian
400	Fishes become abundant		
544	First abundant fossils		
600	Some complex, soft-bodied life	Precambrian	(The Precambrian accounts for the vast majority of geologic time.)
3,500	Earliest single-celled fossils		
4,550	Origin of the Earth		

SUMMARY

Geology is the scientific study of Earth. We benefit from geology in several ways: (1) We need geology to find and maintain a supply of minable commodities and sources of energy; (2) Geology helps protect the environment; (3) Applying knowledge about geologic hazards (such as volcanoes, earthquakes, tsunamis, landslides) saves lives and property; and (4) We have a greater appreciation of rocks and landforms through understanding how they form.

Earth systems are the atmosphere, the hydrosphere, the biosphere, and the geosphere (or solid Earth system). The Earth system is part of the solar system.

Geological investigations indicate that Earth is changing because of internal and surficial processes. Internal processes are driven mostly by temperature differences within Earth's mantle.

Surficial processes are driven by solar energy. Internal forces cause the crust of Earth to move. Plate tectonic theory visualizes the lithosphere (the crust and uppermost mantle) as broken into plates that move relative to each other over the asthenosphere. The plates are moving *away* from divergent boundaries usually located at the crests of mid-oceanic ridges where new crust is being created. Divergent boundaries can develop in a continent and split the continent. Plates move *toward* convergent boundaries. In ocean-continent convergence, lithosphere with oceanic crust is subducted under lithosphere with continental crust. Ocean-ocean convergence involves subduction in which both plates have oceanic crust and the creation of a volcanic island arc. Continent-continent convergence takes place when two continents collide. Plates slide past one another at transform boundaries.

Plate tectonics and isostatic adjustment cause parts of the crust to move up or down.

Erosion takes place at Earth's surface where rocks are exposed to air and water. Rocks that formed under high pressure and temperature inside Earth are out of equilibrium at the surface and tend to alter to substances that are stable at the surface. Sediment is transported to a lower elevation, where it is deposited (commonly on a sea floor in layers). When sediment is cemented, it becomes sedimentary rock.

Although Earth is changing constantly, the rates of change are generally extremely slow by human standards.

Terms to Remember

asthenosphere 13
atmosphere 11
biosphere 11
continental drift 22
continent-continent convergence 15
convergent boundary 15
core 13
crust 13
data 22
divergent boundary 14
Earth system 11
equilibrium 21

erosion 21
geology 4
geosphere (solid Earth system) 11
hydrosphere 11
hypothesis 22
igneous rock 15
isostatic adjustment 21
lithosphere 13
magma 15
mantle 13
metamorphic rock 15
mid-oceanic ridge 14

ocean-continent convergence 15
ocean-ocean convergence 15
plate tectonics 14
scientific method 22
sediment 24
sedimentary rock 24
subduction zone 15
tectonic forces 13
theory 22
transform boundary 19

Testing Your Knowledge

Use the following questions to prepare for exams based on this chapter.

1. What is meant by *equilibrium?* What happens when rocks are forced out of equilibrium?

2. What tectonic plate are you presently on? Where is the nearest plate boundary, and what kind of boundary is it?

3. What is the most likely geologic hazard in your part of the country?

4. What are the three major types of rocks?

5. What are the relationships among the mantle, the crust, the asthenosphere, and the lithosphere?

6. What would the surface of Earth be like if there were no tectonic activity?

7. Explain why cavemen never saw a dinosaur.

8. Plate tectonics is a result of Earth's internal heat engine, powered by (choose all that apply)

   a. the Sun   b. gravity

   c. heat flowing from Earth's interior outward

9. A typical rate of plate motion is

   a. 3–4 meters per year   b. 1 kilometer per year

   c. 1–10 centimeters per year   d. 1,000 kilometers per year

10. Volcanic island arcs are associated with

    a. transform boundaries   b. divergent boundaries

    c. ocean-continent convergence   d. ocean-ocean convergence

11. The division of geology concerned with Earth materials, changes in the surface and interior of the Earth, and the dynamic forces that cause those changes is

    a. physical geology   b. historical geology

    c. geophysics   d. paleontology

12. Which is a geologic hazard?

    a. earthquake   b. volcano

    c. mudflows   d. floods

    e. wave erosion at coastlines   f. landslides

    g. all of the preceding

13. The largest zone of Earth's interior by volume is the

    a. crust   b. mantle

    c. outer core   d. inner core

14. Oceanic and continental crust differ in

    a. composition   b. density

    c. thickness   d. all of the preceding

15. The forces generated inside Earth that cause deformation of rock as well as vertical and horizontal movement of portions of Earth's crust are called

    a. erosional forces   b. gravitational forces

    c. tectonic forces   d. all of the preceding

16. Plate tectonics is a
    a. conjecture   b. opinion
    c. hypothesis   d. theory

17. Which is the type of a plate boundary?
    a. divergent   b. transform
    c. convergent   d. all of the preceding

18. The lithosphere is
    a. the same as the crust   b. the layer beneath the crust
    c. the crust and uppermost mantle   d. only part of the mantle

19. Erosion is a result of Earth's external heat engine, powered by (choose all that apply)
    a. the Sun   b. gravity
    c. heat flowing from Earth's interior outward

Expanding Your Knowledge

1. Why are some parts of the lower mantle hotter than other parts?
2. According to plate tectonic theory, where are crustal rocks created? Why doesn't Earth keep getting larger if rock is continually created?
3. What percentage of geologic time is accounted for by the last century?
4. What would Earth be like without solar heating?
5. What are some of the technical difficulties you would expect to encounter if you tried to drill a hole to the center of Earth?

Exploring Web Resources

www.mhhe.com/plummer11e

This is the dedicated website for this book. You can go to it for new and updated information. The universal resource locators (URL) listed in this book are also given as links on the website, making it easy to go to those websites without typing in the URL. When you visit our Information Center, go to the Student Edition and then to the chapter of interest. Here you will find additional readings and media resources, as well as answers to the Testing Your Knowledge section, more quizzes, animation and video clips, and direct links to the sites listed in this book. Links to additional websites can also be found. We have added questions for some of the sites to allow you to get the most of your exploration of the web. Using the web is an enjoyable way of enhancing your knowledge of geology.

http://pubs.usgs.gov/publications/text/dynamic.html

This Dynamic Earth by the U.S. Geological Survey is an online, illustrated publication explaining plate tectonics. You may want to go to the section "Understanding plate motion." This will help reinforce what you read about plate tectonics in this chapter. It goes into plate tectonics in greater depth, however, covering material that is in chapter 19 of this textbook.

www.uh.edu/~jbutler/anon/anontrips.html

Virtual Field Trips. The site provides access to geologic sites throughout the world. Many are field trips taken by geology classes. Check the alphabetical listing and see if there are any sites near you. Or watch a video clip in one of the Quick Time field trips.

www.usgs.gov

The *U.S. Geological Survey's* home page. Use this as a gateway to a wide range of geologic information.

www.nrcan.gc.ca/gsc/

The *Geological Survey of Canada* home page.

 # Animation

This chapter includes the following animations available on our Online Learning Center at **www.mhhe.com/plummer11e**.

1.9 Divergence of plates at mid-ocean ridge

2

Atoms, Elements, and Minerals

This chapter is the first of six on the material of which Earth is made. The following chapters are mostly about rocks. Nearly all rocks are made of minerals. Therefore, to be ready to learn about rocks, you must first understand what minerals are as well as the characteristics of some of the most common minerals.

In this chapter, you are introduced to some basic principles of chemistry (this is for those of you who have not had a chemistry course). This will help you understand material covered in the chapters on rocks, weathering, and the composition of Earth's crust and its interior. You will discover that each mineral is composed of specific chemical elements, the atoms of which are in a remarkably orderly arrangement. A mineral's chemistry

Crystals of tourmaline (variety: elbaite). Differences in color within each crystal are due to small changes in chemical composition incorporated into the minerals as they grew. *Photo © Parvinder Sethi*

29

and the architecture of its internal structure determine the physical properties used to distinguish it from other minerals. You should learn how to readily determine physical properties and use them to identify common minerals. (Appendix A is a further guide to identifying minerals.)

 ## Relationships to Earth Systems

Minerals are part of the *geosphere* (the solid Earth system). However, many minerals form through interaction with other components of Earth systems (described in chapter 1). Some minerals form in water—the *hydrosphere.* For example, *calcite* (the mineral that makes up the common rock *limestone*) forms when calcium and carbon dioxide are precipitated from seawater. Calcite can also be formed by organisms (the *bio-sphere*) creating shells or other hard parts. Coral, clams, and oysters create hard parts of calcite derived from seawater. Some minerals form from interaction between the *atmosphere* and the hydrosphere. *Halite,* which we know as table salt, forms when salty water is evaporated. Minerals can also be lost to or changed by the atmosphere and hydrosphere. Halite will dissolve when immersed in fresh water. Clay minerals form when water, with dissolved atmospheric gases, reacts with other minerals. A newly formed clay mineral has water incorporated into its crystal structure. Humans (part of the biosphere) are prodigious users of minerals. Most of what we make or use depends on minerals. We make bricks out of clay. Our jewelry may be made from gold as well as gems such as diamonds and emeralds. Steel is made from iron-rich and other metal-bearing minerals.

MINERALS

Introduction

A quick glance at a rock may show some color or pattern, perhaps a speckled appearance that may be attractive to a collector, but otherwise holds no meaning for most people. But these colors and patterns, made up of packages of matter called *minerals,* tell a very important story about the origin of our world, and indeed about all Earth-like planets. The considerable information conveyed by minerals enriches our appreciation for nature, and perhaps gives us more reason to take good care of it.

Minerals are compositionally and physically distinctive substances (figure 2.1). Each mineral type develops in a particular way under natural conditions, guided by the principles of chemistry. Factors such as heat, pressure, oxygen, available atoms, and acid content all play a role in determining how minerals form.

There are about 4,500 kinds of minerals in the world, with only a couple hundred that are really common and only a couple dozen that form the majority of all rocks. Each type of mineral is distinguished by a combination of properties, some of which we can see with the unaided eye, others that are discernable only at the microscopic and atomic levels. Examples of these properties include color, luster, hardness, chemical composition, and the transmission of light under a microscope. Minerals are so important and so easily distinguishable that geologists use them as the basis for classifying almost all rocks.

What most people call *crystals* essentially are just "perfectly formed" minerals. They are mineral specimens whose surfaces consist of faces, edges, and corners that show beautiful geometrical symmetry. For example, the quartz you can see lining the interior of a geode (a natural pocket in certain kinds of volcanic rocks; figure 2.2) forms crystals whose smooth faces sparkle brightly in reflected light. Most mineral grains in rocks do not display the faces that we associate with crystals.

Yet, geologists often call them *crystals* anyway because they have an orderly arrangement of their constituent atoms. It would be less ambiguous to say they are *crystalline* substances. The irregular outlines of most mineral grains seen in rocks results from the fact that these minerals form simultaneously under conditions of close confinement, or grow to fill the gaps left between earlier-forming minerals.

The existence of crystals reflects the fact that the arrangements of atoms within *all* minerals are *orderly and regular.* Detailed X-ray study of minerals shows that there are 230 different kinds of symmetrical atomic arrangements possible in nature and each mineral type exhibits one or several of these 230 arrangements. Minerals are said to be *crystalline* because of this universal internal property. A **crystalline** substance is one in which the atoms are arranged in a three-dimensional, regularly repeating, orderly pattern. The print by M. C. Escher (figure 2.3*A*) vividly expresses what crystallinity is about. You can visualize what crystallinity is in nature by mentally substituting identical clusters of atoms for each fish and imagining the clusters packed together. Figure 2.3*B* is a model of the crystal structure of one mineral.

Putting it all together, then, **minerals** can be defined as a family of naturally occurring, crystalline substances that are physically and chemically distinctive. They are the "building blocks" of rocks. Minerals also are compositionally inorganic; that is, they don't consist of carbon-hydrogen molecules that also form crystalline substances through biological processes (sugars, for example).

When vitamin advertisers and nutritional specialists talk about "minerals," they are not, of course, referring to the strictly geologic definition, but rather to single elements, such as calcium or magnesium, that have certain dietary benefits. Commercial processing of true minerals yields these popular ingredients. In reality, most true minerals are complex assemblages of multiple elements.

The chemical formula of a mineral represents not only the types of atoms in the mineral, but their relative proportions as

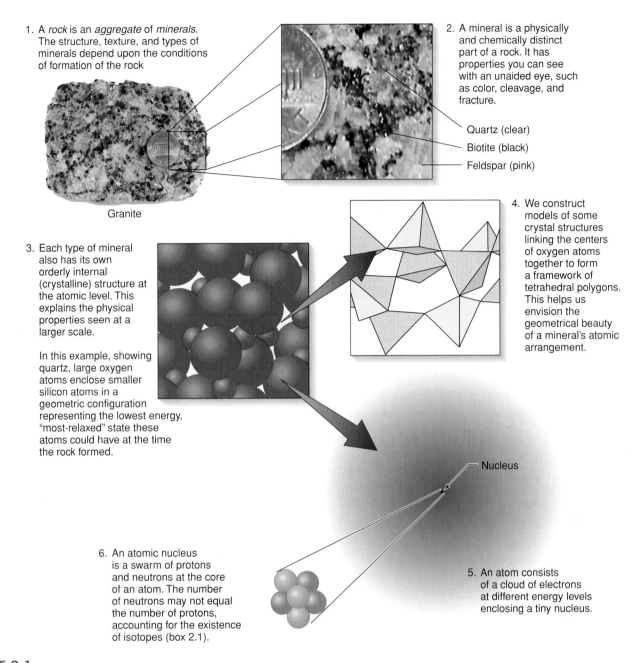

1. A *rock* is an *aggregate* of *minerals*. The structure, texture, and types of minerals depend upon the conditions of formation of the rock

Granite

2. A mineral is a physically and chemically distinct part of a rock. It has properties you can see with an unaided eye, such as color, cleavage, and fracture.

Quartz (clear)

Biotite (black)

Feldspar (pink)

3. Each type of mineral also has its own orderly internal (crystalline) structure at the atomic level. This explains the physical properties seen at a larger scale.

In this example, showing quartz, large oxygen atoms enclose smaller silicon atoms in a geometric configuration representing the lowest energy, "most-relaxed" state these atoms could have at the time the rock formed.

4. We construct models of some crystal structures linking the centers of oxygen atoms together to form a framework of tetrahedral polygons. This helps us envision the geometrical beauty of a mineral's atomic arrangement.

Nucleus

6. An atomic nucleus is a swarm of protons and neutrons at the core of an atom. The number of neutrons may not equal the number of protons, accounting for the existence of isotopes (box 2.1).

5. An atom consists of a cloud of electrons at different energy levels enclosing a tiny nucleus.

FIGURE 2.1

Relationships between granite (a rock), its minerals and their atoms arranged in a crystalline structure. *Photo by C. C. Plummer*

well, expressed in lowest whole number ratios. For example, quartz is made up exclusively of oxygen and silicon atoms. More precisely, quartz contains twice as many oxygen (O) atoms as silicon (Si) atoms. Therefore, the chemical formula for quartz is SiO_2, its specific composition.

Another common mineral is halite (rock salt), whose chemical composition is NaCl. This means that halite is composed of *equal numbers* of sodium (Na) and chlorine (Cl) atoms. These atoms are arranged in an orderly, three-dimensional lattice that resembles a stack of boxes (figure 2.4). This imparts an overall cubic shape to crystals of halite.

Consider the formula of a more common mineral, feldspar—$KAlSi_3O_8$. This reflects not only a more complex composition, but a less symmetric atomic arrangement than that of halite. How do the atoms in a mineral like feldspar stick together? Why are minerals crystalline at all? Science reveals an underlying order to physical reality that is breathtaking and largely hidden from view when we look at the apparent randomness and chaos of the natural world.

FIGURE 2.2

Quartz crystals line the inside of a geode. This purple variety of quartz is known as amethyst. *Photo © Pervinder Sethi*

ATOMS AND ELEMENTS

To answer the questions just posed, we need to take a look at what is happening at an infinitesimally small scale.

Atoms are the smallest, electrically neutral assemblies of energy and matter that we know exist in the universe. It is important to understand what "electrically neutral" means. Many of us have the misfortune of knowing electrical force as a sharp jolt that occurs when we accidentally touch a live wire (or when we touched a wall socket when we were children!). This force results when tiny, charged particles called **electrons** flow from one place to another; for example, along a wire. Physicists say that the electrons carry a negative charge—the electrical force that we exploit to power the world.

Electrons move in directions that allow them to balance out, or "neutralize" their charges. In atoms, electron charges are neutralized as the electrons crowd in around a central core of "positively charged" **protons.** The core, or **nucleus** of the atom, also contains **neutrons,** which are neutrally charged particles of some importance to geology, as you soon will see (figure 2.1).

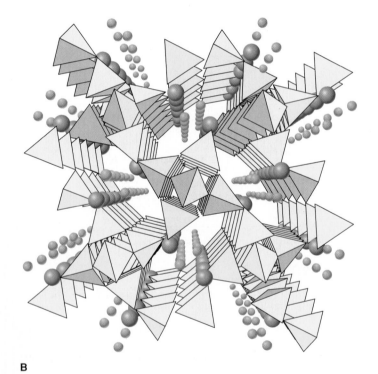

B

A

FIGURE 2.3

(A) *Depth*, print by Dutch artist M. C. Escher. (B) Model of the crystal structure of the mineral natrolite. The small (gray) spheres represent sodium; the large (blue) spheres are water molecules. The "pyramids" are silicon-oxygen tetrahedrons (explained in the text). Figure B from M. Ross, Malcom, Flohr, Marta, J. K., and Ross, Daphne R., 1992, Crystalline Solution Series and order-disorder within the natrolite mineral group. *American Mineralogist 77*, 685–703. Reprinted by permission of the Mineralogical Society of America.

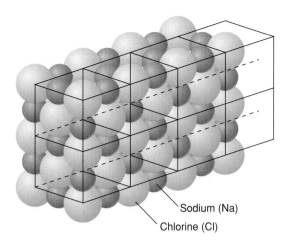

FIGURE 2.4

Model of the atomic structure of halite. The alternating three-dimensional stacking of atoms creates a box-like grid that is expressed in the cubic form of halite crystals seen in hand samples.

A Sodium (Na⁺)

B Chlorine (Cl⁻)

FIGURE 2.5

Diagrammatic representation of (A) sodium and (B) chlorine ions. The dots represent electrons in energy levels within an ion. Sodium has lost the electron that would have made it electrically neutral because a single electron in a higher energy level would be unstable. Chlorine has gained an electron to complete its outer energy level and make it stable.

There are ninety-two different kinds of naturally occurring atoms arranged in order of increasing size and complexity on the periodic table (see appendix D) used by chemists. We call each "species" of atom an *element*. An **element** is defined by the number of protons in its nucleus. For example, oxygen has 8 protons. The number of protons in an atom is that element's *atomic number* (see box 2.1). For example, sodium, potassium, chlorine, and oxygen are all different kinds of elements. In addition to having 8 protons, each atom of oxygen contains 8 electrons and, in its most stable form, 8 neutrons. Chlorine, in contrast, brings together 17 electrons, 17 protons, and 18 neutrons in each atom. Notice that the number of neutrons need not match the numbers of protons and electrons in each atom (see Box 2.1).

The electrons in an atom are continuously on the move, like bees buzzing around a hive. Some are more energetic than others and move farther away from the nucleus as they move in the space around it. Although each electron moves throughout the space surrounding the nucleus, it will spend most its time as part of an *energy level*. (Energy levels used to be shown as concentric spherical shells, but chemists regard that as misleading.) The most stable configuration is to have complete energy levels. The first energy level is complete with 2 electrons. Helium has a complete energy level because it has 2 electrons that balance the 2 protons in its nucleus. The second and third energy levels are each complete with 8 electrons (see figure 2.5). For a more thorough explanation of atomic theory from a chemist's perspective, go to *Understanding Chemistry*, http://www.chemguide.co.uk/atommenu.html#top.

The linking together (**bonding**) of atoms to form minerals largely takes place because most individual atoms in a free state have a deficit or surplus of electrons in their outermost energy levels. They are not fully charge-neutral, in other words, and those with negative net charges are drawn toward those with positive net charges to address this imbalance. An

electrically charged atom is called an **ion.** It is largely the combination of negative and positive ions in a stable, charge-neutral arrangement that makes up the crystalline structure of minerals (but also see box 2.3).

Ions and Crystalline Structures

The most important reason why ions exist is that a typical atom is most stable when each of its energy levels is *completely filled* with electrons—that is, when the electrical energy around the nucleus is compact and concentrated. The innermost energy level in the standard model of an atom is full when it possesses 2 electrons. The second and third orbital energy levels each require 8 electrons for an atom to be nonreactive. (Elements having additional energy levels are more complicated.)

Consider the two ions that bond to form halite—chlorine and sodium. Note that sodium, shown in figure 2.5, has a complete inner energy level with 2 electrons and a second energy level, also filled with 8 electrons. One more electron would neutralize the positive charge of the 11 protons in the nucleus, but an eleventh electron alone in an energy level is unstable, so

Atomic Number, Atomic Mass Number, Isotopes, and Atomic Weight

The number of protons in an atom controls the "behavior" of an element more than does the number of other subatomic particles. The **atomic number** of an element is the number of protons in each atom. As per our earlier definition of an element, each atom of an element has the *same number of protons.*

The **atomic mass number** is the total number of neutrons and protons in an atom. The atomic mass number of the oxygen atom shown in figure 1 is 16 (8 protons plus 8 neutrons) and is indicated by the symbol ^{16}O. Heavier elements have more neutrons and protons than do lighter ones. For example, the heavy element gold has an atomic mass number of 197, whereas helium has an atomic mass number of only 4.

Isotopes of an element are atoms containing different numbers of neutrons but the same number of protons. Isotopes are either stable or unstable. An unstable, or *radioactive, isotope* is one in which protons or neutrons are, over time, spontaneously lost from the nucleus. The subatomic particles that unstable isotopes emit are what Geiger counters detect. This is *radioactivity,* which we know can be hazardous in high doses. Unstable isotopes of uranium and a few other elements are very important to geology because they are used to determine the ages of rocks. These isotopes decay at a known rate and, as described in chapter 8, are used as a kind of geologic stopwatch that starts running at the time some rocks form.

A *stable isotope* is an isotope that will retain all of its protons and neutrons through time. During recent years, stable isotopes have become increasingly important to geology and related sciences. The isotopes most commonly studied in geology are those of carbon, nitrogen, oxygen, sulfur, and hydrogen. Their usefulness in scientific investigations is due to the tendency of isotopes of a given element to partition (distribute preferentially between substances) in different proportions due to their minute weight difference. For instance, oxygen and hydrogen isotopes can be used as a proxy for the surface temperature of the Earth because when water vapor evaporates from liquid water, the vapor will have a slightly higher ratio of lighter to heavier isotopes compared to the isotopes that remain

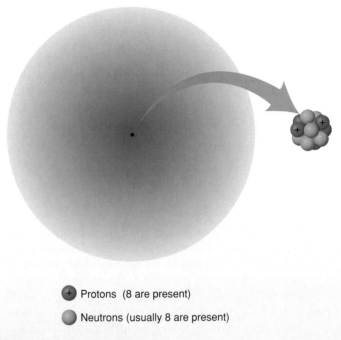

+ Protons (8 are present)

Neutrons (usually 8 are present)

BOX 2.1 ■ FIGURE 1

Model of an oxygen atom and its nucleus.

in the liquid. Box 2.2 describes one of the applications of oxygen isotopes.

An element's atomic weight is closely related to the mass number. **Atomic weight** is the weight of an *average* atom of an element, given in atomic mass units. Because sodium has only one naturally occurring isotope, its atomic mass number and its atomic weight are the same—23. On the other hand, chlorine has two common isotopes, with mass numbers of 35 and 37. The atomic weight of chlorine, which takes into account the abundance of each isotope, is about 35.5 because the lighter isotope is more common than the heavier one.

the sodium atom gives it up if it can be taken up by *other* electron-deficient atoms. In each sodium ion, then, the 11 protons (11^+) and 10 electrons (10^-) add up to a single excess positive charge ($^+1$). Positively charged ions like this are called *cations.* Chemists customarily abbreviate the sodium cation as Na^+.

Chlorine, with an atomic number of 17, has a complete inner energy level with 2 electrons and a complete second energy level of 8 electrons around the inner level. A neutral chlorine atom would have only 7 electrons in the third level, but this level requires 8 electrons, so an extra electron is cap-

tured and incorporated. The chlorine ion then contains 18 electrons and 17 protons, and so have a single excess negative charge (Cl^-). Chlorine is an example of a negatively charged ion, or *anion.*

Several factors explain why ions such as chlorine and sodium bond together to form a crystalline solid. One is the need for different ions of like-charge to be as widely separated as possible. Under ordinary circumstances, like-charged ions repel one another and quickly move apart. They only come close to form a stable mineral structure because they are "glued" into place by bonding with ions of the opposite charge.

EARTH SYSTEMS 2.2

Oxygen Isotopes and Climate Change

Oxygen has three stable isotopes. ^{16}O (the 16 tells us there are 16 protons and neutrons in the nucleus) is most abundant, making up 99.762% of Earth's oxygen. ^{17}O constitutes 0.038%, and ^{18}O, 0.200%. The ratio of ^{18}O to ^{16}O in a substance is determined using very accurate instruments called *mass spectrometers*. The ratio of ^{18}O to ^{16}O is 0.0020:1. If partitioning did not take place, we would expect to find the same ratio of isotopes in any substance containing oxygen. However, there is considerable deviation because of the tendency of lighter and heavier atoms to partition.

Water that evaporates or is respirated by plants or animals will have a slightly higher abundance of the lighter isotope (^{16}O) relative to the heavier isotope (^{18}O) than the water left behind. Colder water will have a higher ratio of ^{18}O to ^{16}O than warmer water.

Oxygen isotope studies have allowed scientists to identify climate changes during relatively recent geologic time by determining the temperature changes of ocean water. As we cannot sample past oceans, we use fossil shells to determine the oxygen isotope ratios at the time the organisms were alive. *Foraminifera* are microscopic and nearly microscopic shells of organisms that live in considerable abundance just beneath an ocean surface. While they are alive, they grow their shells of calcite ($CaCO_3$), incorporating oxygen from the seawater. The oxygen in the shells has the $^{18}O/^{16}O$ ratio that is the same as that of the seawater. The particular isotopic ratio reflects the temperature of the seawater.

When foraminifera die, their shells settle onto the deep ocean floor, where they form a thin layer upon older layers of tiny shells. Deep-sea drilling retrieves cores of these layers of sediment. Foraminifera from each layer are analyzed and the $^{18}O/^{16}O$ ratios determined. The ages of the layers are also determined. From these data, the temperature of the ocean's surface water is inferred for the times the foraminifera were alive. Box figure 1 shows the fluctuation in temperature during the past 800,000 years.

These studies show how an Earth systems approach has been useful in determining knowledge about the atmosphere, the geosphere, the biosphere, and the hydrosphere. We can see that climate warming and cooling are natural occurrences in the con-

BOX 2.2 ■ FIGURE 1

Changes in climate during the last 800,000 years as determined by oxygen isotope content in foraminifera shells found in deep-sea sediment cores. Blue—glacial times; red—interglacial times.

text of geologic time. What the data do not tell us is what effect humans are having on the climate. Is the present climate warming part of a natural cycle, or is the rapid increase in greenhouse gases (notably CO_2) reversing what would be a natural cooling cycle?

In other words, the need to neutralize electrical charges, while at the same time keeping like-charges apart works to create a regular arrangement of atoms.

The field of swarming electrons extends farther out from the atomic nuclei of some elements than it does for others. The "size" of an atom (or an ion) is essentially the radius of its electron field; its *ionic radius*, in other words. Ionic radii play an important role in the arrangement of atoms in a crystalline structure as well. When ions come together they tend to pack as efficiently as possible. No irregular holes may exist in the arrangement. A large number of small *anions* (negatively

charged ions) may crowd around a single, large *cation* (positively charged ion), while only a few, large anions may cluster about a small cation (as in figure 2.6).

Of particular importance in this respect are the crystal structures derived from the two most common elements in the Earth's crust—oxygen and silicon (box 2.4).

Silicon is the element used to make computer chips. **Silica** is a term for *oxygen combined with silicon.* Because silicon is the second most abundant element in the crust, most minerals contain silica. The common mineral quartz (SiO_2) is pure silica that has crystallized. Quartz is one of many minerals that

IN GREATER DEPTH 2.3

Bonding

Ions may be regarded as tiny spheres that behave much like magnets. Positively charged ions attract negatively charged ions so that their electrical charges can be neutralized. In saltwater, equal numbers of sodium ions (Na$^+$) and chlorine ions (Cl$^-$) move about freely. The electrical neutrality of the water is maintained because positive sodium ions exactly balance negative chlorine ions. If the water evaporates, the sodium and chlorine are electrically attracted to each other and crystallize into halite. The crystalline structure is the most orderly way for chlorine and sodium ions to pack themselves together and neutralize their collective charges.

A chlorine ion and a sodium ion are fixed in place by their electrical attraction to each other. This is called **ionic bonding** because it is brought about by an attraction between positively and negatively charged ions (box figure 1).

Ionic bonding is the most common type of bonding in minerals. However, in most minerals the bonds between atoms are not purely ionic. Atoms are also commonly bonded together by **covalent bonding,** or bonding in which adjacent atoms *share* electrons. Diamond is composed exclusively of covalently bonded carbon atoms (box figure 2). Carbon has an atomic number of 6, which means that the innermost energy level is full with 2 electrons. Four more electrons are required to maintain electrical neutrality. In a diamond, each carbon atom has 4 electrons in the outer energy level to maintain neutrality, while the need for 8 electrons in that shell is satisfied by electrons that are shared with adjacent carbon atoms. Neighboring carbon atoms are so close together that each of the outer-shell electrons spends half its time in one atom and half in an adjacent atom. Electrical neutrality is maintained, and each atom, in a sense, has 8 electrons in the outer energy level (even though they are not all there at the same time). Covalent bonds in the diamond are extremely strong, and diamond is the hardest natural substance on Earth. However, covalent bonds are not necessarily stronger than ionic bonds.

Graphite, like diamond, is pure carbon. Graphite is used in pencils and as a lubricant. Amazingly, the hardest mineral and

Sodium (Na$^+$) ion

Chlorine (Cl$^-$) ion

BOX 2.3 ■ FIGURE 1

Ionic bonding between sodium (Na$^+$) and chlorine (Cl$^-$).

one of the softest have the same composition. The distinction is in the bonding.

A third type of bonding, *metallic bonding,* is not as important to geology. In metals, such as iron or gold, the atoms are closely packed and the electrons move freely throughout the crystal so as to hold the atoms together. The ease with which electrons move accounts for the high electrical conductivity of metals.

Finally, after all atoms have bonded together, there may be weak, attractive forces remaining. This is the very weak force that holds adjacent sheets of mica or graphite together.

Recommended Web Investigation

To see how diamond and graphite differ, go to www.minweb. co.uk/Mineral_Web.html to see rotating, crystal structures in 3-D. (First you must download and install Chime software, which is easy to do from the website.) From the nonsilicates pull-down menu, select graphite. Note the rotating crystal structure. The short rods connecting carbon atoms represent strong bonds (ignore the long, thin rods). You can use your mouse to stop the rotation and view the structure from any perspective. Then go to diamond. How do the two crystal structures differ? Can you see why plates of tightly bonded graphite slide easily past one another?

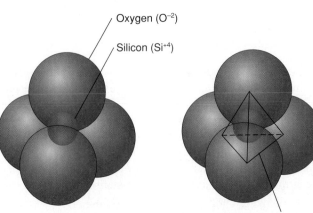

Oxygen (O^{-2})

Silicon (Si^{+4})

FIGURE 2.6

(A) The silicon-oxygen tetrahedron. (B) The silicon-oxygen tetrahedron showing the corners of the tetrahedron coinciding with the centers of oxygen ions.

A Arrangement of atoms in silicon-oxygen tetrahedron

B Diagrammatic representation of a silicon-oxygen tetrahedron

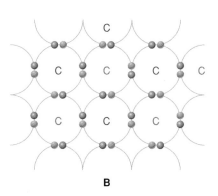

BOX 2.3 ■ FIGURE 2

Covalent bonding in diamond. (*A*) Three-dimensional arrangement of carbon atoms. The rods represent bonds between adjacent carbon atoms. (*B*) Schematic representation of carbon's covalent bonding. Each *C* represents a carbon nucleus with 6 protons and the 2 inner energy level electrons. Dots represent electrons in the outer energy level. For clarity, the color of the "*C*" corresponds to the color of the dots "belonging" to that atom. *Photo © Japack Company/Corbis*

are **silicates,** substances that contain silica (as indicated by their chemical formulas). Most silicate minerals also contain one or more other elements.

The Silicon-Oxygen Tetrahedron

Silicon and oxygen combine to form the atomic framework for most common minerals on Earth. The basic structural unit consists of 4 oxygen atoms (anions) packed together around a single, much smaller silicon atom, as shown in figure 2.6*A*. The four-sided, pyramidal, geometric shape called a *tetrahedron* is used to represent the 4 oxygen atoms surrounding a silicon atom. Each *corner* of the tetrahedron represents the *center* of an oxygen atom (figure 2.6*B*). This basic building block of a

crystal is called a **silicon-oxygen tetrahedron** (also known as a *silica tetrahedron*).

The atoms of the tetrahedron are strongly bonded together. Within a silicon-oxygen tetrahedron, the negative charges exceed the positive charges (see figure 2.7*A*). A single silicon-oxygen tetrahedron is a complex ion with a formula of SiO_4^{-4} because silicon has a charge of $+4$ and the 4 oxygen ions have 8 negative charges (-2 for each oxygen atom).

A silicon-oxygen tetrahedron can either bond with positively charged ions, such as iron or aluminum, or with other silicon-oxygen tetrahedrons. In other words, for the silicon-oxygen tetrahedron to be stable within a crystal structure, it must either (1) be balanced by enough positively charged ions or (2) share oxygen atoms with adjacent tetrahedrons (as

IN GREATER DEPTH 2.4

Elements in the Earth

Estimates of the chemical composition of Earth's crust are based on many chemical analyses of the rocks exposed on Earth's surface. (Models for the composition of the interior of the Earth—the core and the mantle—are based on more indirect evidence.) Table 1 lists the generally accepted estimates of the abundance of elements in the Earth's crust. At first glance, the chemical composition of the crust (and, therefore, the average rock) seems quite surprising.

We think of oxygen as the O_2 molecules in the air we breathe. Yet most rocks are composed largely of oxygen, as it is the most abundant element in the Earth's crust. Unlike the oxygen gas in air, oxygen in minerals is strongly bonded to other elements. By weight, oxygen accounts for almost half the crust, but it takes up 93% of the volume of an average rock. This is because the oxygen atom takes up a large amount of space relative to its weight. (Note how much bigger oxygen atoms are relative to other atoms in figure 2.6 and others.) It is not an exaggeration to regard the crust as a mass of oxygen with other elements occupying positions in crystalline structures between oxygen atoms.

Note that the third most abundant element is aluminum, which is more common in rocks than iron. Knowing this, one might assume that aluminum would be less expensive than iron, but of course this is not the case. Common rocks are not mined for aluminum because it is so strongly bonded to oxygen and other elements. The amount of energy required to break these bonds and separate the aluminum makes the process too costly for commercial production. Aluminum is mined from the uncommon deposits where aluminum-bearing rocks have been weathered, producing compounds in which the crystalline bonds are not so strong.

Collectively, the eight elements listed in table 1 account for more than 98% of the weight of the crust. All the other elements total only about 1.5%. Absent from the top eight elements are such vital elements as hydrogen (tenth by weight) and carbon (seventeenth by weight).

The element copper is only twenty-seventh in abundance, but our industrialized society is highly dependent on this metal. Most of the wiring in electronic equipment is copper, as are many of the telephone and power cables that crisscross the continent. However, the Earth's crust is not homogeneous, and geological processes have created concentrations of elements such as copper in a few places. Exploration geologists are employed by mining companies to discover where (as well as why) ore deposits of copper and other metals occur (see chapter 21).

BOX 2.4 ■ TABLE 1

Crustal Abundance of Elements

Element	Symbol	Percentage by Weight	Percentage by Volume	Percentage of Atoms
Oxygen	O	46.6	93.8	60.5
Silicon	Si	27.7	0.9	20.5
Aluminum	Al	8.1	0.8	6.2
Iron	Fe	5.0	0.5	1.9
Calcium	Ca	3.6	1.0	1.9
Sodium	Na	2.8	1.2	2.5
Potassium	K	2.6	1.5	1.8
Magnesium	Mg	2.1	0.3	1.4
All other elements		1.5	—	3.3

shown in figures 2.7C and D) and therefore reduce the need for extra, positively charged ions. The structures of silicate minerals range from an *isolated silicate structure*, which depends entirely on positively charged ions to hold the tetrahedrons together, to *framework silicates* (quartz, for example), in which all oxygen atoms are shared by adjacent tetrahedrons. The most common types of silicate structures are shown diagrammatically in figure 2.8 and are discussed next.

Isolated Silicate Structure

Silicate minerals that are structured so that none of the oxygen atoms is shared by tetrahedrons have an **isolated silicate structure.** The individual silicon-oxygen tetrahedrons are bonded together by positively charged ions (figure 2.9). The common mineral *olivine*, for example, contains two ions of either magnesium (Mg^{+2}) or iron (Fe^{+2}) for each silicon-oxygen tetrahedron. The formula for olivine is $(Mg,Fe)_2SiO_4$.

Chain Silicates

A **chain silicate structure** forms when two of a tetrahedron's oxygen atoms are shared with adjacent tetrahedrons to form a chain (figures 2.8 and 2.10). Each chain, which extends indefinitely, has a net excess of negative charges. Minerals may have a *single-* or *double-chain structure*. For single-chain silicate structures, the ratio of silicon to oxygen (as figure 2.10 shows) is 1:3; therefore, each mineral in this group (the *pyroxene* group) incorporates SiO_3^{-2} in its formula, and it must be electrically balanced by the positive ions (e.g., Mg^{+2}) that hold the parallel chains together. If a pyroxene has magnesium, as the +2 ions bonding the chains shown in figure 2.10A, it has a formula of $MgSiO_3$.

A double-chain silicate is essentially two adjacent single chains that are sharing oxygen atoms. The *amphibole* group is characterized by two parallel chains in which every other tetrahedron shares an oxygen atom with the adjacent chain's tetra-

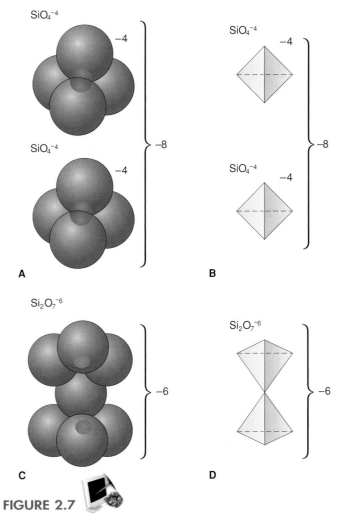

FIGURE 2.7

Two single tetrahedrons (A and B) require more positively charged ions to maintain electrical neutrality than two tetrahedrons sharing an oxygen atom (C and D). B and D are the schematic representations of A and C, respectively.

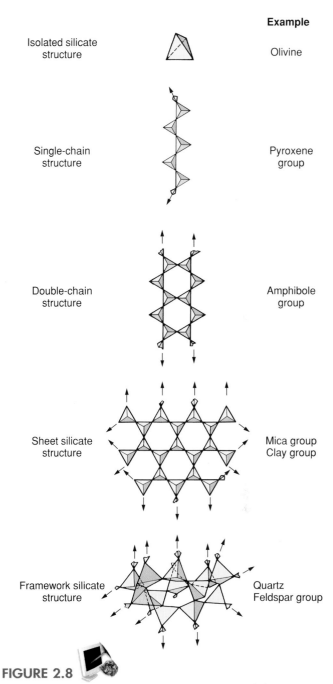

FIGURE 2.8

Common silicate structures. Arrows indicate directions in which structure repeats indefinitely.

hedron (figure 2.8). In even a small amphibole crystal, millions of parallel double chains are bonded together by positively charged ions.

Chain silicates tend to be shaped like columns, needles, or even fibers. The long structure of the external form corresponds to the linear dimension of the chain structure. Fibrous aggregates of certain minerals are called *asbestos* (see box 2.5).

Sheet Silicates

In a **sheet silicate structure** each tetrahedron shares three oxygen atoms to form a sheet (figure 2.8). The *mica* group and the *clay* group of minerals are sheet silicates. The positive ions that hold the sheets together are "sandwiched" between the silicate sheets (box 2.6).

Framework Silicates

When all four oxygen ions are shared by adjacent tetrahedrons, a **framework silicate structure** is formed. *Quartz* is a framework silicate mineral. A *feldspar* is a framework silicate as

well. However, its structure is slightly more complex because aluminum substitutes for some of the silicon atoms in some of the tetrahedrons. The same kind of substitution also takes place in amphiboles and micas, which helps account for the wide variety of silicate minerals.

Nonsilicate Minerals

Although not as abundant in Earth, *nonsilicates,* minerals that do not contain silica, are nevertheless important. The

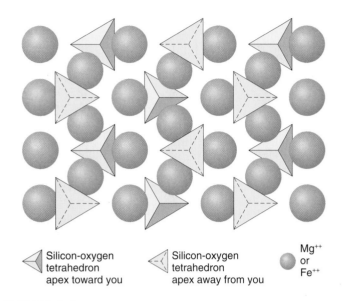

Silicon-oxygen tetrahedron apex toward you   Silicon-oxygen tetrahedron apex away from you Mg^{++} or Fe^{++}

FIGURE 2.9

Diagram of the crystal structure of olivine, as seen from one side of the crystal.

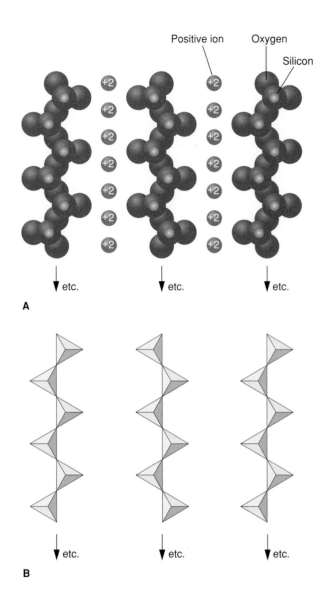

Positive ion   Oxygen   Silicon

A

B

FIGURE 2.10

Single-chain silicate structure. (A) Model of a single-chain silicate mineral. (B) The same chain silicate shown diagrammatically as linked tetrahedrons; positive ions between the chains are not shown.

carbonates have CO_3 in their formulas. Calcite, $CaCO_3$, is a member of this group and is one of the most abundant minerals at the Earth's surface where it occurs mainly in limestone. In dolomite, also a carbonate, magnesium replaces some calcium in the calcite formula. Gypsum is a *sulfate* (containing SO_4). *Sulfides* have S but not O in their formulas (pyrite, FeS_2, is an example). Hematite (Fe_2O_3) is an *oxide*—that is, it contains oxygen not bonded to Si, C, or S. Halite, NaCl, is a member of the *chloride* group. *Native elements* have only one element in their formulas. Some examples are gold (Au), copper (Cu), and the two minerals that are composed of pure carbon (C), diamond and graphite.

VARIATIONS IN MINERAL STRUCTURES AND COMPOSITIONS

It stands to reason that only a limited number of mineral compositions exist in nature because atoms cannot be combined randomly and they can only come together to form a restricted number of crystalline structures. This does not mean, however, that each kind of mineral is compositionally different, or that individual mineral types can't show some internal compositional variation.

Ions of like size and charge may freely substitute for one another in the atomic structures of minerals. Iron (Fe^{2+}) and magnesium (Mg^{2+}), for example, interchangeably substitute to create a range of compositions in the common silicate mineral olivine. This is represented by the parentheses in the formula of olivine—$(Mg,Fe)_2SiO_4$. Olivine is an example of a *solid solution*, with pure magnesium olivine, Mg_2SiO_4, forming the bright green variety forsterite (or peridot, as a gem),

and pure iron olivine forming the jet black variety fayalite, Fe_2SiO_4. The crystal structures of forsterite and fayalite are virtually identical.

Some minerals that show solid solution, like plagioclase feldspar and augite (a pyroxene), also show *compositional zoning,* with the centers of crystals dominated by one type of cation and the rims dominated by another. The grains of plagioclase in certain igneous rocks typically have calcium-rich centers and sodium-rich rims (figure 2.11). The change is due to the cooling of the molten rock from which the plagioclase crystallizes. Calcium-rich plagioclase is more stable at the high temperatures in which the crystals start growing. The crystals then develop sodium-rich rims as the remaining melt crystallizes.

ENVIRONMENTAL GEOLOGY 2.5

Asbestos—How Hazardous?

*A*sbestos is a generic name for fibrous aggregates of minerals (box figure 1). Because it does not ignite or melt in fire, asbestos has a number of valuable industrial applications. Woven into cloth, it may be used to make suits for firefighters. It can also be used as a fireproof insulation for homes and other buildings and has commonly been used in plaster for ceilings. Five of the six commercial varieties of asbestos are amphiboles, known commercially as "brown" and "blue" asbestos. The sixth variety is *chrysotile*, which is not a chain silicate and belongs to the *serpentine* family of minerals, and is more commonly known as "white asbestos." White asbestos is, by far, the most commonly used in North America (about 95% of that used in the United States).

Public fear of asbestos in the United States has resulted in its being virtually outlawed by the federal government. Tens of billions of dollars have been spent (probably unnecessarily) to remove or seal off asbestos from schools and other public buildings.

Asbestos' bad reputation comes from the high death rate among asbestos workers exposed, without protective attire, to extremely high levels of asbestos dust. Some of these workers, who were covered with fibers, were called "snowmen." In Manville, New Jersey, children would catch the "snow" (asbestos particles released from a nearby asbestos factory in their mouths. The high death rates among asbestos workers are attributed to *asbestosis* and lung cancer. Asbestosis is similar to silicosis contracted by miners; essentially, the lungs become clogged with asbestos dust after prolonged heavy exposure. The incidence of cancer has been especially high among asbestos workers who were also smokers. It's not clear that heavy exposure to white asbestos caused cancer among nonsmoking asbestos workers. However, brown and blue (amphibole) varieties, which are not mined in North America, have been linked to cancer for heavy exposure (even if for a short term).

What are the hazards of asbestos to an individual in a building where walls or ceilings contain asbestos? Recent studies from a wide range of scientific disciplines indicate that the risks are minimal to nonexistent, at least for exposure to white asbestos. The largest asbestos mines in the world are at Thetford Mines, Quebec. A study of longtime Thetford Mines residents, whose houses border the waste piles from the asbestos mines, indicated that their incidence of cancer was no higher than that of Canadians overall. Nor have studies in the United States been able to link nonoccupational exposure to asbestos and cancer. One estimate of the risk of death from cancer due to exposure to asbestos dust is one per 100,000 lifetimes. (Compare this to the risk of death from lightning of 4 per 100,000 lifetimes or automobile travel—1,600 deaths per 100,000 lifetimes.)

Following the collapse of New York's World Trade Center towers on September 11, 2001, the dust in the air from destroyed buildings contained high levels of asbestos—much higher than the safety levels set by the Environmental Protection Agency (EPA). Faced with widespread panic and a mass exodus from the city, the EPA reversed itself and declared the air safe. In doing so, the agency admitted that its standards were too stringent and were based on long-term exposure.

In California, a closed-down white asbestos mining site designated for EPA Superfund cleanup is a short distance from where asbestos is being mined cleanly and efficiently. It is packaged and shipped to Japan. It cannot be used in the United States, because the United States is the only industrialized nation whose laws do not distinguish between asbestos types and permit the use of chrysotile.

A reason chrysotile is less hazardous than amphibole asbestos is that chrysotile fibers will dissolve in lungs and amphibole will not. Experiments by scientists at Virginia Polytechnic Institute indicate that it takes about a year for chrysotile fibers to dissolve in lung fluids, whereas, glass fibers of the same size will dissolve only after several hundred years. Yet fiberglass is being used increasingly as a substitute for asbestos.

Additional Resources
Asbestos Institute
- www.asbestos-institute.ca/main.html

National Cancer Institute
- http://cis.nci.nih.gov/fact/3_21.htm

BOX 2.5 ■ FIGURE 1
Chrysotile asbestos. *Photo © Parvinder Sethi*

ENVIRONMENTAL GEOLOGY 2.6

Clay Minerals That Swell

Clay minerals are very common at Earth's surface; they are a major component of soil. There are a great number of different clay minerals. What they all have in common is that they are sheet silicates. They differ by which ions hold sheets together and by the number of sheets "sandwiched" together.

Ceramic products and bricks are made from clay. Surprisingly, some clay minerals are edible; some are used in the manufacturing of pills. *Kaolinite*, a clay mineral, is the main ingredient in Kaopectate, a remedy for intestinal distress. Popular fast-food chains use clay minerals as a thickener for shakes (you can tell which ones, because the chains do not call them "milk shakes"—they do not use milk).

Montmorillonite is one of the more interesting clay minerals. It is better known as *expansive clay* or *swelling clay*. If water is added to the montmorillonite, the water molecules are adsorbed into the spaces between silicate layers (box figure 1). This results in a large increase in volume, sometimes up to several hundred percent. The pressure generated can be up to 50,000 kilograms per square meter. This is sufficient to lift a good-sized building.

If a building is erected on expansive clay that subsequently gets wet, a portion of the building will be shoved upward. In all likelihood, the foundation will break. Some people think that expansive soils have caused more damage than earthquakes and landslides combined. Damage in the United States is estimated to cost $2 billion a year.

On the other hand, swelling clays can be put to use. Montmorillonite, mixed with water, can be pumped into fractured rock or concrete. When the water is adsorbed, swelling clay expands to fill and seal the crack. The technique is particularly useful where dams have been built against fractured bedrock. Sealing the cracks with expansive clays ensures that water will stay in the reservoir behind the dam.

A Dry clay mineral

B Expansion due to adsorption of water

BOX 2.6 ■ FIGURE 1

Expansive clays. (The orange ion represents aluminum in the clay layers and is not drawn to scale.)

FIGURE 2.11

Zoning in plagioclase feldspar, as seen under a polarizing microscope. The concentric color bands each indicate different amounts of Ca and Na in the crystal structure. *Photographed using cross-polarizers and a red-1 (550 nm) retardation plate. Photo by C. C. Plummer*

Some minerals can have the same chemical composition but have different crystalline structures—a phenomenon termed *polymorphism*. For example, calcite and aragonite both have the same formula $CaCO_3$. Their atomic crystal structures differ greatly, however. As you might expect, these two similar, but distinctive mineral types result from separate conditions and processes of formation, with aragonite usually being an indicator of high-pressure crystallization.

Graphite and diamond are another, particularly spectacular example of polymorphism. Both minerals are made up of elemental carbon. They are unusual in that there is no other element involved in their structures. Dull-looking graphite, however, has a sheetlike structure that makes the mineral quite soft—useful as pencil lead; while the structure of shiny diamond is much more compact, making this the hardest substance on Earth. Graphite forms within the crust, while diamond originates much deeper, in the higher-pressure conditions of the mantle.

It is important to note that the physical characteristics of minerals that we can observe without fancy laboratory equipment, such as color, hardness, and luster, are linked closely with the atomic structures and compositions of the minerals.

THE PHYSICAL PROPERTIES OF MINERALS

The best approach to understanding physical properties of minerals is to obtain a sample of each of the most common rock-forming minerals named in table 2.1. The properties described can then be identified in these samples.

To identify an unknown mineral, you should first determine its physical properties, then match the properties with the appropriate mineral, using a mineral identification key or chart such as the ones included in appendix A of this book. With a bit of experience, you may get to know the few diagnostic tests for each common mineral and no longer need to refer to an identification table.

Color

The first thing most people notice about a mineral is its color. For some minerals, color is a useful property. *Muscovite* mica is silvery white or colorless, whereas *biotite* is black or dark brown. Most of the **ferromagnesian minerals** (iron/magnesium-bearing), such as *augite, hornblende,* olivine, and biotite, are either green or black.

Because color is so obvious, beginning students tend to rely too heavily on it as a key to mineral identification. Unfortunately, color is also apt to be the most ambiguous of physical properties (figure 2.12). If you look at a number of quartz crystals, for instance, you may find specimens that are white, pink, black, yellow, or purple. Color is extremely variable in quartz and many other minerals because even minute chemical impurities can strongly influence it. Obviously, it is poor procedure to attempt to identify quartz strictly on the basis of color.

Streak

A pulverized mineral gives a color, called a **streak,** that usually is more reliable than the color of the specimen itself. Scraping the edge of a mineral sample across an unglazed porcelain plate leaves a streak that may be diagnostic of the mineral. For instance, hematite always leaves a reddish brown streak though the sample may be brown or red or silver.

Unfortunately, few of the silicate minerals—the most common minerals—leave an identifying streak because most are harder than the porcelain streak plate.

Luster

The quality and intensity of *light* that is reflected from the surface of a mineral is termed **luster.** (A photograph cannot always show this quality.) The luster of a mineral is described by comparing it to familiar substances.

TABLE 2.1	Minerals of the Earth's Crust	
Name	Chemical Composition	Type of Silicate Structure or Chemical Group
The most common rock-forming minerals (These make up more than 90% of the Earth's crust.)		
Feldspar group		
Plagioclase	Ca and Na Al silicate	Framework silicate
Potassium feldspar (orthoclase, microcline)	K Al silicate	Framework silicate
Pyroxene group (augite most common)	Fe, Mg silicate (some with Al, Na, Ca)	Single-chain silicate
Amphibole group (hornblende most common)	Complex Fe, Mg, Al silicate hydroxide	Double-chain silicate
Quartz	Silica	Framework silicate
Mica group		
Muscovite	K Al silicate hydroxide	Sheet silicate
Biotite	K Fe, Mg Al silicate hydroxide	Sheet silicate
Other common rock-forming minerals		
Silicates		
Olivine	Mg, Fe silicate	Isolated silicate
Garnet group	Complex silicates	Isolated silicate
Clay minerals group	Complex Al silicate hydroxides	Sheet silicate
Nonsilicates		
Calcite	$CaCO_3$	Carbonate
Dolomite	$CaMg(CO_3)_2$	Carbonate
Gypsum	$CaSO_4 \cdot 2H_2O$	Sulfate
Much less common minerals of commercial value		
Halite	NaCl	Chloride
Diamond	C	Native element
Gold	Au (gold)	Native element
Hematite	Iron oxide (Fe_2O_3)	Oxide
Magnetite	Iron oxide (Fe_3O_4)	Oxide
Chalcopyrite	Cu, Fe sulfide	Sulfide
Sphalerite	Zn sulfide	Sulfide
Galena	Pb sulfide	Sulfide

FIGURE 2.12

Why color may be a poor way of identifying minerals. These are all corundum gems, including ruby and sapphire (see box 2.7 for more on gems). *Photo © Parvinder Sethi*

Luster is either *metallic* or *nonmetallic*. A **metallic luster** gives a substance the appearance of being made of metal. Metallic luster may be very shiny, like a chrome car part, or less shiny, like the surface of a broken piece of iron.

Nonmetallic luster is more common. The most important type is **glassy** (also called **vitreous**) **luster,** which gives a substance a glazed appearance, like glass or porcelain. Most silicate minerals have this characteristic. The feldspars, quartz, the micas, and the pyroxenes and amphiboles all have a glassy luster.

Less common is an **earthy luster.** This resembles the surface of unglazed pottery and is characteristic of the various clay minerals. Some uncommon lusters include *resinous* luster (appearance of resin), *silky* luster, and *pearly* luster.

Hardness

The property of "scratchability," or **hardness,** can be tested fairly reliably. For a true test of hardness, the harder mineral or substance must be able to make a groove or scratch on a smooth, fresh surface of the softer mineral. For example, quartz can always scratch calcite or feldspar and is thus said to be harder than both of these minerals. Substances can be compared to **Mohs' hardness scale,** in which ten minerals are designated as standards of hardness (figure 2.13). The softest mineral, talc (used for talcum powder because of its softness), is designated as 1. Diamond, the hardest natural substance on Earth, is 10 on the scale. Mohs' scale is a relative hardness scale. Figure 2.13 shows the absolute hardness for the ten minerals. The absolute hardness is obtained using an instrument that measures how much pressure is required to indent a mineral. Note that the difference in absolute hardness between corundum (9) and diamond (10) is around six times the difference between corundum and topaz (8).

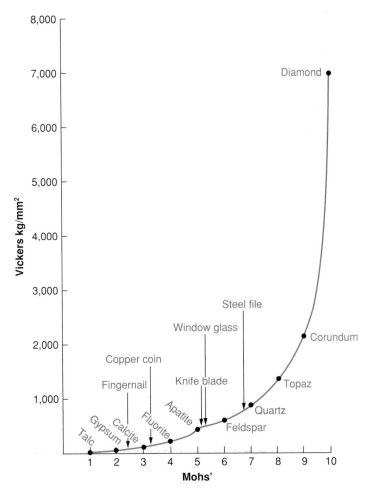

FIGURE 2.13

Mohs' hardness scale plotted against Vickers indentation values (kg/mm²). Indentation values are obtained by an instrument that measures the force necessary to make a small indentation into a substance.

Rather than carry samples of the ten standard minerals, a geologist doing field work usually relies on common objects to test for hardness (figure 2.13). A fingernail usually has a hardness of about 2 1/2. If you can scratch the smooth surface of a mineral with your fingernail, the hardness of the mineral must be less than 2 1/2 (figure 2.14). A copper coin or a penny has a hardness between 3 and 4; however, the brown, oxidized surface of most pennies is much softer, so check for a groove into the coin. A knife blade or a steel nail generally has a hardness slightly greater than 5, but it depends on the particular steel alloy used. A geologist uses a knife blade to distinguish between softer minerals, such as calcite, and similarly appearing harder minerals, such as quartz. Ordinary window glass, usually slightly harder than a knife blade (although some glass, such as that containing lead, is much softer), can be used in the same way as a knife blade for hardness tests. A file (one made of tempered steel for filing metal, not a fingernail file) can be used for a hardness of between 6 and 7. A porcelain streak plate also has a hardness of around 6 1/2.

Precious Gems

Diamond engagement rings are a tradition in our society. The diamond, often a significant financial commitment for the groom-to-be, symbolizes perpetual love. Jewelry with valuable gemstones is glamorous. We expect to see the rich and famous heavily bedecked with gemstones set in gold and other precious metals.

Gemstones are varieties of certain minerals that are valuable because of their beauty. Precious gemstones, or, simply, precious stones, are particularly valuable; semiprecious stones are much less valuable. Diamond, sapphire, ruby, emerald, and aquamarine are regarded as precious stones. What they all have in common is that they are transparent with even coloration and have a hardness greater than quartz (7 on the hardness scale). Their hardness ensures that they are durable.

Diamonds are usually clear, although some tinted varieties are particularly valuable. (The famous Hope diamond is blue.) Diamond's appeal is largely due its unique, brilliant luster (called *adamantine* luster). This results from the way that light reflects from within the crystal and is dispersed into rainbowlike colors. The facets that you see on a diamond have been cut (or, more correctly, ground, using diamond dust) to enhance the gem's brilliance.

Sapphire and ruby are both varieties of corundum (9 on Mohs' scale). Sapphire can be various colors (except red), but blue sapphires are most valuable. Minute amounts of titanium and iron in the crystal structure give sapphire its blue coloration. Rubies are red due to trace amount of chromium in corundum.

Emerald and aquamarine are varieties of beryl (hardness of 7.5). Emerald is the most expensive of these and owes its green color to chromium impurities. Aquamarine's blue color is due to iron impurities in the crystal structure.

Recommended Web Investigation
The Image
- www.theimage.com/

Click on "Gemstone Gallery" and then "Beryl." You can read about the properties of beryl and details about emerald and aquamarine. Below the description, you can access images of these gems. You can go back and click on "Sapphire" for information on sapphire and corundum. A photo of sapphires and a ruby is accessible at the bottom of the text. This site also contains information on how gems are faceted.

FIGURE 2.14

Fingernail (hardness of 2 1/2) easily scratches gypsum (hardness of 2). *Photo © Parvinder Sethi*

External Crystal Form

The **crystal form** of a mineral is a set of faces that have a definite geometric relationship to one another. A well-formed crystal of halite, for example, consists of six faces all square and joined at right angles. The crystal form of halite is a *cube,* in other words. Other kinds of minerals whose crystals commonly consist of single forms include magnetite (octahedron) and garnet (dodecahedron).

Crystals consisting of single forms, like those mentioned previously, are the most highly symmetrical shapes in all of nature other than spheres, which single minerals cannot develop. Crystals more commonly consist of several types of forms combined together to generate the full body of each specimen. As a rule of thumb, if two or more faces on a crystal are identical in shape and size, they belong to the same crystal form (figure 2.15).

Minerals displaying well-developed crystal faces have played an important role in the development of chemistry and physics. Steno, a Danish naturalist of the seventeenth century, first noted that the angle between two adjacent faces of quartz is always exactly the same, no matter what part of the world the quartz sample comes from or the color or size of the quartz. As shown in figure 2.16, the angle between any two adjacent sides of the six-sided "pillar" (which is called a prism by mineralogists) is always exactly 120°, while between a face of the "pillar" and one of the "pyramid" faces (actually part of a rhombohedron) the angle is always exactly 141°45′.

The discovery of such regularity in nature usually has profound implications. When minerals other than quartz were studied, they too were found to have sets of angles for adjacent faces that never varied from sample to sample. This observation became formalized as the *law of constancy of interfacial angles.*

A

B

C

Later the discovery of X-ray beams and their behavior in crystals confirmed Steno's theory about the structure of crystals.

Steno suspected that each type of mineral was composed of many tiny, identical building blocks, with the geometric shape of the crystal being a function of how these building blocks are put together. If you are stacking cubes, you can build a structure having only a limited variety of planar forms. Like-

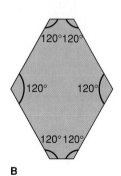

A   B

FIGURE 2.15

Characteristic crystal forms of three common minerals: (A) Cluster of quartz crystals. (B) Crystals of potassium feldspar. (C) Intergrown cubic crystals of fluorite. *Photos © Parvinder Sethi*

FIGURE 2.16

Quartz crystals showing how interfacial angles remain the same in perfectly proportioned (A) and misshapen (B) crystals. Cuts perpendicular to the prisms shows that all angles are exactly 120°. *Photos © Parvinder Sethi*

FIGURE 2.17

Geometric forms built by stacking cubes (*A, B,* and *C*) and rhombohedrons (*D, E, F*). *A* and *D* are from a diagram published in 1801 by Haüy, a French mathematician, *A* and *B* show how cubes can be stacked for cubic and dodecahedral (12-sided) crystal forms. *C* is a photo showing a cube and a dodecahedron of pyrite. *D* and *E* show the relationship of stacked rhombohedrons to a "dog tooth" (scalenohedron) form and a rhombohedral face. *F* is a calcite "dogtooth" without a rhombohedral face. *Photos © Parvinder Sethi*

wise, stacking rhombohedrons in three dimensions limits you to other geometric forms (figure 2.17).

Steno's law was really a precursor of atomic theory, developed centuries later. Our present concept of crystallinity is that atoms are clustered into geometric forms—cubes, bricks, hexagons, and so on—and that a crystal is essentially an orderly, three-dimensional stacking of these tiny geometric forms. Halite, for example, may be regarded as a series of cubes stacked in three dimensions (see figure 2.4). Because of the cubic "building block," its usual crystal form is a cube with crystal faces at 90° angles to each other.

Cleavage

The internal order of a crystal may be expressed externally by crystal faces, or it may be indicated by the mineral's tendency to split apart along certain preferred directions. **Cleavage** is the ability of a mineral to break, when struck or split, along preferred planar directions.

A mineral tends to break along certain planes because the bonding between atoms is weaker there. In quartz, the bonds are equally strong in all directions; therefore, quartz has no cleavage. The micas, however, are easily split apart into sheets (figure 2.18). If we could look at the arrangement of atoms in the crystalline structure of micas, we would see that the individual silicon-oxygen tetrahedrons are strongly bonded to one another within each of the silicate sheets. The bonding *between* adjacent sheets, however, is very weak. Therefore, it is easy to split the mineral apart parallel to the plane of the sheets.

Cleavage is one of the most useful diagnostic tools because it is identical for a given mineral from one sample to another. Cleavage is especially useful for identifying minerals when they are small grains in rocks.

The wide variety of combinations of cleavage and *quality* of cleavage also increases the diagnostic value of this property. Mica has a single direction of cleavage, and its quality is perfect (figures 2.18*A* and 2.19*A*). Other minerals are characterized by one, two, or more cleavage directions; the quality can range from perfect to poor (poor cleavage is very hard for anyone but a well-trained mineralogist to detect).

Three of the most common mineral groups—the feldspars, the amphiboles, and the pyroxenes—have two

A

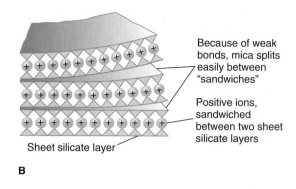

Because of weak bonds, mica splits easily between "sandwiches"

Positive ions, sandwiched between two sheet silicate layers

Sheet silicate layer

B

FIGURE 2.18

(A) Mica cleaves easily parallel to the knife blade. (B) Relationship of mica to cleavage. Mica crystal structure is simplified in this diagram. *Photo by C. C. Plummer*

directions of cleavage (figure 2.19*B* and *C*). In feldspars, the two directions are at angles of about 90° to each other, and both directions are of very good quality. In pyroxenes, the two directions are also at about right angles, but the quality is only fair. In amphiboles (figure 2.20), the quality of the cleavage is very good and the two directions are at an angle of 56° (or 124° for the obtuse angle).

Halite is an example of a mineral with three excellent cleavage directions, all at 90° to each other. This is called *cubic cleavage* (figure 2.19*D*). Halite's cleavage tells us that the

Only direction of cleavage

A

1st direction of cleavage

2nd direction

B

1

2

C

1

3

2

D

1

2

3

E

FIGURE 2.19

Most common types of mineral cleavage. Straight lines and flat planes represent cleavage. (A) One direction of cleavage. Mica is an example. (B) Two directions of cleavage that intersect at 90° angles. Feldspar is an example. (C) Two directions of cleavage that do not intersect at 90° angles. Amphibole is an example. (D) Three directions of cleavage that intersect at 90° angles. Halite is an example. (E) Three directions of cleavage that do not intersect at 90° angles. Calcite is an example. Not shown are the two other possible types of cleavage—four directions (such as in diamond) and six directions (as in sphalerite).

FIGURE 2.20
Amphibole cleavage as seen in a polarizing microscope. *Photo by C. C. Plummer*

FIGURE 2.21
Cleavage fragments of calcite. *Photo © Parvinder Sethi*

bonds are weak in the planes parallel to the cube faces shown in figure 2.4.

Calcite also has three cleavage directions, each excellent. But the angles between them are clearly not right angles. Calcite's cleavage is known as *rhombohedral* cleavage (figures 2.19*E* and figure 2.21).

Some minerals have more than three directions of cleavage. Diamond has very good cleavage in four directions (ironically, the hardest natural substance on Earth can be easily shattered into small cleavage fragments). Sphalerite, the principal ore of zinc, has six directions.

Recognizing cleavage and determining angular relationships between cleavage directions take some practice. Students new to mineral identification tend to ignore cleavage because it is not as immediately apparent to the eye as color. But deter-

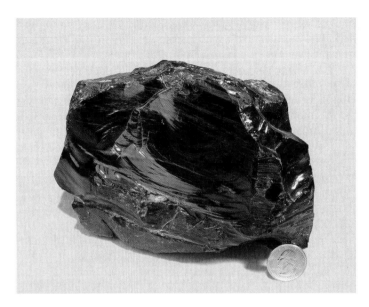

FIGURE 2.22
Conchoidal fracture in volcanic glass (obsidian). *Photo © Parvinder Sethi*

mining cleavage is frequently the key to identifying a mineral, so the small amount of practice needed to develop this skill is worthwhile.

Fracture

Fracture is the way a substance breaks where not controlled by cleavage. Minerals that have no cleavage commonly have an *irregular fracture.*

Some minerals break along curved fracture surfaces known as *conchoidal fractures* (figure 2.22). These look like the inside of a clam shell. This type of fracture is commonly observed in quartz and garnet (but these minerals also show irregular fractures). Conchoidal fracture is particularly common in glass, including obsidian (volcanic glass).

Minerals that have cleavage can fracture along directions other than that of the cleavage. The mica in figure 2.18*A* has irregular edges, which are fractures due to being torn perpendicular to the cleavage direction.

Specific Gravity

It is easy to tell that a brick is heavier than a loaf of bread just by hefting each of them. The brick has a higher **density,** weight per given volume, than the bread. Density is commonly expressed as **specific gravity,** the ratio of a mass of a substance to the mass of an equal volume of water.

Liquid water has a specific gravity of 1. (Ice, being lighter, has a specific gravity of about 0.9.) Most of the common silicate minerals are about two and a half to three times as dense as equal volumes of water: quartz has a specific gravity of 2.65; the feldspars range from 2.56 to 2.76. Special scales are needed to determine specific gravity precisely. However, a person can easily distinguish by hand very dense minerals such as galena (a lead sulfide with a specific gravity of 7.5) from the much less dense silicate minerals.

Gold, with a specific gravity of 19.3, is much heavier than galena. Because of its high density, gold can be collected by "panning." While the lighter clay and silt particles in the pan are sloshed out with the water, the gold lags behind in the bottom of the pan.

Special Properties

Some properties only apply to one or a few minerals. Smell is one. Some clay minerals have a characteristic "earthy" smell when they are moistened. A few minerals have a distinctive taste. If you lick halite, it tastes salty, because it is, of course, table salt.

Plagioclase feldspar commonly exhibits **striations**—straight, parallel lines on the flat surfaces of one of the two cleavage directions (figure 2.23). The lines appear to be etched by a delicate scriber. In plagioclase, they are caused by a systematic change within the pattern of crystalline structure.

The mineral *magnetite* (an iron oxide) owes its name to its characteristic physical property of being attracted to a magnet. Where large bodies of magnetite are found in the Earth's crust, compass needles point toward the magnetite body rather than to magnetic north. Airplanes navigating by compass have become lost because of the influence of large magnetite bodies. Some other minerals are weakly magnetic; their magnetism can only be detected by specialized magnetometers, similar to metal detectors in airports. Magnetism is important to modern civilization. We use magnetic tape (coated with magnetite or other magnetic material) for sound and video recordings as well as for magnetic memory disks in our computers. In later

chapters, you will see how magnetite in igneous rocks has preserved a record of Earth's magnetic field through geologic time; this has been an important part of the verification of plate tectonic theory. Some bacteria create magnetite, and this has been used to support the hypothesis that life has existed on Mars (as described in NASA's Mars micromagnet site, http://science.nasa.gov/headlines/y2000/ast20dec_1.htm).

Quartz has the property of generating electricity when squeezed in a certain crystallographic direction. This property relates to its use in quartz watches (see web box 2.8).

A mineral has numerous other properties, including its melting point, electrical and heat conductivity, and so on. Most are not relevant to introductory geology. Two categories of properties that are important are optical properties and the effects of X rays on minerals.

A clear crystal of calcite exhibits an unusual property. If you place transparent calcite over an image on paper, you will see two images (figure 2.24). This phenomenon is known as

FIGURE 2.23

Plagioclase striations. *Photo by C. C. Plummer*

FIGURE 2.24

Double refraction in calcite. Two images of the letters are seen through the transparent calcite crystal. *Photo © Parvinder Sethi*

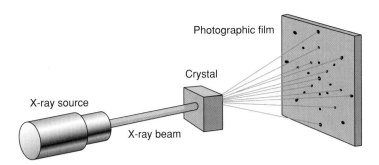

FIGURE 2.25

An X-ray beam passes through a crystal and is deflected by the rows of atoms into a pattern of beams. The dots exposed on the film are an orderly pattern used to identify the particular mineral.

double refraction and is caused by light splitting into two components when it enters some crystalline materials. Each of the components is traveling through the mineral at different velocities. Most minerals possess double refraction, but it is usually slight and can be observed using polarizing filters, notably in polarizing microscopes. Polarizing microscopes are very useful to professional geologists and advanced students for identifying minerals and interpreting how rocks formed. Photomicrographs elsewhere in this book were taken through polarizing microscopes (for example, figures 2.11 and 3.5*B*). Explaining optical phenomena, such as this, is beyond the scope of this book but, if interested, you can go to the Molecular Expressions Microscopy Primer site at http://micro. magnet.fsu.edu/primer/virtual/virtualpolarized.html.

Specialized equipment is needed to determine some properties. Perhaps most important are the characteristic effects of minerals on X rays, which we can explain only briefly here. X rays entering a crystalline substance are deflected by planes of atoms within the crystal. The X rays leave the crystal at precise and measurable angles controlled by the orientation of the planes of atoms that make up the internal crystalline structure (figure 2.25). The pattern of X rays exiting can be recorded on photographic film or by various recording instruments. Each mineral has its own pattern of reflected X rays, which serves as an identifying "fingerprint."

Chemical Tests

One chemical reaction is routinely used for identifying minerals. The mineral calcite, as well as some other carbonate minerals (those containing CO_3^{-2}), reacts with a weak acid to produce carbon dioxide gas. In this test, a drop of dilute hydrochloric acid applied to the sample of calcite bubbles vigorously, indicating that CO_2 gas is being formed. Normally, this is the only chemical test that geologists do during field research.

Chemical analyses of minerals and rocks are done in labs using a wide range of techniques. A chemical analysis can accurately tell us the amount of each element present in a mineral. However, chemical analysis alone cannot be used to conclusively identify a mineral. We also need to know about the mineral's crystalline structure. As we have seen, diamond and graphite have an identical composition but very different crystalline structures.

THE MANY CONDITIONS OF MINERAL FORMATION

Minerals form under an enormously wide variety of conditions—most purely geological; others biological in nature. Some form tens of kilometers beneath the surface; others right at the surface and virtually out of the atmosphere itself.

The most common minerals are silicates, which incorporate the most abundant elements on Earth. Silicate minerals such as quartz, olivine, and the feldspars (plagioclase and potassium feldspar) crystallize primarily from molten rock (magma). They are *precipitates*—products of crystallizing liquid. Other precipitates include the carbonates calcite and aragonite, which grow in spring and cave waters.

Some minerals precipitate due to evaporation (e.g., halite). The very thick salt deposits underlying central Europe and the southern Great Plains exist because of the evaporation of seas millions of years ago.

Ice may be regarded as a very transient mineral at all but the coldest parts of Earth's surface. (Ice is a major crust-forming mineral on planets of the outer solar system, where it cannot melt; box 2.9).

Some minerals result from biological activity; for example, the building of coral reefs creates huge masses of calcite-rich limestone. Many organisms, including human beings, create magnetite within their skull cases. Some researchers believe that birds and whales exploit the properties of this mineral to assist them in migratory navigation (Why do we make magnetite in our own heads?). Bacteria also form huge amounts of sulfur by processing preexisting sulfate minerals. Most of our commercial supply of sulfur, in fact, comes from the mining of these *biogenic* deposits.

Some minerals crystallize directly from volcanic gases around volcanic vents—a process termed *sublimation*. Examples include ordinary sulfur, ralstonite, and thenardite (used as a natural rat poison). Sublimates are much less common than precipitates, though on planets and moons with intense volcanic activity, like Venus and Io, they cover wide swaths of planetary surface in thick beds.

We are able to understand the conditions of formation of most minerals with varying degrees of accuracy and precision using the tools of chemistry, especially with an understanding of thermodynamics and solutions. In fact, as implied at the beginning of this chapter, a good grasp of chemistry is a necessity for any advanced study of minerals.

IN GREATER DEPTH 2.9

Water and Ice—Molecules and Crystals

Earth is often called the *blue planet* because oceans cover 70% of its surface. Ice dominates our planet's polar regions. Perhaps "Aqua" rather than "Earth" would be a more appropriate name for our planet. It is fortunate that water is so abundant because life would be impossible without it. In fact, we humans are made up mostly of water. The nature and behavior of water molecules helps explain why water is vital to life on Earth.

In a water molecule, the two hydrogen atoms are tightly bonded to the oxygen atom. However, the shape of the molecule is asymmetrical, with the two hydrogen atoms on the same side of the atom (box figure 1). This means the molecule is polarized, with a slight excessive positive charge at the hydrogen side of the molecule and a slight excess negative charge at the opposite side. Because of the slight electrical attraction of water molecules, other substances are readily attracted to the molecules and dissolved or carried away by water. Water has been called the universal solvent. Dirt washes out of clothing; water, in blood, carries nutrients to our muscles and transports waste to our kidneys and out of our bodies.

When water is in its liquid state, the molecules are moving about. Because of the polarity, molecules are slightly attracted to one another. For this reason, water molecules are closer together than they are in most other liquids. However, in ice the water molecules are not as tightly packed together as in liquid water.

When water freezes, the hydrogen atoms are attracted to oxygen atoms in adjacent water molecules (box figure 2), resulting in an orderly, three-dimensional pattern that is hexagonal, as in a honeycomb (this explains the hexagonal shape of snowflakes). The openness of the honeycomblike, crystalline structure of ice contrasts with the more closely packed molecules in liquid water. This is the reason ice is less dense than liquid water. This is an unusual solid-liquid relationship. For most substances, the solid is denser than its liquid phase.

The fact that ice is less dense than liquid water has profound implications. Ice floats rather than sinks in liquid water. Icebergs float in the ocean. Lakes freeze from the top down. Ice on a lake surface acts as an insulating layer that retards the freezing of underlying water. If ice sank, lakes would freeze much more readily and thaw much more slowly. Our climate would be very different if ice sank. The Arctic Ocean surface freezes during the winter but only at its surface. If the ice were to sink, more ocean water would be exposed to the cold atmosphere and would freeze and sink. Eventually, the entire Arctic Ocean would freeze and would not thaw during the summer. If this were the case, life, as we know it, probably would not exist.

BOX 2.9 ■ FIGURE 1

Water molecule.

BOX 2.9 ■ FIGURE 2

Hexagonal structure of ice. Small, black dots represent the attraction between hydrogen atoms and oxygen atoms for adjacent water molecules.

When water freezes, it expands. A bottled beverage placed in a freezer breaks its container upon freezing. When water trapped in cracks in rock freezes, it will expand and will help break up the rock (as explained in the chapter on weathering).

Additional Resources

Snow Crystal Research
Nice images taken with an electron microscope.
- www.lpsi.barc.usda.gov/emusnow/

Snow Crystals
Caltech's site. More about ice and nice pictures of snow crystals. Click on "Ice Properties" under "Snowflake Physics" to see a model of the arrangement of oxygen and hydrogen atoms in the crystal structure of ice.
- www.its.caltech.edu/~atomic/snowcrystals/

SUMMARY

Atoms are composed of *protons* (+), *neutrons,* and *electrons* (−). A given element always has the same number of protons. An atom in which the positive and negative electric charges do not balance is an *ion.*

Ions or atoms bond together in very orderly, three-dimensional structures that are *crystalline.*

A *mineral* is a crystalline substance that is naturally occurring and is chemically and physically distinctive. Minerals are the building blocks of rocks.

The two most abundant elements in the Earth's crust are oxygen and silicon. Most minerals are silicates, having the silicon-oxygen tetrahedron as their basic building block.

Minerals are usually identified by their physical properties. Cleavage is perhaps the most useful physical property for identification purposes. Other important physical properties are external crystal form, fracture, hardness, luster, color, streak, and specific gravity.

Terms to Remember

atom 32
atomic mass number 34
atomic number 34
atomic weight 34
bonding 33
chain silicate structure 38
cleavage 47
covalent bonding 36
crystal form 45
crystalline 30
density 49
earthy luster 44
electron 32

element 33
ferromagnesian mineral 43
fracture 49
framework silicate structure 39
glassy (vitreous) luster 44
hardness 44
ion 33
ionic bonding 36
isolated silicate structure 38
isotope 34
luster 43
metallic luster 44
mineral 30

Mohs' hardness scale 44
neutron 32
nonmetallic luster 44
nucleus 32
proton 32
sheet silicate structure 39
silica 35
silicates 37
silicon-oxygen tetrahedron 37
specific gravity 49
streak 43
striations 50

Testing Your Knowledge

Use the following questions to prepare for exams based on this chapter.

1. Compare feldspar and quartz.
   a. How do they differ chemically?
   b. What type of silicate structure does each have?
   c. How would you distinguish between them on the basis of cleavage?
2. How do the crystal structures of pyroxenes and amphiboles differ from one another?
3. How do the various feldspars differ from one another chemically?
4. Distinguish between the following pairs of terms:
   silica/silicate
   silicon/silicon-oxygen tetrahedron
5. What is the distinction between cleavage and external crystal form?

6. How would you distinguish the following on the basis of physical properties? (You might refer to appendix A.)
   feldspar/quartz
   muscovite/feldspar
   calcite/feldspar
   pyroxene/feldspar
7. Using triangles to represent tetrahedrons, start with a single triangle (to represent isolated silicate structure) and, by drawing more triangles, build on the triangle to show a single-chain silicate structure. By adding more triangles, convert that to a double-chain structure. Turn your double-chain structure into a sheet silicate structure.
8. What major factor controls chemical activity between atoms?

9. What are the three most common elements (by number and approximate percentage) in the Earth's crust?

10. What are the next five most common elements?

11. A substance that cannot be broken down into other substances by ordinary chemical methods is a(n)

   a. crystal   b. element

   c. molecule   d. compound

12. The subatomic particle that contributes mass and a single positive electrical charge is the

   a. proton   b. neutron

   c. electron

13. Atoms containing different numbers of neutrons but the same number of protons are called

   a. compounds   b. ions

   c. elements   d. isotopes

14. Atoms with either a positive or negative charge are called

   a. compounds   b. ions

   c. elements   d. isotopes

15. The bonding between Cl and Na in halite is

   a. ionic   b. covalent

   c. metallic   d. male

16. Which is not true of a single silicon-oxygen tetrahedron?

   a. The atoms of the tetrahedron are strongly bonded together.

   b. It has a net negative charge.

   c. The formula is SiO_4.

   d. It has four silicon atoms.

17. Which is not a type of silicate structure?

   a. isolated   b. single chain

   c. double chain   d. sheet

   e. framework   f. pentagonal

18. Which of the common minerals is not a silicate?

   a. quartz   b. calcite

   c. pyroxene   d. feldspar

   e. biotite

19. On Mohs' hardness scale, ordinary window glass has a hardness of about

   a. 2–3   b. 3–4

   c. 5–6   d. 7–8

20. The ability of a mineral to break along preferred directions is called

   a. fracture   b. crystal form

   c. hardness   d. cleavage

21. Striations are associated with

   a. quartz   b. mica

   c. potassium feldspar   d. plagioclase

22. Glass is

   a. atoms randomly arranged   b. crystalline

   c. ionically bonded   d. covalently bonded

23. Crystalline substances are always

   a. ionically bonded   b. minerals

   c. made of repeating patterns of atoms   d. made of glass

Expanding Your Knowledge

1. Why are nonsilicate minerals more common on the surface of the Earth than within the crust?

2. How does oxygen in the atmosphere differ from oxygen in rocks and minerals?

3. What happens to the atoms in water when it freezes? Is ice a mineral? Is a glacier a rock?

4. How would you expect the appearance of a rock high in iron and magnesium to differ from a rock with very little iron and magnesium?

Exploring Web Resources

www.mhhe.com/plummer11e

This is the dedicated website for this book. You can go to it for new and updated information. The universal resource locators (URL) listed in this book are also given as links on the website, making it easy to go to those websites without typing in the URL. When you visit our Online Learning Center, go to the Student Center and then to the chapter of interest. Here you will find additional readings and media resources, as well as answers to the Testing Your Knowledge section, more quizzes, animation and video clips, and direct links to the sites listed in this book. Links to additional websites can also be found. We have added questions for some of the sites to allow you to get the most of your exploration of the web. Using the web is an enjoyable way of enhancing your knowledge of geology.

www.rockhounds.com/

Bob's Rock Shop. Contains a great amount of information for mineral collectors. Find and go to "crystallography and mineral crystal systems" for a more in-depth study of crystallography than presented in this book.

www.minweb.co.uk/

Mineral Web. Crystal structures are displayed in 3-D. The structures rotate, and you can manipulate the rotation using a mouse. You must install Chime, a program for viewing the structures. Chime can be downloaded easily from this site. Once installed, take a look at the crystal structures of diamond, olivine, muscovite, and other minerals. You can also observe the various silicon-oxygen tetrahedron structures.

http://webmineral.com/

Mineral Database. There are descriptions of close to 4,000 mineral species. The descriptions include mineral properties beyond the scope of an introductory geology course; however, there are links to other sites that include pictures of minerals.

www.mindat.org/

The *online mineralogy resource.* Another comprehensive source for mineral information. You can search for a mineral by name or by locality.

www.theimage.com/

The Image. Photos of minerals and gems. Click on Mineral Gallery and choose a mineral to view photos and properties of that mineral. The Gemstone Gallery has photos of gem minerals.

www.webelements.com/

Web elements periodic table. The periodic table of elements. You can click on an element to determine its properties.

www.uky.edu/Projects/Chemcomics/

The comic book periodic table of elements. An entertaining site in which you click on an element and see examples of comic book stories about that element.

 # Animations

This chapter includes the following animation available on our Online Learning Center at **www.mhhe.com/plummer11e**.

2.7–2.8 Silicate mineral structures

Igneous Rocks, Intrusive Activity, and the Origin of Igneous Rocks

A geologist investigating intrusive rocks in northern Victoria Land, Antarctica.
Photo by C. C. Plummer

Chapters 3 and 4 are about igneous rocks and igneous processes. (Either chapter may be read first.) Chapter 4 focuses on volcanoes and igneous activity that takes place at the Earth's surface. Chapter 3 describes igneous processes that take place underground. However, you will learn early in this chapter how volcanic as well as intrusive rocks are classified based on their grain size and mineral content.

We begin the chapter by introducing the rock cycle. This is a conceptual device that shows the interrelationship between igneous, sedimentary, and metamorphic rocks. We then begin focusing on igneous rocks. After the section on igneous rock classification, we describe structural relationships between bodies of intrusive rock and other rocks in the Earth's crust. This is followed by a discussion of how magmas form and are altered. We conclude by discussing various hypotheses that relate igneous activity to plate tectonic theory.

 ## Relationships to Earth Systems

Our atmosphere and hydrosphere are products of intense igneous activity during the very early history of Earth. At that time, over 4 billion years ago, the mantle was largely molten, and as hot magma welled upward, hot gases were released to form the oceans and atmosphere. During the billions of years since the oceans and atmosphere formed, solidifying molten rock has released water (and circulated ground water) that contains dissolved elements. When the water passes through cooler rocks, the elements crystallize into minerals, some of which are vital to civilization. Copper, lead, gold, and other metals are mined from these ore deposits. A unique part of the biosphere thrives at very hot springs along the sea floor where magmatically heated water meets seawater. Volcanic activity relationships to Earth systems are discussed in chapter 4.

THE ROCK CYCLE

A **rock** is naturally formed, consolidated material usually composed of grains of one or more minerals. You will see how some minerals break down chemically and form new minerals when a rock finds itself in a new physical setting. For instance, feldspars that may have formed at high temperatures deep within the Earth can react with surface waters to become clay minerals at the Earth's surface.

As mentioned in chapter 1, the Earth changes because of its internal and external heat engines. If the Earth's internal engine had died (and tectonic forces had therefore stopped operating), the external engine plus gravity would long ago have leveled the continents virtually at sea level. The resulting sediment would have been deposited on the sea floor. Solid Earth would not be changing (except when struck by a meteorite or other extraterrestrial body). The rocks would be at rest. The minerals, water, and atmosphere would be in *equilibrium* (and geology would be a dull subject). But this is not the case. The internal and external forces continue to interact, forcing substances out of equilibrium. Therefore, the Earth has a highly varied and ever-changing surface. And minerals and rocks change as well.

A useful aid in visualizing these changing relationships is the **rock cycle** shown in figure 3.1. The three major rock types—igneous, metamorphic, and sedimentary—are shown. As you see, each may form at the expense of another if it is forced out of equilibrium with its physical or climatic environment by either internal or surficial forces. It is important to be aware that rock moves from deep to shallow, and from high to low in response to tectonic forces and isostasy (covered in chapter 1).

As described in chapter 1, *magma* is molten rock. *Igneous rocks* form when magma solidifies. If the magma is brought to the surface by a volcanic eruption, it may solidify into an *extrusive* igneous rock. Magma may also solidify very slowly

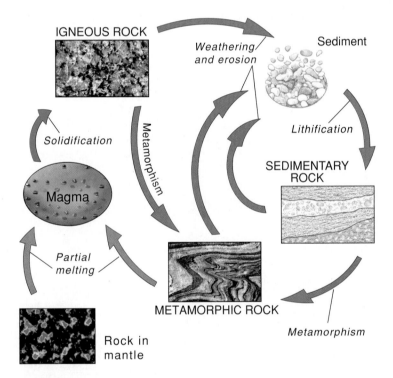

FIGURE 3.1

The rock cycle.

beneath the surface. The resulting *intrusive* igneous rock may be exposed later after uplift and erosion remove the overlying rock (as shown in figure 1.14). The igneous rock, being out of equilibrium, may then undergo *weathering* and *erosion,* and the debris produced is transported and ultimately deposited (usually on a sea floor) as *sediment.* If the unconsolidated sediment becomes *lithified* (cemented or otherwise consolidated into a rock), it becomes a *sedimentary rock.* As the rock is buried by additional layers of sediment and sedimentary rock, heat and pressure increase. Tectonic forces may also increase the temperature and pressure. If the temperature and pressure become high enough, usually at depths greater than several kilometers below the surface, the original sedimentary rock is no longer in equilibrium and recrystallizes. The new rock that forms is called a *metamorphic rock*. If the temperature gets very high, the rock partially melts, producing magma and completing the cycle.

The cycle can be repeated, as implied by the arrows in figure 3.1. However, there is no reason to expect all rocks to go through each step in the cycle. For instance, sedimentary rocks might be uplifted and exposed to weathering, creating new sediment.

We should emphasize that the rock cycle is a conceptual device to help students place the common rocks and how they form in perspective. As such, it is a simplification and does not encompass all geologic processes. For instance, most magma comes from partial melting of the mantle. Note from the diagram that this magma does not come from recycled rocks, so, strictly speaking is not part of a "cycle."

A Plate Tectonic Example

One way of relating the rock cycle to plate tectonics is illustrated by an example from what happens at a convergent plate boundary (figure 3.2). *Magma* is created in the zone of melting above the subduction zone. The magma, being less dense than adjacent rock, works its way upward. A volcanic eruption takes place if magma reaches the surface. The magma solidifies into *igneous rock*. The igneous rock is exposed to the atmosphere and subjected to *weathering* and *erosion*. The resulting *sediment* is transported and then deposited in low-lying areas. In time, the buried layers of sediment solidify into *sedimentary rock*. The sedimentary rock becomes increasingly more deeply buried as more sediment accumulates. After the sedimentary rock is buried to depths of several kilometers, the heat and pressure become too great and the rock recrystallizes into a *metamorphic rock*. As the depth of burial becomes even greater (several tens of kilometers), the metamorphic rock may find itself in a zone of melting. Temperatures are now high enough so that the metamorphic rock partially melts. Magma is created, thus completing the cycle.

FIGURE 3.2

The rock cycle with respect to a convergent plate boundary. Magma solidifies as igneous rock at the volcano. Sediment from the eroded volcano collects in the basin to the right of the diagram. Sediment converts to sedimentary rock as it is buried by more sediment. Deeply buried sedimentary rocks are metamorphosed. The most deeply buried metamorphic rocks partially melt, and the magma moves upward. An alternate way the rock cycle works is shown on the left of the diagram. Sediment from the continent (and volcano) becomes sedimentary rock, some of which is carried down the subduction zone. It is metamorphosed as it descends. It may contribute to the magma that forms in the mantle above the subduction zone.

The rock cycle diagram reappears on the opening pages of chapters 4 through 7. The highlighted portion of the diagram will indicate where the material covered in each chapter fits into the rock cycle.

IGNEOUS ROCKS

If you go to the island of Hawaii, you might observe red hot lava flowing over the land, and, as it cools, solidifying into the fine-grained (the grains are less than 1 millimeter across), black rock we call basalt. Basalt is an **igneous rock,** rock that has solidified from magma. **Magma** is molten rock, usually rich in silica and containing dissolved gases. (**Lava** is magma on the Earth's surface.) Igneous rocks may be either **extrusive** if they form at the Earth's surface (e.g., basalt) or **intrusive** if magma solidifies underground. **Granite,** a coarse-grained (the grains are larger than 1 millimeter) rock composed predominantly of feldspar and quartz, is an intrusive rock. In fact, granite is the most abundant intrusive rock found in the continents.

Unlike the volcanic rock in Hawaii, nobody has ever seen magma solidify into intrusive rock. So what evidence suggests that bodies of granite (and other intrusive rocks) solidified underground from magma?

- Mineralogically and chemically, intrusive rocks are essentially identical to volcanic rocks.
- Volcanic rocks are fine-grained (or glass) due to their rapid solidification; intrusive rocks are generally coarse-grained, which is inferred to mean that the magma crystallized slowly underground.
- Experiments have confirmed that most of the minerals in these rocks can form only at high temperatures. Other experiments indicate that some of the minerals could have formed only under high pressures, implying they were deeply buried. More evidence comes from examining *intrusive contacts,* such as shown in figures 3.3 and 3.4. (A **contact** is a surface separating different rock types. Other types of contacts are described elsewhere in this book.)
- Preexisting solid rock, *country rock,* appears to have been forcibly broken by an intruding liquid, with the magma flowing into the fractures that developed. **Country rock,** incidentally, is an accepted term for any older rock into which an igneous body intruded.
- Close examination of the country rock immediately adjacent to the intrusive rock usually indicates that it appears "baked," or *metamorphosed,* close to the contact with the intrusive rock.

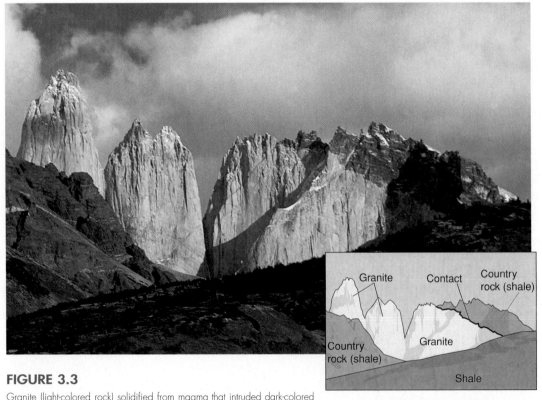

FIGURE 3.3

Granite (light-colored rock) solidified from magma that intruded dark-colored country rock in Torres del Paine, Chile. The dark-colored country rock is shale deposited in a marine environment. The spires are erosional remnants of rock that were once deep underground. *Photo by Kay Kepler*

Geologist's View

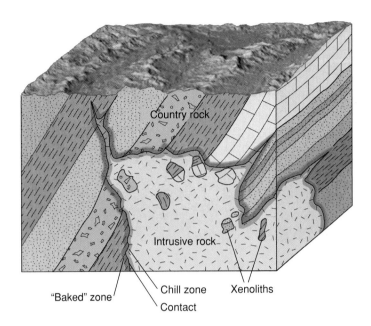

FIGURE 3.4

Igneous rock intruded preexisting rock (country rock) as a liquid. (Xenoliths are usually much smaller than indicated.)

- Rock types of the country rock often match **xenoliths,** fragments of rock that are distinct from the body of igneous rocks in which they are enclosed.

- In the intrusive rock adjacent to contacts with country rock are **chill zones,** finer-grained rocks that indicate magma solidified more quickly here because of the rapid loss of heat to cooler rock.

Laboratory experiments have greatly increased our understanding of how igneous rocks form. However, geologists have not been able to artificially make coarsely crystalline granite. Only very fine-grained rocks containing the minerals of granite have been made from artificial magmas, or "melts." The temperature and pressure at which granite apparently forms can be duplicated in the laboratory—but not the time element. According to calculations, a large body of magma requires over a million years to solidify completely. This very gradual cooling causes the coarse-grained texture of most intrusive rocks. Chemical processes involving silicates are known to be exceedingly slow. Yet another problem in trying to apply experimental procedures to real rocks is determining the role of water and other gases in the crystallization of rocks such as granite. Only a small amount of gases is retained in rock crystallized underground from a magma, but large amounts of gas (especially water vapor) are released during volcanic eruptions. No one has seen an intrusive rock forming; hence, we can only speculate about the role these gases might have played before they escaped. One example shows why gases are important. Laboratory studies have shown that granite can melt at temperatures as low as 650°C if water is present and under high pressure. Without the water, the melting temperature is several hundred degrees higher. Not knowing how much water was present during crystallization makes accurate determination of temperatures difficult and speculative.

Igneous Rock Textures

Texture refers to a rock's appearance with respect to the size, shape, and arrangement of its grains or other constituents. Most (but not all) igneous rocks are *crystalline;* that is, they are made of interlocking crystals (of, for instance, quartz and feldspar). The most significant aspect of texture in igneous rocks is grain (or crystal) size. Extrusive rocks typically are **fine-grained rocks,** in which most of the grains are smaller than 1 millimeter. The grains, if they are crystals, are small because magma cools rapidly at the Earth's surface, and so they have less time to form. Some intrusive rocks are also fine-grained; these occur as smaller bodies that apparently solidified near the surface upon intrusion into relatively cold country rock (probably within a couple kilometers of the Earth's surface). *Basalt, andesite,* and *rhyolite* are the common fine-grained igneous rocks. Igneous rocks that formed at considerable depth—usually more than several kilometers—are called **plutonic rocks** (after Pluto, the Roman god of the underworld). Characteristically, these rocks are coarse-grained, reflecting the slow cooling and solidification of magma. For our purposes, **coarse-grained** (or **coarsely crystalline**) **rocks** are defined as those in which most of the grains are larger than 1 millimeter. The crystalline grains of plutonic rocks are commonly interlocked in a mosaic pattern (figure 3.5). An extremely coarse-grained (grains over 5 centimeters) igneous rock is called a *pegmatite* (see box 3.1).

The crystals or grains of most fine-grained rocks are considerably smaller than 1 millimeter and cannot be distinguished by the unaided eye. So, for practical purposes, if you can discern the individual grains, regard the rock as coarse-grained; if not, consider it fine-grained.

Some rocks are **porphyritic;** that is, large crystals are enclosed in a *groundmass* of finer-grained crystals or glass. An analogy for porphyritic texture is a milk chocolate bar containing whole almonds. If the groundmass is fine-grained, extrusive rock names are used. For instance, figure 3.7 shows a *porphyritic andesite.* Porphyritic extrusive rocks are usually interpreted as having begun crystallizing slowly underground followed by eruption and rapid solidification of the remaining magma at the Earth's surface. Some porphyritic rocks have a coarse-grained groundmass in which the individual grains are over 1 millimeter. The larger crystals enclosed in the groundmass are much bigger, usually two or more centimeters across. *Porphyritic granite* is an example.

Identification of Igneous Rocks

Igneous rock names are based on texture (notably grain size) and mineralogical composition (which reflects chemical composition). Mineralogically (and chemically) equivalent rocks are *granite-rhyolite, diorite-andesite,* and *gabbro-basalt.* The relationships between igneous rocks are shown in figure 3.6.

A   B

FIGURE 3.5

(A) Coarse-grained texture characteristic of plutonic rock. Feldspars are white and pink. Quartz is transparent. Biotite mica is black. A U.S. penny is used for a scale as the "roof" of the monument is 1 millimeter thick and 1 centimeter wide. (B) A similar rock seen through a polarizing microscope. Note the interlocking crystal grains of individual minerals. *Photos by C. C. Plummer*

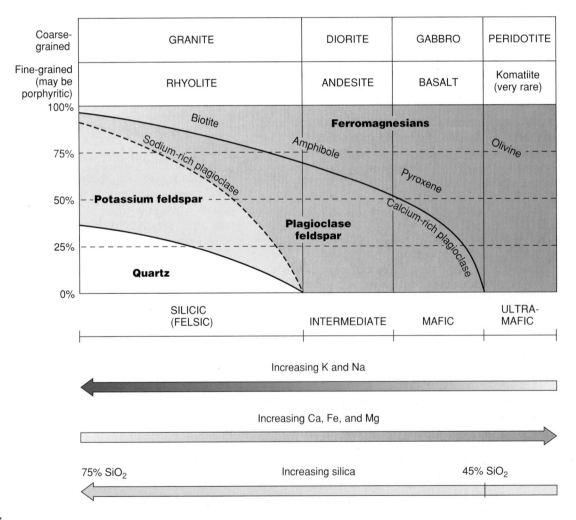

FIGURE 3.6

Classification chart for the most common igneous rocks. Rock names based on special textures are not shown. Sodium-rich plagioclase is associated with silicic rocks, whereas calcium-rich plagioclase is associated with mafic rocks. The names of the particular ferromagnesian minerals (biotite, etc.) are placed in the diagram at the approximate composition of the rocks in which they are most likely to be found.

Pegmatite—A Rock Made of Giant Crystals

Pegmatites are extremely coarse-grained igneous rocks. In some pegmatites, crystals are as large as 10 meters across. Strictly speaking, a pegmatite can be of diorite, gabbro, or granite. However, the vast majority of pegmatites are silicic, with very large crystals of potassium feldspar, sodium-rich plagioclase feldspars, and quartz. Hence, the term *pegmatite* generally refers to a rock of granitic composition (if otherwise, a term such as *gabbroic pegmatite* is used). Pegmatites are interesting as geological phenomena and important as minable resources.

The extremely coarse texture of pegmatites is attributed to both slow cooling and the low viscosity (resistance to flow) of the fluid from which they form. Lava solidifying to rhyolite is very viscous. Magma solidifying to granite, being chemically similar, should be equally viscous.

Pegmatites, however, probably crystallize from a fluid composed largely of water under high pressure. Water molecules and ions from the parent, granitic magma make up a residual magma. Geologists believe the following sequence of events accounts for most pegmatites.

As a granite pluton cools, increasingly more of the magma solidifies into the minerals of a granite. By the time the pluton is well over 90% solid, the residual magma contains a very high amount of silica and ions of elements that will crystallize into potassium and sodium feldspars. Also present are elements that could not be accommodated into the crystal structures of the common minerals that formed during the normal solidification phase of the pluton. Fluids, notably water, that were in the original magma are left over as well. If no fracture above the pluton permits the fluids to escape, they are sealed in, as in a pressure cooker. The watery residual magma has a low viscosity, which allows appropriate atoms to migrate easily toward growing crystals. The crystals add more and more atoms and grow very large.

Pegmatite bodies are generally quite small. Many are pod-like structures, located either within the upper portion of a granite pluton or within the overlying country rock near the contact with granite, the fluid body evidently having squeezed into the country rock before solidifying. Pegmatite dikes are fairly common, especially within granite plutons, where they apparently filled cracks that developed in the already solid granite. Some pegmatites form small dikes along contacts between granite and country rock, filling cracks that developed as the cooling granite pluton contracted.

Most pegmatites contain only quartz, feldspar, and perhaps mica. Minerals of considerable commercial value are found in a few pegmatites. Large crystals of muscovite mica are mined from pegmatites. These crystals are called "books" because the cleavage flakes (tens of centimeters across) look like pages. Because muscovite is an excellent insulator, the cleavage sheets are used in electrical devices, such as toasters, to separate uninsulated

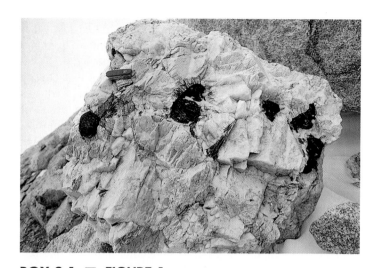

BOX 3.1 ■ FIGURE 1

Pegmatite in northern Victoria Land, Antarctica. The knife is 8 centimeters long. The black crystals are tourmaline. Quartz and feldspar are light colored. *Photo by C. C. Plummer*

electrical wires. Even the large feldspar crystals in pegmatites are mined for various industrial uses, notably the manufacture of ceramics.

Many rare elements are mined from pegmatites. These elements were not absorbed by the minerals of the main pluton and so were concentrated in the residual pegmatitic magma, where they crystallized as constituents of unusual minerals. Minerals containing the element lithium are mined from pegmatites. Lithium becomes part of a sheet silicate structure to form a pink or purple variety of mica (called lepidolite). Uranium ores, similarly concentrated in the residual melt of magmas, are also extracted from pegmatites.

Some pegmatites are mined for gemstones. Emerald and aquamarine, varieties of the mineral beryl, occur in pegmatites that crystallized from a solution containing the element beryllium. A large number of the world's very rare minerals are found only in pegmatites, many of these in only one known pegmatite body. These rare minerals are mainly of interest to collectors and museums.

Hydrothermal veins (described in chapter 7) are closely related to pegmatites. Veins of quartz are common in country rock near granite. Many of these are believed to be caused by water that escapes from the magma. Silica dissolved in the very hot water cakes on the walls of cracks as the water cools while traveling surfaceward. Sometimes valuable metals such as gold, silver, lead, zinc, and copper are deposited with the quartz in veins.

Because of their larger mineral grains, plutonic rocks are easier to identify than extrusive rocks. The physical properties of each mineral in a plutonic rock can be determined more readily. And, of course, knowing what minerals are present makes rock identification a simpler task. For instance, **gabbro** is formed of coarse-grained ferromagnesian minerals and gray, plagioclase feldspar. (Recall from the mineral chapter that ferromagnesian minerals are silicates that contain iron and magnesium—amphibole, pyroxene, olivine, and biotite.) One can positively identify the feldspar on the basis of cleavage and, with practice, verify that no quartz is present. Gabbro's fine-grained counterpart is **basalt,** which is also composed of ferromagnesian minerals and plagioclase. The individual minerals cannot be identified by the naked eye, however, and one must use the less reliable attribute of color—basalt is usually dark gray to black.

As you can see from figure 3.6, granite and **rhyolite** are composed predominantly of feldspars (usually white or pink) and quartz. Granite, being coarse-grained, can be positively identified by verifying that quartz is present. Rhyolite is usually cream-colored, tan, or pink. Its light color indicates that ferromagnesian minerals are not abundant. **Diorite** and **andesite** are composed of feldspars and significant amounts of ferromagnesian minerals (30–50%). The minerals can be identified and their percentages estimated to indicate diorite. Andesite, being fine-grained, can usually be identified by its medium-gray or medium-green color. Its appearance is intermediate between light-colored rhyolite and dark basalt.

Use the chart in figure 3.6 along with table 3.1 to identify common igneous rocks. You may also find it helpful to turn to appendix B, which includes a key for identifying common igneous rocks. (Photos of typical igneous rocks are shown in figure 3.7.)

Varieties of Granite

Granite and rhyolite occupy a larger area in the classification chart than do the other rocks. This reflects their greater variation in composition. Figure 3.8 reproduces the granite-rhyolite field shown in figure 3.6. Rocks X and Y represent two quite different granites (or rhyolites). Figure 3.8 shows the mineral composition of each of these rocks as determined by using the percentage scale on the left of the diagram. Rock X has 34% quartz and 52% potassium feldspar (together, 86% of the rock). Only 8% is sodium-rich plagioclase feldspar and 6% ferromagnesian minerals (most likely biotite mica). This is a very silica-rick (*silicic*) rock. By contrast, granite (or rhyolite) Y is close to the diorite-andesite field and has a composition that has only 11% quartz and 21% potassium feldspar. Plagioclase feldspar is, at 44%, the most abundant mineral and 24% of the rock is ferromagnesian minerals (likely a mixture of biotite and amphibole). Geologists have subdivided the field of granite and named each of the varieties; a rock in the right portion of the field, rock Y, for example, is called granodiorite.

Any classification system is, of course, a human device, and for this reason, classification systems differ somewhat among groups of geologists. We define the boundary between granite and diorite by the presence or absence of quartz; but we could just as easily have placed the boundary slightly to the left, so that a rock with 10% or less quartz would be diorite.

Chemistry of Igneous Rocks

The chemical composition of the magma determines which minerals and how much of each will crystallize when an igneous rock forms. For instance, the presence of quartz in a rock indicates that the magma was enriched in silica (SiO_2). The lower part of figure 3.6 shows the relationship of chemical composition to rock type. Chemical analyses of rocks are reported as weight percentages of oxides (e.g., SiO_2, MgO, Na_2O, etc.) rather than as separate elements (e.g., Si, O, Mg, Na). Figure 3.9 shows the chemical composition of average rocks. For virtually all igneous rocks, SiO_2 (silica) is the most abundant component. The amount of SiO_2 varies from about 45% to 75% of the total weight of common volcanic rocks. The variations between these extremes account for striking differences in the appearance and mineral content of the rocks.

Mafic Rocks

Rocks with a silica content close to 50% (by weight) are considered *silica-deficient*, even though SiO_2 is, by far, the most abundant constituent (figure 3.9). Chemical analyses show that the remainder is composed mostly of the oxides of aluminum (Al_2O_3), calcium (CaO), magnesium (MgO), and iron (FeO and

TABLE 3.1	**Identification of Igneous Rocks**			
Coarse-Grained	Granite	Diorite	Gabbro	Peridotite
Fine-Grained (often porphyritic)	Rhyolite	Andesite	Basalt	—
Mineral Content	Quartz, feldspars (white, light gray, or pink). Minor ferromagnesian minerals.	Feldspars (white or gray) and about 35–50% ferromagnesian minerals. No quartz.	Predominance of ferromagnesian minerals. Rest of rock is plagioclase feldspar (medium to dark gray).	Entirely ferromagnesian minerals (olivine and pyroxene).
Color of Rock (most commonly)	Light-colored	Medium-gray or medium-green	Dark gray to black	Green to black

FIGURE 3.7

Samples of common igneous rocks and their relationship to the classification diagram (figure 3.6). Peridotite is not shown. Do not try to identify real rocks by simply comparing them to photos—use the properties, such as identifying minerals and their amounts. *Photo of gabbro by Larry Davis; all others by C. C. Plummer*

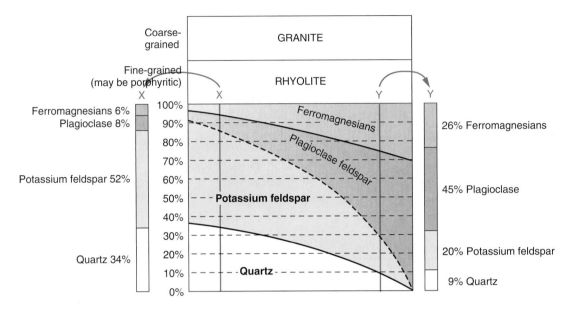

FIGURE 3.8

The compositional variation of granite shown in figure 3.6. Granite x and granite y are quite different in the proportions of minerals present. The bar graphs for each show the percentage of minerals in each rock.

FIGURE 3.9

The average chemical composition of silicic, intermediate, and mafic rocks. Composition is given in weight percent of oxides. Note that as the amount of silica decreases, the oxides of Na and K decrease, and the oxides of Ca, Fe, and Mg increase. Al oxide does not vary significantly.

Fe_2O_3). (These oxides generally combine with SiO_2 to form the silicate minerals as described in chapter 2.) Rocks in this group are called **mafic**—silica-deficient igneous rocks with a relatively high content of magnesium, iron, and calcium. (The term *mafic* comes from *magnesium* and *ferric*.) Basalt and gabbro are, of course, mafic rocks.

Silicic (Felsic) Rocks

At the other extreme, the *silica-rich* (65% or more SiO_2) rocks tend to have only very small amounts of the oxides of calcium, magnesium, and iron. The remaining 25% to 35% of these rocks is mostly aluminum oxide (Al_2O_3) and oxides of sodium (Na_2O) and potassium (K_2O). These are called **silicic** or **felsic** rocks—silica-rich igneous rocks with a relatively high content of potassium and sodium (the *fel* part of the name comes from *feldspar*, which crystallizes from the potassium, sodium, aluminum, and silicon oxides; *si* in *felsic* is for silica). The silicic rocks rhyolite and granite are light-colored because of the low amount of ferromagnesian minerals.

Intermediate Rocks

Rocks with a chemical content between that of felsic and mafic are classified as **intermediate rocks.** *Andesite,* which is usually green or medium gray, is the most common intermediate volcanic rock.

Ultramafic Rocks

An **ultramafic rock** is composed entirely or almost entirely of ferromagnesian minerals. No feldspars are present and, of course, no quartz. **Peridotite,** a coarse-grained rock composed of pyroxene and olivine, is the most abundant ultramafic rock. Chemically, these rocks contain less than 45% silica.

Note from the chart (figure 3.6) that komatiite, the volcanic ultramafic rock, is very rare. Ultramafic extrusive rocks are mostly restricted to the very early history of the Earth. For our purposes, they need not be discussed further.

Some ultramafic rocks form from differentiation (explained later in this chapter) of a basaltic magma at very high temperatures. Most ultramafic rocks come from the mantle, rather than from the Earth's crust. Where we find large bodies of ultramafic rocks, the usual interpretation is that a part of the mantle has traveled upward as solid rock.

INTRUSIVE BODIES

Intrusions, or **intrusive structures,** are bodies of intrusive rock whose names are based on their size and shape, as well as their relationship to surrounding rocks. They are important aspects of the architecture, or *structure,* of the Earth's crust. The various intrusions are named and classified on the basis of the following considerations: (1) Is the body large or small? (2) Does it have a particular geometric shape? (3) Did the rock form at a considerable depth, or was it a shallow intrusion? (4) Does it follow layering in the country rock or not?

Shallow Intrusive Structures

Some igneous bodies apparently solidified near the surface of the Earth (probably at depths of less than 2 kilometers). These bodies appear to have solidified in the subsurface "plumbing systems" of volcanoes or lava flows. Shallow intrusive structures tend to be relatively small compared with those that formed at considerable depth. Because the country rock near the Earth's surface generally is cool, intruded magma tends to chill and solidify relatively rapidly. Also, smaller magma bodies will cool faster than larger bodies, regardless of depth. For both of these reasons, shallow intrusive bodies are likely to be fine-grained.

A **volcanic neck** is an intrusive structure apparently formed from magma that solidified within the throat of a volcano. One of the best examples is Ship Rock in New Mexico (figure 3.10). Here is how geologists interpret the history of this feature. A volcano formed above what is now Ship Rock.

The magma for the volcano moved upward through a more or less cylindrical "plumbing system." Eruptions ceased and the magma underground solidified into what is now Ship Rock. In time, the volcano and its underlying rock—the country rock around Ship Rock—eroded away. The more resistant igneous body eroded more slowly into its present shape. Weathering and erosion are continuing (falling rock has been a serious hazard to rock climbers).

Dikes and Sills

Another, and far more common, intrusive structure can also be seen at Ship Rock. The low, wall-like ridge extending outward from Ship Rock is an eroded dike. A **dike** is a tabular (shaped like a tabletop), discordant, intrusive structure (figure 3.11). *Discordant* means that the body is not parallel to any layering in the country rock. (Think of a dike as cutting across layers of country rock.) Dikes may form at shallow depths and be fine-grained, such as those at Ship Rock, or form at greater depths and be coarser-grained. Dikes need not appear as walls protruding from the ground (figure 3.12). The ones at Ship Rock do so only because they are more resistant to weathering and erosion than the country rock.

A **sill** is also a tabular intrusive structure, but it is *concordant.* That is, sills, unlike dikes, are parallel to any planes or layering in the country rock (figures 3.11 and 3.13). Typically, the country rock bounding a sill is layered sedimentary rock. As magma squeezes into a crack between two layers, it solidifies into a sill.

If the country rock is not layered, a tabular intrusion is regarded as a dike.

A

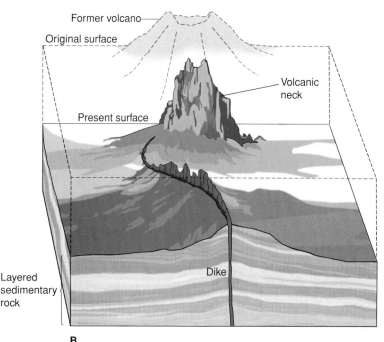

B

FIGURE 3.10

(A) Ship Rock in New Mexico, which rises 420 meters (1,400 feet) above the desert floor. (B) Relationship to the former volcano. *Photo by Frank M. Hanna*

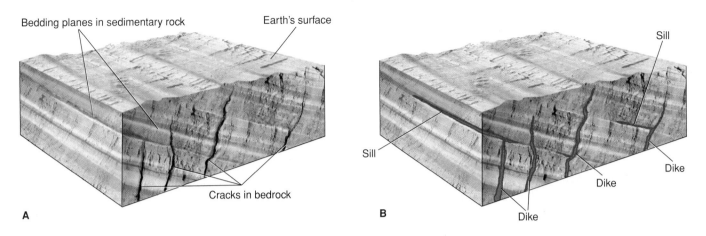

FIGURE 3.11

(A) Cracks and bedding planes are planes of weakness. (B) Concordant intrusions where magma has intruded between sedimentary layers are sills; discordant intrusions are dikes.

Intrusives that Crystallize at Depth

A **pluton** is a body of magma or igneous rock that crystallized at considerable depth within the crust. Where plutons are exposed at the Earth's surface, they are arbitrarily distinguished by size. A **stock** is a small discordant pluton with an outcrop area (i.e., the area over which it is exposed to the atmosphere) of less than 100 square kilometers. If the outcrop area is greater than 100 square kilometers, the body is called a **batholith** (figure 3.14). Most batholiths crop out over areas vastly greater than the minimum 100 square kilometers.

Although batholiths may contain mafic and intermediate rocks, they almost always are predominantly composed of granite. Detailed studies of batholiths indicate that they are formed of numerous, coalesced plutons. Apparently, large blobs of magma worked their way upward through the lower crust and collected 5 to 30 kilometers below the surface, where they solidified (figure 3.15). These blobs of magma,

FIGURE 3.12

Dikes (light-colored rocks) in northern Victoria Land, Antarctica. *Photo by C. C. Plummer*

FIGURE 3.13

A sill (dark layer) intruded between limestone layers, Glacier National Park, Montana. The limestone adjacent to the sill has been contact metamorphosed into light-colored marble (explained in chapter 7). *Photo © William E. Ferguson*

Geologist's View

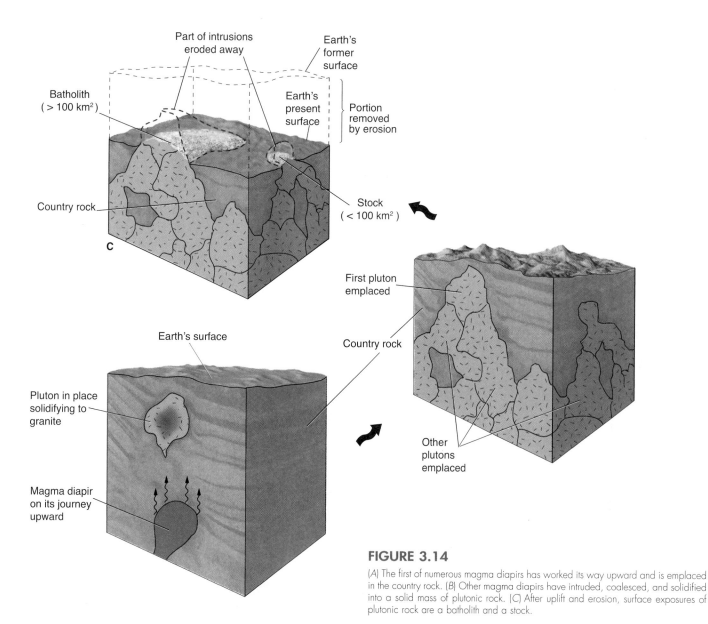

FIGURE 3.14

(A) The first of numerous magma diapirs has worked its way upward and is emplaced in the country rock. (B) Other magma diapirs have intruded, coalesced, and solidified into a solid mass of plutonic rock. (C) After uplift and erosion, surface exposures of plutonic rock are a batholith and a stock.

known as **diapirs,** are less dense than the surrounding rock that is pliable and shouldered aside as the magma rises. Batholiths occupy large portions of North America, particularly in the west. Over half of California's Sierra Nevada mountains (figure 3.16) is a batholith whose individual plutons were emplaced during a period of over 100 million years. An even larger batholith extends almost the entire length of the mountain ranges of Canada's west coast and southeastern Alaska—a distance of 1,800 kilometers. Smaller batholiths are also found in eastern North America in the Piedmont east of the Appalachian Mountains and in New England and the coastal provinces of Canada. (The extent and location of North American batholiths are shown on the geologic map on the inside front cover.)

Granite is considerably more common than rhyolite, its volcanic counterpart. Why is this? Silicic magma is much more *viscous* (that is, more resistant to flow) than mafic magma. Therefore, a silicic magma body will travel upward through the crust more slowly and with more difficulty than mafic or intermediate magma. Unless it is exceptionally hot, a silicic magma will not be able to work its way through the relatively cool and rigid rocks of the upper few kilometers of crust. Instead, it is much more likely to solidify slowly into a pluton.

ABUNDANCE AND DISTRIBUTION OF PLUTONIC ROCKS

Granite is the most abundant igneous rock in mountain ranges. It is also the most commonly found igneous rock in the interior lowlands of continents. Throughout the lowlands

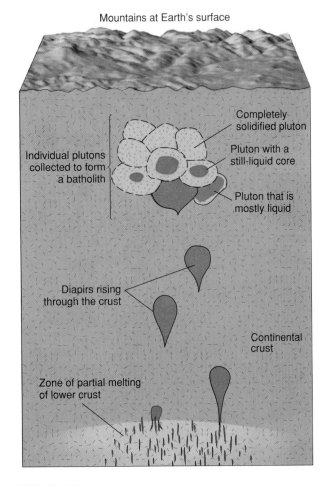

Mountains at Earth's surface

Individual plutons collected to form a batholith

Completely solidified pluton

Pluton with a still-liquid core

Pluton that is mostly liquid

Diapirs rising through the crust

Continental crust

Zone of partial melting of lower crust

FIGURE 3.15

Diapirs of magma travel upward from the lower crust and solidify in the upper crust. (Not drawn to scale.)

of much of Canada, very old plutons have intruded even older metamorphic rock. As explained in chapter 20 on mountains and the continental crust, very old mountain ranges have, over time, eroded and become the stable interior of a continent. Metamorphic and plutonic rocks similar in age and complexity to those in Canada are found in the Great Plains of the United States. Here, however, they are mostly covered by a veneer (a kilometer or so) of younger, sedimentary rock. These "basement" rocks are exposed to us in only a few places. In Grand Canyon, Arizona, the Colorado River has eroded through the layers of sedimentary rock to expose the ancient plutonic and metamorphic basement. In the Black Hills of South Dakota, local uplift and subsequent erosion have exposed similar rocks.

Granite, then, is the predominant igneous rock of the continents. As described in chapter 4, basalt and gabbro are the predominant rocks underlying the oceans. Andesite (usually along continental margins) is the building material of most young volcanic mountains. Underneath the crust, ultramafic rocks make up the upper mantle.

HOW MAGMA FORMS

If a rock is heated sufficiently, it begins melting to form magma. Under ideal conditions, rock can melt and yield a granitic magma at temperatures as low as 650°C. Temperatures over 1,000°C are required to create basaltic magma. However, several factors control the melting temperature of rock. Pressure, amount of gas (particularly water) present, and the particular mix of minerals all influence when melting takes place. These factors are discussed in the following sections.

FIGURE 3.16

Part of the Sierra Nevada batholith. All light-colored rock shown here (including that under the distant snow-covered mountains) is granite. The extent of the Sierra Nevada batholith is shown in the inset. *Photo by C. C. Plummer*

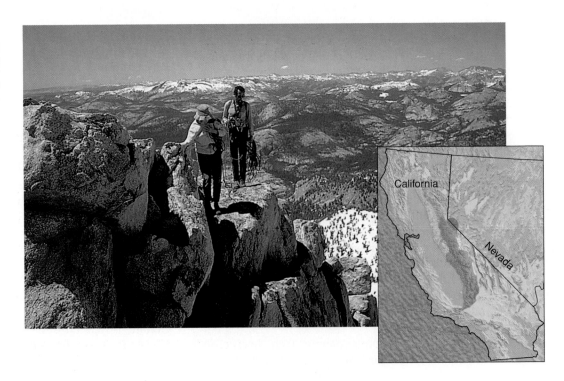

Heat for Melting Rock

Most of the heat that contributes to the generation of magma comes from the very hot Earth's core (where temperatures are estimated to be greater than 5,000°C). Heat is conducted toward the Earth's surface through the mantle and crust. This is comparable to the way heat is conducted through the wall from a hot room into a cooler room or through the metal of a frying pan. Heat is also brought from the lower mantle when part of the mantle flows upward, either through convection (described in chapters 1 and 19) or by hot mantle plumes. The geothermal gradient, described next, is a manifestation of heat transfer in the mantle.

Geothermal Gradient

A miner descending a mine shaft notices a rise in temperature. This is due to the **geothermal gradient,** the rate at which temperature increases with increasing depth beneath the surface. Data show the geothermal gradient, on the average, to be about 3°C for each 100 meters (30°C/km) of depth in the upper part of the crust. The geothermal gradient is not the same everywhere. Figure 3.17 shows geothermal gradients for two regions. The curve for the volcanic region indicates a higher geothermal gradient than that for the continental interior. Temperatures high enough to melt rock would be expected at a relatively shallow depth beneath the volcanic region. You would have to go deeper in the continental interior to reach the same temperature; however, the rock there does not melt because of the increased pressure at that depth.

One reason for a higher geothermal gradient is that deeper, and therefore hotter, mantle rock has worked its way upward closer to the Earth's surface due either to mantle convection or

FIGURE 3.17

Geothermal gradients at two parts of the Earth's crust.

mantle plumes. "Hot spots" in the crust (where the geothermal gradient is locally very high) have been hypothesized to be due to hot **mantle plumes,** which are narrow upwellings of hot material within the mantle. Mantle plumes have been used to explain some igneous activity, notably that which takes place in the interior of tectonic plates, far from a plate boundary. Examples include the long-lasting volcanic activity that built the Hawaiian Islands and the eruptions at Yellowstone National Park in Wyoming. The silicic eruptions at Yellowstone that took place some 600,000 years ago were much larger and more violent than any eruptions that have occurred in historical time. The hypothesis that these eruptions were caused by a plume originating deep in the mantle was, until very recently, widely accepted. However, researchers using new data have questioned the extent, and even the existence, of deep mantle plumes. (For an in-depth discussion on the debate about mantle plumes, go to www.mantleplumes.org/.)

Factors that Control Melting Temperatures

Pressure

The melting point of a mineral generally *increases* with increasing pressure. Pressure increases with depth in the Earth's crust, just as temperature does. So a rock that melts at a given temperature at the surface of the Earth requires a higher temperature to melt deep underground. Rock will not melt where the geothermal gradient is close to that of the plate interior because at all depths, the melting temperature will always be higher than the temperature of the rock. Thus, we need mechanisms that raise the rock to a higher temperature or lower the melting temperature for the rock.

Decompression melting takes place when a body of hot mantle rock moves upward and the pressure is reduced to the extent that the melting point drops to the temperature of the body. Hawaiian volcanic activity, perhaps attributable to an underlying mantle plume, illustrates how *reduced* pressure contributes to the creation of magma. Solid rock that was once deeper in the mantle (and, therefore, very hot) has worked its way upward. Most of its heat has been retained during the upward journey. However, the pressure decreases as the rock body travels upward. As it approaches 50 kilometers or so from the Earth's surface, pressure is sufficiently reduced so that melting takes place.

Water under Pressure

If enough gas, especially water vapor, is present and under high pressure, a dramatic change occurs in the melting process. Water sealed in under high pressure helps break the silicon-oxygen bonds causing the crystal to liquify. A mineral's melting temperature is significantly lowered by water under high pressure (figure 3.18).

Experiments have shown that, under moderately high pressure, water mixed with granite lowers the melting point of granite from over 900°C (when dry) to as low as 650°C when

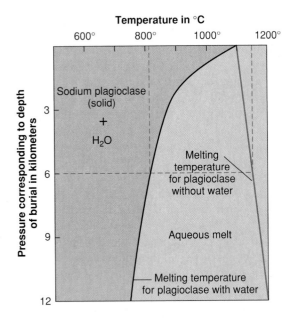

FIGURE 3.18

Melting temperature of a mineral with and without water present. The curve on the left is for melting of plagioclase saturated with water under the pressure corresponding to depth of burial. The line on the right corresponds to melting of dry plagioclase. The dashed, red line indicates that at pressures corresponding to 6 kilometers, plagioclase with water melts at just over 800°C, whereas dry plagioclase melts at around 1150°C. *After Tuttle and Brown, Geologic Society of America, 1958*

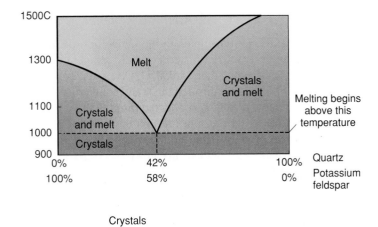

FIGURE 3.19

Melting temperatures for mixtures of quartz and potassium feldspar at atmospheric pressure. *Modified from Schairer and Bowen, 1956. V 254, p. 16. American Journal of Science*

saturated with water under pressure equivalent to that of 10,000 atmospheres, or *bars*. (Pressure at depth is usually expressed in *kilobars;* one kilobar is equal to 1,000 bars.) Ten kilobars corresponds to a depth of approximately 35 kilometers.

Effect of Mixed Minerals

Two metals—as in solder, which, typically, is a mixture of tin and lead—can be mixed in a ratio that lowers their melting temperature far below that of the melting points of the pure metals. Minerals behave similarly. Experiments have shown that in some cases, mixed fragments of two minerals melt at a lower temperature than either mineral alone. Figure 3.19 shows the melting temperatures for quartz and potassium feldspar mixed in various proportions. If the mixture is 42% quartz (58% potassium feldspar), melting takes place at just above 1,000°C. On the other hand, 1,500°C (the top of the diagram) is needed to liquify a mixture of 80% quartz and 20% potassium feldspar. Pure quartz requires even higher temperatures to melt.

HOW MAGMAS OF DIFFERENT COMPOSITIONS EVOLVE

A major topic of investigation for geologists is why igneous rocks are so varied in composition. On a global scale, magma composition is clearly controlled by geologic setting. But why?

Why are basaltic magmas associated with oceanic crust, whereas granitic magmas are common in the continental crust? On a local scale, igneous bodies often show considerable variation in rock type. For instance, individual plutons typically display a considerable range of compositions, mostly varieties of granite, but many also will contain minor amounts of gabbro or diorite. In this section, we describe processes that result in differences in composition of magmas. The final section of this chapter relates these processes to plate tectonics for the larger view of igneous activity.

Sequence of Crystallization and Melting

Early in the twentieth century, N. L. Bowen conducted a series of experiments that determined the sequence in which minerals crystallize in a cooling magma. The sequence became known as **Bowen's reaction series** and is shown in figure 3.20. A simplified explanation of the series and its importance to igneous rocks is presented next. For a more in-depth presentation, go to our website at **www.mhhe.com/plummer11e**.

Bowen's experiments showed that in a cooling magma, certain minerals are stable at higher melting temperatures and crystallize before those stable at lower temperatures. Looking at the *discontinuous branch*, which contains only ferromagnesian minerals, we can see that olivine crystallizes before pyroxene and pyroxene crystallizes before amphibole. A complication is that early formed crystals *react* with the remaining melt and recrystallize as cooling proceeds. For instance, early formed olivine crystals react with the melt and recrystallize to pyroxene when pyroxene's temperature of crystallization is reached. Upon further cooling, pyroxene continues to crystallize until all of the melt is used up or the melting temperature of amphibole is reached. At this point, pyroxene reacts with the

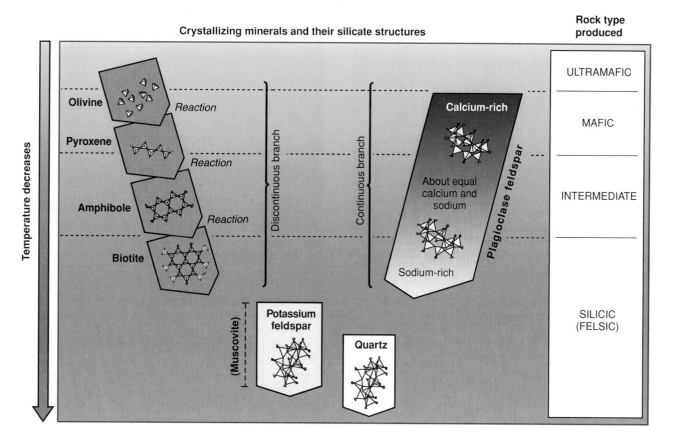

Crystallizing minerals and their silicate structures

Rock type produced

- ULTRAMAFIC
- MAFIC
- INTERMEDIATE
- SILICIC (FELSIC)

FIGURE 3.20

Bowen's reaction series. The reaction series as shown is very generalized. Moreover, it represents Bowen's experiments that involved melting a relatively silica-rich variety of basalt.

remaining melt and amphibole forms at its expense. If all of the iron and magnesium in the melt is used up before all of the pyroxene recrystallizes to amphibole, then the ferromagnesian minerals in the solid rock would be amphibole and pyroxene. (The rock would not contain olivine or biotite.)

Crystallization in the discontinuous and the *continuous branch* takes place at the same time. The continuous branch contains only plagioclase feldspar. Plagioclase is a *solid-solution* mineral (discussed in chapter 2 on minerals) in which either sodium or calcium atoms can be accommodated in its crystal structure, along with aluminum, silicon, and oxygen. The composition of plagioclase changes as magma is cooled and earlier formed crystals react with the melt. The first plagioclase crystals to form as a hot melt cools contain calcium but little or no sodium. As cooling continues, the early formed crystals grow and incorporate progressively more sodium into their crystal structures.

Any magma left after the crystallization is completed along the two branches is richer in silicon than the original magma and also contains abundant potassium and aluminum. The potassium and aluminum combine with silicon to form *potassium feldspar.* (If the water pressure is high, *muscovite* may also form at this stage.) Excess SiO_2 crystallizes as *quartz.*

From Bowen's reaction series, we can derive several important concepts that are necessary to understand igneous rocks and processes:

- A mafic magma will crystallize into pyroxene (with or without olivine) and calcium-rich plagioclase—that is, basalt or gabbro—if the early formed crystals are not removed from the remaining magma. Similarly, an intermediate magma will crystallize into diorite or andesite, if early formed minerals are not removed.

- If minerals are separated from a magma, the remaining magma is more silicic than the original magma. For example, if olivine and calcium-rich plagioclase are removed, the residual melt would be richer in silicon and sodium and poorer in iron and magnesium.

- If you heat a rock, the minerals will melt in reverse order. In other words, you would be going up the series as diagrammed in figure 3.20. Quartz and potassium feldspar would melt first. If the temperature is raised further, biotite and sodium-rich plagioclase would contribute to the melt. Any minerals higher in the series would remain solid unless the temperature is raised further.

Harnessing Magmatic Energy

Buried magma chambers indirectly contribute the heat for today's geothermal electric generating plants. As explained in chapter 11 (Ground Water), water becomes heated in hot rocks. The heat source is usually presumed to be an underlying magma chamber. The rocks containing the hot water are penetrated by drilling. Steam exiting the hole is used to generate electricity.

Why not drill into and tap magma itself for energy? The amount of energy stored in a body of magma is enormous. The U.S. Geological Survey estimates that magma chambers in the United States within 10 kilometers of Earth's surface contain about 5,000 times as much energy as the country consumes each year. Our energy problems could largely be solved if significant amounts of this energy were harnessed.

There are some formidable technical difficulties in drilling into a magma chamber and converting the heat into useful energy. Despite these difficulties, the United States has considered developing magmatic energy. Experimental drilling has been carried out in Hawaii through the basalt crust of a lava lake that formed in 1960.

As drill bits approach a magma chamber, they must penetrate increasingly hotter rock. The drill bit must be made of special alloys to prevent it from becoming too soft to cut rock. The rock immediately adjacent to a basaltic magma chamber is around 1,000°C, even though that rock is solid. Drilling into the magma would require a special technique. One being experimented with is a jet-augmented drill. As the drill enters the magma chamber, it simultaneously cools and solidifies the magma in front of the drill bit. Thus, the drill bit creates a column of rock that extends downward into the magma chamber and simultaneously bores a hole down the center of this column. Once the hollow column is deep enough within the magma chamber, a boiler is placed in the hole. The boiler is protected from the magma by the jacket of the column of rock. Water would be pumped down the hole and turned to water vapor in the boiler by heat from the magma. Steam emerging from the hole would be used to generate electricity.

In principle, the idea is fairly simple, but there are serious technical problems. For one thing, high pressures would have to be maintained on the drill bit during drilling and while the boiler system was being installed; otherwise, gases within the magma might blast the magma out of the drill hole and create a humanmade volcano. (The closest thing to a humanmade volcano occurred in Iceland when a small amount of magma broke into a geothermal steam well and erupted briefly at the well head, showering the area with a few tons of volcanic debris.)

• Bowen's reaction series can be used to show how two important processes that create and modify magma composition work. These are *differentiation* and *partial melting*.

Differentiation

The process by which different ingredients separate from an originally homogenous mixture is **differentiation.** An example is the separation of whole milk into cream and nonfat milk. Differentiation in magmas takes place mainly through **crystal settling,** the downward movement of minerals that are denser (heavier) than the magma from which they crystallized.

If crystal settling takes place in a mafic magma chamber, olivine and, perhaps, pyroxene crystallize and settle to the bottom of the magma chamber (figure 3.21). This makes the remaining magma more silicic. Calcium-rich plagioclase also separates as it forms. The remaining magma is, therefore, depleted of calcium, iron, and magnesium. Because these minerals were economical in using the relatively abundant silica, the remaining magma becomes richer in silica as well as in sodium and potassium.

It is possible that by removing enough mafic components, the residual magma would be silicic enough to solidify into granite (or rhyolite). But it is more likely only enough mafic components would be removed to allow an intermediate residual magma, which would solidify into diorite or andesite. The lowermost portions of some large sills are composed predominantly of olivine and pyroxene, whereas upper levels are considerably less mafic. Even in large sills, however, differentiation has rarely progressed far enough to produce granite within the sill.

Ore Deposits Due to Crystal Settling

Crystal settling accounts for important ore deposits that are mined for chromium and platinum. Most of the world's chromium and platinum come from a huge sill in South Africa. The sill, the famous Bushveldt Complex, is 8 kilometers thick and 500 kilometers long. Layers of chromite (a chromium-bearing mineral) up to 2 meters thick are found, and mined, at the base of the sill. Layers containing platinum overlie the chromite-rich layers.

Partial Melting

As mentioned earlier, progressing upward through Bowen's reaction series (going from cool to hot) gives us the sequence in which minerals in a rock melt. As might be expected, the first portion of a rock to melt as temperatures rise forms a liquid with the chemical composition of quartz and potassium feldspar. The oxides of silicon plus potassium and aluminum "sweated out" of the solid rock could accumulate into a pocket of silicic magma. If higher temperatures prevailed, more mafic magmas would be created. Small pockets of magma could merge and form a large enough mass to rise as a

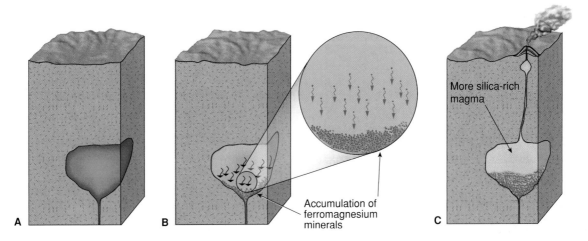

FIGURE 3.21

Differentiation of a magma body. (A) Recently intruded mafic magma is completely liquid. (B) Upon slow cooling, ferromagnesian minerals, such as olivine, crystallize and sink to the bottom of the magma chamber. The remaining liquid is now an intermediate magma. (C) Some of the intermediate magma moves upward to form a smaller magma chamber at a higher level that feeds a volcano.

diapir. In nature, temperatures rarely rise high enough to entirely melt a rock.

Partial melting of the lower continental crust likely produces silicic magma. The magma rises and eventually solidifies at a higher level in the crust into granite, or rhyolite if it reaches Earth's surface.

Geologists generally regard basaltic magma (Hawaiian lava, for example) as the product of partial melting of ultramafic rock in the mantle, at temperatures hotter than those in the crust. The solid residue left behind in the mantle when the basaltic magma is removed is an even more silica-deficient ultramafic rock.

Assimilation

A very hot magma may melt some of the country rock and *assimilate* the newly molten material into the magma (figure 3.22). This is like putting a few ice cubes into a cup of hot coffee. The ice melts and the coffee cools as it becomes diluted. Similarly, if a hot basaltic magma, perhaps generated from the mantle, melts portions of the continental crust, the magma simultaneously becomes richer in silica and cooler. Possibly intermediate magmas such as are associated with circum-Pacific andesite volcanoes may derive from assimilation of some crustal rocks by a basaltic magma.

Mixing of Magmas

Some of our igneous rocks may be "cocktails" of different magmas. The concept is quite simple. If two magmas meet and merge within the crust, the combined magma will be compositionally intermediate (figure 3.23). If you had approximately equal amounts of a granitic magma mixing with a basaltic magma, the resulting magma should crystallize underground as diorite or erupt on the surface to solidify as andesite.

EXPLAINING IGNEOUS ACTIVITY BY PLATE TECTONICS

One of the appealing aspects of the theory of plate tectonics is that it accounts reasonably well for the variety of igneous rocks and their distribution patterns. (Chapter 1 has an overview of plate tectonics.) Divergent boundaries are associated with creation of basalt and gabbro of the oceanic crust. Andesite and granite are associated with convergent boundaries. Table 3.2 summarizes the relationships.

Igneous Processes at Divergent Boundaries

The crust beneath the world's oceans (over 70% of Earth's surface) is mafic volcanic and intrusive rock, covered to a varying extent by sediment and sedimentary rock. Most of this basalt and gabbro was created at mid-oceanic ridges, which also are divergent plate boundaries. Geologists agree that the mafic magma produced at divergent boundaries is due to partial melting of the asthenosphere. The *asthenosphere,* as described in chapter 1, is the plastic zone of the mantle beneath the rigid *lithosphere* (the upper mantle and crust that make up a plate). Along divergent boundaries, the asthenosphere is relatively close (5 to 10 kilometers) to the surface (figure 3.24).

The probable reason the asthenosphere is plastic or "soft" is that temperatures there are only slightly lower than the temperatures required for partial melting of mantle rock.

If extra heat is added, or pressure is reduced, partial melting should take place. The asthenosphere beneath divergent boundaries probably is mantle material that has welled upward from deeper levels of the mantle. As the hot asthenosphere gets close to the surface, decrease in pressure results in partial melting. In other words, *decompression melting* takes place. The

FIGURE 3.22

Assimilation. Magma formed is intermediate in composition between the original magma and the absorbed country rock. (A) Ascending magma breaks off blocks of country rock (the process is called *stoping*). (B) Xenoliths of country rock with melting temperatures lower than the magma melt. (C) The molten country rock blends with the original magma, leaving unmelted portions as inclusions.

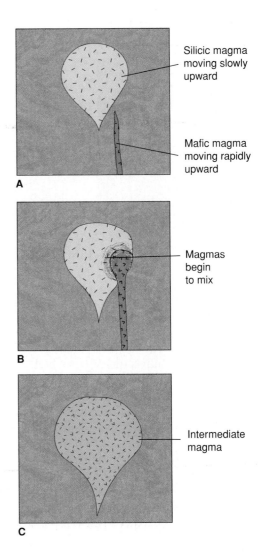

FIGURE 3.23

Mixing of magmas. (A) Two bodies of magma moving surfaceward. (B) The mafic magma catches up with the silicic magma. (C) The two magmas combine and become an intermediate magma.

magma that forms is mafic and will solidify as basalt or gabbro. The portion that did not melt remains behind as a silica-depleted, iron-and-magnesium-enriched ultramafic rock.

Some of the basaltic magma erupts along a submarine ridge to form pillow basalts (described in chapter 4), while some fills near-surface fissures to create dikes. Deeper down, magma solidifies more slowly into gabbro. The newly solidified rock is pulled apart by spreading plates; more magma fills the new fracture and some erupts on the sea floor. The process is repeated, resulting in a continuous production of mafic crust.

The basalt magma that builds the oceanic crust is removed from the underlying mantle, depleting the mantle beneath the ridge of much of its calcium, aluminum, and silicon oxides. The unmelted residue (olivine and pyroxene) becomes depleted mantle, but it is still a variety of ultramafic rock. The rigid ultramafic rock, the overlying gabbro and basalt, and any sediment that may have deposited on the basalt collectively are the lithosphere of an oceanic plate, which moves away from a spreading center over the asthenosphere. (The nature of the oceanic crust is described in more detail in chapter 18.)

Intraplate Igneous Activity

Igneous activity within a plate, a long distance from a plate boundary, is unusual. As described earlier, Hawaii and Yellowstone eruptions represent intraplate igneous activity. The ongo-

TABLE 3.2	Relationships between Rock Types and Their Usual Plate Tectonic Setting			
Rock	Original Magma	Final Magma	Processes	Plate Tectonic Setting
Basalt and gabbro	Mafic	Mafic	Partial melting of mantle (asthenosphere)	1. Divergent boundary—oceanic crust created 2. Intraplate • plateau basalt • volcanic island chains (e.g., Hawaii)
Andesite and diorite	Mafic (usually)	Intermediate	Partial melting of mantle (asthenosphere) followed by: • differentiation or • assimilation or • magma mixing	Convergent boundary
Granite and rhyolite	Silicic	Silicic	Partial melting of lower crust	1. Convergent boundary 2. Intraplate • over mantle plume

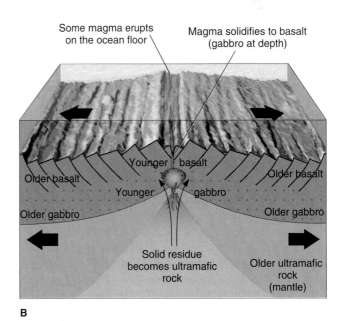

FIGURE 3.24

Schematic representation of how basaltic oceanic crust and the underlying ultramafic mantle rock form at a divergent boundary. The process is more continuous than the two-step diagram implies. (A) Partial melting of asthenosphere takes place beneath a mid-oceanic ridge and magma rises into a magma chamber. (B) The magma squeezes into the fissure system. Solid mafic minerals are left behind as ultramafic rock.

ing volcanic eruptions in Hawaii take place on oceanic crust, whereas eruptions at Yellowstone represent continental intraplate activity.

The huge volume of mafic magma that erupted to form the Columbia plateau basalts of Washington and Oregon (described in chapter 4) is attributed to a past hot mantle plume, according to a recent hypothesis (figure 3.25). In this case, the large volume of basalt is due to the arrival beneath the lithosphere and decompression melting of a mantle plume with a large head on it.

Igneous Processes at Convergent Boundaries

Intermediate and silicic magmas are clearly related to the convergence of two plates and subduction. However, exactly what takes place is debated by geologists. Compared to divergent boundaries, there is less agreement about how magmas are generated at convergent boundaries. The scenarios that follow are currently believed by geologists to be the best explanations of the data.

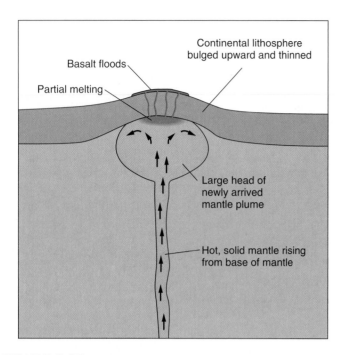

FIGURE 3.25

A hot mantle plume with a large head rises from the lower mantle. When it reaches the base of the lithosphere, it uplifts and stretches the overlying lithosphere. The reduced pressure results in decompression melting, producing basaltic magma. Large volumes of magma travel through fissures and flood the Earth's surface.

The Origin of Andesite

Magma for most of our andesitic composite volcanoes (such as those found along the west coast of the Americas) seems to originate from a depth of about 100 kilometers. This coincides with the depth at which the subducted oceanic plate is sliding under the asthenosphere (figure 3.26). Partial melting of the asthenosphere takes place, resulting in a mafic magma. In most cases, melting occurs because the subducted oceanic crust releases water into the asthenosphere. The water collected in the oceanic crust when it was beneath the ocean and is driven out as the descending plate is heated. The water lowers the melting temperature of the ultramafic rocks in this part of the mantle. Partial melting produces a mafic magma.

But how can we keep producing magma from ultramafic rock after those rocks have been depleted of the constituents of the mafic magma? The answer is that hot asthenospheric rock continues to flow into the zone of partial melting. As shown in figure 3.26, asthenospheric ultramafic rock is dragged downward by the descending lithospheric slab. More ultramafic rock flows laterally to replace the descending material. A continuous flow of hot, "fertile" (containing the constituents of basalt) ultramafic rock is brought into the zone where water, moving upward from the descending slab, lowers the melting temperature. After being depleted of basaltic

magma, the solid, residual, ultramafic rock continues to sink deeper into the mantle.

On its slow journey through the crust, the mafic magma evolves into an intermediate magma by differentiation, assimilation of silicic crustal rocks, and by magma mixing.

Under special circumstances basalt of the descending oceanic crust can partially melt to yield an intermediate magma. In most subduction zones, the basalt remains too cool to melt, even at a depth of over 100 kilometers. But, geologists believe that partial melting of the subducted crust produces the magma for andesitic volcanoes in South America. Here, the oceanic crust is much younger and considerably hotter than normal. The spreading axis where it was created is not far from the trench. Because the lithosphere has not traveled far before being subducted, it is still relatively hot. As can be seen from figure 3.27, subduction is at a shallower angle, because this hotter crust is more buoyant than the usual case (as in figure 3.26).

The reason that partial melting of subducted basalt is unusual is that this kind of subduction and magma generation is, geologically speaking, short-lived. Subduction will end when the overriding plate crashes into the mid-oceanic ridge. Most subduction zones are a long distance from the divergent boundaries of their plates, so steep subduction and magma production from the asthenosphere are the norm.

The Origin of Granite

To explain the great volumes of granitic plutonic rocks, most geologists think that partial melting of the lower continental crust must take place. The continental crust contains the high amount of silica needed for a silicic magma. As the silicic rocks of the continental crust have relatively low melting temperatures (especially if water is present), partial melting of the lower continental crust is likely. However, calculations indicate that the temperatures we would expect from a normal geothermal gradient are too low for melting to take place. Therefore, we need an additional heat source.

Currently, geologists think that the additional heat is provided by mafic magma that was generated in the asthenosphere and moved upward. The process of *magmatic underplating* involves mafic magma pooling at the base of the continental crust, supplying the extra heat necessary to partially melt the overlying, silica-rich crustal rocks (figure 3.28). Mafic magma generated in the asthenosphere rises to the base of the crust. The mafic magma is denser than the overlying silica-rich crust; therefore, it collects as a liquid mass that is much hotter than the crust. The continental crust becomes heated (as if by a giant hotplate). When the temperature of the lower crust rises sufficiently, partial melting takes place, creating silicic magma. The silicic magma collects and forms diapirs, which rise to a higher level in the crust and solidify as granitic plutons (or, on occasion, reach the surface and erupt violently).

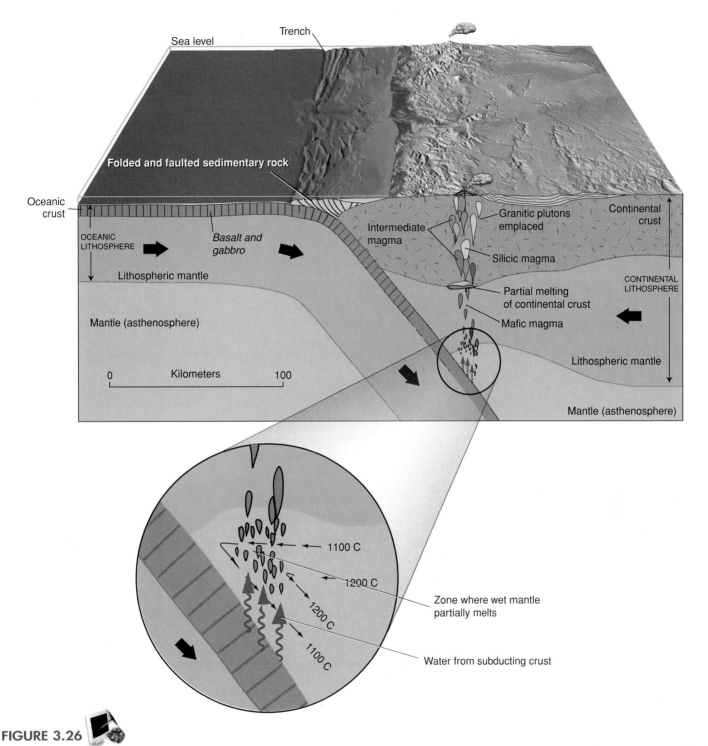

FIGURE 3.26

Generation of magma at a convergent boundary. Mafic magma is generated in the asthenosphere above the subducting oceanic lithosphere, and silicic magma is created in the lower crust. The insert shows the circulation of asthenosphere and lines of equal temperature (isotherms). Partial melting of "wet" ultramafic rock takes place in the zone where it is between 1100 and 1200°C.

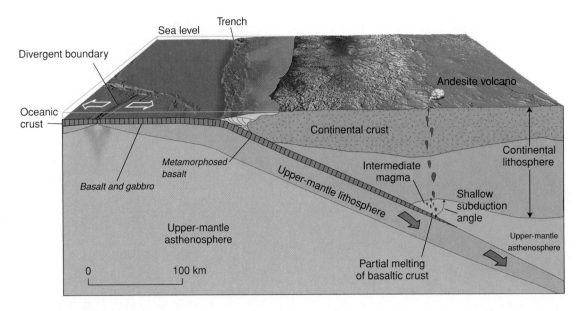

FIGURE 3.27

Young, hot, oceanic lithosphere is buoyant and subducts at a shallower angle than normal. Direct, partial melting of basalt in the subducting slab takes place to form intermediate magma. Basalt partially melts when it is heated further by the overlying asthenosphere.

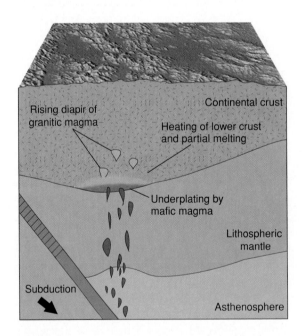

FIGURE 3.28

How mafic magma could add heat to the lower crust and result in partial melting to form a granitic magma. Mafic magma from the asthenosphere rises to underplate the continental crust.

SUMMARY

The interaction between the internal and external forces of the Earth is illustrated by the rock cycle, a conceptual device relating igneous, sedimentary, and metamorphic rocks to each other, to surficial processes such as weathering and erosion, and to internal processes such as tectonic forces. Changes take place when one or more processes force Earth's material out of equilibrium.

Igneous rocks form from solidification of magma. If the rock forms at the Earth's surface it is *extrusive. Intrusive rocks* are igneous rocks that formed underground. Some intrusive rocks have solidified near the surface as a direct result of volcanic activity. Volcanic *necks* solidified within volcanoes. Fine-grained *dikes* and *sills* may also have formed in cracks during local extrusive activity. A sill is *concordant*—parallel to the planes within the country rock. A dike is *discordant*—not parallel to planes in the country rock. Both are tabular bodies. Coarser grains in either a dike or a sill indicate that it probably formed at considerable depth.

Most intrusive rock is *plutonic*—that is, coarse-grained rock that solidified slowly at considerable depth. Most plutonic rock exposed at the Earth's surface is in *batholiths*—large plutonic bodies. A smaller body is called a *stock*.

Silicic (or felsic) rocks are rich in silica, whereas mafic rocks are silica deficient. Most igneous rocks are named on the basis of their mineral content, which in turn reflects the chemical composition of the magmas from which they formed, and on grain sizes. *Granite, diorite,* and *gabbro* are the coarse-grained equivalents of *rhyolite, andesite,* and *basalt,* respectively. *Peridotite* is an *ultramafic* rock made entirely of ferromagnesian minerals and is mostly associated with the mantle.

Basalt and gabbro are predominant in the oceanic crust. Granite predominates in the continental crust. Younger granite batholiths occur mostly within younger mountain belts. Andesite is largely restricted to narrow zones along convergent plate boundaries.

The *geothermal gradient* is the increase in temperature with increase in depth. Hot *mantle plumes* and magma at shallow depths in volcanic regions locally raise the geothermal gradient.

No single process can satisfactorily account for all igneous rocks. In the process of *differentiation,* based on *Bowen's reaction series,* a residual magma more silicic than the original mafic magma is created when the early-forming minerals separate out of the magma. In *assimilation,* a hot, original magma is contaminated by picking up and absorbing rock of a different composition. *Magma mixing* produces a magma whose composition is intermediate, between that of the two types of magma that were mixed.

Partial melting of the mantle usually produces basaltic magma, whereas granitic magma is most likely produced by partial melting of the lower continental crust.

The theory of *plate tectonics* incorporates the preceding concepts. Basalt is generated where hot mantle rock partially melts, most notably along divergent boundaries. The fluid magma rises easily through fissures, if present. The ferromagnesian portion that stays solid remains in the mantle as ultramafic rock. Granite and andesite are associated with subduction. Differentiation, assimilation, partial melting, and mixing of magmas may each play a part in creating the observed variety of rocks.

Terms to Remember

andesite 64
basalt 64
batholith 68
Bowen's reaction series 72
chill zone 61
coarse-grained (coarsely crystalline) rock 61
contact 60
country rock 60
crystal settling 74
decompression melting 71
diapir 69
differentiation 74
dike 67
diorite 64

extrusive rock 60
fine-grained rock 61
gabbro 64
geothermal gradient 71
granite 60
igneous rock 60
intermediate rock or magma 66
intrusion (intrusive structure) 67
intrusive rock 60
lava 60
mafic rock or magma 66
magma 60
mantle plume 71
peridotite 66

pluton 68
plutonic rock 61
porphyritic 61
rhyolite 64
rock 58
rock cycle 58
silicic (felsic) rock or magma 66
sill 67
stock 68
texture 61
ultramafic rock 66
volcanic neck 67
xenolith 61

Testing Your Knowledge

Use the following questions to prepare for exams based on this chapter.

1. Why do mafic magmas tend to reach the surface much more often than silicic magmas?

2. What role does the asthenosphere play in generating magma at (a) a convergent boundary; (b) a divergent boundary?

3. How do batholiths form?

4. How would you distinguish, on the basis of minerals present, among granite, gabbro, and diorite?

5. How would you distinguish andesite from a diorite?

6. What rock would probably form if magma that was feeding volcanoes above subduction zones solidified at considerable depth?

7. Why is a higher temperature required to form magma at the oceanic ridges than in the continental crust?

8. What is the difference between feldspar found in gabbro and feldspar found in granite?

9. What is the difference between a dike and a sill?

10. Describe the differences between the continuous and the discontinuous branches of Bowen's reaction series.

11. A surface separating different rock types is called a
    a. xenolith   b. contact
    c. chill zone   d. none of the preceding

12. The major difference between intrusive igneous rocks and extrusive igneous rocks is
    a. where they solidify   b. chemical composition
    c. type of minerals   d. all of the preceding

13. Which is not an intrusive igneous rock?
    a. gabbro   b. diorite
    c. granite   d. andesite

14. By definition, stocks differ from batholiths in
    a. size   b. shape
    c. chemical composition   d. all of the preceding

15. Which is not a source of heat for melting rock?
    a. geothermal gradient   b. the hotter mantle
    c. mantle plumes   d. water under pressure

16. The geothermal gradient is, on the average, about
    a. 1°C/km   b. 10°C/km
    c. 30°C/km   d. 50°C/km

17. The continuous branch of Bowen's reaction series contains the mineral
    a. pyroxene   b. plagioclase
    c. amphibole   d. biotite

18. The discontinuous branch of Bowen's reaction series contains the mineral
    a. pyroxene   b. amphibole
    c. biotite   d. all of the preceding

19. The most common igneous rock of the continents is
    a. basalt   b. granite
    c. rhyolite   d. ultramafic

20. Granitic magmas are associated with
    a. convergent boundaries and magmatic underplating
    b. divergent boundaries and differentiation
    c. convergent boundaries and decompression melting
    d. divergent boundaries and water release

21. The difference in texture between intrusive and extrusive rocks is primarily due to
    a. different mineralogy
    b. different rates of cooling and crystallization
    c. different amounts of water in the magma

22. Mafic magma is generated at divergent boundaries because of
    a. water under pressure   b. decompression melting
    c. magmatic underplating   d. melting of the lithosphere

23. A change in magma composition due to melting of surrounding country rock is called
    a. magma mixing   b. assimilation
    c. crystal setting   d. partial melting

Expanding Your Knowledge

1. In parts of major mountain belts there are sequences of rocks that geologists interpret as slices of ancient oceanic lithosphere. Assuming that such a sequence formed at a divergent boundary and was moved toward a convergent boundary by plate motion, what rock types would you expect to make up this sequence, going from the top downward?

2. What would happen, according to Bowen's reaction series, under the following circumstances: olivine crystals form and only the surface of each crystal reacts with the melt to form a coating of pyroxene that prevents the interior of olivine from reacting with the melt?

Exploring Web Resources

www.mhhe.com/plummer11e

This is the dedicated website for this book. You can go to it for new and updated information. The universal resource locators (URLs) listed in this book are also given as links on the website, making it easy to go to those websites without typing in the URL. When you visit our Online Learning Center, go to the Student Center and then to the chapter of interest. Here you will find additional readings and media resources, as well as answers to the Testing Your Knowledge section, more quizzes, animation and video clips, and direct links to the sites listed in this book. Links to additional websites can also be found. We have added questions for some of the sites to allow you to get the most of your exploration of the web. Using the web is an enjoyable way of enhancing your knowledge of geology.

http://uts.cc.utexas.edu/~rmr/

Rob's Granite Page. This site has a lot of information on granite and related igneous activity. The site is useful for people new to geology as well as for professionals. There are numerous images of granite. Click on "Did you know that granite is like ice cream?" for an interesting comparison. The page also has photos of various granites and links to other sites that have more images.

www.geolab.unc.edu/Petunia/IgMetAtlas/mainmenu.html

Atlas of Rocks, Minerals, and Textures (from University of North Carolina). This site contains some photomicrographs of plutonic and volcanic rocks. The images are thin sections (slices of rock so thin that most minerals are transparent) seen in a polarizing microscope. Most images are taken from cross-polarized light, which causes many minerals to appear in distinctive, bright colors. For some of the rocks (gabbro, for instance), you can also see what they look like under plain polarized light by clicking the circle with the horizontal, gray lines.

http://seis.natsci.csulb.edu/basicgeo/IGNEOUS_TOUR.html

Igneous Rocks Tour. This site has some hand specimen images of common igneous rocks and should provide a useful review for rock identification.

www.gpc.peachnet.edu/~pgore/stonemtn/stonemountain.html

Stone Mountain, Georgia, Virtual Field Trip. Stone Mountain is an exposure of granite in Georgia that is a famous landmark. Begin by reading the geologic summary, then take the virtual tour.

Animation

This chapter includes the following animations available on our Online Learning Center at **www.mhhe.com/plummer11e**.

3.26 How subduction causes volcanism

Volcanism and Extrusive Rocks

Chapters 3 and 4 cover igneous activity. Either may be read before the other. Chapter 3 emphasizes intrusive activity, but it also covers igneous rock classification and the origin of magmas, which are applicable both to volcanic and intrusive phenomena. Chapter 4 concentrates on volcanoes and related extrusive activity.

Volcanic eruptions, while awesome natural spectacles (figure 4.1), also provide important information on the workings of the Earth's interior. Volcanic eruptions vary in nature and in degree of explosive violence. A strong correlation exists between the chemical composition of

Lava flows into the sea at Hawaii during the eruption of Kilauea that has been ongoing since 1983. *Photo © Parvinder Sethi*

IGNEOUS ROCK

Weathering and erosion

Sediment

Solidification

Metamorphism

Lithification

Magma

SEDIMENTARY ROCK

Partial melting

METAMORPHIC ROCK

Metamorphism

Rock in mantle

magma (or lava), its physical properties, and the violence of an eruption. The size and shape of volcanoes and lava flows and their pattern of distribution on the Earth's surface also correlate with the composition of their lavas.

Our observations of volcanic activity fit nicely into plate-tectonic theory as described in chapter 3. Understanding volcanism also provides a background for theories relating to mountain building, the development and evolution of continental and oceanic crust (topics covered in later chapters).

Landforms are created through volcanic activity and portions of Earth's surface built up. Less commonly, as at Mount St. Helens, landforms are destroyed by violent eruptions (box 4.1).

By studying the magma, gases, and rocks from eruptions, we can infer the chemical conditions as well as the temperatures and pressures within Earth's crust or underlying mantle.

 Relationships to Earth Systems

The *atmosphere* was created by degassing magma during the time following Earth's formation. Even now, gases and dust given off by major volcanic eruptions can profoundly alter worldwide climate.

Condensation of the water vapor during the degassing produced the *hydrosphere.* Volcanic islands occasionally blow up, creating tsunamis, giant sea waves. In 1883, Krakatoa, a volcanic island in Indonesia, was destroyed by one of the most violent eruptions in recorded history. Although nobody lived on Krakatoa, over 36,000 people in the region drowned from the resulting, huge tsunami. (Krakatoa is not very far from where the much more devastating December 16, 2004 tsunami originated.) Eruptions also take place beneath glaciers. In Iceland in 1996, a large volcanic eruption beneath a glacier resulted in large-scale melting of the ice that burst out of the glacier as a flood, destroying three bridges and 10 kilometers of road.

The effect of volcanic activity on the *biosphere* ranges from benign to catastrophic. Volcanic rock in Hawaii has reacted with water and atmospheric gases to form the soil that supports lush, tropical vegetation. Some violent eruptions in other parts of the world have destroyed virtually all living things (including humans) that happened to be in their paths. In the 1980 Mount St. Helens eruption, forests were leveled by a huge lateral blast (box 4.1). Extended periods of major eruptions are believed to have contributed to or caused some of the mass extinctions that have taken place in Earth's history. A mass extinction is a time in which a large number of plant and animal species are wiped out.

PYROCLASTIC DEBRIS AND LAVA FLOWS

The May 18, 1980, eruption of Mount St. Helens (figure 4.1*A* and box 4.1) was a spectacular release of energy from the Earth's interior. The plate-tectonic explanation is that North America is overriding a portion of the Pacific Ocean floor. Melting of previously solid rock takes place at depth, just above the subducting plate. (This is described briefly in chapter 1 and more thoroughly in chapter 3.) Some of the **magma** (molten rock or liquid that is mostly silica) worked its way upward to the Earth's surface to erupt. At Mount St. Helens, magma solidified quickly as it was blasted explosively by gases into the air, producing rock fragments known as **pyroclasts** (from the Greek *pyro,* "fire," and *clast,* "broken"). *Pyroclastic debris* is also known as *tephra.*

In Hawaii, **lava** (magma on Earth's surface) extrudes out of fissures in the ground as *lava flows* (figure 4.1*B*). Pyroclastic debris and rock formed by solidification of lava are collectively regarded as **extrusive rock,** surface rock resulting from volcanic activity.

The most obvious landform created by **volcanism** is a **volcano,** a hill or mountain formed by the extrusion of lava or ejection of rock fragments from a vent. However, volcanoes are not the only volcanic landforms. Very fluid lava may flow out of the Earth and flood an area, solidifying into a nearly horizontal layer of extrusive rock. Successive layers of lava flows may accumulate, building a lava plateau.

LIVING WITH VOLCANOES

Supernatural Beliefs

Not surprisingly, myths and religions relating gods to volcanoes flourish in cultures that live with volcanoes. In Iceland, Loki, of Norse mythology, is regarded as imprisoned underground,

A

B

FIGURE 4.1

Contrasting styles of volcanic eruptions. (A) Mount St. Helens, May 18, 1980. Looking north, we can see the last of the huge lateral explosion from the far side of the volcano. This was followed by vertical eruption of gases and pyroclasts from the top of the volcano. (B) Lava flow in Hawaii, 1969. A lava fountain is at the source of lava cascading over a cliff. *Photo A © Gary Braasch/Corbis. Photo B by D. A. Swanson, U.S. Geological Survey*

ENVIRONMENTAL GEOLOGY 4.1

Mount St. Helens Blows Up

Before 1980, Mount St. Helens, in southern Washington, had not erupted since 1857. On March 27, 1980, ash and steam eruptions began and continued for the next six weeks. These were minor eruptions in which magma was not erupted. Rather, they were due to exploding gas blasting out the volcano's previously formed rock. However, the steam and the pattern of earthquakes indicated magma was working its way upward beneath the volcano.

After several weeks, the peak began swelling—like a balloon being inflated—indicating magma was now inside the volcano. The northern flank of the volcano bulged outward at a rate of 1.5 meters per day. Bulging continued until the surface of the northern slope was displaced outward over a hundred meters from its original position. The bulge was too steep to be stable, and the U.S. Geological Survey warned of another hazard—a mammoth landslide.

On May 18, a monumental blast destroyed the summit and north flank of Mount St. Helens (see figure 4.1). Seconds after the eruption began, an area extending northward 10 kilometers was stripped of all vegetation and soil.

Although the sequence of events was exceedingly rapid, it is now clear what happened (box figure 1). A fairly strong earthquake loosened the bulging north slope, triggering a landslide. The landslide, known as a *debris avalanche*, moved at speeds of over 160 kilometers per hour (100 mph). It was one of the largest landslides ever to occur, but it was eclipsed by the huge eruption that followed. The landslide stripped away the lid on the magma chamber, and because of the reduced pressure, the previously dissolved gases in the magma exploded (figure 4.1A). The violent froth of gas and magma blasted away the mountain's north flank and roared outward at up to 1,000 kilometers per hour (600 mph). The huge lateral blast of hot gas and volcanic rock debris killed everything near the volcano and, beyond the 10-kilometer scorched zone, knocked down every tree in the forest.

For the next 30 hours, exploding gases propelled frothing magma and volcanic ash vertically into the high atmosphere. The mushroom-shaped cloud of ash was blown northeastward by winds. A rain of ash went on for days, causing damage as far away as Montana. Volcanic mudflows caused enormous damage during and after the eruption. The mudflows resulted from water from melted snow and glacier ice mixing with volcanic debris to form a slurry having the consistency of wet cement. Mudflows flowed down river valleys, carrying away steel bridges and other structures (see chapter 9, notably figure 9.13).

Damage was in the hundreds of millions of dollars, and 63 people were killed. The death toll might have been much worse had not scientists warned public officials about the potential hazards, causing them to evacuate the danger zone before the eruption. For comparison, 29,000 people were killed during an eruption of Mount Pelée (described later in this chapter), and 23,000 lives were lost in a 1985 volcanic mudflow in Colombia.

BOX 4.1 ■ FIGURE 1
Sequence of events at Mount St. Helens, May 18, 1980. (A) Just before the eruption. (B) The landslide relieves the pressure on the underlying magma. (C) Magma blasts outward. (D) Full vertical eruption.

Perhaps Mount St. Helens will remain quiescent for decades or a century. Other volcanoes in the Pacific Northwest, however, could erupt and be disastrous to nearby cities. Seattle and Tacoma are close to Mount Rainier. Mount Hood is practically in Portland, Oregon's suburbs. Vancouver, British Columbia, could be in danger if either Mount Garibaldi to the north or Mount Baker in Washington to the south erupt.

Additional Resource
USGS Cascade Volcano Observatory—Mount St. Helens
• http://vulcan.wr.usgs.gov/Volcanoes/MSH/framework.html

blowing steam and lava up through fissures. Pacific Northwest Indians regarded the Cascade volcanoes as warrior gods who would sometimes throw red-hot boulders at each other. They also had a romantic side. Mount Hood and Mount Adams fought over Mount St. Helens, the youngest and prettiest of the volcano gods. In British Columbia, a lava flow killed about 2,000 members of the Nisga'a tribe around A.D. 1700. According to the Nisga'as, children were harassing salmon, including putting flaming sticks in a fish's back and watching the smoking fish swim upstream. Disrespect of fish is a major taboo and was believed to have brought on the lava eruption. In Hawaii, Madame Pele, is regarded as a goddess who controls eruptions. According to legend, Pele and her sister tore up the ocean floor to produce the Hawaiian island chain. Today, many fervently believe that Pele dictates when and where an eruption will take place. In the 1970s, when Kilauea began erupting near a village, residents chartered an airplane and dropped flowers and a bottle of gin into the lava vent to appease Pele.

Volcanism is also relevant to human affairs in very tangible ways. Its effects can be catastrophic or, surprisingly, beneficial.

The Growth of an Island

Although occasionally a highway or village is overrun by outpourings of lava, the overall effects of volcanism have been favorable to humans in Hawaii. Lava flowing into the sea and solidifying adds real estate to the island of Hawaii. Kilauea Volcano has been erupting since 1983, spewing out an average of 325,000 cubic meters of lava a day. This is the equivalent of 40,000 dump truck-loads of material. In twenty years, 2.5 billion cubic meters of lava were produced—enough to build a highway that circles the world over five times. The down side is that during the 1980s and 1990s, 181 houses were destroyed by lava flows.

Were it not for volcanic activity, Hawaii would not exist. The islands are the crests of a series of volcanoes that have been built up from the bottom of the Pacific Ocean over millions of years (the vertical distance from the summit of Mauna Loa Volcano to the ocean floor greatly exceeds the height above sea level of Mount Everest). When lava flows into the sea and solidifies, more land is added to the islands. Hawaii is, quite literally, growing.

In addition to gaining more land, Hawaii benefits in other ways from its volcanoes. Weathered volcanic ash and lava produce excellent fertile soils (think pineapples and papayas). Moreover, Hawaii's periodically erupting volcanoes (which are relatively safe to watch) are great spectacles that attract both tourists and scientists, benefiting the island's economy (figure 4.1B).

Geothermal Energy

In other areas of recent volcanic activity, underground heat generated by igneous activity is harnessed for human needs. Steam or superheated water trapped in layers of hot volcanic rock is tapped by drilling and then piped out of the ground to power turbines that generate electricity. The United States is the biggest producer of geothermal power, followed by the Philippines, Italy, and Mexico. Naturally heated geothermal fluids can also be tapped for space or domestic water heating or industrial use, as in paper manufacturing. (For more information, go to http://geothermal.marin.org/ or chapter 11 on ground water.)

Effect on Climate

Occasionally, a volcano will spew large amounts of fine, volcanic dust and gas into the high atmosphere. Winds can keep fine particles suspended over the Earth for years. The 1991 eruption of Mount Pinatubo in the Philippines produced noticeably more colorful sunsets worldwide (see description in chapter 1). More significantly, it reduced solar radiation that penetrates the atmosphere. Measurements indicated that the worldwide average temperature dropped approximately half a degree Celsius for a couple of years. While this may not seem like much, it was enough to temporarily offset the global warming trend of the past 100 years.

The 1815 eruption of Tambora in Indonesia was the largest, single eruption in a millennium—40 cubic kilometers of material were blasted out of a volcanic island, leaving a 6-kilometer-wide depression. The following year, 1816, became known as "the year without summer." In New England, snow in June was widespread and frosts throughout the summer ruined crops. Parts of Europe suffered famine because of the cold weather effects on agriculture.

Volcanic Catastrophes

While the eruption of Mount St. Helens in 1980 was indeed awesome, its effects were not nearly as disastrous as a number of historical eruptions elsewhere in the world. For instance, the Roman city of Pompeii and at least four other towns near Naples in Italy were destroyed in A.D. 79 when Mount Vesuvius erupted (figure 4.2). Before the eruption, vineyards on the flanks of the apparently "dead" volcano extended to the summit. After it erupted without warning, Pompeii was buried under 5 to 8 meters of hot ash. Seventeen centuries later, the town was rediscovered.

Excavation revealed molds of people suffocated by the ashfall, many with facial expressions of terror. This eruption was not the end of Vesuvius's activity. The volcano was active almost continually from 1631 to 1944, with major twentieth-century eruptions in 1906, 1929, and 1944. Naples is a major city and has expanded onto the lower flanks of Vesuvius. A new eruption could be a disaster.

The island of Krakatoa in Indonesia, composed of three apparently inactive volcanoes, erupted in 1883 with the force of several hydrogen bombs. The eruption took place as an estimated 13 cubic kilometers of rock collapsed into a subsurface magma chamber. Six cubic kilometers of the displaced magma rose to the surface and flashed into gas and pyroclast eruptions.

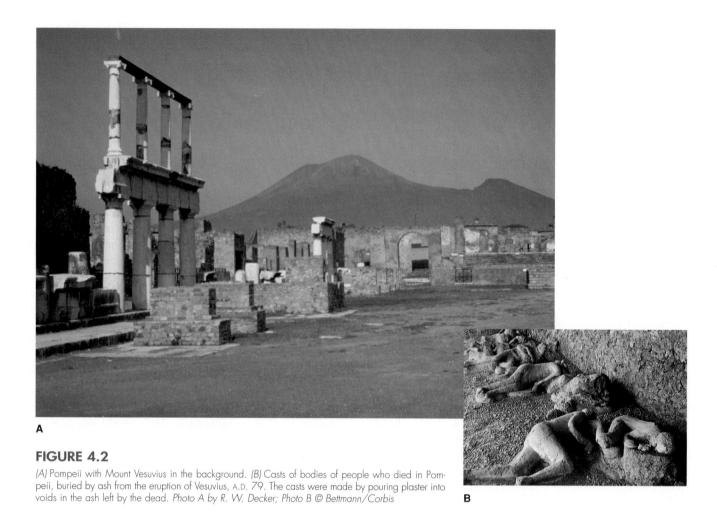

A

B

FIGURE 4.2

(A) Pompeii with Mount Vesuvius in the background. *(B)* Casts of bodies of people who died in Pompeii, buried by ash from the eruption of Vesuvius, A.D. *79.* The casts were made by pouring plaster into voids in the ash left by the dead. *Photo A by R. W. Decker; Photo B © Bettmann/Corbis*

Only a third of the island remained above sea level. The rest, which formerly rose to 800 meters above sea level, became a 300-meter-deep, underwater depression. The huge explosion was heard 5,000 kilometers away. Over 34,000 people died as a result of the giant sea waves (tsunamis) generated by the explosion.

A similar series of explosions in prehistoric time (about 6,600 years ago) created the depression occupied by Crater Lake in Oregon (figure 4.3). Volcanic debris covering more than a million square kilometers in Oregon and neighboring states has been traced to those eruptions. The original volcano, named Mount Mazama (now regarded as a cluster of overlapping volcanoes), is estimated to have been about 2,000 meters higher than the present rim of Crater Lake. For more on Crater Lake and Mount Mazama, go to http://geopubs.wr.usgs.gov/fact-sheet/fs092-02/.

The southern Cascade Mountains, where Crater Lake is located, have been built up by eruptions over the past 30 to 40 million years (figure 4.4; see also the geologic map, inside front cover). Only the youngest peaks (those built within the past 2 million years), such as Mount St. Helens, Mount Rainier, Mount Shasta, and Mount Hood, still stand out as cones. As Mount St. Helens has demonstrated, any of these could again erupt.

The Record of Fatalities

Figure 4.5 shows the results of research at the Smithsonian Institute and Macquarie University, Australia. Note the dramatic increase in fatalities during the recent centuries (figure 4.5*A*). This is not due to increasing volcanic activity but to increasing population and more people living near volcanoes. Figure 4.5*B*, which shows the cumulative number of deaths during the last seven centuries, also shows that most of the fatalities have been caused by seven major eruptions.

Volcanoes can kill in a number of ways. Figure 4.5*C* indicates that pyroclastic flows account for the most fatalities. A *pyroclastic flow,* described in the Extrusive Rocks and Gases section of this chapter, is a mixture of hot gas and pyroclastic debris that rapidly flows down a volcano's flanks. Famine and other indirect causes account for the next greatest number of fatalities. Widespread destruction of crops and farm animals can cause regional famine (as occurred with the eruption of Tambora in 1815). Note that relatively few events (specific eruptions) have caused the large number of deaths attributable to famine.

Pyroclastic fall accounts for the largest number of deadly events; however, few people die in each event, so the total number of deaths is not great. Most of the deaths due to pyroclastic fall are caused by collapse of ash-covered roofs or by being hit by falling rock fragments.

A

B

C

D

FIGURE 4.3

Crater Lake, Oregon. Its development and geologic history. (*A*) Cluster of overlapping volcanoes form. (*B*) Collapse into the partially emptied magma chamber is accompanied by violent eruptions. (*C*) Volcanic activity ceases, but steam explosions take place in the caldera. (*D*) Water fills the caldera to become Crater Lake, and minor renewed volcanism builds a cinder cone (Wizard Island). *Photo © Greg Vaughn/Tom Stack & Associates; Illustration after C. Bacon, U.S. Geological Survey*

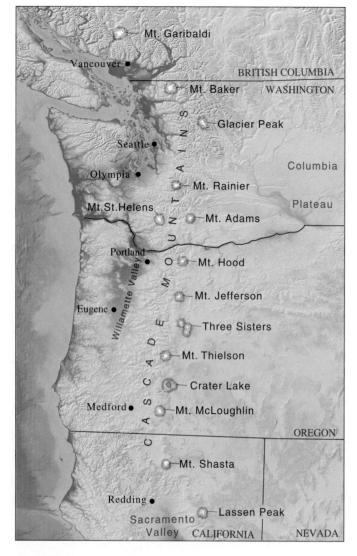

FIGURE 4.4

The Cascade volcanoes. The named volcanoes are ones that have erupted in geologically recent time. *Adapted from U.S. Geological Survey*

Eruptive Violence and Physical Characteristics of Lava

What determines the degree of violence associated with volcanic activity? Why can we state confidently that active volcanism in Hawaii poses only slight danger to humans but we expect violent explosions to occur in the Cascade Mountains? Whether eruptions are very explosive or relatively "quiet" is largely determined by two factors: (1) the amount of gas in the lava or magma and (2) the ease or difficulty with which the gas can escape to the atmosphere. The **viscosity,** or resistance to flow, of a lava determines how easily the gas escapes. The more viscous the lava and the greater the volume of gas trying to escape, the more violent the eruption. Later we will show how these factors not only determine the degree of violence of an eruption but also influence the shape and height of a volcano.

The three factors that influence viscosity are (1) the silica (SiO_2) content of the lava; (2) the temperature of the lava; and (3) gas dissolved in magma—the greater the dissolved gas content, the more fluid the lava. If the lava being extruded is considerably hotter than its solidification temperature, the lava is less viscous (more fluid) than when its temperature is near its solidification point. Temperatures at which lavas solidify range from about 700°C for silicic rocks to 1,200°C for mafic rocks.

Volcanic rocks, and the magma from which they formed, have a silica content that ranges from 45% to 75% by weight. **Silicic (or felsic) rocks** are silica-rich (65% or more SiO_2) rocks. *Rhyolite* is the most abundant silicic volcanic rock. **Mafic rocks** are *silica-deficient* rocks. Their silica content is close to 50%. *Basalt* is the most common mafic rock. **Intermediate rocks** have a chemical content between that of silicic and mafic rocks. The most common intermediate rock is *andesite*. Chapter 3 contains a more complete description of the chemistry of igneous rocks and their relationship to the mineral content of rocks.

FIGURE 4.5

Volcano fatalities. (*A*) Fatal volcano eruptions per century. (*B*) Cumulative volcano fatalities. Note the big jumps with the seven most deadly eruptions. These were eruptions that killed over 10,000 people and account for two-thirds of the total. (*C*) The causes of volcano fatalities. *Reprinted with permission from "Volcano Fatalities" by T. Simkin, L. Siebert, and R. Blong, Science, v. 291: p. 255. Copyright © 2001 American Association for the Advancement of Science*

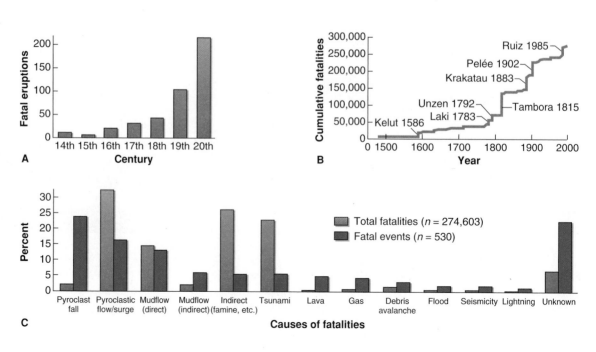

Volcanoes and Flying

There have been several occasions in which jumbo jets have flown into volcanic ash clouds with nearly disastrous results. In 1989, a KLM Boeing 747 unknowingly entered an ash plume over Mount Redoubt, a volcano in Alaska, at an altitude of 8,000 meters (26,000 feet). The pilot applied full power hoping to climb out of the plume. After climbing a thousand meters, all four engines stopped. The plane dropped to an altitude of 4,000 meters (13,000 feet) in eight tension-filled minutes before the flight crew was able to restart the engines. Although the plane landed safely in Anchorage, the cost to repair it was $80 million. Its engines, which had to be replaced, contained glassy coatings that turned out to be melted and resolidified ash. When full power was applied, the engines became very hot—hotter than the melting temperature of the ash. After this discovery, the standard procedure now is to reduce power to keep the engine temperature well below the melting point of volcanic ash and lessen the chances of engine failure. It is, of course, preferable to fly around pyroclastic clouds.

Another, less serious problem is what appears to be extensive scratching of airplane windows. The enormous amount of sulfuric acid aerosol that was belched into the atmosphere by Mount Pinatubo in 1991 caused scratching. Acid attacks the acrylic windows and etches fine lines in them. Although Pinatubo is in the Philippines, airplanes flying above 10 kilometers throughout the Northern Hemisphere have had their windows damaged—an annoyance to passengers and costly to airlines.

Mafic lavas, which are relatively low in SiO_2, tend to flow easily. Conversely, silicic lavas are much more viscous and flow sluggishly. Mafic lava is around 10,000 times as viscous as water, whereas silicic magma is around 100 million times the viscosity of water. Lavas rich in silica are more viscous because even before they have cooled enough to allow crystallization of minerals, silicon-oxygen tetrahedrons have linked to form small, framework structures in the lava. Although too few atoms are involved for the structures to be considered crystals, the total effect of these silicate structures is to make the liquid lava more viscous, much the way that flour or cornstarch thickens gravy.

Because silicic magmas are the most viscous, they are associated with the most violent eruptions. Mafic magmas are the least viscous and commonly erupt as lava flows (such as in Hawaii). Eruptions associated with intermediate magma can be violent or can produce lava flows. The Cascade volcanoes are predominantly composed of intermediate rock.

EXTRUSIVE ROCKS AND GASES

Scientific Investigation of Volcanism

Volcanoes and lava flows, unlike many other geologic phenomena, can be observed directly, and samples can be collected without great difficulty (at least for the quiet, Hawaiian-type of eruption). We can measure the temperature of lava flows, collect samples of gases being given off, observe the lava solidifying into rock, and take newly formed rock samples into the laboratory for analysis and study. By comparing rocks observed solidifying from lava with similar ones from other areas of the world (and even with samples from the Moon) where volcanism is no longer active, we can infer the nature of volcanic activity that took place in the geologic past.

Gases

From active volcanoes we have learned that most of the gas released during eruptions is water vapor, which condenses as steam. Other gases, such as carbon dioxide, sulfur dioxide, hydrogen sulfide (which smells like rotten eggs), and hydrochloric acid, are given off in lesser amounts with the steam.

Surface water introduced into a volcanic system can greatly increase the explosivity of an eruption, as exemplified by the devastation of the island of Krakatoa (described earlier).

Gases and Pyroclasts

During an eruption, expanding, hot gases may propel pyroclasts high into the atmosphere as a column rising from a volcano. At high altitudes, the pyroclasts often spread out into a dark, mushroom cloud. The fine particles are transported by high atmosphere winds. Eventually, debris settles back to Earth under gravity's influence as *pyroclast fall* (often called *ashfall* or *pumice fall*) deposits.

A **pyroclastic flow** is a mixture of gas and pyroclastic debris that is so dense that it hugs the ground as it flows rapidly into low areas (figure 4.6). Pyroclastic flows develop in several ways. Some are associated with volcanic domes (discussed later). An exploding froth of gas and magma can blast out of the side of the dome or viscous plug capping a volcano. A steep-sided dome might collapse, allowing violent release of magma and its gases. For some volcanoes, a pyroclastic flow results from gravitational collapse of a column of gas and pyroclastic debris that was initially blasted vertically into the air. These turbulent masses can travel up to 200 kilometers per hour and are extremely dangerous. In 1991, a pyroclastic flow at Japan's Mount Unzen killed 43 people, including three geologists and famous volcano photographers, Maurice and Katia Krafft. Far worse was the destruction of St. Pierre on the Caribbean island of Martinique

FIGURE 4.6

Pyroclastic flow descending Mayon Volcano, Philippines, in 1984. Ways in which pyroclastic flows can form: (B) Blasting out from under a plug capping a volcano. (C) Collapse of part of a steep-sided dome. (D) Gravitational collapse of an eruptive column. *Photo by Chris Newhall, U.S. Geological Survey*

FIGURE 4.7

The ruins of St. Pierre in 1902. Mount Pelée is in the clouds. *Photo by Underwood & Underwood, courtesy of Library of Congress*

(figure 4.7), where about 29,000 people were killed by a pyroclastic flow in 1902 (see box 4.5).

EXTRUSIVE ROCKS

Most extrusive rocks are named and identified on the basis of their composition and texture. But some names are based solely on texture (e.g., pumice).

Composition

The amount of silica in a lava largely controls not only the viscosity of lava and the violence of eruptions but also which particular rock is formed. Chapter 3 describes how igneous rocks are identified based on the minerals present and their relative abundance in the rock. (For photos and diagrams refer to figures 3.6 and 3.7 on pages 62 and 65.) Because extrusive igneous rocks are generally fine-grained, a specialized microscope is usually needed for precise identification of the component minerals. In most cases, however, we can guess the probable mineral content by noting how dark or light in color an extrusive rock is. Most silicic rocks are light-colored because they contain abundant feldspar and quartz (both of which are silica-rich) and few dark minerals (which contain iron and magnesium and are silica-deficient). Mafic rocks, on the other hand, tend to be dark because of the abundance of ferromagnesian minerals.

Rhyolite, a silicic rock, is usually cream-colored, tan, or pink; it is made up mostly of feldspar but always includes some quartz. Note that the rhyolite (and granite) portion of figures 3.6 and 3.7 is larger than the areas shown for andesite and basalt. Geologists commonly subdivide this portion of the classification system. For example, *dacite,* the rock associated with the 1980 Mount St. Helens eruptions, contains more ferromagnesian minerals and plagioclase but less potassium feldspar and quartz than the average rhyolite. In our classification system, dacite corresponds to the right portion of the area in figure 3.6 assigned to rhyolite.

Andesite, which crystallizes from an intermediate lava, can be recognized by its moderately gray or green color. It is this color because a little over half the rock is light- to medium-gray plagioclase feldspar, while the rest is ferromagnesian minerals (usually *pyroxene* or *amphibole*).

A **basalt** has a relatively low amount (about 50% by weight) of SiO_2. Much of that silica is bonded to iron and magnesium to form ferromagnesian minerals, such as *olivine* or *pyroxene,* which are green or black. The remaining silica plus aluminum is bonded predominantly with calcium to form calcium-rich *plagioclase feldspar* (which tends to be darker gray than the white or pink potassium or sodium feldspars associated with silicic rocks). Basalt does not contain quartz because no silica is left over after the other minerals have formed. Because of the preponderance of dark minerals in basalt, this rock is usually dark gray to black.

Extrusive Textures

Texture refers to a rock's appearance with respect to the size, shape, and arrangement of its grains or other constituents. Table 4.1 is a summary of extrusive rock textures.

Some extrusive rocks (such as obsidian and pumice) are classified solely on the basis of their textures, but most are classified by composition *and* texture. *Grain size* is a rock's most important textural characteristic. For the most part, extrusive rocks are fine-grained or else made of glass.

A **fine-grained rock** is one in which most of the mineral grains are smaller than 1 millimeter. In most, the individual minerals are distinguishable only with a microscope. **Obsidian** (figure 4.8), which is volcanic glass, is one of the few rocks that is not composed of minerals. A fine-grained or glassy texture distinguishes extrusive rocks from most intrusive rocks.

Two critical factors determine grain size during the solidification of igneous rocks: rate of cooling and viscosity. If lava cools rapidly, the atoms have time to move only a short distance; they bond with nearby atoms, forming only small crystals. With extremely rapid or almost instantaneous cooling,

Name	Description
TABLE 4.1	**Summary of Textures in Volcanic Rocks**
Fine-grained (adjective)	Mosaic of interlocking minerals that are smaller than 1 millimeter.
Porphyritic (adjective)	Some crystals, phenocrysts, are larger than 1 millimeter (usually considerably larger). Most grains are smaller than 1 millimeter. Or phenocrysts are enclosed in glass.
Obsidian	Glass. Arrangement of atoms is disordered.
Vesicular (adjective)	Holes (vesicles) in rock due to trapped gas
Pumice	Frothy glass
Tuff	Consolidated, fine pyroclastic material
Volcanic breccia	Consolidated pyroclastic debris that includes blocks or bombs

FIGURE 4.8
Obsidian. *Photo by C. C. Plummer*

individual atoms in the lava are "frozen" in place, forming glass rather than crystals.

Grain size is controlled to a lesser extent by the viscosity of the lava. Atoms in a highly viscous lava cannot move as freely as those in a more fluid lava. Hence, a rock formed from viscous lava is more likely to be obsidian or of finer grains than one formed from more fluid lava. Most obsidian, when chemically analyzed, has a very high silica content and is silicic, the chemical equivalent of rhyolite.

Porphyritic Textures

Extrusive rock that does not have a uniformly fine-grained texture throughout is described as porphyritic. A **porphyritic rock** is one in which larger crystals are enclosed in a *groundmass* of much finer-grained minerals or obsidian. The larger crystals are termed **phenocrysts.** A porphyritic rock looks rather like raisin bread; the groundmass is the bread, the phenocrysts are the raisins. In the porphyritic andesite shown in figure 4.9*A*, phenocrysts of feldspar and ferromagnesian minerals are enclosed in a groundmass of crystals too fine-grained to distinguish with the naked eye but visible under a microscope (figure 4.9*B*).

Porphyritic texture in extrusive rocks usually indicates two stages of solidification. Slow cooling takes place while the magma is underground. Minerals that form at higher temperatures crystallize and grow to form phenocrysts in the still partly fluid magma. If the entire mass is then erupted, the remaining liquid portion cools rapidly and forms the fine-grained groundmass.

Textures Due to Trapped Gas

A magma deep underground is under high pressure, generally high enough to keep all its gases in a dissolved state. On eruption, the pressure is suddenly released and the gases come out of solution. This is analogous to what happens when a bottle of beer or soda is opened. Because the drink was bottled under pressure, the gas (carbon dioxide) is in solution. Uncapping the drink relieves the pressure, and the carbon dioxide separates from the liquid as gas bubbles. If you freeze the newly opened drink very quickly, you have a piece of ice with small, bubble-shaped holes. Similarly, when a lava solidifies while gas is bubbling through it, holes are trapped in the rock, creating a distinctive vesicular texture. **Vesicles** are cavities in extrusive rock resulting from gas bubbles that were in lava, and the texture is called *vesicular.* A vesicular rock has the appearance of Swiss cheese (whose texture is caused by trapped carbon dioxide gas). *Vesicular basalt* is quite common (figure 4.10). *Scoria,* a highly vesicular basalt, actually contains more gas space than rock.

In more viscous lavas, where the gas cannot escape as easily, the lava is churned into a froth (like the head in a glass of beer). When cooled quickly, it forms **pumice** (figure 4.11), a frothy glass with so much void space that it floats in water. Powdered pumice is used as an abrasive because it can scratch metal or glass.

The great eruptions accompanying caldera-forming events (such as Krakatoa in 1883 and Pinatubo in 1991) create

A

B

FIGURE 4.9

Porphyritic andesite. A few large crystals (phenocrysts) are surrounded by a great number of fine grains. (*A*) Hand specimen. Grains in groundmass are too fine to see. (*B*) Photomicrograph (using polarized light) of the same rock. The black-and-white striped phenocrysts are plagioclase, and the green ones are ferromagnesian minerals. *Photo A © Parvinder Sethi. Photo B by C. C. Plummer*

FIGURE 4.10

Vesicular basalt. *Photo © Parvinder Sethi*

huge amounts of pumice pyroclasts of all sizes. The seas near Krakatoa were covered with floating pumice pyroclasts, greatly hindering ship traffic. Baseball- and smaller-sized pumice fragments rained down on people during the eruption of Pinatubo. For some eruptions, most of the pumice fragments are around the size of a fingertip. These are, appropriately, called *popcorn pumice*. The ground east of the Sierra Nevada in California and Nevada near Mono Craters is layered with popcorn pumice from eruptions taking place during the past several thousand years.

Fragmental Textures

Pyroclasts, the fragments formed by volcanic explosion, can be almost any size.

Their size-based names are:

Dust	<1/8 millimeter
Ash	1/8–2 millimeters
Cinder or lapilli	2–64 millimeters
Blocks and bombs	>64 millimeters

Cinder is often used as a less-restricted, general term for smaller pyroclasts. *Lapilli* is used for the 2–64 millimeter particles—a size range that extends from that of a grain of rice to a peach. When solid rock has been blasted apart by a volcanic explosion, the pyroclastic fragments are *angular,* with no rounded edges or corners and are called **blocks.** If lava is ejected into the air, a molten blob becomes streamlined during flight, solidifies, and falls to the ground as a **bomb,** a spindle- or lens-shaped pyroclast (figure 4.12).

When pyroclastic material (ash, bombs, etc.) accumulates and is cemented or otherwise consolidated, the new rock is called *tuff* or *volcanic breccia,* depending on the size of the fragments. A **tuff** (figure 4.13) is a rock composed of fine-grained pyroclastic particles (dust and ash). A **volcanic breccia** is a rock that includes larger pieces of volcanic rock (cinder, blocks, bombs).

TYPES OF VOLCANOES

Volcanic material that is ejected from and deposited around a central vent produces the conical shape typical of volcanoes. The **vent** is the opening through which an eruption takes place. The **crater** of a volcano is a basinlike depression over a vent at the summit of the cone (figure 4.14). Material is not always ejected from the central vent. In a **flank eruption,** lava pours from a vent on the side of a volcano.

A **caldera** is a volcanic depression much larger than the original crater, having a diameter of at least 1 kilometer. (The most famous caldera in the United States is misnamed "Crater Lake.") A caldera can be created when a volcano's summit is blown off by exploding gases or, as in the case of Crater Lake, when a volcano (or several volcanoes) collapses into a partially emptied magma chamber (see figure 4.3).

The three major types of volcanoes (shield, cinder cone, and composite), discussed on pages 98–101 and compared in table 4.2, are markedly distinct from one another in size, shape, and, usually, composition. Note from the scales and the

A

B

mm

FIGURE 4.11

(A) A boulder of pumice can be easily carried because it is mostly air. (B) Seen close up, pumice is a froth of volcanic glass. *Photos by C. C. Plummer*

A

B

FIGURE 4.12

(A) Volcanic bombs. (B) Night-time eruption at Cerro Negro, a cinder cone in Nicaragua. Magma blobs that solidify in the air will land as bombs. If they are still molten upon landing, they will spatter. *Photo A by C. C. Plummer; photo B by R. W. Decker*

relative size diagram that the shield volcano is vastly bigger than the other two and the composite volcano is much bigger than the cinder cone. Although volcanic domes are not cones, they are associated with volcanoes and are also examined in this section.

Shield Volcanoes

Shield volcanoes are broad, gently sloping volcanoes constructed of solidified lava flows. During eruptions, the lava spreads widely and thinly due to its low viscosity. Because the

lava flows from a central vent, without building up much near the vent, the slopes are usually between 2° and 10° from the horizontal, producing a volcano in the shape of a flattened dome or "shield" (figure 4.15).

The islands of Hawaii are essentially a series of shield volcanoes built upward from the ocean floor by intermittent

FIGURE 4.13

Photomicrograph of a tuff. Fragments of different rocks, mainly obsidian and pumice, are angular and variously colored. *Photo by C. C. Plummer*

FIGURE 4.14

Crater and caldera in Kamchatka, Russia. In the foreground is the crater on Karymsky Volcano. In the background is a lake-filled caldera. *Photo by C. Dan Miller, U.S. Geological Survey*

TABLE 4.2 Comparison of the Three Types of Volcanoes

Profile of Volcano	Description	Composition
Shield volcano (10 km, ~4°, 100 km)	**Shield Volcano** Gentle slopes – between 2° and 10°. The Hawaiian example rises 10 kilometers from the sea floor.	Basalt. Layers of solidified lava flows.
Composite volcano (Typically 1,000 to 4,000 meters, ~25°, 1,000–4,000 m)	**Composite Volcano** Slopes less than 33°. Considerably larger than cinder cones.	Layers of pyroclastic fragments and lava flows. Mostly andesite.
Cinder cone (< 500 meters, 33°, < 500 m)	**Cinder Cone** Steep slopes – 33°. Smallest of the three types.	Pyroclastic fragments of any composition. Basalt is most common.

Mauna Loa   Kilauea

Shield volcano: Hawaii

Composite volcano: Mt. Shasta, California   **Cinder cone:** Sunset Crater, Arizona

Profiles drawn to same scale

New lava

Feeders

Layers of basalt

A

B

FIGURE 4.15

(A) Cutaway view of a shield volcano. (B) The top of Mauna Loa, a shield volcano in Hawaii, and its summit caldera. The smaller depressions are pit craters. *Photo © James L. Amos/Corbis Images*

FIGURE 4.16
Flow of lava solidifying to pahoehoe in Hawaii. *Photo © Parvinder Sethi*

eruptions over millions of years (figure 4.15*B*). Although spectacular to observe, the eruptions are relatively nonviolent because the lavas are fairly fluid (low viscosity). By implication, then, the shield volcanoes of the Hawaiian Islands are composed of a series of layers of basalt.

Hawaiian names have been given to two distinctive surfaces of basalt flows. *Pahoehoe* (pronounced *pah-hoy-hoy*) is characterized by a ropy or billowy surface (figure 4.16). The surface is formed by the quick cooling and solidification from the surface downward of a lava flow or pool of lava that was fully liquid. By contrast, basalt that is cool enough to have partially solidified moves as a slow, pasty mass. Its largely solidified front is shoved forward as a pile of rubble. A flow such as this is called *aa* (pronounced *ah-ah*) and has a jagged, rubbly surface (figure 4.17).

A (usually) minor feature called a *spatter cone*, a small, steep-sided cone built from lava sputtering out of a vent (figure 4.18), will occasionally develop on a solidifying lava flow. When a small concentration of gas is trapped in a cooling lava flow, lava is belched out of a vent through the solidified surface of the flow. Falling lava plasters itself onto the developing

FIGURE 4.17
An *aa* flow in Hawaii, 1983. *Photo by J. D. Griggs, U.S. Geological Survey*

FIGURE 4.18

A spatter cone (approximately 1 meter high) erupting in Hawaii. *Photo by J. B. Judd, U.S. Geological Survey*

cone and solidifies. The sides of a spatter cone can be very steep, but they are rarely over 10 meters high. An exception to this is Pu'u 'O'o, the 250-meter-high, combined spatter and cinder cone on the eastern flank of Kilauea shield volcano. It is located at the vent for the ongoing (1983–onward) lava eruptions.

Much of the lava in the ongoing Hawaiian eruptions flows underground in a lava tube, traveling about 7 kilometers from Pu'u 'O'o to the sea. A *lava tube* is a tunnel-like conduit for lava that develops after most of a flow has solidified (figure 4.19). The tube's roof and walls solidified along with the earlier, broader flow. The tube provides insulation so that the rapidly flowing lava loses little heat and remains fluid.

Cinder Cones

A **cinder cone** (less commonly called a *pyroclastic cone*) is a volcano constructed of pyroclastic fragments ejected from a central vent (figure 4.20). Unlike a shield volcano, which is made up of lava flows, a cinder cone is formed exclusively of pyroclasts. In contrast to the gentle slopes of shield volcanoes, cinder cones commonly have slopes of about 30°. Most of the ejected material lands near the vent during an eruption, building up the cone to a peak. The steepness of slopes of accumulating loose material is limited by gravity to about 33°. Cinder cones tend to be very much smaller than shield volcanoes. In fact, cinder cones are commonly found on the flanks and in the calderas of Hawaii's shield volcanoes. Few cinder cones exceed a height of 500 meters.

Cinder cones form by pyroclastic material accumulating around a vent. They form because of a buildup of gases and are independent of composition. Most cinder cones are associated with mafic or intermediate lava. Silicic cinder cones, which are made of fragments of pumice, are also known as pumice cones.

The life span of an active cinder cone tends to be short. The local concentration of gas is depleted rather quickly during the eruptive periods. Moreover, as landforms, cinder cones are temporary features in terms of geologic time. The unconsolidated pyroclasts are eroded relatively easily.

Composite Volcanoes

A **composite volcano** (also called **stratovolcano**) is one constructed of alternating layers of pyroclastic fragments and solidified lava flows (figure 4.21A). The slopes are intermediate in steepness compared with cinder cones and shield volcanoes. Pyroclastic layers build steep slopes as debris collects

A

B

FIGURE 4.19

(A) Lava stream seen through a collapsed roof of a lava tube during a 1970 eruption of Kilauea Volcano, Hawaii. Note the ledges within the tube, indicating different levels of flows. (B) Lava tube at Lava Beds National Monument, California. The narrow, dark shelf on either side of the tube marks the level of the lava stream, indicating where lava solidified against the walls of the tube. *Photo A by J. B. Judd, U.S. Geological Survey. Photo B by C. C. Plummer*

FIGURE 4.20

Cerro Negro, a cinder cone in Nicaragua, erupting. Figure 4.12B shows a night-time eruption of Cerro Negro. *Photo by Mark Hurd Aerial Surveys Corp., courtesy of California Division of Mines and Geology*

near the vent, just as in cinder cones. However, subsequent lava flows partially flatten the profile of the cone as the downward flow builds up the height of the flanks more than the summit area. The solidified lava acts as a protective cover over the loose pyroclastic layers, making composite volcanoes less vulnerable to erosion than cinder cones.

Composite volcanoes are built over long spans of time. Eruption is intermittent, with hundreds or thousands of years of inactivity separating a few years of intense activity. During the quiet intervals between eruptions, composite volcanoes may be eroded by running water, landslides, or glaciers. These surficial processes tend to alter the surface, shape, and form of the cone. But because of their long lives and relative resistance to erosion, composite cones can become very large. Aconcagua, a composite volcano in the Andes, is 6,960 meters (22,835 feet) above sea level and is the highest peak in the Western Hemisphere.

The extrusive material that builds composite cones is predominantly of intermediate composition, although there may be some silicic and mafic eruptions. Therefore, *andesite* is the rock most associated with composite volcanoes. If the lava is especially hot, the relatively low viscosity fluid flows easily from the crater down the slopes. On the other hand, if enough gas pressure exists, an explosion may litter the slopes with pyroclastic andesite, particularly if the lava has fully or partially solidified and clogged the volcano's vent.

The composition as well as eruptive history of individual volcanoes can vary considerably. For instance, Mount Rainier is composed of 90% lava flows and only 10% pyroclastic layers. Conversely, Mount St. Helens was built mostly from pyroclastic eruptions—reflecting a more violent history. As would be expected, the composition of the rocks formed during the 1980 eruptions of Mount St. Helens is somewhat higher in silica than average for Cascade volcanoes.

Distribution of Composite Volcanoes

Nearly all the larger and better known volcanoes of the world are composite volcanoes. They tend to align along two major belts (figure 4.22). The **circum-Pacific belt,** or "Ring of Fire," is the larger. The Cascade Range volcanoes described earlier make up a small segment of the circum-Pacific belt.

Several composite volcanoes in Mexico rise higher than 5,000 meters, including Orizaba (third highest peak in North America) and Popocatépetl (see box 4.4).

The circum-Pacific belt includes many volcanoes in Central America, western South America (including Nevado del Ruiz in Colombia), and Antarctica. Mount Erebus, in Antarctica, is the southernmost active volcano in the world (figure 4.23).

The western portion of the Pacific belt includes volcanoes in New Zealand, Indonesia, the Philippines (with Pinatubo, whose 1991 caldera-forming eruption was the second-largest eruption of the twentieth century), and Japan. The beautifully symmetrical Fujiyama, in Japan, is probably the most frequently painted volcano in the world (figure 4.24), as well as its most climbed mountain. The northernmost part of the circum-Pacific belt includes active volcanoes in Russia (see figure 4.14) and on Alaska's Aleutian Islands. The 1912 eruption of Katmai in Alaska was the world's largest in the twentieth century.

The second major volcanic belt is the **Mediterranean belt,** which includes Mount Vesuvius. An exceptionally violent eruption of Mount Thera, an island in the Mediterranean, may have destroyed an important site of early Greek civilization. (Some archaeologists consider Thera the original "lost continent" of Atlantis.) Mount Etna, on the island of Sicily, is Europe's largest volcano and one of the world's most active volcanoes. The largest eruption in 300 years began in 1991 and lasted for 473 days. Some 250 million cubic meters of lava covered 7 square kilometers of land. A town was saved from the lava by heroic efforts that included building a dam to retain the lava (the lava quickly overtopped it), plugging some natural channels, and diverting the lava into other, newly constructed channels.

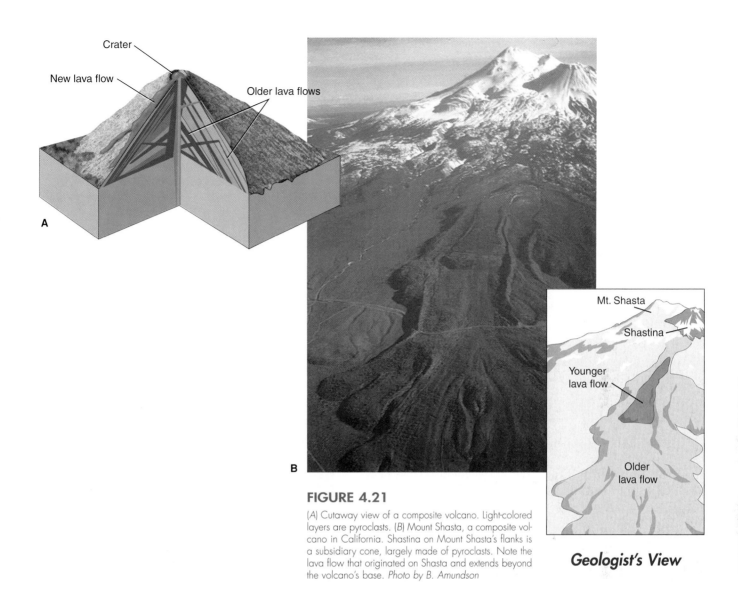

FIGURE 4.21

(A) Cutaway view of a composite volcano. Light-colored layers are pyroclasts. (B) Mount Shasta, a composite volcano in California. Shastina on Mount Shasta's flanks is a subsidiary cone, largely made of pyroclasts. Note the lava flow that originated on Shasta and extends beyond the volcano's base. *Photo by B. Amundson*

Geologist's View

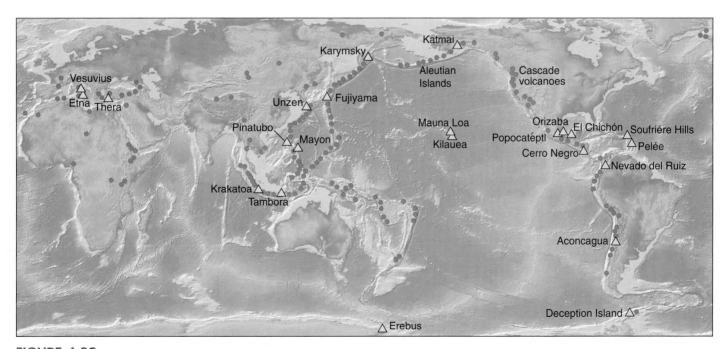

FIGURE 4.22

Map of the world showing recently active major volcanoes. Red dots represent individual volcanoes. Yellow triangles represent volcanoes mentioned in this chapter.

PLANETARY GEOLOGY 4.3

Extraterrestrial Volcanic Activity

Volcanic activity has been a common geologic process operating on the Moon and several other bodies in the solar system. Approximately one-sixth of the Moon's surface consists of nearly circular, dark-colored, smooth, relatively flat lava plains. The lava plains, found mostly on the near side of the Moon, are called *maria* (singular, *mare;* literally, "seas"). They are believed to be huge meteorite impact craters that were flooded with basaltic lava during the Moon's early history. There are also a few extinct shield volcanoes on the Moon.

Elongate trenches or cracklike valleys called *rilles* are found mainly in the smoother portions of the lunar maria. They range in length from a few kilometers to hundreds of kilometers. Some are arc-shaped or crooked and are regarded as drained basaltic lava channels.

Mercury, the innermost planet, also has areas of smooth plains, suspected to be volcanic in origin.

Radar images of Venus show a surface that is young and probably still volcanically active. More than three-fourths of that surface is covered by continuous plains formed by enormous floods of lava. Close examination of these plains reveals extensive networks of lava channels and individual lava flows thousands of kilometers long.

Large shield volcanoes, some in chains along a great fault, have been identified on Venus, and molten lava lakes may exist. In other places, thick lavas have oozed out to form kilometer-high, pancake-shaped domes. Radar studies have shown that some of these domes are composed of a glassy substance mixed with bubbles of trapped gas. Fan-shaped deposits adjacent to some volcanoes may be pyroclastic debris.

Several of Venus's volcanoes emit large amounts of sulfur gases, causing the almost continuous lightning that has been observed by spacecraft. It is strongly suspected that the planet is still volcanically active.

Nearly half of the planet Mars may be covered with volcanic material. There are areas of extensive lava flows similar to the lunar maria and a number of volcanoes, some with associated lava flows.

Mars has at least nineteen large shield volcanoes, probably composed of basalt. The largest one, Olympus Mons (box figure 1), is three times the height of Mount Everest and wider than Arizona. Its caldera is more than 90 kilometers across.

Hundreds of volcanoes have been discovered on Jupiter's moon Io (box figure 2), and some of those have erupted for periods of at least four months. Material rich in sulfur compounds is thrown at least 500 kilometers into space at speeds of up to 3,200 kilometers per hour. This material often forms umbrella-shaped clouds as it spreads out and falls back to the surface. Lakes of very hot silicate lava, perhaps mafic or ultramafic, are common. More than 100 calderas larger than 25 kilometers across have been observed, including one that vents sulfur gases. The energy source for Io's volcanoes may be the gravitational pulls of Jupiter and two of its other larger

BOX 4.3 ■ FIGURE 1

Perspective view of Olympus Mons, the largest volcano and tallest mountain in the solar system. This Martian volcano is over 650 km wide and 24 km high. Note the outline of the state of Arizona for size comparison. *Photo by NASA/MOLA Science Team*

BOX 4.3 ■ FIGURE 2

Two volcanic plumes on Jupiter's moon Io. The plume on left horizon (and upper insert) is 140 kilometers high; the one in the center (and lower insert) is 75 kilometers high. For details go to photojournal.jpl.nasa.gov/catalog/PIA00703. *Photo by JPL/NASA*

satellites, causing Io to heat up much as a piece of wire will do if it is flexed continuously.

Neptune's moon Triton is the third object in the solar system that has active volcanoes. There, "ice volcanoes" erupt what is probably nitrogen frost.

Additional Resource

The Nine Planets
- www.nineplanets.org/

FIGURE 4.23

Mount Erebus, Antarctica, the southernmost active volcano in the world. The photo is taken on sea ice. The summit is 3,794 meters (12,444 feet) above sea level. One of its two summit craters contains a convecting lava lake. *Photo by Philip R. Kyle*

FIGURE 4.24

Mount Fuji, woodblock print by Japanese artist Hiroshige (1797–1858).

Volcanic Domes

Volcanic domes are steep-sided, dome- or spine-shaped masses of volcanic rock formed from viscous lava that solidifies in or immediately above a volcanic vent. A volcanic dome grew within the crater of Mount St. Helens after the climactic eruption of May 1980 (figure 4.25). This was expected because of the

high viscosity of the lava from the eruptions. In 1983 alone, the dome increased its elevation by 200 meters. After years of quiescence, dome growth resumed in October 2004. At that time, lava extrusion shifted and a new dome began growing adjacent to the original dome (figure 4.25 and http://vulcan.wr.usgs.gov/ Volcanoes/MSH/Images/MSH04/framework.html).

Most of the viscous lavas that form volcanic domes are high in silica. Commonly, they solidify as obsidian that is the chemical equivalent of rhyolite (or, less commonly, andesite). If minerals do crystalize, the rock is rhyolite, if from a silicic magma, or andesite, if from an intermediate magma.

Because the thick, pasty lava that squeezes from a vent is too viscous to flow, it builds up a steep-sided dome or spine

FIGURE 4.25

Dome growth in Mt. St. Helens crater after the 1980 cataclysmic eruption that blasted away the top and front of the mountain. The photo, taken November 4, 2004, shows the glow of lava that is part of the dome building event that began a month earlier. The snow-covered "old" dome in the foreground has been volcanically inactive since 1991. That dome has a height of 267 meters (876 feet) above the crater floor. *Photo by Elliot Endo, U.S. Geological Survey*

Geologist's View

ENVIRONMENTAL GEOLOGY 4.4

Popocatépetl—Will It Erupt Big Time?

Popocatépetl, located 55 kilometers east of Mexico City, one of the world's largest cities, and 45 kilometers west of the city of Puebla, began erupting in 1994. Some 30 million people live within view of Popocatépetl (Aztec for "smoking mountain"). A major eruption could endanger hundreds of thousands of those people.

Popocatépetl, affectionately called "Popo," at 5,484 meters (17,991 feet) above sea level, is one of North America's highest mountains. Not only does Popocatépetl provide a majestic scenic presence (box figure 1A), but it figures prominently in Mexico's history, art, and culture. According to Aztec legend, Popocatépetl is a warrior eternally guarding his sleeping lover, the neighboring mountain Ixtaccihuatl (Aztec for "white lady"), whose outline resembles that of a supine woman. Cortez sent his men to climb Popocatépetl during the Spanish conquest of Mexico in 1521. They were lowered into the smoking crater and returned with sulfur used to make gunpowder. (This was the world's first recorded ascent of a major mountain.)

The volcano began awakening from a long period of dormancy in December 1994 with a minor dusting of ash on Puebla. Some 75,000 people living on the eastern flank of the volcano were temporarily evacuated. An extensive monitoring network of instruments was deployed, and teams of Mexican scientists assisted by members of the U.S. Geological Survey Volcanic Disaster Assistance Program began assessing the potential hazards.

The threat of a disaster is taken very seriously because in 1982, an apparently insignificant, jungle-covered, 1,000-meter-high volcano in southern Mexico called El Chichón erupted with a series of violent explosions. Towns near the previously inactive volcano were buried by heavy ashfall or blasted by searing, gas-charged pyroclastic flows. The death toll could only be roughly estimated to be in the thousands.

By determining the size and extent of ancient pyroclastic deposits and dating them, geologists have determined that Popocatépetl produced major explosive eruptions every 1,000 to 3,000 years for the last 10,000 years. Each has produced widespread pumice falls, pyroclastic flows, and mudflows. Pre-

BOX 4.4 ■ FIGURE 1

Popocatépetl in 1960 (A) and during the December 19, 2000, eruption (B) Snowcovered glaciers that stand out in the 1960 photo are now covered with ash. *Photo A by C. C. Plummer; Photo B © AP/Wide World Photos*

(figure 4.26). Some volcanic domes act like champagne corks, keeping gases from escaping. If the plug is removed or broken, the gas and magma escape suddenly and violently, usually as a pyroclastic flow (figure 4.6). Some of the most destructive volcanic explosions known have been associated with volcanic domes (see box 4.5).

LAVA FLOODS

Not all extrusive rocks are associated with volcanoes. Lava that is very nonviscous and flows almost as easily as water does not build a cone around its vents. Such lava is, of course, mafic (low in silica).

Plateau basalts were produced during the geologic past by vast outpourings of lava. The Columbia Plateau area of Washington, Idaho, and Oregon (see inside front cover), for example, is constructed of layer upon layer of basalt (figure 4.27), in places as thick as 3,000 meters. The area covered is over 400,000 square kilometers. Each individual flood of lava added a layer usually between 15 and 100 meters thick and sometimes thousands of square kilometers in extent. The outpourings of lava that built the Columbia Plateau took place from 17.5 to 6 million years ago but 95% erupted between 17 and 15.5 million years ago. Similar huge, lava plateau-building events have not occurred since then. (The hypothesis that these are due to the arrival of huge mantle plumes beneath the lithosphere is described in chapter 3.)

conquest population centers were repeatedly destroyed by these catastrophic eruptions. Since the year 1345, records indicate that there have been some thirty small eruptions before the present activity. Volcanologists consider it one of the world's most dangerous volcanoes. Will the current activity culminate in a colossal event like one of those that takes place every thousand years or so?

If such an event were to occur, pyroclastic flows and mudflows would destroy villages and could kill thousands of people, if they are not evacuated in time. Heavy ashfall would cause further damage. Mexico City is not likely to be affected by pyroclastic flows or mudflows, but if ash is blown over and deposited in the city, there could be serious consequences. Air traffic to and from Mexico City International Airport would be threatened. Water supplies, electrical power grids, and sewer systems could be damaged or destroyed.

Minor steam and ash eruptions from Popocatépetl continued through 1995. On March 29, 1996, a new lava dome was identified in the summit crater of the volcano. A month later, an explosion of the new dome killed five climbers who were at the summit of the volcano. Small ash-producing explosions took place throughout 1996 and into 1997. In April 1997, several slightly larger explosions took place. These generated ash columns that rose as much as 5 kilometers above the mountain. Volcanic bombs showered the flanks of the cone; some started grass fires, scaring rural residents.

In late 2000, activity increased, and on December 18, Popo's largest eruption in over 1,000 years took place with spectacular nighttime displays of incandescent lava expelled from the mountain. By this time, 14,000 people had been evacuated to shelters. Evacuation of high-risk towns had begun days earlier, and by December 21, some 50,000 people had been evacuated. Concern developed over the potential for a large mudflow because of melting of a glacier. On January 31, 2001, a pyroclastic flow descended the volcano to within 8 kilometers of a town.

One author's (Plummer) reflections on the mountain:

One of the indelible memories of growing up in Mexico City is that of the huge, magnificent volcanoes on the eastern skyline—Popocatépetl and Ixtaccihuatl. They were always visible when the sky was clear, which was frequent in the days before Mexico became one of the world's smoggiest cities. At age 15, I was fortunate enough to join some experienced mountaineers and climb the snow-covered mountain. I was stunned by the debilitating effect high altitude has on climbers. I would count out ten steps upward, then collapse over my ice axe, panting for several minutes before taking another ten steps. When we reached the summit, everyone with me felt nauseated and had splitting headaches—altitude sickness. I didn't feel too bad. After the climb, I felt an enormous sense of accomplishment. My life changed. The beauty of mountains and the challenge of climbing them became the focal point of my existence. At college, I became interested in geology (after taking an introductory physical geology course) as a natural extension of my love of mountains. Ultimately, my interest in geology overtook my interest in climbing.

Additional Resource

CENAPRED Volcano Site

Although a daily report can be accessed in English, you can get a wealth of information in Spanish. You can also see the volcano live by clicking on *Tamaño B* under *Imagen del volcán*.

- www.cenapred.unam.mx/mvolcan.html

Even larger basalt plateaus are found in India and Siberia. Their times of eruption coincide with the two largest mass extinctions of life on Earth. The one in Siberia occurred about 250 million years ago, around the time of the largest mass extinction, when over 90% of living species were wiped out. The eruptions are a prime suspect because of the enormous amount of gases that must have been emitted. These would have changed the atmosphere and worldwide climate. The Indian eruptions occurred about 65 million years ago and coincided with the mass extinction in which the last of the dinosaurs died. Although this mass extinction is generally blamed on a large asteroid hitting Earth (see chapter 8), the intense volcanic activity may have been a contributing factor.

Basalt layers give the landscape a striking appearance in most places where they are exposed. Instead of stacked-up slabs or tablets of solid, unbroken rock, the individual layers may appear to be formed of parallel, vertical columns, mostly six-sided. This characteristic of basalt is called **columnar structure** or **columnar jointing** (figure 4.28). The columns can be explained by the way in which basalt contracts as it cools *after* solidifying. Basalt solidifies completely at temperatures below about 1,200°C. The hot layer of rock then continues to cool to temperatures normal for the Earth's surface. Like most solids, basalt contracts as it cools. The layer of basalt is easily able to accommodate the shrinkage in the narrow vertical dimension; but the cooling rock cannot "pull in" its edges, which may be many kilometers away. Instead, the rock contracts toward

A Tale of Two Volcanoes—Lives Lost and Lives Saved in the Caribbean

Montserrat and Martinique are two of the tropical islands that are part of a volcanic island arc (box figure 1). During the twentieth century, both islands had major eruptions that destroyed towns. Violent and deadly pyroclastic flows associated with growth of volcanic domes caused most of the destruction. For one island, the death toll was huge, and for the other, it was minimal.

In 1902, the port city of St. Pierre on the island of Martinique was destroyed after a period of dome growth and pyroclastic flows on Mount Pelée (no relationship to Pele, Hawaii's goddess of volcanoes). A series of pyroclastic flows broke out of a volcanic dome and flowed down the sides of the volcano. Searingly hot pyroclastic flows can travel at up to 200 kilometers per hour and will destroy any living things in their paths. After the pyroclastic flows began, the residents of St. Pierre became fearful and many wanted to leave the island. The authorities claimed there was no danger and prevented evac-

uation. There was an election coming up, and the governor felt that most of his supporters lived in the city. He did not want to lose their votes, but neither the governor nor any of the city's residents would ever vote. The climax came on the morning of May 8, when great fiery, exploding clouds descended like an avalanche down the mountainside, raced down a stream valley, through the port city and onto the harbor. St. Pierre and the ships anchored in the harbor were incinerated (see figure 4.7). Temperatures within the pyroclastic flow were estimated at 700°C. Some of the dead had faces that appeared unaffected by the incinerating storm. However, the backs of their skulls were blasted open by their boiling brains. About 29,000 people were burned to death or suffocated (of the two survivors in St. Pierre, one was a condemned prisoner in a poorly ventilated dungeon).

Ninety-three years later, in July 1995, small steam-ash eruptions began at Soufriére Hills volcano on the neighboring island

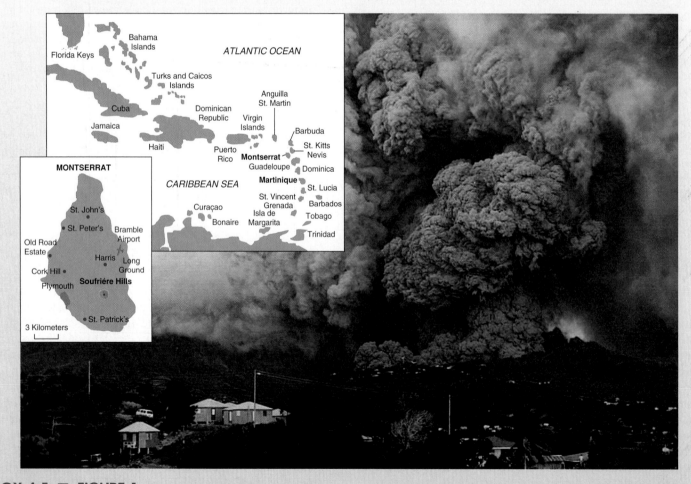

BOX 4.5   ■   FIGURE 1

Eruption of Soufriére Hills volcano on Montserrat, August 4, 1997. An ash cloud billows upward above a ground-hugging pyroclastic flow. Map of the West Indies showing location of Montserrat, Martinique, and Soufriére Hills volcano. *Photo by AP/Kevin West* of Montserrat. As a major eruption looked increasingly likely, teams of volcanologists from France, the United Kingdom, the United States (including members of the U.S. Geological Survey's Volcano Disaster Assistance Team that had successfully predicted the eruption of Mount Pinatubo in the Philippines, as described in chapter 1), and elsewhere flew in to study the volcano and help assess the hazards. An unprecedented array of modern instruments (including seismographs, tiltmeters, and gas analyzers) were deployed around the volcano. In November 1995, viscous, andesitic lava built a dome over the vent. Pyroclastic flows began when the dome collapsed in March 1996. Pyroclastic flows continued with more dome building and collapsing. By 1997, nearly all of the people in the southern part of the island were evacuated, following advice from the scientific teams. In June 1997, large eruptions took place and pyroclastic flows destroyed the evacuated capital city of Plymouth. In contrast to the tragedy of St. Pierre, only nineteen people were killed in the region.

In August 1997, major eruptions resumed. This time, the northern part of the island, previously considered safe, was faced with pyroclastic flows (box figure 1), and more people were evacuated from the island. Activity continued, at least into the mid-2000s, but with decreasing intensity. In May 2004, a volcanic mudflow went through the already uninhabitable town of Plymouth. Up to 6 meters of debris were deposited, partially burying buildings still left in the town.

Additional Resources

Mount Pelée, West Indies (Volcano World site)
This site contains some excellent photos from the 1902 eruptions. The second page has photos of the famous spine that grew in Mount Pelée after the tragic eruption.
- www.volcanoworld.org/vwdocs/volc_images/img_mt_pelee.html

Montserrat Volcano Observatory
Includes up-to-date reports on volcanic activity.
- www.mvo.ms

A Viscous lava wells up into a crater.

B A dome grows as more magma is extruded. The outer part is solid and breaks as the growing dome expands.

C If magma continues to be fed into the steep-sided dome, it may rise above the rim of the crater.

D

FIGURE 4.26

A volcanic dome forming in the crater of a cinder cone (A, B, C). (D) Mono craters, eastern California, is a line of craters with lava domes. The dome in the crater in the foreground has not grown above the level of the crater's rim (like B). Some in the background have overtopped their rims. You can also see some short and steep lava flows, reflecting the very viscous silicic lava that erupted. The Sierra Nevada range is on the skyline. *Photo by C. Dan Miller, U.S. Geologic Survey*

FIGURE 4.27

Basalt layers in the Columbia Plateau, Dry Falls State Park, Washington. *Photo by Cynthia Shaw*

evenly spaced centers of contraction. Tension cracks develop halfway between neighboring centers. A hexagonal fracture pattern is the most efficient way in which a set of contraction centers can share fractures. Although most columns are six-sided, some are five- or seven-sided.

SUBMARINE ERUPTIONS

Basalt plateaus have their counterparts in the oceans. These were unknown until they were discovered through deep-ocean drilling a couple of decades ago. The largest of these *oceanic plateaus* is the Ontang Java Plateau in the western Pacific ocean. This plateau is larger in area than Alaska. A thick sequence of sedimentary rocks covers the huge volume of basalt that formed the plateau around 90 million years ago.

Oceanic plateaus are only a small part of the sea floor. Most of the formation of the sea floor has involved eruptions along mid-oceanic ridges. The eruptions almost always consist of mafic lavas that create basalt. As described in chapter 3, basaltic rock, thought to have been formed from lava erupting along mid-oceanic ridges or solidifying underground beneath the ridges, makes up virtually the entire crust underlying the oceans. In a few places—Iceland, for example—volcanic islands rise above the otherwise submerged system (see box 4.6).

Pillow Basalts

Figure 4.29 shows **pillow structure**—rocks, generally basalt, occurring as pillow-shaped, rounded masses closely fitted together. From observations of submarine eruptions by divers, we know how the pillow structure is produced. Elongate blobs of lava break out of a thin skin of solid basalt over the top of a flow that is submerged in water. Each blob is squeezed out like toothpaste, and its surface is chilled to rock within seconds. A new blob forms as more lava inside breaks out. Each new pillow settles down on the pile, with little space left in between. Some pillow basalt forms in lakes and rivers or where lava flows from land into the sea (as in Hawaii). However, most pillow basalt forms at mid-oceanic ridge crests (figure 4.30).

According to plate-tectonic theory, basalt magma flows up the fracture that develops at a divergent boundary (explained in chapter 3). The magma that reaches the sea floor solidifies as pillow basalt. The rest solidifies in the fracture as a dike. Pillow basalt that is overlying a series of dikes is sometimes found in mountain ranges. These probably formed during seafloor spreading in the distant past followed, much later, by uplift.

Centers of
contraction

FIGURE 4.28
Columnar jointing at Devil's Postpile, California. Insert shows top view with centers of contraction drawn in. (Scratches were caused by glacial erosion as described in chapter 12.) *Photos by C. C. Plummer*

FIGURE 4.29
Pillow basalt in Iceland. *Photo by R. W. Decker*

FIGURE 4.30
Pillow basalt on a mid-oceanic ridge. Photo taken from a submersible vessel. *Courtesy of Woods Hole Oceanographic Institution*

WEB BOX 4.6

Fighting a Volcano in Iceland—and Winning

In 1973, a volcano began erupting on a small island in Iceland (box figure 1). Go to the book's website www.mhhe.com/plummer11e to learn about:

- how a town was almost buried by ash;
- what volunteers did to keep roofs from collapsing under heavy ash deposits;
- a lava flow that threatened to seal off the harbor and end the town's thriving fishing industry;
- an unprecedented effort to halt the lava flow;
- the cleanup and rebuilding of the town;
- how the residents get heat and hot water from the lava flow.

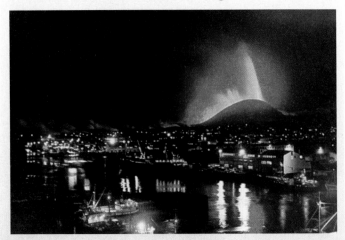

BOX 4.6 ■ FIGURE 1

Lava fountaining at a cinder cone behind the town on Heimaey. The glow behind the town in the left part of the photo is the lava flow advancing to the harbor. *Photo © Solarfilma*

SUMMARY

Lava is molten rock that reaches the Earth's surface, having been formed as *magma* from rock within the Earth's crust or from the uppermost part of the mantle.

More people have been killed by pyroclastic flows and, indirectly, by famine than by other volcanic hazards.

Lava contains 45% to 75% *silica* (SiO_2). The more silica, the more viscous the lava is. Viscosity is also influenced by the temperature and gas content of the lava. Viscous lavas are associated with more violent eruptions than are fluid lavas. *Volcanic domes* form from the extrusion of very viscous lavas.

Collapse of volcanoes into magma chambers forms calderas and results in the most explosive eruptions in which huge amounts of pyroclasts and gas are blasted into the atmosphere.

A *mafic* lava, relatively low in silica, crystallizes into *basalt*, the most abundant extrusive igneous rock. Basalt, which is dark in color, is composed of minerals that are relatively high in iron, magnesium, and calcium.

Rhyolite, a light-colored rock, forms from *silicic* lavas that are high in silica but contain little iron, magnesium, or calcium.

Because potassium and sodium are important elements in rhyolite, its constituent minerals are mostly potassium- and sodium-rich feldspars and quartz.

A lava with a composition between mafic and silicic crystallizes to *andesite*, a moderately dark rock. Andesite contains about equal amounts of ferromagnesian minerals and sodium- and calcium-rich feldspars.

Extrusive rocks are characteristically fine-grained. *Porphyritic* rock contains some larger crystals in an otherwise finer-grained rock. Rocks that solidified too rapidly for crystals to develop form a natural glass called *obsidian*. Gas trapped in rock forms *vesicles*.

Pyroclasts are the result of volcanic explosions. *Tuff* is volcanic ash that has consolidated into a rock. If large pyroclastic fragments have reconsolidated, the rock is a *volcanic breccia*.

A *cinder cone* is composed of loose pyroclastic material that forms steep slopes as it falls from the air back to near the crater. Cinder cones are not as large as the other two major types of cones.

A *shield volcano* is built up by successive eruptions of mafic lava. Its slopes are gentle, but its volume is generally large.

Composite cones are made of alternating layers of pyroclastic material and solidified lava flows. They are not as steep as cinder cones but steeper than shield volcanoes. Young composite volcanoes, predominantly composed of andesite, are aligned along the circum-Pacific belt and, less extensively, in the Mediterranean belt.

Plateau basalts are thick sequences of lava floods. *Columnar jointing* develops in solidified basalt flows. Basalt that erupts underwater forms a *pillow structure.* Pillow basalts commonly form along the crests of mid-oceanic ridges.

Terms to Remember

andesite 95
basalt 95
block 97
bomb 97
caldera 97
cinder cone 101
circum-Pacific belt 102
columnar structure (columnar jointing) 107
composite volcano (stratovolcano) 101
crater 97
extrusive rock 86
fine-grained rock 95
flank eruption 97

intermediate rock 92
lava 86
mafic rock 92
magma 86
Mediterranean belt 102
obsidian 95
phenocryst 96
pillow structure (pillow basalts) 110
plateau basalts 106
porphyritic rock 96
pumice 96
pyroclast 86
pyroclastic flow 93

rhyolite 95
shield volcano 98
silicic (felsic) rock 92
texture 95
tuff 97
vent 97
vesicle 96
viscosity 92
volcanic breccia 97
volcanic dome 105
volcanism 86
volcano 86

Testing Your Knowledge

Use the following questions to prepare for exams based on this chapter.

1. Compare the hazards of lava flows to those of pyroclastic flows.

2. What roles do gases play in volcanism?

3. What do pillow structures indicate about the environment of volcanism?

4. Name the minerals and the approximate percentage of each that you would expect to be present in each of the following rocks: andesite, rhyolite, basalt.

5. What property (or characteristic) of obsidian makes it an exception to the usual geologic definition of *rock?*

6. What determines the viscosity of a lava?

7. What determines whether a series of volcanic eruptions builds a shield volcano, a composite volcano, or a cinder cone? Describe each type of volcanic cone.

8. Explain how a vesicular porphyritic andesite might have formed.

9. Why are extrusive igneous rocks fine-grained?

10. Why don't flood basalts build volcanic cones?

11. Mount St. Helens

    a. last erupted violently in 1980

    b. is part of the Cascade Range

    c. is located in southern Washington

    d. all of the preceding

12. Volcanic eruptions can affect the climate because

    a. they heat the atmosphere

    b. volcanic dust and gas can reduce the amount of solar radiation that penetrates the atmosphere

    c. they change the elevation of the land

    d. all of the preceding

13. Whether volcanic eruptions are very explosive or relatively quiet is largely determined by
    a. the amount of gas in the lava or magma
    b. the ease or difficulty with which the gas escapes to the atmosphere
    c. the viscosity of a lava
    d. all of the preceding

14. Temperatures at which lavas solidify range from about ____°C for silicic rocks to ____°C for mafic rocks.
    a. 100, 200   b. 300, 1,000
    c. 700, 1,200   d. 1,000, 2,000

15. One gas typically not released during a volcanic eruption is
    a. water vapor   b. carbon dioxide
    c. sulfur dioxide   d. hydrogen sulfide
    e. oxygen

16. Mafic rocks contain about ____% silica.
    a. 10   b. 25
    c. 50   d. 65
    e. 80

17. Silicic rocks contain about ____% silica.
    a. 10   b. 25
    c. 50   d. 70
    e. 80

18. Which is not an extrusive igneous rock?
    a. granite   b. rhyolite
    c. basalt   d. andesite

19. Which is not a major type of volcano?
    a. shield   b. cinder cone
    c. composite   d. stratovolcano
    e. spatter cone

20. A typical example of a shield volcano is
    a. Mount St. Helens   b. Kilauea in Hawaii
    c. El Chichón   d. Mount Vesuvius

21. An example of a composite volcano is
    a. Mount St. Helens   b. El Chichón
    c. Mount Vesuvius   d. all of the preceding

22. Which volcano is not usually made of basalt?
    a. shield   b. composite cone
    c. spatter cone   d. cinder cone

23. An igneous rock made of pyroclasts has a texture called
    a. fragmental   b. vesicular
    c. porphyritic   d. fine-grained

Expanding Your Knowledge

1. What might explain the remarkable alignment of the Cascade volcanoes?

2. What would the present-day environmental effects be for an eruption such as that which created Crater Lake?

3. Why are there no active volcanoes in the eastern parts of the United States and Canada?

4. Why are continental igneous rocks richer in silica than oceanic igneous rocks?

Exploring Web Resources

www.mhhe.com/plummer11e
This is the dedicated website for this book. You can go to it for new and updated information. The universal resource locators (URL) listed in this book are also given as links on the website, making it easy to go to those websites without typing in the URL. When you visit our Online Learning Center, go to the Student Center and then to the chapter of interest. Here you will find additional readings and media resources, as well as answers to the Testing Your Knowledge section, more quizzes, animation and video clips, and direct links to the sites listed in this book. Links to additional websites can also be found. We have added questions for some of the sites to allow you to get the most of your exploration of the web. Using the web is an enjoyable way of enhancing your knowledge of geology.

http://volcano.und.ndak.edu/
Volcano World. This is an excellent site to learn about volcanoes. At the home page, you may click on "Volcanoes!" and go to a menu that includes currently active volcanoes, volcano video clips, and Earth's volcanoes.

www.geo.mtu.edu/volcanoes/
Michigan Tech volcanoes page. The focus of this site is on scientific and educational information relative to volcanic hazard mitigation. Clicking on "volcanic humor" will show the lighter side of volcanology.

www.volcanolive.com/contents.html
Volcano Live. This well-organized site is maintained by an Australian volcanologist. You can link to live cameras at most of the volcanoes discussed in this chapter (Mount Fuji, Mount Erebus, Mount Etna, etc.). You can get up-to-date information on what is erupting in the world and much more.

http://hvo.wr.usgs.gov/kilauea/update/
Hawaii Volcano Observatory's Kilauea Update. You can see a summary of present activity as well as photos taken today and during the past. Go to "Kilauea" for more information and data on Kilauea.

http://volcanoes.usgs.gov/Products/sproducts.html#fs
Products and fact sheets of the U.S. Geological Survey's volcanic hazards program. Lists many of the USGS online fact sheets on volcanoes.

Weathering and Soil

In this chapter, you will study several visible signs of weathering in the world around you, including the cliffs and slopes of the Grand Canyon and the rounded edges of boulders. As you study these features, keep in mind that weathering processes made the planet suitable for human habitation. The weathering of rock affects the composition of Earth's atmosphere, helping to maintain a habitable climate. Weathering also produces soils, upon which grow the forests, grasslands, and agriculture of the world.

Differential weathering at Bryce Canyon National Park in Utah has produced spires in the sandstone beds. *Photo © Richard J. Green/Photo Researchers, Inc*

How does rock weather? You learned in chapters 3 and 4 that the minerals making up igneous rocks crystallize at relatively high temperatures and sometimes at high pressures as magma and lava cool. Although these minerals are stable when they form, most of them are not stable during prolonged exposure at the surface. In this chapter, you see how minerals and rocks change when they are subjected to the physical and chemical conditions existing at the surface. Rocks undergo mechanical weathering (physical disintegration) and chemical weathering (decomposition) as they are attacked by air, water, and microorganisms. Your knowledge of the chemical composition and atomic structure of minerals will help you understand the reactions that occur during chemical weathering.

Weathering processes create sediments (primarily mud and sand) and soil. Sedimentary rocks, which form from sediments, are discussed in chapter 6. In a general sense, weathering prepares rocks for erosion and is a fundamental part of the rock cycle, transforming rocks into the raw material that eventually becomes sedimentary rocks. Through weathering, there are important links between the rock cycle and the atmosphere and biosphere.

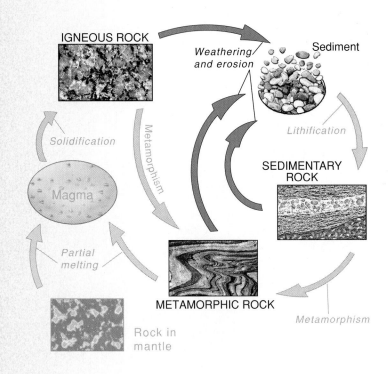

WEATHERING, EROSION, AND TRANSPORTATION

Rocks exposed at Earth's surface are constantly being changed by water, air, varying temperature, and other environmental factors. Granite may seem indestructible, but given time and exposure to air and water, it can decompose and disintegrate into soil. The processes that alter rock are *weathering, erosion,* and *transportation.*

The term **weathering** refers to the group of destructive processes that change the physical and chemical character of rock at or near the surface. For example, if you abandon a car, particularly in a wet climate, eventually the paint will flake off and the metal will rust. The car weathers. Similarly, the tightly bound crystals of any rock can be loosened and altered to new minerals when exposed to air and water during weathering. Weathering breaks down rocks that are either stationary or moving.

Erosion is the picking up or *physical removal* of rock particles by an agent such as running water or glaciers. Weathering helps break down a solid rock into loose particles that are easily eroded. Rainwater flowing down a cliff or hillside removes the loose particles produced by weathering. Similarly, if you sandblast rust off of a car, erosion takes place.

After a rock fragment is picked up (eroded), it is transported. **Transportation** is the movement of eroded particles by agents such as rivers, waves, glaciers, or wind. Weathering processes continue during transportation. A boulder being transported by a stream can be physically worn down and chemically altered as it is carried along by the water. In the car

analogy, transportation would take place when a stream of rust-bearing water flows away from a car in which rust is being hosed off.

WEATHERING AND EARTH SYSTEMS

Solar System

Weathering takes place on Earth because of our atmosphere (which contains oxygen and carbon dioxide) and the abundance of water. Mars has features that indicate water flowed there in the distant past (see box 10.3). Although Mars no longer has surface water, it does have an atmosphere. Winds on Mars, sometimes several times faster than hurricanes on Earth, transport fine-grained material and erode by sandblasting the barren surface (see box 13.3).

Atmosphere

Our atmosphere is crucial to the processes of weathering. Oxygen and carbon dioxide are important for *chemical weathering,* as described later. Water (evaporated from the hydrosphere and distributed as moisture, rain, and snow) is critical to both chemical weathering and *mechanical weathering.* Weathering has also had a dramatic impact on the composition of Earth's atmosphere. Chemical weathering removes carbon dioxide from the atmosphere, allowing it to be transformed into limestone and stored in the crust. Without chemical weathering, the elevated levels of carbon dioxide in the atmosphere would have long ago made Earth too hot to sustain life.

Hydrosphere

Water is necessary for chemical weathering to take place. Oxygen dissolved in water oxidizes iron in rocks. Carbon dioxide mixed with water makes a weak acid that causes most minerals to decompose; this acid is the primary cause of chemical weathering. Running water contributes to weathering and erosion by loosening and removing particles and by abrading rocks during transportation in streams. Ice in glaciers is a very effective agent of erosion as rocks frozen in the base of a glacier grind down the underlying bedrock. Freezing and thawing of water in cracks in rock is also very effective at mechanically breaking them up.

Biosphere

Plants can physically break apart rocks when they grow in cracks. Animals can also contribute to weathering and erosion. You may notice how hillsides have many paths carved by the hooves of grazing cattle or sheep. Humans, of course, are awesome agents of erosion. A single pass by a bulldozer can do more to change a landscape than thousands of years of natural weathering and erosion.

Plants and animals contribute greatly to weathering when they die. When animals and plants decompose, they become mostly water and carbon dioxide. While carbon dioxide dissolved in rain makes the water slightly acidic, carbon dioxide mixed with water in soil with decaying plants produces much more acid. Soil, necessary for plant growth is formed by weathering and includes organic matter from decayed plants. Organic carbon compounds and minerals released by weathering provide the nutrients required for plant growth.

HOW WEATHERING ALTERS ROCKS

Rocks undergo both mechanical weathering and chemical weathering. **Mechanical weathering** (physical disintegration) includes several processes that break rock into smaller pieces. The change in the rock is physical; there is little or no chemical change. For example, water freezing and expanding in cracks can cause rocks to disintegrate physically. **Chemical weathering** is the decomposition of rock from exposure to water and atmospheric gases (principally carbon dioxide, oxygen, and water vapor). As rock is decomposed by these agents, new chemical compounds form.

Mechanical weathering breaks up rock but does not change the composition. A large mass of granite may be broken into smaller pieces by frost action, but its original crystals of quartz, feldspar, and ferromagnesian minerals are unchanged. On the other hand, if the granite is being chemically weathered, some of the original minerals are chemically changed into different

A

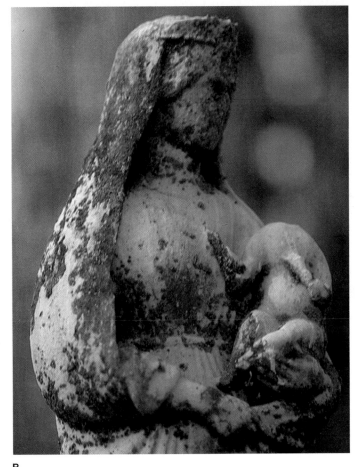

B

FIGURE 5.1

(A) The effects of chemical weathering are obvious in the marble gravestone on the right but not in the slate gravestone on the left, which still retains its detail. Both gravestones date to the 1780s. (B) This marble statue has lost most of the fine detail on the face and the baby's head has been dissolved by chemical weathering. *Photo A by C. C. Plummer; Photo B by David McGeary* minerals. Feldspar, for example, will change into a clay mineral (with a crystal structure similar to mica). In nature, mechanical and chemical weathering usually occur together, and the effects are interrelated.

Weathering is a relatively long, slow process. Typically, cracks in rock are enlarged gradually by frost action or plant growth (as roots pry into rock crevices), and as a result, more surfaces are exposed to attack by chemical agents. Chemical weathering initially works along contacts between mineral grains. Tightly bound crystals are loosened as weathering products form at their contacts. Mechanical and chemical weathering then proceed together, until a once tough rock slowly crumbles into individual grains.

Solid minerals are not the only products of chemical weathering. Some minerals—calcite, for example—dissolve when chemically weathered. We can expect limestone and marble, rocks consisting mainly of calcite, to weather chemically in quite a different way than granite.

EFFECTS OF WEATHERING

The results of chemical weathering are easy to find. Look along the edges or corners of old stone structures for evidence. The inscriptions on statues and gravestones that have stood for several decades may no longer be sharp (figure 5.1). Building blocks of limestone or marble exposed to rain and atmospheric

gases may show solution effects of chemical weathering in a surprisingly short time. Granite and slate gravestones and building materials are much more resistant to weathering due to the strong silicon-oxygen bonds in the silicate minerals. However, after centuries, the mineral grains in granite may be loosened, cracks enlarged, and the surface discolored and dulled by the products of weathering. Surface discoloration is also common on rock *outcrops,* where rock is exposed to view, with no plant or soil cover. That is why field geologists carry rock hammers—to break rocks to examine unweathered surfaces.

We tend to think of weathering as destructive because it mars statues and building fronts. As rock is destroyed, however, valuable products can be created. Soil is produced by rock weathering, so most plants depend on weathering for the soil they need in order to grow. In a sense, then, all agriculture depends on weathering. Weathering products transported to the sea by rivers as dissolved solids make seawater salty and serve as nutrients for many marine organisms. Some metallic ores, such as those of copper and aluminum, are concentrated into economic deposits by chemical weathering.

Many weathered rocks display interesting shapes. **Spheroidal weathering** occurs where rock has been rounded by weathering from an initial blocky shape. It is rounded because chemical weathering acts more rapidly or intensely on the corners and edges of a rock than on the smooth rock faces (figure 5.2).

Differential weathering describes the tendency for different types of rock to weather at different rates. For example,

FIGURE 5.2

(A) Water penetrating along cracks at right angles to one another in an igneous rock produces spheroidal weathering of once-angular blocks. The increase in surface area exposed by the cracks increases chemical weathering. (B) Because of the increased surface area, chemical weathering attacks edges and particularly the corners more rapidly than the flat faces, creating the spheroidal shape shown in (C). (D) Newly eroded granite block with rounded corners contrasted with extensively weathered, spheroidal granite boulder, Acadia National Park, Maine. *Photo by Bret Bennington* shale (composed of soft clay minerals) tends to weather much faster than sandstone (composed of hard quartz mineral). Figure 5.3 shows a striking example of differential weathering. Figure 5.4 illustrates how layers of resistant rock tend to weather to form steep cliffs while softer layers form shallow slopes of eroded rock debris called **talus.**

MECHANICAL WEATHERING

Of the many processes that cause rocks to disintegrate, the most effective are pressure release and frost action.

Pressure Release

The reduction of pressure on a body of rock can cause it to crack as it expands; **pressure release** is a significant type of mechanical weathering. A large mass of rock, such as a batholith, may originally form under great pressure from the weight of several kilometers of rock above it. This batholith is gradually exposed by tectonic uplift of the region followed by erosion of the overlying rock (figure 5.5). The removal of the great weight of rock above the batholith, usually termed *unloading,* allows the granite to expand upward. Cracks called **sheet joints** develop parallel to the outer surface of the rock as the outer part of the rock expands more than the inner part (figures 5.6*A* and *B*). On slopes, gravity may cause the rock between such joints to break loose in concentric slabs from the underlying granite mass. This process of spalling off of rock layers is called **exfoliation;** it is somewhat similar to peeling layers from an onion. **Exfoliation domes** (figure 5.6) are large, rounded landforms developed in massive rock, such as granite, by exfoliation. Some famous examples of exfoliation domes include Stone Mountain in Georgia and Half Dome in Yosemite.

Frost Action

Did you ever leave a bottle of water in the freezer, coming back later to find the water frozen and the bottle burst open? When water freezes at 0°C (32°F), the individual water molecules jumbled together in the liquid align into an ordered crystal structure, forming ice. Because the crystal structure of ice takes up more space than the liquid, water expands 9% in volume when it freezes. This unique property makes water a potent agent of mechanical weathering in any climate where the temperature falls below freezing.

Frost action—the mechanical effect of freezing water on rocks—commonly occurs as frost wedging or frost heaving. In **frost wedging,** the expansion of freezing water pries rock apart. Most rock contains a system of cracks called *joints,* caused by the slow flexing of brittle rock by deep-seated Earth forces (see chapter 15). Water that has trickled into a joint in a rock can freeze and expand when the temperature drops below 0°C (32°F). The expanding ice wedges the rock apart, extending the joint or even breaking the rock into pieces (figure 5.7).

FIGURE 5.3

Pedestal rock near Lees Ferry, Arizona. Resistant sandstone cap protects weak shale pedestal from weathering and erosion. Hammer for scale is barely visible at base of pedestal. *Photo by David McGeary*

FIGURE 5.4

Sedimentary rocks in the Grand Canyon, Arizona. In the foreground, layers of sandstone resist weathering and form steep cliffs. Less resistant layers of shale weather to form gentler slopes of talus between cliffs. *Photo by David McGeary*

A

B

C

FIGURE 5.5

Sheet joints caused by pressure release. A granite batholith (A) is exposed by regional uplift followed by the erosion of the overlying rock (B). Unloading reduces pressure on the granite and causes outward expansion. Sheet joints are closely spaced at the surface where expansion is greatest. Exfoliation of rock layers produces rounded exfoliation domes. (C) Sheet joints in a granite outcrop near the top of the Sierra Nevada, California. The granite formed several kilometers below the surface and expanded outward when it was exposed by uplift and erosion. *Photo by David McGeary*

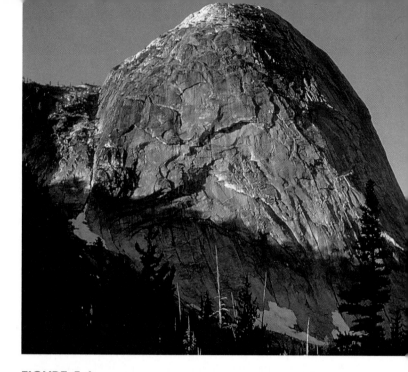

FIGURE 5.6

Exfoliation dome, Yosemite National Park, California. Onionlike layers of rock are peeling off the dome. *Photo by David McGeary*

Frost wedging is most effective in regions with many days of freezing and thawing (mountaintops and midlatitude regions with pronounced seasons). Partial thawing during the day adds new water to the ice in the crack; refreezing at night adds new ice to the old ice.

Frost heaving lifts rock and soil vertically. Solid rock conducts heat faster than soil, so on a cold winter day, the bottom of a partially buried rock will be much colder than soil at the same depth. As the ground freezes in winter, ice forms first under large rock fragments in the soil. The expanding ice layers push boulders out of the ground, a process well known to New England farmers and other residents of rocky soils. Frost heaving bulges the ground surface upward in winter, breaking up roads and leaving lawns spongy and misshapen after the spring thaw.

Other Processes

Several other processes mechanically weather rock but in most environments are less effective than frost action and pressure release. *Plant growth,* particularly roots growing in cracks (figure 5.8A), can break up rocks, as can *burrowing animals.* Such activities help to speed up chemical weathering by enlarging passageways for water and air. The *pressure of salt crystals* formed as water evaporates inside small spaces in rock also helps to disintegrate desert rocks (figure 5.8B). *Extreme changes in temperature,* as in a forest fire, can cause a rock to expand until it cracks. Whatever processes of mechanical weathering are at work, as rocks disintegrate into smaller fragments, the total surface area increases (figure 5.9), allowing more extensive chemical weathering by water and air.

Frost wedging

A

B

A

B

FIGURE 5.7

(A) Frost wedging occurs when water fills joints (cracks) in a rock and then freezes. The expanding ice wedges the rock apart. (B) Frost wedging has broken the rock and sculpted Crawford Mountain in Banff National Park, Alberta, Canada. The broken rock forms cone-shaped piles of debris (talus) at the base of the mountains. *Photo B © Martin G. Miller/Visuals Unlimited*

CHEMICAL WEATHERING

The processes of chemical weathering, or *rock decomposition*, transform rocks and minerals exposed to water and air into new chemical products. A mineral that crystallized deep underground from a water-deficient magma may eventually be exposed at the surface, where it can react with the abundant water there to form a new, different mineral. A mineral containing very little oxygen may react with oxygen in the air, extracting oxygen atoms from the atmosphere and incorporating them into its own crystal structure, thus forming a different mineral. These new minerals are weathering products. They

FIGURE 5.8

(A) Tree roots will pry this rock apart as they grow within the rock joints, Sierra Nevada Mountains, California. (B) This rock is being broken by the growth of salt crystals, which precipitate as water evaporates within the cracks in the rock, Mojave Desert, California. Note the tremendous increase in surface area that results from the rock being split into layers. *Photo A by Diane Carlson; Photo B by Crystal Hootman and Diane Carlson*

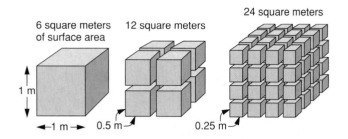

FIGURE 5.9

Mechanical weathering can increase the surface area of a rock, accelerating the rate of chemical weathering. As a cube breaks up into smaller pieces, its volume remains the same, but its surface area increases.

have adjusted to physical and chemical conditions at (or near) Earth's surface. Minerals change gradually at the surface until they come into *equilibrium*, or balance, with the surrounding conditions.

Role of Oxygen

Oxygen is abundant in the atmosphere and quite active chemically, so it often combines with minerals or with elements within minerals that are exposed at Earth's surface.

The rusting of an iron nail exposed to air is a simple example of chemical weathering. Oxygen from the atmosphere combines with the iron to form iron oxide, the reaction being expressed as follows:

$$4Fe^{+3} + 3O_2 \rightarrow 2Fe_2O_3$$
$$\text{iron} + \text{oxygen} \rightarrow \text{iron oxide}$$

Iron oxide formed in this way is a weathering product of numerous minerals containing iron, such as the ferromagnesian group (pyroxenes, amphiboles, biotite, and olivine). The iron in the ferromagnesian silicate minerals must first be separated from the silica in the crystal structure before it can oxidize. The iron oxide (Fe_2O_3) formed is the mineral **hematite,** which has a brick-red color when powdered. If water is present, as it usually is at Earth's surface, the iron oxide combines with water to form **limonite,** which is the name for a group of mostly amorphous, hydrated iron oxides (often including the mineral *goethite*), which are yellowish-brown when powdered. The general formula for this group is $Fe_2O_3 \cdot nH_2O$ (the n represents a small, whole number such as 1, 2, or 3 to show a variable amount of water). The brown, yellow, or red color of soil and many kinds of sedimentary rock is commonly the result of small amounts of hematite and limonite released by the weathering of iron-containing minerals (figure 5.10).

Role of Acids

The most effective agent of chemical weathering is acid. Acids are chemical compounds that give off hydrogen ions (H^+)

FIGURE 5.10

Sandstone has been colored red by hematite, released by the chemical weathering of ferromagnesian minerals, Thermopolis, Wyoming. *Photo by Diane Carlson*

when they dissociate, or break down, in water. Strong acids produce a great number of hydrogen ions when they dissociate, and weak acids produce relatively few such ions.

The hydrogen ions given off by natural acids disrupt the orderly arrangement of atoms within most minerals. Because a hydrogen ion has a positive electrical charge and a very small size, it can substitute for other positive ions (such as Ca^{++}, Na^+, or K^+) within minerals. This substitution changes the chemical composition of the mineral and disrupts its atomic structure. The mineral decomposes, often into a different mineral, when it is exposed to acid.

Some strong acids occur naturally on Earth's surface, but they are relatively rare. Sulfuric acid and hydrofluoric acid are strong acids emitted during many volcanic eruptions. They can kill trees and cause intense chemical weathering of rocks near volcanic vents. The bubbling mud of Yellowstone National Park's mudpots (figure 5.11) is produced by rapid weathering caused by acidic sulfur gases that are given off by some hot springs. Strong acids also drain from some mines as sulfur-containing minerals such as pyrite oxidize and form acids at the surface (figure 5.12). Uncontrolled mine drainage can kill fish and plants downstream and accelerate rock weathering.

The most important natural source of acid for rock weathering at Earth's surface is dissolved carbon dioxide (CO_2) in water. Water and carbon dioxide form *carbonic acid* (H_2CO_3), a weak acid that dissociates into the hydrogen ion and the bicarbonate ion (see equation *A* in table 5.1). Even though carbonic acid is a weak acid, it is so abundant at Earth's surface that it is the single most effective agent of chemical weathering.

Earth's atmosphere (mostly nitrogen and oxygen) contains 0.03% carbon dioxide. Some of this carbon dioxide dissolves in rain as it falls, so most rain is slightly acidic when it hits the

FIGURE 5.11

A mudpot of boiling mud is created by intense chemical weathering of the surrounding rock by the acid gases dissolved in a hot spring, Yellowstone National Park, Wyoming. *Photo by David McGeary*

ground. Large amounts of carbon dioxide also dissolve in water that percolates through soil. The openings in soil are filled with a gas mixture that differs from air. Soil gas has a much higher content of carbon dioxide (up to 10%) than does air, because carbon dioxide is produced by the decay of organic matter and the respiration of soil organisms in the biosphere, such as worms. Rainwater that has trickled through soil is therefore usually acidic and readily attacks minerals in the unweathered rock below the soil (figure 5.13).

Solution Weathering

Some minerals are completely dissolved by chemical weathering. *Calcite,* for instance, goes into solution when exposed to carbon dioxide and water, as shown in equation *B* in table 5.1 and in figure 5.13. The carbon dioxide and water combine to form carbonic acid, which dissociates into the hydrogen ion and the bicarbonate ion, as you have seen, so the equation for the solution of calcite can also be written as equation *C* in table 5.1.

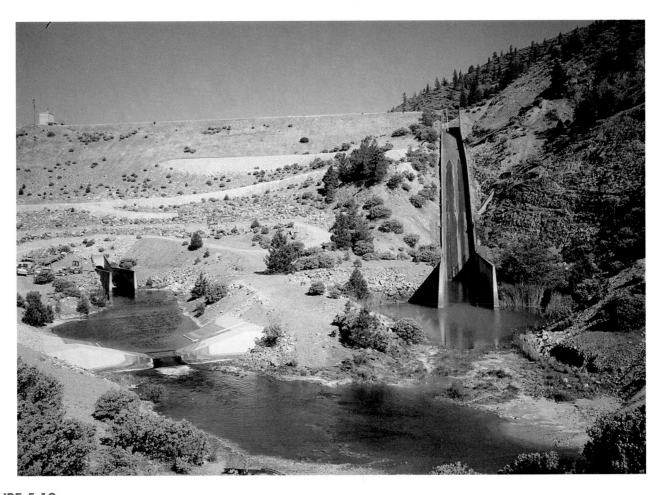

FIGURE 5.12

Spring Creek debris dam collects acid mine drainage from the Iron Mountain Mines Superfund site in northern California. *Photo by Charles Alpers, U.S. Geological Survey*

TABLE 5.1	Chemical Equations Important to Weathering						

A. Solution of Carbon Dioxide in Water to Form Acid

CO_2	+	H_2O	$\rightleftharpoons$	H_2CO_3	$\rightleftharpoons$	H^+	+	HCO_3^-
carbon dioxide		water		carbonic acid		hydrogen ion		bicarbonate ion

B. Solution of Calcite

$CaCO_3$	+	CO_2	+	H_2O	$\rightleftharpoons$	Ca^{++}	+	$2HCO_3^-$
calcite		carbon dioxide		water		calcium ion		bicarbonate ion

C. Solution of Calcite

$CaCO_3$	+	H^+	+	HCO_3^-	$\rightleftharpoons$	Ca^{++}	+	$2HCO_3^-$

D. Chemical Weathering of Feldspar to Form a Clay Mineral

$2KAlSi_3O_8$	+	$2H^+ + 2HCO_3^-$	+	H_2O	$\rightarrow$	$Al_2Si_2O_5(OH)_4$	+	$2K^+ + 2HCO_3^-$	+	$4\,SiO_2$
potassium feldspar		(from CO_2 and H_2O)				clay mineral		(soluble ions)		silica in solution or as fine solid particles

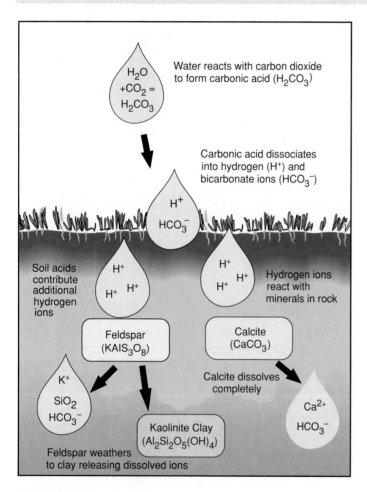

FIGURE 5.13

Chemical weathering of feldspar and calcite by carbonic and soil acids. Water percolating through soil weathers feldspar to clay and completely dissolves calcite. Soluble ions and soluble silica weathering products are washed away.

There are no solid products in the last part of the equation, indicating that complete solution of the calcite has occurred. Caves can form underground when flowing ground water dissolves the sedimentary rock limestone, which is mostly calcite. Rain can discolor and dissolve statues and tombstones carved from the metamorphic rock marble, which is also mostly calcite (see figure 5.1).

Chemical Weathering of Feldspar

The weathering of feldspar is an example of the alteration of an original mineral to an entirely different type of mineral as the weathered product. When feldspar is attacked by the hydrogen ion of carbonic acid (from carbon dioxide and water), it forms clay minerals. In general, a **clay mineral** is a hydrous aluminum silicate with a sheet-silicate structure like that of mica. Therefore, the entire silicate structure of the feldspar crystal is altered by weathering: feldspar is a framework silicate, but the clay mineral product is a sheet silicate, differing both chemically and physically from feldspar.

Let us look in more detail at the weathering of feldspar (equation *D* in table 5.1). Rainwater percolates down through soil, picking up carbon dioxide from the atmosphere and the upper part of the soil. The water, now slightly acidic, comes in contact with feldspar in the lower part of the soil (figure 5.13), as shown in the first part of the equation. The acidic water reacts with the feldspar and alters it to a clay mineral.

The hydrogen ion (H^+) attacks the feldspar structure, becoming incorporated into the clay mineral product. When the hydrogen moves into the crystal structure, it releases potassium (K) from the feldspar. The potassium is carried away in solu-

ENVIRONMENTAL GEOLOGY 5.1

Acid Rain

The burning of coal, oil, and natural gas (the *fossil fuels*) adds a great deal of carbon dioxide to the atmosphere (box figure 1). As you have seen in table 5.1, this carbon dioxide combines with water to form carbonic acid in rain. Coal and oil can also contain nitrogen and sulfur, which are given off as gases (NO_2 and SO_2) when these fuels are burned, forming nitric acid and sulfuric acid in rain. These two acids are much stronger than carbonic acid.

The strength of an acidic solution is measured on the pH scale from 0 to 14 (box figure 2). A solution of pH 7 is chemically neutral, neither acidic nor alkaline. Values below 7 are acidic; the lower the number, the more acidic the solution. Values above 7 are alkaline or basic. The pH scale is logarithmic, so a change

BOX 5.1 ■ FIGURE 2

The pH scale.

of 1 on the scale means a change of 10 in the concentration of H^+ ions that make a solution acidic.

Ordinary rain has a pH of about 5.5 to 6 from the small amount of carbon dioxide in the atmosphere and from natural sources of acidic sulfur gases such as volcanoes and coastal marshes. Ordinary rain is about as acidic as milk—hardly a strong acid.

In cities and downwind of industrial smokestacks (often for hundreds of miles), the increased amount of acid gases can reduce the pH of rain to 4, 3, or even 2 (the pH of lemon juice or vinegar). This is the environmental problem termed "acid rain" (although all rain is really acid). This rain in turn can lower the pH of streams, lakes, and soils. Such low pH values are hard on organisms; fish may die in streams and lakes polluted by acid rain, and forests suffer under acid rain. This is particularly a problem where rocks and soil do not buffer the acid. For example, areas with exposed limestone are least affected.

Chemical weathering is accelerated by acid rain. Statues and stone buildings in cities weather many times faster than stone structures in rural areas free of acid rain (see figure 5.1).

Additional Resources

- http://minerals.er.usgs.gov/acid1.html
- http://bqs/usgs.gov/acidrain/index.htm

BOX 5.1 ■ FIGURE 1

The burning of fossil fuels releases carbon dioxide and other acid gases, such as sulfur dioxide and nitrous oxide, that combine with water to produce acid rain. *Photo by David McGeary*

---

tion as a dissolved ion (K^+). The bicarbonate ion from the original carbonic acid does not enter into the reaction; it reappears on the right side of the equation. The soluble potassium and bicarbonate ions are carried away by water (ground water or streams).

All the silicon from the feldspar cannot fit into the clay mineral, so some is left over and is carried away as silica (SiO_2) by the moving water. This excess silica may be carried in solution or as extremely small solid particles.

The weathering process is the same regardless of the type of feldspar: K-feldspar forms potassium ions; Na-feldspar and Ca-feldspar (plagioclase) form sodium ions and calcium ions, respectively. The ions that result from the weathering of Cafeldspar are calcium ions (Ca^{++}) and bicarbonate ions

(HCO_3^-), both of which are very common in rivers and underground water, particularly in humid regions.

Chemical Weathering of Other Minerals

The weathering of ferromagnesian or dark minerals is much the same as that of feldspars. Two additional products are found on the right side of the equation—magnesium ions and iron oxides (hematite, limonite, and goethite).

The susceptibility of the rock-forming minerals to chemical weathering is dependent on the strength of the mineral's chemical bonding within the crystal framework. Because of the strength of the silicon-oxygen bond, quartz is quite resistant to chemical weathering. Thus, quartz (SiO_2) is the rock-forming mineral least susceptible to chemical attack at Earth's surface. Ferromagnesian minerals such as olivine, pyroxene, and amphibole include other positively charged ions such as Al, Fe, Mg, and Ca. The presence of these positively charged ions in the crystal framework makes these minerals vulnerable to chemical attack due to the weaker chemical bonding between these ions and oxygen, as compared to the much stronger silicon-oxygen bonds. For example, olivine—$(Fe, Mg)_2SiO_4$—weathers rapidly because its isolated silicon-oxygen tetrahedra are held together by relatively weak ionic bonds between oxygen and iron and magnesium. These ions are replaced by H^+ ions during chemical weathering similar to that described for the feldspars.

Weathering and Diamond Concentration

Diamond is the hardest mineral known and is also extremely resistant to weathering. This is due to the very strong covalent bonding of carbon, as described in chapter 2. But diamonds are often concentrated by weathering, as illustrated in figure 5.14. Diamonds are brought to the surface of Earth in *kimberlite pipes,* columns of brecciated or broken ultramafic rock that have risen from the upper mantle. Diamonds are widely scattered in diamond pipes when they form. At the surface, the ultramafic rock in the pipe is preferentially weathered and eroded away. The diamonds, being more resistant to weathering, are left behind, concentrated in rich deposits on top of the pipes. Rivers may redistribute and reconcentrate the diamonds, as in South Africa and India. In Canada, kimberlite pipes have been eroded by glaciers, and diamonds may be found widely scattered in glacial deposits.

Weathering Products

Table 5.2 summarizes weathering products for the common minerals. Note that quartz and clay minerals commonly are left after complete chemical weathering of a rock. Sometimes other solid products, such as iron oxides, also are left after weathering.

The solution of calcite supplies substantial amounts of calcium ions (Ca^{++}) and bicarbonate ions (HCO_3^-) to underground water. The weathering of Ca-feldspars (plagioclase)

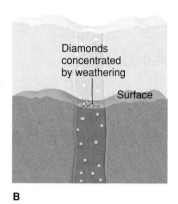

FIGURE 5.14

Residual concentration by weathering. (*A*) Cross-sectional view of diamonds widely scattered within kimberlite pipe. (*B*) Diamonds concentrated on surface by removal of rock by weathering and erosion.

into clay minerals can also supply Ca^{++} and HCO_3^- ions, as well as silica (SiO_2), to water. Under ordinary chemical circumstances, the dissolved Ca^{++} and HCO_3^- can combine to form solid $CaCO_3$ (calcium carbonate), the mineral calcite. Dissolved silica can also precipitate as a solid from underground water. This is significant because calcite and silica are the most common materials precipitated as *cement,* which binds loose particles of sand, silt, and clay into solid sedimentary rock (see chapter 6). The weathering of calcite, feldspars, and other minerals is a likely source for such cement.

If the soluble ions and silica are not precipitated as solids, they remain in solution and may eventually find their way into a stream and then into the ocean. Enormous quantities of dissolved material are carried by rivers into the sea (one estimate is 4 billion tons per year). This is the main reason seawater is salty.

Factors Affecting Weathering

The intensity of both mechanical and chemical weathering is affected by a variety of factors. Chemical weathering is largely a function of the availability of liquid water. Rock chemically

TABLE 5.2	Weathering Products of Common Rock-Forming Minerals			
Original Mineral	**Under Influence of CO₂ and H₂O**	**Main Solid Product**		**Other Products (Mostly Soluble)**
Feldspar	→	Clay mineral	+	Ions (Na^+, Ca^{++}, K^+), SiO_2
Ferromagnesian minerals (including biotite mica)	→	Clay mineral	+	Ions (Na^+, Ca^{++}, K^+, Mg^{++}), SiO_2, Fe oxides
Muscovite mica	→	Clay mineral	+	Ions (K^+), SiO_2
Quartz	→	Quartz grains (sand and silt)		
Calcite	→	—		Ions (Ca^{++}, HCO_3^-)

weathers much faster in humid climates than in arid climates. Limestone, which is extremely susceptible to dissolution, weathers quickly and tends to form valleys in wet regions such as the Appalachian Mountains. However, in the arid west, limestone is a resistant rock that forms ridges and cliffs. Temperature is also a factor in chemical weathering. The most intense chemical weathering occurs in the tropics, which are both wet and hot. Polar regions experience very little chemical weathering because of the frigid temperatures and the absence of liquid water. Mechanical weathering intensity is also related to climate (temperature and humidity), as well as to slope. Temperate climates, where abundant water repeatedly freezes and thaws, promote extensive frost weathering. Steep slopes cause rock to fall and break up under the influence of gravity. The most intense mechanical weathering probably occurs in high mountain peaks where the combination of steep slopes, precipitation, freezing and thawing, and flowing glacial ice rapidly pulverize the solid rock.

SOIL

In civil engineering and construction, soil is the usual name for any kind of loose, unconsolidated Earth material; but most geologists commonly use the term **soil** for a layer of *weathered,* unconsolidated material on top of bedrock (a general term for rock beneath soil). Soil scientists further restrict the term *soil* to horizons of weathered, unconsolidated material that contains organic matter and is capable of supporting plant growth. (If this definition is used, then the term *regolith* can be applied to any loose surface sediment; soil would be the upper part of the regolith.) A mature, fertile soil is the product of centuries of mechanical and chemical weathering of rock, combined with the addition and decay of plant and other organic matter. In terms of Earth systems, soil forms an important interface between the solid Earth (geosphere), atmosphere, hydrosphere, and biosphere.

The term **loam** refers to a soil of approximately equal amounts of sand, silt, and clay. (*Clay-sized* particles usually consist of *clay minerals.*) Loamy soils are often well-drained, may contain organic matter, and are often very fertile. *Topsoil* is the upper part of the soil and is more fertile than the underlying *subsoil,* which is often stony and lacks organic matter.

Clay minerals and quartz, the two minerals usually remaining after complete weathering of rock (table 5.2), have important roles in soil development and plant growth. Quartz crystals form sand grains that help keep soil loose and aerated, allowing good water drainage. (Partially weathered crystals of feldspar and other minerals can also form sand-sized grains.)

Clay minerals help to hold water and plant nutrients in a soil. Clay minerals occur as microscopic plates. Because of ion substitution within their sheet-silicate structure, most clay minerals have a negative electrical charge on the flat faces of the plates. This negative charge attracts water and nutrient ions to the clay mineral.

The water molecule, made up of two hydrogen atoms and one oxygen atom, is neutral in charge but has a positive end and a negative end. The negative charge on the flat faces of the clay mineral attracts the positive ends of the water molecules to the clay flake (figure 5.15). The clay holds the water loosely enough that most of it is available for uptake by plant roots.

Plant nutrients, such as Ca^{++} and K^+, commonly supplied by the weathering of minerals such as feldspar, are also held loosely on the surface of clay minerals. A plant root is able to release H^+ from organic acids and exchange it for the Ca^{++} and K^+ that the plant needs for healthy growth (figure 5.16).

Soil Horizons

Most soils take a long time to form. The rate of soil formation is controlled by rainfall, temperature, slope, and to some extent the type of bedrock that weathers to form soil. High temperature and abundant rainfall speed up soil formation, but in most places, a fully developed soil that can support plant growth takes hundreds or thousands of years to form.

As soils mature, distinct layers appear in them (figure 5.17). Soil layers are called **soil horizons** and can be distinguished from one another by appearance and chemical composition. Boundaries between soil horizons are usually transitional rather than sharp. By observing a vertical cross section, or *soil profile,* various horizons can be identified.

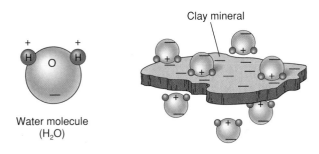

FIGURE 5.15
Negative charges on a clay mineral attract positive ends of water molecules.

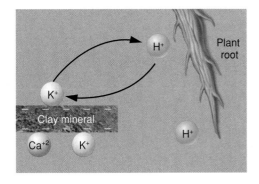

FIGURE 5.16
Ion exchange between plant root and clay mineral.

Weathering, the Carbon Cycle, and Global Climate

Weathering has affected the long-term climate of Earth by changing the carbon dioxide content of the atmosphere through the inorganic carbon cycle (see box figure 1). Carbon dioxide is a "greenhouse gas" that traps solar heat near the surface, warming the Earth. The planet Venus has a dense atmosphere composed mostly of CO_2, which traps so much solar heat that the surface temperature averages a scorching 480°C (about 900°F—see chapter 22). Earth has comparatively very little CO_2 in its atmosphere (see box table 1)—enough to keep most of the surface above freezing but not too hot to support life. However, when Earth first formed, its atmosphere was probably very much like that of Venus, with much more CO_2. What happened to most of the original carbon dioxide in Earth's atmosphere? Geologists believe that a quantity of CO_2 equal to approximately 65,000 times the mass of CO_2 in the present atmosphere lies buried in the crust and upper mantle of Earth. Some of this CO_2 was used to make organic molecules during photosynthesis and is now trapped as buried organic matter and fossil fuels in sedimentary rocks. However, the majority of the missing CO_2 was converted to bicarbonate ion (HCO_3^-) during chemical weathering and is locked away in carbonate minerals (primarily $CaCO_3$) that formed layers of limestone rock.

The inorganic carbon cycle helps to regulate the climate of Earth because CO_2 is a greenhouse gas, chemical weathering accelerates with warming, and the formation of limestone occurs mostly in warm, tropical oceans. When Earth's climate is warm, chemical weathering and the formation of limestone increase, drawing CO_2 from the atmosphere, which cools the climate. When the global climate cools, chemical weathering and limestone formation slow down, allowing CO_2 to accumulate in the atmosphere from volcanism, which warms the Earth. An increase in chemical weathering can also lead to global cooling by removing more CO_2 from the atmosphere. For example, the Cenozoic uplift and weathering of large regions of high mountains such as the Alps and the Himalaya may have triggered the global cooling that culminated in the glaciations of the Pleistocene epoch.

Additional Resources

- http://earthobservatory.nasa.gov/Library/CarbonCycle/

BOX 5.2 ■ TABLE 1			
Carbon Dioxide in the Atmospheres of Earth, Mars, and Venus			
	Earth	Mars	Venus
CO_2%	0.33	95.3	96.5
Total surface pressure, bars	1.0[a]	.006	92

[a]Approximately 50 bars of CO_2 is buried in the crust of the Earth as limestone and organic carbon.

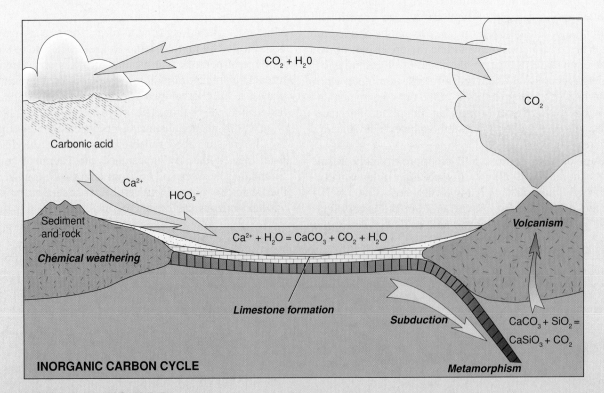

BOX 5.2 ■ FIGURE 1

Carbon dioxide dissolves in water to form carbonic acid in the atmosphere. Carbonic acid reacts with sediment and rocks during chemical weathering, releasing calcium ions and bicarbonate ions (HCO_3^-), which are carried by rivers into the sea. The precipitation of $CaCO_3$ mineral in the oceans (see chapter 6) forms layers of limestone rock. Deep burial of limestone leads to metamorphism, which reacts silica and calcite to form calcium silicate minerals and carbon dioxide. The CO_2 remains trapped in Earth's interior until it is released during volcanic eruptions.

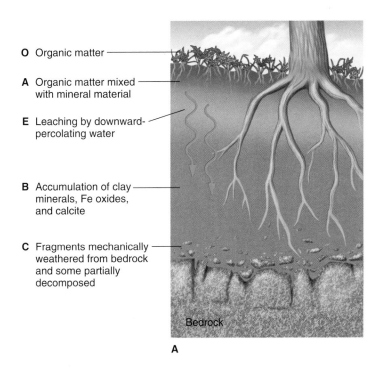

O Organic matter

A Organic matter mixed with mineral material

E Leaching by downward-percolating water

B Accumulation of clay minerals, Fe oxides, and calcite

C Fragments mechanically weathered from bedrock and some partially decomposed

Bedrock

A

B

FIGURE 5.17

(A) Horizons (O, A, E, B, and C) in a soil profile that form in a humid climate. (B) Soil profile that shows the A horizon stained dark by humus. The E horizon is lighter in color, sandy, and crumbly. The clayey B horizon is stained red by hematite, leached downward from the E horizon. *Photo by United States Department of Agriculture*

The **O horizon** is the uppermost layer that consists entirely of nondecomposed and highly decomposed organic material. For example, fallen leaves and needles along with ground vegetation would constitute the O horizon in a forested area.

The **A horizon** is the dark-colored soil layer that is rich in organic material and forms just below surface vegetation. This horizon contains decomposed plant material, or *humus,* and contributes to the formation of organic acids that accelerate leaching in the *E horizon.* The **E horizon,** or **zone of leaching,** is characterized by the downward movement of water and removal (or *eluviation*) of fine-grained soil components such as clay. Part of the rain falling on the ground percolates downward through the soil. This tends to leach or carry dissolved chemicals downward to lower levels in the soil profile. In a humid (wet) climate, iron oxides and dissolved calcite are most typically leached downward; clays are also transported downward. Leaching may make the E horizon pale and sandy.

The **B horizon,** or **zone of accumulation,** is a soil layer characterized by the accumulation of material leached downward from the E horizon above. This layer is often quite clayey and stained red or brown by hematite and limonite. Calcite may also build up in B horizons.

The **C horizon** is incompletely weathered parent material lying below the B horizon. The parent material is commonly the underlying bedrock, which is subjected to mechanical and chemical weathering from frost action, roots, plant acids, and other agents. In such a case, the C horizon is transitional between unweathered bedrock below and developing soil above.

Soil Classification

The Soil Conservation Service of the U.S. Department of Agriculture has developed a soil classification system to group soils with similar properties so that soils can be mapped in a systematic way. There are twelve large groups called *orders* that are distinguished by the characteristics of the horizons present in soil profiles. Brief descriptions of the orders are given in table 5.3, and their distribution in the continental United States is shown on the map in figure 5.18. Each order can be further subdivided into many subdivisions, or suborders, which are defined by even more specific diagnostic physical and chemical properties observed in the soil.

Residual and Transported Soils

A **residual soil** is one that develops from weathering of the rock directly beneath it. Figure 5.17A is a diagram of a residual soil developing in a humid climate from a bedrock source.

TABLE 5.3	Soil Orders

Soil Orders (meaning of name)	Description
Gelisols (frozen soils)	Soils with permafrost within 2 meters of the surface
Histosols (organic soils)	Wet, organic soils such as peat in swamps and marshes
Spodosols (ashy soils)	Acid soils low in plant nutrient ions with subsurface accumulation of organic matter and compounds of aluminum and iron; cool, humid forests
Andisols (volcanic ash)	Soils formed in volcanic ash
Oxisols (oxide soils)	Heavily weathered soils low in plant nutrient ions and rich in aluminum and iron oxides; tropical, usually moist
Vertisols (inverted soils)	Clayey soils that swell when wet and shrink when dry, forming wide, deep cracks
Aridisols (arid soils)	Dry, desert soils low in organic matter and with carbonate horizons
Ultisols (ultimate soils)	Strongly weathered soils low in plant nutrient ions with clay accumulation in the subsurface; usually moist
Mollisols (soft soils)	Nearly black surface horizon rich in organic matter and plant nutrient ions; subhumid to subarid grasslands
Alfisols (pedalfers)	Gray to brown surface horizon, subsurface horizon of clay accumulation; medium to high in plant nutrient ions, usually moist, as in humid forests
Inceptisols (beginning soils)	Very young soils that have weak horizons; usually moist
Entisols (recent soils)	Soils that have no horizons

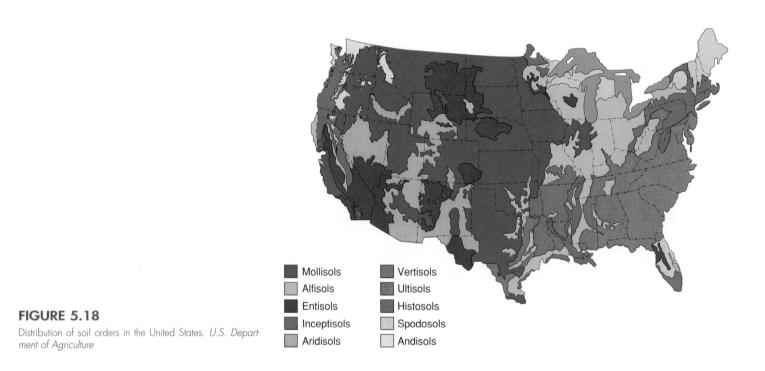

FIGURE 5.18

Distribution of soil orders in the United States. *U.S. Department of Agriculture*

Although this is a typical situation, a number of important agricultural regions in the United States and elsewhere have developed on **transported soils,** which did not form from the local rock but from regolith brought in from some other region. Transported soils usually form on sediment deposited by running water, wind, or glacial ice. For example, mud deposited by a river during times of flooding can form an excellent agricultural soil next to the river after floodwaters recede. The soil-forming mud was not weathered from the rock beneath its present location but was carried downstream from regions perhaps hundreds of kilometers away. Transported wind deposits called *loess* (see chapter 13) are the parent material for some of the most valuable food-producing soils in the Midwest and the Pacific Northwest.

Soils, Parent Material, Time, and Slope

The character of a soil depends partly on the parent material from which it develops. A soil developing on weathering granite will be sandy, as sand-sized particles of quartz and partially weathered feldspar are released from the granite. As time passes, the partially weathered feldspar grains weather completely, forming fine-grained clay minerals. The quartz does not chemically weather, so the resulting soil has both sand and clay (and perhaps silt) in it.

A soil forming on basalt may never be sandy, even in its early stages of development (this depends on the relative rates of chemical weathering versus mechanical weathering). The fine-grained feldspars in the basalt weather to fine-grained clay minerals. Since the parent rock had no coarse-grained minerals and no quartz to start with, the resulting soil may lack sand. Such a soil may not drain well, although it can be quite fertile.

Note that the character of a soil changes with time. A soil developing from granite begins as a sandy soil and becomes more clayey with time. Over very long periods, the type of parent rock becomes less and less important. Given enough time, soils forming from many different kinds of igneous, metamorphic, and sedimentary rocks can become quite similar (in the same climate). The presence or absence of coarse grains of quartz in the parent rock becomes the only characteristic of the parent rock to have long-term significance.

With time, soils tend to become thicker (figure 5.19A); most modern soils have taken centuries to form. A new deposit of volcanic ash, which is very fine-grained and rich in plant nutrients, may be covered with grass and other low plants in just a few years, but a new lava flow, which weathers much more slowly than ash, may not have enough soil to support grass for many decades. It may take centuries for the lava-flow soil to thicken enough to support shrubs and thousands of years to support trees. The fertile agricultural soils of the Canadian plains and the northern United States took more than 10,000 years to develop on glacial deposits after the thick continental ice sheets melted.

Another factor controlling soil thickness is the slope of the land surface (figure 5.19B). Soils tend to be thick on flat land where erosion is slow and water can collect, and thin on steep slopes where gravity pulls water and soil particles downhill.

Organic Activity

The biosphere plays an important role in the development of soil. Plant roots break up rocks, and burrowing organisms such as ants, worms, and rodents bring soil particles to the surface and create passageways for water and air to get underground, thus speeding up chemical weathering. Respiration of soil organisms and decay of plant and animal material add carbon dioxide gas to soil, creating carbonic acid. Plants and humus also release organic acids that increase chemical weathering. Once soil begins to develop on a newly exposed rock, it attracts plants and soil organisms that increase chemical weathering, accelerating the rate of soil development. Partially decayed organic matter provides plant nutrients, increasing soil fertility.

Soils and Climate

Climate affects soil thickness and character. Soils in wet climates, as in Europe, most of Canada, and the eastern United States, tend to be thick and are generally characterized by downward movement of water through Earth materials (figure 5.17 shows such a soil). In general, these soils tend to have a high content of aluminum and iron oxides, and are marked by effective downward leaching due to high rainfall and to the acids produced by the decay of abundant humus.

In arid (dry) climates, as in many parts of the western United States, soils tend to be thin and are characterized by little leaching, scant humus, and the *upward* movement of soil water beneath the land surface. The water is drawn up by subsurface evaporation and capillary action.

The evaporation of water beneath the land surface can cause the precipitation of salts within the soil. These salts are usually calcium salts such as calcite (figure 5.20). An extreme example of salt buildup can be found in desert *alkali soils,* in which heavy concentrations of toxic sodium salts may prevent plant growth.

Hardpan

"Hardpan" is a general term for a hard layer of Earth material that is difficult to dig or drill. Geologists usually restrict the term to a hard, often clayey, layer of cemented soil particles.

FIGURE 5.19

Soil thickness. (A) Soil thickens with time. Oldest basalt flow was exposed to soil-forming processes for a longer time and has a thicker soil developed on it than younger flows. The soils developed below the younger basalt flow are examples of a buried soil. (B) Steep slopes have thin soil.

FIGURE 5.20

Soil profile marked by upward-moving ground water that evaporates underground in a drier climate, precipitating calcium carbonate within the soil, sometimes forming a light-colored layer. *Photo by D. Yost, U.S. Agriculture Department Soil Conservation Service*

Accumulation of a calcite layer and nodules as upward-moving ground water evaporates

FIGURE 5.21

Laterite soil (oxisol) develops in very wet climates, where intense, downward leaching carries away all but iron and aluminum oxides. Many laterites are a rusty orange to deep red color from the oxidation of the iron oxides. *Photo by USDA Natural Resources Conservation Service*

Such a layer may be too hard for even backhoes to dig through; planting a tree in a lawn with a hardpan layer may require a jackhammer. Hardpan layers in wet climates are usually formed of clay minerals, silica, and iron compounds that have accumulated in the B horizon from eluviation of the overlying E horizon. In arid climates, a different type of hardpan forms from the cementing of soil by calcium carbonate and other salts that precipitate in the soil as water evaporates. Both types of hardpan are really layers of rock within loose soil. A hardpan layer can break plows, prevent water drainage through the soil, and act as a barrier to plant roots. Tree roots may grow laterally along rather than down through hardpan; such shallow-rooted trees are easily uprooted by wind.

Laterites

In tropical regions where temperatures are high and rainfall is abundant, highly leached soils called **laterites** (*oxisols*) form. Under such conditions, weathering is deep and intense. Laterites are usually red and are composed almost entirely of iron and aluminum oxides, generally the least soluble products of rock weathering in tropical climates (figure 5.21). If the soil is rich in hematite, it can be mined as iron ore, but tropical rainfall usually hydrates the hematite to limonite, which is seldom rich enough to mine. However, aluminum is sometimes found in nearly pure layers of *bauxite* ($Al_2O_3 \cdot nH_2O$, the principal ore of aluminum), particularly in laterites formed by the weathering of aluminum-rich volcanic tuffs.

Under tropical conditions of high rainfall and high temperature, most weathering products are soluble—even silica. The least soluble product is aluminum oxide, which remains on top of the weathering rocks, forming bauxite in a soil very rich in aluminum. Like the diamonds, the aluminum has been concentrated residually by the removal of everything else. The aluminum ores may be redistributed slightly by running water (figure 5.22).

Because bauxite forms under conditions of tropical weathering, the United States has very little aluminum ore and depends almost entirely on recycling and tropical countries for its aluminum supply. A small percentage of the U.S. aluminum supply has come from bauxite deposits near Little Rock, Arkansas, that formed on igneous rock with a high aluminum content approximately 50 million years ago when the region had a tropical climate.

FIGURE 5.22

Bauxite forms by intense tropical weathering of an aluminum-rich source rock such as a volcanic tuff.

FIGURE 5.23

Soil erosion caused by clear cutting of rain forest, north of Kuantan, Malaysia. *Photo © George Loun/Visuals Unlimited*

Laterites are relatively nonproductive soils. This may seem strange when you think of the lush jungle growth that often exists on tropical lateritic soils. Jungle vegetation, though, is nourished largely by a layer of humus on top of the soil. If the jungle and the humus layer are cleared away or burned—an increasingly common practice in tropical regions—the laterite quickly becomes incapable of sustaining plant growth, making tropical agriculture very difficult (figure 5.23). Laterite exposed to the sun is apt to bake into a permanent, bricklike layer that makes digging nearly impossible. This hard layer can be quarried, however, and makes a durable building material.

Buried Soils

A soil may become buried by volcanic ash, wind-blown dust, glacial deposits, other sediment, or lava (see figure 5.19). A buried soil is called a *paleosol* (*paleo* = ancient). Such soils may be distinctive and traceable over wide regions and may contain buried organic remains, making them useful for dating rocks and sediments, and for interpreting past climates and topography.

SUMMARY

When rocks that formed deep in Earth become exposed at the Earth's surface, they are altered by *mechanical* and *chemical weathering.*

Weathering processes produce *spheroidal weathering, differentially weathered* landforms, *sheet joints,* and *exfoliation domes.*

Mechanical weathering, largely caused by *frost action* and *pressure release* after unloading, disintegrates (breaks) rocks into smaller pieces.

By increasing the exposed surface area of rocks, mechanical weathering helps speed chemical weathering.

Chemical weathering results when a mineral is unstable in the presence of water and atmospheric gases. As chemical weathering proceeds, the mineral's components recombine into new minerals that are more in equilibrium.

Weak acid, primarily from the solution of carbon dioxide in water, is the more effective agent of chemical weathering.

Calcite dissolves when it is chemically weathered. Most of the silicate minerals form *clay minerals* when they chemically weather. Quartz is very resistant to chemical weathering.

Soil develops by chemical and mechanical weathering of a parent material. Some definitions of soil require that it contain organic matter and be able to support plant growth.

Soils, which can be *residual* or *transported,* usually have distinguishable layers, or *horizons,* caused in part by water movement within the soil.

Climate is the most important factor determining soil type. Other factors in soil development are parent material, time, slope, and organic activity.

Laterites form under conditions of intense tropical weathering; they are usually red from concentrated iron oxides. Bauxite, the ore of aluminum, may be found in laterites.

Terms to Remember

A horizon 131
B horizon (zone of accumulation) 131
chemical weathering 119
C horizon 131
clay mineral 126
differential weathering 120
E horizon (zone of leaching) 131
erosion 118
exfoliation 121
exfoliation dome 121

frost action 121
frost heaving 122
frost wedging 121
hematite 124
laterite 134
limonite 124
loam 129
mechanical weathering 119
O horizon 131
pressure release 121

residual soil 131
sheet joints 121
soil 129
soil horizon 129
spheroidal weathering 120
talus 121
transportation 118
transported soil 132
weathering 118

Testing Your Knowledge

Use the following questions to prepare for exams based on this chapter.

1. Why are some minerals stable several kilometers underground but unstable at Earth's surface?

2. Describe what happens to each mineral within granite during the complete chemical weathering of granite in a humid climate. List the final products for each mineral.

3. Explain what happens chemically when calcite dissolves. Show the reaction in a chemical equation.

4. Why do stone buildings tend to weather more rapidly in cities than in rural areas?

5. Describe at least three processes that mechanically weather rock.

6. How can mechanical weathering speed up chemical weathering?

7. Name at least three natural sources of acid in solution. Which one is most important for chemical weathering?

8. What is the difference between a residual soil and a transported soil?

9. What is a laterite (oxisol) and how does it form?

10. What are soil horizons?

11. What are the soil horizons? How do they form?

12. Physical disintegration of rock into smaller pieces is called
    a. chemical weathering   b. transportation
    c. deposition   d. mechanical weathering

13. The decomposition of rock from exposure to water and atmospheric gases is called
    a. chemical weathering   b. transportation
    c. deposition   d. mechanical weathering

14. Which is not a type of mechanical weathering?
    a. frost wedging   b. frost heaving
    c. pressure release   d. oxidation

15. The single most effective agent of chemical weathering at Earth's surface is
    a. carbonic acid H_2CO_3 b. water H_2O
    c. carbon dioxide CO_2 d. hydrochloric acid HCl

16. The most common end product of the chemical weathering of feldspar is
    a. clay minerals   b. pyroxene
    c. amphibole   d. calcite

17. The most common end product of the chemical weathering of quartz is
    a. clay minerals   b. pyroxene
    c. amphibole   d. calcite
    e. quartz does not usually weather chemically

18. Soil with approximately equal amounts of sand, silt, and clay along with a generous amount of organic matter is called
    a. loam   b. inorganic
    c. humus   d. caliche

19. Which is characteristic of soil horizons?
    a. they can be distinguished from one another by appearance and chemical composition
    b. boundaries between soil horizons are usually transitional rather than sharp
    c. they are classified by letters
    d. all of the preceding

20. The soil horizon containing only organic material is the
    a. A horizon   b. B horizon
    c. C horizon   d. O horizon

21. Hardpan forms in the
    a. A horizon   b. B horizon
    c. C horizon   d. E horizon

22. Tropical soils are typically
    a. rich in organic material   b. very fertile
    c. deeply leached   d. easily replenished

Expanding Your Knowledge

1. Which mineral weathers faster—hornblende or quartz? Why?

2. Compare and contrast the weathering rate and weathering products for Ca-rich plagioclase in the following localities:

   a. central Pennsylvania with 40 inches of rain per year;

   b. Death Valley with 2 inches of rain per year;

   c. an Alaskan mountaintop where water is frozen year-round.

3. The amount of carbon dioxide gas has been increasing in the atmosphere for the past 40 years as a result of the burning of fossil fuels. What effect will the increase in CO_2 have on the rate of chemical weathering? The increase in CO_2 may cause global warming in the future. What effect would a warmer climate have on the rate of chemical weathering? Give the reasons for your answers.

4. In a humid climate, is a soil formed from granite the same as one formed from gabbro? Discuss the similarities and possible differences with particular regard to mineral content and soil color.

Exploring Web Resources

www.mhhe.com/plummer11e

This is the dedicated website for this book. You can go to it for new and updated information. The universal resource locators (URL) listed in this book are also given as links on the website, making it easy to go to those websites without typing in the URL. When you visit our Online Learning Center, go to the Student Center and then to the chapter of interest. Here you will find additional readings and media resources, as well as answers to the Testing Your Knowledge section, more quizzes, animation and video clips, and direct links to the sites listed in this book. Links to additional websites can also be found. We have added questions for some of the sites to allow you to get the most of your exploration of the web. Using the web is an enjoyable way of enhancing your knowledge of geology.

http://soils.ag.uidaho.edu/soilorders/

University of Idaho Soil Science Division. Web page contains photos, descriptions, and surveys of the twelve major soil orders.

http://res.agr.ca/cansis/_overview.html

Canadian Soil Information System provides links to detailed soil surveys and land inventories.

Sediment and Sedimentary Rocks

The rock cycle is a conceptual model of the constant recycling of rocks as they form, are destroyed, and then reform. We began our discussion of the rock cycle with igneous rock (chapters 3 and 4), and we now discuss sedimentary rocks. In this chapter, we first describe sediment and sedimentary rock, and then discuss sedimentary structures and fossils. We also consider the

Eroded sandstone formations formed from ancient sand dunes, Colorado Plateau, Arizona. *Photo © R-F Masterfile*

139

importance of sedimentary rocks for interpreting the history of Earth and their tremendous economic importance. Metamorphic rocks, the third major rock type, are the subject of chapter 7.

You saw in chapter 5 how weathering produces sediment. In this chapter, we explain more about sediment origin, as well as the erosion, transportation, sorting, deposition, and eventual transformation of sediments to sedimentary rock. Because they have such diverse origins, sedimentary rocks are difficult to classify. We divide them into detrital, chemical, and organic sedimentary rocks, but this classification does not do justice to the great variety of sedimentary rock types. Furthermore, despite their great variety, only three sedimentary rocks are very common—shale, sandstone, and limestone.

Sedimentary rocks contain sedimentary structures such as ripple marks, crossbeds, and mud cracks, as well as the fossilized remains of extinct organisms. These features, combined with knowledge of the sediment types within the rock and the sequence of rock layers, allow geologists to interpret the environments in which the rocks were deposited. About three-fourths of the surface of the continents is blanketed by sedimentary rock, providing geologists with the information they need to reconstruct a detailed history of the surface of Earth and its biosphere.

Sedimentary rocks are also economically important. Most building materials such as stone, concrete, silica (glass), gypsum (plaster), and iron are quarried and mined from sedimentary rock. Salt is also a sedimentary product and, in many places in the world, supplies of fresh water are pumped from sedimentary layers. Coal, crude oil, and natural gas, the fossil fuels that drove the industrial revolution and that power our technological society, are all formed within and extracted from sedimentary rock.

Relationships to Earth Systems

Sediment and sedimentary rocks are important components of the solid Earth system. They are especially important at the surface of Earth, where sedimentary rocks account for the majority of exposed bedrock. The atmosphere, hydrosphere, and biosphere are deeply intertwined in the creation of sediment and in it becoming sedimentary rock.

Atmosphere

Most sediment is the product of weathering of rocks exposed to air; the important role that the atmosphere plays was described in chapter 5. Wind is one of the agents by which sediment is transported. Sand is skipped along the ground, moving in ripples and often accumulating into sand dunes. Dust, which col-

lects relentlessly on almost everything, is deposited by the atmosphere. Finer sedimentary particles are carried as dust by the air and may travel great distances before settling out on land or sea.

Hydrosphere

Water plays a role in the making of nearly all sedimentary rocks. Typically, sediment is created during weathering with water being a vital ingredient in the process. Sediment is further modified during transportation by streams and ocean currents. In colder regions, glaciers (frozen water) move sediment. Most sediment is ultimately deposited on a sea floor.

The conversion of sediment to sedimentary rock usually involves water. Water carrying dissolved material flows between grains of sediment. Precipitation of the dissolved substances onto the grains cements them together, and sediment is turned into sedimentary rock.

Biosphere

Most sedimentary rocks contain fossils and part or all of many sedimentary rocks are made by organisms. Limestone, for example, is often made from the shells or remains of other hard parts of animals and algae. Plants can partially decompose and be converted into coal (which is a rock). Our civilization depends on crude oil for our principal source of energy. Crude oil is found in sedimentary rocks and is formed through the partial decay of organic matter.

SEDIMENT

Sediment is the collective name for loose, solid particles of mineral that originate from:

1. Weathering and erosion of preexisting rocks (detrital sediments).

2. Precipitation from solution, including secretion by organisms in water (chemical sediments).

Sediment includes such particles as sand on beaches, mud on a lake bottom, boulders frozen into glaciers, pebbles in streams, and dust particles settling out of the air. An accumulation of clam shells on the sea bottom offshore is sediment, as are coral fragments broken from a reef by large storm waves.

These particles usually collect in layers on Earth's surface. An important part of the definition is that the particles are loose. Sediments are said to be *unconsolidated,* which means that the grains are separate, or unattached to one another.

Detrital sediment particles are classified and defined according to the size of individual fragments. Table 6.1 shows the precise definitions of particles by size.

Gravel includes all rounded particles coarser than 2 millimeters in diameter, the thickness of a U.S. nickel. (Angular fragments of this size are called *rubble.*) *Pebbles* range from 2 to 64 millimeters (about the size of a tennis ball). *Cobbles* range from 64 to 256 millimeters (about the size of a basketball), and *boulders* are coarser than 256 millimeters.

Sand grains are from 1/16 millimeter (about the thickness of a human hair) to 2 millimeters in diameter. Grains of this size are visible and feel gritty between the fingers. **Silt** grains are from 1/256 to 1/16 millimeter. They are too small to see without a magnifying device, such as a geologist's hand lens. Silt does not feel gritty between the fingers, but it does feel gritty between the teeth (geologists often bite sediments to test their grain size). **Clay** is the finest sediment, at less than 1/256 millimeter, too fine to feel gritty to fingers or teeth. *Mud* is a term loosely used for a mixture of silt and clay.

Note that we have two different uses of the word *clay*—a *clay-sized particle* (table 6.1) and a *clay mineral.* A clay-sized

FIGURE 6.1

These boulders have been rounded by abrasion as wave action rolled them against one another on this beach. *Photo by David McGeary*

particle can be composed of any mineral at all provided its diameter is less than 1/256 millimeter. A clay mineral, on the other hand, is one of a small group of silicate minerals with a sheet-silicate structure. Clay minerals usually form in the clay-size range.

Quite often the composition of sediment in the clay-size range turns out to be mostly clay minerals, but this is not always the case. Because of its resistance to chemical weathering, quartz may show up in this fine-size grade. (Most silt is quartz.) Intense mechanical weathering can break down a wide variety of minerals to clay size, and these extremely fine particles may retain their mineral identity for a long time if chemical weathering is slow. The great weight of glaciers is particularly effective at grinding minerals down to the silt- and clay-size range, producing "rock flour," which gives a milky appearance to glacial meltwater streams (see chapter 12).

Weathering, erosion, and transportation are some of the processes that affect the character of sediment. Both mechanically weathered and chemically weathered rock and sediment can be eroded, and weathering continues as erosion takes place. Sand being transported by a river also can be actively weathered, as can mud on a lake bottom. The character of sediment can also be altered by *rounding* and *sorting* during transportation, and after eventual *deposition.*

Transportation

Most sediment is **transported** some distance by gravity, wind, water, or ice before coming to rest and settling into layers. During transportation, sediment continues to weather and change in character in proportion to the distance the sediment is moved.

Rounding is the grinding away of sharp edges and corners of rock fragments during transportation. Rounding occurs in sand and gravel as rivers, glaciers, or waves cause particles to hit and scrape against one another (figure 6.1) or against a rock

TABLE 6.1	Sediment Particles and Detrital Sedimentary Rocks		
Diameter (mm)	Sediment		Sedimentary Rock
256	Boulder		**Breccia** (angular particles) or **conglomerate** (rounded particles)
64	Cobble	Gravel	
	Pebble		
2	Sand		**Sandstone**
1/16	Silt		Siltstone (mostly silt)
1/256	Clay	"Mud"	**Shale** or mudstone (mostly clay)

Sandstone and shale are quite common; the others are relatively rare.

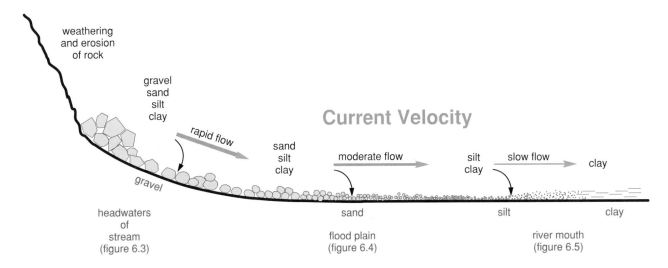

FIGURE 6.2

Sorting of sediment by a river. The coarse sediment is deposited first, and the finest sediment is carried the farthest.

surface, such as a rocky streambed. Boulders in a stream may show substantial rounding in less than 1 kilometer of travel.

Sorting is the process by which sediment grains are selected and separated according to grain size (or grain shape or specific gravity) by the agents of transportation, especially by running water. Because of their high viscosity and manner of flow, glaciers are poor sorting agents. Glaciers deposit all sediment sizes in the same place, so glacial sediment usually consists of a mixture of clay, silt, sand, and gravel. Such glacial sediment is considered *poorly sorted*. Sediment is considered *well-sorted* when the grains are nearly all the same size. A river, for example, is a good sorting agent, separating sand from gravel, and silt and clay from sand. Sorting takes place because of the greater weight of larger particles. Boulders weigh more than pebbles and are more difficult for the river to transport, so a river must flow more rapidly to move boulders than to move pebbles. Similarly, pebbles are harder to move than sand, and sand is harder to move than silt and clay.

Figure 6.2 shows the sorting of sediment by a river as it flows out of steep mountains onto a gentle flood plain, where the water loses energy and slows down. As the river loses energy, the heaviest particles of sediment are deposited. The boulders come to rest first (figure 6.3). As the river continues to slow and becomes less turbulent, cobbles and then pebbles are deposited. Sand comes to rest as the river loses still more energy (figure 6.4). Finally, the river is carrying only the finest sediment—silt and clay (figure 6.5). The river has sorted the original sediment mix by grain size.

Deposition

When transported material settles or comes to rest, **deposition** occurs. Sediment is deposited when running water, glacial ice, waves, or wind loses energy and can no longer transport its load.

Deposition also refers to the accumulation of chemical or organic sediment, such as clam shells on the sea floor or plant material on the floor of a swamp. Such sediments may form as organisms die and their remains accumulate, perhaps with no transportation at all. Deposition of salt crystals can take place as seawater evaporates. A change in the temperature, pressure, or chemistry of a solution may also cause precipitation—hot springs may deposit calcite or silica as the warm water cools.

The **environment of deposition** is determined by the location in which deposition occurs. A few examples of environments of deposition are the deep-sea floor, a desert valley, a

FIGURE 6.3

Coarse gravel (boulder and cobble size) is deposited first along a river's course as the river sorts out the various sediment sizes, Oak Creek Canyon near Sedona, Arizona. *Photo © Parvinder Sethi*

FIGURE 6.4
Deposition of sand occurs as a river loses energy as it flows across the flood plain. *Photo by David McGeary*

FIGURE 6.5
The river on the right is carrying silt and clay as it enters the clear river on the left. This fine sediment may come to rest at the mouth of a river where it enters a lake or the sea. *Photo by C. W. Montgomery*

river channel, a coral reef, a lake bottom, a beach, and a sand dune. Each environment is marked by characteristic physical, chemical, and biological conditions. You might expect mud on the sea floor to differ from mud on a lake bottom. Sand on a beach may differ from sand in a river channel. Some differences are due to varying sediment sources and transporting agents, but most are the result of conditions in the environments of deposition themselves.

One of the most important jobs of geologists studying sedimentary rocks is to try to determine the ancient environment of deposition of the sediment in which the rock formed. Factors that can help in determining this are a detailed knowledge of modern environments, the vertical sequence of rock layers in the field, the fossils and sedimentary structures found within the rock, the mineral composition of the rock, and the size, shape, and surface texture of the individual sediment grains. Later in this chapter, we give a few examples of interpreting environments of deposition.

Preservation

Not all sediments are preserved as sedimentary layers. Gravel in a river may be deposited when a river is low but then may be eroded and transported by the next flood on the river. Many sediments on land, particularly those well above sea level, are easily eroded and carried away, so they are not commonly preserved. Sediments on the sea floor are easier to preserve. In general, continental and marine sediments are most likely to be preserved if they are deposited in a *subsiding* (sinking) *basin* and if they are covered or *buried by later sediments.*

Lithification

Lithification is the general term for the processes that convert loose sediment into sedimentary rock. Most sedimentary rocks are lithified by a combination of *compaction,* which packs loose sediment grains tightly together, and *cementation,* in which the precipitation of cement around sediment grains binds them into a firm, coherent rock. *Crystallization* of minerals

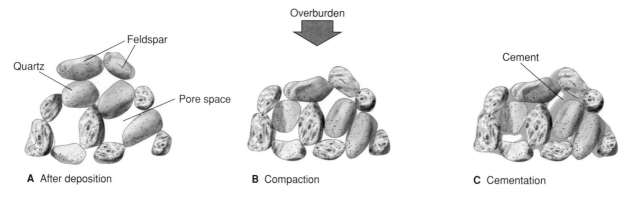

A After deposition   **B** Compaction   **C** Cementation

FIGURE 6.6

Lithification of sand grains to become sandstone. (A) Loose sand grains are deposited with open pore space between the grains. (B) The weight of overburden compacts the sand into a tighter arrangement, reducing pore space. (C) Precipitation of cement in the pores by ground water binds the sand into the rock sandstone, which has a clastic texture.

from solution, without passing through the loose-sediment stage, is another way that rocks may be lithified. Some layers of sediment persist for tens of millions of years without becoming fully lithified. Usually, layers of *partially lithified sediment* have been buried deep enough to become compacted, but have not experienced the conditions required for cementation.

As sediment grains settle slowly in a quiet environment such as a lake bottom, they form an arrangement with a great deal of open space between the grains (figure 6.6*A*). The open spaces between grains are called *pores,* and in a quiet environment, a deposit of sand may have 40% to 50% of its volume as open **pore space.** (If the grains were traveling rapidly and impacting one another just before deposition, the percentage of pore space will be less.) As more and more sediment grains are deposited on top of the original grains, the increasing weight of this *overburden* packs the original grains together, reducing the amount of pore space. This shift to a tighter packing, with a resulting decrease in pore space, is called **compaction** (figure 6.6*B*). As pore space decreases, some of the interstitial water that usually fills sediment pores is driven out of the sediment.

As underground water moves through the remaining pore space, solid material called **cement** can precipitate in the pore space and bind the loose sediment grains together to form a solid rock. The cement attaches very tightly to the grains, holding them in a rigid framework. As cement partially or completely fills the pores, the total amount of pore space is further reduced (figure 6.6*C*), and the loose sand forms a hard, coherent sandstone by **cementation.**

Sedimentary rock cement is often composed of the mineral calcite or of other carbonate minerals. Dissolved calcium and bicarbonate ions are common in surface and underground waters. If the chemical conditions are right, these ions may recombine to form solid calcite, as shown in the following reaction.

$$\underbrace{Ca^{++} + 2HCO_3^-}_{\substack{\text{dissolved} \\ \text{ions}}} \rightarrow \underbrace{CaCO_3}_{\text{calcite}} + H_2O + CO_2$$

Silica is another common cement. Iron oxides and clay minerals can also act as cement but are less common than calcite and silica. The dissolved ions that precipitate as cement originate from the chemical weathering of minerals such as feldspar and calcite. This weathering may occur within the sediments being cemented, or at a very distant site, with the ions being transported tens or even hundreds of kilometers by water before precipitating as solid cement.

A sedimentary rock that consists of sediment grains bound by cement into a rigid framework is said to have a **clastic texture.** Usually such a rock still has some pore space because cement rarely fills the pores completely (figure 6.6*C*).

Some sedimentary rocks form by **crystallization,** the development and growth of crystals by precipitation from solution at or near Earth's surface (the term is also used for igneous rocks that crystallize as magma cools). These rocks have a **crystalline texture,** an arrangement of interlocking crystals that develops as crystals grow and interfere with each other.

FIGURE 6.7

Crystalline dolostone as seen through a polarizing microscope. Note the interlocking crystals of dolomite mineral that grew as they precipitated during recrystallization. Such crystalline sedimentary textures have no cement or pore spaces. *Photo by Bret Bennington*

Crystalline rocks lack cement. They are held together by the interlocking of crystals. Such rocks have minimal pore space because the crystals typically grow until they fill all available space. Some sedimentary rocks with a crystalline texture are the result of *recrystallization,* the growth of new crystals that form from and then destroy the original clastic grains of a rock that has been buried (figure 6.7).

TYPES OF SEDIMENTARY ROCKS

Sedimentary rocks are formed from (1) eroded mineral grains, (2) minerals precipitated from solution, or (3) consolidation of the organic remains of plants. These different types of sedimentary rocks are called, respectively, *detrital, chemical,* and *organic* rocks.

Most sedimentary rocks are **detrital sedimentary rocks,** formed from cemented sediment grains that are fragments of preexisting rocks. The rock fragments can be either identifiable pieces of rock, such as pebbles of granite or shale, or individual mineral grains, such as sand-sized quartz and feldspar crystals loosened from rocks by weathering and erosion. Clay minerals formed by chemical weathering are also considered fragments of preexisting rocks. During transportation the grains may have been rounded and sorted. Table 6.1 shows the detrital rocks, such as conglomerate, sandstone, and shale, and how these rocks vary in grain size.

Chemical sedimentary rocks are deposited by precipitation of minerals from solution. An example of inorganic precipitation is the formation of *rock salt* as seawater evaporates. Chemical precipitation can also be caused by organisms. The sedimentary rock *limestone* is often formed from the cementation of broken pieces of seashell and fragments of calcite mineral produced by corals and algae. Such a rock is called a *bioclastic* limestone.

Not all chemical sedimentary rocks accumulate as sediment. Some limestones are crystallized as solid rock by corals and coralline algae in reefs. Chert crystallizes in solid masses within some layers of limestone. Rock salt may crystallize directly as a solid mass or it may form from the crystallization of individual salt crystals that behave as sedimentary particles until they grow large enough to interlock into solid rock.

Organic sedimentary rocks are rocks that are composed of organic carbon compounds. *Coal* is an organic rock that forms from the compression of plant remains, such as moss, leaves, twigs, roots, and tree trunks.

Appendix B describes and helps you identify the common sedimentary rocks. The standard geologic symbols for these rocks (such as dots for sandstone, and a "brick-wall" symbol for limestone) are shown in appendix F and will be used in the remainder of the book.

DETRITAL ROCKS

Detrital sedimentary rocks are formed from the weathered and eroded remains (detritus) of bedrock. Detrital rocks are also

FIGURE 6.8

Breccia is characterized by coarse, angular fragments. The cement in this rock is colored by hematite. The wide black and white bars on the scale are 1 centimeter long, the small divisions are 1 millimeter. Note that most grains exceed 2 millimeters (table 6.1). *Photo by David McGeary*

often referred to as *terrigenous clastic rocks* because they are composed of *clasts* (broken pieces) of mineral derived from the erosion of the land.

Breccia and Conglomerate

Sedimentary breccia is a coarse-grained sedimentary rock formed by the cementation of coarse, angular fragments of rubble (figure 6.8). Because grains are rounded so rapidly during transport, it is unlikely that the angular fragments within breccia have moved very far from their source. Sedimentary breccia might form from fragments that have accumulated at the base of a steep slope of rock that is being mechanically weathered. Landslide deposits also might lithify into sedimentary breccia. This type of rock is not particularly common.

Conglomerate is a coarse-grained sedimentary rock formed by the cementation of rounded gravel. It can be distinguished from breccia by the definite roundness of its particles (figure 6.9). Because conglomerates are coarse-grained,

FIGURE 6.9

An outcrop of a poorly sorted conglomerate. Note the rounding of cobbles, which vary in composition and size. Long scale bar is 10 centimeters; short bars are 1 centimeter. *Photo by David McGeary*

FIGURE 6.10

Types of sandstone. (*A*) Quartz sandstone; more than 90% of the grains are quartz. (*B*) Arkose; the grains are mostly feldspar and quartz. (*C*) Graywacke; the grains are surrounded by dark, fine-grained matrix. (Small scale divisions are 1 millimeter; most of the sand grains are about 1 millimeter in diameter.) *Photos by David McGeary* the particles may not have traveled far; but some transport was necessary to round the particles. Angular fragments that fall from a cliff and then are carried a few kilometers by a river or pounded by waves crashing in the surf along a beach are quickly rounded. Gravel that is transported down steep submarine canyons or carried by glacial ice, however, can be transported tens or even hundreds of kilometers before deposition.

Sandstone

Sandstone is formed by the cementation of sand grains (figure 6.10). Any deposit of sand can lithify to sandstone. Rivers deposit sand in their channels, and wind piles up sand into dunes. Waves deposit sand on beaches and in shallow water. Deep-sea currents spread sand over the sea floor. As you might imagine, sandstones show a great deal of variation in mineral composition, degree of sorting, and degree of rounding.

Quartz sandstone is a sandstone in which more than 90% of the grains are quartz (figure 6.10A). Because quartz is resistant to chemical weathering, it tends to concentrate in sand deposits as the less-resistant minerals such as feldspar are weathered away. The quartz grains in a quartz sandstone are usually well-sorted and well-rounded because they have been transported for great distances (figure 6.11A). Most quartz sandstone was deposited as beach sand or dune sand.

A sandstone with more than 25% of the grains consisting of feldspar is called *arkose* (figure 6.10B). Because feldspar grains are preserved in the rock, the original sediment obviously did not undergo severe chemical weathering, or the feldspar would have been destroyed. Mountains of granite in a desert could be a source for such a sediment, for the rapid erosion associated with rugged terrain would allow feldspar to be mechanically weathered and eroded before it is chemically weathered (a dry climate slows chemical weathering). Most arkoses contain coarse, angular grains (figure 6.11B), so transportation distances were probably short. An arkose may have been deposited within an **alluvial fan,** a large, fan-shaped pile of sediment that usually forms where a stream emerges from a narrow canyon onto a flat plain at the foot of a mountain range (figure 6.12).

Sandstones may contain a substantial amount of **matrix** in the form of fine-grained silt and clay in the space between larger sand grains (figure 6.13). A matrix-rich sandstone is poorly sorted and often dark in color. It is sometimes called a "dirty sandstone."

Graywacke (pronounced "gray-wacky") is a type of sandstone in which more than 15% of the rock's volume consists of fine-grained matrix (figures 6.10C and 6.11C). Graywackes are often tough and dense, and are generally dark gray or green. The sand grains may be so coated with matrix that they are hard to see, but they typically consist of quartz, feldspar, and sand-sized fragments of other fine-grained sedimentary, volcanic, and metamorphic rocks. Most graywackes probably formed from sediments transported by **turbidity currents** (see figure 6.31).

A

B

C

FIGURE 6.11

Detrital sedimentary rocks viewed through a polarizing microscope. (*A*) Quartz sandstone; note the well-rounded and well-sorted grains. (*B*) Arkose; large feldspar grain in center surrounded by angular quartz grains. (*C*) Graywacke; quartz grains surrounded by brownish matrix of mud. *Photos by Bret Bennington*

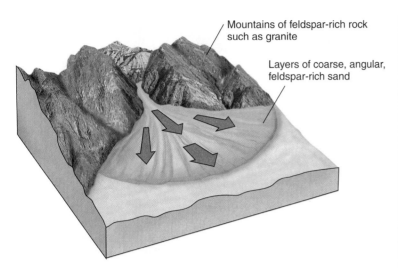

FIGURE 6.12

Feldspar-rich sand (arkose) may accumulate from the rapid erosion of feldspar-containing rock such as granite. Steep terrain accelerates erosion rates so that feldspar may be eroded before it is completely chemically weathered into clay minerals.

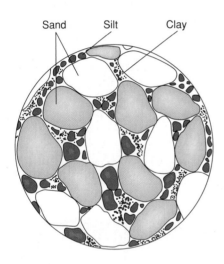

FIGURE 6.13

A poorly sorted sediment of sand grains surrounded by a matrix of silt and clay grains. Lithification of such a sediment would produce a "dirty sandstone."

FIGURE 6.14

(A) An outcrop of shale from Hudson Valley in New York. Note how this fine-grained rock tends to split into very thin layers. (B) Shale pieces; note the very fine grain (scale in centimeters), very thin layers (laminations) on the edge of the large piece, and tendency to break into small, flat pieces (fissility). *Photo A © John Buitenkant/Photo Researchers Inc.; Photo B by David McGeary*

The Fine-Grained Rocks

Rocks consisting of fine-grained silt and clay are called *shale, siltstone, claystone,* and *mudstone.*

Shale is a fine-grained sedimentary rock notable for its ability to split into layers (called *fissility*). Splitting takes place along the surfaces of very thin layers (called *laminations*) within the shale (figure 6.14). Most shales contain both silt and clay (averaging about two-thirds clay-sized clay minerals and one-third silt-sized quartz) and are so fine-grained that the surface of the rock feels very smooth. The silt and clay deposits that lithify as shale accumulate on lake bottoms, at the ends of rivers in deltas, on river flood plains, and on quiet parts of the deep-ocean floor.

Fine-grained rocks such as shale typically undergo pronounced compaction as they lithify. Figure 6.15 shows the role of compaction in the lithification of shale from wet mud. Before compaction, as much as 80% of the volume of the wet mud may have been pore space filled with water. The flakelike clay minerals were randomly arranged within the mud. Pressure from overlying material packs the sediment grains together and reduces the overall volume by squeezing water out of the pores. The clay minerals are reoriented perpendicular to the pressure, becoming parallel to one another like a deck of cards. The fissility of shale is due to weaknesses between these parallel clay flakes.

Compaction by itself does not generally lithify sediment into sedimentary rock. It does help consolidate clayey sedi-

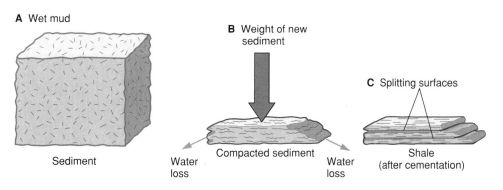

A Wet mud

Sediment

B Weight of new sediment

Compacted sediment

Water loss   Water loss

C Splitting surfaces

Shale (after cementation)

FIGURE 6.15

Lithification of shale from the compaction and cementation of wet mud. (A) Randomly oriented silt and clay particles in wet mud. (B) Particles reorient, water is lost, and pore space decreases during compaction caused by the weight of new sediment deposited on top of the wet mud. (C) Splitting surfaces in cemented shale form parallel to the oriented mineral grains.

ments by pressing the microscopic clay minerals so closely together that attractional forces at the atomic level tend to bind them together. Even in shale, however, the primary method of lithification is cementation.

A rock consisting mostly of silt grains is called *siltstone.* Somewhat coarser-grained than most shales, siltstones lack the fissility and laminations of shale. *Claystone* is a rock composed predominately of clay-sized particles but lacking the fissility of shale. *Mudstone* contains both silt and clay, having the same grain size and smooth feel of shale but lacking shale's laminations and fissility. Mudstone is massive and blocky, while shale is visibly layered and fissile.

CHEMICAL SEDIMENTARY ROCKS

Chemical sedimentary rocks are precipitated from an aqueous environment. Chemical sedimentary rocks are precipitated either directly by inorganic processes or by the actions of organisms. Chemical rocks include *carbonates, chert,* and *evaporates.*

Carbonate Rocks

Carbonate rocks contain the CO_3^{2-} ion as part of their chemical composition. The two main types of carbonates are limestone and dolomite.

Limestone

Limestone is a sedimentary rock composed mostly of calcite ($CaCO_3$). Limestones are precipitated either by the actions of organisms or directly as the result of inorganic processes. Thus, the two major types of limestone can be classified as either *biochemical* or *inorganic limestone.*

Biochemical limestones are precipitated through the actions of organisms. Most biochemical limestones are formed on continental shelves in warm, shallow water. Biochemical limestone may be precipitated directly as a solid rock in

the core of a reef by corals, encrusting algae, or other shell-forming organisms (figure 6.16). Such a rock would have a crystalline texture and would contain the fossil remains of organisms preserved in growth position.

The great majority of limestones are biochemical limestones formed of wave-broken fragments of algae, corals, and shells. The fragments may be of any size (gravel, sand, silt, and clay) and are often sorted and rounded as they are transported by waves and currents across the sea floor (figure 6.17). The action of these waves and currents and subsequent cementation of these fragments into rock give these limestones a clastic texture. These *bioclastic* (or *skeletal*) *limestones* take a great variety of appearances. They may be relatively coarse-grained with recognizable fossils (figure 6.18) or uniformly fine-grained and

FIGURE 6.16

Corals precipitate calcium carbonate to form limestone in a reef. Water depth about 8 meters (25 feet), San Salvador Island, Bahamas. *Photo by David McGeary*

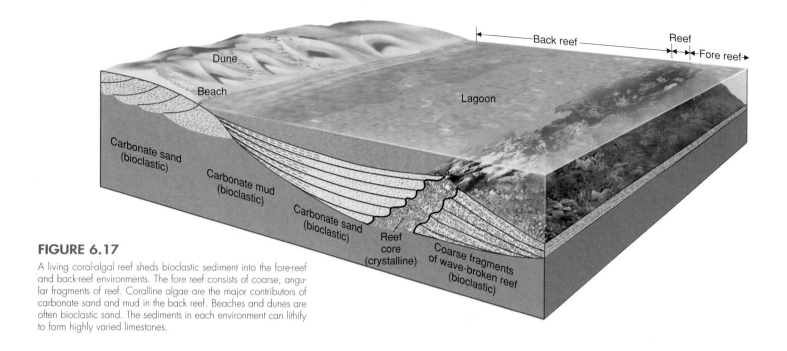

FIGURE 6.17

A living coral-algal reef sheds bioclastic sediment into the fore-reef and back-reef environments. The fore reef consists of coarse, angular fragments of reef. Coralline algae are the major contributors of carbonate sand and mud in the back reef. Beaches and dunes are often bioclastic sand. The sediments in each environment can lithify to form highly varied limestones.

Labels in figure:
Dune
Beach
Back reef
Reef
Fore reef
Lagoon
Carbonate sand (bioclastic)
Carbonate mud (bioclastic)
Carbonate sand (bioclastic)
Reef core (crystalline)
Coarse fragments of wave-broken reef (bioclastic)

FIGURE 6.18

Bioclastic limestones. The two on the left are coarse-grained and contain visible fossils of corals and shells. The limestone on the right consists of fine-grained carbonate mud formed by coralline algae. *Photo by David McGeary*

FIGURE 6.19

Coralline algae on the sea floor in 3 meters of water on the Bahama Banks. The "shaving brush" alga is *Penicillus*, which produces great quantities of fine-grained carbonate mud. *Photo by David McGeary*

FIGURE 6.21

Chalk is a fine-grained variety of bioclastic limestone formed of the remains of microscopic marine organisms that live near the sea surface. *Photo by David McGeary*

FIGURE 6.20

Coquina, a variety of bioclastic limestone, is formed by the cementation of coarse shells. *Photo by David McGeary*

dense from the accumulation of microscopic fragments of coralline algae (figures 6.18 and 6.19). A variety of limestone called *coquina* forms from the cementation of shells and shell fragments that accumulated on the shallow sea floor near shore (figure 6.20). It has a clastic texture and is usually coarse-grained, with easily recognizable shells and shell fragments in it. *Chalk* is a light-colored, porous, very fine-grained variety of bioclastic limestone that forms from the sea-floor accumulation of microscopic marine organisms that drift near the sea surface (figure 6.21).

Inorganic limestones are precipitated directly as the result of inorganic processes. *Oolitic limestone* is a distinctive variety of inorganic limestone formed by the cementation of sand-

sized *ooids,* small spheres of calcite inorganically precipitated in warm, shallow seawater (figure 6.22). Strong tidal currents roll the oolites back and forth, allowing them to maintain a nearly spherical shape as they grow. Wave action may also contribute to their shape.

Oolitic limestone has a clastic texture. *Tufa* and *travertine* are inorganic limestones that form from fresh water. Tufa is precipitated from solution in the water of a continental spring or lake, or from percolating ground water. Travertine may form in caves when carbonate-rich water loses CO_2 to the cave atmosphere. Tufa and travertine both have a crystalline texture; however, tufa is generally more porous, cellular, or open than travertine, which tends to be more dense.

Limestones are particularly susceptible to **recrystallization,** the process by which new crystals, often of the same mineral composition as the original grains, develop in a rock. Calcite grains recrystallize easily, particularly in the presence of water and under the weight of overlying sediment. The new crystals that form are often large and can be easily seen in a rock as light reflects off their broad, flat faces. Because recrystallization often destroys the original clastic texture and fossils of a rock, replacing them with a new crystalline texture, the geologic history of such a rock can be very difficult to determine.

Dolomite

The term **dolomite** (table 6.2) is used to refer to both a sedimentary rock and the mineral that composes it, $CaMg(CO_3)_2$. (Some geologists use *dolostone* for the rock.) Dolomite often forms from limestone as the calcium in calcite is partially replaced by magnesium, usually as water solutions move through the limestone.

$$Mg^{++} + 2CaCO_3 \rightarrow CaMg(CO_3)_2 + Ca^{++}$$

magnesium   calcite   dolomite   calcium
in solution   in solution

A

B

FIGURE 6.22

(A) Aerial photo of underwater dunes of ooids chemically precipitated from seawater on the shallow Bahama Banks, south of Bimini. Tidal currents move the dunes. (B) An oolitic limestone formed by the cementation of ooids (small spheres). Small divisions on scale are 1 millimeter wide. *Photos by David McGeary*

TABLE 6.2	Chemical Sedimentary Rocks		
Inorganic Sedimentary Rocks			
Rock	**Composition**	**Texture**	**Origin**
Limestone	$CaCO_3$	Crystalline Oolitic	May be precipitated directly from seawater. Cementation of oolites (ooids) precipitated chemically from warm shallow seawater (*oolitic* limestone). Also forms in caves as *travertine* and in springs, lakes, or percolating ground water as *tufa*
Dolomite	$CaMg(CO_3)_2$	Crystalline	Alteration of limestone by Mg-rich solutions (usually)
Evaporites			Evaporation of seawater or a saline lake
Rock salt	$NaCl$	Crystalline	
Rock gypsum	$CaSO_4 \cdot 2H_2O$	Crystalline	
Biochemical Sedimentary Rocks			
Rock	**Composition**	**Texture**	**Origin**
Limestone	$CaCO_3$ (calcite)	Clastic or crystalline	Cementation of fragments of shells, corals, and coralline algae (*bioclastic limestone* such as coquina and chalk). Also precipitated directly by organisms in reefs
Chert	SiO_2 (silica)	Crystalline (usually)	Cementation of microscopic marine organisms; rock usually recrystallized

FIGURE 6.23

(A) Chert nodules in limestone near Bluefield, West Virginia. (B) Bedded chert from the coast ranges, California. Camera lens cap (5.5 centimeters) for scale. *Photo A © Parvinder Sethi; Photo B by David McGeary*

Regionally extensive layers of dolomite are thought to form in one of two ways:

1. As magnesium-rich brines created by solar evaporation of seawater trickle through existing layers of limestone.

2. As chemical reactions take place at the boundary between fresh underground water and seawater; the Mg ions could migrate through layers of limestone as sea level rises or falls.

The dolomitization process causes recrystallization of the preexisting limestone, resulting in dolomite rock that is hard and very finely crystalline. Original features such as grain size, fossils, and sedimentary structures are often destroyed during recrystallization, making it difficult to interpret the environment of deposition of the original limestone.

Chert

A hard, compact, fine-grained sedimentary rock formed almost entirely of silica, **chert** occurs in two principal forms—as irregular, lumpy nodules within other rocks and as layered deposits like other sedimentary rocks (figure 6.23). The nodules, often found in limestone, probably formed from inorganic precipitation as underground water replaced part of the original rock with silica. The layered deposits typically form from the accumulation of delicate, glass-like shells of microscopic marine organisms on the sea floor.

Microscopic fossils composed of silica are abundant in some cherts. But because chert is susceptible to recrystallization, the original fossils are easily destroyed, and the origin of many cherts remains unknown.

Evaporites

Rocks formed from crystals that precipitate during evaporation of water are called **evaporites.** They form from the evaporation of seawater or a saline lake (figure 6.24), such as Great Salt Lake in Utah. *Rock gypsum,* formed from the mineral gypsum ($CaSO_4 \cdot 2H_2O$), is a common evaporite. *Rock salt,* composed of the mineral halite (NaCl), may also form if evaporation continues. Other less common evaporites include the borates, potassium salts, and magnesium salts. All evaporites have a crystalline texture. Extensive deposits of rock salt and rock gypsum have formed in the past where shallow, continental seas existed in hot, arid climates. Similarly, modern evaporite deposits are forming in the Persian Gulf and in the Red Sea.

FIGURE 6.24

Salt deposited on the floor of a dried-up desert lake, Bonneville salt flats, Utah. *Photo by Diane Carlson*

Valuable Sedimentary Rocks

Many sedimentary rocks have uses that make them valuable. *Limestone* is widely used as building stone and is also the main rock type quarried for crushed rock for road construction. Pulverized limestone is the main ingredient of cement for mortar and concrete and is also used to neutralize acid soils in the humid regions of the United States. *Coal* is a major fuel, used widely for generating electrical power and for heating. Plaster and plasterboard for home construction are manufactured from *gypsum,* which is also used to stabilize the shrink-swell characteristics of clay-rich soils in some areas. Huge quantities of *rock salt* are consumed by industry, primarily for the manufacture of hydrochloric acid. More familiar uses of rock salt are for table salt and melting ice on roads.

Some *chalk* is used in the manufacture of blackboard chalk, although most classroom chalk is now made from pulverized limestone. The filtering agent for beer brewing and for swimming pools is likely to be made of *diatomite,* an accumulation of the siliceous remains of microscope diatoms.

Clay from *shale* and other deposits supplies the basic material for ceramics of all sorts, from hand-thrown pottery and fine porcelain to sewer pipe. *Sulfur* is used for matches, fungicides, and sulfuric acid; and *phosphates* and *nitrates* for fertilizers are extracted from natural occurrences of special sedimentary rocks (although other sources also are used). Potassium for soap manufacture comes largely from *evaporites,* as does boron for heat-resistant cookware and fiberglass, and sodium for baking soda, washing soda, and soap. *Quartz sandstone* is used in glass manufacturing and for building stone.

Many *metallic ores,* such as the most common iron ores, have a sedimentary origin. The pore space of sedimentary rocks acts as a reservoir for ground water (chapter 11), crude oil, and natural gas. In chapter 21, we take a closer look at these resources and other useful Earth materials.

ORGANIC SEDIMENTARY ROCKS

Coal

Coal is a sedimentary rock that forms from the compaction of plant material that has not completely decayed (figure 6.25). Rapid plant growth and deposition in water with a low oxygen content are needed, so shallow swamps or bogs in a temperate or tropical climate are likely environments of deposition. The plant fossils in coal beds include leaves, stems, tree trunks, and stumps with roots often extending into the underlying shales, so apparently most coal formed right at the place where the plants grew. Coal usually develops from *peat,* a brown, lightweight, unconsolidated or semiconsolidated deposit of plant remains that accumulate in wet bogs. Peat is transformed into coal largely by compaction after it has been buried by sediments.

Partial decay of the abundant plant material uses up any oxygen in the swamp water, so the decay stops and the remaining organic matter is preserved. Burial by sediment compresses the plant material, gradually driving out any water or other volatile compounds. The coal changes from brown to black as the amount of carbon in it increases. Several varieties of coal are recognized on the basis of the type of original plant material and the degree of compaction (see chapter 21).

THE ORIGIN OF OIL AND GAS

Oil and natural gas seem to originate from organic matter in marine sediment. Microscopic organisms, such as diatoms and other single-celled algae, settle to the sea floor and accumulate in marine mud. The most likely environments for this are restricted basins with poor water circulation, particularly on continental shelves. The organic matter may partially decompose, using up the dissolved oxygen in the sediment. As soon as the oxygen is gone, decay stops and the remaining organic matter is preserved.

FIGURE 6.25

Coal bed in the Black Warrior Coal Basin, Alabama. Note the fossil tree stump preserved in place at the top of the coal. *Photo by Bret Bennington*

Continued sedimentation buries the organic matter and subjects it to higher temperatures and pressures, which convert the organic matter to oil and gas. As muddy sediments compact, the gas and small droplets of oil may be squeezed out of the mud and may move into more porous and permeable sandy layers nearby. Over long periods of time, large accumulations of gas and oil can collect in the sandy layers. Both oil and gas are less dense than water, so they generally tend to rise upward through water-saturated rock and sediment. Natural gas represents the end point in petroleum maturation.

Details of the origin of coal, oil, and gas are discussed in chapter 21.

SEDIMENTARY STRUCTURES

Sedimentary structures are features found within sedimentary rock. They usually form during or shortly after deposition of the sediment but before lithification. Structures found in sedimentary rocks are important because they provide clues that help geologists determine the means by which sediment was transported and also its eventual resting place, or environment of deposition. Sedimentary structures may also reveal the orientation, or upward direction, of the deposit, which helps geologists unravel the geometry of rocks that have been folded and faulted in tectonically active regions.

One of the most prominent structures, seen in most large bodies of sedimentary rock, is **bedding,** a series of visible layers within rock (figure 6.26). Most bedding is horizontal because the sediments from which the sedimentary rocks formed were originally deposited as horizontal layers. The principle of **original horizontality** states that most waterlaid sediment is deposited in horizontal or near-horizontal layers that are essentially parallel to Earth's surface. In many cases, this is also true for sediments deposited by ice or wind. If each new layer of sediment buries previous layers, a stack of horizontal layers will develop with the oldest layer on the bottom and the layers becoming younger upward. This is the principle of **superposition.** Sedimentary rocks formed from such sediments preserve the horizontal layering in the form of beds (figure 6.26). A **bedding plane** is a nearly flat surface of deposition separating two layers of rock. A change in the grain size or composition of the particles being deposited, or a pause during deposition, can create bedding planes.

In sandstone, a thicker bed of rock will often consist of a series of thinner, inclined beds called **cross-beds** (figure 6.27). Cross-beds form because in flowing air and water, sand grains move as migrating ripples and dunes. Sand is pushed up the

FIGURE 6.26

Bedding in sandstone and shale, Utah. The horizontal layers formed as one type of sediment buried another in the geologic past. The layers get younger upwards. *Photo by David McGeary*

Sedimentary Rocks: The Key To Mars' Past

Sedimentary rocks on Mars will one day allow planetary geologists to decipher its early history and determine if Mars was once a warmer, wetter planet. Currently, the atmosphere on Mars is too thin and its surface too cold to allow liquid water to exist (see chapter 22). But was Mars wet enough to host lakes and seas long ago? New observations from robotic spacecraft exploring Mars show evidence for extensive deposits of water-lain sedimentary rock. In orbit around Mars, the *Mars Global Surveyor* and *Mars Express* spacecraft have taken thousands of high-resolution photographs, many of which reveal widespread, laterally continuous layers that appear to be sedimentary rock. For example, hundreds of layers of rock are exposed in parts of the walls of the Valles Marineris, a large chasm on Mars that resembles the Grand Canyon but is almost 4,000 kilometers (2,700 miles) long! Recently, robotic rovers have explored two regions of Mars that may have once been covered by liquid water. The Mars Exploration Rover named *Opportunity* landed inside of a small crater with exposures of layered rock and later traversed the Martian surface to enter a larger crater with more layered rock (box figure 1). Detailed photographic and spectrographic analysis of these layered rocks has revealed sedimentary features such as cross-beds, hematite mineral concretions, and the presence of minerals such as jarosite that typically form in water. Halfway around the planet in Gusev Crater, the Mars Exploration Rover named *Spirit* has discovered exposures of bedrock in hills near its landing site. Although most rocks in this area appear to be of volcanic origin, some have characteristics that indicate they have been chemically and texturally altered by exposure to liquid water.

Although the accomplishments of these robotic geologists are remarkable, confirmation of the presence of water-deposited sedimentary rock on Mars will have to wait until samples can be returned to Earth for detailed study or until human geologists arrive on Mars.

Scientists are particularly excited by the prospect of Mars being a wet planet early in its history because of the possibility that life could have evolved there. If, as now seems likely, the early history of Mars is recorded in layers of sedimentary rock, perhaps fossils of Martian microbes remain to be discovered. Mars continues to be the most promising place to look for evidence of extraterrestrial life in our solar system.

Additional Resources

M. C. Malin and K. S. Edgett. 2000. Sedimentary rocks of early Mars. *Science* 290 (5498): 1927–37.

Information about the Mars exploration program at NASA, including images and updates from ongoing missions such as the Mars Global Surveyor and the Mars Exploration Rovers, is available from the Jet Propulsion Laboratory/NASA Mars Program website.

- http://marsprogram.jpl.nasa.gov/

Visit the European Space Agency's Mars Express website for information about this ongoing mission.

- http://www.esa.int/SPECIALS/Mars_Express/

Visit the Malin Space Science Systems website for an extensive collection of archived Mars Orbiter Camera images.

- www.msss.com

BOX 6.2 ■ FIGURE 1

Layers of sedimentary rock exposed inside the rim of Endurance Crater photographed by the Mars Exploration Rover *Opportunity. Photo by NASA/JPL/Cornell*

FIGURE 6.27

Cross-bedded sandstone in Zion National Park, Utah. Note how the thin layers have formed at an angle to the more extensive bedding planes (also tilted) in the rock. This cross-bedding was formed in sand dunes deposited by the wind. *Photo by David McGeary*

shallow side of the ripple to the crest, where it then avalanches down the steep side, forming a cross-bed. Cross-beds form one after the other as the ripple migrates downstream (figure 6.28). Ripples can also be preserved on the surface of a bed of sandstone, forming **ripple marks,** if they are buried by another layer of sediment (figure 6.29). Ripple marks produced by currents flowing in a single direction are asymmetrical (as discussed previously and in figures 6.29*B* and *D*). In waves, water moves back and forth, producing symmetrical wave ripples (figures 6.29*A* and *C*). Ripple marks and cross-beds can form in conglomerates, sandstones, siltstones, and limestones, and in environments such as deserts, river channels, river deltas, and shorelines.

A **graded bed** is a layer with a vertical change in particle size, usually from coarse grains at the bottom of the bed to progressively finer grains toward the top (figure 6.30). A single bed may have gravel at its base and grade upward through sand and silt to fine clay at the top. A graded bed may build up as

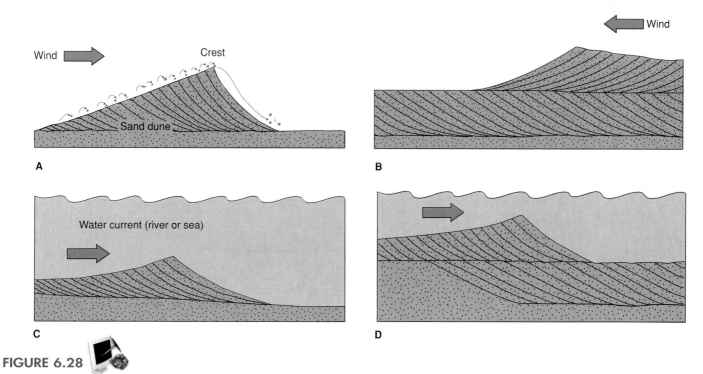

FIGURE 6.28

The development of cross-beds in wind-blown sand (*A* and *B*) and water-deposited sand (*C* and *D*). (*A*) Sand grains migrate up the shallow side of the dune and avalanche down the steep side, forming cross-beds. (*B*) Second layer of cross-beds forms as wind shifts and a dune migrates from the opposite direction. (*C*) Underwater current deposits cross-beds as ripple migrates downstream. (*D*) Continued deposition and migration of ripples produces multiple layers of cross-beds.

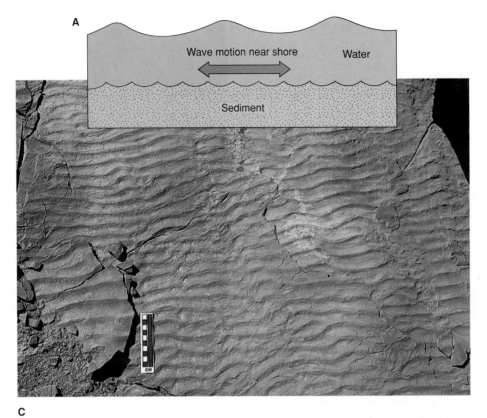

A

Wave motion near shore

Water

Sediment

C

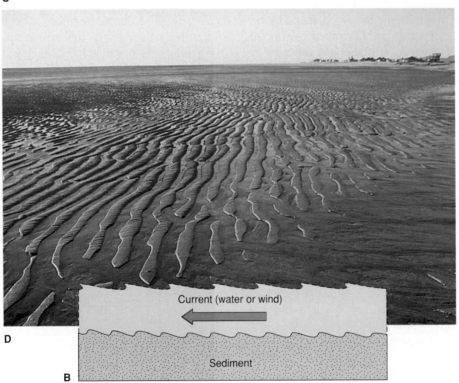

Current (water or wind)

D

Sediment

B

FIGURE 6.29

Development of ripple marks in loose sediment. (*A*) Symmetric ripple marks form beneath waves. (*B*) Asymmetric ripple marks, forming beneath a current, are steeper on their down-current sides. (*C*) Ripple marks on a bedding plane in sandstone, Capitol Reef National Park, Utah. Scale in centimeters. (*D*) Current ripples in wet sediment of a tidal flat, Baja California. *Photo C by David McGeary; Photo D by Frank M. Hanna*

FIGURE 6.30

A graded bed has coarse grains at the bottom of the bed and progressively finer grains toward the top. Coin for scale. *Photo by David McGeary*

sediment is deposited by a gradually slowing current. This seems particularly likely to happen during deposition from a *turbidity current* on the deep-sea floor. Figure 6.31 shows the development of a graded bed by turbidity-current deposition.

Mud cracks are a polygonal pattern of cracks formed in very fine-grained sediment as it dries (figure 6.32). Because drying requires air, mud cracks form only in sediment exposed above water. Mud cracks may form in lake-bottom sediment as the lake dries up; in flood-deposited sediment as a river level drops; or in marine sediment exposed to the air, perhaps temporarily by a falling tide. Cracked mud can lithify to form shale, preserving the cracks. The filling of mud cracks by sand can form casts of the cracks in an overlying sandstone.

FOSSILS

Fossils are the remains of organisms preserved in sedimentary rock. Most sedimentary layers contain some type of fossil and some limestones are composed entirely of fossils. Most fossils are preserved by the rapid burial in sediment of bones, shells, or teeth, which are the mineralized hard parts of animals most resistant to decay (figure 6.33). The original bone or shell is rarely preserved unaltered; the original mineral is often recrystallized or replaced by a different mineral such as pyrite or silica. Bone and wood may be *petrified* as organic material is replaced and pore spaces filled with mineral. Shells entombed within rock are commonly dissolved away by pore waters, leaving only impressions or *molds* of the original fossil. Leaves and undecayed organic tissue can also be preserved as a thin film of carbon (figure 6.34*A*). *Trace fossils* are a type of sedimentary structure produced by the impact of an organisms's activities on the sediment. Footprints, trackways, and burrows are the most common trace fossils (figure 6.34*B*).

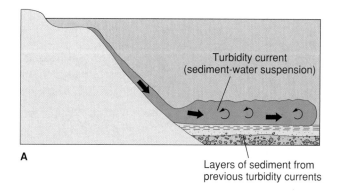

A

Turbidity current (sediment-water suspension)

Layers of sediment from previous turbidity currents

Sediment-laden turbidity current flows beneath clear water

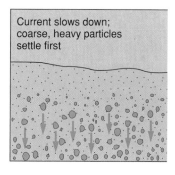

Current slows down; coarse, heavy particles settle first

Main body of current comes to rest

Fine-grained "tail" of turbidity current continues to flow, adding fine-grained sediment to top of deposit

Progressively finer sediments settle on top of coarse particles

B

A graded bed

FIGURE 6.31

Formation of a graded bed by deposition from a turbidity current. (*A*) Slurry of sediment and water moves downslope along the sea floor. (*B*) As the turbidity flow slows down, larger grains are deposited first, followed by progressively finer grains, to produce a graded bed.

Many *paleontologists* study fossils to learn about the evolution of life on Earth, but fossils are also very useful for interpreting depositional environments and for reconstructing the climates of the past. Fossils can be used to distinguish fresh water from marine environments and to infer the water depth at which a particular sedimentary layer was deposited. Tropical, temperate, and arid climates can be associated with distinctive

A

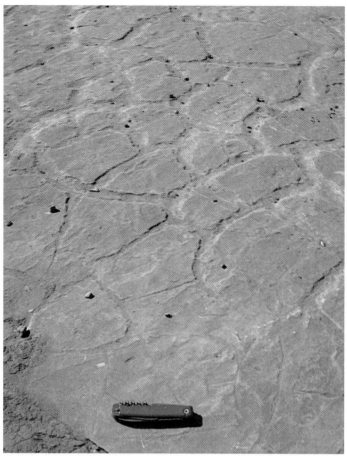

B

FIGURE 6.32

(A) Mud cracks in recently dried mud. (B) Mud cracks preserved in shale; they have been partially filled with sediment. *Photos by David McGeary*

FIGURE 6.33

Fossil clams, brachiopods, and trilobites in the Hamilton Shale of New York. Some of the fossils have their original shell material, other fossils are preserved as impressions. *Photo by Bret Bennington*

types of fossil plants. Marine *microfossils,* the tiny shells produced by ocean-dwelling plankton, can be analyzed to determine the water temperature that surrounded the shell when it formed. Much of our detailed knowledge of Earth's climate changes over the last 150 million years has come from the study of microfossils extracted from layers of mud deposited on the deep-ocean floor.

FORMATIONS

A **formation** is a body of rock of considerable thickness that is large enough to be mappable, and with characteristics that distinguish it from adjacent rock units. Although a formation is usually composed of one or more beds of sedimentary rock, units of metamorphic and igneous rock are also called formations. It is a convenient unit for mapping, describing, or interpreting the geology of a region.

Formations are often based on rock type. A formation may be a single thick bed of rock such as sandstone. A sequence of several thin sandstone beds could also be called a formation, as could a sequence of alternating limestone and shale beds.

The main criterion for distinguishing and naming a formation is some visible characteristic that makes it a recognizable

B

FIGURE 6.34

(A) Fossil fish in a rock from western Wyoming. (B) Dinosaur footprint in shale, Tuba City, Arizona. Scale in centimeters. *Photo A by U.S. Geological Survey; Photo B by David McGeary*

unit. This characteristic may be rock type or sedimentary structures or both. For example, a thick sequence of shale may be overlain by basalt flows and underlain by sandstone. The shale, the basalt, and the sandstone are each a different formation. Or a sequence of thin limestone beds, with a total thickness of many tens of meters, may have recognizable fossils in the lower half and distinctly different fossils in the upper half. The limestone sequence is divided into two formations on the basis of its fossil content.

Formations are given proper names: the first name is often a geographic location where the rock is well exposed, and the second the name of a rock type, such as Navajo Sandstone, Austin Chalk, Baltimore Gneiss, Onondaga Limestone, or Chattanooga Shale. If the formation has a mixture of rock types, so that one rock name does not accurately describe it, it is called simply "formation," as in the Morrison Formation or the Martinsburg Formation.

A **contact** is the boundary surface between two different rock types or ages of rocks. In sedimentary rock formations, the contacts are usually bedding planes.

Figure 6.35 shows the three formations that make up the upper part of the canyon walls in Grand Canyon National Park in Arizona. The contacts between formations are also shown.

INTERPRETATION OF SEDIMENTARY ROCKS

Sedimentary rocks are important in interpreting the geologic history of an area. Geologists examine sedimentary formations to look for clues such as fossils; sedimentary structures; grain shape, size, and composition; and the overall shape and extent of the formation. These clues are useful in determining the source area of the sediment, environment of deposition, and the possible plate-tectonic setting at the time of deposition.

Source Area

The **source area** of a sediment is the locality that eroded and provided the sediment. The most important things to determine about a source area are the type of rocks that were exposed in it and its location and distance from the site of eventual deposition.

The *rock type* exposed in the source area determines the character of the resulting sediment. The composition of a sediment can indicate the source area rock type, even if the source area has been completely eroded away. A conglomerate may contain cobbles of basalt, granite, and chert; these rock types were obviously in its source area. An arkose containing coarse feldspar, quartz, and biotite may have come from a granitic source area. Furthermore, the presence of feldspar indicates the source area was not subjected to extensive chemical weathering and that erosion probably took place in an arid environment with high relief. A quartz sandstone containing well-rounded quartz grains, on the other hand, probably represents the erosion and deposition of quartz grains from preexisting sandstone. Quartz is a hard, tough mineral very resistant to rounding by abrasion, so if quartz grains are well-rounded, they have undergone many cycles of erosion, transportation, and deposition, probably over tens of millions of years.

Sedimentary rocks are also studied to determine the *direction* and *distance* to the source area. Figure 6.36 shows how several characteristics of sediment may vary with distance from a source area. Many sediment deposits get thinner away from the source, and the sediment grains themselves usually become finer and more rounded.

Sedimentary structures often give clues about the directions of ancient currents (*paleocurrents*) that deposited sediments. Refer back to figure 6.27 and notice how cross-beds slope downward in the direction of current flow. Ancient

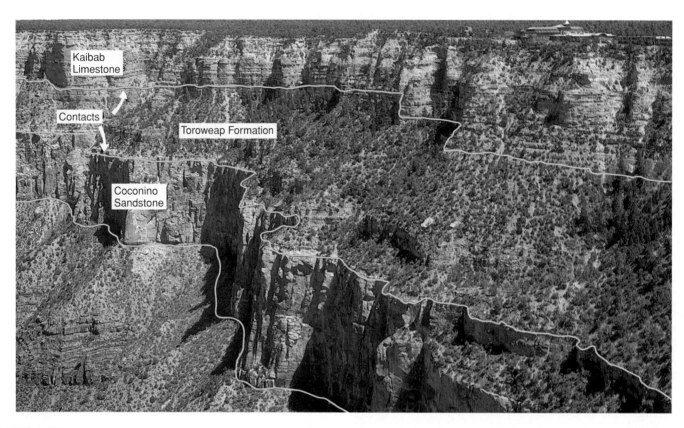

FIGURE 6.35

The upper three formations in the cliffs of the Grand Canyon, Arizona. The Kaibab Limestone and the Coconino Sandstone are resistant in the dry climate and form cliffs. The Toroweap Formation contains some shale and is less resistant, forming slopes. The tan lines are drawn to show the boundaries between the formations. *Photo by David McGeary*

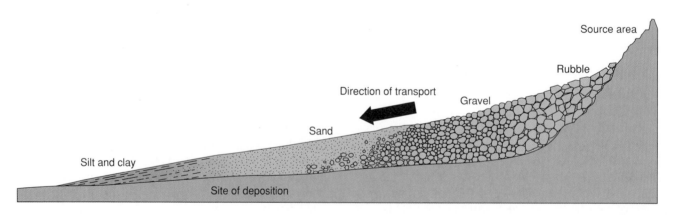

FIGURE 6.36

Sediment deposits often become thinner away from the source area, and sediment grains usually become finer and more rounded. The rocks that form from these sediments would change with distance from the source area from breccia to conglomerate to sandstone to shale. See appendix F for rock symbols.

current direction can also be determined from asymmetric ripple marks (figure 6.29C and D).

Figure 6.37 shows how three of these characteristics were used to determine the location of the source area for a particular rock unit in the southwestern United States. The unit is the Salt Wash Member of the Morrison Formation (a *member* is a subdivision of a formation). It is an important rock unit, for it contains a great deal of uranium, deposited within the rock by ground water long after the rock formed. The unit thickens and coarsens to the southwest, and cross-beds show that the old currents that deposited the sediment came from rivers that flowed from the southwest. These three facts strongly suggest that the source area was to the southwest. This information helps exploration geologists search for uranium within the Salt

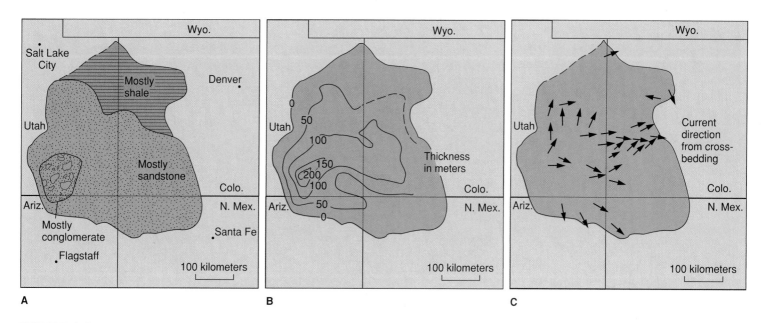

FIGURE 6.37

Characteristics of the Salt Wash Member of the Morrison Formation that help locate its source area. (A) The sediment grains become coarser to the southwest. (B) The deposit becomes thicker to the southwest. The contour lines show the thickness of the Salt Wash Member in meters. (C) Cross-bedding shows that the depositing currents came mostly from the southwest (arrows point down current). *Redrawn and simplified from Craig and others, 1955, U.S. Geological Survey Bulletin*

Wash Member. The Morrison Formation is also world famous for its dinosaur fossils.

Environment of Deposition

Figure 6.38 shows the common environments in which sediments are deposited. Geologists study modern environments in great detail so that they can interpret ancient rocks. Clues to the ancient environment of deposition come from a rock's composition, the size and sorting and rounding of the grains, the sedimentary structures and fossils present, and the vertical sequence of the sedimentary layers.

Continental environments include alluvial fans, river channels, flood plains, lakes, and dunes. Sediments deposited on land are subject to erosion, so they often are destroyed. The great bulk of sedimentary rocks comes from the more easily preserved shallow marine environments, such as deltas, beaches, lagoons, shelves, and reefs. The characteristics of major environments are covered in detail in later chapters (10, 12–14, 18). In this section, we describe the main sediment types and sedimentary structures found in each environment.

Glacial Environments

Glacial ice often deposits narrow ridges and layers of sediment in valleys and widespread sheets of sediment on plains. Glacial sediment (*till*) is an unsorted mix of unweathered boulders, cobbles, pebbles, sand, silt, and clay. The boulders and cobbles may be scratched from grinding over one another under the great weight of the ice.

Alluvial Fan

As streams emerge from mountains onto flatter plains, they deposit broad, fan-shaped piles of sediment. The sediment often consists of coarse, arkosic sandstones and conglomerates, marked by coarse cross-bedding and lens-like channel deposits (figure 6.39).

River Channel and Flood Plain

Rivers deposit elongate lenses of conglomerate or sandstone in their channels (figure 6.40). The sandstones may be arkoses or may consist of sand-sized fragments of fine-grained rocks. River channel deposits typically contain cross-beds and current ripple marks. Broad, flat flood plains are covered by periodic floodwaters, which deposit thin-bedded shales characterized by mud cracks and fossil footprints of animals. Hematite may color flood-plain deposits red.

Lake

Thin-bedded shale, perhaps containing fish fossils, is deposited on lake bottoms. If the lake periodically dries up, the shales will be mud-cracked and perhaps interbedded with evaporites such as gypsum or rock salt.

Delta

A delta is a body of sediment deposited when a river flows into standing water, such as the sea or a lake. Most deltas contain a great variety of subenvironments but are generally made up of thick sequences of siltstone and shale, marked by low-angle cross-bedding and cut by coarser channel deposits. Delta

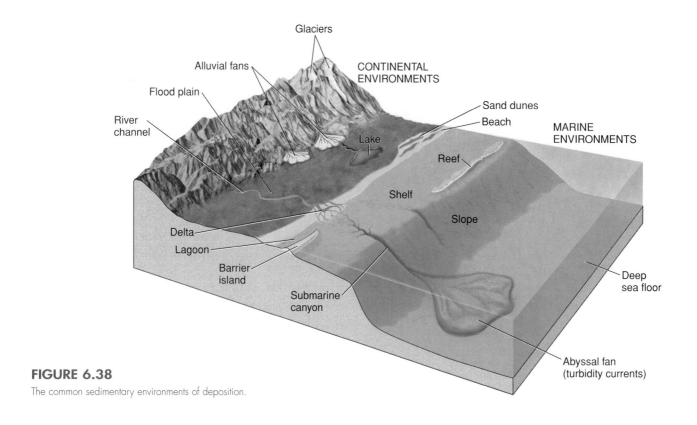

FIGURE 6.38

The common sedimentary environments of deposition.

FIGURE 6.39

Alluvial fan deposits, Baja California. A channel deposit of conglomerate occurs within the coarse-grained sequence. *Photo by David McGeary*

Geologist's View

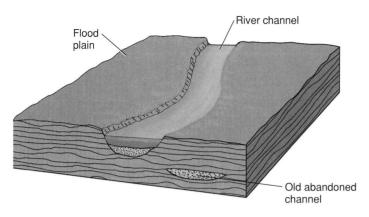

Flood plain

River channel

Old abandoned channel

FIGURE 6.40

A river deposits an elongate lens of sand and gravel in its channel. Fine-grained silt and clay are deposited beside the channel on the river's flood plain.

sequences may contain beds of peat or coal, as well as marine fossils such as clam shells.

Beach, Barrier Island, Dune

A barrier island is an elongate bar of sand built by wave action. Well-sorted quartz sandstone with well-rounded grains is deposited on beaches, barrier islands, and dunes. Beaches and barrier islands are characterized by cross-bedding (often low-angle) and marine fossils. Dunes have both high-angle and low-angle cross-bedding and occasionally contain fossil footprints of land animals such as lizards. All three environments can also contain carbonate sand in tropical regions, thus yielding cross-bedded clastic limestones.

Lagoon

A semienclosed, quiet body of water between a barrier island and the mainland is a lagoon. Fine-grained dark shale, cut by tidal channels of coarse sand and containing fossil oysters and other marine organisms, is formed in lagoons. Limestones may also form in lagoons adjacent to reefs (see figure 6.17).

Shallow Marine Shelves

On the broad, shallow shelves adjacent to most shorelines, sediment grain size decreases offshore. Widespread deposits of sandstone, siltstone, and shale can be deposited on such shelves. The sandstone and siltstone contain symmetrical ripple marks, low-angle cross-beds, and marine fossils such as clams and snails. If fine-grained *tidal flats* near shore are alternately covered and exposed by the rise and fall of tides, mud-cracked marine shale will result.

Reefs

Massive limestone is deposited in reef cores, with steep beds of limestone breccia forming seaward of the reef, and horizontal beds of sand-sized and finer-grained limestones forming landward (see figure 6.17). All these limestones are full of

fossil fragments of corals, coralline algae, and numerous other marine organisms.

Deep Marine Environments

On the deep-sea floor are deposited shale and graywacke sandstones. The graywackes are deposited by turbidity currents (figure 6.31) and typically contain graded bedding and current ripple marks.

WEB BOX 6.3

Transgression and Regression

View and understand how sequences of marine sedimentary rocks are deposited with the rise (transgression) and fall (regression) of sea level.

For the full story, go to:
www.mhhe.com/plummer11e

Plate Tectonics and Sedimentary Rocks

The dynamic forces that move plates on Earth are also responsible for the distribution of many sedimentary rocks. As such, the distribution of sedimentary rocks often provides information that helps geologists reconstruct past plate-tectonic settings.

In tectonically active areas, particularly along *convergent plate boundaries,* the thickening of the crust that forms a mountain belt also cause the adjacent crust to subside, forming basins (figure 6.41). Rapid erosion of the rising mountains produces enormous quantities of sediment that are transported by streams and turbidity currents to the adjacent basins. Continued subsidence of the basins results in the formation of great thicknesses of sedimentary rock that record the history of uplift and erosion in the mountain belt. For example, uplift of the ancestral Sierra Nevada and Klamath mountain ranges in California is recorded by the thick accumulation of turbidite deposits preserved in basins to the west of the mountains. There, graywacke sandstone deposited by turbidity currents contains mainly volcanic clasts in the lower part of the sedimentary sequence and abundant feldspar clasts in the upper part of the sequence. This indicates that a cover of volcanic rocks was first eroded from the ancestral mountains, and then, as uplift and erosion continued, the underlying plutonic rocks were exposed and eroded. Other eroded mountains, such as the Appalachians, have left similar records of uplift and erosion in the sedimentary record.

It is not uncommon for rugged mountain ranges, such as the Canadian Rockies, European Alps, and Himalayas, that stand several thousand meters above sea level to contain

FIGURE 6.41

Sedimentary basins associated with convergent plate boundary include a forearc basin on the oceanward side that contains mainly clastic sediments deposited by streams and turbidity currents from an eroding magmatic arc. Toward the craton (continent), a foreland basin also collects clastic sediment derived from the uplifted mountain belt and craton.

sedimentary rocks of marine origin that were originally deposited below sea level. The presence of marine sedimentary rocks such as limestone, chert, and shale containing marine fossils at high elevations attests to the tremendous uplift associated with mountain building at convergent plate boundaries (see chapter 20).

Transform plate boundaries are also characterized by rapid rates of erosion and deposition of sediments as fault-bounded basins open and subside rapidly with continued plate motion. Because of the rapid rate of deposition and burial of organic material, fault-bounded basins are good places to explore for petroleum. Many of the petroleum occurrences in California are related to basins that formed as the San Andreas transform fault developed.

A *divergent plate boundary* may result in the splitting apart of a continent and formation of a new ocean basin. In the initial stages of continental divergence, a rift valley forms and fills with thick wedges of gravel and coarse sand along its fault-bounded margins; lake bed deposits and associated evaporite rocks may form in the bottom of the rift valley (figure 6.42). In the early stages, continental rifts will have extensive volcanics that contribute to the sediments in the rift. The Red Sea and adjacent East African Rift Zone have good examples of the features and sedimentary rocks formed during the initial stages of continental rifting.

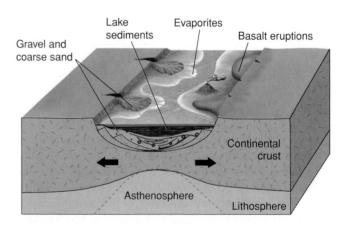

FIGURE 6.42

Divergent plate boundary showing thick wedges of gravel and coarse sand along fault-bounded margins of developing rift valley. Lake bed deposits and evaporite rocks are located on the floor of the rift valley. Refer to figure 19.25 for more detail of faulted margin and sediments deposited along a rifted continental margin.

SUMMARY

Sediment forms by the weathering and erosion of preexisting rocks and by chemical precipitation, sometimes by organisms.

Gravel, sand, silt, and *clay* are sediment particles defined by grain size.

The composition of sediment is governed by the rates of chemical weathering, mechanical weathering, and erosion. During transportation, grains can become rounded and sorted.

Sedimentary rocks form by *lithification* of sediment, by *crystallization* from solution, or by consolidation of remains of organisms. Sedimentary rocks may be *detrital, chemical,* or *organic.*

Detrital sedimentary rocks form mostly by *compaction* and *cementation* of grains. *Matrix* can partially fill the *pore space* of clastic rocks.

Conglomerate forms from coarse, rounded sediment grains that often have been transported only a short distance by a river or waves. *Sandstone* forms from sand deposited by rivers, wind, waves, or turbidity currents. *Shale* forms from river, lake, or ocean mud.

Limestone consists of calcite, formed either as a chemical precipitate in a reef or, more commonly, by the cementation of shell and coral fragments or of ooids. *Dolomite* usually forms from the alteration of limestone by magnesium-rich solutions.

Chert consists of silica and usually forms from the accumulation of microscopic marine organisms. *Recrystallization* often destroys the original texture of chert (and some limestones).

Evaporites, such as rock salt and gypsum, form as water evaporates. *Coal,* a major fuel, is consolidated plant material.

Sedimentary rocks are usually found in *beds* separated by *bedding planes* because the original sediments are deposited in horizontal layers.

Cross-beds and *ripple marks* develop as moving sediment forms ripples and dunes during transport by wind, underwater currents, and waves.

A *graded bed* forms as coarse particles fall from suspension before fine particles due to decreasing water flow velocity (perhaps in a turbidity current).

Mud cracks form in drying mud.

Fossils are the traces of an organism's hard parts or tracks preserved in rock.

A *formation* is a convenient rock unit for mapping and describing rock. Formations are lithologically distinguishable from adjacent rocks; their boundaries are *contacts.*

Geologists try to determine the *source area* of a sedimentary rock by studying its grain size, composition, and sedimentary structures. The source area's rock type and location are important to determine.

The *environment of deposition* of a sedimentary rock is determined by studying bed sequence, grain composition and rounding, and sedimentary structures. Typical environments include alluvial fans, river channels, flood plains, lakes, dunes, deltas, beaches, shallow marine shelves, reefs, and the deep-sea floor.

Plate tectonics plays an important role in the distribution of sedimentary rocks; the occurrence of certain types of sedimentary rocks is used by geologists to construct past plate-tectonic settings.

Terms to Remember

alluvial fan 147	deposition 142	recrystallization 151
bedding 155	detrital sedimentary rock 145	ripple mark 157
bedding plane 155	dolomite 151	rounding 141
cement 144	environment of deposition 142	sand 141
cementation 144	evaporite 153	sandstone 147
chemical sedimentary rock 145	formation 160	sediment 141
chert 153	fossil 159	sedimentary breccia 145
clastic texture 144	graded bed 157	sedimentary rock 145
clay 141	gravel 141	sedimentary structure 155
coal 154	limestone 149	shale 148
compaction 144	lithification 143	silt 141
conglomerate 145	matrix 147	sorting 142
contact 161	mud crack 159	source area 161
cross-beds 155	organic sedimentary rock 145	superposition 155
crystalline texture 144	original horizontality 155	transportation 141
crystallization 144	pore space 144	turbidity current 147

Testing Your Knowledge

Use the following questions to prepare for exams based on this chapter.

1. Quartz is a common mineral in sandstone. Under certain circumstances, feldspar is common in sandstone, even though it normally weathers rapidly to clay. What conditions of climate, weathering rate, and erosion rate could lead to a feldspar-rich sandstone? Explain your answer.

2. Describe with sketches how wet mud compacts before it becomes shale.

3. What do mud cracks tell about the environment of deposition of a sedimentary rock?

4. How does a graded bed form?

5. List the detrital sediment particles in order of decreasing grain size.

6. How does a sedimentary breccia differ in appearance and origin from a conglomerate?

7. Describe three different origins for limestone.

8. How does dolomite usually form?

9. What is the origin of coal?

10. Sketch the cementation of sand to form sandstone.

11. How do evaporites form? Name two evaporites.

12. Name the three most common sedimentary rocks.

13. What is a formation?

14. Explain two ways that cross-bedding can form.

15. Particles of sediment from 1/16 to 2 millimeters in diameter are of what size?
    a. gravel   b. sand
    c. silt   d. clay

16. Rounding is
    a. the rounding of a grain to a spherical shape
    b. the grinding away of sharp edges and corners of rock fragments during transportation
    c. a type of mineral
    d. none of the preceding

17. Compaction and cementation are two common processes of
    a. erosion   b. transportation
    c. deposition   d. lithification

18. Which is not a chemical or organic sedimentary rock?
    a. rock salt   b. shale
    c. limestone   d. gypsum

19. The major difference between breccia and conglomerate is
    a. size of grains   b. rounding of the grains
    c. composition of grains   d. all of the preceding

20. Which is not a type of sandstone?
    a. quartz sandstone   b. arkose
    c. graywacke   d. coal

21. Shale differs from mudstone in that
    a. shale has larger grains
    b. shale is visibly layered and fissile; mudstone is massive and blocky
    c. shale has smaller grains
    d. there is no difference between shale and mudstone

22. The chemical element found in dolomite not found in limestone is
    a. Ca   b. Mg
    c. C   d. O
    e. Al

23. In a graded bed, the particle size decreases
    a. upward
    b. downward
    c. in the direction of the current
    d. particle size stays the same

24. A body or rock of considerable thickness with characteristics that distinguish it from adjacent rock units is called a/an
    a. formation   b. contact
    c. bedding plane   d. outcrop

25. If sea level drops or the land rises, what is likely to occur?
    a. a flood   b. a regression
    c. a transgression   d. no geologic change will take place

26. Thick accumulations of graywacke and volcanic sediments can indicate an ancient
    a. divergent plate boundary
    b. convergent boundary
    c. transform boundary

27. A sedimentary rock made of fragments of preexisting rocks is
    a. organic   b. chemical
    c. clastic

28. Clues to the nature of the source area of sediment can be found in
    a. the composition of the sediment
    b. sedimentary structures
    c. rounding of sediment
    d. all of the preceding

Expanding Your Knowledge

1. How might graded bedding be used to determine the tops and bottoms of sedimentary rock layers in an area where sedimentary rock is no longer horizontal? What other sedimentary structures can be used to determine the tops and bottoms of tilted beds?

2. Which would weather faster in a humid climate, a quartz sandstone or an arkose? Explain your answer.

3. A cross-bedded quartz sandstone may have been deposited as a beach sand or as a dune sand. What features could you look for within the rock to tell if it had been deposited on a beach? On a dune?

4. Why is burial usually necessary to turn a sediment into a sedimentary rock?

5. Why are most beds of sedimentary rock formed horizontally?

6. Discuss the role of sedimentary rocks in the rock cycle, diagramming the rock cycle as part of your answer. What do sedimentary rocks form from? What can they turn into?

Exploring Web Resources

www.mhhe.com/plummer11e

This is the dedicated website for this book. You can go to it for new and updated information. The universal resource locators (URL) listed in this book are also given as links on the website, making it easy to go to those websites without typing in the URL. When you visit our Online Learning Center, go to the Student Center and then to the chapter of interest. Here you will find additional readings and media resources, as well as answers to the Testing Your Knowledge section, more quizzes, animation and video clips, and direct links to the sites listed in this book. Links to additional websites can also be found. We have added questions for some of the sites to allow you to get the most of your exploration of the web. Using the web is an enjoyable way of enhancing your knowledge of geology.

http://darkwing.uoregon.edu/~dogsci/dorsey/SedResources.html

Web Resources for Sedimentary Geology site contains a comprehensive listing of resources available on the worldwide web.

www.lib.utexas.edu/Libs/GEO/FolkReady/TitlePage.html

Online version of *Petrology of Sedimentary Rocks* by Professor Robert Folk at the University of Texas at Austin.

http://walrus.wr.usgs.gov/seds/

Visit the *U.S. Geological Survey Bedform and Sedimentology* site for computer and photographic images and movies of sedimentary structures.

http://zircon.geology.union.edu/Gildner/stack.html

For a virtual field trip of the Ordovician-age rocks in the Mohawk Valley of New York state, visit Union College's geology website.

www.palaeo.de/edu/JRP/index.html

For a virtual field trip of the Jurassic Park Reef in Germany. Learn how fossil reefs preserved in sedimentary rocks are used to interpret past ecosystems.

 # Animations

This chapter includes the following animations available on our Online Learning Center at www.mhhe.com/plummer11e.

6.28 Migration of sand grains to form ripples, dunes, and crossbeds

6.31 Formation of graded bed

Metamorphism, Metamorphic Rocks, and Hydrothermal Rocks

This chapter on metamorphic rocks, the third major category of rocks in the rock cycle, completes our description of Earth materials (rocks and minerals). The information on igneous and sedimentary processes in previous chapters should help you understand metamorphic rocks, which form from *preexisting* rocks.

Photo taken through a polarizing microscope of a metamorphic rock that was once shale. Micas (brightly colored crystals) grew while the rock was being folded during regional metamorphism. The area shown is approximately 2 centimeters wide. *Photo by C. C. Plummer*

After reading chapter 5 on weathering, you know how rocks are altered when exposed at Earth's surface. *Metamorphism* (a word from Latin and Greek that means literally "changing of form") also involves alterations, but the changes are due to deep burial, tectonic forces, and/or high temperature rather than surface conditions.

Because nearly all metamorphic rocks form deep within the Earth's crust, they provide geologists with many clues about conditions at depth. Therefore, understanding metamorphism will help you when we consider geologic processes involving Earth's internal forces. Metamorphic rocks are a feature of the oldest exposed rocks of the continents and of major mountain belts. They are especially important in providing evidence of what happens during subduction and plate convergence.

We also discuss hydrothermally deposited rocks and minerals, which are usually found in association with both igneous and metamorphic rocks. Hydrothermal ore deposits, while not volumetrically significant, are of great importance to the world's supply of metals.

Relationships to Earth Systems

Metamorphism takes place at depth in the solid Earth, so it involves no interaction between the Earth systems at the surface of Earth. However, water is important in metamorphic processes. Water from the hydrosphere seeps through cracks and pores in rocks and may penetrate to at least the shallower depths where metamorphism is taking place. Water is also incorporated in minerals that form during igneous and sedimentary processes. When rock containing these minerals is subducted or otherwise carried to depth and heated, the water is driven out of these minerals, affecting the metamorphic process. Cold water going downward and heated cycles upward as hot water. Hot water rising through rock is important for creating important metallic ore deposits, the hydrothermal deposits discussed toward the end of this chapter. Copper, lead, gold, and other metals are mined from these ore deposits and profoundly affect the *biosphere*, most notably the human part of the biosphere. But the mining, processing, and disposal of mined metals can adversely affect other living things.

The *atmosphere* may have been altered by metamorphism in the geologic past. When the word's highest mountain chain, the Himalaya, began forming about 60 million years ago, huge quantities of carbon dioxide were released during metamorphism, according to one hypothesis. The Himalayan mountain belt is a product of collision of India with Asia (described in chapter 20). Before the collision, great thicknesses of limestone and other sedimentary rocks built up on the ocean floors separating the landmasses. Upon collision, the sedimentary layers crumpled and portions were deeply buried. Heat and pressure metamorphosed these rocks. Calcite reacted with quartz and other silicate minerals to produce new minerals as well as carbon dioxide gas. It is estimated that several hundred million tons of CO_2 per year were released into the atmosphere over 10 million years. The amount of CO_2 added to the atmosphere would have contributed greatly to the greenhouse effect and would account for the warmer climate inferred for that part of Earth's history.

INTRODUCTION

From your study so far of Earth materials and the rock cycle, you know that rocks change, given enough time, when their physical environment changes radically. In chapter 3, you saw how deeply buried rocks melt (or partially melt) to form magma when temperatures are high enough. What happens to rocks that are deeply buried but are not hot enough to melt? They become metamorphosed. **Metamorphism** refers to changes to rocks that take place in Earth's interior. The changes may be new textures, new mineral assemblages, or both. Trans-

formations occur in the *solid state* (meaning the rock does not melt). The new rock is a **metamorphic rock.**

The conversion of a slice of bread to toast is a solid-state process analogous to metamorphism of rock. When the bread (think "sedimentary rock") is heated, it converts to toast (think "metamorphic rock"). The toast is texturally and compositionally different from its parent material, bread. Although the rock remains solid during metamorphism, it is important to recognize that fluids, notably water, often play a significant role in the metamorphic process.

As most metamorphism takes place at moderate to great depths in Earth's crust, metamorphic rocks provide us with a window to processes that take place deep underground, beyond our direct observation. Metamorphic rocks are exposed over large regions because of erosion of mountain belts and its accompanying uplift due to isostatic adjustment (the vertical movement of a portion of Earth's crust to achieve balance, described in chapter 1). In fact, the cores of the continents are largely metamorphic rocks and granitic plutons. As described in chapter 20 on mountain belts and the continental crust, these form the stable interior of North America, the central lowlands between the Appalachians and the Rocky Mountains, and other ranges of western North America. Very ancient (Precambrian) complexes of metamorphic and intrusive igneous rocks are exposed over much of Canada (known as the *Canadian Shield*). The inside front cover shows the Canadian Shield as the region underlain by Precambrian rocks. In the Great Plains of the United States, similar rocks form the *basement* underlying a veneer of younger sedimentary rocks (see the brown area on the inside front cover map that the legend indicates is "Platform deposits on Precambrian basement"). Similar ancient metamorphic and plutonic rocks form the stable cores, or *cratons,* of the other continental landmasses (e.g., Africa, Antarctica, Australia) as well.

In nearly all cases, a metamorphic rock has a texture clearly different from that of the *original* rock, or **parent rock.** When limestone is metamorphosed to marble, for example, the fine grains of calcite coalesce and recrystallize into larger calcite crystals. The calcite crystals are interlocked in a mosaic pattern that gives marble a texture distinctly different from that of the parent limestone. If the limestone is composed entirely of calcite, then metamorphism into marble involves no new minerals, only a change in texture.

More commonly, the various elements of a parent rock react chemically and crystallize into new minerals, thus making the metamorphic rock distinct both mineralogically and texturally from the parent rock. This is because the parent rock is unstable in its new environment. The old minerals recrystallize into new ones that are at *equilibrium* in the new environment. For example, clay minerals form at Earth's surface (see chapter 5). Therefore, they are stable at the low temperature and pressure conditions both at and just below Earth's surface. When subjected to the temperatures and pressures deep within Earth's crust, the clay minerals of a shale can recrystallize into coarse-grained mica. Another example is that under appropriate temperature and pressure conditions, a quartz sandstone with a calcite cement metamorphoses as follows:

$$CaCO_3 \;+\; SiO_2 \;\rightarrow\; CaSiO_3 \;+\; CO_2$$

calcite	quartz	wollastonite	carbon
		(a mineral)	dioxide

No one has observed metamorphism taking place, just as no one has ever seen a granite pluton form. What, then, leads us to believe that metamorphic rocks form in a solid state (i.e., without melting) at high pressure and temperature? Many metamorphic rocks found on Earth's surface exhibit contorted layering

FIGURE 7.1

Metamorphic rock from Greenland. Metamorphism took place 3,700 million years ago—it is one of the oldest rocks on Earth. *Photo by C. C. Plummer*

(figure 7.1). The layering can be demonstrated to have been either caused by metamorphism or inherited from original, flat-lying sedimentary bedding (even though the rock has since recrystallized). These rocks, now hard and brittle, would shatter if smashed with a hammer. But they must have been **ductile** (or **plastic**), capable of being bent and molded under stress, to have been folded into such contorted patterns. In a laboratory, we can reproduce high pressure and temperature conditions and demonstrate such ductile behavior of rocks on a small scale. Therefore, a reasonable conclusion is that these rocks formed at considerable depth, where such conditions exist. Moreover, crystallization of a magma would not produce contorted layering.

FACTORS CONTROLLING THE CHARACTERISTICS OF METAMORPHIC ROCKS

A metamorphic rock owes its characteristic texture and particular mineral content to several factors, the most important being (1) the composition of the parent rock before metamorphism,

(2) temperature and pressure during metamorphism, (3) the effects of tectonic forces, and (4) the effects of fluids, such as water.

Composition of the Parent Rock

Usually no new elements or chemical compounds are added to the rock during metamorphism, except perhaps water. (Metasomatism, discussed later in this chapter, does involve the addition of other elements.) Therefore, the mineral content of the metamorphic rock is controlled by the chemical composition of the parent rock. For example, a basalt always metamorphoses

into a rock in which the new minerals can collectively accommodate the approximately 50% silica and relatively high amounts of the oxides of iron, magnesium, calcium, and aluminum in the original rock. On the other hand, a limestone, composed essentially of calcite ($CaCO_3$), cannot metamorphose into a silica-rich rock.

Temperature

Heat, necessary for metamorphic reactions, comes primarily from the outward flow of geothermal energy from Earth's deep interior. The deeper a rock is beneath the surface, the hotter it will be. The particular temperature for rock at a given depth depends on the local *geothermal gradient* (described in chapter 3). Additional heat could be derived from magma, if magma bodies are locally present.

A mineral is said to be *stable* if, given enough time, it does not react with another substance or convert to a new mineral or substance. Any mineral is stable only within a given temperature range. The stability temperature range of a mineral varies with factors such as pressure and the presence or absence of other substances. Some minerals are stable over a wide temperature range. Quartz, if not mixed with other minerals, is stable at atmospheric pressure (i.e., at Earth's surface) up to about

A

FIGURE 7.2

Confining pressure. (A) The diver's suit is pressurized to counteract hydrostatic pressure. Object (cube) has a greater volume at low pressure than at high pressure. (B) Both styrofoam cups were identical. The shrunken cup was carried to a depth of 2,250 meters by the submersible ALVIN in a biological sampling dive to the Juan de Fuca Ridge, off the coast of Washington state. *Photo courtesy of the National Science Foundation-funded REVEL Project, University of Washington*

B

800°C. At higher pressures, quartz remains stable to even higher temperatures. Other minerals are stable over a temperature range of only 100° or 200°C.

By knowing (from results of laboratory experiments) the particular temperature range in which a mineral is stable, a geologist may be able to deduce the temperature of metamorphism for a rock that includes that mineral.

Minerals stable at higher temperatures tend to be less dense (or have a lower specific gravity) than chemically identical minerals (polymorphs) stable at lower temperatures. (An example, discussed later in this chapter, is sillimanite, which forms at higher temperature and is less dense than andalusite.) As temperature increases, the atoms vibrate more within their sites in the crystal structure. A more open (less tightly packed) crystal structure, such as high-temperature minerals tend to have, allows greater vibration of atoms. (If the heat and resulting vibrations become too great, the bonds between atoms in the crystal break and the substance becomes liquid.)

The upper limit on temperature in metamorphism overlaps the temperature of partial melting of a rock. If partial melting takes place, the component that melts becomes a magma; the solid residue remains a metamorphic rock. Temperatures at which the igneous and metamorphic realms can coexist vary considerably. For an ultramafic rock (containing only ferromagnesian silicate minerals), the temperature will be over 1,200°C. For a metamorphosed shale under high water pressure, a granitic melt component can form in the metamorphic rock at temperatures as low as 650°C.

Pressure

Usually, when we talk about pressure, we mean **confining pressure;** that is, pressure applied equally on all surfaces of a substance as a result of burial or submergence. A diver senses confining pressure (known as *hydrostatic pressure*) proportional to the weight of the overlying water (figure 7.2). The pressure uniformly squeezes the diver's entire body surface. Likewise, an object buried deeply within Earth's crust is compressed by strong confining pressure, called *lithostatic pressure,* which forces grains closer together and eliminates pore space. For metamorphism, pressure is usually given in *kilobars*. A kilobar is 1,000 bars. A *bar* is very close (0.99) to standard atmospheric pressure, so that, for all practical purposes, a kilobar is the pressure equivalent of a thousand times the pressure of the atmosphere at sea level. The *pressure gradient,* the increase in lithostatic pressure with depth, is approximately 1 kilobar per each 3.3 kilometers of burial in crustal rock.

Any new mineral that has crystallized under high-pressure conditions tends to occupy less space than did the mineral or minerals from which it formed. The new mineral is denser than its low-pressure counterparts because the pressure forces atoms closer together into a more closely packed crystal structure.

But what if pressure and temperature both increase, as is commonly the case with increasing depth into the Earth? If the effect of higher temperature is greater than the effect of higher pressure, the new mineral will likely be less dense. A denser new mineral is likely if increasing pressure effects are greater than increasing temperature effects.

Differential Stress

Most metamorphic rocks show the effects of tectonic forces. When forces are applied to an object, the object is under **stress,** force per unit area. If the forces on a body are stronger or weaker in different directions, a body is subjected to **differential stress.** Differential stress tends to deform objects into oblong or flattened forms. If you squeeze a rubber ball between your thumb and forefinger, the ball is under differential stress. If you squeeze a ball of dough (figure 7.3A), it will remain flattened after you stop squeezing, because dough is ductile (or plastic). To illustrate the difference between confining pressure and differential stress, visualize a drum filled with water. If you place a ball of putty underwater in the bottom of the drum, the ball will not change its shape (its volume will decrease slightly due to the

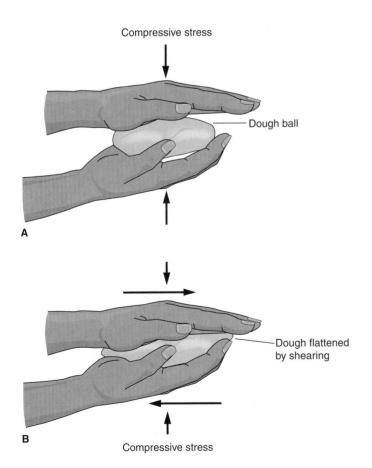

Compressive stress

Dough ball

A

Dough flattened by shearing

B   Compressive stress

FIGURE 7.3

(A) Compressive stress exerted on a ball of putty by two hands. More force is exerted in the direction of arrows than elsewhere on the putty. (B) Shearing takes place as two hands move parallel to each other at the same time that some compressive force is exerted perpendicular to the flattening putty.

weight of the overlying water). Now take the putty ball out of the water and place it under the drum. The putty will be flattened into the shape of a pancake due to the differential stress. In this case, the putty is subjected to *compressive differential stress* or, more simply, **compressive stress** (as is the dough ball shown in figure 7.3*A*).

Differential stress is also caused by **shearing,** which causes parts of a body to move or slide relative to one another across a plane. An example of shearing is when you spread out a deck of cards on a table with your hand moving parallel to the table. Shearing often takes place perpendicular to, or nearly perpendicular to, the direction of compressive stress. If you put a ball of putty between your hands and slide your hands while compressing the putty, as shown in figure 7.3*B,* the putty flattens parallel to the shearing (the moving hands) as well as perpendicular to the compressive stress.

Some rocks can be attributed exclusively to shearing during faulting (movement of bedrock along a fracture, described in chapter 15) in a process sometimes called *dynamic metamorphism.* Rocks in contact along the fault are broken and crushed when movement takes place. A *mylonite* is an unusual rock that is formed from pulverized rock in a fault zone. The rock is streaked out parallel to the fault in darker and lighter components due to shearing. Mylonites are believed to form at a depth of around a kilometer or so, where the rock is still cool and brittle (rather than ductile), but the pressure is sufficient to compress the pulverized rock into a compact, hard rock. Where found, they occupy zones that are only about a meter or so wide.

Foliation

Differential stress has a very important influence on the texture of a metamorphic rock because it forces the constituents of the rock to become parallel to one another. For instance, the pebbles in the metamorphosed conglomerate shown in figure 7.4 were originally more spherical but have been flattened by differential stress. When a rock has a planar texture, it is said to be *foliated.* **Foliation** is manifested in various ways. If a platy mineral (such as mica) is crystallizing within a rock that is undergoing differential stress, the mineral grows in such a way that it remains parallel to the direction of shearing or perpendicular to the direction of compressive stress (figure 7.5). Any platy mineral attempting to grow against shearing is either ground up or forced into alignment. Minerals that crystallize in needlelike shapes (for example, hornblende) behave similarly, growing with their long axes parallel to the plane of foliation. The three very different textures described next (from lowest to highest degree of metamorphism) are all variations of foliation and are important in classifying metamorphic rocks:

1. If the rock splits easily along nearly flat and parallel planes, indicating that preexisting, microscopic, platy minerals were realigned during metamorphism, we say the rock is *slaty,* or that it possesses *slaty cleavage.*

FIGURE 7.4

Metamorphosed conglomerate in which the pebbles have been flattened (sometimes called a stretched pebble conglomerate). Compare to the inset photo of a conglomerate (this is figure 6.9). *Background photo by C. C. Plummer; Inset photo by David McGeary*

2. If visible platy or needle-shaped minerals have grown essentially parallel to a plane due to differential stress, the rock is *schistose* (figure 7.6).

3. If the rock became very ductile and the new minerals separated into distinct (light and dark) layers or lenses, the rock has a layered or *gneissic* texture, such as in figure 7.13.

Fluids

Hot water (as vapor) is the most important fluid involved in metamorphic processes, although other gases, such as carbon dioxide, sometimes play a role. The water may have been trapped in a parent sedimentary rock or given off by a cooling pluton. Water may also be given off from minerals that have water in their crystal structure (e.g., clay, mica). As temperature rise during metamorphism and a mineral becomes unstable, its water is released.

Water is thought to help trigger metamorphic chemical reactions. Water, moving through fractures and along grain margins, is a sort of intrarock rapid transit for ions. Under high pressure, it moves between grains, dissolves ions from one mineral, and then carries these ions elsewhere in the rock where they can react with the ions of a second mineral. The new mineral that forms is stable under the existing conditions.

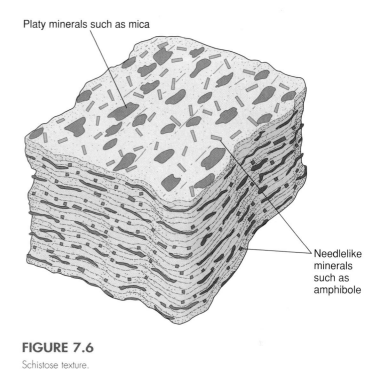

Platy minerals such as mica

Needlelike minerals such as amphibole

FIGURE 7.6

Schistose texture.

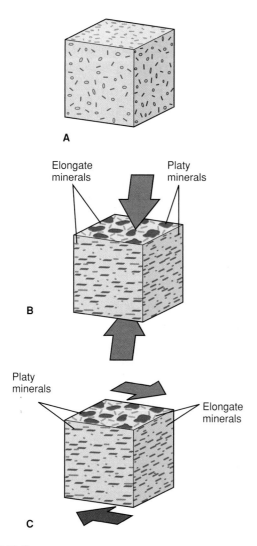

A

Elongate minerals   Platy minerals

B

Platy minerals   Elongate minerals

C

FIGURE 7.5

Orientation of platy and elongate minerals in metamorphic rock. (A) Platy minerals randomly oriented (e.g., clay minerals before metamorphism). No differential stress involved. (B) Platy minerals (e.g., mica) and elongate minerals (e.g., amphibole) have crystallized under the influence of compressive stress. (C) Platy and elongate minerals developed with shearing as the dominant stress.

Time

The effect of time on metamorphism is hard to comprehend. Most metamorphic rocks are composed predominantly of silicate minerals, and silicate compounds are notorious for their sluggish chemical reaction rates. Garnet crystals taken from a metamorphic rock collected in Vermont were analyzed, and scientists calculated a growth rate of 1.4 millimeters per million years. The garnets' growth was sustained over a 10.5-million-year period. Many laboratory attempts to duplicate metamorphic reactions believed to occur in nature have been frustrated by the time element. The several million years during which a particular combination of temperature and pressure may have prevailed in nature are impossible to duplicate.

CLASSIFICATION OF METAMORPHIC ROCKS

As we noted before, the kind of metamorphic rock that forms is determined by the metamorphic environment (primarily the particular combination of pressure, stress, and temperature) and by the chemical constituents of the parent rock. Many kinds of metamorphic rocks exist because of the many possible combinations of these factors. These rocks are classified based on broad similarities. (Appendix B contains a systematic procedure for identifying common metamorphic rocks.) The relationship of texture to rock name is summarized in table 7.1.

First, consider the texture of a metamorphic rock. Is it *foliated* or *nonfoliated* (figure 7.7)?

Nonfoliated Rocks

If the rock is nonfoliated, it is named on the basis of its composition. The two most common nonfoliated rocks are marble and quartzite, composed, respectively, of calcite and quartz.

Marble, a coarse-grained rock composed of interlocking calcite crystals (figure 7.8), forms when limestone recrystallizes during metamorphism. If the parent rock is dolomite, the recrystallized rock is a *dolomite marble.* Marble has long been valued as a building material and as a material for sculpture (figure 7.8*B*), partly because it is easily cut and polished and partly because it reflects light in a shimmering pattern, a result of the excellent cleavage of the individual calcite crystals.

A   **B**

FIGURE 7.7

Photomicrographs taken through a polarizing microscope of metamorphic rocks. (A) Nonfoliated rock and (B) Foliated rock. *Photos by C. C. Plummer*

| TABLE 7.1 | Classification and Naming of Metamorphic Rocks (Based Primarily on Texture) |

Nonfoliated

Name Based on Mineral Content of Rock

Usual Parent Rock	Rock Name	Predominant Minerals	Identifying Characteristics
Limestone	Marble	Calcite	Coarse interlocking grains of calcite (or, less commonly, dolomite)
Dolomite	Dolomite marble	Dolomite	Calcite (or dolomite) has rhombohedral cleavage; hardness intermediate between glass and fingernail. Calcite effervesces in weak acid
Quartz sandstone	Quartzite	Quartz	Rock composed of interlocking small granules of quartz. Has a sugary appearance and vitreous luster; scratches glass
Shale	Hornfels	Fine-grained micas	A fine-grained, dark rock that generally will scratch glass. May have a few coarser minerals present
Basalt	Hornfels	Fine-grained ferromagnesian minerals, plagioclase	

Foliated

Name Based Principally on Kind of Foliation Regardless of Parent Rock. Adjectives Describe the Composition (e.g., biotite-garnet schist)

Texture	Rock Name	Typical Characteristic Minerals	Identifying Characteristics
Slaty	Slate	Clay and other sheet silicates	A very fine-grained rock with an earthy luster. Splits easily into thin, flat sheets
Intermediate between slaty and schistose	Phyllite	Mica	Fine-grained rock with a silky luster. Generally splits along wavy surfaces
Schistose	Schist	Biotite and muscovite amphibole	Composed of visible platy or elongated minerals that show planar alignment. A wide variety of minerals can be found in various types of schist (e.g., garnet-mica schist, hornblende schist, etc.).
Gneissic	Gneiss	Feldspar	Light and dark minerals are found in separate, parallel layers or lenses. Commonly, the dark layers include biotite and hornblende; the light-colored layers are composed of feldspars and quartz. The layers may be folded or appear contorted

FIGURE 7.8

(A) Hand specimen of marble. Inset is a photomicrograph showing interlocking crystals of calcite. Each crystal is approximately 2 millimeters across. (B) Michelangelo's unfinished sculpture *Bound Slave* in a block of marble quarried in Carrara, Italy. *Photo A by C. C. Plummer; Photo B by Nimatallah/Art Resource, NY*

Marble is, however, highly susceptible to chemical weathering (see chapter 5).

Quartzite (figure 7.9) is produced when grains of quartz in sandstone are welded together while the rock is subjected to high temperature. This makes it as difficult to break along grain boundaries as through the grains. Therefore, quartzite, being as hard as a single quartz crystal, is difficult to crush or break. It is the most durable of common rocks used for construction, both because of its hardness and because quartz is not susceptible to chemical weathering.

Hornfels is a very fine-grained, nonfoliated, metamorphic rock whose parent rock is either shale or basalt. If it forms from shale, characteristically only microscopically visible micas form from the shale's clay minerals. Sometimes a few minerals grow large enough to be seen with the naked eye; these are minerals that are especially capable of crystallizing under the particular temperature attained during metamorphism. If hornfels forms from basalt, amphibole, rather than mica, is the predominant fine-grained mineral produced.

Foliated Rocks

If the rock is foliated, you need to determine the type of foliation to name the rock. For example, a schistose rock is called a

FIGURE 7.9

Quartzite. Inset shows photomicrograph taken using a polarizing microscope. Interlocking quartz crystals are about ½ millimeter across. *Photos by C. C. Plummer schist*. But this name tells us nothing about what minerals are in this rock, so we add adjectives to describe the composition—for example, *garnet-mica schist*. The following are the most common foliated rocks progressing from lower grade (they usually form at lower temperatures) to higher grade:

Slate is a very fine-grained rock that splits easily along flat, parallel planes (figure 7.10). Although some slate forms from volcanic ash, the usual parent rock is shale. Slate develops under temperatures and pressures only slightly greater than those found in the sedimentary realm. The temperatures are not high enough for the rock to thoroughly recrystallize. The important controlling factor is differential stress. The original clay minerals partially recrystallize into equally fine-grained, platy minerals. Under differential stress, the old and new platy minerals are aligned, creating slaty cleavage in the rock. A slate indicates that a relatively cool and brittle rock has been subjected to intense tectonic activity.

Because of the ease with which it can be split into thin, flat sheets, slate is used for making chalkboards, pool tables, and roofs.

Phyllite is a rock in which the newly formed micas are larger than the platy minerals in slate but still cannot be seen with the naked eye. This requires a further increase in temperature over that needed for slate to form. The very fine-grained

FIGURE 7.11

Phyllite, exhibiting a crinkled, silky-looking surface. *Photo by C. C. Plummer*

mica imparts a satin sheen to the rock, which may otherwise closely resemble slate (figure 7.11). But the slaty cleavage may be crinkled in the process of conversion of slate to phyllite.

A **schist** is characterized by megascopically visible, approximately parallel-oriented minerals. Platy or elongate minerals that crystallize from the parent rock are clearly visible to the naked eye. Which minerals form depends on the particular combination of temperature and pressure prevailing during recrystallization as well as the composition of the parent rock. Two, of several, schists that form from shale are *mica schist* and *garnet-mica schist* (figure 7.12). Although they both have the same parent rock, they form under different combinations of temperature and pressure. If the parent rock is basalt, the schists that form are quite different. If the predominant ferromagnesian mineral that forms during metamorphism of basalt is amphibole, it is an *amphibole schist* (also called an *amphibolite*). At a lower grade, the predominant mineral is chlorite, a green micaceous mineral, in a *chlorite schist* (or *greenschist*).

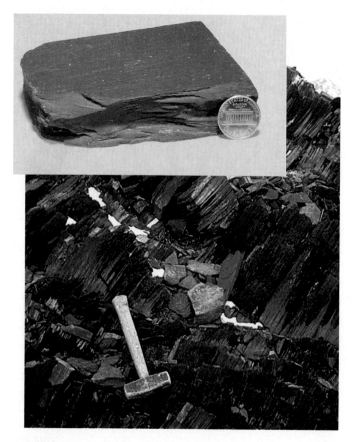

FIGURE 7.10

Slate outcrop in Antarctica. Inset is hand specimen of slate. *Background photo by P. D. Rowley, U.S. Geological Survey; Inset photo © Parvinder Sethi*

FIGURE 7.12

Garnet-mica schist. Small, subparallel flakes of muscovite mica reflect light. Garnet crystals give the rock a "raisin bread" appearance. *Photo by C. C. Plummer*

180

FIGURE 7.13
Gneiss. *Photo by C. C. Plummer*

Gneiss is a rock consisting of light and dark mineral layers or lenses. The highest temperatures and pressures have changed the rock so that minerals have separated into layers. Platy or elongate minerals (such as mica or amphibole) in dark layers alternate with layers of light-colored minerals of no particular shape. Within the light-colored layers, coarse feldspars have crystallized. In composition, a gneiss may resemble granite or diorite, but it is distinguishable from those plutonic rocks by its foliation (figure 7.13).

Temperature conditions under which a gneiss develops approach those at which granite solidifies. It is not surprising, then, that the same minerals are found in gneiss and in granite. In fact, a previously solidified granite can be converted to a gneiss under appropriate pressure and temperature conditions and if the rock is under differential stress.

TYPES OF METAMORPHISM

The two most common types of metamorphism are contact metamorphism and regional metamorphism. Hydrothermal processes, in which hot water plays a major role during metamorphism, are discussed later in this chapter.

Contact Metamorphism

Contact metamorphism (also known as *thermal* metamorphism) is metamorphism in which high temperature is the dominant factor. Confining pressure may influence which new minerals crystallize; however, the confining pressure is usually relatively low. This is because contact metamorphism mostly takes place not too far beneath Earth's surface (less than 10 kilometers). Contact metamorphism occurs adjacent to

a pluton when a body of magma intrudes relatively cool country rock. The process can be thought of as the "baking" of country rock adjacent to an intrusive contact; hence, the term *contact metamorphism.* The zone of contact metamorphism (also called an *aureole*) is usually quite narrow—generally from 1 to 100 meters wide. Differential stress is rarely significant. Therefore, the most common rocks found in an aureole are the nonfoliated rocks: marble when igneous rock intrudes limestone; quartzite when quartz sandstone is metamorphosed; hornfels when shale is scorched.

Marble and quartzite also form under conditions of regional metamorphism. When grains of calcite or quartz recrystallize, they tend to be equidimensional, rather than elongate or platy. For this reason, marble and quartzite do not usually exhibit foliation, even though subjected to differential stress during metamorphism.

Regional Metamorphism

The great majority of the metamorphic rocks found on Earth's surface are products of **regional metamorphism,** which is metamorphism that takes place at considerable depth underground (generally greater than 5 kilometers). Regional metamorphic rocks are almost always foliated, indicating differential stress during recrystallization (for this reason, regional metamorphism is sometimes referred to as *dynamothermal* metamorphism). Metamorphic rocks are prevalent in the most intensely deformed portions of mountain ranges. They are visible where once deeply buried cores of mountain ranges are exposed by erosion. Furthermore, large regions of the continents are underlain by metamorphic rocks, thought to be the roots of ancient mountains long since eroded down to plains or rolling hills.

Temperatures during regional metamorphism vary widely. Usually, the temperatures are in the range of 300 to 800°C. Temperature at a particular place depends to a large extent on depth of burial and the geothermal gradient of the region. Locally, temperature may also increase because of heat given off by nearby magma bodies. The high confining pressure is due to burial under 5 or more kilometers of rock. The differential stress is due to tectonism; that is, the constant movement and squeezing of the crust during mountain-building episodes.

Temperatures and pressures during metamorphism can be estimated through the results of laboratory experimental studies of minerals. In many cases, we can estimate temperature and pressure by determining the conditions under which an assemblage of several minerals can coexist. In some instances, a single mineral, or *index mineral,* suffices for determining the pressure and temperature combination under which a rock recrystallized (box 7.2).

Depending on the pressure and temperature conditions during metamorphism, a particular parent rock may recrystallize into one of several metamorphic rocks. For example, if basalt is metamorphosed at relatively low temperatures and pressures, it will recrystallize into a *greenschist,* a schistose rock containing chlorite (a green sheet-silicate), actinolite (a green amphibole), and sodium-rich plagioclase. Or it will

PLANETARY GEOLOGY 7.1

Impact Craters and Shock Metamorphism

The spectacular collision of the comet Shoemaker-Levy with Jupiter in 1994 served to remind us that asteroids and comets occasionally collide with a planet. Earth is not exempt from collisions. Large meteorites have produced impact craters when they have collided with Earth's surface. One well-known meteorite crater is Meteor Crater in Arizona, which is a little more than a kilometer in diameter (box figure 1). Many much larger craters are known in Canada, Germany, Australia, and other places.

Impact craters display an unusual type of metamorphism called *shock metamorphism*. The sudden impact of a large extraterrestrial body results in brief but extremely high pressures. Quartz may recrystallize into the rare SiO_2 minerals coesite and stishovite. Quartz that is not as intensely impacted suffers damage (detectable under a microscope) to its crystal lattice.

The impact of a meteorite also may generate enough heat to locally melt rock. Molten blobs of rock are thrown into the air and become streamlined in the Earth's atmosphere before solidifying into what are called *tektites*. Tektites may be found hundreds of kilometers from the point of meteorite impact.

A large meteorite would blast large quantities of material high into the atmosphere. According to theory, the change in global climate due to a meteorite impact around 65 million years ago caused extinctions of many varieties of creatures (see box 8.2 on the extinction of dinosaurs). Evidence for this impact includes finding tiny fragments of shock metamorphosed quartz and tektites in sedimentary rock that is 65 million years old.

Shock metamorphosed rock fragments are much more common on the Moon than on Earth. There may be as many as 400,000 craters larger than a kilometer in diameter on the Moon. Mercury's surface is remarkably similar to that of the Moon. Our two neighboring planets, Venus and Mars, are not as extensively

BOX 7.1 ■ FIGURE 1

Meteor Crater in Arizona. Diameter of the crater is 1.2 kilometers. *Photo by Frank M. Hanna*

cratered as is the Moon. This is because these planets, like Earth, have been tectonically active since the time of greatest meteorite bombardment, about 4 billion years ago. If Earth had not been tectonically active and if we didn't have an atmosphere driving erosion, Earth would have around sixteen times the number of meteorite craters as the Moon and would appear just as pockmarked with craters.

Additional Resource

Meteor Crater

See an animation of the meteorite impact. Go to "reference information" for details about the meteorite impact.

• www.meteorcrater.com/

---

recrystallize into a *greenstone*, a rock that has similar minerals but is not foliated. (A greenstone would indicate that the tectonic forces were not strong enough to induce foliation while the basalt was recrystallizing.) At higher temperatures and pressures, the same basalt would recrystallize into an *amphibole schist* (also called *amphibolite*), a rock composed of hornblende, plagioclase feldspar, and, perhaps, garnet. Metamorphism of other parent rocks under conditions similar to those that produce amphibole schist from basalt should produce the metamorphic rocks shown in table 7.2.

The minerals present in a rock indicate its *metamorphic grade*. Low-grade rocks formed under relatively cool temperatures and high-grade rocks at high temperatures, whereas medium-grade rocks recrystallized at around the middle of the range of metamorphic temperatures. Greenschist and greenstone are regarded as low-grade rocks, while amphibole schist is regarded as a medium-grade rock.

Prograde Metamorphism

When a rock becomes buried to increasingly greater depths, it is subjected to increasingly greater temperatures and pressures and will undergo *prograde metamorphism*—that is, it recrystallizes into a higher-grade rock. To show how rocks are changed by regional metamorphism, we look at what happens to shale during prograde metamorphism as progressively greater pressure and temperature act on a rock type with increasing depth in Earth's crust (figure 7.14).

Slate, which looks quite similar to the shale from which it forms, is the lowest-grade rock in progressive metamorphism. Its slaty cleavage develops as a result of differential stress during incipient recrystallization of clay minerals to other platy minerals. As described earlier, phyllite is a rock that is transitional between slate and schist and, as such, we expect it to have formed at a depth between where slate and schist form.

IN GREATER DEPTH 7.2

Index Minerals

Certain minerals can only form under a restricted range of pressure and temperature. Stability ranges of these minerals have been determined in laboratories. When found in metamorphic rocks, these minerals can help us infer, within limits, what the pressure and temperature conditions were during metamorphism. For this reason, they are known as *index minerals*. Among the best known are *andalusite, kyanite,* and *sillimanite*. All three have an identical chemical composition (Al_2SiO_5) but different crystal structures (they are *polymorphs*). They are found in metamorphosed shales that have an abundance of aluminum. Box figure 1 is a phase diagram showing the pressure-temperature fields in which each is stable. Box figure 2 is a map showing metamorphic patterns across the Grenville Province of the Canadian Shield. These patterns were established using the minerals andalusite-sillimanite-kyanite.

If andalusite is found in a rock, this indicates that pressures and temperatures were relatively low. Andalusite is often found in contact metamorphosed shales (hornfels). Kyanite, when found in schists, is regarded as an indicator of high pressure; but note

BOX 7.2 ■ FIGURE 2

Regional metamorphic patterns across the Grenville Province of the Canadian Shield. Colored bands represent reconstructed burial temperatures based on minerals present in the metamorphic rocks. Higher grades of metamorphism occur in the west of the Grenville Province and indicate deeper burial and higher temperatures in that area. *Courtesy of Nick Eyles*

BOX 7.2 ■ FIGURE 1

Phase diagram showing the stability relationships for the Al_2SiO_5 minerals. *M. J. Holdaway, 1971,* American Journal of Science, v. 271. *Reprinted by permission of* American Journal of Science *and Michael J. Holdaway*

that the higher the temperature of the rock, the greater the pressure needed for kyanite to form. Sillimanite is an indicator of high temperature and can be found in some contact metamorphic rocks adjacent to very hot intrusions as well as in regionally metamorphosed schists and gneisses that formed at considerable depths.

Note that if you find all three minerals in the same rock and could determine that they were mutually stable, you could infer that the temperature was close to 500°C and the confining pressure was almost 4 kilobars during metamorphism.

TABLE 7.2	Regional Metamorphic Rocks That Form under Approximately Similar Pressure and Temperature Conditions	
Parent Rock	**Rock Name**	**Predominant Minerals**
Basalt	Amphibole schist (amphibolite)	Hornblende, plagioclase, garnet
Shale	Mica schist	Biotite, muscovite, quartz, garnet
Quartz sandstone	Quartzite	Quartz
Limestone or dolomite	Marble	Calcite or dolomite

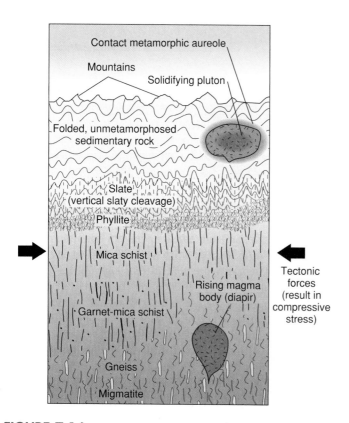

FIGURE 7.14

Schematic cross section representing an approximately 30-kilometer portion of Earth's crust during metamorphism. Rock names given are those produced from shale.

metamorphism. But why doesn't a rock recrystallize to one stable at lower temperature and pressure conditions during its long journey to the surface, where we now find it? The answer is that water is usually available during prograde metamorphism and the rock is relatively dry after reaching its peak temperatures. The absence of water means that chemical reaction will be prohibitively slow at the cooler temperatures. Substantial *retrograde metamorphism* only occurs if additional water is introduced to the rock after peak metamorphism. Tectonic forces at work during the peak of metamorphism fracture the rock extensively and permit water to get to the mineral grains. After tectonic forces are relaxed, the rocks move upward as a large block as isostatic adjustment takes place. It is unusual to find rocks that indicate retrograde metamorphism. These are rocks that recrystallized under lower temperature and pressure conditions than during the peak of metamorphism. They were fractured during their ascent, permitting water to trigger reactions to new, lower-grade minerals.

Pressure and Temperature Paths in Time

Index minerals and mineral assemblages in a rock can be used to determine the approximate temperature and pressure conditions that prevailed during metamorphism. Precise determination of the chemical composition of some minerals can determine the temperature or pressure present during the

Schist forms at higher temperatures and usually higher pressures than does phyllite. However, schist with shale as a parent rock forms over a wide range of temperatures and pressures. Figure 7.14 indicates the metamorphic setting for two varieties of schist (there are a number of others) that form from shale. *Mica schist* indicates a grade of metamorphism slightly higher than that of phyllite. Garnet requires higher temperatures to crystallize in a schist, so the *garnet-mica schist* probably formed at a deeper level than that of mica schist.

If schist is subjected to high enough temperatures, its constituents become more mobile and the rock recrystallizes into gneiss. The constituents of feldspar migrate (probably as ions) into planes of weakness caused by differential stress where feldspars, along with quartz, crystallize to form light-colored layers. The ferromagnesian minerals remain behind as the dark layers.

If the temperature is high enough, partial melting of rock may take place, and a magma collects in layers within the foliation planes of the solid rock. After the magma solidifies, the rock becomes a **migmatite,** a mixed igneous and metamorphic rock (figure 7.15). A migmatite can be thought of as a "twilight zone" rock that is neither fully igneous nor entirely metamorphic.

The metamorphic rocks that we see usually have minerals that formed at or near the highest temperature reached during

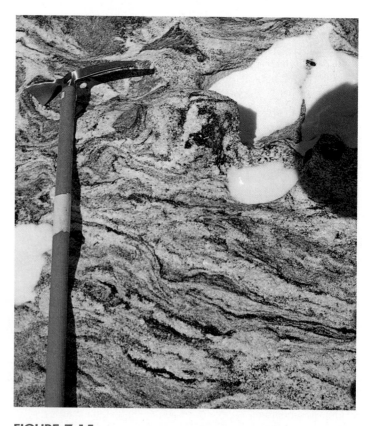

FIGURE 7.15

Migmatite in the Daniels Range, Antarctica. *Photo by C. C. Plummer* growth of a particular mineral. The usual basis for determining temperature (*geothermometry*) or pressure (*geobarometry*) during mineral growth is the ratio of pairs of elements (e.g., Fe and Al) within the crystal structure of the mineral.

Modern techniques allow us to determine chemical compositional changes across a grain of a mineral in a rock. An *electron microprobe* is a microscope that allows the user to focus on a tiny portion of a mineral in a rock, then shoot a very narrow beam of electrons into that point in the mineral. The extent and manner in which the beam is absorbed by the mineral are translated (by computer) into the precise chemical composition of the mineral at that point. If the mineral is *zoned* (that is, the chemical composition changes within the mineral, as described in chapter 2), the electron microprobe will indicate the differing composition within the mineral grain.

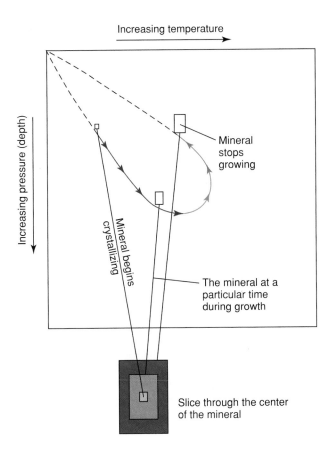

FIGURE 7.16

Pressure-temperature-time path for growth of a mineral during metamorphism. An electron microprobe is used to determine the precise chemical composition of the concentric zones of the mineral. The data are used to determine the pressure and temperate during the growth of the mineral. Three stages during the growth of the mineral are correlated to the graph—beginning of growth (center of crystal), an arbitrary point during its growth, and the end of crystallization (the outermost part of the crystal).

The green segment of the path indicates increasing pressure and temperature during metamorphism. The orange segment indicates that pressure was decreasing while temperature continued to rise. The blue segment indicates temperature and pressure were both decreasing. The decrease in pressure is likely to be the result of uplift and erosion at the surface. The dashed lines are inferred pressure and temperature paths before and after metamorphism.

A mineral will grow from the center outward, adding layers of atoms as it becomes larger. If pressure and temperature conditions change as the mineral grows, the concentric zoning will reflect those changes. Figure 7.16 shows the results of one such study. The diagram shows the changes of temperature and pressure in time, with the line showing the temperature-pressure-time path. If pressure and temperature are both increasing, this indicates the rock is being buried deeper while becoming hotter. If temperature and pressure are both decreasing, the rock is cooling down at the same time that pressure is being reduced because of erosion at Earth's surface.

PLATE TECTONICS AND METAMORPHISM

Studies of metamorphic rocks have provided important information on conditions and processes within the lithosphere and have aided our understanding of plate tectonics. Conversely, plate tectonic theory has provided models that allow us to explain many of the observed characteristics of metamorphic rocks.

Foliation and Plate Tectonics

Figure 7.17 shows an oceanic-continental boundary (oceanic lithosphere is subducted beneath continental lithosphere). One of the things the diagram shows is where differential stress that is responsible for foliation is taking place. Shearing takes place in the subduction zone where the oceanic crust slides beneath continental lithosphere. For here, we infer that the sedimentary rocks and some of the basalt becomes foliated, during metamorphism, roughly parallel to the subduction zone (parallel to the lines in the diagram).

Within the thickest part of the continental crust shown in figure 7.17, flowage of rock is indicated by the purple arrows. The crust is thickest here beneath a growing mountain belt. The thickening is due to the compression caused by the two colliding plates. Within this part of the crust, rocks flow downward and then outward (as indicated by the arrows) in a process (described in chapter 20 on mountains) of *gravitational collapse and spreading.* Under this concept, the central part of a mountain belt becomes too high after plate convergence and is gravitationally unstable. This forces the rock downward and outward. Regional metamorphism takes place throughout and we expect foliation in the recrystallizing rocks to be approximately parallel to the arrows.

Pressure-Temperature Regimes

Before the advent of plate tectonics, geologists were hard-pressed to explain how some rocks apparently were metamorphosed at relatively cool temperatures yet high pressures. We expect rocks to be hotter as they become more deeply buried. How could rocks stay cool, yet be deeply buried?

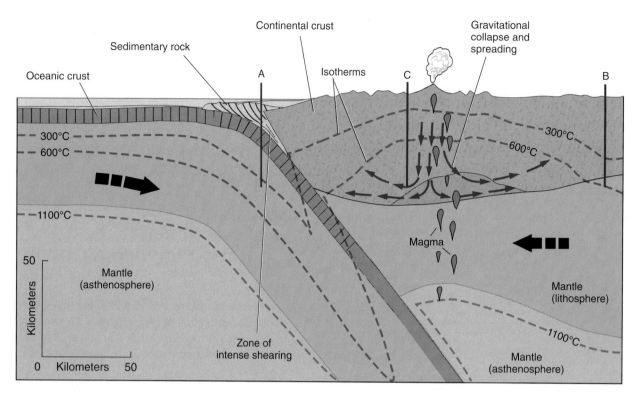

FIGURE 7.17

Metamorphism across a convergent plate boundary. All rock that is hotter than 300° or deeper than 5 kilometers is likely to be undergoing metamorphism. *Modified from W. G. Ernst.* Metamorphism and Plate Tectonic Regimes. *Stroudsberg, Pa.: Dowden, Hutchinson & Ross, 1975; p. 425. Reprinted by permission of the publisher*

Figure 7.18 shows experimentally determined stability fields for a few metamorphic minerals. Line *x* indicates a common geothermal gradient during metamorphism. At the appropriate pressure and temperature, kyanite begins to crystallize in the rock. If it is buried deeper, its pressure and temperature would change along line *x*. Eventually, it would cross the stability boundary and sillimanite would crystallize rather than kyanite. By contrast, if a rock contains glaucophane (an amphibole), rather than hornblende, the rock must have formed under high pressure but abnormally low temperature for its depth of burial. Line *y* represents a possible geothermal gradient that must have been very low and the increase in temperature was small with respect to the increase in pressure.

If we return to figure 7.17, we can use it to see how plate tectonics explains these very different pressure-temperature regimes at a convergent boundary. Confining pressure is directly related to depth. For this reason, we expect the same pressure at any given depth. For example, the pressure corresponding to 20 kilometers is the same under a hot volcanic area as it is within the relatively cool rocks of a plate's interior. Temperature, however, is quite variable as indicated by the dashed red lines. Each of these lines is an **isotherm,** a line connecting points of equal temperature.

Each of the three places (*A, B,* and *C*) in figure 7.17 would have a different geothermal gradient. If you were somehow

able to push a thermometer through the lithosphere, you would find the rock is hotter at shallower depths in areas with higher geothermal gradients than at places where the geothermal gradient is low. As indicated in figure 7.17, the geothermal gradient is higher progressing downward through an active volcanic-plutonic complex (for instance, the Cascade Mountains of Washington and Oregon) than it is in the interior of a plate (beneath the Great Plains of North America, for example). The isotherms are bowed upward in the region of the volcanic-plutonic complex because magma created at lower levels works its way upward and brings heat from the asthenosphere into the mantle and crust of the continental lithosphere. At point *C* we would expect the metamorphism that takes place to result in minerals that reflect the high temperature relative to pressure conditions such as those along line *x* in figure 7.18.

If we focus our attention at the line at *A* in figure 7.17, we can understand how minerals can form under high pressure but relatively low temperature conditions. You may observe that the bottom of line *A* is at a depth of about 50 kilometers, and if a hypothetical thermometer were here, it would read just over 300° because it would be just below the 300° isotherm. Compare this to vertical line *C* in the volcanic-plutonic complex. The confining pressure at the base of this line would be the same as at the base of line *A,* yet the temperature at the base of

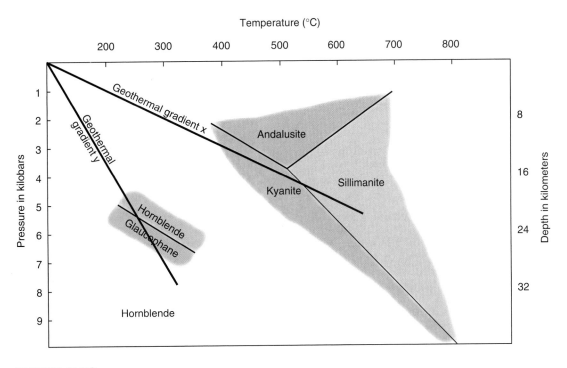

FIGURE 7.18

Stability fields for a few minerals. (Many more mineral stability fields can be used for increased accuracy.) The fields are based on laboratory research. Prograde metamorphism taking place with a geothermal gradient x involves a high temperature increase with increasing pressure. Prograde metamorphism under conditions of geothermal gradient y involves low temperature increase with increasing pressure.

line *C* would be well over 600°. The minerals that could form at the base of line *A* would not be the same as those that could form at line *C*. Therefore, we would expect quite different metamorphic rocks in the two places, even if the parent rock had been the same (box 7.3).

So when we find high-pressure/low-temperature minerals (such as glaucophane) in a rock, we can infer that metamorphism took place while subduction carried basalt and overlying sedimentary rocks downward. Thus, plate tectonics accounts for the abnormally high-pressure/low-temperature geothermal gradients (such as line *y* in figure 7.18).

HYDROTHERMAL PROCESSES

Rocks that have precipitated from hot water or have been altered by hot water passing through are hard to classify. As described earlier, hot water is involved to some extent in most metamorphic processes. Beyond metamorphism, hot water also plays an important role creating new rocks and minerals. These form entirely by precipitation of ions derived from hydrothermal solutions. *Hydrothermal minerals* can form in void spaces or between the grains of a host rock. An aggregate of hydrothermal minerals, a **hydrothermal rock,** may crystallize within a preexisting fracture in a rock to form a hydrothermal **vein.**

WEB BOX 7.3

Metamorphic Facies and the Relationship to Plate Tectonics

If the mineral assemblages in rocks of a given parent rock are the same, they are regarded as having formed under similar pressure and temperature conditions. If you find a rock of the same parentage but with a different assemblage of minerals, you could infer that metamorphism took place under different pressure and temperature conditions. Early in the twentieth century, geologists assigned metamorphosed basalts to *metamorphic facies* based on the assemblage of minerals present in the rocks. For instance, metabasalts that are mostly hornblende and plagioclase feldspar belong to the amphibolite facies. If a rock of the same chemical composition is composed largely of actinolite (an amphibole), chlorite (a green sheet silicate mineral) and sodium-rich feldspar, it belongs to the greenschist facies. Field relationships indicated that the greenschist facies represents metamorphism under lower pressure and temperature conditions than those of the amphibole facies.

To learn more, including how the various facies are used to infer the plate tectonic setting of metamorphism, go to the box in the Learning Center www.mhhe.com/plummer11e.

TABLE 7.3	**Hydrothermal Processes**
Role of Water	**Name of Process or Product**
Water transports ions between grains in a rock. Some water may be incorporated into crystal structures.	Metamorphism
Water brings ions from outside the rock, and they are added to the rock during metamorphism. Other ions may be dissolved and removed.	Metasomatism
Water passes through cracks or pore spaces in rock and precipitates minerals on the walls of cracks and within pore spaces.	Hydrothermal rocks

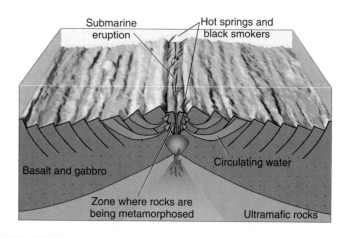

FIGURE 7.19

Cross section of a mid-oceanic ridge (divergent plate boundary). Water descends through fractures in the oceanic crust, is heated by magma and hot igneous rocks, and rises.

Hydrothermal processes are summarized in table 7.3. As we have seen, water is important for metamorphic processes not only because water transports ions from one mineral to another but because many of the minerals (the micas, for instance) that crystallize during metamorphism incorporate water into their crystal structures.

Hydrothermal Activity at Divergent Plate Boundaries

Hydrothermal processes are particularly important at mid-oceanic ridges (which are also divergent plate boundaries). As shown in figure 7.19, cold seawater moves downward through cracks in the basaltic crust and is cycled upward by heat from magma beneath the ridge crest. Very hot water returns to the ocean at submarine hot springs (*hydrothermal vents*).

Hot water traveling through the basalt and gabbro of the oceanic lithosphere helps metamorphose these rocks while they are close to the divergent boundary. This is sometimes called *seafloor metamorphism.* During metamorphism, the ferromagnesian igneous minerals, olivine and pyroxene, become converted to *hydrous* (water-bearing) minerals such as amphibole. An important consequence of this is that the hydrous minerals may eventually contribute to magma generation at convergent boundaries. After oceanic crust is subducted, the minerals are dehydrated deep in a subduction zone (see figure 7.20). The water produced moves upward into the overlying asthenosphere and contributes to melting and magma generation, as described in chapter 3.

Ore Deposits at Divergent Plate Boundaries

As the seawater moves through the crust, it dissolves metals and sulfur from the crustal rocks and magma. When the hot, metal-rich solutions contact cold seawater, metal sulfides are precipitated in a mound around the hydrothermal vent. This process has been filmed in the Pacific, where some springs spew clouds of fine-grained ore minerals that look like black smoke (figure 7.21). (To see a video clip of a "black smoker" or learn more about hydrothermal vents, go to the seafloor geology page of *Voyage to the Deep,* www.ocean.udel.edu/deepsea/level/geology/geology.html.)

The metals in rift-valley hot springs are predominantly iron, copper, and zinc, with smaller amounts of manganese, gold, and silver. Although the mounds are nearly solid metal

FIGURE 7.20

Water at a convergent boundary. Seawater trapped in the oceanic crust is carried downward and released upon heating at various depths within the subduction zone.

FIGURE 7.21

"Black smoker" or submarine hot spring on the crest of the mid-oceanic ridge in the Pacific Ocean near 21° North Latitude. The "smoke" is a hot plume of metallic sulfide minerals being discharged into cold seawater from a chimney 0.5 meters high. The large mounds around the chimney are metal deposits. The instruments in the foreground are attached to the small submersible from which the picture was taken. *Photo © WHOI, D. Foster/Visuals Unlimited*

sulfide, they are usually small and widely scattered on the sea floor, so commercial mining of them may not be practical.

Water at Convergent Boundaries

Water that percolates from the surface into the ground becomes *ground water*. Ground water seeps downward through pores and fractures in rocks. However, the depth to which surface-derived water can penetrate is quite limited.

Plate tectonics can account for water at deeper levels in the lithosphere as seawater trapped in the oceanic crust can be carried to depths of up to 100 kilometers through subduction (figure 7.20). Water trapped in sediment and in sedimentary rocks lying on basalt may be carried down with the descending crust. It is driven out by pressure at depths up to around 30 kilometers. However, studies indicate that most of the water is carried by hydrous minerals (amphibole, for example) in the basaltic crust. When the rocks get hot enough, the hydrous minerals recrystallize, releasing water. The water vapor works its way upward through the overlying continental lithosphere through fissures. In the process of ascending, water assists in the metamorphism of rocks, dissolves minerals, and carries the ions to interact during metasomatism, or it deposits quartz and other minerals in fissures as veins. The water can also lower the melting points of rocks at depth, allowing magma to form (as described in chapter 3 on igneous rocks).

Metasomatism

Metasomatism is metamorphism coupled with the introduction of *ions* from an external source. The ions are brought in by water from outside the immediate environment and are incorporated into the newly crystallizing minerals. Often, metasomatism involves ion exchange. Newly crystallizing minerals

replace preexisting ones as water simultaneously dissolves and replaces ions.

When metasomatism takes place during regional metamorphism, very hot water travels through a rock while gneiss or schist is crystallizing. Ions (typically K^+, Na^+, and SiO_4^{-4}) are carried by the water and participate in metamorphic reactions. Large feldspar crystals may grow in schist due to the addition of potassium or sodium ions.

If metasomatism is associated with contact metamorphism, the ions are introduced from a cooling magma. Some important commercially mined deposits of metals such as iron, tungsten, copper, lead, zinc, and silver are attributed to metasomatism. Figure 7.22 shows how magnetite (iron oxide) ore bodies have formed through metasomatism. Ions of the metal are transported by water and react with minerals in the host rock. Elements within the host rock are simultaneously dissolved out of the host rock and replaced by the metal ions

A

B

FIGURE 7.22

Development of a contact metasomatic deposit of iron (magnetite). (A) Magma intrudes country rock (limestone), and marble forms along contact. (B) As magma solidifies, gases bearing ions of iron leave the magma, dissolve some of the marble, and deposit iron as magnetite.

brought in by the fluid. Because of the solubility of calcite, marble commonly serves as a host for metasomatic ore deposits.

Hydrothermal Rocks and Minerals

Quartz veins (figure 7.23) are especially common where igneous activity has occurred. These can form from hot water given off by a cooling magma. They also are produced by ground water heated by a pluton and circulated by convection, as shown in figure 7.24. Where the water is hottest, the most material (notably silica) is dissolved. As water vapor continues upward through increasingly cooler rocks during its journey toward Earth's surface, pressure decreases and heat is lost. Fewer ions can be carried in solution, and so the silicon and oxygen leave the water and cake onto the walls of the crack as silica (SiO_2), forming a quartz vein.

Veins consisting only of quartz are the most widespread, although some quartz veins contain other minerals. Veins with no quartz are not as common and are composed of calcite or some other minerals.

Hydrothermal veins are very important economically. In them, we find most of the world's great deposits of zinc, lead, silver, gold, tungsten, tin, mercury, and, to some extent, copper. Ore minerals containing these metals are usually found in quartz veins. Veins containing commercially extractable amounts of metals are by no means common.

Some ore-bearing solutions percolate upward between the grains of the rock and deposit very fine grains of ore mineral throughout. These are called *disseminated ore deposits.* Usually, metallic sulfide ore minerals are distributed in very low concentration through large volumes of rock, both above and

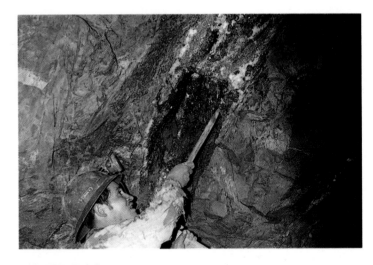

FIGURE 7.23

A wide vein that contains masses of sphalerite (dark), pyrite and chalcopyrite (both shiny yellow) in the Casapalca mine in Peru. It was mined for zinc and copper. *Photo © Brian Skinner*

within a pluton. The ore in the pluton is in the upper part, which solidified earliest. As crystallization continued in the underlying magma, hydrothermal solutions were given off, and ore minerals crystallized in the tiny fractures and between grains in the overlying rock. Most of the world's copper comes from disseminated deposits, also called *porphyry copper deposits,* because the associated pluton is usually porphyritic (see box 7.4). Along with the copper are deposited many other metals, such as lead, zinc, molybdenum, silver, and gold (and iron, though not in commercial quantities). Some very large gold mines are also in disseminated ore deposits.

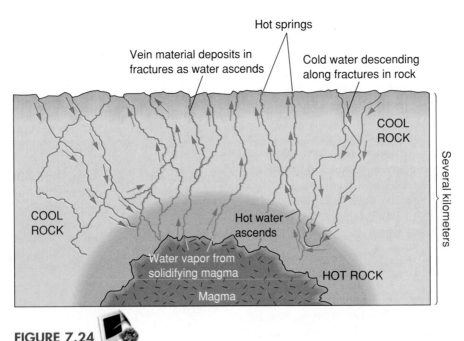

FIGURE 7.24

How veins form. Cold water descends, is heated, dissolves material, ascends, and deposits material as water cools and pressure drops upon ascending.

ENVIRONMENTAL GEOLOGY 7.4

The World's Largest Humanmade Hole—
The Bingham Canyon Copper Mine

The Bingham Canyon mine near Salt Lake City, Utah, is thought to be the biggest single humanmade hole in the world. (The Morenci mine in Arizona is volumetrically larger, but is not a single pit.) The 800-meter (½-mile) deep open pit mine is 4 kilometers (2½ miles) wide at the top and continues to be enlarged. The reason for this hole is copper.

About 40,000 kilograms of explosives are used per day to blast apart over 60,000 tons of ore (copper-bearing rock) and an equal amount of waste rock. An 8-kilometer-long conveyor belt system moves up to 10,000 tons of crushed rock per hour through a tunnel out of the pit for processing.

Mining began here as a typical underground operation in 1863. The shafts and tunnels of the mine followed a series of veins. Originally, ores of silver and lead were mined. Later, it was discovered that fine-grained, copper-bearing minerals (chalcopyrite and other copper sulfide minerals) were disseminated in tiny veinlets throughout a granite stock. Although the percentage of copper in the rock was small, the total volume of copper was recognized as huge. With efficient earth-moving techniques, large volumes of ore-bearing rock can be moved and processed. Today, mining is still going on, and the company is able to make a profit even though only 0.6% of the rock being mined is copper. Since 1904, over 12 million tons of copper have been mined, processed, and sold. The mine has also produced impressive amounts of gold, silver, and molybdenum.

Such an operation is not without environmental problems. Some people regard the huge hole in the mountains as an eyesore (but it is a popular tourist attraction). Disposing of the waste—over 99% of the rock material mined—creates problems. Wind stirs up dust storms from the piles of finely crushed waste rock unless it is kept wet. The nearby smelter that extracts the pure copper from the sulfide minerals has created a toxic smoke containing sulfuric acid fumes. During most of the twentieth century, the toxic smoke was released into the atmosphere; occasionally, wind blew polluted air to Salt Lake City. Now, over 99% of the sulfur fumes are removed at the smelter.

Additional Resource

Mining Technology—Bingham Canyon
- www.mining-technology.com/projects/bingham/

BOX 7.4 ■ FIGURE 1

Bingham Canyon copper mine in Utah. *Photo courtesy of Kennecott Copper Company*

SUMMARY

Metamorphic rocks form from other rocks that are subjected to high temperature, generally accompanied by high confining pressure. Although recrystallization takes place in the solid state, water, which is usually present, aids metamorphic reactions. Foliation in metamorphic rocks is due to *differential stress* (either *compressive stress* or *shearing*). Slate, phyllite, schist, and gneiss are foliated rocks that indicate increasing grade of regional metamorphism. They are distinguished from one another by the type of foliation.

Contact metamorphic rocks are produced during metamorphism usually without significant differential stress but with high temperature. Contact metamorphism occurs in rocks immediately adjacent to intruded magmas.

Regional metamorphism, which involves heat, confining pressure, and differential stress, has created most of the metamorphic rock of Earth's crust. Different parent rocks as well as widely varying combinations of pressure and temperature result in a large variety of metamorphic rocks. Combinations of minerals in a rock can indicate what the pressure and temperature conditions were during metamorphism. Extreme metamorphism, where the rock partially melts, can result in *migmatites.*

Hydrothermal veins form when hot water precipitates material that crystallizes into minerals. During *metasomatism,* hot water introduces ions into a rock being metamorphosed, changing the chemical composition of the metasomatized rock from that of the parent rock.

Plate-tectonic theory accounts for the features observed in metamorphic rocks and relates their development to other activities in Earth. In particular, plate tectonics explains (1) the deep burial of rocks originally formed at or near Earth's surface; (2) the intense squeezing necessary for the differential stress, implied by foliated rocks; (3) the presence of water deep within the lithosphere; and (4) the wide variety of pressures and temperatures believed to be present during metamorphism.

Terms to Remember

compressive stress 176	hydrothermal rock 187	phyllite 180
confining pressure 175	isotherm 186	quartzite 179
contact metamorphism 181	marble 177	regional metamorphism 181
differential stress 175	metamorphic rock 172	schist 180
ductile (plastic) 173	metamorphism 172	shearing 176
foliation 176	metasomatism 189	slate 180
gneiss 181	migmatite 184	stress 175
hornfels 179	parent rock 173	vein 187

Testing Your Knowledge

Use the following questions to prepare for exams based on this chapter.

1. What are the effects on metamorphic minerals and textures of temperature, confining pressure, and differential stress?

2. What are the various sources of heat for metamorphism?

3. How do regional metamorphic rocks commonly differ in texture from contact metamorphic rocks?

4. Why is such a variety of combinations of pressure and temperature environments possible during metamorphism?

5. How would you distinguish
   a. schist and gneiss?   b. slate and phyllite?
   c. quartzite and marble?   d. granite and gneiss?

6. Why is an edifice built with blocks of quartzite more durable than one built of marble blocks?

7. Which is not regarded as a low-grade metamorphic rock?
   a. greenschist   b. phyllite
   c. slate   d. gneiss

8. Shearing is a type of
   a. compressive stress   b. confining pressure
   c. lithostatic pressure d. differential stress

9. Metamorphic rocks with a planar texture (the constituents of the rock are parallel to one another) are said to be
   a. concordant   b. foliated
   c. discordant   d. nonfoliated

10. Metamorphic rocks are classified primarily on
    a. texture—the presence or absence of foliation
    b. mineralogy—the presence or absence of quartz
    c. environment of deposition
    d. chemical composition

11. Which is not a foliated metamorphic rock?
    a. gneiss   b. schist
    c. quartzite   d. slate

12. Limestone recrystallizes during metamorphism into
    a. hornfels   b. marble
    c. quartzite   d. schist

13. Quartz sandstone is changed during metamorphism into
    a. hornfels   b. marble
    c. quartzite   d. schist

14. The correct sequence of rocks that are formed when shale undergoes prograde metamorphism is
    a. slate, gneiss, schist, phyllite   b. phyllite, slate, schist, gneiss
    c. slate, phyllite, schist, gneiss   d. schist, phyllite, slate, gneiss

15. The major difference between metamorphism and metasomatism is
    a. temperature at which each takes place
    b. the minerals involved
    c. the area or region involved
    d. metasomatism is metamorphism coupled with the introduction of ions from an external source

16. Ore bodies at divergent plate boundaries can be created through
    a. contact metamorphism   b. regional metamorphism
    c. hydrothermal processes

17. A schist that developed in a high-pressure, low-temperature environment likely formed
    a. in the lower part of the continental crust
    b. in a subduction zone
    c. in a mid-oceanic ridge
    d. near a contact with a magma body

18. A metamorphic rock that has undergone partial melting to produce a mixed igneous-metamorphic rock is a
    a. gneiss   b. hornfels
    c. schist   d. migmatite

Expanding Your Knowledge

1. Should ultramafic rocks in the upper mantle be regarded as metamorphic rocks rather than igneous rocks?
2. Where were the metals before they were concentrated in hydrothermal vein ore deposits?
3. What happens to originally horizontal layers of sedimentary rock when they are subjected to the deformation associated with regional metamorphism?
4. Where in Earth's crust would you expect most migmatites to form?

Exploring Web Resources

www.mhhe.com/plummer11e
This is the dedicated website for this book. You can go to it for new and updated information. The universal resource locators (URL) listed in this book are also given as links on the website, making it easy to go to those websites without typing in the URL. When you visit our Online Learning Center, go to the Student Center and then to the chapter of interest. Here you will find additional readings and media resources, as well as answers to the Testing Your Knowledge section, more quizzes, animation and video clips, and direct links to the sites listed in this book. Links to additional websites can also be found. We have added questions for some of the sites to allow you to get the most of your exploration of the web. Using the web is an enjoyable way of enhancing your knowledge of geology.

www.geol.ucsb.edu/faculty/hacker/geo102C/lectures/part1.html
University of California Santa Barbara's *Metamorphic Petrology* website. This site is meant for a course on metamorphic rock. It is well-illustrated and can be used for in-depth learning of particular topics.

www.geolab.unc.edu/Petunia/IgMetAtlas/mainmenu.html
University of North Carolina's *Atlas of Rocks, Minerals, and Textures*. Click on "Metamorphic microtextures." Click on terms covered in this chapter (e.g., foliation, gneiss, phyllite, marble, quartzite, slate) to see excellent photomicrographs taken through a polarizing microscope.

 Animation

This chapter includes the following animations available on our Online Learning Center at **www.mhhe.com/plummer11e**.

7.24 Hydrothermal ore vein formation

Kaibab Limestone

Coconino Sandston[e]

Supai Formation

Redwall Limesto[ne]

Bright Angel Shale

Tapeats Sandstone

Vishnu Schist

Grand Canyon Series

Time and Geology

The immensity of geologic time is hard for humans to perceive. It is unusual for someone to live a hundred years, but a person would have to live 10,000 times that long to observe a geologic process that takes a million years. In this chapter, we try to help you develop a sense of the vast amounts of time over which geologic processes have been at work.

Geologists working in the field or with maps or illustrations in a laboratory are concerned with relative time—unraveling the sequence in which geologic events occurred. For instance, a geologist looking at the photo of Arizona's Grand Canyon on the facing page can determine that the tilted sedimentary rocks are older than the horizontal sedimentary rocks and that the lower layers of the horizontal sedimentary rocks are older than the layers above them. But this tells us nothing about how long ago any of the rocks formed. To determine how many years ago rocks formed, we need the specialized techniques of radioactive isotope dating. Through isotopic dating, we have been able to determine that the rocks in the lowermost part of the Grand Canyon are well over a billion years old.

Grand Canyon, Arizona. Horizontal Paleozoic beds (top of photo) overlie tilted Precambrian beds (Grand Canyon Series) and older, Precambrian metamorphic rock (Vishnu Schist). *Photo © Craig Aurness/Corbis Images*

This chapter explains how to apply several basic principles to decipher a sequence of events responsible for geologic features. These principles can be applied to many aspects of geology—as, for example, in understanding geologic structures (chapter 15). Understanding the complex history of mountain belts (chapter 20) also requires knowing the techniques for determining relative ages of rocks.

Determining age relationships between geographically widely separated rock units is necessary for understanding the geologic history of a region, a continent, or the whole Earth. Substantiation of the plate-tectonics theory depends on inter-continental correlation of rock units and geologic events, piecing together evidence that the continents were once one great body.

Widespread use of fossils led to the development of the standard geologic time scale. Originally based on relative age relationships, the subdivisions of the standard geologic time scale have now been assigned numerical ages in thousands, millions, and billions of years through isotopic dating. Think of the geologic time scale as a sort of calendar to which events and rock units can be referred. Its major subdivisions are referred to elsewhere in this book.

THE KEY TO THE PAST

Until the 1800s, almost all people living in Western culture accepted the church concept of Earth being only a few thousand years old. On the other hand, Chinese and Hindu cultures believed the age of Earth was vast beyond comprehension—more in line with what has now been determined scientifically. In the Christendom of the seventeenth and eighteenth centuries, formation of all rocks and other geologic events were placed into a biblical chronology. This required that features we observe in rocks and landscapes were created supernaturally and catastrophically. The sedimentary rocks with marine fossils (clams, fish, etc.) that we find in mountains thousands of meters above sea level were believed to have been deposited by a worldwide flood (Noah's flood) that inundated all of Earth, including its highest mountains, in a matter of days. Because no known physical laws could account for such events, they were attributed to divine intervention. In the eighteenth century, however, James Hutton, a Scotsman often regarded as the father of geology, realized that geologic features could be explained through present-day processes. He recognized that our mountains are not permanent but have been carved into their present shapes and will be worn down by the slow agents of erosion now working on them. He realized that the great thicknesses of sedimentary rock we find on the continents are products of sediment removed from land and deposited as mud and sand in seas. The time required for these processes to take place had to be incredibly long. Hutton broke from conventional thinking that Earth is less than 6,000 years old when he wrote in 1788, "We find no sign of a beginning—no prospect for an end." His writings were not widely read, but a few people realized the logic of his thesis and how important it was for understanding Earth. In the early 1800s, his ideas were given widespread attention by Charles Lyell in a landmark book, *Principles of Geology*. Hutton's concept of geological processes requiring vast amounts of time led to the development of evolutionary theory and the revolutionizing of biological sciences. Charles Darwin was among those influenced by Lyell's writing. His evolutionary theory involving survival of the fittest, published in the mid-1800s, required the great amount of time that Hutton envisioned. So Hutton's ideas became not only a foundation for geology but for the life sciences as well.

Hutton's concept that geologic processes operating at present are the same processes that operated in the past eventually became known as the principle of **uniformitarianism.** The principle is stated more succinctly as "The present is the key to the past." The term *uniformitarianism* is a bit unfortunate, because it suggests that changes take place at a uniform *rate*. Hutton recognized that sudden, violent events, such as a major, short-lived volcanic eruption, also influence Earth's history. Many geologists prefer **actualism** in place of uniformitarianism. The term *actualism* comes closer to conveying Hutton's principle that the same processes and natural laws that operated in the past are those we can actually observe or infer from observation as operating at present. It is based on the assumption, central to the sciences, that physical laws are independent of time and location. Under present usage, uniformitarianism has the same meaning as actualism for most geologists.

We now realize that geology involves time periods much greater than a few thousand years. But how long? For instance, were rocks near the bottom of the Grand Canyon (chapter opening photo) formed closer to 10,000 or 100,000 or 1,000,000 or 1,000,000,000 years ago? What geologists needed was some "clock" that began running when rocks formed. Such a "clock" was found shortly after radioactivity was discovered. Dating based on radioactivity (discussed later in this chapter) allows us to determine a rock's **numerical age** (also known as *absolute age*)—age given in years or some other unit of time. Geologists working in the field or in a laboratory with maps, cross sections, and photographs are more often concerned with **relative time,** the *sequence* in which events took place, rather than the number of years involved.

These statements show the difference between numerical age and relative time:

"The American Revolutionary War took place after the signing of the Magna Carta but before World War II." This statement gives the time of an event (the Revolutionary War) relative to other events.

But in terms of numerical age, we could say: "The Revolutionary War took place about two and a half centuries ago." Note that a numerical age does not have to be an *exact* age, merely age given in units of time. Because most geologic problems are concerned with the sequence of events, we discuss relative time first.

RELATIVE TIME

The geology of an area may seem, at first glance, to be hopelessly complex. A nongeologist might think it impossible to decipher the sequence of events that created such a geologic pattern; however, a geologist has learned to approach seemingly formidable problems by breaking them down to a number of simple problems. (In fact, a geologic education trains students in a broad spectrum of problem-solving techniques, useful for a wide variety of applications and career opportunities.) As an example, the geology of the Grand Canyon, shown diagrammatically in figure 8.16 and in the chapter opening photo, can be analyzed in four parts: (1) horizontal layers of rock; (2) inclined layers; (3) rock underlying the inclined layers (plutonic and metamorphic rock); and (4) the canyon itself, carved into these rocks.

After you have studied the following section, return to the photo of the Grand Canyon and see if you can determine the sequence of geologic events that took place.

Principles Used to Determine Relative Age

Most of the individual parts of the larger problem are solved by applying several simple principles while studying the exposed rock. In this way, the sequence of events or the relative time involved can be determined. Contacts are particularly useful for deciphering the geologic history of an area. (**Contacts,** as described in previous chapters, are the surfaces separating two different rock types or rocks of different ages.) To explain various principles, we will use a fictitious place that bears some resemblance to the Grand Canyon. We will call this place, represented by the block diagram of figure 8.1, Minor Canyon. The formation names are also fictitious. (**Formations,** as described in chapter 6, are bodies of rock of considerable thickness with recognizable characteristics that make each distinguishable from adjacent rock units. They are named after local geographic features, such as towns or landmarks. Grand Canyon's formation names are shown on the chapter's opening photo.) Note the contacts between the tilted formations, the horizontal formations, the granite, and the dike. What sequence of events might be responsible for the geology of Minor Canyon? (You might briefly study the block diagram and see how much of the geologic history of the area you can decipher before reading further.)

Our interpretations are based mainly on layered rock (sedimentary or volcanic). The subdiscipline of geology that uses interrelationships between layered rock (mostly) or sediment to interpret the history of an area or region is known as *stratigraphy*. Stratigraphy uses four principles to determine the geologic history of a locality or a region. These are the principles of (1) original horizontality, (2) superposition, (3) lateral continuity, and (4) cross-cutting relationships. These principles will be used in interpreting figure 8.1.

FIGURE 8.1

Block diagram representing the Minor Canyon area.

Original Horizontality

The principle of **original horizontality** states that beds of sediment deposited in water formed as horizontal or nearly horizontal layers (as described in chapter 6). (The sedimentary rocks in figure 8.1 were originally deposited in a marine environment.)

Note in figure 8.1 that the Larsonton Formation and overlying rock units (Foster City Formation, Hamlinville Formation, and Skinner Gulch Limestone) are horizontal. Evidently, their original horizontal attitude has not changed since they were deposited. However, the Lutgrad, Birkland, Tarburg, and Leet Junction Formations must have been tilted after they were deposited as horizontal layers. By applying the principle of original horizontality, we have determined that a geologic event—tilting of bedrock—occurred after the Leet Junction, Tarburg, Birkland, and Lutgrad Formations were deposited on a sea floor. We can also see that the tilting event did not affect the Larsonton and overlying formations. (A reasonable conclusion is that tilting was accompanied by uplift and erosion, all before renewed deposition of younger sediment.)

Superposition

The principle of **superposition** states that within a sequence of undisturbed sedimentary or volcanic rocks, the layers get younger going from bottom to top.

Obviously, if sedimentary rock is formed by sediment settling onto the sea floor, then the first (or bottom) layer must be there before the next layer can be deposited on top of it. The principle of superposition also applies to layers formed by multiple lava flows, where one lava flow is superposed on a previously solidified flow.

Applying the principle of superposition, we can determine that the Skinner Gulch Limestone is the youngest layer of sedimentary rock in the Minor Canyon area. The Hamlinville Formation is the next oldest formation, and the Larsonton Formation is the oldest of the still-horizontal sedimentary rock units. Similarly, we assume that the inclined layers were originally horizontal (by the first principle). By mentally restoring them to their horizontal position (or "untilting" them), we can see that the youngest formation of the sequence is the Leet Junction Formation and that the Tarburg, Birkland, and Lutgrad Formations are progressively older.

Lateral Continuity

The principle of **lateral continuity** states that an original sedimentary layer extends laterally until it tapers or thins at its edges. This is what we expect at the edges of a depositional environment, or where one type of sediment interfingers laterally with another type of sediment as environments change. In figure 8.1, the bottom bed of the Hamlinville Formation tapers as we would expect from this principle. We are not seeing any other layers taper, either because we are not seeing their full extent within the diagram or because they have been truncated (cut off abruptly) due to later events.

Cross-Cutting Relationships

The fourth principle can be applied to determine the remaining age relationships at Minor Canyon. The principle of **cross-cutting relationships** states that a disrupted pattern is older than the cause of disruption. A layer cake (the pattern) has to be baked (established) before it can be sliced (the disruption).

To apply this principle, look for disruptions in patterns of rock. Note that the valley in figure 8.1 is carved into the horizontal rocks as well as into the underlying tilted rocks. The sedimentary beds on either side of the valley appear to have been sliced off, or *truncated,* by the valley. (The principle of lateral continuity tells us that sedimentary beds normally become thinner toward the edges rather than stop abruptly.) So the event that caused the valley must have come after the sedimentation responsible for deposition of the Skinner Gulch Limestone and underlying formations. That is, the valley is younger than these layers. We can apply the principle of cross-cutting relationships to contacts elsewhere in figure 8.1, with the results shown in table 8.1.

We can now describe the geological history of the Minor Canyon area represented in figure 8.1 on the basis of what we have learned through applying the principles. Figures 8.2 through 8.11 show how the area changed over time, progressing from oldest to youngest events.

By *superposition,* we know that the Lutgrad Formation, the lowermost rock unit in the tilted sequence, must be the oldest of the sedimentary rocks as well as the oldest rock unit in the diagram. From the principle of *original horizontality,* we infer that these layers must have been tilted after they formed. Figure 8.2 shows initial sedimentation of the Lutgrad Forma-

TABLE 8.1	Relative Ages of Features in Figure 8.1 Determinable by Cross-Cutting Relationships	
Feature	**Is Younger Than**	**But Older Than**
Valley (canyon)	Skinner Gulch Limestone	
Foster City Formation	Dike	Hamlinville Formation
Dike	Larsonton Formation	Foster City Formation
Larsonton Formation	Leet Junction Formation and granite	Dike
Granite	Tarburg Formation	Larsonton Formation

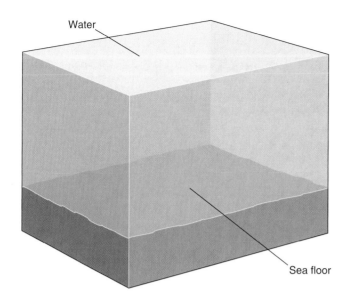

FIGURE 8.2

The area during deposition of the initial sedimentary layer of the Lutgrad Formation.

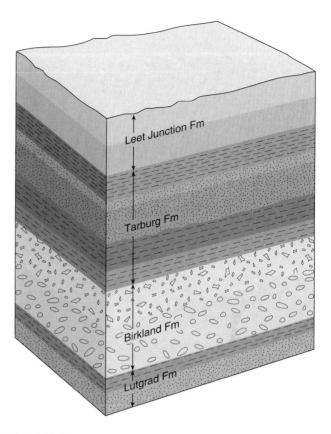

FIGURE 8.3

The area before intrusion of the granite.

tion taking place. If the entire depositional basin were shown, the layer would be tapered at its edges, according to the principle of *lateral continuity*.

Superposition indicates that the Birkland Formation was deposited on top of the Lutgrad Formation. Deposition of the Tarburg and Leet Junction Formations followed in turn (figure 8.3).

The truncation of bedding in the Lutgrad, Birkland, and Tarburg Formations by the granite tells us that the granite intruded sometime after the Tarburg Formation was formed (this is an *intrusive contact*). Although figure 8.4 shows that the granite was emplaced before tilting of the layered rock, we cannot determine from looking at figure 8.1 whether the granite intruded the sedimentary rocks before or after tilting. We can, however, determine through *cross-cutting relationships* that tilting and intrusion of the granite occurred before deposition of the Larsonton Formation. Figure 8.5 shows the rocks in the area have been tilted and erosion has taken place. Sometime later, sedimentation was renewed, and the lowermost layer of the Larsonton Formation was deposited on the erosion surface, as shown in figure 8.6. Contacts representing buried erosion surfaces such as these are called *unconformities* and are discussed in more detail in the Unconformities section of this chapter.

After the Larsonton Formation was deposited, an unknown additional thickness of sedimentary layers was deposited, as shown in figure 8.7. This can be determined through application of cross-cutting relationships. The dike is truncated by the Foster City Formation; therefore, it must have extended into some rocks that are no longer present, such as shown in figure 8.8. Figure 8.9 shows the area after the erosion that truncated the dike took place.

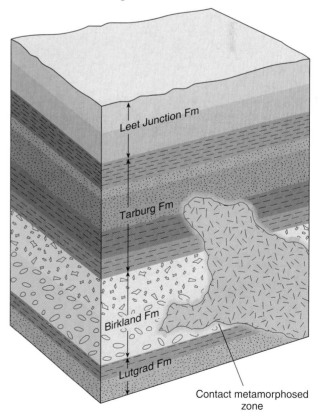

FIGURE 8.4

The area before layers were tilted.

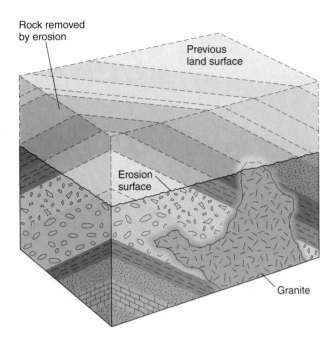

FIGURE 8.5

The area before deposition of the Larsonton Formation. Dashed lines show rock probably lost through erosion.

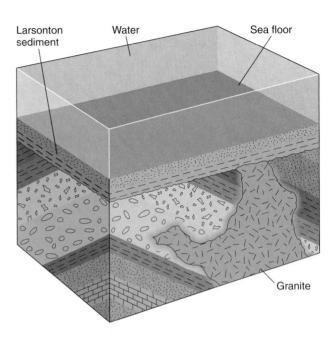

FIGURE 8.6

The area at the time the Larsonton Formation was being deposited.

FIGURE 8.7

Area before intrusion of dike. Thickness of layers above the Larsonton Formation is indeterminate.

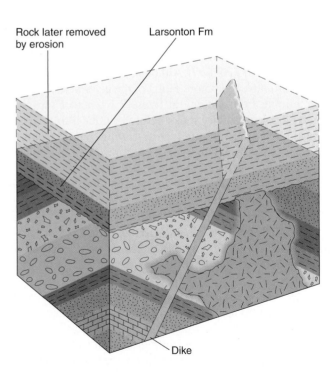

FIGURE 8.8

Dike intruded into the Larsonton Formation and preexisting, overlying layers of indeterminate thickness.

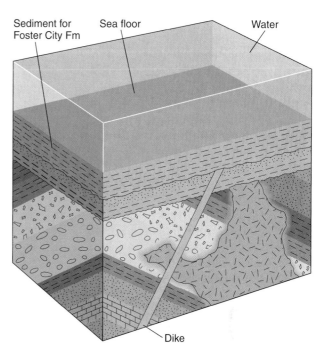

FIGURE 8.9

The area after rock overlying the Larsonton Formation, along with part of the dike, was removed by erosion.

FIGURE 8.10

Sediment being deposited that will become part of the Foster City Formation.

Once again, sedimentation took place as the lowermost layer of the Foster City Formation blanketed the erosion surface (figure 8.10). Sedimentation continued until the uppermost layer (top of the Skinner Gulch Limestone) was deposited. At some later time, the area was raised above sea level, and the stream began to carve the canyon (figure 8.11). Because the valley sides truncated the youngest layers of rock, we can determine from figure 8.1 that the last event was the carving of the valley.

Note that there are limits on how precisely we can determine the relative age of the granite body. It definitely intruded

before the Larsonton Formation was deposited and *after* the Tarburg Formation was deposited. As no contacts can be observed between the Leet Junction Formation and the granite, we cannot say whether the granite is younger or older than the Leet Junction Formation. Nor, as mentioned earlier, can we determine whether the granite formed before, during, or after the tilting of the lower sequence of sedimentary rocks.

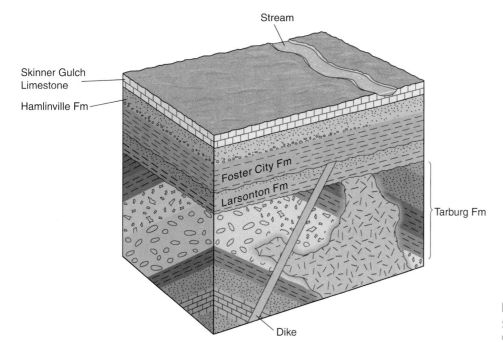

FIGURE 8.11

Same area as shown in figure 8.1 but before valley was carved into rock.

We must also point out some special circumstances under which these principles do not apply. For example, intense deformation by tectonic forces can overturn or disrupt beds so much that the principle of superposition cannot be used (see chapter 15). A geologist must avoid being dogmatic in applying principles.

Now, if you take another look at the chapter opening photo of the Grand Canyon (and figure 8.16), you should be able to determine the sequence of events. The sequence (going from older to younger) is as follows. Regional metamorphism took place resulting in the Vishnu Schist of the lower part of the Grand Canyon (you cannot tell these are schists from the photograph). Erosion followed and leveled the land surface. Sedimentation followed, resulting in the Grand Canyon Series rocks. These sedimentary layers were subsequently tilted (they were also faulted, although this is not evident in the photograph). Once again, erosion took place. The lowermost of the presently horizontal layers of sedimentary rock was deposited (the Tapeats Sandstone followed by the Bright Angel Shale). Subsequently, each of the layers progressively higher up the sequence formed. Finally, the stream (the Colorado River) eroded its way through the rock, carving the Grand Canyon.

Other Time Relationships

Other characteristics of geology can be applied to help determine relative ages (figure 8.12). The tilted layers in figure 8.12 immediately adjacent to the granite body have been *contact metamorphosed* (or "baked"). This indicates that the Tarburg Formation and older formations shown in figure 8.1 had to be there before intrusion of the hot, granite magma. The base of

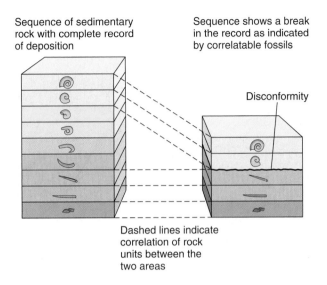

FIGURE 8.13

Schematic representation of a disconformity. The disconformity is in the block on the right.

the Larsonton Formation in contact with the granite would not be contact metamorphosed because it was deposited after the granite had cooled (and exposed by erosion).

The principle of **inclusion** states that fragments included in a host rock are older than the host rock. In figure 8.12, the granite contains inclusions of the tilted sedimentary rock. Therefore, the granite is younger than the tilted rock. The rock overlying the granite has granite pebbles in it. Therefore, the granite is older than the horizontal sedimentary rock.

Unconformities

In this and earlier chapters, we noted the importance of *contacts* for deciphering the geologic history of an area. In chapters 3 and 6 we described intrusive contacts and sedimentary contacts. Faults (described in chapter 15) are a third type of contact. The final important type of contact is an *unconformity*. Each type of contact has a very different implication about what took place in the geologic past.

An **unconformity** is a surface (or contact) that represents a *gap in the geologic record,* with the rock unit immediately above the contact being considerably younger than the rock beneath. Most unconformities are buried erosion surfaces. Unconformities are classified into three types—disconformities, angular unconformities, and nonconformities—with each type having important implications for the geologic history of the area in which it occurs.

Disconformities

In a **disconformity,** the contact representing missing rock strata separates beds that are parallel to one another. Probably what has happened is that older rocks were eroded away parallel to the bedding plane; renewed deposition later buried the erosion surface (figure 8.13).

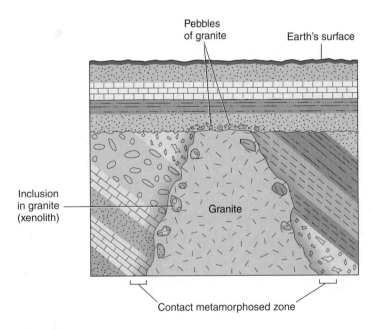

FIGURE 8.12

Age relationships indicated by contact metamorphism, inclusions (xenoliths) in granite, and pebbles of granite.

Because it often appears to be just another sedimentary contact (or bedding plane) in a sequence of sedimentary rock, a disconformity is the hardest type of unconformity to detect in the field. Rarely, a telltale weathered zone is preserved immediately below a disconformity. Usually, the disconformity can be detected only by studying fossils from the beds in a sequence of sedimentary rocks. If certain fossil beds are absent, indicating that a portion of geologic time is missing from the sedimentary record, it can be inferred that a disconformity is present in the sequence. Although it is most likely that some rock layers are missing because erosion followed deposition, in some instances neither erosion nor deposition took place for a significant amount of geologic time.

Angular Unconformities

An **angular unconformity** is a contact in which younger strata overlie an erosion surface on tilted or folded layered rock. It implies the following sequence of events, from oldest to youngest: (1) deposition and lithification of sedimentary rock (or solidification of successive lava flows if the rock is volcanic); (2) uplift accompanied by folding or tilting of the layers; (3) erosion; and (4) renewed deposition (usually preceded by subsidence) on top of the erosion surface (figure 8.14). Figures 8.1 and 8.12 also show angular unconformities but with simple tilting rather than folding of the older beds.

Nonconformities

A **nonconformity** is a contact in which an erosion surface on plutonic or metamorphic rock has been covered by younger sedimentary or volcanic rock (figure 8.15). A nonconformity generally indicates deep or long-continued erosion before subsequent burial, because metamorphic or plutonic rocks form at considerable depths in Earth's crust.

The geologic history implied by a nonconformity, shown in figure 8.15, is (1) crystallization of igneous or metamorphic rock at depth; (2) erosion of at least several kilometers of overlying rock (the great amount of erosion further implies considerable

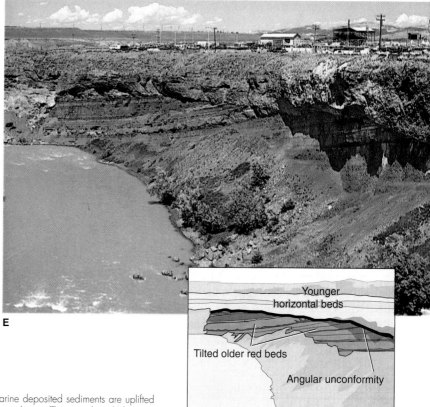

FIGURE 8.14

A particular sequence of events (*A–D*) producing an angular unconformity. Marine deposited sediments are uplifted and folded (probably during plate-tectonic convergence). Erosion removes the upper layers. The area drops below sea level (or sea level rises) and renewed sedimentation takes place. (An angular unconformity can also involve terrestrial sedimentation.) (*E*) is an angular unconformity at Cody, Wyoming. *Photo by C. C. Plummer*

Geologist's View

FIGURE 8.15

(*A–F*) Sequence of events implied by a nonconformity underlain by metamorphic and plutonic rock. (*G*) A nonconformity in Grand Canyon, Arizona. Paleozoic sedimentary rocks overlie vertically foliated Precambrian metamorphic rocks. *Photo by C. C. Plummer*

uplift of this portion of Earth's crust); and (3) deposition of new sediment, which eventually becomes sedimentary rock, on the ancient erosion surface. Figures 8.1 and 8.12 also show nonconformities; however, these represent erosion to a relatively shallow depth as the rocks intruded by the pluton have not been regionally metamorphosed, as was the case for those in figure 8.15.

Correlation

In geology, **correlation** usually means determining time equivalency of rock units. Rock units may be correlated within a region, a continent, and even between continents. Various

methods of correlation are described along with examples of how the principles we described earlier in this chapter are used to determine whether rocks in one area are older or younger than rocks in another area.

Physical Continuity

Finding **physical continuity**—that is, being able to trace physically the course of a rock unit—is one way to correlate rocks between two different places. The prominent white layer of cliff-forming rock in figure 8.16 is the Coconino Sandstone, exposed along the upper part of the Grand Canyon. It can be seen all the way across the photograph. You can physically fol-

FIGURE 8.16

Schematic cross section through part of the Colorado Plateau showing the relationship of the Coconino Sandstone, the white cliff-forming unit in the left photo, in Grand Canyon, to the Navajo Sandstone, white unit in the right photo, at Zion National Park. *Photos by C. C. Plummer*

low this unit for several tens of kilometers, thus verifying that, wherever it is exposed in the Grand Canyon, it is the same rock unit. The Grand Canyon is an ideal location for correlating rock units by physical continuity. However, it is not possible to follow this rock unit from the Grand Canyon into another region because it is not continuously exposed. We usually must use other methods to correlate rock units between regions.

Similarity of Rock Types

Under some circumstances, correlation between two regions can be made by assuming that similar rock types in two regions formed at the same time. This method must be used with extreme caution, especially if the rocks being correlated are common ones.

To show why correlation by similarity of rock type does not always work, we can try to correlate the white, cliff-forming Coconino Sandstone in the Grand Canyon with a rock unit of similar appearance in Zion National Park about 100 kilometers away (figure 8.16). Both units are white sandstone. Cross-bedding indicates that both were once a series of sand dunes. It is tempting to correlate them and conclude that both formed at the same time. But if you were to drive or walk from the rim of

the Grand Canyon (where the Coconino Sandstone is *below* you), you would get to Zion by ascending a series of layers of sedimentary rock stacked on one another. In other words, you would be getting into progressively younger rock, as shown diagrammatically in figure 8.16. In short, you have shown through *superposition* that the sandstone in Zion (called the Navajo Sandstone) is younger than the Coconino Sandstone.

Correlation by similarity of rock types is more reliable if a very unusual sequence of rocks is involved. If you find in one area a layer of green shale on top of a red sandstone that, in turn, overlies basalt of a former lava flow and then find the same sequence in another area, you probably would be correct in concluding that the two sequences formed at essentially the same time.

When the hypothesis of continental drift was first proposed (see chapter 1), important evidence was provided by correlating a sequence of rocks (figure 8.17) consisting of glacially deposited sedimentary rock (tillites, described in chapter 12 on glaciation), overlain by continental sandstones, shales, and coal beds. These strata are in turn overlain by basalt flows. The sequence is found in parts of South America, Australia, Africa, Antarctica, and India. It is very unlikely that an identical sequence of rocks could have formed on each of the continents

FIGURE 8.17

Rock sequences similar to this are found in India, Africa, South America, Australia, and Antarctica. The rocks in each of these localities contain the fossil plant *Glossopteris*.

if they were widely separated, as they are at present. Therefore, the continents on which the sequence is found are likely to have been part of a single, super-continent on which the rocks were deposited. Fossils found in these rocks further strengthened the correlation.

In some regions, a *key bed,* a very distinctive layer, can be used to correlate rocks over great distances. An example is a layer of volcanic ash produced from a very large eruption and distributed over a significant portion of a continent.

Correlation by Fossils

Fossils are common in sedimentary rock, and their presence is important for correlation. Plants and animals that lived at the time the rock formed were buried by sediment, and their fossil remains are preserved in sedimentary rock. Most of the fossil species found in rock layers are now extinct—99.9% of all species that ever lived are extinct. (The concept of *species* for fossils is similar to that in biology.)

In a thick sequence of sedimentary rock layers, the fossils nearer the bottom (that is, in the older rock) are more unlike today's plants and animals than are those near the top. As early as the end of the eighteenth century, naturalists realized that the fossil remains of creatures of a series of "former worlds" were preserved in Earth's sedimentary rock layers. In the early nineteenth century, a self-educated English surveyor named William Smith realized that different sedimentary layers are characterized by distinctive fossil species and that *fossil species succeed one another through the layers in a predictable order.* Smith's discovery of this principle of **faunal succession** allowed rock layers in different places to be correlated based on their fossils. We now understand that faunal succession works because there is an evolutionary history to life on Earth. Species evolve, exist for a time, and go extinct. Because the same species never evolves twice (extinction is forever), any period of time in Earth history can be identified by the species that lived at that time. *Paleontologists,* specialists in the study of fossils, have patiently and meticulously over the years identified many thousands of species of fossils and determined the time sequence in which

they existed. Therefore, sedimentary rock layers anywhere in the world can be assigned to their correct place in geologic history by identifying the fossils they contain.

Ideally, a geologist hopes to find an **index fossil,** a fossil from a very short-lived, geographically widespread species known to exist during a specific period of geologic time. A single index fossil allows the geologist to correlate the rock in which it is found with all other rock layers that contain that fossil.

Many fossils are of little use in time determination because the species thrived during too large a portion of geologic time. Sharks, for instance, have been in the oceans for a long time, so discovering an ordinary shark's tooth in a rock is not very helpful in determining the rock's relative age.

A geologist is likely to find a **fossil assemblage,** several different fossil species in a rock layer. A fossil assemblage is generally more useful for dating rocks than a single fossil is, because the sediment must have been deposited at a time when all the species represented existed (figure 8.18).

Some fossils are restricted in geographic occurrence, representing organisms adapted to special environments. But many former organisms apparently lived over most of the Earth, and fossil assemblages from these may be used for worldwide correlation. Fossils in the lowermost horizontal layers of the Grand Canyon are comparable to ones collected in Wales, Great Britain, and many other places in the world (the trilobites in figure 8.19 are an example). We can, therefore, correlate these rock units and say they formed during the same general span of geologic time.

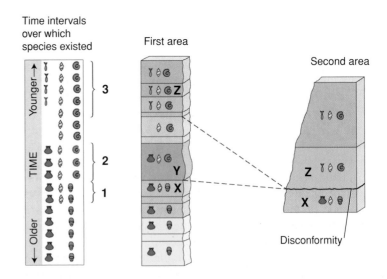

FIGURE 8.18

The use of fossil assemblages for determining relative ages. Rock X contains 🐚 🌿 🐚. Therefore it must have formed during time interval 1. Rock Y contains 🐚 🌿 🐚. Therefore, it must have formed during time interval 2. Rock Z contains 🐚 🌿 🐚. Therefore, it must have formed sometime during time interval 3. In the second area, fossils of time interval 2 are missing. Therefore, the surface between X and Z is a disconformity.

FIGURE 8.19

Elrathia kingii trilobites from the Middle Cambrian Wheeler Formation of Utah. The larger one is 10 mm in diameter. *Photo by Robert R. Gaines*

The Standard Geologic Time Scale

Geologists can use fossils in rock to refer the age of the rock to the **standard geologic time scale,** a worldwide relative time scale. Based on fossil assemblages, the geologic time scale subdivides geologic time. On the basis of fossils found, a geologist can say, for instance, that the rocks of the lower portion of horizontal layers in the Grand Canyon formed during the *Cambrian Period.* This implicitly correlates these rocks with certain rocks in Wales (in fact, the period takes its name from *Cambria,* the Latin name for Wales) and elsewhere in the world where similar fossils occur.

The geologic time scale, shown in a somewhat abbreviated form in table 8.2, has had tremendous significance as a unifying concept in the physical and biological sciences. The working out of the evolutionary chronology by successive generations of geologists and other scientists has been a remarkable human achievement. The geologic time scale, representing an extensive fossil record, consists of three **eras,** which are divided into **periods,** which are, in turn, subdivided into **epochs.** (Remember that this is a relative time scale.)

Precambrian denotes the vast amount of time that preceded the Paleozoic Era (which begins with the Cambrian Period). The **Paleozoic Era** (meaning "old life") began with the appearance of complex life (trilobites, for example), as indicated by fossils. Rocks older than Paleozoic contain few fossils. This is because creatures with shells or other hard parts, which are easily preserved as fossils, did not evolve until the beginning of the Paleozoic.

The **Mesozoic Era** (meaning "middle life") followed the Paleozoic. On land, dinosaurs became the dominant animals of the Mesozoic. We live in the **Holocene** (or **Recent**) **Epoch** of

the **Quaternary Period** of the **Cenozoic Era** (meaning "new life"). The Quaternary also includes the most recent ice ages, which were part of the **Pleistocene Epoch.**

It is noteworthy that the fossil record indicates mass extinctions, in which a large number of species became extinct, occurred a number of times in the geologic past. The two greatest mass extinctions define the boundaries between the three eras (see boxes 8.1 and 8.2).

Fossils have been used to determine ages of the horizontal rocks in Grand Canyon. All are Paleozoic. The lowermost horizontal formations (chapter opening photo) are Cambrian, above which are Devonian, Mississippian, Pennsylvanian, and Permian rock units. By referring to the geologic time scale (table 8.2), we can see that Ordovician and Silurian rocks are not represented. Thus, an unconformity (buried erosion surface) is present within the horizontally layered rocks of Grand Canyon.

NUMERICAL AGE

Counting annual growth rings in a tree trunk will tell you how old a tree is. Similarly, layers of sediment deposited annually in

TABLE 8.2	Geologic Time Scale		
Era	**Period**		**Epoch**
Cenozoic	*Quaternary*		Holocene (Recent)
			Pleistocene
	***Tertiary**	Neogene	Pilocene
			Miocene
		Paleogene	Oligocene
			Eocene
			Paleocene
Mesozoic	*Cretaceous*		
	Jurassic		
	Triassic		
Paleozoic	*Permian*		
	Pennsylvanian	} Carboniferous*	
	Mississippian		
	Devonian		
	Silurian		
	Ordovician		
	Cambrian		
Precambrian Time			

*Outside of North America, Carboniferous Period is used rather than Pennsylvanian and Mississippian.

**In 2003, the International Commission on Stratigraphy recommended dropping Tertiary and replacing it with Paleogene and Neogene (shown in red, along with their boundaries). It will probably take the larger geologic community a long time, if ever, to adapt to the change.

EARTH SYSTEMS 8.1

Highlights of the Evolution of Life through Time

The history of the biosphere is preserved in the fossil record. Through fossils, we can determine their place in the evolution of plants and animals as well as get clues as to how extinct creatures lived. The oldest readily identifiable fossils found are prokaryotes—microscopic, single-celled organisms that lack a nucleus. These date back to around 3.5 billion years (b.y.) ago, so life on Earth is at least that old. It is likely that even more primitive organisms date back further in time but are not preserved in the fossil record. Fossils of much more complex, single-celled organisms that contained a nucleus (eukaryotes) are found in rocks as old as 1.4 b.y. These are the earliest living creatures to have reproduced sexually. Colonies of unicellular organisms likely evolved into multicellular organisms. Multicellular algae fossils date back at least a billion years.

Imprints of larger multicellular creatures appear in rocks of late Precambrian age, about 700 to 550 million years ago (m.y.). These resemble jellyfish and worms.

Sedimentary rocks from the Paleozoic, Mesozoic, and Cenozoic Eras have abundant fossils. Large numbers of fossils appeared early in the Cambrian Period. Trilobites (see figure 8.19) evolved into many species and were particularly abundant during the Cambrian. Trilobites were arthropods that crawled on muddy sea floors and are the oldest fossils with eyes. They became less significant later in the Paleozoic, and finally, all trilobites became extinct by the end of the Paleozoic.

The most primitive fishes, the first vertebrates, date back to late in the Cambrian. Fishes similar to presently living species (including sharks) flourished during the Devonian (named after Devonshire, England). The Devonian is often called the "age of fishes." Amphibians evolved from air-breathing fishes late in the Devonian. These were the first land vertebrates. However, invertebrate land animals date back to the latest Cambrian, and land plants first appeared in the Ordovician. Reptiles and early ancestors of mammals evolved from amphibians in Pennsylvanian time or perhaps earlier.

The Paleozoic ended with the greatest mass extinction ever to occur on Earth. Over 95% of species that existed died out.

During the Mesozoic, new creatures evolved to occupy ecological domains left vacant by extinct creatures. Dinosaurs and mammals evolved from the animal species that survived the great extinction. Dinosaurs became the dominant group of land animals. Birds likely evolved from dinosaurs in the Mesozoic. Large, now extinct, marine reptiles lived in Mesozoic seas. Ichtyosaurs, for example, were up to 20 meters long, had dolphinlike bodies, and were probably fast swimmers. Flying reptiles, pterosaurs, some of which had wingspans of almost 10 meters, soared through the air.

The Cretaceous Period (and Mesozoic Era) ended with the second-largest mass extinction (around 75% of species were wiped out).

The Cenozoic is often called the age of mammals. Mammals, which were small, insignificant creatures during the Mesozoic, evolved into the many groups of mammals (whales, bats, canines, cats, elephants, primates, and so forth) that occupy Earth at present. Many species of mammals evolved and became extinct throughout the Cenozoic. Hominids (modern humans and our extinct ancestors) have a fossil record dating back 6 m.y. and likely evolved from a now extinct ancestor common to hominids, chimpanzees, and other apes.

We tend to think of mammals' evolution as being the great success story (because we are mammals); mammals, however, pale in comparison to insects. Insects have been around far longer than mammals and now account for an estimated 1 million species.

Additional Resources

University of California Museum of Paleontology
Find the fossils mentioned here.

- www.ucmp.berkeley.edu/

The Paleontology Portal
Another site to find out about fossils. You can search by type of creature, by time, or by location.

- www.paleoportal.org/

---

glacial lakes can be counted to determine how long those lakes existed (*varves,* as these deposits are called, are explained in chapter 12). But only within the few decades following the discovery of radioactivity in 1896 have scientists been able to determine numerical ages of rock units. We have subsequently been able to assign numerical values to the geologic time scale and determine how many years ago the various eras, periods, and epochs began and ended: The Cenozoic Era began some 65 million years ago, the Mesozoic Era started about 250 million years ago, and the Precambrian ended (or the Paleozoic began) about 545 million years ago. The Precambrian includes most of geologic time, because the age of Earth is commonly regarded as about 4.5 to 4.6 billion years.

The oldest rock found on Earth is a gneiss from northwestern Canada (its location is indicated on the inside front cover) that has been dated at 4.03 billion years old. In 2001, the oldest known mineral was dated at 4.4 billion years old, which is much older than the oldest rock dated so far. The mineral, a zircon crystal from Australia, was likely originally in a granite. Scientists who have studied this mineral think that its chemical makeup indicates that the granite formed from a magma that had a component of melted sedimentary rock. This would indicate that seas existed much earlier than geologists had previously thought possible.

Isotopic Dating

Radioactivity provides a "clock" that begins running when radioactive elements are sealed into newly crystallized minerals. The rates at which radioactive elements decay have been

EARTH SYSTEMS 8.2

Demise of the Dinosaurs—Was It Extraterrestrial?

The story of the rise and fall of dinosaurs involves the biosphere (the dinosaurs and their ecosystem), the solar system (extraterrestrial objects), the atmosphere (which changed abruptly), and the hydrosphere (part of an ocean was vaporized). Dinosaurs dominated the continents during the Mesozoic Era. Now they prey on the imaginations of children of all ages and are featured in media ranging from movies to cartoons. It's hard to accept that beings as powerful and varied as dinosaurs existed and were wiped out. But the fossil record is clear—when the Mesozoic came to a close, dinosaurs became extinct. Not a single of the numerous dinosaur species survived into the Cenozoic Era. Not only did the dinosaurs go, but about 75% of all plant and animal species, marine as well as terrestrial, were extinguished. This was one of Earth's "great dyings"—an even "greater dying" was when the Paleozoic Era ended with the extinction of over 95% of all species. Most major extinctions have been gradual, and scientists usually have attributed them to climate changes.

A couple decades ago, geologist Walter Alvarez, his father, physicist Luis Alvarez, and two other scientists proposed a hypothesis that the dinosaur extinction was caused by the impact of an asteroid. This was based on the chemical analysis of a thin layer of clay marking the boundary between the Mesozoic and Cenozoic Eras (usually referred to as the K-T boundary—it separates the Cretaceous [K] and Tertiary [T] Periods). The K-T boundary clay was found to have about 30 times the amount of the rare element iridium as is normal for crustal rocks. Iridium is relatively abundant in meteorites and other extraterrestrial objects such as comets, and the scientists suggested that the iridium was brought in by an extraterrestrial body.

A doomsday scenario is visualized in which an asteroid 10 kilometers in diameter struck Earth. The asteroid would have blazed through the atmosphere at astonishing speed and, likely, impacted at sea. Part of the ocean would have been vaporized and a crater created on the ocean floor. There would have been an earthquake much larger than any ever felt by humans. Several-hundred-meter-high waves would crisscross the oceans, devastating life anywhere near shorelines. The lower atmosphere would have become intolerably hot, at least for a short period of time. The atmosphere worldwide would have been altered and the climate cooled because of the increased blockage of sunlight by dust particles suspended in the upper atmosphere.

For a while, the hypothesis was hotly debated. Other scientists hypothesized that the extinctions were caused by exceptionally large volcanic activity. Further evidence supporting the asteroid hypothesis accumulated. K-T layers throughout the world were found to have grains of quartz that had been subjected to shock metamorphism (see box 7.1). Microscopic spheres of glass that formed when rock melted from the impact and droplets were thrown high into the air were also found in the K-T layers. Sedi-

ment that appeared to have been deposited by giant sea waves was found in various locations.

The asteroid hypothesis advocates predicted that a large meteorite crater should be found someplace on Earth that could be dated as having formed around 65 million years ago, when the Mesozoic ended.

In 1990, the first evidence for the "smoking gun" crater was found. The now-confirmed crater is over 200 kilometers in diameter and centered along the coast of Mexico's Yucatan peninsula at a place called Chicxulub. The crater at Chicxulub, now buried beneath younger sedimentary rock, is the right size to have been formed by a 10-kilometer asteroid.

The existence of the crater was confirmed by geologists going over Mexican oil company records compiled during drilling for oil at Yucatan and finding breccias of the right age buried in the Chicxulub area. Breccias, due to meteorite impact, are common at known meteorite craters. The evidence for an asteroid impact is overwhelming. However, the impact may not be entirely to blame for dinosaur extinction.

As the Mesozoic was ending, huge lava floods were taking place in India over a 2- to 3-million-year period and building a basalt plateau larger than the Columbia Plateau described in chapter 4. Gas emissions very likely had an effect on the atmosphere and worldwide climate. It is noteworthy that Earth's largest mass extinction at the end of the Paleozoic has often been blamed on even larger lava floods that occurred in Siberia. However, recently found evidence strongly indicates that a major asteroid impact also occurred at that time. (Recently, geologists reported that they found evidence for an appropriate crater buried beneath sedimentary rocks off the coast of Australia. Other geologists remain skeptical.) It seems to many geologists that having the two largest mass extinctions associated with an impact and lava floods is more than coincidence. Some have suggested that the lava floods were somehow triggered by the asteroid impacts, even though they were in different parts of the world.

We will probably never know exactly how big a role either the asteroid or the lava floods played in the unfortunate extinction of dinosaurs. But "unfortunate" is from the perspective of dinosaurs, not humans. The only mammals in the Cretaceous were inconsequential, rat-sized creatures. They survived the K-T extinction and, with dinosaurs no longer dominating the land, evolved into the many mammal species that populate Earth today, including humans.

Additional Resource
Walking with Dinosaurs
Visit *Tyrannosaurus rex* and other famous dinosaurs and read more about the K-T extinction.
- www.bbc.co.uk/dinosaurs/

measured and duplicated in many different laboratories. Therefore, if we can determine the ratio of a particular radioactive element and its decay products in a mineral, we can calculate how long ago that mineral crystallized.

Determining the age of a rock through its radioactive elements is known as **isotopic dating** (previously, and somewhat inaccurately, called *radiometric dating*). Geologists who specialize in this important field are known as *geochronologists*.

Isotopes and Radioactive Decay

As discussed in chapter 2, every atom of a given element possesses the same number of protons in its nucleus. The number of neutrons, however, need not be the same in all atoms of the same element. The **isotopes** of a given element have different numbers of neutrons but the same number of protons.

Uranium, for example, commonly occurs as two isotopes, uranium-238 (^{238}U) and uranium-235 (^{235}U). The former has 238 protons and neutrons in its nucleus, whereas the latter has 235. (^{238}U is, by far, the most abundant of naturally occurring uranium isotopes. Only 0.72% of uranium is ^{235}U; however, this is the isotope used for nuclear weapons and power generators.) For both isotopes, 92 (the atomic number of uranium) nuclear particles must be protons and the rest neutrons.

Radioactive decay is the spontaneous nuclear change of isotopes with unstable nuclei. Energy is produced with radioactive decay. Emissions from radioactive elements can be detected by a Geiger counter or similar device and, in high concentrations, can damage or kill humans (see box 8.3).

Nuclei of radioactive isotopes change primarily in three ways (figure 8.20). An *alpha* (α) *emission* is the ejection of 2 protons and 2 neutrons from a nucleus. When an alpha emission takes place, the atomic number of the atom is reduced by 2, and its atomic mass number is reduced by 4. After an alpha emission, ^{238}U becomes ^{234}Th (thorium), which has an atomic number of 90. The original isotope (^{238}U) is referred to as the *parent isotope*. The new isotope (^{234}Th) is the *daughter product*.

Beta (β) *emissions* involve the release of an electron from a nucleus. To understand this, we need to explain that electrons, which have virtually no mass and are usually in orbit around the nucleus, are also in the nucleus as part of a neutron. A neutron is a proton with an electron inside of it; thus, it is electrically neutral. If an electron is emitted from a neutron during radioactive decay, the neutron becomes a proton and the atom's atomic number is increased by one. For example, when ^{234}Th undergoes a beta emission, it becomes ^{234}Pa, an element with an atomic number of 91. Note that the atomic mass number has not changed. This is because the weight of an electron is negligible.

The third mode of change is *electron capture*, whereby a proton in the nucleus captures an orbiting electron. The proton becomes a neutron. The atom becomes a different element having an atomic number one less than its parent isotope. An example of this is the potassium-argon system in table 8.3, in

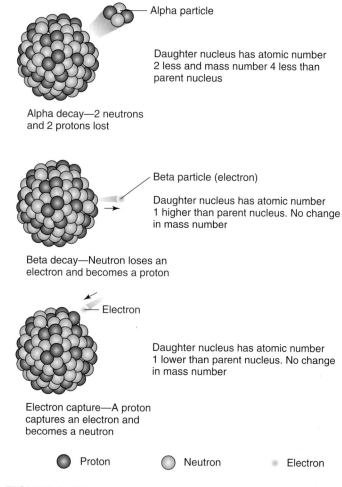

Alpha particle

Daughter nucleus has atomic number 2 less and mass number 4 less than parent nucleus

Alpha decay—2 neutrons and 2 protons lost

Beta particle (electron)

Daughter nucleus has atomic number 1 higher than parent nucleus. No change in mass number

Beta decay—Neutron loses an electron and becomes a proton

Electron

Daughter nucleus has atomic number 1 lower than parent nucleus. No change in mass number

Electron capture—A proton captures an electron and becomes a neutron

● Proton   ○ Neutron   • Electron

FIGURE 8.20

Three modes of radioactive decay.

which ^{40}K becomes ^{40}Ar. The parent isotope, potassium, has an atomic number of 19, and the atomic number of argon, the daughter product, is 18.

Figure 8.21 shows how ^{238}U decays to ^{206}Pb (lead-206) in a series of alpha and beta emissions. The important point is not the intermediate steps but the starting and ending isotopes. In the process, ^{238}U loses 10 protons, so that the daughter product has an atomic number of 82 (which is lead) and a total of 32 protons and neutrons, so the new atomic mass number is 206. ^{206}Pb can only be produced by the decay of ^{238}U.

To understand how isotopic dating works, it is important to recognize that if a large number of atoms of a given radioactive isotope are present in a rock or mineral, the *proportion* (or percentage) of those atoms that will radioactively decay over a given time span is constant. For example, if you have 100,000 atoms of isotope X and over a period of a million years, a quarter of those atoms (25,000) radioactively decay (the proportion would be 1 in 4). You would have the same proportion of 1 in 4 if you started out with 300,000 atoms: after a million years, 75,000 of the atoms would have decayed. The proportional

FIGURE 8.21

Uranium 238 decays to lead 206. The different intermediate steps in the process are shown below the models of the nuclei of ^{238}U and ^{206}Pb. Refer to appendix C or the periodic table of elements in appendix D for names of the elements shown.

amount of atoms that decay in time is unaffected by chemical reactions or by the high pressures and high temperatures of Earth's interior.

The rate of proportional decay for isotopes is expressed as **half-life,** the time it takes for a given amount of a radioactive isotope to be reduced by one-half. (The other half disintegrates into daughter products and energy.) The half-lives of some isotopes created in nuclear reactors are in fractions of a second. Naturally occurring isotopes used to date rocks have very long half-lives (table 8.3). ^{40}K has a half-life of 1.3 billion years. If you began with 1 milligram of ^{40}K, 1.3 billion years later one-half milligram of ^{40}K would remain. After another 1.3 billion years, there would be one-fourth of a milligram, and after another half-life, only one-eighth of a milligram. (Note that two half-lives do not equal a whole life.)

TABLE 8.3	Radioactive Istopes Commonly Used for Determining Ages of Earth's Materials		
Parent Isotope	**Half-Life**	**Daughter Product**	**Effective Dating Range (years)**
K-40 ^{40}K	1.3 billion years	^{40}Ar	100,000–4.6 billion
U-238 ^{238}U	4.5 billion years	^{206}Pb	10 million–4.6 billion
U-235 ^{235}U	713 million years	^{207}Pb	10 million–4.6 billion
Th-232 ^{232}Th	14.1 billion years	^{208}Pb	10 million–4.6 billion
Rb-87 ^{87}Rb	49 billion years	^{87}Sr	10 million–4.6 billion
C-14 ^{14}C	5,730 years	^{14}N	100–40,000

Radon, a Radioactive Health Hazard

Radon is an odorless, colorless gas. Every time you breathe outdoors, you inhale a harmless, minute amount of radon. If the concentration of radon that you breathe in a building is too high, however, you could, over time, develop lung cancer. It is one of the intermediate daughter products in the radioactive disintegration of ^{238}U to ^{206}Pb. It has a half-life of only 3.8 days.

Concentrations of radon are highest in areas where the bedrock is granite, gneiss, limestone, black shale, or phosphate-rich rock—rocks in which uranium is relatively abundant. Concentrations are also high where glacial deposits are made of fragments of these rocks. Even in these areas, radon levels are harmless in open, freely circulating air. Radon may dissolve in ground water or build up to high concentrations in confined air spaces.

The U.S. Environmental Protection Agency (EPA) regards 5 million American homes to have unacceptable radon levels in the air. Scientists outside of EPA have concluded that the standards the EPA is using are too stringent. They think that a more reasonably defined danger level means that only 50,000 homes have radon concentrations that pose a danger to their occupants.

Radon was first recognized in the 1950s as a health hazard in uranium mines, where the gas would collect in poorly ventilated air spaces. Radon lodges in the respiratory system of an individual, and as it deteriorates into daughter products, the subatomic particles given off damage lung tissue. Three-quarters of the uranium miners studied were smokers. Thus, it is difficult to determine the extent to which smoking or radon induced lung cancer. (All studies show, however, that smoking and exposure to high radon levels are more likely to cause lung cancer than either alone.)

Interpolating the high rates of cancer incidence from the uranium miners to the population exposed to the very much lower radium levels in homes, as the EPA has done, is scientifically questionable.

What should you do if you are living in a high radon area? First, have your house checked to see what the radon level is. Then, read up on what acceptable standards should be. In most buildings with a high radon level, the gas seeps in from the underlying soil through the building's foundation. If a building's windows are kept open and fresh air circulates freely, radon concentrations cannot build up. But houses are often kept sealed for air conditioning during the summer and heating during the winter. Air circulation patterns are such that a slight vacuum sucks the gases from the underlying soil into the house. Thus, radon concentrations might build up to dangerous levels.

The problem may be solved in several ways (aside from leaving windows open winter and summer). Basements can be made air tight so that gases cannot be sucked into the house from the soil. Air circulation patterns can be altered so that gases are not sucked in from underlying soil or are mixed with sufficient fresh, outside air.

If you are purchasing a new house, it would be a good idea to have it tested for radon before buying, particularly if the house is in an area of high-uranium bedrock or soil.

Additional Resource

Radon in Earth, Air, and Water

Check the extent of radon hazard for any part of the United States.

- http://sedwww.cr.usgs.gov/radon/radonhome.html

To determine the age of a rock by using ^{40}K, the amount of ^{40}K in that rock must first be determined by chemical analysis. The amount of ^{40}Ar (the daughter product) must also be determined. Adding the two values gives us how much ^{40}K was present when the rock formed. By knowing how much ^{40}K was originally present in the rock and how much is still there, we can calculate the age of the rock on the basis of its half-life mathematically (see box 8.4). The graph in figure 8.22A applies the mathematical relationship between a radioactively decaying isotope and time and can be used to easily determine an isotopic age.

Radiocarbon Dating

Because of its short half-life of 5,730 years, radiocarbon dating is useful only in dating things and events accurately back to about 40,000 years—about seven half-lives. The technique is most useful in archaeological dating and for very young geologic events (Holocene, or Recent, volcanic and glacial features for instance). It is also used to date historical artifacts. For instance, the Dead Sea Scrolls, the oldest of the surviving bib-lical manuscripts, were radiocarbon dated and their ages ranged from the third century B.C. to 68 A.D. These ages are consistent with estimates previously made by archaeologists and other scholars.

Radiocarbon dating is fundamentally different from the parent-daughter systems described previously in that ^{14}C is being created continuously in the atmosphere. Carbon (atomic number 6) is in the air as part of CO_2. It is mostly the stable isotope ^{12}C. However, ^{14}C is created in the atmosphere when cosmic radiation bombards nitrogen (N), atomic number 7. A neutron strikes and is captured by an ^{14}N atom. A proton is expelled from the nucleus and the atom becomes ^{14}C. The nucleus of the newly created carbon atom is unstable and will, sooner or later, through a beta emission, revert to ^{14}N. The electron is emitted from the atom as radiation. The rate of production of ^{14}C approximately balances the rate at which ^{14}C reverts to ^{14}N so that the level of ^{14}C remains essentially constant in the atmosphere.

Living matter incorporates ^{12}C and ^{14}C into its tissues; the ratios of ^{12}C and ^{14}C in the new tissues are usually the same as in the atmosphere. On dying, the plant or animal ceases to

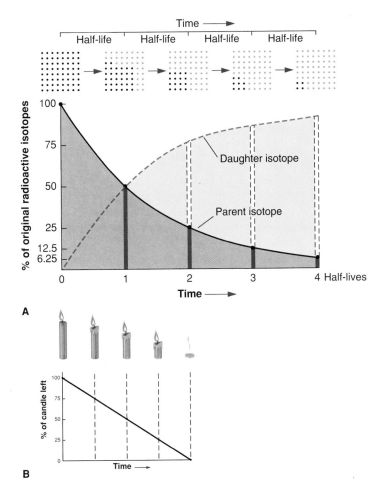

FIGURE 8.22

(A) The curve used to determine the age of a rock by comparing the percentage of radioactive isotope remaining in time to the original amount. Dark-blue bars show the amount left after each half-life. Dashed red curve shows the amount disintegrated into daughter product and lost nuclear particles. The numbers of dots in the squares above the graph are proportional to the numbers of atoms. (B) For comparison, a candle burns at a linear rate.

build new tissue. The ^{14}C disintegrates radioactively at the fixed rate of its half-life (5,730 years). The ratio of ^{12}C to radioactive ^{14}C in organic remains is determined in a laboratory. Using the ratio, the time elapsed since the death of the organism is calculated.

Cosmogenic Isotope Dating

During the past couple decades, another dating technique has been added to geologists' numerical age determination arsenal. *Cosmogenic isotope dating,* or *surface exposure dating,* uses the effects of constant bombardment by neutron radiation coming from deep space (cosmogenic) of material at Earth's surface. The high-energy particles hit atoms in minerals and alter their nuclei. For instance, when the atoms in quartz are hit, oxygen is converted to beryllium-10 (^{10}Be) and silicon is changed to aluminum-26 (^{26}Al). The concentrations of these isotopes increase at a constant rate once a rock surface is exposed to the atmosphere because the influx of cosmogenic radiation is uni-

form over time. The length of time a rock surface has been exposed can be calculated by knowing the rate of increase of a cosmogenic isotope and determining the amount of that isotope in a mineral at a rock's surface.

One application of cosmogenic dating has been to determine how long ago boulders were deposited by advancing glaciers during the geologically recent ice ages (see chapter 12). However, dates obtained are minimum ages, because snow that covered the boulders for part of the year reduced their exposure to cosmogenic radiation.

Uses of Isotopic Dating

When we are dating a rock, we are usually attempting to determine how long ago that rock formed. But exactly what is being dated depends on the type of rock and the isotopes analyzed. For a metamorphic rock, we are likely to be dating a time during the millions of years of the cooling of that rock rather than the peak of high temperature during metamorphism. Some techniques determine isotopic ratios for a whole rock, while others use single minerals within a rock. Usually, an isotopic date determines how long ago the rock or mineral became a closed system. That is, how long ago it was sealed off so that neither parent nor daughter isotopes could enter or leave the mineral or rock. Different isotopic pairs have different closure temperatures; when a rock cools below that temperature, the system is closed and the "clock" starts. For instance, the ^{40}K ^{40}Ar isotopic pair has closure temperatures ranging from 150°C to 550°C, depending on the mineral. (Ar is a gas and gets trapped in different crystal structures at different temperatures.)

Generally, the best dates are obtained from igneous rocks. For a lava flow, which cools and solidifies rapidly, the age determined is the precise time at which the rock formed. On the other hand, plutonic rocks, which may take over a million years to solidify, will not necessarily yield the time of intrusion but the time at which a mineral cooled below the closure temperature. Dating metamorphic rocks usually means determining when closure temperatures for particular minerals are reached during cooling. Sedimentary rocks are difficult to date reliably.

For an isotopic age determination to be accurate, several conditions must be met. To ensure that the isotopic system has remained closed, the rock collected must show no signs of weathering or hydrothermal alteration. Second, one should be able to infer there were no daughter isotopes in the system at the time of closure or make corrections for probable amounts of daughter isotopes present before the "clock" was set. Third, there must be sufficient parent and daughter atoms to be measurable by the instrument (a mass spectrometer) being used. And, of course, technicians and geochronologists must be highly skilled at working sophisticated equipment and collecting and processing rock specimens.

Whenever possible, geochronologists will use more than one isotope pair for a rock. The two U-Pb systems (table 8.3) can usually be used together and provide an internal cross-check

Calculating the Age of a Rock

(NOTE: This box is not intended for the mathematically challenged.)

The relationship between time and radioactive decay of an isotope is expressed by the following equation (which is used to plot curves such as shown in figure 8.22).

$$N = N_0 e^{-\lambda t}$$

N is the number of atoms of the isotope at time t, the time elapsed. N_0 is the number of atoms of that isotope present when the "clock" was set. The mathematical constant e has a value of 2.718. λ is a decay constant—a proportionality constant that relates the rate of decay of an isotope to the number of atoms of that isotope remaining.

The relationship between λ and the half-life (t_{hl}) is

$$\lambda = \frac{\ln 2}{t_{hl}} = \frac{0.693}{t_{hl}}$$

Replacing λ in the first equation and converting that equation to natural logarithmic (to the base e) form, we get

$$t = \frac{t_{hl}}{.693} \ln \frac{N}{N_0}$$

N/N_0 is the ratio of parent atoms at present to the original number of parent atoms.

As an example, we will calculate the age of a mineral using ^{235}U decaying to ^{207}Pb. Table 8.3 indicates that the half-life is 713 million years. A laboratory determines that, at present, there are 440,000 atoms of ^{235}U and that the amount of ^{207}Pb indicates that when the mineral crystallized, there were 1,200,000 atoms of ^{235}U. (We assume that there was no ^{207}Pb in the mineral at the time the mineral crystallized.) Plugging these values into the formula, we get

$$t = \frac{713,000,000}{.693} \ln \frac{440,000}{1,200,000}$$

Solving this gives us 1,032,038,250 years. Rounded off, we can say the mineral formed 1.032 billion years ago.

on the age determination. Because of their high closure temperatures, U-Pb systems are usually more realistic of crystallization ages of rocks than K-Ar or Rb-Sr results.

How Reliable Is Isotopic Dating?

Half-lives of radioactive isotopes, whether short-lived, such as used in medicine, or long-lived, such as used in isotopic dating, have been found not to vary beyond statistical expectations. The half-life of each of the isotopes we use for dating rocks has not changed with physical conditions or chemical activity, nor could the rates have been different in the distant past. It would violate laws of physics for decay rates (half-lives) to have been different in the past. Moreover, when several isotopic dating systems are painstakingly done on a single ancient igneous rock, the same age is obtained, or we understand the reason for differences in ages; it confirms that the decay constants for each system are indeed constant.

Comparing isotopic ages with relative age relationships confirms the reliability of isotopic dating. For instance, a dike that crosscuts rocks containing Cenozoic fossils gives us a relatively young isotopic age (less than 65 million years old), whereas a pluton truncated by overlying sedimentary rocks with earliest Paleozoic fossils yields a relatively old age (greater than 250 million years). Many thousands of similar determinations have confirmed the reliability of radiometric dating.

COMBINING RELATIVE AND NUMERICAL AGES

Radiometric dating can provide numerical time brackets for events whose relative ages are known. Figure 8.23 adds isotopic dates for each of the two igneous bodies in the fictitious Minor Canyon area of figure 8.1. The date obtained for the granite is 540 million years B.P. (before present), while the dike formed 78 million years ago. We can now state that the Tarburg Formation and older tilted layers formed before 540 million years ago (though we cannot say how much older they are). We still do not know whether the Leet Junction Formation is older or younger than the granite because of the lack of cross-cutting relationships. The Larsonton Formation's age is bracketed by the age of the granite and the age of the dike. That is, it is between 540 and 78 million years old. The Foster City and overlying formations are younger than 78 million years old; how much younger we cannot say.

Isotopic dates from volcanic ash layers or lava flows interlayered between fossiliferous sedimentary rocks have been used to assign numerical ages to the geologic time scale (figure 8.24). Isotopic dating has also allowed us to extend the time scale back into the Precambrian. There is, of course, a margin of uncertainty in each of the given dates. The beginning of the Paleozoic, for instance, was regarded until recently to be 570 million

FIGURE 8.23
The Minor Canyon area as shown in figure 8.1 but with isotopic dates for igneous rocks indicated.

years ago but with an uncertainty of ±30 million years. Recent work has fixed the age as 544 ± 1 million years. There are inherent limitations on the dating techniques as well as problems in finding the ideal rock for dating. For instance, if you wanted to obtain the date for the end of the Paleozoic Era and the beginning of the Mesozoic Era, the ideal rock would be found where there is no break in deposition of sediments between the two eras, as indicated by fossils in the rocks. But the difficulties in dating sedimentary rock mean you would be unlikely to date such rocks. Therefore, you would need to date volcanic rocks interlayered with sedimentary rocks found as close as possible to the transitional sedimentary strata. Alternatively, isotopically dated intrusions, such as dikes, whose cross-cutting relationships indicate that the age of intrusion is close to that of the transitional sedimentary layers, could be used to approximate the numerical age of the transition.

Isotopic dating has shown that the Precambrian took up most of geologic time (87%). Obviously, the Precambrian needed to be subdivided. The three major subdivisions of the Precambrian are the **Hadean** (*Hades* is, in Greek mythology, the underground place where the dead live; the name alludes to the hell-like nature of Earth's early surface), the **Archean,** and the **Proterozoic** (Greek for "beginning life"). Each is regarded as an **eon,** the largest unit of geological time. A fourth, and youngest, eon is the **Phanerozoic** (Greek for "visible life"). The Phanerozoic Eon is all of geologic time with an abundant fossil record; in other words, it is made up of the three eras that followed the Precambrian.

AGE OF THE EARTH

In 1625, Archbishop James Ussher determined that Earth was created in the year 4004 B.C. His age determination was made by counting back generations in the Bible. This would make Earth 6,000 years old at present. That very young age of Earth was largely taken for granted by Western countries. By contrast, Hindus at the time regarded Earth as very old. According to an ancient Hindu calendar, the year A.D. 2000 would be year 1,972,949,101.

With the popularization of uniformitarianism in the early 1800s, Earth scientists began to realize that Earth must be very old—at least in the hundreds of millions of years. They were dealt a setback by the famous English physicist, Lord Kelvin. Kelvin, in 1866, calculated from the rate at which Earth loses heat that Earth must have been entirely molten between 20 and 100 million years ago. He later refined his estimate to between 20 and 40 million years. He was rather arrogant in scoffing at Earth scientists who believed that uniformitarianism indicated a much older age for Earth. The discovery of radioactivity in 1896 invalidated Kelvin's claim because it provided a heat source that he had not known about. When radioactive elements decay, heat is given off and that heat is added to the heat already in Earth. The amount of radioactive heat given off at present approximates the heat Earth is losing. So, for all practical purposes, Earth is not getting cooler.

The discovery of radioactivity also provided the means to determine how old Earth is. In 1905, the first crude isotopic

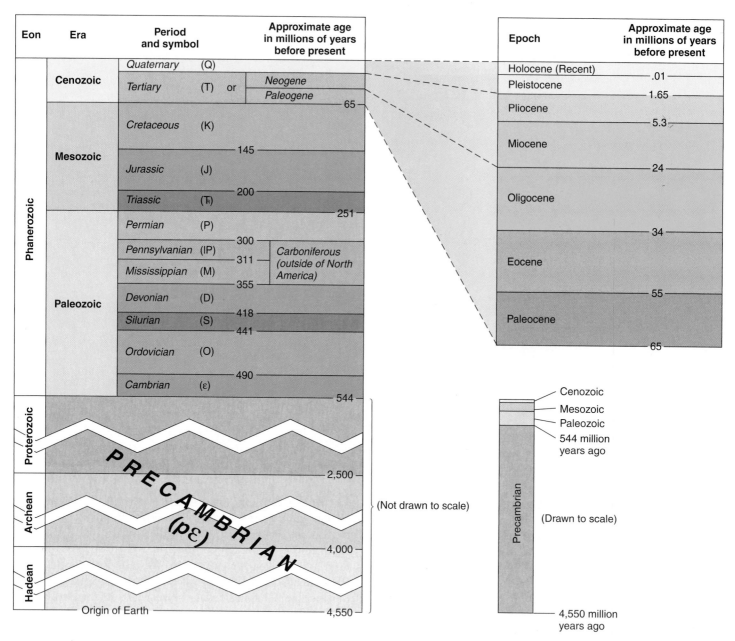

FIGURE 8.24

The geologic time scale. The small diagram to the right shows the Precambrian and the three eras at the same scale. Note that the Precambrian accounts for almost 90% of geologic time. *After A. V. Okulitch, 1999, Geological Survey of Canada, Open File 3040*

dates were done and indicated an age of 2 billion years. But since then, we have dated rocks on Earth that are twice that age.

Earth is now regarded as between 4.5 and 4.6 billion years old—much older than the oldest rock found. Because erosion and tectonic activity have recycled the original material at Earth's surface, we cannot determine Earth's age from its rocks. The age determination comes primarily from dates obtained from meteorites and lunar rocks. Most meteorites are regarded as fragments of material that did not coalesce into a planet. The oldest dates obtained from meteorites and lunar rocks are in the 4.5 to 4.6 billion-year range. It is highly likely that the planets and other bodies of the solar system, including Earth, formed at approximately the same time.

Comprehending Geologic Time

The vastness of geologic time (sometimes called deep time) is difficult for us to comprehend. One way of visualizing deep time is to imagine driving from Los Angeles to New York, a distance of approximately 4,500 kilometers, where each kilometer represents 1 million years—this is a very, very slow trip. The highlights of the trip corresponding to Earth's history are shown in figure 8.25. Note that if you live to be 100, your life is represented by less than the width of a curb at the edge of a sidewalk.

Another way to get a sense of geologic time is to compare it to a motion picture. A movie is projected at a rate of thirty-two frames per second; that is, each image is flashed on the

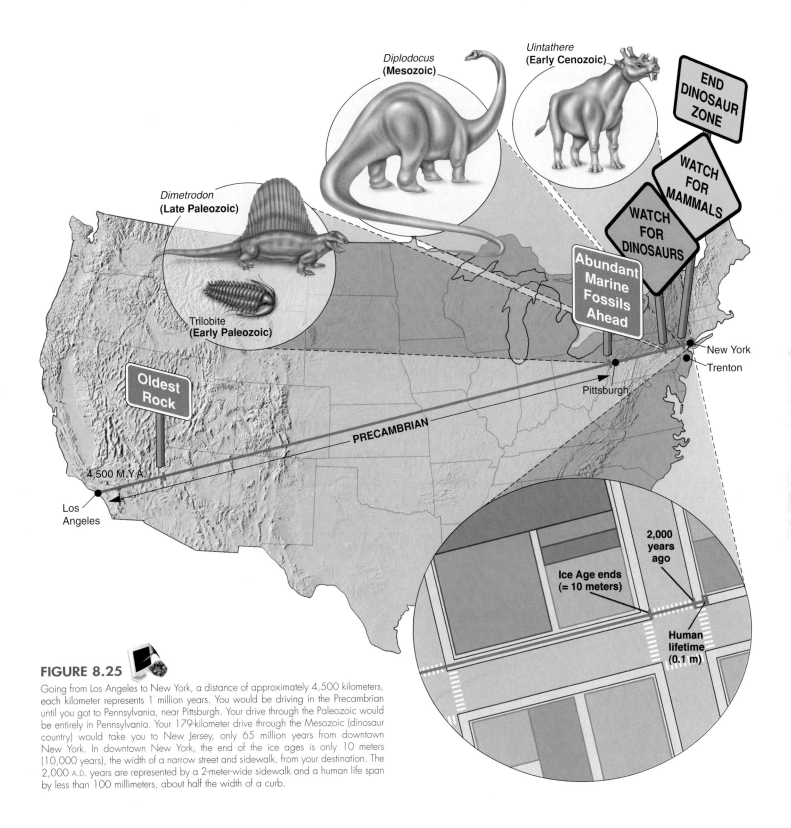

Labels on image:
- *Diplodocus* **(Mesozoic)**
- *Uintathere* **(Early Cenozoic)**
- *Dimetrodon* **(Late Paleozoic)**
- Trilobite **(Early Paleozoic)**
- END DINOSAUR ZONE
- WATCH FOR MAMMALS
- WATCH FOR DINOSAURS
- **Abundant Marine Fossils Ahead**
- New York
- Trenton
- Pittsburgh
- **Oldest Rock**
- PRECAMBRIAN
- 4,500 M.Y.A.
- Los Angeles
- **2,000 years ago**
- **Ice Age ends (= 10 meters)**
- **Human lifetime (0.1 m)**

FIGURE 8.25

Going from Los Angeles to New York, a distance of approximately 4,500 kilometers, each kilometer represents 1 million years. You would be driving in the Precambrian until you got to Pennsylvania, near Pittsburgh. Your drive through the Paleozoic would be entirely in Pennsylvania. Your 179-kilometer drive through the Mesozoic (dinosaur country) would take you to New Jersey, only 65 million years from downtown New York. In downtown New York, the end of the ice ages is only 10 meters (10,000 years), the width of a narrow street and sidewalk, from your destination. The 2,000 A.D. years are represented by a 2-meter-wide sidewalk and a human life span by less than 100 millimeters, about half the width of a curb.

screen for only 1/32 of a second, giving the illusion of continuous motion. But suppose that each frame represented 100 years. If you lived 100 years, one frame would represent your whole lifetime.

If we were able to show the movie on a standard projector, each 100 years would flash by in 1/32 of a second. It would take only 1/16 of a second to go back to the signing of the Declaration of Independence. The 2,000-year-old Christian era would be on screen for 3/4 of a second. A section showing all time back to the last major ice age would only be less than seven seconds long. However, you would have to sit through almost six hours of film to view a scene at the close of the Mesozoic Era (perhaps you would see the last dinosaur die). And to give a complete record from the beginning of the Paleozoic Era, this epic film would have to run continuously for two days. You would have to spend over two weeks (sixteen days) in the theater, without even a popcorn break between reels, to see a movie entitled *The Complete Story of Earth, from Its Birth to Modern Civilization.*

Thinking of our lives as taking less than a frame of such a movie can be very humbling. From the perspective of being stuck in that one last frame, geologists would like to know what the whole movie is like or, at least, get a synopsis of the most dramatic parts of the film.

SUMMARY

The principle of *uniformitarianism* (or *actualism,*) a fundamental concept of geology, states that the present is the key to the past.

Relative time, or the sequence in which geologic events occur in an area, can be determined by applying the principles of *original horizontality, superposition, lateral continuity,* and *cross-cutting relationships.*

Unconformities are buried erosion surfaces that help geologists determine the relative sequence of events in the geologic past. Beds above and below a *disconformity* are parallel, generally indicating less intense activity in Earth's crust. An *angular unconformity* implies that folding or tilting of rocks took place before or around the time of erosion. A *nonconformity* implies deep erosion because metamorphic or plutonic rocks have been exposed and subsequently buried by younger rock.

Rocks can be correlated by determining the physical continuity of rocks between the two areas (generally, this works only for a short distance). A less useful means of correlation is similarity of rock types (which must be used cautiously).

The principle of *faunal succession* states that fossil species succeed one another in a definite and recognizable order. Fossils are used for worldwide correlation of rocks. Sedimentary rocks are assigned to the various subdivisions of the *geologic time scale* on the basis of fossils they contain, which are arranged according to the principle of faunal succession.

Numerical age—how many years ago a geologic event took place—is generally obtained by using *isotopic dating* techniques. Isotopic dating is accomplished by determining the ratio of the amount of a radioactive isotope presently in a rock or mineral being dated to the amount originally present. The time it takes for a given amount of an isotope to decay to half that amount is the *half-life* for that isotope. Numerical ages have been determined for the subdivisions of the geologic time scale. The scientifically determined age of Earth is 4.5 to 4.6 billion years.

Terms to Remember

actualism 196
angular unconformity 203
Archean Eon 215
Cenozoic Era 207
contacts 197
correlation 204
cross-cutting relationship 198
disconformity 202
eon 215
epoch 207
era 207
faunal succession 206
formation 197
fossil assemblage 206

Hadean Eon 215
half-life 211
Holocene (or Recent) Epoch 207
inclusion 202
index fossil 206
isotope 210
isotopic dating 210
lateral continuity 198
Mesozoic Era 207
nonconformity 203
numerical age 196
original horizontality 198
Paleozoic Era 207
period 207

Phanerozoic Eon 215
physical continuity 204
Pleistocene Epoch 207
Precambrian 207
Proterozoic Eon 215
Quaternary Period 207
radioactive decay 210
relative time 196
standard geologic time scale 207
superposition 198
unconformity 202
uniformitarianism 196

Testing Your Knowledge

Use the following questions to prepare for exams based on this chapter.

1. Why is it desirable to find an index fossil in a rock layer? In the absence of index fossils, why is it desirable to find several fossils in a rock unit to determine relative age?

2. Suppose you had a radioactive isotope X whose half-life in disintegrating to daughter product Y is 120,000 years. By calculating how much it took to make the present amount of Y, you determine that, the rock originally contained 8 grams of isotope X. At present, only 1/4 gram of X is in the rock. How many half-lives have gone by? How old is the rock?

3. By applying the various principles, draw a cross section of an area in which the following sequence of events occurred. The relative time relationship for all events should be clear from your single cross section that shows what the geology looks like at present.

   a. Metamorphism took place during the Archean. During later Precambrian time, uplift and erosion reduced the area to a plane.

   b. Three layers of marine sedimentary rock were deposited on the plain during Ordovician through Devonian time.

   c. Although sedimentation may have taken place during the Mississippian through Permian, there are presently no sedimentary rocks of that age in the area.

   d. A vertical dike intruded all rocks that existed here during the Permian.

   e. A layer of sandstone was deposited during the Triassic.

   f. All of the rocks were tilted 45° during the early Cretaceous. This was followed by erosion to a planar surface.

   g. The area dropped below sea level, and two layers of Tertiary sedimentary rock were deposited on the erosion surface.

   h. Uplift and erosion during the Quaternary resulted in a slightly hilly surface.

   i. Following erosion, a vertical dike fed a small volcano.

4. Name as many types of contacts (e.g., intrusive contact) as you can.

5. Using information from box 8.4, calculate the age of a feldspar. At present, there are 1.2 million atoms of ^{40}K. The amount of ^{40}Ar in the mineral indicates that originally, there were 1.9 million ^{40}K atoms in the rock. Use a half-life of 1.3 billion years. (Hint: The answer is 862 million years.)

6. "Geological processes operating at present are the same processes that have operated in the past" is the principle of

   a. correlation   b. catastrophism
   c. uniformitarianism   d. none of the preceding

7. "Within a sequence of undisturbed sedimentary rocks, the layers get younger going from bottom to top" is the principle of

   a. original horizontality b. superposition
   c. crosscutting   d. none of the preceding

8. If rock A cuts across rock B, then rock A is _____ rock B.

   a. younger than   b. the same age as
   c. older than

9. Which is a method of correlation?

   a. physical continuity b. similarity of rock types
   c. fossils   d. all of the preceding

10. Eras are subdivided into

    a. periods   b. eons
    c. ages   d. epochs

11. Periods are subdivided into

    a. eras   b. epochs
    c. ages   d. time zones

12. Which division of geologic time was the longest?

    a. Precambrian   b. Paleozoic

    c. Mesozoic   d. Cenozoic

13. Which is a useful radioactive decay scheme?

    a. $^{238}U\ ^{206}Pb$ b. $^{235}U\ ^{207}Pb$

    c. $^{40}K\ ^{40}Ar$ d. $^{87}Rb\ ^{87}Sr$

    e. all of the preceding

14. C-14 dating can be used on all of the following except

    a. wood   b. shell

    c. the Dead Sea Scrolls d. granite

    e. bone

15. Concentrations of radon are highest in areas where the bedrock is

    a. granite   b. gneiss

    c. limestone   d. black shale

    e. phosphate-rich rock f. all of the preceding

16. Which is not a type of unconformity?

    a. disconformity   b. angular unconformity

    c. nonconformity   d. triconformity

17. A geologist could use the principle of inclusion to determine the relative age of

    a. fossils   b. metamorphism

    c. shale layers   d. xenoliths

18. The oldest abundant fossils of complex multicellular life with shells and other hard parts date from the

    a. Precambrian   b. Paleozoic

    c. Mesozoic   d. Cenozoic

19. A contact between parallel sedimentary rock that records missing geologic time is

    a. a disconformity   b. an angular unconformity

    c. a nonconformity   d. a sedimentary contact

Expanding Your Knowledge

1. How much of the ^{238}U originally part of Earth is still present?

2. As indicated by fossil records, why have some ancient organisms survived through very long periods of time whereas others have been very short-lived?

3. To what extent would a composite volcano (see chapter 4) be subject to the three principles described in this chapter?

4. Suppose a sequence of sedimentary rock layers was tilted into a vertical position by tectonic forces. How might you determine (a) which end was originally up and (b) the relative ages of the layers?

5. Note that in table 8.2, the epochs are given only for the Cenozoic Era (as is commonly done in geology textbooks). Why are the epochs for the Mesozoic and Paleozoic considered less important and not given?

6. Why would you not be able to use the principle of superposition to determine the age of a sill (defined in chapter 3)?

Exploring Web Resources

www.mhhe.com/plummer11e

This is the dedicated website for this book. You can go to it for new and updated information. The universal resource locators (URL) listed in this book are also given as links on the website, making it easy to go to those websites without typing in the URL. When you visit our Online Learning Center, go to the Student Center and then to the chapter of interest. Here you will find additional readings and media resources, as well as answers to the Testing Your Knowledge section, more quizzes, animation and video clips, and direct links to the sites listed in this book. Links to additional websites can also be found. We have added questions for some of the sites to allow you to get the most of your exploration of the web. Using the web is an enjoyable way of enhancing your knowledge of geology.

www.ucmp.berkeley.edu/exhibit/exhibits.html

Paleontology without Walls. University of California Museum of Paleontology virtual exhibit. Click on "Geologic Time."

http://vearthquake.calstatela.edu/VirtualDating/

Virtual Dating. This site provides an excellent, interactive way of learning how isotopic dating works. You can change data presented and watch graphs and other illustrations change accordingly. Quizzes help you understand the material.

www.talkorigins.org/origins/faqs.html

Talk Origins. This is an excellent site for in-depth information on geologic time. Click on "Age of the Earth." Topics include isotopic dating, the geologic time scale, and changing views of the age of Earth. The site includes in-depth presentations of arguments for a young Earth and the scientific rebuttals to them.

www.asa3.org/ASA/resources/Wiens.html

Radiometric Dating: A Christian Perspective. At this website, you can get a very thorough knowledge of isotopic dating, how it works, and how it has been used to determine the age of Earth and other events. The author addresses concerns of people who feel that an old Earth is incompatible with their religious beliefs.

 Animation

This chapter includes the following animations available on our Online Learning Center at **www.mhhe.com/plummer11e**.

8.25 The geologic history of the Earth scaled to a single year

Mass Wasting

Surficial Processes

Plate tectonics explains how rock is deformed and why we have mountains. This chapter and chapters 10 through 14 are concerned with surficial processes, the interaction of rock, air, and water in response to gravity at or near the Earth's surface. Nearly all of the features we see as landforms—rounded or rugged mountains, river valleys, cliffs and beaches along seashores, caves, sand dunes, and so on—are products of surficial processes. Surficial processes involve weathering, erosion, transportation, and deposition. Subsequent chapters address the work of running water, ground water (water that is beneath the surface), glaciers, wind, and ocean waves. This chapter is about the downward

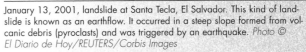

January 13, 2001, landslide at Santa Tecla, El Salvador. This kind of landslide is known as an earthflow. It occurred in a steep slope formed from volcanic debris (pyroclasts) and was triggered by an earthquake. *Photo © El Diario de Hoy/REUTERS/Corbis Images*

223

movement of masses of rock or loose material. Mass wasting mostly involves landsliding (a very general term).

 Relationships to Earth Systems

The "spheres" of Earth systems interact to play vital roles in surficial processes. The geosphere, of course, provides the solid rock to be sculpted, weathered, and altered by the various agents of erosion. Water from the hydrosphere is vital to almost all of the surficial processes. Running water plays an important role in weathering and in carving landscapes in desert as well as in wet climates. Water is a common contributor to landslides. Ice, frozen water, is a very efficient agent of erosion and transportation in glaciated areas. Water flowing underground is

an important water resource and can result in distinctive landscapes where caves are carved in limestone. The force of crashing waves shapes our coastlines. Wind, motion of the atmosphere, causes the waves that sculpt coastlines. Wind blowing over the land plays a lesser role than running water, but it is responsible for sand dunes and dust storms. The atmosphere provides the gases, notably carbon dioxide, that mix with water to form acid for chemical weathering. The biosphere would, of course, not exist if it were not for the other "spheres." But plants and animals also play a role in stabilizing or destabilizing slopes. For instance, plant roots help prevent soil erosion. A beaver dam may change the course of a stream. Humans with heavy equipment alter the normal rate of change of a landscape on a massive scale.

INTRODUCTION TO MASS WASTING

You may recall from previous chapters that mountains are products of tectonic forces. Most mountains are associated with present or past convergent plate boundaries. If tectonism were not at work, the surfaces of the continents would long ago have been reduced to featureless plains due to weathering and erosion. We consider the material on mountain slopes or hillsides to be out of equilibrium with respect to gravity. Because of the force of gravity, the various agents of erosion (moving water, ice, and wind) work to make slopes gentler and therefore increasingly more stable. In this chapter, we discuss the process of mass wasting.

Mass wasting is movement in which bedrock, rock debris, or soil moves downslope in bulk, or as a mass, because of the pull of gravity. Mass wasting includes movement so slow that it is almost imperceptible (called *creep*) as well as **landslides,** a general term for the slow to very rapid descent of rock or soil. The term *landslide* tells us nothing about the processes

involved. As you will see, terms such as *earthflow* and *rockslide* are far more descriptive than *landslide*.

Mass wasting affects people in many ways. Its effects range from the devastation of a killer landslide (such as the debris avalanche described in box 9.1) to the nuisance of having a fence slowly pulled apart by soil creep. The cost in lives and property from landslides is surprisingly high. Damage and casualty reports for landslides are often overlooked because they are part of a larger disaster, such as an earthquake or heavy rain from a hurricane. According to the U.S. Geological Survey, more people in the United States died from landslides during the last three months of 1985 than were killed during the last twenty years by all other geologic hazards, such as earthquakes and volcanic eruptions. Over time, landslides have cost Americans triple the combined costs of earthquakes, hurricanes, floods, and tornadoes. On average, the annual cost of landslides in the United States has been $1.5 billion and twenty-five lost lives. In many cases of mass wasting, a little knowledge of geology, along with appropriate preventive action, could have averted destruction.

TABLE 9.1	Some Types of Mass Wasting[1]			
	Slowest ⟶	Increasing Velocities ⟶		Fastest
Type of Movement	Less than 1 centimeter/year	1 millimeter/day to 1 kilometer/hour	1 to 5 kilometer/hour	*Velocities generally greater than 4 kilometers/hour*
Flow	Creep (soil) ⟶	◀— Debris Flow —▶ Earthflow ⟶	Mudflow ⟶	Debris avalanche (debris) ⟶ Rock avalanche (bedrock)
Slide		◀— Debris slide or earthslide —▶ ◀— Rockslide (bedrock) —▶		
Fall				Rockfall (bedrock)
		◀— "Landslides" —▶		

1. *The type of material at the start of movement is shown in parentheses. Rates given are typical velocities for each type of movement.*

CLASSIFICATION OF MASS WASTING

A number of systems are used by geologists, engineers, and others for classifying mass wasting, but none has been universally accepted. Some are very complex and useful only to the specialist.

The classification system used here and summarized in table 9.1 is based on (1) rate of movement, (2) type of material, and (3) nature of the movement.

Rate of Movement

A landslide (debris avalanche) like the one in Peru (box 9.1) clearly involves rapid movement. Just as clearly, movement of soil at a rate of less than a centimeter a year is slow movement. There is a wide range of velocities between these two extremes.

Type of Material

Mass wasting processes are usually distinguished on the basis of whether the descending mass started as bedrock (as in a rockslide) or as unconsolidated material. For this discussion, we call any unconsolidated or weakly consolidated material at the Earth's surface, regardless of particle size or composition, **soil** (also called *engineering soil*—see chapter 5 for other definitions of soil). Soil can be debris, earth, or mud. **Debris** implies that coarse-grained fragments predominate in the soil. If the material is predominantly fine-grained (sand, silt, clay) it is called **earth** (not capitalized). **Mud** as the name suggests, has a high content of water, clay, and silt.

The amount of water (or ice and snow) in a descending mass strongly influences the rate and type of movement.

Type of Movement

In general, the type of movement in mass wasting can be classified as mainly flow, slide, or fall (figure 9.1). A **flow** implies that the descending mass is moving downslope as a viscous fluid. **Slide** means the descending mass remains relatively intact, moving along one or more well-defined surfaces. A **fall** occurs when material free-falls or bounces down a cliff.

Flow

Fall

Slide

Translational slide

Rotational slide (slump)

FIGURE 9.1

Flow, slide, and fall.

ENVIRONMENTAL GEOLOGY 9.1

Disaster in the Andes

As a result of a tragic combination of geologic conditions and human ignorance of geologic hazards, one of the most devastating landslides (a debris avalanche) in history destroyed the town of Yungay in Peru in 1970. Yungay was one of the most picturesque towns in the Santa River Valley, which runs along the base of the highest peaks of the Peruvian Andes. Heavily glaciated Nevado Huascarán, 6,768 meters (22,204 feet) above sea level, rises steeply above the populated, narrow plains along the Santa River.

In May 1970, a sharp earthquake occurred. The earthquake was centered offshore from Peru about 100 kilometers from Yungay. Although the tremors in this part of the Andes were no stronger than those that have done only light damage to cities in the United States, many poorly constructed homes collapsed. Because of the steepness of the slopes, thousands of small rockfalls and rockslides were triggered.

The greatest tragedy began when a slab of glacier ice about 800 meters wide, perched near the top of Huascarán, was dislodged by the shaking. (A few years earlier, American climbers returning from the peak had warned that the ice looked highly unstable. The Peruvian press briefly noted the danger to the towns below, but the warning was soon forgotten.)

The mass of ice rapidly avalanched down the extremely steep slopes, breaking off large masses of rock debris and scooping out small lakes and loose rock that laid in its path. Eyewitnesses described the mass as a rapidly moving wall the size of a ten-story building. The sound was deafening. More than 50 million cubic meters of muddy debris traveled 3.7 kilometers (12,000 feet) vertically and 14.5 kilometers (9 miles) horizontally in less than four minutes, attaining speeds between 200 and 435 kilometers per hour (125 to 270 miles per hour). The main mass of material traveled down a steep valley until it came to rest, blocking the Santa River and burying about 1,800 people in the small village of Ranrahirca (box figure 1). A relatively small part of the mass of mud and debris that was moving especially rapidly shot up the valley sidewall at a curve and overtopped a ridge. The mass was momentarily airborne before it fell on the town of Yungay, completely burying it under several meters of mud and loose rock. Only the top of the church and tops of palm trees were visible, marking where the town center was buried (box figure 2). Ironically, the cemetery was not buried because it occupied the high ground. The few survivors were people who managed to run to the cemetery.

The estimated death toll at Yungay was 17,000. This was considerably more than the town's normal population, because it was Sunday, a market day, and many families had come in from the country.

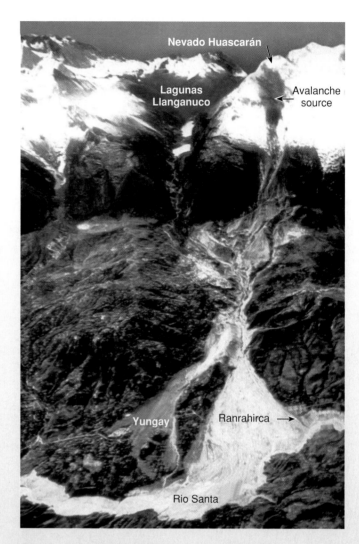

BOX 9.1 ■ FIGURE 1

Air photo showing the 1970 debris avalanche in Peru, which buried Yungay. The main mass of debris destroyed the small village of Ranrahirca. *Photo by Servicio Aerofotografico de Peru, courtesy of U.S. Geological Survey*

For several days after the slide, the debris was too muddy for people to walk on, but within three years, grass had grown over the site. Except for the church steeple and the tops of palm trees that still protrude above the ground, and the crosses erected by families of those buried in the landslide, the former site of Yun-

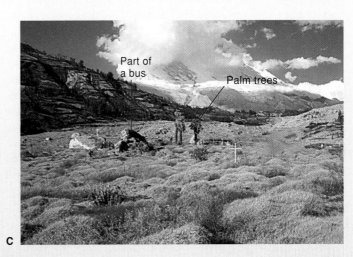

BOX 9.1 ■ FIGURE 2

(A) Yungay is completely buried, except for the cemetery and a few houses on the small hill in the lower right of the photograph. (B) Behind the palm trees is the top of a church buried under 5 meters of debris at Yungay's central plaza. (C) Three years later. *Photos A and B by George Plafker, U.S. Geological Survey; Photo C by C. C. Plummer*

gay appears to be a scenic meadow overlooking the Santa River. The U.S. Geological Survey and Peruvian geologists found evidence that Yungay itself had been built on top of debris left by an even bigger slide in the recent geologic past. More slides will almost surely occur here in the future.

Additional Resource

G. E. Ericksen, G. Plafker, and J. Fernandez Concha. 1970. *Preliminary report on the geologic events associated with the May 31, 1970, Peru earthquake.* U.S. Geological Survey Circular 639.

Two kinds of slip are shown in figure 9.1. In a **translational slide,** the descending mass moves along a plane approximately parallel to the slope of the surface. A **rotational slide** (also called a **slump**) involves movement along a curved surface, the upper part moving downward while the lower part moves outward.

CONTROLLING FACTORS IN MASS WASTING

Table 9.2 summarizes the factors that influence the likelihood and the rate of movement of mass wasting. The table makes apparent some of the reasons why the landslide (a debris avalanche) in Peru (box 9.1) occurred and why it moved so rapidly. (1) The slopes were exceptionally steep, and (2) the **relief** (the vertical distance between valley floor and mountain summit) was great, allowing the mass to pick up speed and momentum. (3) Water and ice not only added weight to the mass of debris but made it more fluid. (4) Abundant loose rock and debris were available in the course of the slide. (5) Where the slide began, there were no plants with roots to anchor loose material on the slope. Finally, (6) the region has earthquakes. Although the slide would have occurred eventually even without one, it was triggered by an earthquake.

Other factors influence susceptibility to mass wasting as well as its rate of movement. The orientation of planes of weakness in bedrock (bedding planes, foliation planes, etc.) is important if the movement involves bedrock rather than debris. Fractures or bedding planes oriented so that slabs of rock can slide easily along these surfaces greatly increase the likelihood of mass wasting.

Climatic controls inhibit some types of mass wasting and aid others (table 9.2). Climate influences how much and what kinds of vegetation grow in an area and what type of weather-

ing occurs. Infrequent but heavy rainfall aids mass wasting because it quickly saturates debris that lacks the protective vegetation found in wetter climates. By contrast, a climate in which rain drizzles intermittently much of the year will have vegetation that tends to inhibit mass wasting. In cold climates, freezing and thawing contribute to downslope movement.

Gravity

The driving force for mass wasting is gravity. Figures 9.2A–C show gravity acting on a block on a slope. The length of the red, vertical arrow is proportional to the force—the heavier the material, the longer the arrow. The effect of gravity is resolvable into two component forces, indicated by the black arrows. One, the *normal force,* is perpendicular to the slope and is the component of gravity that tends to hold the block in place. The greater its length, the more force is needed to move the block. The other, called the **shear force,** is parallel to the slope and indicates the block's ability to move. The length of the arrows is proportional to the strength of each force. The steeper the slope (and the heavier the block), the greater the shear force and the tendency of the block to slide. Friction counteracts the shear force. *Shear resistance* (represented by the brown arrow) is the force that would be needed to move the block. If that arrow is larger than the arrow representing shear force (as in figure 9.2A), the block will not move. The magnitude of the shear resistance (and the length of the brown arrow) is a function of friction and the size of the normal force. The brown arrow will be shorter (and the shear resistance lower) if water or ice reduces the friction beneath the block. If the shear resistance becomes lower than the shear force, the block will slide (figure 9.2B). Similar forces act on soil on a hillside (figure 9.2D). The resistance to movement or deformation of that soil is its **shear strength.** Shear strength is controlled by fac-

TABLE 9.2	**Summary of Controls of Mass Wasting**	
Driving Force: Gravity		
Contributing Factors	**Most Stable Situation**	**Most Unstable Situation**
Slope angle	Gentle slopes or horizontal surface	Steep or vertical
Local relief	Low	High
Thickness of soil over bedrock	Slight thickness (usually)	Great thickness
Orientation of planes of weakness in bedrock	Planes at right angles to hillside slopes	Planes parallel to hillside slopes
Climatic factors:		
Ice	Temperature stays above freezing	Freezing and thawing for much of the year
Water in soil or debris	Film of water around fine particles	Saturation of soil with water
Precipitation	Frequent but light rainfall or snow	Long periods of drought with rare episodes of heavy precipitation
Vegetation	Heavily vegetated	Sparsely vegetated
Triggering Mechanisms: (1) earthquakes; (2) weight added to upper part of a slope; (3) undercutting of bottom of slope; (4) heavy rainfall		

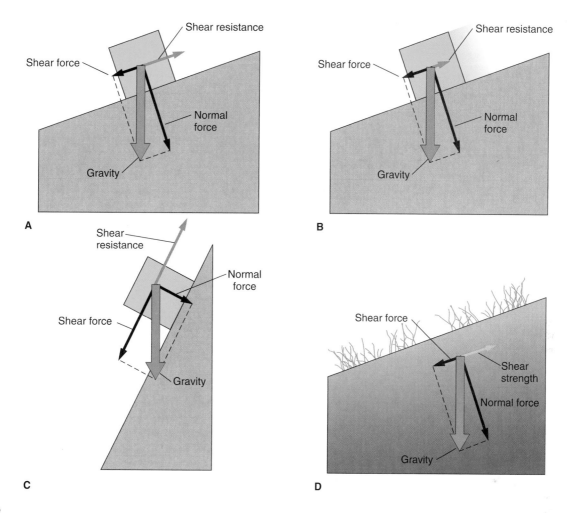

FIGURE 9.2

Relationship of shear force and normal force to gravity. (A) A block on a gently inclined slope in which the shear resistance (brown arrow) is greater than the shear force; therefore, the block will not move. (B) The same situation as in A, except that the shear resistance is less than the shear force; therefore, the block will be moving. (C) A block on a steep slope. Note how much greater the shear force is and how much larger the shear resistance has to be to prevent the block from moving. (D) Forces acting at a point in soil. Shear strength is represented by a yellow arrow. If that arrow is longer than the one represented by shear force, soil at that point will not slide or be deformed.

tors such as the cohesiveness of the material, friction between particles, pore pressure of water, and the anchoring effect of plant roots. Shear strength is also related to the normal force. The larger the normal force, the greater the shear strength is. If the shear strength is greater than the shear force, the soil will not move or be deformed. On the other hand, if shear strength is less than shear force, the soil will flow or slide.

Building a heavy structure high on a slope demands special precautions. To prevent movement of both the slope and the building, pilings may have to be sunk through the soil, perhaps even into bedrock. Developers may have to settle for fewer buildings than planned if the weight of too many structures will make the slope unsafe.

Water

Water is a critical factor in mass wasting. When soil is saturated with water (as from heavy rain or melting snow), it becomes heavier and less viscous, and is more likely to flow downslope. The added gravitational shear force from the increased weight is usually less important than the reduction in shear strength. This is due to increased *pore pressure* in which water forces soil grains apart.

Paradoxically, a small amount of water in soil can actually prevent downslope movement. When water does not completely fill the pore spaces between the grains of soil, it forms a thin film around each grain (as shown in figure 9.3). Loose grains adhere to one another because of the *surface tension* created by the film of water, and shear strength increases. Surface tension of water between sand grains is what allows you to build a sand castle. The sides of the castle can be steep or even vertical because surface tension holds the moist sand grains in place. Dry sand cannot be shaped into a sand castle because the sand grains slide back into a pile that generally slopes at an angle of about 30° to 35° from the horizontal. On the other hand, an experienced sand castle builder also knows that it is impossible to build anything with sand that is too wet. In this case, the water completely occupies the pore space between sand grains,

FIGURE 9.3

The effect of water in sand. (A) Unsaturated sand held together by surface tension of water. (B) Saturated sand grains forced apart by water; mixture flows easily. (C) A sand castle in Acapulco, Mexico. *Photo by C. C. Plummer*

forcing them apart and allowing them to slide easily past one another. When the tide comes in or someone pours a pail of water on your sand castle, all you have is a puddle of wet sand.

Similarly, as the amount of water in soil increases, rate of movement tends to increase. Damp soil may not move at all, whereas moderately wet soil moves slowly downslope. Slow types of mass wasting, such as creep, are generally characterized by a relatively low ratio of water to earth. Mudflows always have high ratios of water to earth. A mudflow that continues to gain water eventually becomes a muddy stream.

Triggering Mechanisms

A sudden event may trigger mass wasting of a hillside that is unstable. Eventually, movement would occur without the triggering if conditions slowly became more unstable.

Earthquakes commonly trigger landslides. The 1970 debris avalanche in Peru (see box 9.1) was one of thousands of landslides, mostly small ones, triggered by a quake. The worst damage from California's 1989 Loma Prieta earthquake was in the nearly flat areas of San Francisco's Marina District (where fires from broken gas mains ravaged buildings) and across the bay in Oakland. In Oakland, ground failure occurred beneath a two-tiered freeway, the upper level of which collapsed onto the lower level, crushing many vehicles (see chapter 16). Without the earthquake, movement may not have taken place for decades or centuries, and then mass wasting would have been slow enough so that corrective measures could have been taken to stabilize the ground.

Landslides often are triggered by heavy rainfall. The sudden influx of voluminous water quickly adds weight and increases pore pressure in material. Rainstorms in normally dry southern California caused numerous landslides in January 2005 and parts of the two previous years. These are discussed in this chapter as are other examples of water-triggered landslides, notably the Gros Ventre, Wyoming, and Vaiont, Italy, rockslides, described in this chapter.

Construction sometimes triggers mass wasting. The extra weight of buildings on a hillside can cause a landslide, as can bulldozing a road cut at the base of a slope.

COMMON TYPES OF MASS WASTING

The types of mass wasting are shown in table 9.1. Here we will describe the most common ones in detail.

Creep

Creep (or **soil creep**) is very slow, downslope movement of soil. Shear forces, over time, are only slightly greater than shear strengths. The rate of movement is usually less than a centimeter per year and can be detected only by observations taken over months or years. When conditions are right, creep can take place along nearly horizontal slopes. Some indicators of creep are illustrated in figures 9.4 and 9.5.

Two factors that contribute significantly to creep are water in the soil and daily cycles of freezing and thawing. As we have said, water-saturated ground facilitates movement of soil downhill. What keeps downslope movement from becoming more rapid in most areas is the presence of abundant grass or other plants that anchor the soil. (Understandably, overgrazing can severely damage sloping pastures.)

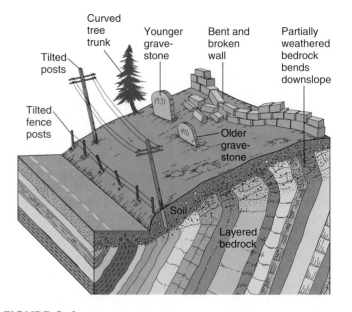

FIGURE 9.4

Indicators of creep. *After C. F. S. Sharpe*

Several processes contribute to soil creep. Particles are displaced in cycles of wetting and drying. The soil tends to swell when wet and contract when dry so that movement takes place in a manner similar to that of a freeze-thaw cycle. Burrowing worms and other creatures "stir" the soil and facilitate movement under gravity's influence. The process is more active where the soil freezes and thaws during part of the year. During the winter in regions such as the northeastern United States, the temperature may rise above and fall below freezing once a day. When there is moisture in the soil, each freeze-thaw cycle moves soil particles a minute amount downhill, as shown in figure 9.6.

FIGURE 9.5

(A) Tilted gravestones in a churchyard at Lyme Regis, England (someone probably straightened the one upright gravestone). Grassy slope is inclined gently to the left. (B) Soil and partially weathered, nearly vertical sedimentary strata have crept downslope. (C) As a young tree grew, it grew vertically but was tilted by creeping soil. As it continued to grow, its new, upper part would grow vertically but in turn would be tilted. *Photo A by C. C. Plummer; Photo B by Frank M. Hanna; Photo C © Parvinder Sethi*

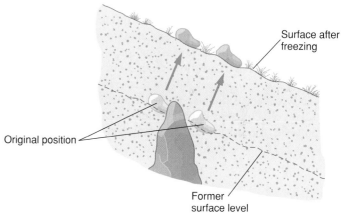

FIGURE 9.6

Downslope movement of soil, illustrated by following two sand grains (each less than a millimeter in size) during a freeze-thaw cycle. Movement downward might not be precisely vertical if adjacent grains interfere with each other.

Creep is not as dramatic as landsliding. However, it can be a costly nuisance. If you have a home on creeping ground, you likely will have doors that stick, cracks in walls, broken pipes, and a driveway that will need repaving often. You will find that you are spending more time and money on repairs than does a person who lives on stable ground.

Flow

Flow occurs when motion is taking place within a moving mass of unconsolidated or weakly consolidated material. Grains move relative to adjacent grains, or motion takes place along closely spaced, discrete fractures. The common varieties of flow—earthflow, debris flow, mudflow, and debris avalanche—are described in this section.

Earthflow

In an **earthflow,** earth moves downslope as a viscous fluid; the process can be slow or rapid. Earthflows usually occur on hillsides that have a thick cover of soil in which finer grains are predominant, often after heavy rains have saturated the soil. Typically, the flowing mass remains covered by a blanket of vegetation, with a *scarp* (steep cut) developing where the moving debris has pulled away from the stationary upper slope.

A landslide may be entirely an earthflow, as in figure 9.7, with soil particles moving past one another roughly parallel to the slope. Commonly, however, rotational sliding (slumping) takes place above the earthflow, as in figure 9.8. This example is a *rotational slide* (upper part) and an *earthflow* (lower part) and can be called a *slump-earthflow.* In such cases, soil remains in a relatively coherent block or blocks that rotate downward and outward, forcing the soil below to flow.

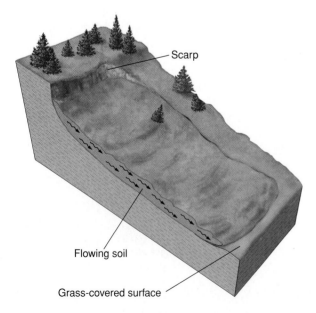

FIGURE 9.7

Earthflow. Soil flows beneath a blanket of vegetation.

Labels in figure: Scarp; Flowing soil; Grass-covered surface

A hummocky lobe usually forms at the toe or front of the earthflow where soil has accumulated. An earthflow can be active over a period of hours, days, or months; in some earthflows, intermittent movement continues for years.

In March 1995, following an extraordinarily wet year, a slump-earthflow destroyed or severely damaged fourteen homes in the southern California coastal community of La Conchita (figure 9.8*B*). In January 2005, following 15 days of record-breaking rainfall, around 15% of the 1995 landslide remobilized (figure 9.8*C*). Rapidly moving flow of soil killed 10 people and severely damaged or destroyed 36 houses. Because future landslides are likely, the town of La Conchita was abandoned. For details of the La Conchita landslides go to http://pubs.usgs.gov/of/2005/1067/pdf/OF2005-1067.pdf.

People can trigger earthflows by adding too much water to soil from septic tank systems or by overwatering lawns. In one case, in Los Angeles, a man departing on a long trip forgot to turn off the sprinkler system for his hillside lawn. The soil became saturated, and both house and lawn were carried downward on an earthflow whose lobe spread out over the highway below.

Earthflows, like other kinds of landslides, can be triggered by undercutting at the base of a slope. The undercutting can be caused by waves breaking along shorelines or streams eroding and steepening the base of a slope. Along coastlines, mass wasting commonly destroys buildings. Entire housing developments and expensive homes built for a view of the ocean are lost. A home buyer who knows nothing of geology may not realize that the sea cliff is there because of the relentless erosion of waves along the shoreline. Nor is the person likely to be aware that a steepened slope creates the potential for landslides.

Bulldozers can undercut the base of a slope more rapidly than wave erosion, and such oversteepening of slopes by human activity has caused many landslides. Unless careful engineering measures are taken at the time a cut is made, road cuts or platforms carved into hillsides for houses may bring about disaster (figure 9.20).

Solifluction and Permafrost

One variety of earthflow is usually associated with colder climates. **Solifluction** is the flow of water-saturated soil over impermeable material. Because the impermeable material beneath the soil prevents water from draining freely, the soil between the vegetation cover and the impermeable material becomes saturated (figure 9.9). Even a gentle slope is susceptible to movement under these conditions.

The impermeable material beneath the saturated soil can be either impenetrable bedrock or, as is more common, **permafrost,** ground that remains frozen for many years. Most solifluction takes place in areas of permanently frozen ground, such as in Alaska and northern Canada. Permafrost occurs at depths ranging from a few centimeters to a few meters beneath the surface. The ice in permafrost is a cementing agent for the soil. Permafrost is as solid as concrete.

Above the permafrost is a zone that, if the soil is saturated, is frozen during the winter and indistinguishable from the underlying permafrost. When this zone thaws during the

FIGURE 9.8

Earthflow with rotational sliding (slumping). (A) Soil in the upper part of the diagram remained mostly intact as it rotated downward in blocks. Soil in the lower portion flowed. (B) A slump-earthflow destroyed several houses in March, 1995, at La Conchita, California. (C) After heavy rains in January 2005 part of the landslide remobilized and flowed into the town of La Conchita. *Photo B by Robert L. Schuster, U.S. Geological Survey; Photo C © Kevork Djansezian/AP/Wide World Photos*

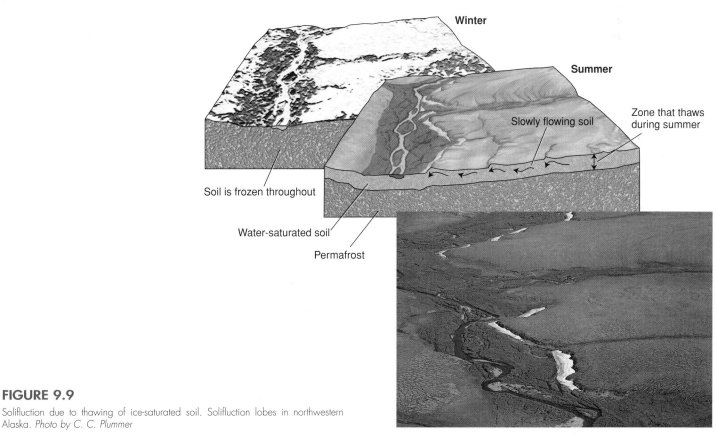

FIGURE 9.9

Solifluction due to thawing of ice-saturated soil. Solifluction lobes in northwestern Alaska. *Photo by C. C. Plummer*

FIGURE 9.10

A railroad built on permafrost terrain in Alaska. *Photo by Lynn A. Yehle, U.S. Geological Survey*

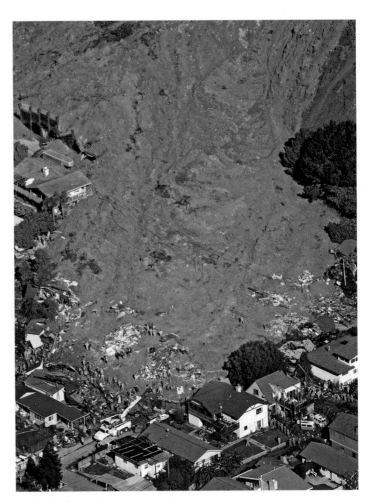

FIGURE 9.11

The portion of the La Conchita landslide that remobilized and killed ten people in January 2005 (see also figure 9.8C). The U.S. Geological Survey classified this as a debris flow, because of the abundance of coarse soil mixed with mud. Note that the flow overtopped the metal wall (upper left) put up to protect the town. *Photo © Kevork Djansezian/Wide World Photos*

summer, the water, along with water from rain and runoff, cannot percolate downward through the permafrost, and so the slopes become susceptible to solifluction.

As solifluction movement is not rapid enough to break up the overlying blanket of vegetation into blocks, the water-saturated soil flows downslope, pulling vegetation along with it and forming a wrinkled surface. Gradually, the soil collects at the base of the slope, where the vegetated surface bulges into a hummocky lobe.

Solifluction is not the only hazard associated with permafrost. Great expanses of flat terrain in Arctic and subarctic climates become swampy during the summer because of permafrost, making overland travel very difficult. Building and maintaining roads is an engineering headache (figure 9.10). In the preliminary stages of planning the Alaska pipeline, a road was bulldozed across permafrost terrain during the winter, removing the vegetation from the rock-hard ground. It was an excellent truck route during the winter, but when summer came, the road became a quagmire several hundred kilometers long. The strip can never be used by vehicles as planned, nor will the vegetation return for many decades. Building structures on permafrost terrain presents serious problems. For instance, heat from a building can melt underlying permafrost; the building then sinks into the mud. To learn more about permafrost, go to *Permafrost Research at the Geological Survey of Canada,* http://sts.gsc.nrcan.gc.ca/permafrost/.

Debris Flow

A **debris flow** is flow involving soil in which coarse material (gravel, boulders) is predominant. A debris flow can be like an earthflow and travel relatively short distances to the base of a slope or, if there is a lot of water, a debris flow can behave like a mudflow and flow rapidly, traveling considerable distance in a channel. Rapidly moving debris flows can be extremely devastating.

The steep mountains that rise above Los Angeles and other southern California urban centers are sources of sometimes catastrophic debris flows (and mudflows). In December 2003, dozens of debris flows took place in the San Bernadino Mountains (figure 9.11). The debris flows followed a typical scenario that began during the hot, dry summer. Widespread forest fires scorched southern California, killing trees and ground cover. The anchoring effect of vegetation was gone. Geologists predicted that steep slopes underlain with thick debris were ripe for producing debris flows. Heavy rains would saturate the soil and trigger the debris flows. Heavy rains did indeed come in late December. On Christmas day, one of many debris flows destroyed a church camp, killing fourteen people.

Debris flows (and mudflows) illustrate the interplay between Earth systems: Soil is produced through weathering—interaction of the atmosphere and the geosphere. Vegetation (the biosphere) grows in the soil and adds shear strength to stabilize the hillside. The atmosphere heats and dries the vegetation and produces thunderstorms, which ignite forest fires. Part of the biosphere is destroyed. Atmospheric conditions bring heavy rain and part of the hydrosphere mixes with the soil to produce the debris flows.

Mudflow

A **mudflow** is a flowing mixture of soil and water, usually moving down a channel (figure 9.12). It can be visualized as a stream with the consistency of a thick milkshake. Most of the solid particles in the slurry are clay and silt (hence, the muddy appearance), but coarser sediment commonly is part of the mixture. A slurry of soil and water forms after a heavy rainfall or other influx of water and begins moving down a slope. Most mudflows quickly become channeled into valleys. They then move downvalley like a stream except that, because of the heavy load of sediment, they are more viscous. Mud moves more slowly than a stream but, because of its high viscosity, can transport boulders, automobiles, and even locomotives. Houses in the path of a mudflow will be filled with mud, if not broken apart and carried away.

Mudflows are most likely to occur in places where soil is not protected by a vegetative cover. For this reason, mudflows are more likely to occur in arid regions than in wet climates. A hillside in a desert environment, where it may not have rained for many years, may be covered with a blanket of loose material. With sparse desert vegetation offering little protection, a sudden thunderstorm with drenching rain can rapidly saturate the soil and create a mudflow in minutes.

Mudflows frequently occur on young volcanoes that are littered with ash. Water from heavy rains mixes with pyroclastic debris, as at Mount Pinatubo in 1991 (see chapter 1). For over a decade after the big eruption, mudflows near Mount Pinatubo continue to cost lives and destroy property. Water also can come from glaciers that are melted by lava or hot pyroclastic debris, as occurred at Mount St. Helens in 1980 (figure 9.13) and at Colombia's Nevado del Ruiz in 1985, which cost 23,000 lives (described in chapter 1). Like debris flows, mudflows also occur after forest fires destroy slope vegetation that normally anchors soil in place.

The year 1978 was particularly bad for debris flows and mudflows in southern California. One flow roared through a Los Angeles suburb carrying almost as many cars as large boulders. A sturdily built house withstood the onslaught but began filling with muddy debris. Two of its occupants were pinned to the wall of a bedroom and could do nothing as the room filled slowly with mud. The mud stopped rising just as it was reaching their heads. Hours later they were rescued. (John McPhee's *The Control of Nature,* listed in *Exploring Web Resources* on this book's website [www.mhhe.com/plummer11e], has a highly readable account of the 1978 flows in southern California.)

FIGURE 9.12

A dried mudflow in the Peruvian Andes. *Photo by C. C. Plummer*

FIGURE 9.13

Man examining a 75-meter-long bridge on Washington state highway 504, across the North Fork of the Toutle River. The bridge was washed out by mudflow during the May 18, 1980, eruption of Mount St. Helens. The steel structure was carried about 0.5 kilometers downstream and partially buried by the mudflow. *Photo by Robert L. Schuster, U.S. Geological Survey*

Debris Avalanche

The fastest variety of debris flow is a **debris avalanche,** a very rapidly moving, turbulent mass of debris, air, and water. The best modern example is the one that buried Yungay (described in box 9.1).

Some geologists have suggested that in very rapidly moving rock avalanches, air trapped under the rock mass creates an air cushion that reduces friction. This could explain why some landslides reach speeds of several hundred kilometers per hour.

ENVIRONMENTAL GEOLOGY 9.2

Los Angeles, A Mobile Society*

The following satirical newspaper column was written by humorist Art Buchwald in 1978, a year, in which southern California had many landslides because of unusually wet weather.

Los Angeles—I came to Los Angeles last week for rest and recreation, only to discover that it had become a rain forest.

I didn't realize how bad it was until I went to dinner at a friend's house. I had the right address, but when I arrived there was nothing there. I went to a neighboring house where I found a man bailing out his swimming pool.

I beg your pardon, I said. Could you tell me where the Cables live?

"They used to live above us on the hill. Then, about two years ago, their house slid down in the mud, and they lived next door to us. I think it was last Monday, during the storm, that their house slid again, and now they live two streets below us, down there. We were sorry to see them go—they were really nice neighbors."

I thanked him and slid straight down the hill to the new location of the Cables' house. Cable was clearing out the mud from his car. He apologized for not giving me the new address and explained, "Frankly, I didn't know until this morning whether the house would stay here or continue sliding down a few more blocks."

Cable, I said, you and your wife are intelligent people, why do you build your house on the top of a canyon, when you know that during a rainstorm it has a good chance of sliding away?

"We did it for the view. It really was fantastic on a clear night up there. We could sit in our Jacuzzi and see all of Los Angeles, except of course when there were brush fires.

"Even when our house slid down two years ago, we still had a great sight of the airport. Now I'm not too sure what kind of view we'll have because of the house in front of us, which slid down with ours at the same time."

But why don't you move to safe ground so that you don't have to worry about rainstorms?

"We've thought about it. But once you live high in a canyon, it's hard to move to the plains. Besides, this house is built solid and has about three more good mudslides in it."

Still, it must be kind of hairy to sit in your home during a deluge and wonder where you'll wind up next. Don't you ever have the desire to just settle down in one place?

"It's hard for people who don't live in California to understand how we people out here think. Sure we have floods, and fire and drought, but that's the price you have to pay for living the good life. When Esther and I saw this house, we knew it was a dream come true. It was located right on the tippy top of the hill, way up there. We would wake up in the morning and listen to the birds, and eat breakfast out on the patio and look down on all the smog.

"Then, after the first mudslide, we found ourselves living next to people. It was an entirely different experience. But by that time we were ready for a change. Now we've slid again and we're in a whole new neighborhood. You can't do that if you live on solid ground. Once you move into a house below Sunset Boulevard, you're stuck there for the rest of your life.

"When you live on the side of a hill in Los Angeles, you at least know it's not going to last forever."

Then, in spite of what's happened, you don't plan to move out?

"Are you crazy? You couldn't replace a house like this in L.A. for $500,000."

What happens if it keeps raining and you slide down the hill again?

"It's no problem. Esther and I figure if we slide down too far, we'll just pick up and go back to the top of the hill, and start all over again; that is, if the hill is still there after the earthquake."

Additional Resource

John McPhee's *The Control of Nature* contains a factual, and highly readable, account of 1978 landslides in southern California.

*Reprinted by permission of the author

But other geologists have contended that the rock mass is too turbulent to permit such an air cushion to form.

Rockfalls and Rockslides

Rockfall

In May 3, 2003, New Hampshire lost its beloved symbol, the Old Man of the Mountain (figure 9.14), to rockfall. The Granite State's citizens associated resolute individualism with the rugged features outlined by the face high on a cliff. But the relentless work of water and frost-wedging enlarged the cracks in the granite until the overhanging rock broke apart.

When a block of bedrock breaks off and falls freely or bounces down a cliff, it is a **rockfall** (figure 9.15). Cliffs may form naturally by the undercutting action of a river, wave action, or glacial erosion. Highway or other construction projects may also oversteepen slopes. Bedrock commonly has cracks (joints) or other planes of weakness such as foliation (in metamorphic rocks) or sedimentary bedding planes. Blocks of rock will break off along these planes. In colder climates, rock is effectively broken apart by frost wedging (as explained in chapter 5).

Commonly, an apron of fallen rock fragments, called **talus,** accumulates at the base of a cliff (figure 9.16).

A spectacular rockfall took place in Yosemite National Park in the summer of 1996, killing one man and injuring several other people. The rockfall originated from near Glacier Point (the place where the photo for figure 12.1 was taken). Two huge slabs (weighing approximately 80,000 tons) of an

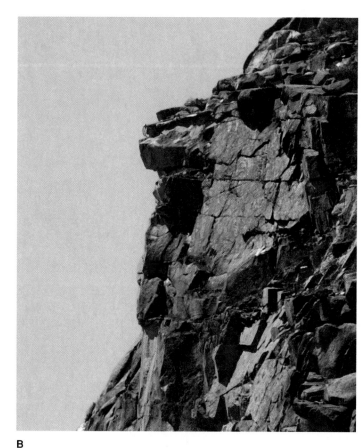

A

B

FIGURE 9.14

The Old Man of the Mountain in New Hampshire. (A) The profile of the face of a man was a product of weathering and erosion controlled largely by subhorizontal joints in granite. This is the profile that appeared on license plates and New Hampshire publications. (B) After succumbing to continuing erosion, features of the Old Man broke apart and became a rock fall on May 3, 2003. *Photos © Jim Cole/AP/Wide World Photos*

overhanging arch broke loose just seconds apart. (The arch was a product of exfoliation and broke loose along a sheet joint; see chapter 5.) The slabs slid a short distance over steep rock from which they were launched outward, as if from a ski jump, away from the vertical cliffs. The slabs fell free for around 500 meters (1,700 feet) and hit the valley floor 30 meters out from the base of the cliff (you would not have been hit if you were standing at the base of the cliff). They shattered upon impact and created a dust cloud (figure 9.17) that obscured visibility for hours. A powerful air blast was created as air between the rapidly falling rock, and the ground was compressed. The debris-laden wind felled a swath of trees between the newly deposited talus and a nature center building. In 1999, another rockfall in the same area killed one rock climber and injured three others. For an excellent and thorough report, go to the U.S. Geological Survey site *Rockfall in Yosemite*, http:// greenwood.cr.usgs.gov/pub/open-file-reports/ofr-99-0385/.

Rockslide and Rock Avalanche

A **rockslide** is, as the term suggests, the rapid sliding of a mass of bedrock along an inclined surface of weakness, such as a bedding plane, a major fracture in the rock, or a foliation plane (box 9.3). Once sliding begins, a rock slab usually breaks up into rubble. Like rockfalls, rockslides can be caused by undercutting at the base of the slope from erosion or construction.

Some rockslides travel only a few meters before halting at the base of a slope. In country with high relief, however, a rockslide may travel hundreds or thousands of meters before reaching a valley floor. If movement becomes very rapid, the rockslide may break up and become a rock avalanche. A **rock avalanche** is a very rapidly moving, turbulent mass of broken-up bedrock. Movement in a rock avalanche is flowage on a grand scale. The only difference between a rock avalanche and a debris avalanche is that a rock avalanche begins its journey as bedrock.

Ultimately, a rockslide or rock avalanche comes to rest as the terrain becomes less steep. Sometimes the mass of rock fills the bottom of a valley and creates a natural dam. If the rock mass suddenly enters a lake or bay, it can create a huge wave that destroys lives and property far beyond the area of the original landslide.

An example is a disastrous rock avalanche that took place in northern Italy in 1963. A huge layer (1.8 kilometers long

FIGURE 9.15

Two examples of rockfall.

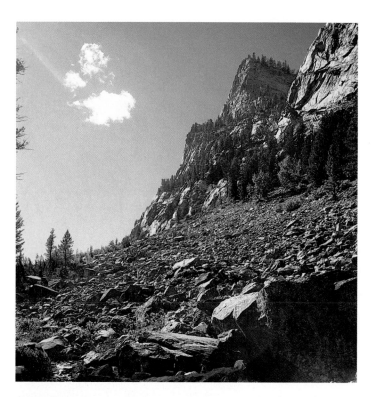

FIGURE 9.16

Talus. *Photo by C. C. Plummer*

FIGURE 9.17

Small dust clouds linger high above Yosemite Valley where rock slabs broke loose and fell to the valley floor, creating upon impact, the debris-laden blast of air climbing up the other side of the valley. The photo was taken by a rock climber on a nearby cliff. *Photo by Ed Youmans*

and 1 kilometer wide) of limestone broke loose parallel to its bedding planes. The translational slide involved around 270 million cubic meters that moved up to 100 kilometers per hour into the Vaiont Reservoir, creating a giant wave. The 175-meter (almost two football fields) high wave overtopped the Vaiont Dam. It was the world's highest dam, rising 265 meters (870 feet) above the valley floor. Three thousand people were killed in the villages that the water flooded in the valleys below. The dam was not destroyed (figure 9.18), a tribute to excellent engineering, but the men in charge of the building project were convicted of criminal negligence for ignoring the landslide hazards. The chief engineer committed suicide.

As in slower mass movements, water can play an important role in causing a rockslide. In 1925, exceptionally heavy

FIGURE 9.18

Part of the 270 million cubic meters of rock from the rock avalanche that filled the former reservoir behind the Vaiont Dam. The dam, at the right of the photo, was the world's highest when built. Most of the dam face is buried under the debris. Part of the bedding plane over which sliding took place can be seen on the mountainside. The inset photo shows the upper portion of the downstream face of the dam within a steep, narrow valley. It was taken in 2004 from Longarone, one of the largest towns along the Piave River destroyed by the flood. Visualize the 175-meter wall of water overtopping the dam. *Large photo by Earle F. McBride; inset by C. C. Plummer*

Geologist's View

rains in the Gros Ventre Mountains of Wyoming caused water to seep into a layer of sandstone, wetting the underlying layer of shale and greatly reducing its shear strength (figure 9.19). The layers of sedimentary rock were inclined roughly parallel to the hillside. With the wet shale acting as a lubricant, the overlying sedimentary rock and its soil cover slid into the valley, blocking the river. A rancher on horseback saw the ground beneath him begin to move and had to gallop to safe ground. The slide itself merely created a lake, but the natural dam broke two years later, and the resulting flood destroyed the small town of Kelly several kilometers downstream. Several residents who were standing on a bridge watching the floodwaters come down the valley were killed.

UNDERWATER LANDSLIDES

The steeper parts of the ocean floors sometimes have very large landslides. Prehistoric ones are indicated by large masses of jumbled debris on the deep-ocean floor. One, off the coast of the Hawaiian Islands, is much larger than any landslide mass on land. The debris from what is called the Nuuanu debris avalanche covers an area of 5,000 square kilometers, larger than all of the present Hawaiian Islands combined, and includes volcanic rock blocks several kilometers across.

One very large landslide evidently took place off the coast of northeastern Canada in 1929 following an earth-

FIGURE 9.19

(A) Photo of Gros Ventre slide. (B) and (C) Diagram of the Gros Ventre, Wyoming, slide. *Photo by D. A. Rahm, courtesy of Rahm Memorial Collection, Western Washington University. B and C after W. C. Alden, U.S. Geological Survey*

quake. It systematically cut a series of trans-Atlantic telephone and telegraph lines. The existence and extent of the event were inferred decades later by analyzing the timing of the telephone conversation cutoffs and the distance of cables from the earthquake's epicenter. The underwater debris avalanche, described as a *turbidity current* in chapter 18 (see figure 18.11), traveled over 700 kilometers in thirteen hours at speeds from 15 to 60 kilometers per hour. The lengths of the sections of cable carried away indicate that the debris flow was up to 100 kilometers wide.

Scientists have recently found that a very large area of thick sediment off of the central part of the East Coast of the United States is unstable and could become a giant submarine landslide. If it does go, it very likely will generate a giant *tsunami* (discussed in chapter 16) that could be disas-

trous to coastal communities in Europe as well as North America.

PREVENTING LANDSLIDES

Preventing Mass Wasting of Soil

Mass movements of soil usually can be prevented. Proper engineering is essential when the natural environment of a hillside is altered by construction. As shown in figure 9.20, construction generally makes a slope more susceptible to mass wasting of soil in several ways: (1) the base of the slope is undercut, removing the natural support for the upper part of the slope; (2) vegetation is removed during construction; (3) buildings constructed on the

ENVIRONMENTAL GEOLOGY 9.3

Failure of the St. Francis Dam—A Tragic Consequence of Geology Ignored

In 1928, the St. Francis Dam near Los Angeles, California, broke, only a year after it had been completed (box figure 1). The concrete dam was about 60 meters (200 feet) high, and the wall of water that roared down the valley killed about 400 people in two counties.

The failure of this ill-conceived dam destroyed the reputation of its chief designer, William Mulholland. Mulholland and his cronies had earlier in the 1900s become rich by building an unsurpassed but controversial system of aqueducts to bring (some say, steal) water from eastern California and the Colorado River to transform previously arid southern California. This led to the booming growth of Los Angeles and San Diego as well as unprecedented agricultural productivity.

The eastern edge of the dam had been built against a metamorphic rock with foliation planes parallel to the sides of the valley. Landslide scars in the valley should have been ample warning to the builders that the metamorphic rock moved even under only the force of gravity. A competent engineer worries as much about the stability of the rock against which a dam is built as about the strength of the dam itself. Water pressure at the base of the dam exerted a force of 5.7 tons per square foot against the dam. With pressure such as this, the dam and part of the bordering foliated rock could easily slide. Movement would be parallel to the weak foliation planes, just as if the dam had been anchored against a giant deck of cards.

Ironically, investigators never found out for sure whether this was what caused the failure of the dam. Many other blunders had been made in construction, and any one of them could have caused the dam to break. The base of the dam was on a fault with ground-up rock; and, incredibly, the other side of the dam was built against rock that disintegrates in water. This is but one of many instances in which ignorance of geology cost lives and money. Had professional geologic advice been sought, the dam probably would not have been built in that spot.

Additional Resource

St. Francis Dam virtual field trip.

- http://seis.natsci.csulb.edu/VIRTUAL_FIELD/ Francesquito_Dam/franmain.htm

BOX 9.3 ■ FIGURE 1

The St. Francis Dam, California, after its failure. The former water level in the reservoir is visible on the hills in the background. *Photo by H. T. Stearns, U.S. Geological Survey*

upper part of a slope add weight to the potential slide; and (4) extra water may be allowed to seep into the soil.

Some preventive measures can be taken during construction. A retaining wall is usually built where a cut has been made in the slope, but this alone is seldom as effective a deterrent to downslope movement as people hope. If, in addition, drain pipes are put through the retaining wall and into the hillside, water can percolate through and drain away rather than collecting in the soil behind the wall (figure 9.21). Without drains, excess water results in decreased shear strength and the whole soggy mass can easily burst through the wall.

Another practical preventive measure is to avoid oversteepening the slope. The hillside can be cut back in a series of terraces rather than in a single steep cut. This reduces not only the slope angle but also the shear force by removing much of the overlying material. It also prevents loose material (such as boulders dislodged from the top of the cut) from rolling to the base. Road cuts constructed in this way are usually reseeded with rapidly growing grass or plants whose roots

Before construction

Sprinkler adds water to soil

Building adds weight to slope

Fill

Vegetation removed

Steepening of slope for road cut

FIGURE 9.20

A hillside becomes vulnerable to mass wasting due to construction activities.

Water trapped in soil causes movement, pushing down retaining wall.

Water drains through pipe, allowing wall to keep slope from moving.

FIGURE 9.21

Use of drains to help prevent mass wasting.

help anchor the slope. A vegetation cover also minimizes erosion from running water.

Some roads and railroads in steep, mountainous areas are covered by sheds with sloping, reinforced concrete roofs (figure 9.22). Sliding debris and snow avalanches pass safely overhead rather than block a road and endanger lives.

Preventing Rockfalls and Rockslides on Highways

Rockslides and rockfalls are a major problem on highways built through mountainous country. Steep slopes and cliffs are created when road cuts are blasted and bulldozed into mountain sides. If the bedrock has planes of weakness (such as joints, bedding planes, or foliation planes), the orientation of these planes relative to the road cut determines whether there is a rockslide hazard (as in figure 9.23A). If the planes of weakness are inclined into the hill, there is no chance of a rockslide. On the other hand, where the planes of weakness are approximately parallel to the slope of the hillside, a rockslide may occur.

Various techniques are used to prevent rockslides. By doing a detailed geologic study of an area before a road is built, builders might avoid a hazard by choosing the least dangerous route for the road. If a road cut must be made through bedrock that appears prone to sliding, all of the rock that might slide could be removed (sometimes at great expense), as shown in figure 9.23B.

In some instances, slopes prone to rock sliding have been "stitched" in place by the technique shown in figure 9.24.

FIGURE 9.22

A concrete shed with a sloping roof protects a highway from avalanches in the Canadian Rockies. *Photo by C. W. Montgomery*

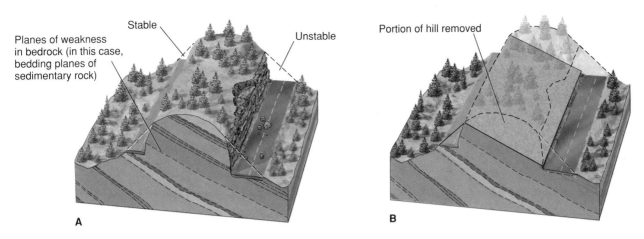

FIGURE 9.23

(A) Cross section of a hill showing a relatively safe road cut on the left and a hazardous road cut on the right. (B) The same hazardous road cut after removal of rock that might slide.

Spraying a roadside exposure with cement may retard a landslide in some instances (figure 9.24*E*). Fences or railing on the side of a road can keep minor rockfalls from blocking the road.

Radio-transmitted, real-time monitoring of areas where mass wasting is active is valuable for predicting when a mass movement is about to speed up and be dangerous. Five sites along U.S. Highway 50 in California's Sierra Nevada are monitored by means of instruments placed at steep mass wasting sites by the U.S. Geologic Survey, and the data is immediately available on the Internet. The instruments include buried pore pressure gauges as well as motion sensors that can tell when a slow-moving mass is starting to move faster.

The best way to avoid mass wasting damage to or destruction of your house is to get information on the susceptibility of the land to mass wasting before building or buying it. This way, you can determine that the house is constructed so as to not cause movement of the terrain, if possible. If not, you had best avoid building or buying the structure. A starting place, if you live in the forty-eight contiguous states, is the U.S. Geological Survey's landslide overview map at http://landslides .usgs.gov/html_files/landslides/ nationalmap/national.html.

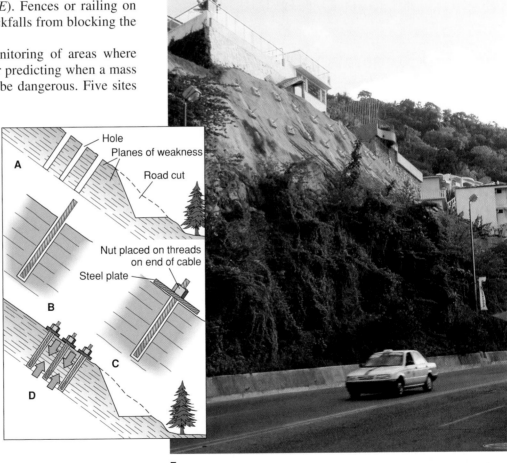

FIGURE 9.24

"Stitching" a slope to keep bedrock from sliding along planes of weakness. (A) Holes are drilled through unstable layers into stable rock. (B) Expanded view of one hole. A cable is fed into the hole and cement is pumped into the bottom of the hole and allowed to harden. (C) A steel plate is placed over the cable and a nut tightened. (D) Tightening all the nuts pulls unstable layers together and anchors them in stable bedrock. (E) Road cut in Acapulco, Mexico stabilized by "stitching" and sprayed concrete. *Photo by C. C. Plummer*

SUMMARY

Mass wasting is the movement of a mass of soil or bedrock toward the base of a slope. *Soil,* as used in this chapter, is unconsolidated or weakly consolidated material, regardless of particle size. If soil is predominantly fine material, it is *earth;* if predominantly coarse, it is *debris.* Movement can take place as a flow, slide, or fall. Gravity is the driving force. The component of gravitational force that propels mass wasting is the *shear force,* which occurs parallel to the slopes. The resistance to that force is the *shear strength* of rock or soil. If shear force exceeds shear strength, mass wasting takes place. Water is usually an important factor in mass wasting.

A number of other factors determine whether movement will occur and, if it does, the rate of movement.

The slowest type of movement, *creep,* occurs mostly on relatively gentle slopes, usually aided by water in the soil. In colder climates, repeated freezing and thawing of water within the soil contributes to creep. *Landsliding* is a general term for more rapid mass wasting of rock, soil, or both. *Flows* include earthflows, debris flows, mudflows, and debris avalanches. *Earthflows,* in which finer-grained material is predominant, vary greatly in velocity, although they are not as rapid as *debris avalanches,* which are turbulent masses of debris, water, and air. *Solifluction,* a special variety of earthflow, usually takes place in Arctic or subarctic climates, where ground is permanently frozen (*permafrost*). *Debris flows* involve coarser material than present in earthflows. Typically, they travel farther than earthflows and, if a lot of water is present, travel long distances in channels and behave similarly to mudflows. A *mudflow* is a slurry of mostly clay, silt and water. Most mudflows flow in channels much as streams do.

Rockfall is the fall of broken rock down a vertical or near-vertical slope. A *rockslide* is a slab of rock sliding down a less-than-vertical surface.

Landslides also take place underwater. The larger ones of these are vastly bigger than any that have occurred on land.

Terms to Remember

creep (soil creep) 230	mass wasting 224	shear force 228
debris 225	mud 225	shear strength 228
debris avalanche 235	mudflow 235	slide 225
debris flow 234	permafrost 232	slump 228
earth 225	relief 228	soil 225
earthflow 232	rock avalanche 237	solifluction 232
fall 225	rockfall 236	talus 236
flow 225	rockslide 237	translational slide 228
landslide 224	rotational slide (slump) 228	

Testing Your Knowledge

Use the following questions to prepare for exams based on this chapter.

1. Describe the effect on shear strength of the following:

   a. thickness of soil;   b. orientation of planes of weakness;

   c. water in soil; and   d. vegetation.

2. Compare the shear force to the force of gravity (drawing diagrams similar to figure 9.2) for the following situations:

   a. a vertical cliff;   b. a flat horizontal plane; and

   c. a 45° slope.

3. How does a rotational slide differ from a translational slide?

4. What role does water play in each of the types of mass wasting?

5. Why is solifluction more common in colder climates than in temperate climates?

6. List and explain the key factors that control mass wasting.

7. What is the slowest type of mass wasting process?

   a. debris flow   b. rockslide

   c. creep   d. rockfall

   e. avalanche

8. The largest landslide has taken place

   a. on the sea floor   b. in the Andes

   c. on active volcanoes d. in the Himalaya

9. A descending mass moving downslope as a viscous fluid is referred to as a

   a. fall   b. landslide

   c. flow   d. slide

10. The driving force behind all mass wasting processes is
    a. gravity   b. slope angle
    c. type of bedrock material   d. presence of water
    e. vegetation

11. The resistance to movement or deformation of soil is its
    a. mass   b. shear strength
    c. shear force   d. density

12. Flow of water-saturated soil over impermeable material is called
    a. solifluction   b. flow
    c. slide   d. fall

13. A flowing mixture of soil and water, usually moving down a channel is called a
    a. mudflow   b. slide
    c. fall   d. earthflow

14. An apron of fallen rock fragments that accumulates at the base of a cliff is called
    a. bedrock   b. sediment
    c. soil   d. talus

15. How does construction destabilize a slope?
    a. adds weight to the top of the slope
    b. decreases water content of the slope
    c. adds weight to the bottom of the slope
    d. increases the shear strength of the slope

16. How can landslides be prevented during construction? (choose all that apply)
    a. retaining walls   b. cut steeper slopes
    c. install water drainage systems   d. add vegetation

Expanding Your Knowledge

1. Why do people fear earthquakes, hurricanes, and tornadoes more than they fear landslides?

2. If you were building a house on a cliff, what would you look for to ensure that your house would not be destroyed through mass wasting?

3. Why isn't the land surface of Earth flat after millions of years of erosion by mass wasting as well as by other erosional agents?

4. Can any of the indicators of creep be explained by processes other than mass wasting?

Exploring Web Resources

www.mhhe.com/plummer11e

This is the dedicated website for this book. You can go to it for new and updated information. The universal resource locators (URL) listed in this book are also given as links on the website, making it easy to go to those websites without typing in the URL. When you visit our Online Learning Center, go to the Student Center and then to the chapter of interest. Here you will find additional readings and media resources, as well as answers to the Testing Your Knowledge section, more quizzes, animation and video clips, and direct links to the sites listed in this book. Links to additional websites can also be found. We have added questions for some of the sites to allow you to get the most of your exploration of the web. Using the web is an enjoyable way of enhancing your knowledge of geology.

http://landslides.usgs.gov/

Geologic hazards, landslides, U.S. Geological Survey. You can get to several useful sites from here. Reports on recent landslides can be accessed by clicking on the ones listed. Click on "National Landslide Information

Center" for photos of landslides, including some described in this chapter. Watch animation of a landslide. You can access sources of information on landslides and other geologic features for any state, usually from a state's geologic survey.

http://landslides.nrcan.gc.ca/

Landslides. Geological Survey of Canada's site has generalized descriptions of significant Canadian landslides.

http://seis.natsci.csulb.edu/VIRTUAL_FIELD/Palos_Verdes/pvportuguese.htm

Portuguese Bend Landslide. Note the photo showing the scarps and the diagrams that illustrate how a rockslide takes place along planes of weakness approximately parallel to the hillside. How did the people in the city of Rolling Hills contribute to this slide? What role did the mineral *bentonite* play? Look through the remaining photos. What affect does the ongoing sliding have on roads?

 # Animation

This chapter includes the following animation available on our Online Learning Center at **www.mhhe.com/plummer11e**.

9.1 Types of Earth movements

Streams and Floods

Running water, aided by mass wasting, is the most important geologic agent in eroding, transporting, and depositing sediment. Almost every landscape on Earth shows the results of stream erosion or deposition. Although other agents—ground water, glaciers, wind, and waves—can be locally important in sculpturing the land, stream action and mass wasting are the dominant processes of landscape development.

The Gooseberry River flows over a resistant ledge of Precambrian-age lava flows at Gooseberry Falls State Park, Minnesota. *Photo © Ray Coleman/ Photo Researchers*

We begin by examining the relationship of running water to other water in the Earth systems. The first part of this chapter also deals with the various ways that streams erode, transport, and deposit sediment. The second part describes landforms produced by stream action, such as val-leys, flood plains, deltas, and alluvial fans, and shows how each of these is related to changes in stream characteristics. The chapter also includes a discussion of the causes and effects of flooding, and various measures used to control flooding.

EARTH SYSTEMS—THE HYDROLOGIC CYCLE

The interrelationship of the hydrosphere, geosphere, biosphere, and atmosphere is easy to visualize through the **hydrologic cycle,** the movement and interchange of water between the sea, air, and land (figure 10.1). Solar radiation provides the necessary energy for *evaporation* of water vapor from the land and sea. When air becomes saturated with water (100% relative humidity), rises, and cools in the atmosphere, liquid droplets condense to form clouds. These droplets grow larger as more water leaves the gaseous state to form rain or snow, depending on the temperature. When rain (or snow) falls on the land surface as *precipitation,* more than half the water returns rather rapidly to the atmosphere by evaporation or *transpiration* from plants. Some of the water is held as ice in glaciers and snow

FIGURE 10.1

The hydrologic cycle. Water vapor evaporates from the land and sea, condenses to form clouds, and falls as precipitation (rain and snow). Water falling on land runs off over the surface as streams or infiltrates into the ground to become ground water. It returns to the atmosphere again by evaporation and transpiration (the loss of water to the air by plants) *Visit http://observe.nasa.gov/nasa/earth/hydrocycle/hydro1.html*

pack. The remainder either flows over the land surface as *runoff* in streams, is held temporarily in lakes, or soaks into the ground by *infiltration* to form ground water. Ground water (the subject of chapter 11) moves, usually very slowly, underground and may flow back onto the surface a long distance from where it seeped into the ground. Most water eventually reaches the sea, where ongoing evaporation completes the cycle.

Only about 15% to 20% of rainfall normally ends up as surface runoff in rivers, although the amount of runoff can range from 2% to more than 25% with variations in climate, steepness of slope, soil and rock type, and vegetation. Steady, continuous rains can saturate the ground and the atmosphere, however, and lead to floods as runoff approaches 100% of rainfall.

RUNNING WATER

A **stream** is a body of running water that is confined in a channel and moves downhill under the influence of gravity. In some parts of the country, *stream* implies size: rivers are large,

streams somewhat smaller, and brooks or creeks even smaller. Geologists, however, use *stream* for any body of running water, from a small trickle to a huge river.

Figure 10.2A shows a *longitudinal profile* of a typical stream viewed from the side. The stream begins in steep mountains and flows out across a gentle plain into the sea. The *headwaters* of a stream are the upper part of the stream near its source in the mountains. The *mouth* is the place where a stream enters the sea, a lake, or a larger stream. A *cross section* of a stream in steep mountains is usually a V-shaped valley cut into solid rock, with the stream channel occupying the narrow bottom of the valley; there is little or no flat land next to the stream on the valley bottom (figure 10.2B). Near its mouth a stream usually flows within a broad, flat-floored valley. The stream channel is surrounded by a flat *flood plain* of sediment deposited by the stream (figure 10.2C).

A stream normally stays in its **stream channel,** a long, narrow depression eroded by the stream into rock or sediment. The stream *banks* are the sides of the channel; the *streambed* is the bottom of the channel. During a flood, the waters of a stream

A Longitudinal profile of a stream beginning in mountains and flowing across a plain into the sea.

B Cross section of the stream at B-B'. The channel is at the bottom of a V-shaped valley cut into rock.

C Cross section at C-C'. The channel is surrounded by a broad flood plain of sediment.

FIGURE 10.2

Longitudinal profile and cross sections of a typical stream.

A

B

FIGURE 10.3

A stream normally stays in its channel, but during a flood it can spill over its banks onto the adjacent flatland (flood plain) as shown in these three-dimensional satellite images. (A) Before flooding image (August 14, 1991) of Missouri River (bottom), Mississippi River (upper left), and Illinois River (upper right). Vegetation is shown in green and red indicates recently plowed fields (bare soil). (B) Image taken on November 7 after the huge floods of 1993 showing how the rivers spilled over their banks onto the flat flood plains. *Photos © NASA/GSFC/Photo Researchers*

may rise and spill over the banks onto the flat flood plain of the valley floor (figure 10.3).

Not all water that moves over the land surface is confined to channels. Sometimes, particularly during heavy rains, water runs off as **sheetwash,** a thin layer of unchanneled water flowing downhill. Sheetwash is particularly common in deserts, where the lack of vegetation allows rainwater to spread quickly over the land surface. It also occurs in humid regions during heavy thunderstorms when water falls faster than it can soak into the ground. A series of closely spaced storms can also promote sheetwash; as the ground becomes saturated, more water runs over the surface.

Sheetwash, along with the violent impact of raindrops on the land surface, can produce considerable *sheet erosion,* in which a thin layer of surface material, usually topsoil, is removed by the flowing sheet of water. This gravity-driven movement of sediment is a process intermediate between mass wasting and stream erosion.

Overland sheetwash becomes concentrated in small channels, forming tiny streams called *rills.* Rills merge to form small streams, and small streams join to form larger streams. Most regions are drained by networks of coalescing streams.

DRAINAGE BASINS

Each stream, small or large, has a **drainage basin,** the total area drained by a stream and its tributaries (a **tributary** is a small stream flowing into a larger one). A drainage basin can be outlined on a map by drawing a line around the region drained by all the tributaries to a river (figure 10.4). The Mis-

sissippi River's drainage basin, for example, includes all the land area drained by the Mississippi River itself and by all its tributaries, including the Ohio and Missouri Rivers. This great drainage system includes more than one-third the land area of the contiguous 48 states.

A ridge or strip of high ground dividing one drainage basin from another is termed a **divide** (figure 10.4). The best known in the United States is the Continental Divide, a line separating streams that flow to the Pacific Ocean from those that flow to the Atlantic and the Gulf of Mexico. The Continental Divide, which extends from the Yukon Territory down into Mexico, crosses Montana, Idaho, Wyoming, Colorado, and New Mexico in the United States. Road signs indicating the crossing of the Continental Divide have been placed at numerous points where major highways intersect the divide.

DRAINAGE PATTERNS

The arrangement, in map view, of a river and its tributaries is a **drainage pattern.** A drainage pattern can, in many cases, reveal the nature and structure of the rocks underneath it.

Most tributaries join the main stream at an acute angle, forming a V (or Y) pointing downstream. If the pattern resembles branches of a tree or nerve dendrites, it is called **dendritic** (figures 10.4 and 10.5A). Dendritic drainage patterns develop on uniformly erodible rock or regolith and are the most common type of pattern. A **radial pattern,** in which streams diverge outward like spokes of a wheel, forms on high conical mountains, such as composite volcanoes and domes (figure 10.5B). A **rectangular pattern,** in which tributaries have frequent 90° bends and tend to join other streams at right angles, develops on

FIGURE 10.4

The drainage basin of the Mississippi River is the land area drained by the river and all its tributaries, including the Ohio and Missouri Rivers; it covers more than 1.6 million square kilometers. Heavy rain in any part of the basin can cause flooding on the lower Mississippi River in the states of Mississippi and Louisiana. The Continental Divide separates rivers that flow into the Pacific from rivers that flow into the Atlantic and the Gulf of Mexico.

regularly fractured rock (figure 10.5*C*). A network of fractures meeting at right angles forms pathways for streams because fractures are eroded more easily than unbroken rock. A **trellis pattern** consists of parallel main streams with short tributaries meeting them at right angles (figure 10.5*D*). A trellis pattern forms in a region where tilted layers of resistant rock such as sandstone alternate with nonresistant rock such as shale. Erosion of such a region results in a surface topography of parallel ridges and valleys.

FACTORS AFFECTING STREAM EROSION AND DEPOSITION

Stream erosion and deposition are controlled primarily by a river's *velocity* and, to a lesser extent, its *discharge*. Velocity is largely controlled by the stream *gradient,* channel shape, and channel roughness.

Velocity

The distance water travels in a stream per unit time is called the **stream velocity.** A moderately fast river flows at about 5 kilometers per hour (3 miles per hour). Rivers flow much faster during flood, sometimes exceeding 25 kilometers per hour (15 miles per hour).

The cross-sectional views of a stream in figure 10.6 show that a stream reaches its maximum velocity near the middle of the channel. When a stream goes around a curve, the region of maximum velocity is displaced by inertia toward the outside of the curve. Velocity is the key factor in a stream's ability to erode, transport, and deposit. High velocity (meaning greater energy) generally results in erosion and transportation; low velocity causes sediment deposition. Slight changes in velocity can cause great changes in the sediment load carried by the river.

Figure 10.7 shows the stream velocities at which sediments are eroded, transported, and deposited. For each grain size, these velocities are different. The upper curve represents the minimum velocity needed to erode sediment grains. This curve shows the velocity at which previously stationary grains are first picked up by moving water. The lower curve represents the velocity below which deposition occurs, when moving grains come to rest. Between the two curves, the water is moving fast enough to transport grains that have already been eroded. Note that it takes a higher stream velocity to erode grains (set them in motion) than to transport grains (keep them in motion).

Point A on figure 10.7 represents fine sand on the bed of a stream that is barely moving. The vertical red arrows represent a flood with gradually increasing stream velocity. No sediment moves until the velocity is high enough to intersect the *upper* curve and move into the area marked "erosion." As the flood recedes, the velocity drops below the upper curve and into the transportation area. Under these conditions, the sand that was already eroded continues to be transported, but no new sand is eroded. As the velocity falls below the lower curve, all the sand is deposited again, coming to rest on the streambed.

The right half of the diagram shows that coarser particles require progressively higher velocities for erosion and transportation, as you might expect (boulders are harder to move than sand grains). The erosion curve also rises toward the left of the diagram, however. This shows that fine-grained silt and clay are actually harder to erode than sand. The reason is that molecular forces tend to bind silt and clay into a smooth, cohesive mass that resists erosion. Once silt and clay are eroded, however, they are easily transported. As you can see from the lower curve, the silt and clay in a river's suspended load are not deposited until the river virtually stops flowing.

A Dendritic

B Radial

C Rectangular

D Trellis

FIGURE 10.5

Drainage patterns can reveal something about the rocks underneath. (A) Dendritic pattern develops on uniformly erodible rock. (B) A radial pattern develops on a conical mountain or dome. (C) A rectangular pattern develops on regularly fractured rock. (D) A trellis pattern develops on alternating ridges and valleys caused by the erosion of resistant and nonresistant tilted rock layers.

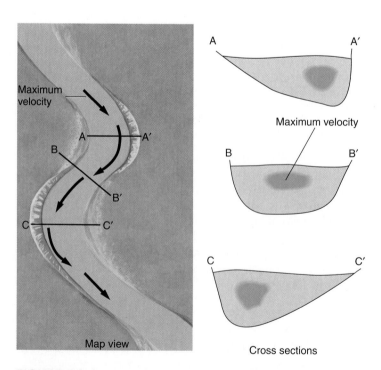

FIGURE 10.6

Regions of maximum velocity in a stream. Arrows on the map show how the maximum velocity shifts to the outside of curves. Sections show maximum velocity on outside of curves and in the center of the channel on a straight stretch of stream.

Gradient

One factor that controls a stream's velocity is the **stream gradient,** the downhill slope of the bed (or of the water surface, if the stream is very large). A stream gradient is usually measured in feet per mile in the United States, because these units are used on U.S. maps (elsewhere, gradients are expressed in meters per kilometer). A gradient of 5 feet per mile means that the river drops 5 feet vertically for every mile that it travels horizontally. Mountain streams may have gradients as steep as 50 to 200 feet per mile (10 to 40 meters per kilometer). The lower Mississippi River has a very gentle gradient, 0.5 foot per mile (0.1 meter per kilometer) or less.

A stream's gradient usually decreases downstream. Typically, the gradient is greatest in the headwater region and decreases toward the mouth of the stream (see figure 10.2). Local increases in the gradient of a stream are usually marked by rapids.

Channel Shape and Roughness

The *shape of the channel* also controls stream velocity. Flowing water drags against the stream banks and bed, and the resulting friction slows the water down. In figure 10.8, the streams in *A* and *B* have the same cross-sectional area, but stream *B* flows slower than *A* because the wide, shallow channel in *B* has more surface for the moving water to drag against.

A stream may change its channel width as it flows across different rock types. Hard, resistant rock is difficult to erode, so a stream may have a relatively narrow channel in such rock. As a result, it flows rapidly (figure 10.9*A*). If the stream flows onto a softer rock that is easier to erode, the channel may widen, and the river will slow down because of the increased surface area dragging on the flowing water. Sediment may be deposited as the velocity decreases.

The width of a stream may be controlled by factors external to the stream. A landslide may carry debris onto a valley floor, partially blocking a stream's channel (figure 10.9*B*). The constriction causes the stream to speed up as it flows past the slide, and the increased velocity may quickly erode the landslide debris, carrying it away downstream. Human interference with a river can promote erosion and deposition. Construction of a culvert or bridge can partially block a channel, increasing the stream's velocity (figure 10.9*C*). If the bridge was poorly designed, it may increase velocity to the point where erosion may cause the bridge to collapse.

The *roughness of the channel* also controls velocity. A stream can flow rapidly over a smooth channel, but a rough, boulder-strewn channel floor creates more friction and slows the flow (see figure 10.8*C*). Coarse particles increase the roughness more than fine particles, and a rippled or wavy sand bottom is rougher than a smooth sand bottom.

Discharge

The **discharge** of a stream is the volume of water that flows past a given point in a unit of time. It is found by multiplying the cross-sectional area of a stream by its velocity (or width × depth × velocity). Discharge can be reported in cubic feet per second (cfs), which is standard in the United States, or in cubic meters per second (m³/sec).

Discharge (cfs) = average stream width (ft)
× average depth (ft)
× average velocity (ft/sec)

A stream 100 feet wide and 15 feet deep flowing at 4 miles per hour (6 ft/sec) has a discharge of 9,000 cubic feet per second

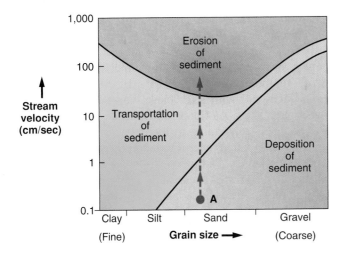

FIGURE 10.7

Logarithmic graph showing the stream velocities at which erosion and deposition of sediment occur. These velocities vary with the grain size of the sediment. See text for a discussion of point A and the dashed red line above it.

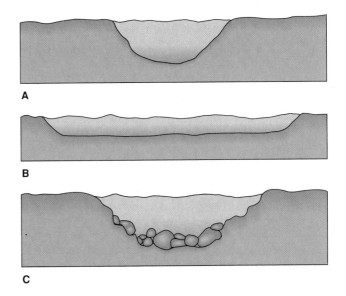

FIGURE 10.8

Channel shape and roughness influence stream velocity. (A) Semicircular channel allows stream to flow rapidly. (B) Wide, shallow channel increases friction, slowing river down. (C) Rough, boulder-strewn channel slows river.

FIGURE 10.9

Channel width variations caused by rock type and obstructions. Length of arrow indicates velocity. (A) A channel may widen in soft rock. Deposition may result as stream velocity drops. (B) Landslide may narrow a channel, increasing stream velocity. Resulting erosion usually removes landslide debris. (C) Bridge piers (or other obstructions) will increase velocity and sometimes erosion next to the piers.

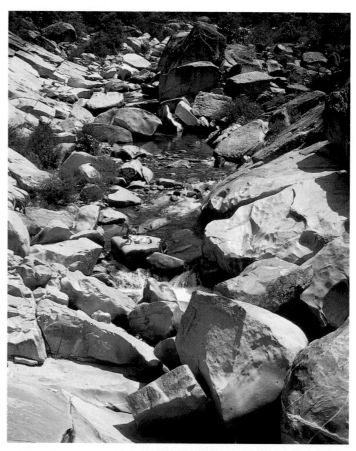

FIGURE 10.10

These large boulders of granite in a mountain stream are moved only during floods. Note the rounding of the boulders and the scoured high-water mark of floods on the valley walls. Note people for scale. *Photo by David McGeary*

(cfs). In streams in humid climates, discharge increases downstream for two reasons: (1) water flows out of the ground into the river through the streambed; and (2) small tributary streams flow into a larger stream along its length, adding water to the stream as it travels.

To handle the increased discharge, these streams increase in width and depth downstream. Some streams surprisingly increase slightly in velocity downstream, as a result of the increased discharge (the increase in discharge and channel size,

and the typical downstream smoothness of the channel override the effect of a lessening gradient).

During floods, a stream's discharge and velocity increase, usually as a result of heavy rains over the stream's drainage basin. Flood discharge may be 50 to 100 times normal flow. Stream erosion and transportation generally increase enormously as a result of a flood's velocity and discharge. Swift mountain streams in flood can sometimes move boulders the size of automobiles (figure 10.10). Flooded areas may be intensely scoured, with river banks and adjacent lawns and fields washed away. As floodwaters recede, both velocity and discharge decrease, leading to the deposition of a blanket of sediment, usually mud, over the flooded area.

In a dry climate, a river's discharge can decrease in a downstream direction as river water evaporates into the air and soaks into the dry ground (or is used for irrigation). As the discharge decreases, the load of sediment is gradually deposited.

STREAM EROSION

A stream usually erodes the rock and sediment over which it flows. In fact, streams are one of the most effective sculptors of the land. Streams cut their own valleys, deepening and widening them over long periods of time and carrying away the sediment that mass wasting delivers to valley floors. The particles of rock and sediment that a stream picks up are carried along to be deposited farther downstream. Streams erode rock and sediment in three ways—*hydraulic action, solution,* and *abrasion.*

Hydraulic action refers to the ability of flowing water to pick up and move rock and sediment (figure 10.11). The force of running water swirling into a crevice in a rock can crack the rock and break loose a fragment to be carried away by the stream. Hydraulic force can also erode loose material from a stream bank on the outside of a curve. The pressure of flowing water can roll or slide grains over a streambed, and a swirling eddy of water may exert enough force to lift a rock fragment above a streambed. The great force of falling water makes hydraulic action particularly effective at the base of a waterfall, where it may erode a deep plunge pool. You may be able to hear the results of hydraulic action by standing beside a swift mountain stream and listening to boulders and cobbles hitting one another as they tumble along downstream.

From what you have learned about weathering, you know that some rocks can be dissolved by water. **Solution,** although ordinarily slow, can be an effective process of weathering and erosion (weathering because it is a response to surface chemical conditions; erosion because it removes material). A stream flowing over limestone, for example, gradually dissolves the rock, deepening the stream channel. A stream flowing over other sedimentary rocks, such as sandstone, can dissolve calcite cement, loosening grains that can then be picked up by hydraulic action.

The erosive process that is usually most effective on a rocky streambed is **abrasion,** the grinding away of the stream channel by the friction and impact of the sediment load. Sand and gravel tumbling along near the bottom of a stream wear away the streambed much as moving sandpaper wears away wood. The abrasion of sediment on the streambed is generally much more effective in wearing away the rock than hydraulic action alone. The more sediment a stream carries, the faster it is likely to wear away its bed.

The coarsest sediment is the most effective in stream erosion. Sand and gravel strike the streambed frequently and with great force, while the finer-grained silt and clay particles weigh so little that they are easily suspended throughout the stream and have little impact when they hit the channel.

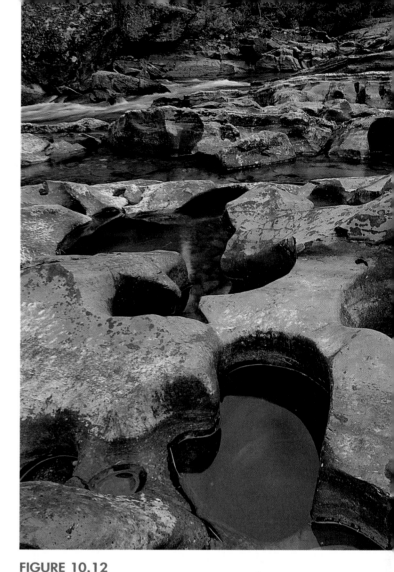

FIGURE 10.12

Potholes scoured along bed of McDonald River in Glacier National Park, Montana. *Photo © Joe McDonald/Visuals Unlimited*

Potholes are depressions that are eroded into the hard rock of a streambed by the abrasive action of the sediment load (figure 10.12). During high water when a stream is full, the swirling water can cause sand and pebbles to scour out smooth, bowl-shaped depressions in hard rock. Potholes tend to form in spots where the rock is a little weaker than the surrounding rock. Although potholes are fairly uncommon, you can see them on the beds of some streams at low water level. Potholes may contain sand or an assortment of beautifully rounded pebbles.

STREAM TRANSPORTATION OF SEDIMENT

The sediment load transported by a stream can be subdivided into *bed load, suspended load,* and *dissolved load.* Most of a stream's load is carried in suspension and in solution.

The **bed load** is the large or heavy sediment particles that travel on the streambed (figure 10.13). Sand and gravel, which form the usual bed load of streams, move by either *traction* or *saltation.*

FIGURE 10.11

Hydraulic action can loosen, roll, and lift grains from the streambed.

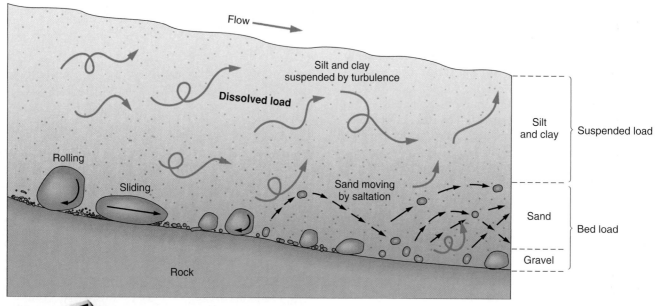

FIGURE 10.13

A stream's bed load consists of sand and gravel moving on or near the streambed by traction and saltation. Finer silt and clay form the suspended load of the stream. The dissolved load of soluble ions is invisible.

Large, heavy particles of sediment, such as cobbles and boulders, may never lose contact with the streambed as they move along in the flowing water. They roll or slide along the stream bottom, eroding the streambed and each other by abrasion. Movement by rolling, sliding, or dragging is called **traction.**

Sand grains move by traction, but they also move downstream by **saltation,** a series of short leaps or bounces off the bottom (see figure 10.13). Saltation begins when sand grains are momentarily lifted off the bottom by *turbulent* water (eddying, swirling flow). The force of the turbulence temporarily counteracts the downward force of gravity, suspending the grains in water above the streambed. The water soon slows down because the velocity of water in an eddy is not constant; then gravity overcomes the lift of the water, and the sand grain once again falls to the bed of the stream. While it is suspended,

the grain moves downstream with the flowing water. After it lands on the bottom, it may be picked up again if turbulence increases, or it may be thrown up into the water by the impact of another falling sand grain. In this way, sand grains saltate downstream in leaps and jumps, partly in contact with the bottom and partly suspended in the water.

The **suspended load** is sediment that is light enough to remain lifted indefinitely above the bottom by water turbulence (see figure 10.13). The muddy appearance of a stream during a flood or after a heavy rain is due to a large suspended load. Silt and clay usually are suspended throughout the water, while the coarser bed load moves on the stream bottom. Suspended load has less effect on erosion than the less visible bed load, which causes most of the abrasion of the streambed. Vast quantities of sediment, however, are transported in suspension.

FIGURE 10.14

Sand and gravel bars deposited along the banks and middle of a stream. Burnside River Nunavut, Canada.
Photo © Brian Sytnyk/Masterfile

FIGURE 10.15

A flood can wash away bars in a stream, depositing new bars as the water recedes. (*A*) Normal water flow with sand and gravel bar. (*B*) Increased discharge and velocity during flood moves all sediment downstream. Channel deepens and widens if banks erode easily. (*C*) New bars are deposited as water level drops and stream slows down.

Soluble products of chemical weathering processes can make up a substantial **dissolved load** in a stream. Most streams contain numerous ions in solution, such as bicarbonate, calcium, potassium, sodium, chloride, and sulfate. The ions may precipitate out of water as evaporite minerals if the stream dries up, or they may eventually reach the ocean. Very clear water may in fact be carrying a large load of material in solution, for the dissolved load is invisible. Only if the water evaporates does the material become visible as crystals begin to form.

One estimate is that rivers in the United States carry about 250 million tons of solid load and 300 million tons of dissolved load each year. (It would take a freight train eight times as long as the distance from Boston to Los Angeles to carry 250 million tons.)

STREAM DEPOSITION

The sediments transported by a stream are often deposited temporarily along the stream's course (particularly the bed load sediments). Such sediments move sporadically downstream in repeated cycles of erosion and deposition, forming *bars* and *flood-plain deposits*. At or near the end of a stream, sediments may be deposited more permanently in a *delta* or an *alluvial fan*.

Bars

Stream deposits may take the form of a **bar**, a ridge of sediment, usually sand and gravel, deposited in the middle or along the banks of a stream (figure 10.14). Bars are formed by deposition when a stream's discharge or velocity decreases. During a flood, a river can move all sizes of sediment, from silt and clay up to huge boulders, because the greatly increased volume of water is moving very rapidly. As the flood begins to recede, the water level in the stream falls and the velocity drops. With the stream no longer able to carry all its sediment load, the larger boulders drop down on the streambed, slowing the water locally even more. Finer gravel and sand are deposited between the boulders and downstream from them. In this way, deposition builds up a sand and gravel bar that may become exposed as the water level falls.

The next flood on the river may erode most of the sediment in this bar and move it farther downstream. But as the flood slows, it may deposit new gravel in approximately the same place, forming a new bar (figure 10.15). After each flood, river anglers and boat operators must relearn the size and position of the bars. Sometimes gold panners discover fresh gold in a mined-out river bar after a flood has shifted sediment downstream. A dramatic example of the shifting of sandbars occurred during the planned flood on the Colorado River downstream from the Glen Canyon Dam (box 10.1).

Placer Deposits

Placer deposits are found in streams where the running water has mechanically concentrated heavy sediment. The heavy sediment is concentrated in the stream where the velocity of the water is high enough to carry away lighter material but not the heavy sediment. Such places include river bars on the inside of meanders, plunge pools below waterfalls, and depressions on a streambed (figure 10.16). Grains concentrated in this manner include gold dust and nuggets, native platinum, diamonds and other gemstones, and worn pebble or sand grains composed of the heavy oxides of titanium and tin.

A Map view

B Side view

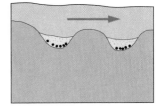
C Side view

FIGURE 10.16

Types of placer deposits. (*A*) Stream bar. (*B*) Below waterfall. (*C*) Depressions on streambed. Valuable grains shown in black.

ENVIRONMENTAL GEOLOGY 10.1

A Controlled Flood in the Grand Canyon: A Bold Experiment to Restore Sediment Movement in the Colorado River

On March 26, 1996, one of the largest experiments ever conducted on a river took place along the Colorado River below the Glen Canyon Dam (box figure 1). For six days, the discharge from the Glen Canyon Dam was increased from 8,000 cfs to 45,000 cfs (a spike flow) to emulate the effects of a flood on the Colorado River (box figure 2). One of the main goals of this controlled flooding experiment was to determine whether the higher flows would result in bed scour and redeposition of sandbars and beaches along the sides of the channel. Another goal was to measure and observe how rocks move along the bed of the river bed with increasing discharge and velocity of floodwaters.

The Colorado River had not experienced its usual summertime floods since the Glen Canyon Dam was completed in 1963. The construction of the dam controlled peak discharges or flows on the Colorado River, which resulted in sand being deposited mainly along the bed or bottom of the river and erosion of beaches along the banks of the river. The Glen Canyon Dam cuts off a significant percentage of the sand supply to the lower Colorado River such that most of the downstream sand is supplied by two tributary streams, the Paria and Little Colorado Rivers. In August 1992, the Paria River flooded and deposited 330,000 tons of sand into the Colorado River, and in January 1993, a flood on the Little Colorado River deposited 10 million tons of sediment below its confluence with the Colorado River. The influx of sediment, coupled with the relatively low discharges from the dam (8–20,000 cfs), resulted in sand being concentrated along the bed of the Colorado River.

One of the main predictions of the experiment proved true. That is, sand caught in deep pools in the bottom of the main channel was scoured and carried in suspension downstream, where it was redeposited along the river banks as beaches (box figure 3). Deposition of sand along the banks occurs due to back eddies that upwell and move upstream along the river banks and decrease the velocity of the downstream flow so that deposition can occur. Most of the scouring and deposition occurred in the first three days of the experiment; however, when flows were reduced back to 8,000 cfs, beaches began to erode and redeposition occurred once again in the deep pools on the bottom of the river.

Downstream at Lava Falls, an experiment was set up to determine how and if large boulders deposited in the main channel from a debris flow would move with the increased discharge and velocity of the floodwater. Holes were drilled into 150 basalt boulders and radio tags were inserted (box figure 4) so their movement could be monitored and correlated with the increase in discharge and velocity of the river. Surface velocity measurements were taken by kayaking the river and charting the speed at which floating balls moved. The surface velocities were used to calculate the velocity of the water close to the riverbed where the boulders were positioned. Dye was also injected into the river at peak flows to determine the average velocity of the water. The dye indicated that the velocity of the water increased downstream, particularly at the Lava Falls debris flow. This is because the floodwater accelerated as it flowed downstream, pushing the river water in front of it, which increased the downstream velocity. The first crest of water actually arrived behind Hoover Dam at Lake Mead a day ahead of the floodwater marked with a red dye.

The experiment was deemed a success, and for the first time, a flood was studied as it happened. The experimental flood, even though smaller than a naturally occurring flood (box figure 2), showed that beaches could be restored below a dam and that boulders could be moved out of rapids much like that which occurs on an undammed river during a seasonal flood. It is proposed that other dammed rivers would benefit from periodic floods to help restore their natural conditions and thus minimize the adverse effects of damming a river.

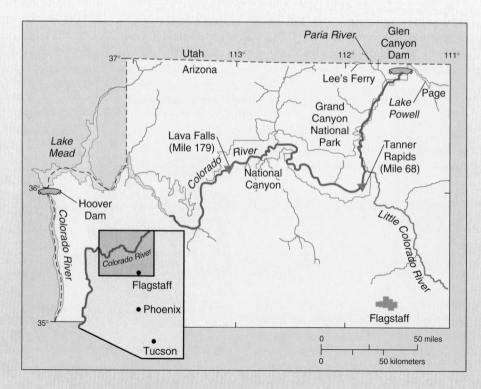

BOX 10.1 ■ FIGURE 1

Location map of the Grand Canyon controlled-flood experiment.
U.S. Geological Survey

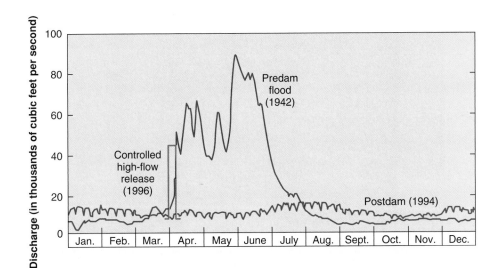

BOX 10.1 ■ FIGURE 2

Graph of annual variation in discharges before (black) and after Glen Canyon Dam (blue) and the 1996 controlled high-flow release (red). *U.S. Geological Survey*

BOX 10.1 ■ FIGURE 3

Cross section of the channel downstream of the confluence with the Little Colorado River. Increased flows have scoured the bottom sediment and redeposited it as a beach along the river bank. *U.S. Geological Survey*

Additional Resources

Flooding in Grand Canyon. *Scientific American*, January 1997, pp. 82–89.

Grand Canyon Flood! *Nova* video, 1997.

R. H. Webb, J. C. Schmidt, G. R. Marzolf, and R. A. Valdez, eds. 1999. *The Controlled Flood in Grand Canyon*. Geophysical Monograph Series 110.

For an overview and details of the specific experiments conducted during the planned flood, visit the following websites:

- http://water.usgs.gov/wid/FS_089–96/FS_089–96.html and www.pbs.org/kuat/grandcanyonflood

BOX 10.1 ■ FIGURE 4

(A) Hole being drilled into a basalt boulder and (B) radio tag installed to track the movement of boulders as the discharge and velocity of the Colorado River increase at the Lava Falls debris flow locality. *Photos courtesy of KUAT-TV, University of Arizona, Dan Duncan*

FIGURE 10.17

(A) A midchannel bar can divert a stream around it, widening the stream. (B) Braided stream occurs where there is an excess of sediment. Bars split main channel into many smaller channels, greatly widening the stream. (C) A braided stream carrying a heavy suspended load of sand and gravel from melting glaciers. Denali National Park, Alaska. Photo © Michael Giannechini/Photo Researchers, Inc.

Braided Streams

Deposition of a bar in the center of a stream (a *midchannel bar*) diverts the water toward the sides, where it washes against the stream banks with greater force, eroding the banks and widening the stream (figure 10.17A). A stream heavily loaded with sediment may deposit many bars in its channel, causing the stream to widen continually as more bars are deposited. Such a stream typically goes through many stages of deposition, erosion, deposition, and erosion, especially if its discharge fluctuates. The stream may fill its main channel with sediment and become a **braided stream,** flowing in a network of interconnected rivulets around numerous bars (figures 10.17B and C). A braided stream characteristically has a wide, shallow channel.

A stream tends to become braided when it is heavily loaded with sediment (particularly bed load) and has banks that are easily eroded. The braided pattern develops in deserts as a sediment-laden stream loses water through evaporation and percolation into the ground. In meltwater streams flowing off glaciers, braided patterns tend to develop when the discharge from the melting glaciers is low relative to the great amount and ranges of size of sediment the stream has to carry.

Meandering Streams and Point Bars

Rivers that carry fine-grained silt and clay in suspension tend to be narrow and deep and to develop pronounced, sinuous curves called **meanders** (figure 10.18). In a long river, sediment tends to become finer downstream, so meandering is common in the lower reaches of a river.

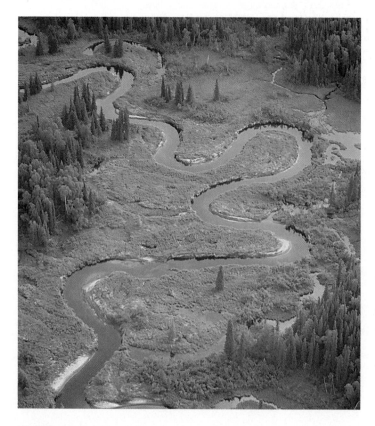

FIGURE 10.18

Meanders in a stream. These sinuous curves develop because a stream's velocity is highest on the outside of curves, promoting erosion there. Photo © Glenn M. Oliver/Visuals Unlimited

You have seen in figure 10.6 that a river's velocity is higher on the outside of a curve than on the inside. This high velocity can erode the river bank on the outside of a curve, often rapidly (figure 10.19).

The low velocity on the inside of a curve promotes sediment deposition. The sandbars in figure 10.20 have been deposited on the inside of curves because of the lower velocity there. Such a bar is called a **point bar** and usually consists of a series of arcuate ridges of sand or gravel.

The simultaneous erosion on the outside of a curve and deposition on the inside can deepen a gentle curve into a hairpin-like meander (see figure 10.20). Meanders are rarely fixed in position. Continued erosion and deposition cause them to migrate back and forth across a flat valley floor, as well as downstream, leaving scars and arcuate point bars to mark their former positions.

At times, particularly during floods, a river may form a **meander cutoff,** a new, shorter channel across the narrow neck of a meander (figure 10.21). The old meander may be abandoned as sediment separates it from the new, shorter channel. The cutoff meander becomes a crescent-shaped **oxbow lake** (figure 10.22). With time, an oxbow lake may fill with sediment and vegetation.

Flood Plains

A **flood plain** is a broad strip of land built up by sedimentation on either side of a stream channel. During floods, flood plains may be covered with water carrying suspended silt and clay

A

B

FIGURE 10.19

River erosion on the outside of a curve. Newaukum River, Washington. Pictures were taken in (A) January and (B) March 1965. *Photos by P. A. Glancy, U.S. Geological Survey*

Cross section

Corkscrew water motion on a curve helps cause erosion and deposition

FIGURE 10.20

Development of river meanders and point bars by erosion and deposition on curves.

FIGURE 10.21

Creation of an oxbow lake by a meander neck cutoff. Old channel is separated from river by sediment deposition.

FIGURE 10.22

An oxbow lake marks the former position of a river meander, Blackfoot River near Vallet, Montana. *Photo © James Steinberg/Photo Researchers*

FIGURE 10.23

River flood plains. Flooded flood plain of the Animas River, Colorado. *Photo by D. A. Rahm, courtesy of Rahm Memorial Collection, Western Washington University*

(figure 10.23). When the floodwaters recede, these fine-grained sediments are left behind as a horizontal deposit on the flood plain.

Some flood plains are constructed almost entirely of horizontal layers of fine-grained sediment, interrupted here and there by coarse-grained channel deposits (figure 10.24*A*). Other flood plains are dominated by meanders shifting back and forth over the valley floor and leaving sandy point bar deposits on the inside of curves. Such a river will deposit a characteristic fining-upward sequence of sediments: coarse channel deposits are gradually covered by medium-grained point bar deposits, which in turn are overlain by fine-grained flood deposits (figure 10.24*B*).

As a flooding river spreads over a flood plain, it slows down. The velocity of the water is abruptly decreased by friction as the water leaves the deep channel and moves in a thin sheet over the flat valley floor. The sudden decrease in velocity of the water causes the river to deposit most of its sediment near the main channel, with progressively less sediment deposited away from the channel (figure 10.25). A series of floods may build up **natural levees**—low ridges of flood-deposited sediment that form on either side of a stream channel and thin away from the channel. The sediment near the

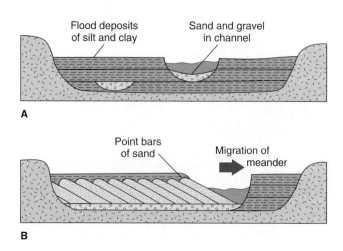

FIGURE 10.24

Flood plains. (*A*) Horizontal layers of fine-grained flood deposits with lenses of coarse-grained channel deposits. (*B*) A fining-upward sequence deposited by a migrating meander. Channel gravel is overlain by sandy point bars, which are overlain by fine-grained flood deposits.

A

B

C

FIGURE 10.25

Natural levee deposition during a flood. Levees are thickest and coarsest next to the river channel and build up from many floods, not just one. (Relief of levees is exaggerated.) (A) Normal flow. (B) Flood. (C) After flood.

river is coarsest, often sand and silt, while the finer clay is carried farther from the river into the flat, lowland area (the backswamp).

Deltas

Most streams ultimately flow into the sea or large lakes. A stream flowing into quiet water usually builds a **delta,** a body of sediment deposited at the mouth of a river when the river's velocity decreases (figure 10.26).

The surface of most deltas is marked by **distributaries**—small, shifting channels that carry water away from the main river channel and distribute it over the surface of the delta (figure 10.27). Sediment deposited at the end of a distributary tends to block the water flow, causing distributaries and their sites of sediment deposition to shift periodically.

The shape of a marine delta in map view depends on the balance between sediment supply from the stream and the erosive power of waves and tides (figure 10.28). Some deltas, like that of the Nile River, are broadly triangular; this delta's resemblance to the Greek letter *delta* (Δ) is the origin of the name.

The Nile Delta is a *wave-dominated delta* that contains barrier islands along its oceanward side (figure 10.28A); the barrier islands form by waves actively reworking the deltaic sediments. Some deltas form along a coast that is dominated by strong tides, and the sediment is reshaped into tidal bars that

FIGURE 10.26

Delta of Tongariro River, New Zealand. Note the sediment-laden water and how the land is being built outward by river sedimentation. A river typically divides into several channels (distributaries) on a delta. *Photo © G. R. "Dick" Roberts/Natural Sciences Image Library*

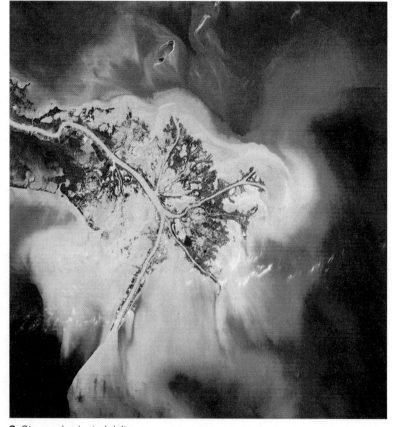

River

Distributaries

Marshy delta surface

Topset beds

Foreset beds

Lake

Bottomset beds

FIGURE 10.27

Internal construction of a small delta.

are aligned parallel to a tidal current (figure 10.28*B*). The Ganges-Brahmaputra Delta in Bangladesh is a good example of a *tide-dominated delta.* Other deltas, including that of the Mississippi River, are created when very large amounts of sediment are carried into relatively quiet water. Partly because dredging has kept the major distributary channels (locally called "passes") fixed in position for many decades, the Mississippi's distributaries have built long fingers of sediment out into the sea. The resulting shape has been termed a *birdfoot delta.* Because of the dominance of stream sedimentation that forms the fingerlike distributaries, birdfoot deltas like the Mississippi's are also referred to as *stream-dominated deltas* (figure 10.28*C*).

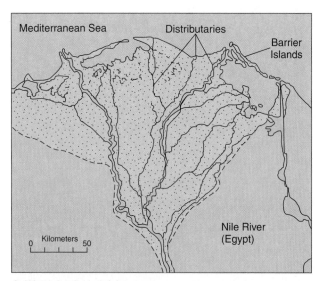

Mediterranean Sea

Distributaries

Barrier Islands

Kilometers
0   50

Nile River (Egypt)

A Wave-dominated delta

Bangladesh

Tidal current

Kilometers
0   15

Tidal sandbars

B Tidal-dominated delta

C Stream-dominated delta

FIGURE 10.28

The shape of a delta depends on the amount of sediment being carried by the river and on the vigor of waves and tides in the sea. (*A*) The Nile Delta is a wave-dominated delta with prominent barrier islands. (*B*) The Ganges-Brahmaputra Delta in Bangladesh contains tidal sandbars formed by strong tidal currents. (*C*) A Landsat image of the Mississippi River Delta. Note how sediment (light yellow) is carried by both river and ocean currents. *After R. G. Walker, "Facies Models," Geoscience Canada, fig. 7, p. 109, 1979; Photo by NASA*

Many deltas, particularly small ones in freshwater lakes, are built up from three types of deposits, shown in the diagram in figure 10.27. *Foreset beds* form the main body of the delta. They are deposited at an angle to the horizontal. This angle can be as great as 20° to 25° in a small delta where the foreset beds are sandy or less than 5° in large deltas with fine-grained sediment. On top of the foreset beds are the *topset beds*, nearly horizontal beds of varying grain size formed by distributaries shifting across the delta surface. Out in front of the foreset beds are the *bottomset beds*, deposits of the finest silt and clay carried out into the lake by the river water flow or by sediments sliding downhill on the lake floor. Many of the world's great deltas in the ocean are far more complex than the simplified diagram shown in the figure. Shifting river mouths, wave energy, currents, and other factors produce many different internal structures.

The persistence of large deltas as relatively "dry" land depends on a balance between the rate of sedimentation and the rates of tectonic subsidence and compaction of water-saturated sediment. Many deltas are sinking, with seawater encroaching on once-dry land. The Mississippi Delta in Louisiana is sinking, as upstream dams catch sediment, reducing the delta's supply, and as extraction of oil and gas from beneath the delta accelerates subsidence. The flat surface of a delta is a risky place to live or farm, particularly in regions threatened by the high waves and raised sea level of hurricanes, such as the U.S. Gulf Coast and the country of Bangladesh (the Ganges-Brahmaputra Delta) on the Indian Ocean.

Alluvial Fans

Some streams, particularly in dry climates, do not reach the sea or any other body of water. They build alluvial fans instead of deltas. An **alluvial fan** is a large, fan- or cone-shaped pile of sediment that usually forms where a stream's velocity decreases as it emerges from a narrow mountain canyon onto a flat plain (figure 10.29). Alluvial fans are particularly well developed and exposed in the southwestern desert of the United States and in other desert regions, but they are by no means limited to arid regions.

An alluvial fan builds up its characteristic fan shape gradually as streams shift back and forth across the fan surface and deposit sediment, usually in a braided pattern. Deposition on an alluvial fan in the desert is discontinuous because streams typically flow for only a short time after the infrequent rainstorms. When rain does come, the amount of sediment to be moved is often greater than the available water and material is moved as a debris flow before it comes to rest and is deposited.

The sudden loss of velocity when a stream flows from narrow mountain canyons onto a broad plain causes the sediment to deposit on an alluvial fan. The loss of velocity is due to the widening or branching of the channel as it leaves the narrow canyon. The gradual loss of water as it infiltrates into the fan also promotes sediment deposition. On large fans, deposits are graded in size within the fan, with the coarsest sediment dropped nearest the mountains and the finer material deposited progressively farther away. Small fans do not usually show such grading.

FIGURE 10.29

(A) An alluvial fan at the mouth of a desert canyon. (B) This alluvial fan formed on the salt-encrusted floor of Death Valley, California, as sediment was washed out of the canyon by thunderstorms. *Photo by Frank M. Hanna*

FLOODING

Many of the world's cities, such as Pittsburgh, St. Louis, and Florence, Italy, are built beside rivers and therefore can be threatened by floods. Rivers are important transportation routes for ships and barges, and flat flood plains have excellent agricultural soil and offer attractive building sites for houses and industry.

Flooding does not occur every year on every river, but flooding is a natural process on all rivers; those who live in river cities and towns must be prepared. Heavy rains and the rapid melting of snow in the spring are the usual causes of floods. The rate and volume of rainfall and the geographic path of rainstorms often determine whether flooding will occur.

Floods are described by *recurrence interval,* the average time between floods of a given size. A "100-year flood" is one that can occur, on the *average,* every 100 years (box 10.2). A 100-year flood has a 1-in-100, or 1%, chance of occurring in any given year. It is perfectly possible to have two 100-year floods in successive years—or even in the same year. If a 100-year flood occurs this year on the river you live beside, you should not assume that there will be a 99-year period of safety before the next one.

Flood erosion is caused by the high velocity and large volume of water in a flood. Although relatively harmless on an uninhabited flood plain, flood erosion can be devastating to a city. As a river undercuts its banks, particularly on the outside of curves where water velocity is high, buildings, piers, and bridges may fall into the river. As sections of flood plain are washed away, highways and railroads are cut.

High water covers streets and agricultural fields, and invades buildings, shorting out electrical lines and backing up sewers. Water-supply systems may fail or be contaminated. Water in your living room will be drawn upward in your walls by capillary action in wall plasterboard and insulation, creating a soggy mess that has to be torn out and replaced. High water on flat flood plains often drains away very slowly; street travel

may be by boat for weeks. If floodwaters are deep enough, houses may float away (figure 10.30).

Flood deposits are usually silt and clay. A new layer of wet mud on a flood plain in an agricultural region can be beneficial in that it renews the fields with topsoil from upstream, as used to be the case with the Nile River until the Aswan Dam was built. The same mud in a city will destroy lawns, furniture, and machinery. Cleanup is slow; imagine shoveling 4 inches of worm-filled mud that smells like sewage out of your house.

Urban Flooding

Urbanization contributes to severe flooding. Paved areas and storm sewers increase the amount and rate of surface runoff of water. This is due to their inhibiting infiltration of rainwater into the ground and their rapid delivery of the resulting increased runoff to the channels, making river levels higher during storms (figure 10.31). Such rapid increases in runoff or discharge to a river are called a "flashy" discharge. Storm sewers are usually designed for a 100-year storm; however, large storms that drop a lot of rain in a short period of time (cloudburst) may overwhelm sewer systems and cause localized flooding. Rising river levels may block storm sewer outlets and add to localized flooding problems.

Bridges, docks, and buildings built on flood plains can also constrict the flow of floodwaters, increasing the water height and velocity and promoting erosion.

Flash Floods

Some floods occur rapidly and die out just as quickly. *Flash floods* are local, sudden floods of large volume and short duration, often triggered by heavy thunderstorms. A startling example occurred in 1976 in north-central Colorado along the Big Thompson River (figure 10.32). Strong winds from the east pushed moist air up the front of the Colorado Rockies, causing thunderstorms in the steep mountains. The storms were unusually stationary, allowing as much as 30 centimeters (12 inches) of rain to fall in two days. Some areas received 19 centimeters in just over

FIGURE 10.30

January 1997 floodwaters were so deep in the Arboga, California, area that a house floated off its foundation. *Photo courtesy of Robert A. Eplett, Governor's Office of Emergency Services*

FIGURE 10.31

The presence of a city can increase the chance of floods. The blue curve shows the normal increase in a river's discharge following a rainstorm (black bar). The red curve shows the great increase in runoff rate and amount caused by pavement and storm sewers in a city.

FIGURE 10.32

(A) Location map of the 1976 flash flood on the Big Thompson River in Colorado and the 1997 flash flood in Fort Collins. (B) A cabin sits crushed against a bridge following the Big Thompson Canyon flash flood of 1976. (C) Devastation along Spring Creek after the 1997 Fort Collins flash flood. *Photos by W. R. Hansen, U.S. Geological Survey*

A

B

C

IN GREATER DEPTH 10.2

Estimating the Size and Frequency of Floods

Because people have encroached on the flood plains of many rivers, flooding is one of the most universally experienced geologic hazards. To minimize flood damage and loss of life, it is useful to know the potential size of large floods and how often they might occur. This is often a difficult task because of the lack of long-term records for most rivers. The U.S. Geological Survey monitors the stage (water elevation) and discharge of rivers and streams throughout the United States to collect data that can be used to attempt to predict the size and frequency of flooding and to make estimates of water supply.

Hydrologists designate floods based on their *recurrence interval*, or *return period*. For example, a 100-year flood is the largest flood expected to occur within a period of 100 years. This does not mean that a 100-year flood occurs once every century but that there is a 1-in-100 chance, or a 1% probability, each year that a flood of this size will occur. Usually, flood control systems are built to accommodate a 100-year flood because that is the minimum margin of safety required by the federal government if an individual wants to obtain flood insurance subsidized by the Federal Emergency Management Agency (FEMA).

To calculate the recurrence interval of flooding for a river, the annual peak discharges (largest discharge of the year) are collected and ranked according to size (box figure 1 and table 1). The largest annual peak discharge is assigned a rank (m) of 1, the second a 2, and so on until all of the discharges are assigned a rank number. The *recurrence interval* **(R)** of each annual peak discharge is then calculated by adding 1 to the *number of years of record* **(n)** and dividing by its *rank* **(m)**.

$$R = \frac{n + 1}{m}$$

For example, the Cosumnes River in California has 90 years of record **(n = 90)**, and in 1907, the second-largest peak discharge **(m = 2)** of 71,000 cfs occurred. The recurrence inter-

	BOX 10.2 ■ TABLE 1		

Annual Peak Discharges and Recurrence Intervals in Rank Order for the Cosumnes River at Michigan Bar, California

Year	Peak Discharge (cfs)	Magnitude Rank (m)	Recurrence Interval
1997	93,000	1	91.0
1907	71,000	2	45.5
1986	45,100	3	30.33
1956	42,000	4	22.75
1963	39,400	5	18.42
1909	28,400	10	9.20
1943	22,900	20	4.60
1970	16,800	30	3.07
1960	11,200	40	2.30
1971	8,590	50	1.84
1991	6,670	60	1.53

Source: Preliminary data from Richard Hunrichs, hydrologist, U.S. Geological Survey and U.S. Geological Survey Water-Data Report, CA-97-3

val **(R)**, or expected frequency of occurrence, for a discharge this large is 45.5 years:

$$R = \frac{90 + 1}{2} = 45.5$$

That is, there is a 1-in-45.5, or 2%, chance each year of a peak discharge of 71,000 cfs or greater occurring on the Cosumnes River.

BOX 10.2 ■ FIGURE 1

Annual peak discharge for the Cosumnes River. *After U.S. Geological Survey Water-Data Report, CA-97-3*

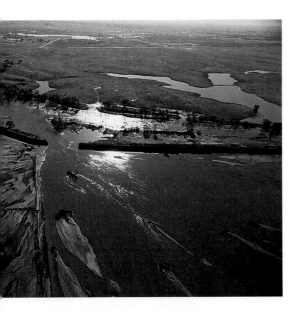

BOX 10.2 ■ FIGURE 2

Levee break along the Cosumnes River. *Courtesy of Robert A. Eplett, Governor's Office of Emergency Services*

BOX 10.2 ■ FIGURE 3

Flood-frequency curves for the Cosumnes River. Data from Richard Hunrichs, hydrologist, U.S. Geological Survey and U.S. Geological Survey Water-Data Report, CA-97-3

The flood of record (largest recorded discharge) occurred on January 2, 1997, when heavy, unseasonably warm rains rapidly melted snow in the Sierra Nevada and caused flooding in much of northern California. A peak discharge of 93,000 cfs in the Cosumnes River resulted in levee breaks and widespread flooding of homes and agricultural areas (box figure 2). The recurrence interval for the 1997 flood (93,000 cfs) is 91 years:

$$R = \frac{90 + 1}{1} = 91 \text{ years}$$

A *flood-frequency curve* can be useful in providing an estimate of the discharge and the frequency of floods. The flood frequency curve is generated by plotting the annual peak discharges against the calculated recurrence intervals (box figure 3). Because most of the data points defining the curve plot in the lower range of discharge and recurrence interval, there is some uncertainty in projecting larger flood events. Two flood frequency curves are drawn in box figure 3; the red line represents the best-fit curve for all of the data, whereas the dashed blue line excludes the 1997 flood of record. Notice that the curve has a steeper slope when the 1997 data is included and that the size of the 100-year flood has increased from 73,000 cfs to 82,000 cfs based on the one additional year of record. Because large floods do not occur as often as small floods, the rare large flood can have a dramatic effect on the shape of the flood-frequency curve and the estimate

of a 100-year event. This is particularly true for a river like the Cosumnes that has had only two large events, one in 1907 and the other 90 years later in 1997.

The 100-year flood plain is based on the estimate of the discharge of the 100-year flood and on careful mapping of the flood plain. Changes in the estimated size of the 100-year flood could result in property that no longer has 100-year flood protection. In this case, property owners may be prevented from getting flood insurance or money to rebuild from FEMA unless new flood-control structures are built to provide additional protection or houses are raised or even relocated out of the flood plain.

Additional Resources

Water Resources Data for California, Water Year 1997. U.S. Geological Survey Water-Data Report CA-97-3.

H. C. Riggs. 1968. Frequency curves. *Techniques of Water-Resources Investigations of the U.S. Geological Survey.* Book 4, Hydrologic Analysis and Interpretation.

To find data sets to calculate the recurrence interval for rivers throughout the United States, access the U.S. Geological Survey Water Data Retrieval Website:

• http://water.usgs.gov/usa/nwis/

an hour. Little of this torrential rainfall could soak into the ground. The volume of water in the Big Thompson River swelled to four times the previously recorded maximum, and the river's velocity rose to an impressive 25 kilometers per hour for a few hours on the night of July 31. By the next morning, the flood was over, and the appalling toll became apparent—139 people dead, 5 missing, and more than $35 million in damages (figure 10.32*B*).

On July 29, 1997, just two days before the twenty-first anniversary of the Big Thompson River flood, Fort Collins, Colorado, was struck by a flash flood when 20 centimeters of rain fell in only five hours. A 4-meter-high wall of water rushed down Spring Creek, a tributary to the Cache la Poudre River, nearly devastating two trailer parks. Five people lost their lives and forty were injured when a 5-meter-high railroad embankment that had temporarily dammed Spring Creek broke and sent the wall of water into the trailer park (figure 10.32*C*). Unlike the Big Thompson Canyon flood, the heavy rains fell over the city of Fort Collins rather than upstream in the steep mountain canyons.

Controlling Floods

Flood-control structures can partially reduce the dangers of floodwaters and sedimentation to river cities (figure 10.33). Upstream dams can trap water and release it slowly after the storm. (A dam also catches sediment, which eventually fills its reservoir and ends its life as a flood-control structure.) Artificial levees are embankments built along the banks of a river channel to contain floodwaters within the channel. Protective walls of stone (riprap) or concrete are often constructed along river banks, particularly on the outside of curves, to slow erosion. Floodwalls, walls of concrete, may be used to protect cities from flooding; however, these flood-control structures may constrict the channel and cause water to flow faster with more erosive power downstream. Bypasses are also used along the Mississippi and other rivers to reduce the discharge in the main channel by diverting water through gates or weirs into designated basins in the flood plain. The bypasses serve to give part of the natural flood plain back to the river.

Dams and levees are designed to control certain specified floods. If the flood-control structures on your river were designed for 75-year floods, then a much larger 100-year flood will likely overtop these structures and may destroy your home. The disastrous floods along the Missouri and Mississippi Rivers and their tributaries north of Cairo, Illinois, in 1993 resulted from many such failures in flood control.

Wise land-use planning and zoning for flood plains should go hand in hand with flood control. Wherever possible, buildings should be kept out of areas that might someday be flooded by 100-year floods.

The Great Flood of 1993

In the spring of 1993, heavy rains soaked the ground in the upper Midwestern United States. In June and July, a stationary weather pattern created a shifting band of thunderstorms that dumped as much as 10 centimeters (4 inches) of rain in a single day on localities such as Bismarck, North Dakota; Cedar Rapids, Iowa; and Manhattan, Kansas (figure 10.34). These torrential rains falling on already-saturated ground created the worst flood disaster in U.S. history, as swollen rivers flooded

FIGURE 10.33

Examples of flood-control structures.

A

B

C

D

FIGURE 10.34

The Great Flood of 1993. (*A*) Area of flood. (*B*) 1993 rainfall at Cedar Rapids, Iowa, compared to normal rainfall. (*C*) Discharge of the Mississippi River at St. Louis compared to 100-year flood. (*D*) Mississippi River water pours through a broken levee onto a farm near Columbia, Illinois, 1993. *Data from U.S. Geological Survey. Photo © St. Louis Post-Dispatch*

FIGURE 10.35

Downcutting, mass wasting, and sheet erosion shape canyons and valleys. (*A*) Downcutting can create slot canyons in resistant rock, particularly where downcutting is rapid during flash floods and fractures in the rock are favorably oriented. (*B*) Stream erosion has cut this unusual slot canyon through porous sandstone, Zion National Park, Utah. (*C*) Downslope movement of rock and soil on valley walls widens most canyons into V-shaped valleys. (*D*) The waterfall and rapids on the Yellowstone River in Wyoming indicate that the river is ungraded and actively downcutting. Note the V-shaped cross-profile and lack of flood plain due to the downslope movement of volcanic rock. *Photo B by Allen Hagood, Zion Natural History Association; Photo D by David McGeary*

6.6 million acres in nine states, killing thirty-eight people and causing $12 billion in damage to houses and crops.

River discharges exceeded 100-year discharges on many rivers including the Mississippi, Missouri, Iowa, Platte, and Raccoon. At St. Louis, the Mississippi River discharge was greater than 1 million cfs on August 1 (six times normal flow), and the river crested 20 feet above flood stage (and 23 feet above flood stage farther downstream at Chester, Illinois). At Hannibal, Missouri, where the 500-year flood height had been determined to be 30 feet, the new, 31-feet-high flood-control levee was completed in April. The river crested at 32 feet. Some rivers remained above flood stage for months.

The high water in major rivers such as the Mississippi and Missouri caused many smaller tributary streams to back up, flooding numerous small towns. Many levees broke as water overtopped them or just physically pushed the saturated levee sediment aside (figure 10.34D). After the rains stopped, in some places the floodwaters took weeks or even months to recede. Some entire towns have been relocated to drier ground.

STREAM VALLEY DEVELOPMENT

Valleys, the most common landforms on the Earth's surface, are usually cut by streams. By removing rock and sediment from the stream channel, a stream deepens, widens, and lengthens its own valley.

Downcutting and Base Level

The process of deepening a valley by erosion of the streambed is called **downcutting.** If a stream removes rock from its bed, it can cut a narrow *slot canyon* down through rock (figures 10.35A and B). Such narrow canyons do not commonly form because mass wasting and sheet erosion usually remove rock from the valley walls. These processes widen the valley from a narrow, vertical-walled canyon to a broader, open, V-shaped canyon (figures 10.35C and D). Slot canyons persist, however, in very

resistant rock with favorably oriented fractures or in regions where downcutting is rapid.

Downcutting cannot continue indefinitely because the headwaters of a stream cannot cut below the level of the streambed at the mouth. If a river flows into the ocean, sea level becomes the lower limit of downcutting. The river cannot cut below sea level, or it would have to flow uphill to get to the sea. For most streams, sea level controls the level to which the land can be eroded.

The limit of downcutting is known as **base level;** it is a theoretical limit for erosion of the Earth's surface (figure 10.36A). Downcutting will proceed until the streambed reaches base level. If the stream is well above base level, downcutting can be quite rapid; but as the stream approaches base level, the rate of downcutting slows down. For streams that reach the ocean, base level is close to sea level, but since streams need at least a gentle gradient to flow, base level slopes gently upward in an inland direction.

During the glacial fluctuations of the Pleistocene Epoch (see chapter 12), sea level rose and fell as water was removed from the sea to form the glaciers on the continents and returned to the sea when the glaciers melted. This means that base level rose and fell for streams flowing into the sea. As a result, the lower reaches of such rivers alternated between erosion (caused by low sea level) and deposition (caused by high sea level). Since the glaciers advanced and retreated several times, the cycle of erosion and deposition was repeated many times, resulting in a complex history of cutting and filling near the mouths of most old rivers.

Base levels for streams that do not flow into the ocean are not related to sea level. In Death Valley in California (figure 10.36B), base level for in-flowing streams corresponds to the lowest point in the valley, 86 meters (282 feet) *below* sea level (the valley has been dropped below sea level by tectonic movement along faults). On the other hand, base level for a stream above a high reservoir or a mountain lake can be hundreds or even thousands of meters above sea level. The surface of the lake or reservoir serves as temporary base level for all the water upstream (figure 10.36C). The base level of a tributary stream is governed by the level of its

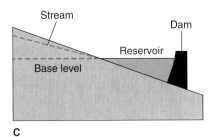

FIGURE 10.36

Base level is the lowest level of downcutting.

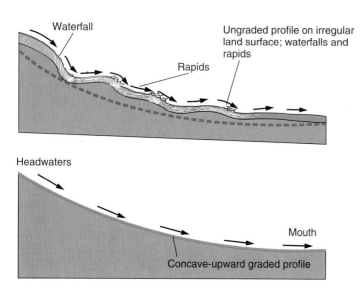

FIGURE 10.37

An ungraded stream has an irregular longitudinal profile with many waterfalls and rapids. A graded stream has smoothed out its longitudinal profile to a smooth, concave-upward curve.

FIGURE 10.38

Lateral erosion can widen a valley by undercutting and eroding valley walls.

junction with the main stream. A ledge of resistant rock may act as a temporary base level if a stream has difficulty eroding through it.

The Concept of a Graded Stream

As a stream begins downcutting into the land, its longitudinal profile is usually irregular with rapids and waterfalls along its course (see figure 10.37). Such a stream, termed *ungraded,* is using most of its erosional energy in downcutting to smooth out these irregularities in gradient.

As the stream smooths out its longitudinal profile to a characteristic concave-upward shape, it becomes graded. A **graded stream** is one that exhibits a delicate balance between its transporting capacity and the sediment load available to it. This balance is maintained by cutting and filling any irregularities in the smooth longitudinal profile of the stream.

In this chapter's section on Factors Affecting Stream Erosion and Deposition, you learned how changes in a stream's gradient can cause changes in its sediment load. An increase in gradient causes an increase in a stream's velocity, allowing the stream to erode and carry more sediment. A balance is maintained—the greater load is a result of the greater transporting capacity caused by the steeper gradient.

The relationship also works in reverse—a change in sediment load can cause a change in gradient. For example, a decrease in sediment load may bring about erosion of the stream's channel, thus lowering the gradient. Because dams trap sediment in the calm reservoirs behind them, most streams are almost completely sediment-free just downstream from dams. In some

streams, this loss of sediment has caused severe channel erosion below a dam, as the stream adjusts to its new, reduced load.

A river's energy is used for two things—transporting sediment and overcoming resistance to flow. If the sediment load decreases, the river has more energy for other things. It may use this energy to erode more sediment, deepening its valley. Or it may change its channel shape or length, increasing resistance to flow, so that the excess energy is used to overcome friction. Or the river may increase the roughness of its channel, also increasing friction. The response of a river is not always predictable, and construction of a dam can sometimes have unexpected and perhaps harmful results.

FIGURE 10.39

Headward erosion is lengthening this stream channel. Note the dendritic drainage pattern that is developing in the headwaters of the streams, New Plymouth, New Zealand. *Photo © G. R. "Dick" Roberts/Natural Sciences Image Library*

Lateral Erosion

A graded stream can be deepening its channel by downcutting while part of its energy is also widening the valley by **lateral erosion,** the erosion and undercutting of a stream's banks and valley walls as the stream swings from side to side across its valley floor. The stream channel remains the same width as it moves across the flood plain, but the valley widens by erosion, particularly on the outside of curves and meanders where the stream impinges against the valley walls (figure 10.38). The valley widens as its walls are eroded by the stream and as its walls retreat by mass wasting triggered by stream undercutting. As a valley widens, the stream's flood plain increases in width also.

Headward Erosion

Building a delta or alluvial fan at its mouth is one way a river can extend its length. A stream can also lengthen its valley by **headward erosion,** the slow uphill growth of a valley above its original source through gullying, mass wasting, and sheet erosion (figure 10.39). This type of erosion is particularly difficult to stop. When farmland is being lost to gullies that are eroding headward into fields and pastures, farmers must divert sheet flow and fill the gully heads with brush and other debris to stop, or at least retard, the loss of topsoil.

STREAM TERRACES

Stream terraces are steplike landforms found above a stream and its flood plain (figure 10.40). Terraces may be benches cut in rock (sometimes sediment-covered), or they may be steps formed in sediment by deposition and subsequent erosion.

Figure 10.41 shows how a terrace forms as a river cuts downward into a thick sequence of its own flood-plain deposits. Originally, the river deposited a thick section of flood-plain sediments. Then the river changed from deposition to erosion and cut into its old flood plain, parts of which remain as terraces above the river.

Why might a river change from deposition to erosion? One reason might be regional uplift, raising a river that was once meandering near base level to an elevation well above base level. Uplift would steepen a river's gradient, causing the river to speed up and begin erosion. But there are several other reasons a river might change from deposition to erosion. A change from a dry to a wet climate may increase discharge and

FIGURE 10.40
Stream terraces near Jackson Hole, Wyoming. The stream has cut downward into its old flood plain. *Photo by Diane Carlson*

A

B

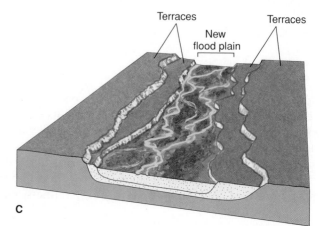

C

FIGURE 10.41
Terraces formed by a stream cutting downward into its own flood-plain deposits. (A) Stream deposits thick, coarse, flood-plain deposits. (B) Stream erodes its flood plain by downcutting. Old flood-plain surface forms terraces. (C) Lateral erosion forms new flood plain below terraces.

cause a river to begin eroding. A drop in base level (such as lowering of sea level) can have the same effect. A situation like that shown in figure 10.41 can develop in a recently glaciated region. Thick valley fill such as glacial outwash (see chapter 12) may be deposited in a stream valley and later, after the glacier stops producing large amounts of sediment, be dissected into terraces by the river.

Terraces can also develop from erosion of a bedrock valley floor. Bedrock benches are usually capped by a thin layer of flood-plain deposits.

INCISED MEANDERS

Incised meanders are meanders that retain their sinuous pattern as they cut vertically downward below the level at which they originally formed. The result is a meandering *valley* with essentially no flood plain, cut into the land as a steep-sided canyon (figure 10.42A).

Some incised meanders may be due to the profound effects of a change in base level. They may originally have been formed as meanders in a laterally eroding river flowing over a flat flood plain, perhaps near base level. If regional uplift elevated the land high above base level, the river would begin downcutting and might be able to maintain its characteristic meander pattern while deepening its valley (figure 10.42B). A drop in base level without land uplift (possibly because of a lowering of sea level) could bring about the same result.

Although uplift may be a key factor in the formation of many incised meanders, it may not be *required* to produce them. Lateral erosion certainly seems to become more prominent as a river approaches base level, but some meandering can occur as soon as a river develops a graded profile. A river flowing on a flat surface high above base level may develop meanders early in its erosional history, and these meanders may become incised by subsequent downcutting. In such a case, uplift is not necessary.

A

B

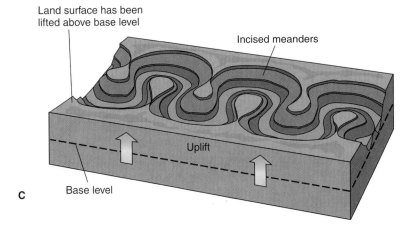

C

FIGURE 10.42

(A) Incised meanders of the Colorado River ("The Loop"), Canyonlands National Park, southwestern Utah. (B) Meandering river flowing over a flat plain cut to base level. (C) Regional uplift of land surface allows river to downcut and incise its meanders. *Photo by Frank M. Hanna*

PLANETARY GEOLOGY 10.3

Stream Features on the Planet Mars

There is probably no liquid water on the surface of Mars today. With the present surface temperatures, atmospheric pressures, and water content in the Martian atmosphere, any liquid water would immediately evaporate. Recent evidence collected from the 2003 Mars Exploration Rovers, *Spirit,* and *Opportunity,* indicate that conditions may have been different in the past and that liquid water existed on Mars, at least temporarily.

Certain features on Mars, called *channels,* closely resemble certain types of stream channels on Earth. They have tributary systems and meanders and are sometimes braided. The channels trend downslope and tend to get wider toward their mouths. These Martian channels are restricted to certain areas and appear to have been formed by intermittent episodes of erosion.

One type of channel on Mars appears to have formed by large flooding events and is similar in appearance to those observed in the Channeled Scablands of Washington. The Channeled Scablands were formed by extensive flooding during the Pleistocene glacial ages, when a naturally formed ice dam broke and released water from a large lake. The mouth of Ares Vallis, an ancient Martian flood channel similar to those observed in the Channeled Scablands, was selected for the July 4, 1997, landing of the *Pathfinder* spacecraft and Sojourner Rover. It was postulated that a variety of rock types should be present in the mouth of an ancient flood channel. The first photos from the Mars *Pathfinder* Lander and *Sojourner* Rover, a "robotic field geologist," revealed a variety of rock types (box figure 1) in what does appear to be an ancient outflow channel.

A second kind of Martian channel (box figure 2), a meandering streamlike feature, occurs on the older surfaces of Mars (more than 3.5 billion years old) and may indicate that early in the history of Mars, temperature and atmospheric conditions were such that rainfall could have occurred and long-lived river systems could have existed. On June 8, 1998, the Mars Orbiter Camera aboard the Mars *Global Surveyor* spacecraft captured an image of what appears to be a meandering stream channel

BOX 10.3 ■ FIGURE 1

View from the Mars Pathfinder Lander showing the Sojourner Rover and a variety of rocks from Ares Vallis. *Photo courtesy of Jet Propulsion Laboratory/NASA*

BOX 10.3 ■ FIGURE 2

Meandering channel and flat terraces within the Nanedi Vallis canyon, which resemble stream features cut by running water on Earth. *Photo courtesy of NASA* and stream terraces inside the Nanedi Vallis canyon. The 2.5-kilometer-wide canyon is located approximately 1,600 kilometers (1,000 miles) south of where the *Pathfinder* landed. The presence of the narrow, streamlike channel and associated terraces within the Nanedi Vallis canyon strongly suggest that a river of water repeatedly flowed down the canyon. But the lack of smaller tributary channels may argue that the channel, like many others on Mars, may have instead formed by surface collapse caused by frozen water underground.

Additional Resources

M. P. Golombek. 1998. The Mars Pathfinder Mission. *Scientific American 7.*

For more information on the possibility of water on Mars, visit the NASA Goddard Institute for Space Studies research site:

- www.giss.nasa.gov/research/intro/gornitz.03/

Information about the Mars Global Surveyor can be found at the Jet Propulsion Laboratory/NASA site:

- http://mars.jpl.nasa.gov/mgs

SUMMARY

Normally, stream *channels* are eroded and shaped by the streams that flow in them. Unconfined sheet flow can cause significant erosion.

Drainage basins are separated by *divides.*

A river and its tributaries form a *drainage pattern.* A *dendritic* drainage pattern develops on uniform rock, a *rectangular* pattern on regularly jointed rock. A *radial* pattern forms on conical mountains, while a *trellis* pattern usually indicates erosion of folded sedimentary rock.

Stream *velocity* is the key factor controlling sediment erosion, transportation, and deposition. Velocity is in turn controlled by several factors.

An increase in a stream's *gradient* increases the stream's velocity. *Channel shape* and *roughness* affect velocity by increasing or lessening friction. As tributaries join a stream, the stream's *discharge* increases downstream. Floods increase stream discharge and velocity.

Streams erode by *hydraulic action, abrasion,* and *solution.* They carry coarse sediment by *traction* and *saltation* as *bed load.* Finer-grained sediment is carried in *suspension.* A stream can also have a substantial *dissolved load.*

Streams create features by erosion and deposition. *Potholes* form by abrasion of hard rock on a streambed. *Bars* form in the middle of streams or on stream banks, particularly on the inside of curves where velocity is low (*point*

bars). A *braided pattern* can develop in streams with a large amount of bed load.

Meanders are created when a laterally eroding stream shifts across the flood plain, sometimes creating cutoffs and oxbow lakes.

A *flood plain* develops by both lateral and vertical deposition. *Natural levees* are built up beside streams by flood deposition.

A *delta* forms when a stream flows into standing water. The shape and internal structure of deltas are governed by river deposition and wave and current erosion. *Alluvial fans* form, particularly in dry climates, at the base of mountains as a stream's channel widens and its velocity decreases.

Rivers deepen their valleys by *downcutting* until they reach *base level,* which is either sea level or a local base level.

A *graded stream* is one with a delicate balance between its transporting capacity and its available load.

Lateral erosion widens a valley after the stream has become graded.

A valley is lengthened by both *headward erosion* and sediment deposition at the mouth.

Stream terraces can form by erosion of rock benches or dissection of thick valley deposits during downcutting.

Incised meanders form as (1) river meanders are cut vertically downward following uplift or (2) lateral erosion and downcutting proceed simultaneously.

Terms to Remember

abrasion 255
alluvial fan 265
bar 257
base level 273
bed load 255
braided stream 260
delta 263
dendritic pattern 250
discharge 253
dissolved load 257
distributary 263
divide 250
downcutting 273
drainage basin 250

drainage pattern 250
flood plain 261
graded stream 274
headward erosion 275
hydraulic action 255
hydrologic cycle 248
incised meander 277
lateral erosion 275
meander 260
meander cutoff 261
natural levee 262
oxbow lake 261
point bar 261
pothole 255

radial pattern 250
rectangular pattern 250
saltation 256
sheetwash 250
solution 255
stream 249
stream channel 249
stream gradient 253
stream terrace 275
stream velocity 251
suspended load 256
traction 256
trellis pattern 251
tributary 250

Testing Your Knowledge

Use the following questions to prepare for exams based on this chapter.

1. What factors control a stream's velocity?

2. Describe how bar deposition creates a braided stream.

3. In what part of a large alluvial fan is the sediment the coarsest? Why?

4. What does a trellis drainage pattern tell about the rocks underneath it?

5. Describe one way that incised meanders form.

6. How does a meander neck cutoff form an oxbow lake?

7. How does a natural levee form?

8. Describe how stream terraces form.

9. Describe three ways in which a river erodes its channel.

10. Name and describe the three main ways in which a stream transports sediment.

11. How does a stream widen its valley?

12. What is base level?

13. The total area drained by a stream and its tributaries is called the
    a. hydrologic cycle   b. tributary area
    c. divide   d. drainage basin

14. Stream erosion and deposition are controlled primarily by a river's
    a. velocity   b. discharge
    c. gradient   d. channel shape
    e. channel roughness

15. What is the gradient of a stream that drops 10 vertical feet over a 2-mile horizontal distance?
    a. 20 feet per mile   b. 10 feet per mile
    c. 5 feet per mile   d. 2 feet per mile

16. What are typical units of discharge?
    a. miles per hour   b. cubic meters
    c. cubic feet per second d. meters per second

17. Hydraulic action, solution, and abrasion are all examples of stream
    a. erosion   b. transportation
    c. deposition

18. Cobbles are more likely to be transported in a stream's
    a. bed load   b. suspended load
    c. dissolved load   d. all of the preceding

19. A river's velocity is _____ on the outside of a meander curve compared to the inside.
    a. higher   b. equal
    c. lower

20. Sandbars deposited on the inside of meander curves are called
    a. dunes   b. point bars
    c. cutbanks   d. none of the preceding

21. Which is not a drainage pattern?
    a. dendritic   b. radial
    c. rectangular   d. trellis
    e. none of the preceding

22. The broad strip of land built up by sedimentation on either side of a stream channel is
    a. a flood plain   b. a delta
    c. an alluvial fan   d. a meander

23. The average time between floods of a given size is
    a. the discharge   b. the gradient
    c. the recurrence interval d. the magnitude

24. A platform of sediment formed where a stream flows into standing water is
    a. an alluvial fan   b. a delta
    c. a meander   d. a flood plain

Expanding Your Knowledge

1. Several rivers have been set aside as "wild rivers" on which dams cannot be built. Give at least four arguments against building dams on rivers. Give at least four arguments in favor of building dams.

2. Discuss the similarities between deltas and alluvial fans. Describe the differences between them.

3. How is the recurrence interval for a flood determined? How may new data affect the flood-frequency curve?

4. What affect would global warming have on the overall water budget in the hydrologic cycle? How might this influence the dynamics of a stream?

Exploring Web Resources

www.mhhe.com/plummer11e

This is the dedicated website for this book. You can go to it for new and updated information. The universal resource locators (URL) listed in this book are also given as links on the website, making it easy to go to those websites without typing in the URL. When you visit our Online Learning Center, go to the Student Center and then to the chapter of interest. Here you will find additional readings and media resources, as well as answers to the Testing Your Knowledge section, more quizzes, animation and video clips, and direct links to the sites listed in this book. Links to additional websites can also be found. We have added questions for some of the sites to allow you to get the most of your exploration of the web. Using the web is an enjoyable way of enhancing your knowledge of geology.

http://water.usgs.gov/

Contains extensive information on water issues throughout the United States and many links to U.S. Geological Survey data and online publications.

http://water.usgs.gov/public/realtime.html

Contains real-time streamflow data from U.S. Geological Survey gaging stations throughout the United States.

http://water.usgs.gov/usa/nwis/

Contains historical streamflow data from U.S. Geological Survey gaging stations throughout the United States.

www.dartmouth.edu/artsci/geog/floods/

The *Dartmouth Flood Observatory* website contains information on flood detection and satellite images of floods and flood damage from around the world.

http://vcourseware.calstatela.edu/VirtualRiver/

California State University, Los Angeles *Virtual River* exercise. Analyze streamflow data and observe animations of flowing streams.

 # Animations

This chapter includes the following animations available on our Online Learning Center at **www.mhhe.com/plummer11e**.

10.13 Modes of sediment transport

10.20 River meander development

Ground Water

How much of the hydrosphere is ground water? Compared to the oceans, not much. Approximately 0.6% of the world's water is ground water, whereas over 97% is ocean water. If we look at fresh water alone, we find that the amount of ground water is 35 times that of all rivers and lakes in the world. (However, the amount of fresh water stored in glaciers is 3.5 times greater than the amount of ground water.)

Ground water is a tremendously important resource. Managing our water resources becomes increasingly difficult as demands increase. Growing cities in arid climates are removing ground water faster than it can be replenished. Pollution of ground water by industrial wastes, agricultural pesticides, and other means can render the water unfit for human consumption. Growing population and improvements in lifestyles increasingly impact our water supply.

Hot ground water is released at fairly regular intervals from Old Faithful in Yellowstone National Park as pressure builds underground. *Photo © Brian Yarvin/Photo Researchers*

How water gets underground, where it is stored, how it moves while underground, how we look for it, and, perhaps most important of all, why we need to protect it are the main topics of this chapter.

Also important is how ground water is related to surface rivers and springs. Ground water can form distinctive geologic features, such as caves, sinkholes, and petrified wood. It also can appear as hot springs and geysers. Hot ground water can be used to generate geothermal energy.

Distribution of Water in the Hydrosphere (%)

Oceans	97.2
Glaciers and other ice	2.15
Ground water	.61
Lakes	
Fresh	.009
Saline	.008
Soil moisture	.005
Atmosphere	.001
Rivers	.0001

INTRODUCTION

Many communities obtain the water they need from rivers, lakes, or reservoirs, sometimes using aqueducts or canals to bring water from distant surface sources. Another source of water lies directly beneath most towns. This resource is **ground water,** the water that lies beneath the ground surface, filling the pore space between grains in bodies of sediment and clastic sedimentary rock, and filling cracks and crevices in all types of rock.

Ground water is a major economic resource, particularly in the dry western areas of the United States and Canada, where surface water is scarce. Many towns and farms pump great quantities of ground water from drilled wells. Even cities next to large rivers may pump their water from the ground because ground water is commonly less contaminated and more economical to use than surface water.

As we saw from the hydrologic cycle described in chapter 10 (see figure 10.1), some of the water that precipitates from the atmosphere as rain and snow infiltrates the geosphere and becomes ground water. How much precipitation soaks into the ground is influenced by climate, land slope, soil and rock type, and vegetation. In general, approximately 15% of the total precipitation ends up as ground water, but that varies locally and regionally from 1% to 20%.

POROSITY AND PERMEABILITY

Porosity, the percentage of rock or sediment that consists of voids or openings, is a measurement of a rock's ability to hold water. Most rocks can hold some water. Some sedimentary rocks, such as sandstone, conglomerate, and many limestones, tend to have a high porosity and therefore can hold a considerable amount of water. A deposit of loose sand may have a porosity of 30% to 50%, but this may be reduced to 10% to 20% by compaction and cementation as the sand lithifies (table 11.1). A sandstone in which pores are nearly filled with

TABLE 11.1   Porosity and Permeability of Sediments and Rocks

Sediment	Porosity (%)	Permeability
Gravel	25 to 40	Excellent
Sand (clean)	30 to 50	Good to excellent
Silt	35 to 50	Moderate
Clay	35 to 80	Poor
Glacial till	10 to 20	Poor to moderate
Rock		
Conglomerate	10 to 30	Moderate to excellent
Sandstone		
Well-sorted, little cement	20 to 30	Good to very good
Average	10 to 20	Moderate to good
Poorly sorted, well-cemented	0 to 10	Poor to moderate
Shale	0 to 30	Very poor to poor
Limestone, dolomite	0 to 20	Poor to good
Cavernous limestone	up to 50	Excellent
Crystalline rock		
Unfractured	0 to 5	Very poor
Fractured	5 to 10	Poor
Volcanic rocks	0 to 50	Poor to excellent

cement and fine-grained matrix material may have a porosity of 5% or less. Crystalline rocks, such as granite, schist, and some limestones, do not have pores but may hold some water in joints and other openings.

Although most rocks can hold some water, they vary a great deal in their ability to allow water to pass through them. **Permeability** refers to the capacity of a rock to transmit a fluid such as water or petroleum through pores and fractures. In other words, permeability measures the relative ease of water flow and indicates the degree to which openings in a rock interconnect. The distinction between porosity and permeability is important. A rock that holds much water is called *porous;* a rock that allows water to flow easily through it is described as *permeable.* Most sandstones and conglomerates are both porous and permeable. An *impermeable* rock is one that does not allow water to flow through it easily. Unjointed granite and schist are impermeable. Shale can have substantial porosity, but it has low permeability because its pores are too small to permit easy passage of water.

THE WATER TABLE

Responding to the pull of gravity, water percolates down into the ground through the soil and through cracks and pores in the rock. The rate of groundwater flow tends to decrease with depth because sedimentary rock pores tend to be closed by increasing amounts of cement and the weight of the overlying rock. Moreover, sedimentary rock overlies igneous and metamorphic crystalline basement rock, which usually has very low porosity.

The subsurface zone in which all rock openings are filled with water is called the **saturated zone** (figure 11.1*A*). If a well were drilled downward into this zone, ground water would fill the lower part of the well. The water level inside the well marks the upper surface of the saturated zone; this surface is the **water table.**

Most rivers and lakes intersect the saturated zone. Rivers and lakes occupy low places on the land surface, and ground water flows out of the saturated zone into these surface depressions. The water level at the surface of most lakes and rivers coincides with the water table. Ground water also flows into mines and quarries cut below the water table (figure 11.1*B*).

Above the water table is a zone where not all the rock openings are filled with water and is referred to as the **unsaturated zone** (figure 11.1*A*). Within the unsaturated zone, surface tension causes water to be held above the water table. The *capillary fringe* is a transition zone with higher moisture content at the base of the unsaturated zone just above the water table. Some of the water in the capillary fringe has been drawn or wicked upward from the water table (much like water rising up a paper towel if the corner is dipped in water). The capillary fringe is generally less than a meter thick but may be much thicker in fine-grained sediments and thinner in coarse-grained sediments such as sand and gravel.

Plant roots generally obtain their water from the belt of soil moisture near the top of the unsaturated zone, where fine-

A

B

FIGURE 11.1

(A) The water table marks the top of the saturated zone in which water completely fills the rock pore space. Above the water table is the unsaturated zone in which rock openings typically contain both air and water. (B) Ground water fills lakes that extend below the water table. The surface of the lakes represent the top of the groundwater table. *Photo reproduced with the permission of the Minister of Public Works and Government Services Canada, 2005 and courtesy of Natural Resources Canada, Geological Survey of Canada*

grained clay minerals hold water and make it available for plant growth. Most plants "drown" if their roots are covered by water in the saturated zone; plants need both water and air in soil pores to survive. (The water-loving plants of swamps and marshes are an exception.)

IN GREATER DEPTH 11.1

Darcy's Law and Fluid Potential

In 1856, Henry Darcy, a French engineer, found that the velocity at which water moves depends on the *hydraulic head* of the water and on the permeability of the material that the water is moving through.

The hydraulic head of a drop of water is equal to the elevation of the drop plus the water pressure on the drop:

$$\text{Hydraulic head} = \text{elevation} + \text{pressure}$$

In box figure 1*A*, the points A and B are both on the water table, so the pressure at both points is zero (there is no water above points A and B to create pressure). Point A is at a higher elevation than B, so A has a higher hydraulic head than B. The difference in elevation is equal to the difference in head, which is labeled **h.** Water will move from point A to point B (as shown by the dark blue arrow), because water moves from a region of high hydraulic head to a region of low head. The distance the water moves from A to B is labeled **L.** The *hydraulic* gradient is the difference in head between two points divided by the distance between the two points:

$$\text{Hydraulic gradient} = \frac{\text{difference in head}}{\text{distance}} = \frac{\Delta h}{L}$$

In box figure 1*B*, the two points have equal elevation, but the pressure on point C is higher than on point D (there is more water to create pressure above point C than point D). The head is higher at point C than at point D, so the water moves from C to D. In box figure 1*C*, point F has a lower elevation than point G, but F also has a higher pressure than G. The difference in pressure is greater than the difference in elevation, so F has a higher head than G, and water moves from F to G. Note that

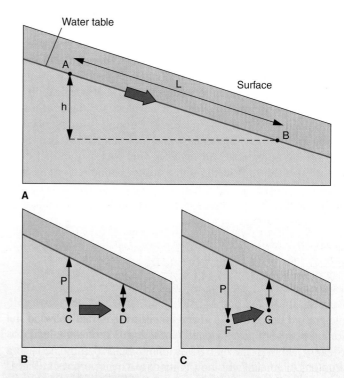

A

B   **C**

BOX 11.1 ■ FIGURE 1

Ground water moves in response to hydraulic head (elevation plus pressure). Water movement shown by dark blue arrows. (*A*) Points A and B have the same pressure, but A has a higher elevation; therefore, water moves from A to B. (*B*) Point C has a higher pressure (arrow marked P) than D; therefore, water moves from C to D at the same elevation. (*C*) Pressure also moves water upward from F to G.

A **perched water table** is the top of a body of ground water separated from the main water table beneath it by a zone that is not saturated (figure 11.2). It may form as ground water collects above a lens of less permeable shale within a more permeable rock, such as sandstone. If the perched water table intersects the land surface, a line of springs can form along the upper contact of the shale lens. The water perched above a shale lens can provide a limited water supply to a well; it is an unreliable long-term supply.

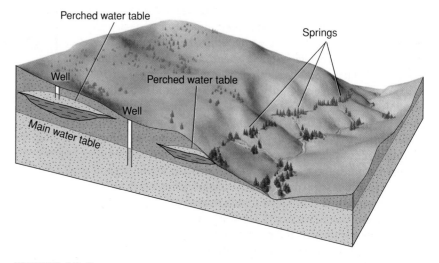

THE MOVEMENT OF GROUND WATER

Compared to the rapid flow of water in surface streams, most ground water moves relatively slowly

FIGURE 11.2

Perched water tables above lenses of less permeable shale within a large body of sandstone. Downward percolation of water is impeded by the less permeable shale.

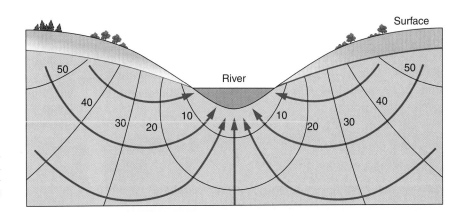

BOX 11.1 ■ FIGURE 2

Dark blue arrows are flow lines, which show direction of groundwater flow. Flow is perpendicular to equipotential lines (black lines with numbers), which show regions of equal hydraulic head. Ground water generally flows from hilltops toward valleys, emerging from the ground as springs into stream beds and banks, lakes, and swamps.

underground water may move downward, horizontally, or upward in response to differences in head but that it always moves in the direction of the downward slope of the water table above it. One of the first goals of groundwater geologists, particularly in groundwater contamination investigations, is to find the slope of the local water table in order to determine the direction (and velocity) of groundwater movement.

The velocity of groundwater flow is controlled by both the permeability of the sediment or rock and the hydraulic gradient. *Darcy's Law* states that the velocity equals the permeability multiplied by the hydraulic gradient. This gives the Darcian velocity (or the velocity of water flowing through an open pipe). To determine the actual velocity of ground water, since ground water only flows through the openings in sediment or rock, the Darcian velocity must be divided by the porosity.

Groundwater velocity = permeability/porosity × hydraulic gradient

$$V = \frac{K}{n} \frac{\Delta h}{L}$$

(Darcy called **K** the hydraulic conductivity; it is a measure of permeability and is specific to a particular aquifer. The porosity is represented by **n** in the equation.)

Groundwater movement is shown in diagrams in relation to *equipotential lines* (lines of constant hydraulic head). Ground water moves from regions of high head to regions of low head. Box figure 2 illustrates how *flow lines,* which show groundwater movement, cross equipotential lines at right angles as water moves from high to low head.

through rock underground. Because it moves in response to differences in water pressure and elevation, water within the upper part of the saturated zone tends to move downward following the slope of the water table (figure 11.3). See box 11.1 for Darcy's Law.

The circulation of ground water in the saturated zone is not confined to a shallow layer beneath the water table. Ground water may move hundreds of feet vertically downward before rising again to discharge as a spring or seep into the beds of rivers and lakes at the surface (figure 11.3) due to the combined effects of gravity and the slope of the water table.

The *slope of the water table* strongly influences groundwater velocity. The steeper the slope of the water table, the faster ground water moves. Water-table slope is controlled largely by topography—the water table roughly parallels the land surface (particularly in humid regions),

as you can see in figure 11.3. Even in highly permeable rock, ground water will not move if the water table is flat.

How fast ground water flows also depends on the *permeability* of the rock or other materials through which it passes. If

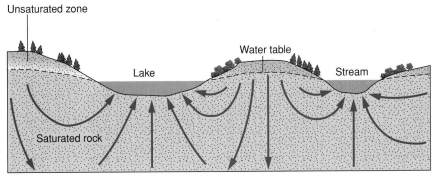

FIGURE 11.3

Movement of ground water beneath a sloping water table in uniformly permeable rock. Near the surface, ground water tends to flow parallel to the sloping water table.

rock pores are small and poorly connected, water moves slowly. When openings are large and well connected, the flow of water is more rapid. One way of measuring groundwater velocity is to introduce a tracer, such as a dye, into the water and then watch for the color to appear in a well or spring some distance away. Such experiments have shown that the velocity of ground water varies widely, averaging a few centimeters to many meters a day. Nearly impermeable rocks may allow water to move only a few centimeters per *year*, but highly permeable materials, such as unconsolidated gravel or cavernous limestone, may permit flow rates of hundreds or even thousands of meters per day.

AQUIFERS

An **aquifer** is a body of saturated rock or sediment through which water can move easily. Aquifers are both highly permeable and saturated with water. A well must be drilled into an aquifer to reach an adequate supply of water (figure 11.4). Good aquifers include sandstone, conglomerate, well-jointed limestone, bodies of sand and gravel, and some fragmental or fractured volcanic rocks such as columnar basalt (table 11.1). These favorable geologic materials are sought in "prospecting" for ground water or looking for good sites to drill water wells.

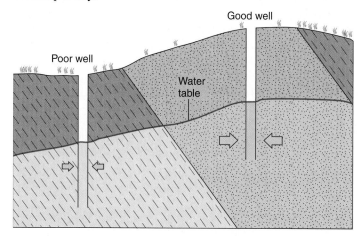

FIGURE 11.4

A well must be installed in an aquifer to obtain water. The saturated part of the highly permeable sandstone is an aquifer, but the less permeable shale is not. Although the shale is saturated, it will not readily transmit water.

FIGURE 11.5

Wells can obtain some water from fractures in crystalline rock. Wells must intersect fractures to obtain water.

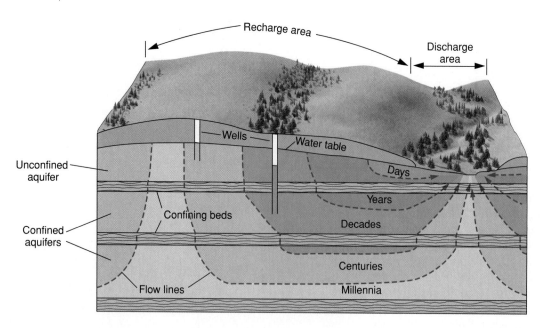

FIGURE 11.6

An unconfined aquifer is exposed to the surface and is only partly filled with water; water in a shallow well will rise to the level of the water table. A confined aquifer is separated from the surface by a confining bed and is completely filled with water under pressure; water in wells rises above the aquifer. Flow lines show direction of groundwater flow. Days, years, decades, centuries, and millennia refer to the time required for ground water to flow from the recharge area to the discharge area. Water enters aquifers in recharge areas and flows out of aquifers in discharge areas.

Wells drilled in shale beds are not usually very successful because shale, although sometimes quite porous, is relatively impermeable (figure 11.4). Wet mud may have a porosity of 80% to 90% and, even when compacted to form shale, may still have a high porosity of 30%. Yet the extremely small size of the pores, together with the electrostatic attraction that clay minerals have for water molecules (see chapter 5), prevents water from moving readily through the shale into a well.

Because they are not very porous, crystalline rocks such as granite, gabbro, gneiss, schist, and some types of limestone are not good aquifers. The porosity of such rocks may be 1% or less. (Shale and crystalline rocks are sometimes called *aquitards* because they retard the flow of ground water.) Crystalline rocks that are highly fractured, however, may be porous and permeable enough to provide a fairly dependable water supply to wells (figure 11.5).

Figure 11.6 shows the difference between an **unconfined aquifer,** which has a water table because it is only partly filled with water, and a **confined aquifer,** which is completely filled with water under pressure and is usually separated from the surface by a relatively impermeable confining bed, or aquitard, such as shale. An unconfined aquifer is recharged by precipitation, has a rising and falling water table during wet and dry seasons, and has relatively rapid movement of ground water

through it (figure 11.7). A confined aquifer is recharged slowly through confining shale beds. With very slow movement of ground water, a confined aquifer may have no response at all to wet and dry seasons.

WELLS

A **well** is a deep hole, generally cylindrical, that is dug or drilled into the ground to penetrate an aquifer within the saturated zone (see figure 11.4). Usually water that flows into the well from the saturated rock must be lifted or pumped to the surface. As figure 11.7 shows, a well dug in a valley usually has to go down a shorter distance to reach water than a well dug on a hilltop. During dry seasons, the water table falls as water flows out of the saturated zone into springs and rivers. Wells not deep enough to intersect the lowered water table go dry, but the rise of the water table during the next rainy season normally returns water to the dry wells. The addition of new water to the saturated zone is called **recharge.**

When water is pumped from a well, the water table is typically drawn down around the well into a depression shaped like an inverted cone known as a **cone of depression** (figure 11.8). This local lowering of the water table, called

A

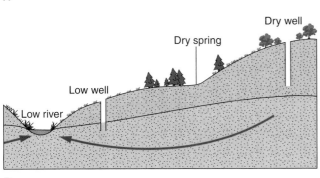

B

FIGURE 11.7

The water table in an unconfined aquifer rises in wet seasons and falls in dry seasons as water drains out of the saturated zone into rivers. (A) Wet season: water table and rivers are high; springs and wells flow readily. (B) Dry season: water table and rivers are low; some springs and wells dry up.

A

B

FIGURE 11.8

Pumping well lowers the water table into a cone of depression. If well is heavily pumped, surrounding shallow wells may go dry.

FIGURE 11.9

The Dakota Sandstone in South Dakota is a relatively unusual type of confined aquifer because it is tilted and exposed to the surface by erosion. Water in most wells rose above the land surface when the aquifer was first tapped in the 1800s.

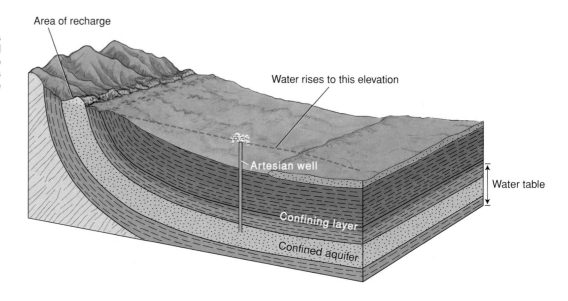

Area of recharge

Water rises to this elevation

Artesian well

Water table

Confining layer

Confined aquifer

FIGURE 11.10

Artesian well spouts water above land surface in South Dakota, early 1900s. Heavy use of this aquifer has reduced water pressure so much that spouts do not occur today. *Photo by N. H. Darton, U.S. Geological Survey*

drawdown, tends to change the direction of groundwater flow by changing the slope of the water table. In lightly used wells that are not pumped, drawdown does not occur and a cone of depression does not form. In a simple, rural well with a bucket lowered on the end of a rope, water cannot be extracted rapidly enough to significantly lower the water table. A well of this type is shown in figure 11.1*A*.

In unconfined aquifers, water rises in shallow wells to the level of the water table. In confined aquifers, the water is under pressure and rises in wells to a level above the top of the aquifer (see figure 11.6). Such a well is called an **artesian well** and confined aquifers are also called *artesian aquifers.*

In some artesian wells, the water rises above the land surface, producing a flowing well that spouts continuously into the air unless it is capped (figure 11.9). Flowing wells used to occur in South Dakota, when the extensive Dakota Sandstone aquifer was first tapped (figure 11.10), but continued use has lowered the water pressure surface below the ground surface in most parts of the state. Water still rises above the aquifer but does not reach the land surface.

SPRINGS AND STREAMS

A **spring** is a place where water flows naturally from rock onto the land surface (figure 11.11). Some springs discharge where the water table intersects the land surface, but they also occur where water flows out from caverns or along fractures, faults, or rock contacts that come to the surface (figure 11.12).

Climate determines the relationship between stream flow and the water table. In rainy regions, most streams are **gaining streams;** that is, they receive water from the saturated zone

FIGURE 11.11

A large spring flowing from limestone in Vaesy's Paradise, Marble Canyon, Arizona.
Photo © Alissa Crandall/Corbis Images

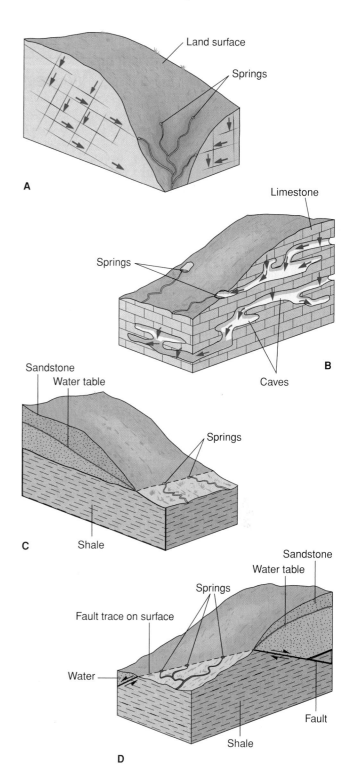

FIGURE 11.12

Springs can form in many ways. (A) Water moves along fractures in crystalline rock and forms springs where the fractures intersect the land surface. (B) Water enters caves along joints in limestone and exits as springs at the mouths of caves. (C) Springs form at the contact between a permeable rock such as sandstone and an underlying less permeable rock such as shale. (D) Springs can form along faults when permeable rock has been moved against less permeable rock. Arrows show relative motion along fault.

(figure 11.13A). The surface of these streams coincides with the water table. Water from the saturated zone flows into the stream through the streambed and banks that lie below the water table. Because of the added ground water, the discharge of these streams increases downstream. Where the water table intersects the land surface over a broad area, ponds, lakes, and swamps are found.

In drier climates, rivers tend to be **losing streams;** that is, they are losing water to the saturated zone (figure 11.13B). The water percolating into the ground beneath a losing stream causes the water table to slope away from the stream. In very dry climates, such as in a desert, a losing stream may be separated or *disconnected* from the underlying saturated zone and a groundwater mound remains beneath the stream even if the streambed is dry (figure 11.13C).

ENVIRONMENTAL GEOLOGY 11.2

Prospecting for Ground Water

Many wells are drilled or dug without any effort to locate a good aquifer. Many of these wells are successful (especially if only small amounts of water are needed) because most rocks hold some water, which flows into wells that intersect the water table. If a large and dependable supply of water is needed—as for a city water system—specialists in groundwater geology (hydrogeologists) may be called in to locate a promising well site. Hydrogeologists use many methods to locate aquifers. A detailed knowledge of the local rocks is necessary. Therefore, a geologist may map the surface rocks and use electrical, magnetic, and seismic surveys to study subsurface rocks to determine the presence of possible aquifers. Sometimes a small-diameter test well is drilled before the larger, more expensive supply well is sunk. Hydrogeologists look for potentially high-producing aquifers rather than searching for water directly. In some regions, however, the presence of certain plants may be a useful guide to locating water, particularly the depth of the water table.

Some people search for water by water witching, or *dowsing*, with a divining rod (also sometimes used to search for metals or lost objects). Usually the dowser holds a forked stick horizontally in the hands while walking over an area. The stick is supposed to deflect or twist downward of its own accord when the dowser passes over water. This method has been tried for centuries, the only modification being that a twisted metal rod, often made of a coat hanger, now may be substituted for the stick. Carefully controlled tests conducted by workers in psychic research have shown that water witchers' "success" is equal to or less than pure chance, while geologists' results are superior both to witching and to chance. Records kept on thousands of wells in Australia in the early 1900s show that of wells that were not divined, more than 83% produced flows of 100 gallons per hour and 7.4% were failures, finding no water at all. Of wells that were divined by water witchers, only about 70% produced more than 100 gallons per hour, and 14.7% were dry. In the early part of this past century, the U.S. Geological Survey concluded that any future testing of the results of water witching would be a misuse of public funds.

Despite such findings, many people believe strongly in dowsing. Water witchers themselves devoutly believe they can find water, and in some regions of the United States, almost no wells are drilled without a witcher's advice. Dowsers are helped in locating water by the fact that most rocks hold some water, and dowsers often have a longstanding knowledge of a particular region and its potential water resources. This is not to say that dowsers are deliberate frauds. Many are convinced that they perform a valuable public service, and some do not charge for their services. Scientists see no reason for dowsing to work and are skeptical about dowsers' "success." Geologists would almost unanimously urge you not to pay a fee for a dowser's service.

A Gaining stream

B Losing stream

FIGURE 11.13

Gaining and losing streams. (A) Stream gaining water from saturated zone. (B) Stream losing water through streambed to saturated zone. (C) Water table can be close to the land surface, but disconnected from the surface water beneath a streambed that intermittently contains water.

C Losing stream (disconnected)

CONTAMINATION OF GROUND WATER

Ground water in its natural state tends to be relatively free of contaminants in most areas. Because it is a widely used source of drinking water, the contamination of ground water can be a very serious problem.

Pesticides and *herbicides* (such as diazinon, atrazine, DEA, and 2,4-D) applied to agricultural crops (figure 11.14*A*) can find their way into ground water when rain or irrigation water leaches the contaminants downward into the soil. *Fertilizers* are also a concern. Nitrate, which forms from one of the most widely used fertilizers, is harmful in even small quantities in drinking water.

Rain can also leach pollutants from city landfills into groundwater supplies (figure 11.14*B*). Consider for a moment some of the things you threw away last year. A partially empty aerosol can of ant poison? The can will rust through in the land-fill, releasing the poison into the ground and into the saturated zone below. A broken thermometer? The toxic mercury may eventually find its way to the groundwater supply. A half-used can of oven cleaner? The dried-out remains of a can of lead-base paint? *Heavy metals* such as mercury, lead, chromium, copper, and cadmium, together with household chemicals and poisons, can all be concentrated in groundwater supplies beneath landfills (figure 11.15). This is particularly a concern in older, unlined landfills than in newer landfills that are engineered to contain contaminants with impermeable layers of clay and synthetic liners.

Liquid and solid wastes from septic tanks, sewage plants, and animal feedlots and slaughterhouses may contain *bacteria, viruses,* and *parasites* that can contaminate ground water (figure 11.14*C*). Liquid wastes from industries (figure 11.14*D*) and military bases can be highly toxic, containing high concentrations of heavy metals and compounds such as cyanide and *PCBs* (polychlorinated biphenyls), which are widely used

A

B

C

D

FIGURE 11.14

Some sources of groundwater pollution. (*A*) Pesticides. (*B*) Household garbage. (*C*) Animal waste. (*D*) Industrial toxic waste. *Photo A by Doug Wilson, USDA Photography Center; Photo B by Frank M. Hanna; Photos C and D by U.S. Department of Agriculture Soil Conservation Service*

A Cross section

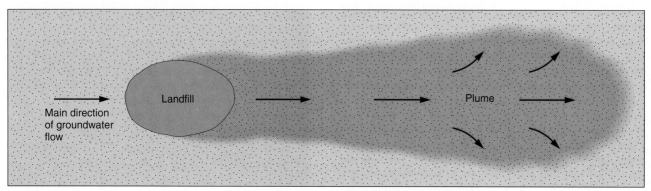

B Map view of contaminant plume. Note how it grows in size with distance from the pollution source.

FIGURE 11.15

Waste piled on the land surface creates a groundwater mound beneath it because the landfill forms a hill and because the waste material is more porous and permeable than the surrounding soil and rock. Rain leaches pollutants into the saturated zone. A plume of contaminated water will spread out in the direction of groundwater flow.

in industry. A degreaser called *TCE* (trichloroethylene) has been increasingly found to pollute both surface and underground water in numerous regions. Toxic liquid wastes are often held in surface ponds or pumped down deep disposal wells. If the ponds leak, ground water can become polluted. Deep wells may be safe for liquid waste disposal if they are deep enough, but contamination of drinking water supplies and even surface water has resulted in some localities from improper design of the disposal wells.

Acid mine drainage from coal and metal mines can contaminate both surface and ground water. It is usually caused by sulfuric acid formed by the oxidation of sulfur in pyrite and other sulfide minerals when they are exposed to air by mining activity. Fish and plants are often killed by the acid waters draining from long-abandoned mines.

Radioactive waste is both an existing and a very serious potential source of groundwater contamination. The shallow burial of *low-level* solid and liquid radioactive wastes from the nuclear power industry has caused contamination of ground water, particularly as liquid waste containers leak into the saturated zone and as the seasonal rise and fall of the water table at some sites periodically covers the waste with ground water. The search for a permanent disposal site for solid, *high-level* radioactive waste (now stored temporarily on the surface) is a major national concern for the United States. The permanent site will be deep underground and must be isolated from groundwater circulation for thousands of years. Salt beds, shale, glassy tuffs, and crystalline rock deep beneath the surface have all been studied, particularly in arid regions where the water table is hundreds of meters below the land surface. The site selected for disposal of high-level waste, primarily spent fuel from nuclear reactors, is Yucca Mountain, Nevada, 180 kilometers northwest of Las Vegas. The site will be deep underground in volcanic tuff well above the current (or predicted future) water table and in a region of very low rainfall. The U.S. Congress, under intense political pressure from other candidate states that did not want the site, essentially chose Nevada in late 1988 by eliminating the funding for the study of all alternative sites. After much controversy over the ultimate safety of the site and objections from Nevada, President George W. Bush approved the Yucca Mountain site in 2002.

Not all groundwater contaminants form plumes within the saturated zone, as shown in figure 11.15. *Gasoline,* which leaks from gas station storage tanks at tens of thousands of U.S. locations, is less dense than water and floats upon the water table (figure 11.16). Some liquids such as TCE are heavier than water and sink to the bottom of the saturated zone, perhaps traveling in unpredicted directions on the surface of an impermeable layer (figure 11.16). Determining the extent and flow direction of groundwater pollution is a lengthy process requiring the drilling of tens, or even hundreds, of costly wells for each contaminated site.

Not all sources of groundwater contamination are man-made. Naturally occurring *minerals within rock and soil* may contain elements such as arsenic, selenium, mercury, and other toxic metals. Circulating ground water can leach these elements out of the minerals and raise their concentrations to harmful levels within the water. Not all spring water is safe to drink. Like a "bad waterhole" depicted in a Western movie, some springs contain such high levels of toxic elements that the water can sicken or kill humans and animals that drink it. Many desert springs contain such high concentrations of sodium chloride or other salts that their water is undrinkable.

Soil and rock filter some contaminants out of ground water. This filtering ability depends on the permeability and mineral composition of rock and soil. Under ideal conditions, human sewage can be purified by only 30 to 45 meters of travel through a sandy loam soil (a mixture of clay minerals, sand, and organic humus). The sewage is purified by filtration, ion absorption by clay minerals and humus, and decomposition by soil organisms in the biosphere (figure 11.17). On the other hand, extremely permeable rock, such as highly fractured granite or cavernous limestone, has little purifying effect on sewage. Ground water flows so rapidly through such rocks that it is not purified even after hundreds of meters of travel. Some pesticides and toxic chemicals are not purified at all by passage through rock and soil, even soil rich in humus and clay minerals.

Contaminated ground water is extremely difficult to clean up. Networks of expensive wells may be needed to pump contaminated water out of the ground and replace it with clean water. Because of the slow movement and large volume of ground water, the cleanup process for a large region can take decades and tens of millions of dollars to complete.

Groundwater contamination can be largely prevented with careful thought and considerable expense. A city landfill can be sited high above the water table and possible flood levels or located in a region of groundwater discharge rather than recharge. A site can be sealed below by impermeable (and expensive) clay barriers and plastic liners, and sealed off from rainfall by an impermeable cover. Dikes can prevent surface runoff through or from the site. Although sanitary landfills are expensive, they are much cheaper than groundwater cleanup.

Pumping wells can cause or aggravate groundwater contamination (figure 11.18). Well drawdown can increase the slope of the water table locally, thus increasing the rate of groundwater flow and giving the water less time to be purified underground before it is used (figure 11.18A). Drawdown can even reverse the original slope of the water table, perhaps contaminating wells that were pure before pumping began (figure 11.18B). Heavily pumped wells near a coast can be contaminated by *saltwater intrusion* (figure 11.18C and *D*). Saltwater intrusion is becoming a serious problem as the demand for drinking water increases in rapidly growing coastal communities.

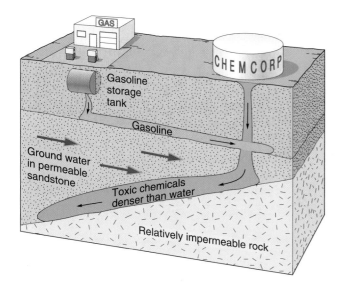

FIGURE 11.16

Not all contaminants move within the saturated zone, as shown in figure 11.15. Gasoline floats on water; many dense chemicals move along impermeable rock surfaces below the saturated zone.

FIGURE 11.17

Rock type and distance control possible sewage contamination of neighboring wells. (A) As little as 30 meters of movement can effectively filter human sewage in sandstone and some other rocks and sediments. (B) If the rock has large open fractures, contamination can occur many hundreds of meters away.

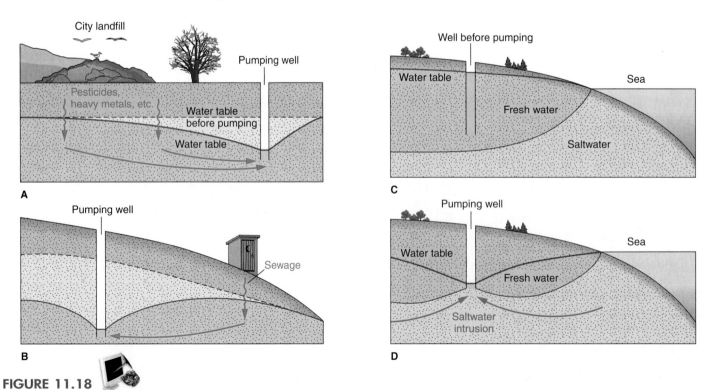

FIGURE 11.18

Groundwater pollution problems caused or aggravated by pumping wells. (A) Water table steepens near a landfill, increasing the velocity of groundwater flow and drawing contaminants into a well. (B) Water-table slope is reversed by pumping, changing direction of the groundwater flow and contaminating the well. (C) Well near a coast (before pumping). Fresh water floats on saltwater. (D) Well in C begins pumping, thinning the freshwater lens and drawing saltwater into the well.

ENVIRONMENTAL GEOLOGY 11.3

Hard Water and Soapsuds

"Hard water" is water that contains relatively large amounts of dissolved calcium (often from the chemical weathering of calcite or dolomite) and magnesium (from the ferromagnesian minerals or dolomite). Water taken from the groundwater supply or from a stream for home use may contain enough of these ions to prevent soap from lathering. Calcium and magnesium ions in hard water form gray curds with soap. The curd continues to form until all the calcium and magnesium ions are removed from the water and bound up in the curd. Only then will soap lather and clean laundry. Cleaning laundry in hard water, therefore, takes an excessively large amount of soap.

Hard water may also precipitate a scaly deposit inside teakettles and hot-water tanks and pipes (box figure 1). The entire hot-water piping system of a home in a hard-water area eventually can become so clogged that the pipes must be replaced.

"Soft water" may carry a substantial amount of ions in solution but not the ions that prevent soap from lathering. Water softeners in homes replace calcium and magnesium ions with sodium ions, which do not affect lathering or cause scale. But water containing a large amount of sodium ions, whether from a softener or from natural sources, may be harmful if used as drinking water by persons on a "salt-free" (low-sodium) diet for health reasons.

BOX 11.3 ■ FIGURE 1

Scale caused by hard water coats the inside of this hot-water pipe. *Photo © Sheila Terry/SPL/Photo Researchers*

BALANCING WITHDRAWAL AND RECHARGE

A local supply of ground water will last indefinitely if it is withdrawn for use at a rate equal to or less than the rate of recharge to the aquifer. If ground water is withdrawn faster than it is being recharged, however, the supply is being reduced and will one day be gone.

Heavy use of ground water causes a regional water table to drop. In parts of western Texas and eastern New Mexico, the pumping of ground water from the Ogallala aquifer has caused the water table to drop 30 meters over the past few decades. The lowering of the water table means that wells must be deepened and more electricity must be used to pump the water to the surface. Moreover, as water is withdrawn, the ground surface may settle because the water no longer supports the rock and sediment. Mexico City has subsided more than 7 meters and portions of California's Central Valley 9 meters because of extraction of ground water (figure 11.19). Such *subsidence* can crack building foundations, roads, and pipelines. Overpumping of ground water also causes compaction and porosity loss in rock and soil, and can permanently ruin good aquifers.

To avoid the problems of falling water tables, subsidence, and compaction, many towns use *artificial recharge* to increase the natural rate of recharge. Natural floodwaters or treated industrial or domestic wastewaters are stored in infiltration ponds on the surface to increase the rate of water percolation into the ground. Reclaimed, clean water from sewage treatment plants is commonly used for this purpose. In some cases, especially in areas where ground water is under confined conditions, water is actively pumped down into the ground to replenish the groundwater supply. This is more expensive than filling surface ponds, but it reduces the amount of water lost through evaporation.

EFFECTS OF GROUNDWATER ACTION
Caves, Sinkholes, and Karst Topography

Caves (or **caverns**) are naturally formed, underground chambers. Most caves develop when slightly acidic ground water

FIGURE 11.19

Subsidence of the land surface caused by the extraction of ground water, near Mendota, San Joaquin Valley, California. Signs on the pole indicate the positions of the land surface in 1925, 1955, and 1977. The land sank 9 meters in 52 years. Since the late 1970s, subsidence decreased to less than a meter due to reduced groundwater pumping and increased use of surface water for irrigation. *Photo by Richard O. Ireland, U.S. Geological Survey*

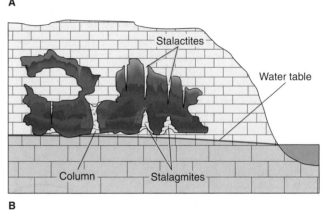

FIGURE 11.20

Solution of limestone to form caves. (*A*) Water moves along fractures and bedding planes in limestone, dissolving it to form caves below the water table. (*B*) Falling water table allows cave system, now greatly enlarged, to fill with air. Calcite precipitation forms stalactites, stalagmites, and columns above the water table.

dissolves limestone along joints and bedding planes, opening up cavern systems as calcite is carried away in solution (figure 11.20). Natural ground water is commonly slightly acidic because of dissolved carbon dioxide (CO_2) from the atmosphere or from soil gases (see chapter 5).

Geologists disagree whether limestone caves form above, below, or at the water table. Most caves probably are formed by ground water circulating below the water table, as shown in figure 11.20. If the water table drops or the land is elevated above the water table, the cave may begin to fill in again by calcite precipitation. The following equation can be read from left to right for calcite solution and from right to left for the calcite precipitation reaction (see also table 5.1).

Ground water with a high concentration of calcium (Ca^{++}) and bicarbonate (HCO_3^-) ions may drip slowly from the ceiling of an air-filled cave. As a water drop hangs on the ceiling of the cave, some of the dissolved carbon dioxide (CO_2) may be lost into the cave's atmosphere. The CO_2 loss causes a small amount of calcite to precipitate out of the water onto the cave ceiling. When the water drop falls to the cave floor, the impact may cause more CO_2 loss, and another small amount of calcite may precipitate on the cave floor. A falling water drop, therefore, can precipitate small amounts of calcite on both the cave ceiling and the cave floor, and each subsequent drop adds more calcite to the first deposits.

$$H_2O \quad + \quad CO_2 \quad + \quad CaCO_3 \quad \rightleftarrows \quad Ca^{++} \quad + \quad 2HCO_3^-$$

H_2O	CO_2	$CaCO_3$	Ca^{++}	$2HCO_3^-$
water	carbon dioxide	calcite in limestone	calcium ion	bicarbonate ion

→ development of caves (solution)

← development of flowstone and dripstone (precipitation)

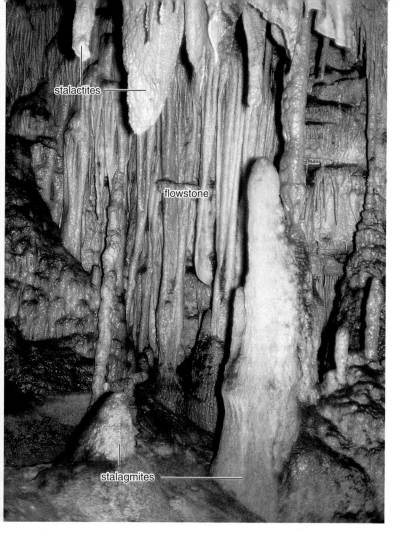

stalactites

flowstone

stalagmites

FIGURE 11.21

Stalactites, stalagmites, and flowstone in Great Onyx Cave, Kentucky. *Photo courtesy of Stanley Fagerlin*

A

B

FIGURE 11.22

(A) Sinkholes formed in limestone near Timaru, New Zealand. (B) A collapse sinkhole that formed suddenly in Winter Park, Florida, in 1981. *Photo A © G. R. "Dick" Roberts/Natural Sciences Image Library; Photo B © AP/Wide World Photos*

Deposits of calcite (and, rarely, other minerals) built up in caves by dripping water are called *dripstone* or **speleothems.** **Stalactites** are iciclelike pendants of dripstone hanging from cave ceilings (figure 11.20*B*). They are generally slender and are commonly aligned along cracks in the ceiling, which act as conduits for ground water. **Stalagmites** are cone-shaped masses of dripstone formed on cave floors, generally directly below stalactites. Splashing water precipitates calcite over a large area on the cave floor, so stalagmites are usually thicker than the stalactites above them. As a stalactite grows downward and a stalagmite grows upward, they may eventually join to form a *column* (figure 11.20*B*). Figure 11.21 shows some of the intriguing features formed in caves.

In parts of some caves, water flows in a thin film over the cave surfaces rather than dripping from the ceiling. Sheetlike or ribbonlike *flowstone* deposits develop from calcite that is precipitated by flowing water on cave walls and floors.

The floors of most caves are covered with sediment, some of which is *residual clay,* the fine-grained particles left behind as insoluble residue when a limestone containing clay dissolves. (Some limestone contains only about 50% calcite.) Other sediment, including most of the coarse-grained material found on cave floors, may be carried into the cave by streams, particularly when surface water drains into a cave system from openings on the land surface.

Solution of limestone underground may produce features that are visible on the surface. Extensive cavern systems can undermine a region so that roofs collapse and form depressions in the land surface above. **Sinkholes** are closed depressions found on land surfaces underlain by limestone (figure 11.22). They form either by the *collapse* of a cave roof or by *solution* as descending water enlarges a crack in limestone. Limestone regions in Florida, Missouri, Indiana, and Kentucky are heavily dotted with sinkholes. Sinkholes can also form in regions underlain by gypsum or rock salt, which are also soluble in water.

FIGURE 11.23

Karst topography is marked by underground caves and numerous surface sinkholes. A major river may cross the region, but small surface streams generally disappear down sinkholes.

An area with many sinkholes and cave systems beneath the land surface is said to have **karst topography** (figure 11.23). Karst areas are characterized by a lack of surface streams, although one major river may flow at a level lower than the karst area.

Streams sometimes disappear down sinkholes to flow through caves beneath the surface. In this specialized instance, a true *underground stream* exists. Such streams are quite rare, however, as most ground water flows very slowly through pores and cracks in sediment or rock. You may hear people with wells describe the "underground stream" that their well penetrates, but this is almost never the case. Wells tap ground water in the rock pores and crevices, not underground streams. If a well did tap a true underground river in a karst region, the water would probably be too polluted to drink, especially if it had washed down from the surface into a cavern without being filtered through soil and rock.

Other Effects

Ground water is important in the preservation of *fossils* such as **petrified wood,** which develops when porous buried wood is either filled in or replaced by inorganic silica carried in by ground water (figure 11.24). The result is a hard, permanent rock, commonly preserving the growth rings and other details of the wood. Calcite or silica carried by ground water can also replace the original material in marine shells and animal bones.

Sedimentary rock *cement,* usually silica or calcite, is carried into place by ground water. When a considerable amount of cementing material precipitates locally in a rock, a hard, rounded mass called a **concretion** develops, typically around an organic nucleus such as a leaf, tooth, or other fossil (figure 11.25).

FIGURE 11.24

Petrified log in the Painted Desert, Arizona. The log was replaced by silica carried in solution by ground water. Small amounts of iron and other elements color the silica in the log. *Photo © Eric and David Hosking/Corbis*

Geodes are partly hollow, globe-shaped bodies found in some limestones and locally in other rocks. The outer shell is amorphous silica, and well-formed crystals of quartz, calcite, or other minerals project inward toward a central cavity (figure 11.26). The origin of geodes is complex but clearly related to ground water. Crystals in geodes may have filled original cavities or have replaced fossils or other crystals.

In arid and semiarid climates, *alkali soil* may develop because of the precipitation of great quantities of sodium salts

FIGURE 11.25

Concretions that have weathered out of shale. Concretions contain more cement than the surrounding rock and therefore are very resistant to weathering. *Photo by David McGeary*

FIGURE 11.26

Concentric layers of amorphous silica are lined with well-formed amethyst (quartz) crystals growing inward toward a central cavity in a geode. *Photo © Martin Land/LANDM/Bruce Coleman*

by evaporating ground water. Such soil is generally unfit for plant growth. Alkali soil generally forms at the ground surface in low-lying areas (see chapter 5).

HOT WATER UNDERGROUND

Hot springs are springs in which the water is warmer than human body temperature. Water can gain heat in two ways while it is underground. First, and more commonly, ground

water may circulate near a magma chamber or a body of cooling igneous rock. In the United States, most hot springs are found in the western states, where they are associated with relatively recent volcanism. The hot springs and pools of Yellowstone National Park in Wyoming are of this type.

Ground water can also gain heat if it circulates unusually deeply in the Earth, perhaps along joints or faults. As discussed in chapter 3, the normal geothermal gradient (the increase in temperature with depth) is 25°C/kilometer (about 75°F/mile). Water circulating to a depth of 2 or 3 kilometers is warmed substantially above normal surface water temperature. The famous springs at Warm Springs, Georgia; Hot Springs, Arkansas; and Sulfur Springs, West Virginia, have all been warmed by deep circulation. Warm water, regardless of its origin, is lighter than cold water and readily rises to the surface.

A **geyser** is a type of hot spring that periodically erupts hot water and steam. The water is generally near boiling (100°C). Eruptions may be caused by a constriction in the underground "plumbing" of a geyser, which prevents the water from rising and cooling. The events thought to lead to a geyser eruption are illustrated in figure 11.27. Water gradually seeps into a partially emptied geyser chamber and heat supplied from below slowly warms the water. Bubbles of water vapor and other gases then begin to form as the temperature of the water rises. The bubbles may clog the constricted part of the chamber until the upward pressure of the bubbles pushes out some of the water above in a gentle surge, thus lowering the pressure on the water in the lower part of the chamber. This drop in pressure causes the chamber water, now very hot, to flash into vapor. The expanding vapor blasts upward out of the chamber, driving hot water with it and condensing into visible steam. The chamber, now nearly empty, begins to fill again, and the cycle is repeated. The entire cycle may be quite regular, as it is in Yellowstone's Old Faithful geyser, which averages about 79 minutes between eruptions (though it varies from about 45 to 105 minutes, depending on the amount of water left in the chamber after an eruption). Many geysers, however, erupt irregularly, some with weeks or months between eruptions.

As hot ground water comes to the surface and cools, it may precipitate some of its dissolved ions as minerals. *Travertine* is a deposit of *calcite* that often forms around hot springs (figure 11.28), while dissolved *silica* precipitates as *sinter* (called *geyserite* when deposited by a geyser, as shown in figure 11.29). The composition of the subsurface rocks generally determines which type of deposit forms, although sinter can indicate higher subsurface temperatures than travertine because silica is harder to dissolve than calcite. Both deposits can be stained by the pigments of bacteria that thrive in the hot water. These thermophilic bacteria are some of the most primitive of living bacteria in the biosphere and suggest that life may have arisen near hot springs.

A *mudpot* is a special type of hot spring that contains thick, boiling mud. Mudpots are usually marked by a small amount of water and strongly sulfurous gases, which combine to form strongly acidic solutions. The mud probably results from

FIGURE 11.27

Eruptive history of a typical geyser in (A) through (D). Photo shows the eruption of Old Faithful geyser in Yellowstone National Park, Wyoming. See text for explanation. *Photo © Hal Beral/Visuals Unlimited*

FIGURE 11.28

Precipitation of calcite in the form of travertine around a hot spring (Mammoth Hot Springs, Yellowstone National Park). Thermophilic bacteria living in the hot water provide the color. *Photo by Diane Carlson*

intense chemical weathering of the surrounding rocks by these strong acids (see figure 5.13).

Geothermal Energy

Electricity can be generated by harnessing naturally occurring steam and hot water in areas that are exceptionally hot underground. In such a *geothermal area,* wells can tap steam (or superheated water that can be turned into steam) that is then piped to a powerhouse, where it turns a turbine that spins a generator, creating electricity.

Geothermal energy production requires no burning of fuel, so the carbon dioxide emissions of power plants that burn coal, oil, or natural gas are not produced. Although geothermal energy is relatively clean, it has some environmental problems. Workers need protection from toxic hydrogen sulfide gas in the steam, and the hot water commonly contains dissolved ions and metals, such as lead and mercury, that can kill fish and plants if discharged on the surface. Geothermal fluids are often highly corrosive to equipment, and their extraction can cause land subsidence. Pumping the cooled wastewater underground can help reduce subsidence problems.

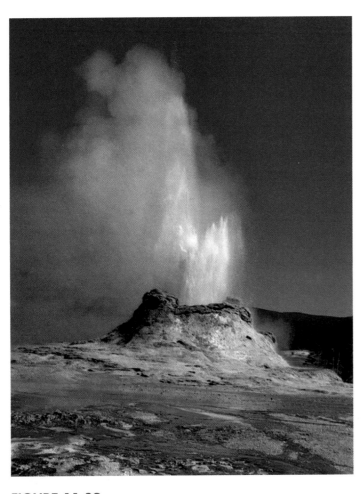

FIGURE 11.29

Geyserite deposits around the vent of Castle geyser, Yellowstone National Park. *Photo by David McGeary*

FIGURE 11.30

Geothermal power plant at The Geysers, California. Underground stream, piped from wells to the power plant, is discharging from the cooling towers and surrounding wells. *Photo © Roger Ressmeyer/Corbis*

Geothermal fields can be depleted. The largest field in the world is at The Geysers in California (figure 11.30), 120 kilometers north of San Francisco. The Geysers field decreased its capacity in recent years to just under 1,000 megawatts of electricity (enough for 1 million people), as the field began running out of steam. As production declined, innovative solutions such as injecting wastewater from nearby communities has increased the steam capacity.

Nonelectric uses of geothermal energy include space heating (in Boise, Idaho; Klamath Falls, Oregon; and Reykjavik, the capital of Iceland), as well as paper manufacturing, ore processing, and food preparation.

SUMMARY

About 15% of the water that falls on land percolates underground to become ground water. Ground water fills pores and joints in rock, creating a large reservoir of usable water in most regions.

Porous rocks can hold water. *Permeable* rocks permit water to move through them.

The *water table* is the top surface of the *saturated zone* and is overlain by the *unsaturated zone*.

Local variations in rock permeability may develop a *perched water table* above the main water table.

Groundwater velocity depends on rock permeability and the slope of the water table.

An *aquifer* is porous and permeable, and can supply water to wells. A *confined aquifer* holds water under pressure, which can create *artesian wells*.

Gaining streams, springs, and lakes form where the water table intersects the land surface. *Losing streams* contribute to the ground water in dry regions.

Ground water can be contaminated by city landfills, agriculture, industry, or sewage disposal. Some pollutants can be filtered out by passage of the water through moderately permeable geologic materials.

A pumped well causes a *cone of depression* that in turn can cause or aggravate groundwater pollution. Near a coast, it can cause *saltwater intrusion*.

Artificial *recharge* can help create a balance between withdrawal and recharge of groundwater supplies and help prevent subsidence.

Solution of limestone by ground water forms *caves, sinkholes,* and *karst topography*. Calcite precipitating out of ground water forms *speleothems* such as *stalactites* and *stalagmites* in caves.

Precipitation of material out of solution by ground water helps form petrified wood, other fossils, sedimentary rock cement, concretions, geodes, and alkali soils.

Geysers and *hot springs* occur in regions of hot ground water. *Geothermal energy* can be tapped to generate electricity.

Terms to Remember

Testing Your Knowledge

Use the following questions to prepare for exams based on this chapter.

1. What conditions are necessary for an artesian well?

2. What distinguishes a geyser from a hot spring? Why does a geyser erupt?

3. What is karst topography? How does it form?

4. What chemical conditions are necessary for caves to develop in limestone? For stalactites to develop in a cave?

5. What causes a perched water table?

6. Describe several ways in which ground water can become contaminated.

7. Discuss the difference between porosity and permeability.

8. What is the water table? Is it fixed in position?

9. Sketch four different origins for springs.

10. What controls the velocity of groundwater flow?

11. Name several geologic materials that make good aquifers. Define *aquifer.*

12. How does petrified wood form?

13. What happens to the water table near a pumped well?

14. How does a confined aquifer differ from an unconfined aquifer?

15. Porosity is
    a. the percentage of a rock's volume that is openings
    b. the capacity of a rock to transmit a fluid
    c. the ability of a sediment to retard water
    d. none of the preceding

16. Permeability is
    a. the percentage of a rock's volume that is openings
    b. the capacity of a rock to transmit a fluid
    c. the ability of a sediment to retard water
    d. none of the preceding

17. The subsurface zone in which all rock openings are filled with water is called the
    a. saturated zone   b. water table
    c. unsaturated zone   d. aquiclude

18. An aquifer is
    a. a body of saturated rock or sediment through which water can move easily
    b. a body of rock that retards the flow of ground water
    c. a body of rock that is impermeable

19. Which rock type would make the best aquifer?
    a. shale   b. mudstone
    c. sandstone   d. all of the preceding

20. Which of the following determines how quickly ground water flows?
    a. elevation   b. water pressure
    c. permeability   d. all of the preceding

21. Ground water flows
    a. always downhill
    b. from areas of high hydraulic head to low hydraulic head
    c. from high elevation to low elevation
    d. from high permeability to low permeability

22. The drop in the water table around a pumped well is the
    a. drawdown   b. hydraulic head
    c. porosity   d. fluid potential

Expanding Your Knowledge

1. Describe any difference between the amounts of water that would percolate downward to the saturated zone beneath a flat meadow in northern New York and a rocky hillside in southern Nevada. Discuss the factors that control the amount of percolation in each case.

2. Where should high-level nuclear waste from power plants be stored? If your state or community uses nuclear power, where is your local waste stored?

3. Should all contaminated ground water be cleaned up? How much money has been set aside by the federal government for cleaning polluted ground water? Who should pay for groundwater cleanup if the company that polluted the water no longer exists? Should some aquifers be deliberately left contaminated if there is no current use of the water or if future use could be banned?

4. Why are most of North America's hot springs and geysers in the western states and provinces?

Exploring Web Resources

www.mhhe.com/plummer11e

This is the dedicated website for this book. You can go to it for new and updated information. The universal resource locators (URL) listed in this book are also given as links on the website, making it easy to go to those websites without typing in the URL. When you visit our Online Learning Center, go to the Student Center and then to the chapter of interest. Here you will find additional readings and media resources, as well as answers to the Testing Your Knowledge section, more quizzes, animation and video clips, and direct links to the sites listed in this book. Links to additional websites can also be found. We have added questions for some of the sites to allow you to get the most of your exploration of the web. Using the web is an enjoyable way of enhancing your knowledge of geology.

http://toxics.usgs.gov/toxics/

Various sites and information about cleanup of toxics in surface and ground water.

http://water.usgs.gov/public/wid/html/bioremed.html

Information about using bioremediation to clean up toxic substances in the soil, on the surface, and in ground water.

http://water.wr.usgs.gov/gwatlas/index.html

Ground Water Atlas for the United States. Good general information about aquifers.

http://water.usgs.gov/

Good general website that has a lot of links to water topics in the United States from the U.S. Geological Survey.

www.caves.org/

Home page of the *National Speleological Society* contains links to web pages of local interest and to the society's bookstore.

Animations

This chapter includes the following animations available on our Online Learning Center at **www.mhhe.com/plummer11e**.

11.7 Basic dynamics of groundwater movement
11.18a Landfill and cone of depression
11.18b, c, d Cone of depression

C H A P T E R

12

Glaciers and Glaciation

In chapters 9, 10, and 11, you have seen how the surface of the land is shaped by mass wasting, running water and, to some extent, ground water. Running water is regarded as the erosional agent most responsible for shaping Earth's land surface. Where glaciers exist, however, they are far more effective agents of erosion, transportation, and deposition. Geologic features characteristic of glaciation are distinctly different from the features formed by running water. Once recognized, they lead one to appreciate the great extent of glaciation during the recent geologic past (that age popularly known as the Ice Age).

Immense and extensive glaciers, covering as much as a third of Earth's land surface, had a profound effect on the landscape and our present civilization. Moreover,

Geologists' camp on a glacier in the Daniels Range, northern Victoria Land, Antarctica. *Photo by C. C. Plummer* worldwide climatic changes during the glacial ages distinctively altered landscapes in areas far from the glacial boundaries.

These episodes of glaciation took place within only the last couple million years, ending about 10,000 years ago. Preserved in the rock record, however, is evidence of extensive older glaciations. The record of a late Paleozoic glacial age is used as evidence for continental drift, as described in chapter 19 on plate tectonics.

 ## Relationships to Earth Systems

Glaciers, along with oceans, lakes, and rivers, are part of the hydrosphere. Most people are surprised to learn that most of the world's fresh water (approximately 75%) is in glaciers. Glaciers (and frozen sea ice) are part of a subsystem of the hydrosphere known as the *cryosphere*. Although glaciers exist in temperate climates, nearly all of the cryosphere is in polar regions.

The cryosphere affects the geosphere in that glaciers are very effective at eroding and transporting rock. Much of this chapter is about the unique landscapes produced by glacial activity. The cryosphere also has a profound influence on the atmosphere. Our climate and weather patterns are heavily influenced by air cooled by the ice in polar regions. Shrinking glaciers are often an indication of a warming climate.

The cryosphere also affects other parts of the hydrosphere. For instance, deep-ocean circulation takes place because dense, cold water from Antarctic oceans sinks beneath warmer waters from equatorial regions. Sea level changes when the world's glaciers grow or melt. During the height of the ice ages, sea level was several hundred meters lower than at present. Sea level rises when global warming takes place. If the current global warming continues, low-lying land will be submerged. (Almost all of Florida will be under water if a significant part of the Antarctic glaciers melts.)

The cryosphere also influences the biosphere. Water kept cool from melted glaciers and sea ice along the coast of Antarctica has the highest per-volume concentration of living organisms in the world. This is because colder water holds higher amounts of dissolved oxygen and other atmospheric gases, and a very diverse fauna (including penguins, whales, and tiny shrimp) have evolved into a unique ecosystem.

INTRODUCTION

A **glacier** is a large, long-lasting mass of ice, formed on land, that moves under its own weight. It develops as snow is compacted and recrystallized. Glaciers can develop any place where, over a period of years, more snow accumulates than melts away or is otherwise lost.

There are two types of *glaciated* terrain on the Earth's surface. **Alpine glaciation** is found in mountainous regions, while **continental glaciation** exists where a large part of a continent (thousands of square kilometers) is covered by glacial ice. In both cases, the moving masses of ice profoundly and distinctively change the landscape.

The spectacularly scenic areas in many North American national parks owe much of their beauty to glacial action. Yosemite Valley in California might have been another nondescript valley if glaciers had not carved it into its present shape (figure 12.1). Unlike stream-carved valleys, Yosemite is straight for long stretches. Its sides are steep and the valley floor is flat (it is U-shaped rather than the characteristic V-shape of a stream-carved valley). The sediment beneath the vegetation in the valley floor is poorly sorted debris, unlike the sorted sediment deposited by a stream. All of these things are evidence that the Yosemite landscape has been carved by a glacier. But there is no glacier in Yosemite Valley. Yosemite indicates, as does overwhelming evidence elsewhere in the world, that glaciation was more extensive in the geologically recent past—that is, during the glacial ages.

Our lives and environment today have been profoundly influenced by the effects of past glaciation. For example, much of the fertile soil of the northern Great Plains of the United States developed on the loose debris transported and deposited

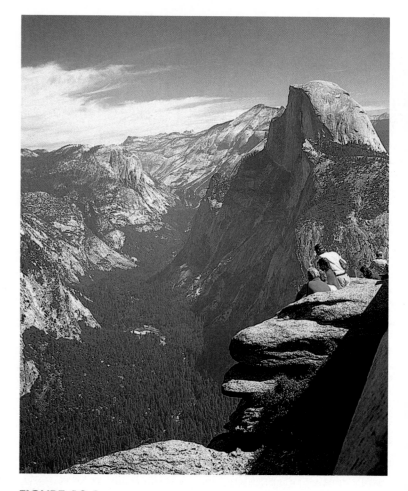

FIGURE 12.1

Yosemite Valley, as seen from Glacier Point, Yosemite National Park, California. Its U-shaped cross profile is typical of glacially carved valleys. *Photo by C. C. Plummer* by glaciers that moved southward from northern Canada. The thick blankets of sediment left in the Midwest store vast amounts of ground water. The Great Lakes and the thousands of lakes in Minnesota and neighboring states and provinces are the products of past glaciation.

Before we can understand how a continental glacier was responsible for much of the soil in the Midwest or how a glacier confined to a valley could carve a Yosemite, we must learn something about present-day glaciers.

GLACIERS—WHERE THEY ARE, HOW THEY FORM AND MOVE

Distribution of Glaciers

Glaciers occur in temperate as well as polar climates. They are found where more snow falls during the cold time of year than can be melted during warm months.

Washington has more glaciers than any other state except Alaska, because of the extensively glaciated mountains of western Washington. Washington's mountains have warmer winters but much more precipitation in the higher elevations than do the Rocky Mountains. There is more snow left after summer melting in Washington than in states to the east of it. Glaciers are common even near the equator in the very high mountains of South America and Africa because of the low temperatures at high altitudes.

Glaciation is most extensive in polar regions, where little melting takes place at any time of year. At present, about one-tenth of the land surface on Earth is covered by glaciers (compared with about one-third during the peak of the glacial ages). Approximately 85% of the present-day glacier ice is on the Antarctic continent, covering an area larger than the combined areas of western Europe and the United States; 10% is in Greenland. All the remaining glaciers of the world amount to only about 5% of the world's freshwater ice. This means that Antarctica is in fact storing most of Earth's fresh water in the form of ice. Some have suggested that ice from the Antarctic, towed as icebergs, could be brought to areas of dry climate to alleviate water shortages. It is worth noting that if all of Antarctica's ice were to melt, sea level around the world would rise over 70 meters (230 feet). This would flood the world's coastal cities and significantly decrease the land surface available for human habitation.

Types of Glaciers

A simple criterion—whether or not a glacier is restricted to a valley—is the basis for classifying glaciers by form. A **valley glacier** is a glacier that is confined to a valley and flows from a higher to a lower elevation. Like streams, small valley glaciers may be tributaries to a larger trunk system. Valley glaciers

FIGURE 12.2

Valley glacier on the flanks of Mount Logan, Canada's highest mountain. *Photo by C. C. Plummer*

are prevalent in areas of alpine glaciation. As might be expected, most glaciers in the United States and Canada, being in mountains, are of the valley type (figure 12.2).

In contrast, an **ice sheet** is a mass of ice that is not restricted to a valley but covers a large area of land (over 50,000 square kilometers). Ice sheets are associated with continental glaciation. Only two places on Earth now have ice sheets: Greenland and Antarctica. A similar but smaller body is called an **ice cap.** Ice caps (and valley glaciers as well) are found in a few mountain highlands in Iceland and on islands in the Arctic Ocean, off Canada, Russia, and Scandinavia. An ice cap or ice sheet flows downward and outward from a central high point, as figure 12.3 shows.

Formation and Growth of Glaciers

Snow converts to glacier ice in somewhat the same way that sediment turns into a sedimentary rock and then into metamorphic rock; figure 12.4 shows the process. A snowfall can be compared to sediment settling out of water. A new snowfall may be in the form of light "powder snow," which consists mostly of air trapped between many six-pointed snowflakes. In a short time, the snowflakes settle by compaction under their own weight, and much of the air between them is driven out. Meanwhile, the sharp points of the snowflakes are destroyed as flakes reconsolidate into granules. In warmer climates, partial thawing and refreezing result in coarse granules—the "corn snow" of spring skiing. In colder climates where little or no melting takes place, the snowflakes will recrystallize into fine granules. After the granular snow is buried by a new snowpack, usually during the following winter, the granules are

FIGURE 12.3

Diagrammatic cross section of an ice sheet. Vertical scale is highly exaggerated.

FIGURE 12.4

(A) Conversion of snow to glacier ice. (B) Thin slice of an ice core from a glacier (a core is shown in Box 12.3, figure 1). The ice is between sheets of polarizing filters. In polarized light, the colors of individual ice grains vary depending on their crystallographic orientation. Without the polarizing filters, the ice would be transparent and clear. The ruler shows that many of the ice grains are over a centimeter in length. *Photo by C. C. Plummer* compacted and weakly "cemented" together by ice. The compacted mass of granular snow, transitional between snow and glacier ice, is called *firn*. Firn is analogous to a sedimentary rock such as sandstone.

Through the years, the firn becomes more deeply buried as more snow accumulates. More air is expelled, the remaining pore space is greatly reduced, and granules forced together recrystallize into the tight, interlocking mosaic of *glacier ice* (figure 12.4B). The recrystallization process involves little or no melting and is comparable to metamorphism. Glacier ice is texturally similar to the metamorphic rock, quartzite.

Under the influence of gravity, glacier ice moves downward and is eventually **ablated,** or lost. For glaciers in all but the coldest parts of the world, ablation is due mostly to melting, although some ice evaporates directly into the atmosphere. If a moving glacier reaches a body of water, blocks of ice break off (or *calve*) and float free as **icebergs** (figure 12.5). In most of the Antarctic, ablation takes place largely through calving of icebergs and direct evaporation. Only along the coast does melting take place, and there for only a few weeks of the year.

Glacial Budgets

If, over a period of time, the amount of snow a glacier gains is greater than the amount of ice and water it loses, the glacier's budget is *positive* and it expands. If the opposite occurs, the glacier decreases in volume and is said to have a *negative budget*. Glaciers with positive budgets push outward and downward at their edges; they are called **advancing glaciers.** Those with negative budgets grow smaller and their edges melt back; they are **receding glaciers.** Bear in mind that the glacial ice moves downvalley, as shown in figure 12.6, whether the glacier is advancing or receding. In a receding glacier, however, the rate of flow of ice is insufficient to replace all of the ice lost in the lower part of the glacier. If the amount of snow retained by the glacier equals the amount of ice and water lost, the glacier has a *balanced budget* and is neither advancing nor receding.

FIGURE 12.5
An iceberg in southern Chile. *Photo by C. C. Plummer*

The upper part of a glacier, called the **zone of accumulation,** is the part of the glacier with a perennial snow cover (figure 12.6). The lower part is the **zone of ablation,** for there ice is lost, or ablated, by melting, evaporation, and calving.

The boundary between these two altitudinal zones of a glacier is an irregular line called the **equilibrium line** (sometimes called the *snow line* or the *firn line*), which marks the highest point at which the glacier's winter snow cover is lost during a melt season (figure 12.7).

The equilibrium line may shift up or down from year to year, depending on whether there has been more accumulation or more ablation. Its location therefore indicates whether a glacier has a positive or negative budget. An equilibrium line migrating upglacier over a period of years is a sign of a negative budget, whereas an equilibrium line migrating downglacier indicates that the glacier has a positive budget. If an

FIGURE 12.6
A valley glacier as it would appear at the end of a melt season. Below the equilibrium line, glacier ice and snow have been lost during the melting season. In the zone of accumulation above that line, firn is added to the glacier from the previous winter snowfall.

A

B

C

equilibrium line remains essentially in the same place year after year, the glacier has a balanced budget.

The **terminus** is the lower edge of a glacier. Its position reflects the glacier's budget. For a valley glacier, a positive budget results in the terminus moving downvalley. In a receding glacier, the terminus melts back upvalley. Because glacial ice moves slowly, migration of the terminus tends to lag several years behind a change in the budget.

If the terminus of an ice sheet is on land, it will advance or retreat in response to a positive or negative budget, just as for a valley glacier. If the terminus is at the continent's shoreline (as it is for our Antarctic and Greenland ice sheets), a positive budget results in a greater volume of icebergs calving into the sea.

Advancing or receding glaciers are significant and sensitive indicators of climatic change. However, an advancing glacier does not necessarily indicate that the climate is getting colder. It may mean that the climate is getting wetter, more precipitation is falling during the winter months, or the summers are cloudier. It is estimated that a worldwide decrease in the mean annual temperature of about 5°C could bring about a new ice age. Conversely, global climate warming can significantly reduce the size and numbers of glaciers. In general, valley glaciers around the world have been receding during the past century. At present, glaciers at Glacier National Park in Montana and Mount Kilamanjaro in Africa are receding at a rate that, if sustained, will lead them to disappear in a few years. (Will Montana's park then be renamed Glacier-Free National Park?)

Movement of Valley Glaciers

Valley glaciers move downslope under the influence of gravity at a variable rate, generally ranging from less than a few millimeters to 15 meters a day. Sometimes a glacier will move much faster for a brief period of time (see box 12.2). A glacier will flow faster where it is steeper. Also, the thicker parts of a glacier will flow faster than where it is thinner. The upper part of a glacier tends to be steeper than at lower levels. If a glacier has an even gradient, the glacier will be thickest near the equilibrium line. So, except for locally steeper stretches, we expect the fastest moving ice to be near the equilibrium line. Below the equilibrium line, the glacier usually becomes progressively thinner and slower.

Glaciers in temperate climates—where the temperature of the glacier is at or near the melting point for ice—tend to move faster than those in colder regions—where the ice temperature stays well below freezing.

FIGURE 12.7

South Cascade glacier, Washington. If the photos were taken at the end of the melt season, the equilibrium line would be the boundary between white snow and darker glacier ice. Photo (A) was taken in 1957; note that the glacier extended into the lake and that small icebergs calved from it. Photo (B) was taken in 1980; notice that the glacier has shrunk and receded. During the 23-year interval, the glacier lost approximately 7.5 meters of ice averaged over its surface, or the equivalent of 18.7 million cubic meters of water for the entire glacier. Photo (C), taken in October 2000, shows that the glacier continued to recede. *Photos by U.S. Geological Survey*

Glaciers as a Water Resource

Few people think of glaciers as frozen reservoirs supplying water for irrigation, hydroelectric power, recreation, and industrial and domestic use. Yet, glacially-derived water is an important resource in places such as Iceland, Norway, British Columbia, and the state of Washington. In Washington, streamflow from the approximately 800 glaciers there amounts to about 470 billion gallons of water during a summer, according to the U.S. Geological Survey. More water is stored in glacier ice in Washington than in all of the state's lakes, reservoirs, and rivers.

One important aspect of glacier-derived water is that it is available when needed most. Snow accumulates on glaciers during the wet winter months. During the winter, streams at lower elevations, where rain rather than snow falls, are full and provide plenty of water. During the summer, the climate in the Pacific Northwest is hotter and drier. Demand for water increases, especially for irrigation of crops. Streams that were fed by rainwater may have dried up. Yet, in the heat of summer, the period of peak demand, snow and ice on glaciers are melting, and streams draining glaciers are at their highest level.

Paradoxically, the greater the snowfall on a glacier during a winter, the smaller the amount of meltwater during the summer. A larger blanket of white snow reflects the sun's radiation more effectively than the darker, bare glacier ice, which absorbs more of the heat of the sun. Experiments have shown that melting can be greatly increased by darkening the snow surface, for instance, by sprinkling coal dust on it. Similarly, the melting of a glacier can be slowed artificially by covering it with highly reflective material. Such means of controlling glacial meltwater have been proposed to benefit power generating stations or to provide additional irrigation.

These ideas are appealing from a shortsighted point of view. However, the long-term effect of tampering with a glacier's natural regime can adversely affect the overall environment. It is conceivable, for example, that we could melt a glacier out of existence.

Additional Resource

U.S. Geological Survey. 1973. *Glaciers, a water resource.* U.S. Geological Survey Information Pamphlet.

Velocity also varies within the glacier itself (figure 12.8). The central portion of a valley glacier moves faster than the sides (as water does in a stream), and the surface moves faster than the base. How ice moves within a valley glacier has been demonstrated by studies in which holes are drilled through the glacier ice and flexible pipes inserted. Changes in the shape and position of the pipes are measured periodically. The results of these studies are shown diagrammatically in figure 12.8.

Note in the diagram that the base of the pipe has moved downglacier. This indicates **basal sliding,** which is the sliding of the glacier as a single body over the underlying rock. A thin film of meltwater that develops along the base from the pressure of the overlying glacier facilitates basal sliding. Think of a large bar of wet soap sliding down an inclined board.

Note that the lower portion of the pipe is bent in a downglacier direction. The bent pipe indicates **plastic flow** of

FIGURE 12.8

Movement of a glacier. Markers on the glacier indicate the center of the glacier moves faster than its side. Cross-sectional view shows movement within the glacier.

FIGURE 12.9

(A) Crevasses on a glacier, looking down from Mount Logan, Yukon Territory, Canada. (B) Crevasses along the course of a glacier. *Photo by C. C. Plummer*

Glacier without snow cover

Crevasses

Snow - covered glacier

Sagging snow bridge on crevasse

Crevasses

Geologist's View

B

Crevasses on outside of curve

Crevasses

Closed crevasses

ice, movement that occurs within the glacier due to the plastic or "deformable" nature of the ice itself. Visualize two neighboring grains of ice within the glacier, one over the other. Both are moving, carried along by the ice below them; however, the higher of the two ice grains slides over its underlying neighbor a bit further. The reason the pipe is bent more sharply near the base of the glacier is that pressure from overlying ice results in greater flowage with increasing depth. Deep in the glacier, ice grains are sliding past their underlying neighbors farther than similar ice grains higher up, where the pipe is less bent. We should point out that a glacier flows not only because ice grains slide past one another but also because ice grains deform and recrystallize.

In the **rigid zone,** or upper part of the glacier, the pipe has been moved downglacier; however, it has remained unbent. The ice nearer the top apparently rides along passively on the plastically moving ice closer to the base. In the rigid zone, grains of ice do not move relative to their neighbors.

Crevasses

Along its length, a valley glacier moves at different rates in response to changes in the steepness of the underlying rock. Typically, a valley glacier rides over a series of rock steps. Where the glacier passes over a steep part of the valley floor, it moves faster. The upper rigid zone of ice, however, cannot stretch to move as rapidly as the underlying plastic-flowing ice. Being brittle, the ice of the rigid zone is broken by the tensional forces. Open fissures, or **crevasses,** develop (figure 12.9). Crevasses also form along the margins of glaciers in places where the path is curved, as shown in part of

figure 12.9. This is because ice (like water) flows faster toward the outside of the curve. For glaciers in temperate climates, a crevasse should be no deeper than about 40 meters, the usual thickness of the rigid zone. If you are falling down a crevasse, it may be of some consolation that, as you are hurtling to death or injury, you realize on the way down that you will not fall more than 40 meters.

After the ice has passed over a steep portion of its course, it slows down, and compressive forces close the crevasses.

Movement of Ice Sheets

An ice sheet or ice cap moves like a valley glacier except that it moves downward and outward from a central high area toward the edges of the glacier (as shown in figure 12.3).

Glaciological research in Antarctica has determined how ice sheets grow and move. Antarctica has two ice sheets: the West Antarctic Ice Sheet is separated by the Transantarctic Mountains from the much larger East Antarctic Ice Sheet (figure 12.10). The two ice sheets join in the low areas between mountain ranges. Both are nearly completely within a zone of accumulation because so little melting takes place (ablation is largely by calving of icebergs) and because occasional snowfalls nourish their high central parts. The ice sheets mostly overlie interior lowlands but also completely bury some mountain ranges. Much of the base of the West Antarctic Ice Sheet is on bedrock that is below sea level. At least one active volcano underlies the West Antarctic Ice Sheet (resulting in a depression in the ice sheet). Where mountain ranges are higher than the ice sheet, the ice flows through as valley glaciers.

ENVIRONMENTAL GEOLOGY 12.2

Water Beneath Glaciers: Floods, Giant Lakes, and Galloping Glaciers

A Galloping Glacier

Glacial motion is often used as a metaphor for slowness ("The trial proceeded at a glacial pace"). But, some glaciers will *surge*—that is, move very rapidly for short periods following years of barely moving at all. The most extensively documented surge (or "galloping glacier") was that of Alaska's Bering Glacier in 1993–94. The Bering Glacier is the largest glacier in continental North America, and it surges on a 20–30 year cycle. After its previous surge in 1967, its terminus retreated 10 kilometers. In August 1993, the latest surge began. Ice traveled at velocities up to 100 meters per day for short periods of time and sustained velocities of 35 meters per day over a period of several months. The terminus advanced 9 kilometers by the time the surge ended in November 1994. When glaciers surge, the previously slow moving, lower part of a glacier breaks into a chaotic mass of blocks (box figure 1). Surges are usually attributed to a buildup of water beneath part of a glacier, floating it above its bed. In July 1994, a large flood of water burst from Bering Glacier's terminus, carrying with it blocks of ice up to 25 meters across.

A Flood

Glacial outburst floods are not always associated with surges. In October 1996, a volcano erupted beneath a glacier in Iceland. The glacier, which is up to 500 meters thick, covers one-tenth of Iceland. Emergency teams prepared for the flood that geologists predicted would follow the eruption. The expected flood took place early in November with a peak flow of 45,000 cubic meters per second (over 1.5 million cubic feet per second)! The flood lasted only a few hours; however, it caused between $10 and 15 million worth of damage. Three major bridges were destroyed or damaged, and 10 kilometers of roads were washed away. Because people had been kept away from the expected flood path, there were no casualties.

A Giant Lake

One of the world's largest lakes was only recently discovered. But don't expect to take a dip in it or go windsurfing on it. It lies below the thickest part of the East Antarctic Ice Sheet and is named after the Russian research station, Vostok, which is 4,000 meters above the lake at the coldest and most remote part of Antarctica. Lake Vostok was discovered in the 1970s through ice-penetrating radar; however, its extent was unknown until 1996, when satellite-borne radar revealed how large it is.

Studies indicate that the lake is 200 kilometers long and 50 kilometers wide—about the size of Lake Ontario. At its deepest, it is 510 meters, placing it among the ten deepest lakes in the world. Recently, more, but smaller, lakes beneath the East Antarctic Ice Sheet have been discovered.

The lake has been sealed off from the rest of the world for around a million years, and it likely contains organisms, such as microbes, dating back to that time. These organisms (and their genes) would not have been affected by modern pollution or nuclear bomb fallout. By coincidence, the world's deepest ice hole (over 3 kilometers) was being drilled from Vostok Station above the lake when the size of Lake Vostok was being determined. The ice core from this hole should add to the findings from the Greenland drilling projects (box 12.3) and provide an even greater picture of Earth's climate during the ice ages. When the hole was completed in 1997, drilling was halted short of reaching the lake due to fear of contaminating it and harming whatever living organisms might be in the very old water. Study of the lake and its organisms is curtailed until a future generation of scientists can devise means of sampling the waters without altering its ecosystem.

BOX 12.2 ■ FIGURE 1

Part of a glacier after a surge (lower part of photo). The debris-covered ice has been broken up into a chaotic mass of blocks. In the background is a small glacier that has retreated up its valley. Photo taken near the Canadian-Alaskan border. *Photo by C. C. Plummer*

Geologist's View

FIGURE 12.10

The Antarctic continent and its ice sheets. Vostok is at the highest part of the East Antarctic Ice Sheet. (False coloring is used to show variations among snow, ice, blue ice, and exposed rock.) *Photo by U.S. Geological Survey/NASA*

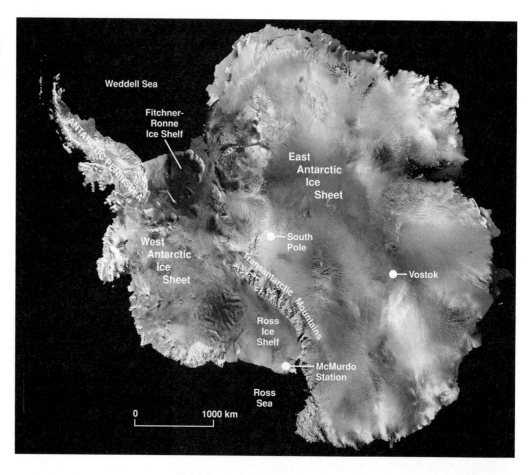

At the South Pole (figures 12.10 and 12.11)—neither the thickest part nor the center of the East Antarctic Ice Sheet—the ice is 2,700 meters thick. The thickest part of the East Antarctic Ice Sheet is 4,776 meters. Research at ice sheets has yielded important information regarding past climates (see box 12.3).

Most of the movement of the East Antarctic Ice Sheet is by means of plastic flow. It has been thought that most of the ice sheet is frozen to the underlying rocks and basal sliding takes place only locally. But the recent discovery of a giant lake and other lakes beneath the thickest part of the East Antarctic Ice Sheet (see box 12.2) indicates that liquid water at its base is more widespread and basal sliding might be more important than previously regarded.

GLACIAL EROSION

Wherever basal sliding takes place, the rock beneath the glacier is abraded and modified. As meltwater works into cracks in bedrock and refreezes, pieces of the rock are broken loose and frozen into the base of the moving glacier, a process known as *plucking*. While being dragged along by the moving ice, the rock within the glacier grinds away at the underlying rock (figure 12.12). The thicker the glacier, the more pressure on the rocks and the more effective the grinding and crushing.

Pebbles and boulders that are dragged along are *faceted*, that is, given a flat surface by abrasion. Bedrock underlying a glacier is *polished* by fine particles and *striated* (scratched) by

FIGURE 12.11

The South Pole. Actually, the true South Pole is several kilometers from here. The moving ice sheet has carried the striped pole away from the site of the true South Pole, where the pole was erected in 1956. *Photo by C. C. Plummer*

sharp-edged, larger particles. Striations and grooves on bedrock indicate the direction of ice movement (figure 12.13).

The grinding of rock across rock produces a powder called **rock flour.** Rock flour is composed largely of very fine (silt- and clay-sized) particles of unaltered minerals (pulverized from chemically unweathered bedrock). When *meltwater* washes rock flour from a glacier, the streams draining the gla-

FIGURE 12.12

Plucking and abrasion beneath a glacier.

FIGURE 12.13

Striated and polished bedrock surface in south Australia. Unlike glacial striations commonly found in North America, these were caused by late Paleozoic glaciation. *Photo by C. C. Plummer*

cier appear milky, and lakes into which glacial meltwater flows often appear a milky green color.

Not all glacier-associated erosion is caused directly by glaciers. Mass wasting takes place on steep slopes created by downcutting glaciers. Frost wedging breaks up bedrock ridges and cliffs above a glacier, causing frequent rockfalls. Snow avalanches bring down loose rocks onto the glacier. If rocks collect in the zone of accumulation, they may be incorporated into the body of the glacier. If rock falls onto the zone of ablation, they will ride on the glacial ice surface. Debris may also fall into crevasses to be transported within or at the base of a glacier, as shown in figure 12.19.

Erosional Landscapes Associated with Alpine Glaciation

We are in debt to glaciers for the rugged and spectacular scenery of high mountain ranges. Figure 12.14 shows how glaciation has radically changed a previously unglaciated mountainous

A

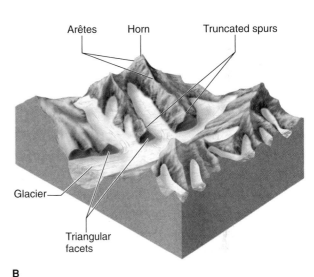

B

C

FIGURE 12.14

(A) A stream-carved mountain landscape before glaciation. *(B)* The same area during glaciation. Ridges and peaks become sharper due to frost wedging. *(C)* The same area after glaciation.

EARTH SYSTEMS 12.3

Global Warming and Glaciers

Most of Earth's glaciers have been receding for a century (see figure 12.7). This is generally regarded as a consequence of global warming. That Earth's climate is warming is now clearly established. But some questions arise with regard to global warming: How does it compare to past episodes of warming? Is it part of a natural cycle? How much of it is anthropogenic (caused by humans)? What are the consequences of continued global warming on Earth systems? Can we do anything to reduce the global warming?

Glaciers, particularly the Antarctic and Greenland ice sheets, provide us with a means to answer these questions. Glaciers preserve records of precipitation, air temperatures, atmospheric dust, volcanic ash, carbon dioxide, and other atmospheric gases.

When snow becomes converted to glacier ice, some of the air that was mixed with the snowflakes becomes bubbles trapped in the glacier ice. By analyzing the air in these bubbles, we are analyzing the air that prevailed when an ancient ice layer formed. Drilling into glaciers and retrieving ice cores allows scientists to sample the environment at the time of ancient snowfalls. A cylindrical core of ice is extracted from a hollow drill after it has penetrated a glacier. The layers in an ice core represent the different layers of snow that converted to glacier ice (box figure 1). Each layer, when analyzed, can reveal information about conditions of the atmosphere at the time the snow accumulated and turned into ice.

As of 2004, the most ambitious drilling project in Antarctica had extracted the largest-ever ice core, representing a record going back 740,000 years. The international team hopes to extract a million years of core before reaching the base of the ice sheet. The next-largest core was drilled at Vostok (described in box 12.2) in Antarctica in the 1990s.

The Vostok core reached a depth of over 3 kilometers and yielded a climate and atmospheric history of the past 420,000 years. Graphs derived from the research project are shown in box figure 2. The temperature variation is relative to the ice sheet's temperature during the past millennium. The team determined the temperature by studying hydrogen isotope variation (see chapter 2) within the ice layers. Methane and carbon dioxide are greenhouse gases. Note how the greenhouse gases correlate closely with the temperature variations. Also note the five periods when the temperature was warmest. These are the *interglacial* periods during which the North American and European ice sheets disappeared. Two of the five warm periods are emphasized for

BOX 12.3 ■ FIGURE 1

An ice core being examined in a cold laboratory. *Photo by Mark Twickler, University of New Hampshire/National Oceanic and Atmospheric Administration Paleoclimatology Program/Department of Commerce*

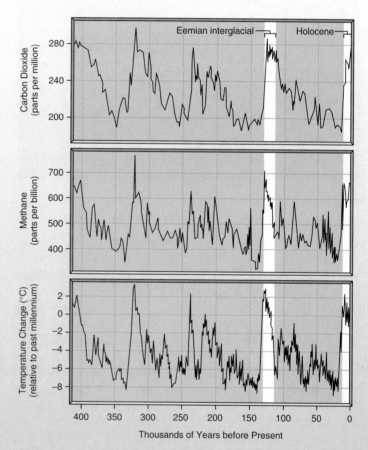

BOX 12.3 ■ FIGURE 2

Temperature, carbon dioxide, and methane content of air at Vostok on the East Antarctic Ice Sheet for the last 420,000 years. See text for explanation. Also note the rapid rising of temperature at the beginning of interglacial periods.

comparison—the Holocene interglacial epoch (which began about 12,000 years ago and is ongoing) and the previous interglacial (the Eemian).

Compare the Holocene temperature pattern to that of the Eemian. From this, you can understand why scientists infer that we should be in a period of declining temperatures leading into the next glacial age. Instead, we have ongoing global warming. Further, there is now strong evidence that the global warming is due to anthropogenic contribution of greenhouse gases to our atmosphere. The two biggest culprits are methane and carbon dioxide. If these gases are not controlled, their levels will rise to the naturally derived levels reached during the warmer Eemian interglacial. Sea level worldwide during that time was several meters higher than at present. (This was caused by the expansion of the ocean waters due to heating, as well as melting of polar ice at a higher rate than at present.)

If computer models projecting higher rates of global warming are correct, we can expect the hydrosphere to be affected by gases in the atmosphere, and sea level will rise significantly. One group of researchers calculated that sea level could rise several tens of centimeters during this century. But this prediction is based mainly on expansion of the oceans due to being heated. If the higher temperatures also trigger disintegration of large parts of the ice sheets, sea levels could rise even higher. A higher sea level affects the biosphere, notably humans. A major proportion of the world's population lives at or close to a coastline. Houses would be destroyed by coastal erosion. Major cities, such as New York, would have to erect dikes to keep water out of buildings that are below sea level.

Can anything be done to stop or slow down global warming? James Hansen (see Additional Resources) thinks so. The rate at which methane and carbon dioxide are produced would have to be reduced. The rate of carbon dioxide production, mainly from burning of fossil fuels, is rising. However, the rate of production of methane has been declining for two decades. Methane is a fuel and it makes economic sense to capture it where it is produced at landfills, mines, and oil fields. Reducing the production of carbon dioxide is more problematic. Additional things that could be done include using more fuel-efficient vehicles and moving toward alternate forms of energy. Cutting back the production of carbon dioxide and other pollutants has the additional benefit of decreasing health risks.

Additional Resources

For more on ice sheet drilling, go to the Online Center at

- www.mhhe.com/plummer11e.

For more on the topic of global warming and its relationship to glaciation, read James Hansen, Defusing the Global Warming Time Bomb. *Scientific American*, March 2004, pp. 68–77. For a more detailed version of that article, go to

- http://www.sciam.com/media/pdf/hansen.pdf.

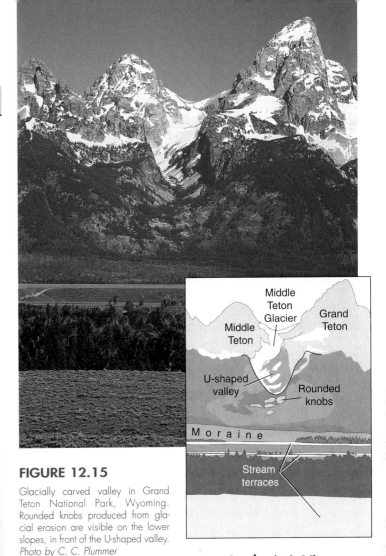

FIGURE 12.15

Glacially carved valley in Grand Teton National Park, Wyoming. Rounded knobs produced from glacial erosion are visible on the lower slopes, in front of the U-shaped valley. *Photo by C. C. Plummer*

Geologist's View

region. The striking and unique features associated with mountain glaciation, described next, are due to the erosional effects of glaciers as well as frost wedging on exposed rock.

Glacial Valleys

Glacially carved valleys are easy to recognize. A **U-shaped valley** (in cross profile) is characteristic of glacial erosion (figure 12.15), just as a V-shaped valley is characteristic of stream erosion.

The thicker a glacier is, the more erosive force it exerts on the underlying valley floor, and the more bedrock is ground away. For this reason, a large trunk glacier erodes downward more rapidly and carves a deeper valley than do the smaller tributary glaciers that join it. After the glaciers disappear, these tributaries remain as **hanging valleys** high above the main valley (figure 12.16).

Valley glaciers, which usually occupy valleys formerly carved by streams, tend to straighten the curves formed by running water. This is because the mass of ice of a glacier is too sluggish and inflexible to move easily around the curves. In the process of carving the sides of its valley, a glacier erodes or "truncates" the lower ends of ridges that extended

319

to the valley. **Truncated spurs** are ridges that have *triangular facets* produced by glacial erosion at their lower ends (figure 12.14*B*).

Although a glacier tends to straighten and smooth the side walls of its valley, ice action often leaves the surface of the underlying bedrock carved into a series of steps. This is due to the variable resistance of bedrock to glacial erosion. Figure 12.17 shows what happens when a glacier abrades a relatively weak rock with closely spaced fractures. Water seeps into cracks in the bedrock, freezes there, and enlarges fractures or makes new ones. Rock frozen into the base of the glacier grinds and loosens more pieces. After the ice has melted back, a chain of **rock-basin lakes** (also known as **tarns**) may occupy the depressions carved out of the weaker rock. A series of such lakes, reminiscent of a string of prayer beads, is sometimes called *paternoster lakes*.

Areas where the bedrock is more resistant to erosion stand out after glaciation as *rounded knobs* (see figures 12.15 and

FIGURE 12.16

A hanging valley in Yosemite National Park, California. *Photo by C. C. Plummer*

FIGURE 12.17

Development of rock steps. (*A*) Valley floor before glaciation. (*B*) During glaciation. (*C*) Rock steps and rock-basin lakes. Sierra Nevada, California. *A and B after F. E. Matthes, 1930, U.S. Geological Survey; Photo by C. C. Plummer*

12.22), usually elongated parallel to the direction of glacier flow. These are also known as *roches moutonnées.* (In French, *roche* is rock and *moutonnée* means fleecy or curled.*)

Cirques, Horns, and Arêtes

A **cirque** is a steep-sided, half-bowl-shaped recess carved into a mountain at the head of a valley carved by a glacier (figure 12.18). In this unique, often spectacular, topographic feature, a large percentage of the snow accumulates that eventually converts to glacier ice and spills over the threshold as the valley glacier starts its downward course.

A cirque is not entirely carved by the glacier itself but is also shaped by the weathering and erosion of the rock walls above the surface of the ice. Frost wedging and avalanches break up the rock and steepen the slopes above the glacier. Broken rock tumbles onto the valley glacier and becomes part of its load, and some rock may fall into a crevasse that develops where the glacier is pulling away from the cirque wall (figure 12.19).

The headward erosional processes that enlarge a cirque also help create the sharp peaks and ridges characteristic of glaciated mountain ranges. A **horn** is the sharp peak that remains after cirques have cut back into a mountain on several sides (figure 12.20).

---

*The term was first used in the 1780s to describe an assemblage of rounded knobs in the Swiss Alps. It alluded to a fleecy wig called a *moutonné,* popular at that time, that was slicked down with sheep tallow. Later, the term came to refer to individual rounded knobs that resembled sheep—*mouton* is French for sheep.

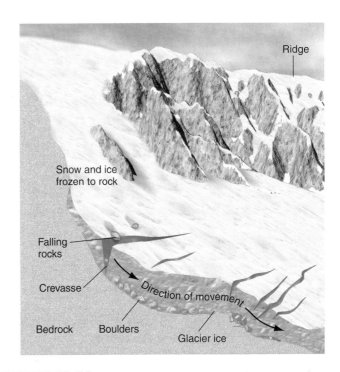

FIGURE 12.19

Cutaway view of a cirque.

FIGURE 12.18

A cirque occupied by a small glacier in the Canadian Rocky Mountains. The glacier was much larger during the ice ages. *Photo by C. C. Plummer*

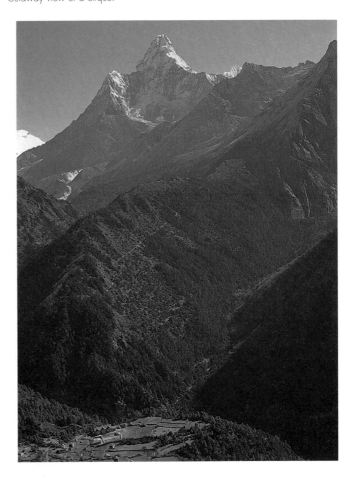

FIGURE 12.20

Ama Dablan, a horn in the Mount Everest region of the Himalaya in Nepal. Note the cirque below the peak. *Photo by C. C. Plummer*

Frost wedging works on the rock exposed above the glacier, steepening and cutting back the side walls of the valley. Sharp ridges called **arêtes** separate adjacent glacially carved valleys (figure 12.21).

Erosional Landscapes Associated with Continental Glaciation

In contrast to the rugged and angular nature of glaciated mountains, an ice sheet tends to produce rounded topography. The rock underneath an ice sheet is eroded in much the same way as the rock beneath a valley glacier; however, the weight and thickness of the ice sheet may produce more pronounced effects. Rounded knobs are common (figure 12.22), as are grooved and striated bedrock. Some grooves are actually channels several meters deep and many kilometers long. The orientation of grooves and striations indicates the direction of movement of a former ice sheet.

An ice sheet may be thick enough to bury mountain ranges, rounding off the ridges and summits and perhaps streamlining them in the direction of ice movement. Much of northeastern Canada, with its rounded mountains and grooved and striated bedrock surface, shows the erosional effects of ice sheets that formerly covered that part of North America (figure 12.23).

GLACIAL DEPOSITION

The rock fragments scraped and plucked from the underlying bedrock and carried along at the base of the ice make up most of the load carried by an ice sheet but only part of a valley glacier's load. Much of a valley glacier's load comes from rocks broken from the valley walls.

Most of the rock fragments carried by glaciers are angular, as the pieces have not been tumbled around enough for the edges and corners to be rounded. The debris is unsorted, and clay-sized to boulder-sized particles are mixed together (figure 12.24). The unsorted and unlayered rock debris carried or deposited by a glacier is called **till.**

Glaciers are capable of carrying virtually any size of rock fragment, even boulders as large as a house. An **erratic** is an ice-transported boulder that has not been derived from underlying bedrock (figure 12.22). If its bedrock source can be found, the erratic indicates the direction of movement of the glacier that carried it.

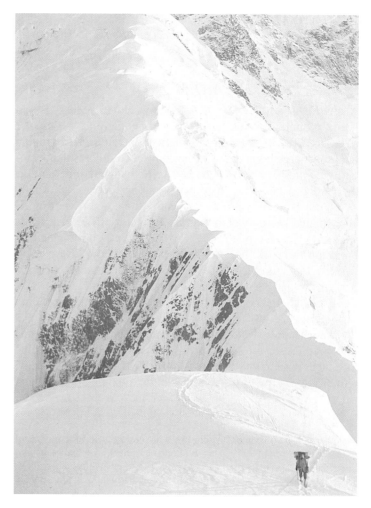

FIGURE 12.21

An arête on Mount Logan, Yukon Territory, Canada. *Photo by C. C. Plummer*

Erratic   Rounded knob

FIGURE 12.22

Rounded knob (*roche moutonnée*) with an erratic (the boulder) on it in Central Park, New York City. *Photo by Charles Merguerian*

FIGURE 12.23

Glacially scoured terrain near Baker Lake, Northwest Territories, Canada. *Photo © 2005 GlobeXplorer*

Moraines

When till occurs as a body of unsorted and unlayered debris either on a glacier or left behind by a glacier, the body is regarded as one of several types of **moraines. Lateral moraines** are elongate, low mounds of till that form along the sides of a valley glacier (figures 12.24, 12.25, 12.26, and 12.27). Rockfall debris from the steep cliffs that border valley glaciers accumulates along the edges of the ice to form lateral moraines.

Where tributary glaciers come together, the adjacent lateral moraines join and are carried downglacier as a single long ridge of till known as a **medial moraine.** In a large trunk glacier that has formed from many tributaries, the numerous medial moraines give the glacier the appearance from the air of a multilane highway (figures 12.25 and 12.26).

An actively flowing glacier brings debris to its terminus. If the terminus remains stationary for a few years or advances, a distinct **end moraine,** a ridge of till, piles up along the front edge of the ice. Valley glaciers build end moraines that are crescent-shaped or sometimes horseshoe-shaped (figures 12.25

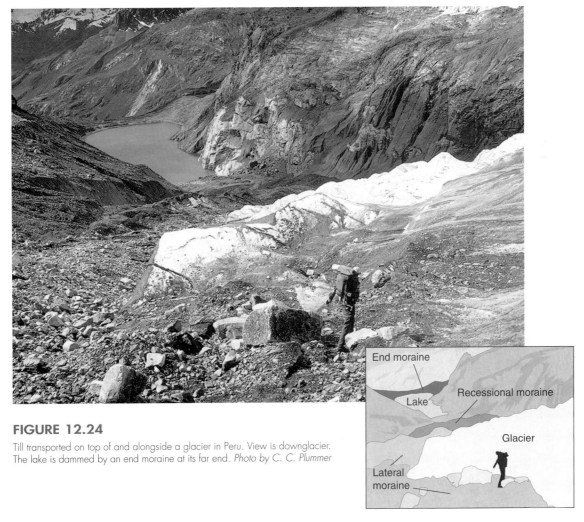

FIGURE 12.24

Till transported on top of and alongside a glacier in Peru. View is downglacier. The lake is dammed by an end moraine at its far end. *Photo by C. C. Plummer*

Geologist's View

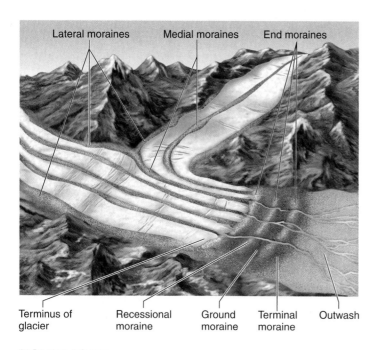

Lateral moraines   Medial moraines   End moraines

Terminus of   Recessional   Ground   Terminal   Outwash
glacier   moraine   moraine   moraine

FIGURE 12.25

Moraines associated with valley glaciers.

and 12.27). The end moraine of an ice sheet takes a similar lobate form but is much longer and more irregular than that of a valley glacier (figure 12.28).

Geologists distinguish two special kinds of end moraines. A *terminal moraine* is the end moraine marking the farthest advance of a glacier. A *recessional moraine* is an end moraine built while the terminus of a receding glacier remains temporarily stationary. A single receding glacier can build several recessional moraines (as in figures 12.24, 12.25, 12.27, and 12.28).

As ice melts, rock debris that has been carried by a glacier is deposited to form a **ground moraine,** a fairly thin, extensive layer or blanket of till (figures 12.25 and 12.28). Very large areas that were once covered by an ice sheet now have the gently rolling surface characteristic of ground moraine deposits.

In some areas of past continental glaciation, there are bodies of till shaped into streamlined hills called **drumlins** (figures 12.28 and 12.29). A drumlin is shaped like an inverted spoon aligned parallel to the direction of ice movement of the former glacier. Its gentler end points in the downglacier direction. Because we cannot observe drumlins forming beneath present ice sheets, we are not certain how till becomes shaped into these streamlined hills.

FIGURE 12.26

Medial and lateral moraines on valley glaciers, Yukon Territory, Canada. Ice is flowing toward viewer and to lower right. *Photo by C. C. Plummer*

FIGURE 12.27

End moraines (recessional moraines) in the foreground, curve into two long, lateral moraines. The two lateral moraines extend back to a glacially carved valley in the Sierra Nevada, California. *Photo by C. C. Plummer*

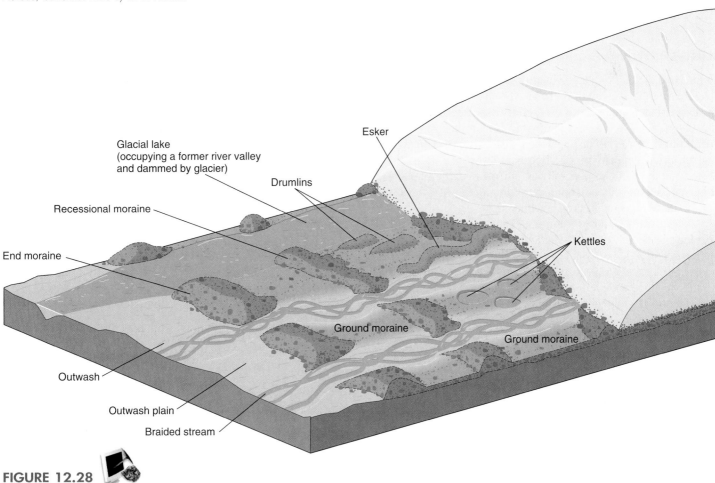

FIGURE 12.28

Depositional features in front of a receding ice sheet.

FIGURE 12.29

Drumlin in New York state. Ice flowed from top right to bottom left of photo. *Photo by Ward's Natural Science Est., Inc., Rochester, N.Y.*

Outwash

In the zone of ablation, large quantities of meltwater usually run over, beneath, and away from the ice. The material deposited by the debris-laden meltwater is called **outwash.** Because it has the characteristic layering and sorting of stream-deposited sediment, outwash can be distinguished easily from the unlayered and unsorted deposits of till. Because outwash is fairly well sorted and the particles generally are not chemically weathered, it is an excellent source of aggregate for building roadways and for mixing with cement to make concrete.

An outwash feature of unusual shape associated with former ice sheets and some very large valley glaciers is an **esker,** a long, sinuous ridge of water-deposited sediment (figures 12.28 and 12.30). Eskers can be up to 10 meters high and are formed of cross-bedded and well-sorted sediment. Evidently eskers are deposited in tunnels within or under glaciers, where meltwater loaded with sediment flows under and out of the ice.

As meltwater builds thick deposits of outwash alongside and in front of a retreating glacier, blocks of stagnant ice may be surrounded and buried by sediment. When the ice block finally melts (sometimes years later), a depression called a **kettle** forms (figures 12.28 and 12.31). Many of the small scenic lakes in the upper middle west of the United States are kettle lakes. A *kame* is a low mound or irregular ridge formed of outwash deposits on a stagnating glacier. Sediment accumulates in depressions or troughs on a glacier's surface. When the ice melts, the sediment

remains as irregular, moundlike hills. The irregular, bumpy landscape of hills and depressions associated with many moraines is known as *kame and kettle topography.*

The streams that drain glaciers tend to be very heavily loaded with sediment, particularly during the melt season. As

FIGURE 12.30

An esker in northeastern Washington. *Photo by D. A. Rahm, courtesy of Rahm Memorial Collection, Western Washington University* they come off the glacial ice and spread out over the outwash deposits, the streams form a braided pattern (see chapter 10 on streams).

The large amount of rock flour that these streams carry in suspension settles out in quieter waters. In dry seasons or drought, the water may dry up, and the rock flour deposits may be picked up by the wind and carried long distances. Some of the best agricultural soil in the United States has been formed by rock flour that has been redeposited by wind. Such fine-grained, wind-blown deposits of dust are called *loess* (see chapter 13).

Glacial Lakes and Varves

Lakes often occupy depressions carved by glacial erosion but can also form behind dams built by glacial deposition. Commonly, a lake forms between a retreating glacier and an earlier end moraine (see figure 12.24).

In the still water of the lake, clay and silt settle on the bottom in two thin layers—one light-colored, one dark—that are characteristic of glacial lakes. Two layers of sediment representing one year's deposition in a lake are called a **varve** (figure 12.32). The light-colored layer consists of slightly coarser sediment (silt) deposited during the warmer part of the year when the nearby glacier is melting and sediment is transported to the lake. The silt settles within a few weeks or so after reaching the lake. The dark layer is finer sediment (clay)—material that sinks down more slowly during the winter after the lake surface freezes and the supply of fresh, coarser sediment stops due to lack of meltwater. The dark color is attributed to fine organic matter mixed with the clay.

Because each varve represents a year's deposit, varves are like tree rings and indicate how long a glacial lake existed.

THE THEORY OF GLACIAL AGES

In the early 1800s, the hypothesis of past extensive continental glaciation of Europe was proposed. Among the many people who regarded the hypothesis as outrageous was the Swiss naturalist Louis Agassiz. But, after studying the evidence in Switzerland, he changed his mind. In 1837, he published a discourse arguing that Switzerland was, in the past, entirely glaciated. Subsequently, he, along with the eminent English geologist William Buckland, found the same evidence in the British Isles for past glaciation that was found in Switzerland. Presently, there are no glaciers in Britain or Ireland. Agassiz, after further studies in northern Europe, concluded that a great glacier had covered most or all of Europe. Agassiz had to overcome skepticism over past climates being quite different from those of today. At the time, the hypothesis seemed to many

FIGURE 12.32

Varves from a former glacial lake. Each pair of light and dark layers represents a year's deposition. *Photo © Nick Eyles, University of Toronto, Scarborough*

FIGURE 12.31

A kettle, a kame, and outwash (*background* and *left*) from a glacier, Yukon Territory, Canada. Stagnant ice underlies much of the till. *Photo by C. C. Plummer*

PLANETARY GEOLOGY 12.4

Mars on a Glacier

Meteorites are extraterrestrial rocks—fragments of material from space that have managed to penetrate Earth's atmosphere and land on Earth's surface. They are of interest not only to astronomers but to geologists, for they help us date Earth (chapter 8) and give us clues to what Earth's interior is like (see chapter 17) because many of the meteorites are thought to represent fragments of destroyed minor planets. Meteorites are rarely found; they usually do not look very different from Earth's rocks with which they are mixed.

The international Antarctic meteorite program has recovered 30,000 specimens during the last three decades. This far exceeds the total collected elsewhere in the past two centuries. Over a thousand meteorites have been collected from one small area where the ice sheet abuts against the Transantarctic Mountains. The reason for this heavy concentration is that meteorites landing on the surface of the ice over a vast area have been incorporated into the glacier and transported to where ablation takes place. The process is illustrated in box figure 1.

A few of the meteorites are especially intriguing. Some almost certainly are rocks from the moon, while several others appar-

ently came from Mars. Their chemistry and physical properties match what we would expect of a Martian rock. But how could a rock escape Mars and travel to Earth? Scientists suggest that a meteorite hit Mars with such force that fragments of that planet were launched into space. Eventually, some of the fragments reached Earth.

In 1996, researchers announced that they found what could be signs of former life on Mars in one of the meteorites collected twelve years earlier in Antarctica. The evidence included carbon-containing molecules that might have been produced by living organisms as well as microscopic blobs that could be fossil alien bacteria. But there are alternate explanations for each line of evidence, and a hot debate has ensued between scientists with opposing viewpoints.

Additional Resource
Antarctic Meteorite Program
- www-curator.jsc.nasa.gov/curator/antmet/program.htm

BOX 12.4 ■ FIGURE 1
Diagram showing the way in which meteorites are concentrated in a narrow zone of wastage along the Transantarctic Mountains. Two meteorites are shown as well as the paths they would have taken from the time they hit the ice sheet until they reached the zone of ablation. The vertical scale is greatly exaggerated. *Source: Antarctic Journal of the United States*

geologists to be a violation of the principle of uniformitarianism. Agassiz later came to North America and worked with American geologists who had found similar indications of large-scale past glaciation on this continent.

As more evidence accumulated, the hypothesis became accepted as a theory that today is seldom questioned. The **theory of glacial ages** states that at times in the past, colder climates prevailed during which much more of the land surface of Earth was glaciated than at present.

As the glacial theory gained general acceptance during the latter part of the nineteenth century, it became clear that much of northern Europe and the northern United States as well as most of Canada had been covered by great ice sheets during the so-called Ice Age. It also became evident that even areas not covered by ice had been affected because of the changes in climate and the redistribution of large amounts of water.

We now know that the last of the great North American ice sheets melted away from Canada less than 10,000 years ago. In many places, however, till from that ice sheet overlies older tills, deposited by earlier glaciations. The older till is distinguishable from the newer till because the older till was deeply weathered during times of warmer climate between glacial episodes.

Geologists can reconstruct with considerable accuracy the last episode of extensive glaciation, which covered large parts of North America and Europe and was at its peak about 18,000 years ago. There has not been enough time for weathering and erosion to alter significantly the effects of glaciation. Less evidence is preserved for each successively older glacial episode, because (1) weathering and erosion occurred during warm interglacial periods and (2) later ice sheets and valley glaciers overrode and obliterated many of the features of earlier glaciation. However, from piecing together the evidence, geologists can see that earlier glaciers covered approximately the same region as the more recent ones.

Ironically, we know more about when the numerous glacial ages began and ended not from glacial deposits but from deep-ocean sediment. As described in chapter 2, oxygen isotope studies ($^{18}O/^{16}O$ ratios in microfossil shells) have delineated changes in temperature of near-surface ocean water. The changes of temperature of the seas have been correlated with periods of extensive worldwide glaciation and intervening, interglacial warm climates. Studies of ice cores from Antarctica and Greenland have also provided important information on climate change (see box 12.3).

Although the glacial ages are generally associated with the Pleistocene Epoch (see chapter 8), cooling actually began earlier. Recent work indicates that worldwide climate changes necessary for northern continental glaciation probably began at least 3 million years ago, late in the Tertiary Period, at least a million years before the

Pleistocene. Moreover, Antarctica has been glaciated for 14 million years.

Earth has undergone episodic changes in climate during the last 2 to 3 million years. Actually, the climate changes necessary for a glacial age to occur are not so great as one might imagine. During the height of a glacial age, the worldwide average of annual temperatures was probably only about 5°C cooler than at present. Some of the intervening interglacial periods were probably a bit warmer worldwide than present-day average temperatures.

Direct Effects of Past Glaciation in North America

Moving ice abraded vast areas of northern and eastern Canada during the growth of the North American ice sheets (figure 12.33). Most of the soil and sedimentary rock was scraped off, and underlying crystalline bedrock was scoured. Many thousands of future lake basins were gouged out of the bedrock.

FIGURE 12.33

Maximum glaciation during the Pleistocene in the Northern Hemisphere. Small glaciated areas are in mountains. Note that ice sheets extended beyond present continental shorelines. This is because sea level was lower than at present. Note that the North Pole (center of map), which is in the Arctic Ocean, was not glaciated. *After J. Ehlers and Gibbard, 2004, Quaternary glaciations—extent and chronology, Parts I-III. Elsevier.*

Causes of Glacial Ages

Go to the book's website at www.mhhe.com/plummer11e for a more in-depth presentation of this summary.

The question of what caused the glacial ages has not been completely answered since the theory of glacial ages was accepted over a century ago. Only in the last few decades have climatologists thought they were beginning to provide acceptable answers.

The primary control on the Pleistocene glacial and interglacial episodes seems to be variations in Earth's orbit and inclination to the Sun. The amount of heat from solar radiation received by any particular portion of Earth is related to the angle of the incoming Sun's rays and, to a lesser degree, the distance to the Sun. The angle of Earth's poles relative to the plane of Earth's orbit about the Sun also changes periodically. Variations in orbital relationships and "wobble" of Earth's axis are largely responsible for glacial and interglacial episodes. These provide variations in incoming solar radiation cycles of 21,000, 41,000, and 100,000 years, as calculated by Milutin Milankovitch, a Serbian mathematician, in 1921. Support for Milankovitch's cycles came from cores of deep sediment taken by oceanographic research ships. Deep-sea sediment provides a fairly precise record of climatic variations over the past few hundred thousand years. The cycles of cooling and warming determined from the marine sediments closely match the times predicted by Milankovitch.

But the theory fails to explain the absence of glaciation over most of geologic time. Thus, one or more of the other mechanisms in the following list (and described in the website) may have contributed to climate change resulting in glacial ages.

- **Changes in the atmosphere.** These changes include the amount of carbon dioxide in the atmosphere. Carbon dioxide has a "greenhouse effect," whereby the more of the gas in the atmosphere, the warmer the global climate. Large volcanic eruptions are known to lower temperature worldwide by placing SO_2 gas and fine dust in the high atmosphere. A series of large, volcanic eruptions might help trigger an ice age.

- **Changes in the positions of continents.** Plate-tectonic movement of continents closer to the poles increases the likelihood of glaciation. Movement of Northern Hemisphere continents closer to the North Pole has placed landmasses in a position more favorable for glaciation.

- **Changes in circulation of sea water.** The surface of the Arctic Ocean freezes during the winter because bordering landmasses block its circulation with the warmer Atlantic Ocean. One hypothesis speculates that an ice age might have begun at a time when there was free circulation between the oceans. The Arctic Ocean surface would not have frozen over, and the increased moisture in the air would have resulted in greater snowfall to the adjacent continents.

The directions of ice flow can be determined from the orientation of striations and grooves in the bedrock and from elongate, rounded knobs. The largest ice sheet (the Laurentide Ice Sheet) moved outward from the general area now occupied by Hudson Bay, which is where the ice sheet was thickest. The present generally barren surface of the Hudson Bay area contrasts markedly with the Great Plains surface of southern Canada and northern United States, where vast amounts of till were deposited.

Most of the till was deposited as ground moraine, which, along with outwash deposits, has partially weathered to yield excellent soil for agriculture. Rock flour that originally washed out of ice sheets has been redistributed by wind, as *loess,* over large parts of the Midwest and eastern Washington to contribute to especially good agricultural land (see chapter 13). In many areas along the southern boundaries of land covered by ground moraines, broad and complex end moraines extend for many kilometers, indicating that the ice margin must have been close to stationary for a long time (figure 12.34). Numerous drumlins are preserved in areas such as Ontario, New England, and upstate New York. New York's Long Island is made of terminal and recessional moraines and outwash deposits. Erratics there come from metamorphic rock in New England. Cape Cod in Massachusetts was also formed from moraines.

Glaciers have a tremendous capability for forming lakes through both erosion and deposition. Most states and provinces that were glaciated have thousands of lakes. By contrast, Virginia, which was not glaciated, has only two natural lakes. Minnesota bills itself as "the land of 10,000 lakes." Most of those lakes are kettle lakes. The Finger Lakes in New York (figure 12.35) are in long, north-south glacially modified valleys that are dammed by recessional moraines at their southern ends. The Great Lakes are, at least in part, a legacy of continental glaciation. Former stream valleys were widened by the ice sheet eroding weak layers of sedimentary rock into the present lake basins. End moraines border the Great Lakes, as shown in figure 12.34. Large regions of Manitoba, Saskatchewan, North Dakota, and Minnesota were covered by ice-dammed lakes. The largest of these is called Lake Agassiz. The former lake beds are now rich farmland.

Alpine glaciation was much more extensive throughout the world during the glacial ages than it is now. For example, small glaciers in the Rocky Mountains that now barely extend beyond their cirques were then valley glaciers 10, 50, or 100 kilometers in length. Yosemite Valley, which is no longer glaciated, was filled by a glacier about a kilometer thick. Its terminus was at an elevation of about 1,300 meters above sea level. Furthermore, cirques and other features typical of valley glaciers can be found in regions that at present have no glaciers, such as the northern Appalachians—notably in the White Mountains of New Hampshire.

Indirect Effects of Past Glaciation

As the last continental ice sheet wasted away, what effects did the tremendous volume of meltwater have on American rivers? Rivers that now contain only a trickle of water were huge in the glacial ages. Other river courses were blocked by the ice sheet or clogged with morainal debris. Large dry stream channels have been found that were preglacial tributaries to the Mississippi and other river systems.

Pluvial Lakes

During the glacial ages, the climate in North America, even beyond the glaciated parts, was more humid than it is now. Most of the presently arid regions of the western United States had moderate rainfall, as traces or remnants of numerous lakes indicate. These **pluvial lakes** (formed in a period of abundant rainfall) once existed in Utah, Nevada, and eastern California (figure 12.34). Some may have been fed by meltwater from mountain glaciers, but most were simply the result of a wetter climate.

Great Salt Lake in Utah is but a small remnant of a much larger body of fresh water called Lake Bonneville, which, at its maximum size, was nearly as large as Lake Michigan is today. Ancient beaches and wave-cut terraces on hillsides indicate the depth and extent of ancient Lake Bonneville. As the climate became more arid, lake levels lowered, outlets were cut off, and the water became salty, eventually leaving behind the Bonneville salt flats and the present very saline Great Salt Lake (see figure 6.24).

FIGURE 12.34

End moraines in the contiguous United States and Canada (shown by brown lines), and glacial Lake Agassiz and pluvial lakes in the western United States (purple). *After C. S. Denny, U.S. Geological Survey, and the Geological Map of North America, Geological Society of America, and the Geological Survey of Canada*

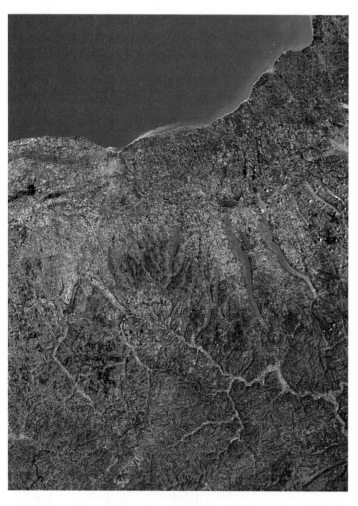

FIGURE 12.35

Satellite image of Finger Lakes in New York. Part of Lake Ontario is at the top. *Photo © Advanced Satellite Productions, Inc.*

Even Death Valley in California—now the driest and hottest place in the United States—was occupied by a deep lake during the Pleistocene. The salt flats that were left when this lake dried include rare boron salts that were mined during the pioneer days of the American West.

Lowering of Sea Level

All of the water for the great glaciers had to come from somewhere. The water was "borrowed" from the oceans, such that sea level worldwide was lower than it is today—at least 130 meters lower, according to scientific estimates.

Recall that if today's ice sheets were to melt, sea level worldwide would rise by over 60 meters, and shorelines would be considerably further inland. It's important to realize that our present shorelines are not fixed and are very much controlled by climate changes. We should also realize that we are still in a cooler than usual (relative to most of Earth's history) time, perhaps the lingering effects of the last ice age.

What is the evidence for lower sea level? Stream channels have been charted in the present continental shelves, the gently inclined, now submerged edges of the continents (described in chapter 18 on the sea floor). These submerged channels are continuations of today's major rivers and had to have been above sea level for stream erosion to take place. Bones and teeth from now-extinct mammoths and mastodons have been dredged up from the Atlantic continental shelf, indicating that these relatives of elephants roamed over what must have been dry land at the time.

A **fiord** (also spelled fjord) is a coastal inlet that is a drowned glacially carved valley (figure 12.36). Fiords are common along the mountainous coastlines of Alaska, British Columbia, Chile, New Zealand, and Norway. Surprisingly, the lower reach of the Hudson River, just north of New York City,

FIGURE 12.36

Lysenfjord, a fiord in Norway. Underwater it is a deep, U-shaped valley. *Photo © Jan Stromme/Alamy*

IN GREATER DEPTH 12.6

The Channeled Scablands

In chapter 4, we described how the Columbia plateau in the Pacific Northwest (see figure 4.27) was built by a series of successive lava floods. The northeastern part of the plateau features a unique landscape, known as the channeled scablands, where the basalt bedrock has been carved into a series of large, interweaving valleys. From the air, the pattern looks like that of a giant braided stream. The channels, however, which range up to 30 kilometers wide and are usually 15 to 30 meters deep, are mostly dry.

The scablands are believed to have been carved by gigantic floods of water. Huge ripples in gravel bars (box figure 1) support this idea. To create these ripples, a flood would have to be about 10 times the combined discharge of all the world's rivers. This is much larger than any flood in recorded history.

What seems to have occurred is that, during the ice ages, a lobe of the ice sheet extended southward into northern Washington, Montana, and Idaho, blocking the head of the valley occupied by the Clark Fork River. The ice provided a natural dam for what is now known as Glacial Lake Missoula. Lake Missoula drowned a system of valleys in western Montana that extended hundreds of kilometers into the Rocky Mountains.

Ice is not ideal for building dams. Upon failure of the glacial dam, the contents of Lake Missoula became the torrential flood that scoured the Columbia plateau. There were dozens of giant floods. Advancing ice from Canada would reestablish the dam, only to be destroyed after the reservoir refilled.

Mars has what appear to be giant outflow channels that are similar to those of the channeled scablands. At present, Mars has no liquid water, but the channels suggest that there must have been a huge amount of water in the distant past.

Additional Resource

Ice Age Flood Home Site

- www.iceagefloodsinstitute.org/

BOX 12.6 ■ FIGURE 1
Giant ripples of gravel from the draining of Lake Missoula, Montana. For scale, see farm buildings in lower middle of photo.
Photo by P. Weiss, U.S. Geological Survey

is a fiord. Fiords are evidence that valleys eroded by past glaciers were later partly submerged by rising sea level.

Crustal Rebound

The weight of an ice sheet several thousand meters thick depresses the crust of Earth much as the weight of a person depresses a mattress. A land surface bearing the weight of a continental ice sheet may be depressed several hundred meters.

Once the glacier is gone, the land begins to rebound slowly to its previous height (see chapter 17 and figures 17.13 and 17.14). Uplifted and tilted shorelines along lakes are an indication of this process. The Great Lakes region is still rebounding as the crust slowly adjusts to the removal of the last ice sheet.

Evidence for Older Glaciation

Throughout most of geologic time, the climate has been warmer and more uniform than it is today. We think that the late Cenozoic Era is unusual because of the periodic fluctuations of climate and the widespread glaciations. However, glacial ages are not restricted to the late Cenozoic.

Evidence of older glaciation comes from rocks called tillites. A **tillite** is lithified till (figure 12.37). Unsorted rock particles, including angular, striated, and faceted boulders, have been consolidated into a sedimentary rock. In some places, tillite layers overlie surfaces of older rock that have

been polished and striated. Tillites of the late Paleozoic and tillites representing a minor part of the late Precambrian crop out in parts of the southern continents. (The striated surface in Australia, shown in figure 12.13, is overlain in places by late Paleozoic tillite.)

The oldest glaciation, for which we have evidence, appears to have taken place in what is now Ontario around 2.3 billion years ago.

Support is growing for the idea that a late Precambrian Ice Age was so extensive that the surfaces of the world's oceans were frozen. Although the concept was first proposed in the early twentieth century, scientists in the 1990s began taking it seriously and called it the *snowball Earth hypothesis*. Evidence for the hypothesis includes tillites that must, at the time, have been deposited near the equator. The hypothesis proposes that the extreme cold was due to the Sun being weaker at the time and the absence of carbon dioxide and other greenhouse gases in the atmosphere.

Paleozoic glaciation provides strong support for plate tectonics. The late Paleozoic tillites in the southern continents (South Africa, Australia, Antarctica, South America) indicate that these landmasses were once joined (see chapter 19 on plate tectonics). Directions of striations show that an ice sheet flowed onto South America from what is now the South Atlantic Ocean. Because an ice sheet can build up only on land, it is reasonable to conclude that the former ice sheet was centered on the ancient supercontinent.

FIGURE 12.37

The Dwyka tillite in South Africa. It is of Permian (late Paleozoic) age. Similar Permian tillites found in South America, Australia, and India are used as evidence for the existence of a super continent and continental drift (see figure 19.4). *Photo by Robert J. Stull*

SUMMARY

A *glacier* is a large, long-lasting mass of ice that forms on land and moves under its own weight. A glacier can form wherever more snow accumulates than is lost. *Ice sheets* and *valley glaciers* are the two most common types of glaciers. Glaciers move downward from where the most snow accumulates toward where the most ice is wasted.

A glacier moves by both basal sliding and internal flow. The upper portion of a glacier tends to remain rigid and is carried along by the ice moving beneath it.

Glaciers advance and recede in response to changes in climate. A receding glacier has a *negative budget,* and an advancing one has a *positive budget.* A glacier's budget for the year can be determined by noting the relative position of the *equilibrium line.*

Snow recrystallizes into firn, which eventually becomes converted to glacier ice. Glacier ice is lost (or ablated) by melting, breaking off as icebergs, and direct evaporation of the ice into the air.

A glacier erodes by plucking and the grinding action of the rock it carries. The grinding produces rock flour and faceted and polished rock fragments. Bedrock over which a glacier moves is generally polished, striated, and grooved.

A mountain area showing the erosional effects of alpine glaciation possesses relatively straight valleys with U-shaped cross profiles. The floor of a glacial valley usually has a *cirque* at its head and descends as a series of rock steps. Small *rock-basin lakes* are commonly found along the steps and in cirques.

A *hanging valley* indicates that a smaller tributary joined the main glacier. A *horn* is a peak between several cirques. *Arêtes* usually separate adjacent glacial valleys.

A glacier deposits unsorted rock debris or *till,* which contrasts sharply with the sorted and layered deposits of glacial *outwash.* Till forms various types of *moraines.*

Fine silt and clay may settle as *varves* in a lake in front of a glacier, each pair of layers representing a year's accumulation.

Multiple till deposits and other glacial features indicate several major episodes of glaciation during the late Cenozoic Era. During each of these episodes, large ice sheets covered most of northern Europe and northern North America, and glaciation in mountain areas of the world was much more extensive than at present. At the peak of glaciation, about a third of Earth's land surface was glaciated (in contrast to the 10% of the land surface presently under glaciers). Warmer climates prevailed during interglacial episodes.

The glacial ages also affected regions never covered by ice. Because of wetter climate in the past, large lakes formed in now-arid regions of the United States. Sea level was considerably lower.

Glacial ages also occurred in the more distant geologic past, as indicated by late Paleozoic and Precambrian tillites.

Terms to Remember

ablation 311
advancing glacier 311
alpine glaciation 308
arête 322
basal sliding 313
cirque 321
continental glaciation 308
crevasse 314
drumlin 324
end moraine 323
equilibrium line 311
erratic 322
esker 326
fiord 332
glacier 308

ground moraine 324
hanging valley 319
horn 321
iceberg 311
ice cap 309
ice sheet 309
kettle 326
lateral moraine 323
medial moraine 323
moraine 323
outwash 326
plastic flow 313
pluvial lake 331
receding glacier 311
rigid zone 314

rock-basin lake 320
rock flour 316
tarns 320
terminus 312
theory of glacial ages 329
till 322
tillite 334
truncated spur 320
U-shaped valley 319
valley glacier 309
varve 327
zone of ablation 311
zone of accumulation 311

Testing Your Knowledge

Use the following questions to prepare for exams based on this chapter.

1. How do erosional landscapes formed beneath glaciers differ from those that developed in rock exposed above glaciers?

2. How do features caused by stream erosion differ from features caused by glacial erosion?

3. How does material deposited by glaciers differ from material deposited by streams?

4. Why is the North Pole not glaciated?

5. How do arêtes, cirques, and horns form?

6. How does the glacial budget control the migration of the equilibrium line?

7. How do recessional moraines differ from terminal moraines?

8. Alpine glaciation

   a. is found in mountainous regions

   b. exists where a large part of a continent is covered by glacial ice

   c. is a type of glacier

   d. none of the preceding

9. Continental glaciation

   a. is found in mountainous regions

   b. exists where a large part of a continent is covered by glacial ice

   c. is a glacier found in the subtropics of continents

   d. none of the preceding

10. At present, about _____% of the land surface of the Earth is covered by glaciers.

    a. 1/2   b. 1

    c. 2   d. 10

    e. 33   f. 50

11. Which is not a type of glacier?

    a. valley glacier   b. ice sheet

    c. ice cap   d. sea ice

12. The boundary between the zone of accumulation and the zone of ablation of a glacier is called the

    a. firn   b. equilibrium line

    c. ablation zone   d. moraine

13. In a receding glacier

    a. ice flows from lower elevations to higher elevations

    b. the terminus moves upvalley

    c. the equilibrium line moves to a lower elevation

    d. all of the preceding

14. Recently, geologists have been drilling through ice sheets for clues about

    a. ancient mammals   b. astronomical events

    c. extinctions   d. past climates

15. Glacially carved valleys are usually _____ shaped.

    a. V   b. U

    c. Y   d. all of the preceding

16. Which is not a type of moraine?

    a. medial   b. end

    c. terminal   d. recessional

    e. ground   f. esker

17. The last episode of extensive glaciation in North America was at its peak about _____ years ago.

    a. 2,000   b. 5,000

    c. 10,000   d. 18,000

18. How fast does the central part of a valley glacier move compared to the sides of the glacier?

    a. faster   b. slower

    c. at the same rate

19. During the Ice Ages, much of Nevada, Utah, and eastern California was covered by

    a. ice   b. huge lakes

    c. deserts   d. the sea

Expanding Your Knowledge

1. How might a warming trend cause increased glaciation?

2. How do, or do not, the Pleistocene glacial ages fit in with the principle of uniformitarianism?

3. Is ice within a glacier a mineral? Is a glacier a rock?

4. Could a rock that looks like a tillite have been formed by any agent other than glaciation?

5. What is the likelihood of a future glacial age? What effect might human activity have on causing or preventing a glacial age?

Exploring Web Resources

www.mhhe.com/plummer11e

This is the dedicated website for this book. You can go to it for new and updated information. The universal resource locators (URL) listed in this book are also given as links on the website, making it easy to go to those websites without typing in the URL. When you visit our Online Learning Center, go to the Student Center and then to the chapter of interest. Here you will find additional readings and media resources, as well as answers to the Testing Your Knowledge section, more quizzes, animation and video clips, and direct links to the sites listed in this book. Links to additional websites can also be found. We have added questions for some of the sites to allow you to get the most of your exploration of the web. Using the web is an enjoyable way of enhancing your knowledge of geology.

http://dir.yahoo.com/science/earth_sciences/geology_and_ geophysics/glaciology/

Glaciers and Glaciology—list of sites. This site provides links and descriptions of numerous icy websites.

www.glacier.rice.edu/

Glacier. Explore Antarctica on Rice University's site. Go to "Ice." There are many topics you can go to for information that expands on that covered in this book. Examples are "How Do Glaciers Move?" "How Do Glaciers Change the Land?" and "What Causes Ice Ages?"

www.crevassezone.org/

Glacier movement studies on the Juneau Icefield, Alaska. Go to "Photo Gallery" to view photos of glacial features and other aspects of the project.

www.museum.state.il.us/exhibits/ice_ages/

Ice Ages. Illinois State Museum's virtual ice ages exhibit. The site features a tape clip showing the retreat of glaciers during the last ice age. You can download the video clip by going to:

www.museum.state.il.us/exhibits/ice_ages/laurentide_ deglaciation.html

http://nsidc.org/cryosphere/index.html

The Cryosphere. General information on snow and ice. You can link to pages on glaciers, avalanches, and icebergs.

 # Animations

This chapter includes the following animations available on our Online Learning Center at **www.mhhe.com/plummer11e**.

12.6   Dynamics of glacial advance and retreat
12.9   Crevasse formation in glaciers
12.28   Formation of glacial features by deposition at a wasting ice front
12.33   Glacial maximum and deglaciation

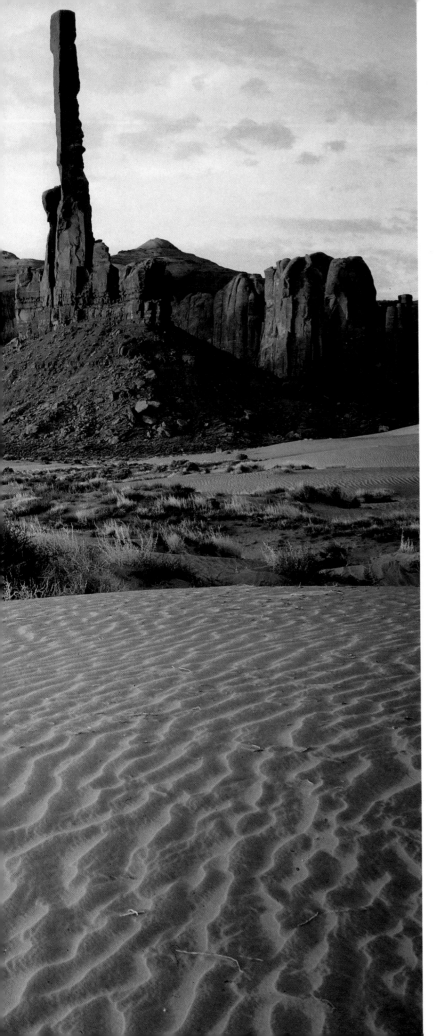

Deserts and Wind Action

Distribution of Deserts
Some Characteristics of Deserts
Desert Features in the Southwestern United
    States
Wind Action
    Wind Erosion and Transportation
    Wind Deposition
Summary

In chapters 9 through 12, you have seen how the land is sculptured by mass wasting, streams, ground water, and glaciers. Here, we discuss the fifth agent of erosion and deposition: wind. Deserts and wind action are discussed together because of the wind's particular effectiveness in dry regions. But wind erosion and deposition can be very significant in other climates as well.

Deserts have a distinctive appearance because a dry climate controls erosional and depositional processes and the rates at which they operate. Although it seldom rains in the desert, running water is actually the dominant agent of land sculpture. Flash floods cause most desert erosion and deposition, even though they are rare events.

The word *desert* may suggest shifting sand dunes. Although moviemakers usually film sand dunes to represent deserts, only small portions of most deserts are covered with dunes. Actually, a **desert** is any region with low rainfall. A region is usually classified as a desert if it has

Totem Pole Buttes, Monument Valley Tribal Park, Navajo Nation, Arizona.
Photo © Craig Blacklock/Larry Ulrich Stock

a dry or *arid climate* with less than 25 centimeters of rain per year. The biosphere of deserts reflects the dryness of the air and the infrequency of precipitation. Few plants can tolerate low rainfall, so most deserts look barren.

Some specialized types of plants, however, grow well in desert climates despite the dryness. These plants are generally salt-tolerant, and they have extensive root systems to conserve water, so they often are widely spaced (figure 13.1). The leaves are usually very small, minimizing water loss by transpiration; they may even drop off the plants between rainstorms. During much of the year, many desert plants look like dead, dry sticks. When rain does fall on the desert, the plants become green, and many will bloom.

The lack of vegetation affects the geosphere because debris between plants is not anchored by roots, making loose material susceptible to mechanical weathering and erosion.

FIGURE 13.1

A scene from the Mojave Desert in southern California showing widely spaced plants that have adapted to less than 25 centimeters of rain per year. *Photo by Diane Carlson*

DISTRIBUTION OF DESERTS

Obviously, deserts are related to the atmosphere, because the climate is caused by circulating air that usually is dry. Deserts, the atmosphere, and the hydrosphere are interrelated in various ways. Water from the hydrosphere must evaporate to become part of the atmosphere. Air over oceans tends to be moist. If conditions cause that air to become dry before it circulates to an arid region, precipitation will not take place.

The location of most deserts is related to descending air. The global pattern of air circulation is shown in simplified form in figure 13.2. The equator receives the Sun's heat more directly than the rest of the Earth. Air warms and rises at the equator, then moves both northward and southward to sink near 30° North latitude and 30° South latitude. The world's best-known deserts lie in a belt 10°–15° wide centered on 30° North and South latitude (figure 13.3).

Air sinking down through the atmosphere is compressed by the weight of the air above it. As air compresses, it warms up, and as it warms, it is able to hold more water vapor. Evaporation of water from the land surface into the warm, dry air is so great under belts of sinking air that moisture seldom falls back to Earth in the form of rain. The two belts at 30° North and South latitude characteristically have clear skies, much sunshine, little rain, and high evaporation.

In contrast to the belts centered on 30°, the equator is marked by rising air masses that expand and cool as they rise. In cooling, the air loses its moisture, causing cloudy skies and heavy precipitation. Thus, a belt of high rainfall at the equator separates the two major belts of deserts.

Not all deserts lie near the 30° latitude belts. Interaction with the geosphere also helps control the location of deserts. Some of the world's deserts are the result of the **rain shadow**

effect of mountain ranges (figure 13.4). As moist air is forced up to pass over a mountain range, it expands and cools, losing moisture by condensation as it rises. The dry air coming down the other side of the mountain compresses and warms, bringing high evaporation with little or no rainfall to the downwind side of the range. This dry region downwind of mountains is the *rain shadow zone*. Parts of the southwestern United States desert in Nevada and northern Arizona are largely the result of the rain shadow effect of the Sierra Nevada range in eastern California.

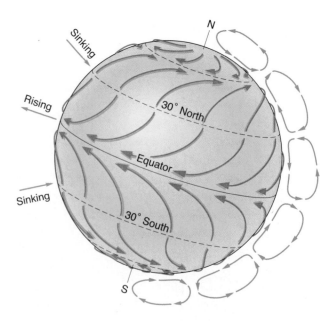

FIGURE 13.2

Global air circulation. Red arrows show surface winds. Blue arrows show vertical circulation of air. Air sinks at 30°N and 30°S latitude (and at the poles).

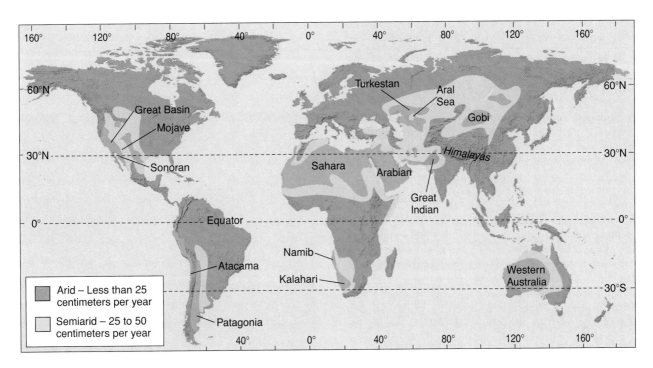

FIGURE 13.3

World distribution of nonpolar deserts. Most deserts lie in two bands near 30°N and 30°S. *Map adapted from U.S. Department of Agriculture*

Great distance from the ocean is another factor that can create deserts, since most rainfall comes from water evaporated from the sea. The dry climate of the large arid regions in China, well north of 30° North latitude, is due to their location in a continental interior and to the rain shadow effect of mountains such as the Himalayas.

Deserts also tend to develop on tropical coasts next to *cold ocean currents*. Cold currents run along the western edges of continents, cooling the air above them. The cold marine air warms up as it moves over land, causing high evaporation and little rain on the coasts. This effect is particularly pronounced on the Pacific coast of South America and the Atlantic coast of Africa and to a lesser extent western Australia.

Not all deserts are hot. The cold, descending air near the North and South Poles (figure 13.2) creates *polar deserts* that have an arid climate along with a snow or ice cover. The entire continent of Antarctica is a desert, as are most of Greenland and the northernmost parts of Alaska, Canada, and Siberia.

SOME CHARACTERISTICS OF DESERTS

Because of their low rainfall, deserts have characteristic drainage and topography that differ from those of humid regions. Desert streams usually flow intermittently. Water runs over the surface after storms, but during most of the year, stream beds are dry. As a result, most deserts *lack through-flowing streams.* The Colorado River in the southwestern United States and the Nile River in Egypt are notable exceptions. Both are fed by heavy rainfall in distant mountains. The runoff is great enough to sustain streamflow across dry regions with high evaporation.

Many desert regions have *internal drainage;* the streams drain toward landlocked basins instead of toward the sea. The surface of an enclosed basin acts as a local base level. Because each basin is generally filled to a different level than the neighboring basins, desert erosion may be controlled by many different *local base levels.* As a basin fills with sediment, its

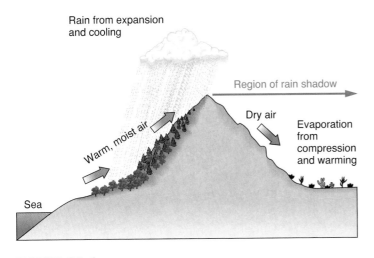

FIGURE 13.4

Rain shadow causes deserts on the downwind side of mountain ranges. Prevailing winds are from left to right.

ENVIRONMENTAL GEOLOGY 13.1

Expanding Deserts

Many geologists and geographers use a two-part definition of *desert*. A desert must have less than 25 centimeters of rain per year or must be so devoid of vegetation that few people can live there. Many dry regions have supported marginally successful agriculture and moderate human populations in the past but are being degraded into barren deserts today by overgrazing, overpopulation, and water diversions. The expansion of barren deserts into once-populated regions is called *desertification.*

Limited numbers of people can exist in dry regions through careful agricultural practices that protect water sources and limit grazing of sparse vegetation. Overuse of the land by livestock and humans, however, can strip it bare and make it uninhabitable. The large desert in northern Africa, shown in figure 13.3, is the Sahara Desert, and the semiarid region (25 to 50 centimeters of rain per year) to the south of it is the Sahel. In the early 1960s, a series of abnormally wet years encouraged farmers in the Sahel to expand their herds and grazing lands.

A severe drought throughout the 1970s and 1980s caused devastation of the plant life of the region as starving livestock searched desperately for food, and humans gathered the last remaining sticks for firewood (box figure 1). Vast areas that were once covered with trees and sparse grass became totally barren, and an acute famine began, killing more than 100,000 people. The desert expanded southward, advancing in some places as much as 50 kilometers per year. The denuded soil in many regions became susceptible to wind erosion, leading to choking dust storms and new, advancing dune fields (some even migrating into cities).

BOX 13.1 ■ FIGURE 2

Fishing boats marooned in a sea of sand as the Aral Sea decreases in size due to diversion of water for irrigation. *Photo © David Turnley/Corbis Images*

Some of the same problems afflicted the midwestern United States in the 1930s, as intense land cultivation coupled with a prolonged drought produced the barren Dust Bowl during the time of the Great Depression. Renewed rains and improved soil-conservation practices have reversed the trend in the United States, but the area is still vulnerable to a future drought, and the possibility of future dust storms in prairie states is very real.

Drought accelerates desertification but is not necessary for it to occur. Overloading the land with livestock and humans can strip marginal regions of vegetation even in wet years.

Diversion of rivers for agricultural use can also cause desertification. Such is the case in the Turkestan Desert where two rivers feeding the Aral Sea (figure 13.3) were diverted to provide irrigation water for agriculture. The Aral Sea, once the fourth-largest inland water body in the world, has decreased in size by nearly half since 1960. Fishing boats are now marooned in a sea of sand as the shoreline has migrated tens of kilometers with the sea continuing to shrink (box figure 2).

Additional Resource

• http://pubs.usgs.gov/gip/deserts/desertification/

Good overview of desertification around the world. Includes photographs.

BOX 13.1 ■ FIGURE 1

Desertification in Africa after a period of drought has left inhabitants desperately searching for water. *Photo © Walt Anderson/Visuals Unlimited*

FIGURE 13.5

This van was swept away in a flash flood that occurred in Death Valley National Park on August 15, 2004 after a cloudburst dropped .33 inches of rain in only 20 minutes. Two people died in the flash flood and the storm caused extensive damage to roads and water and sewer lines, and closed the park for over a week. *Photo Courtesy Caltrans*

FIGURE 13.6

Badlands topography (sharp ridges and V-shaped channels) eroded on shale in a dry region where plants are scarce. Badlands National Park, South Dakota. *Photo by Diane Carlson*

surface rises, leading to a *rising base level,* which is a rare situation in humid (wet) regions.

The limited rainfall that does occur in deserts often comes from violent thunderstorms, with a high volume of rain falling in a very short time. Desert thunderstorms may dump more than 13 centimeters of rain in one hour. Such a large amount of rain cannot soak readily into the sun-baked hardpan soil, so the water runs rapidly over the land surface, particularly where vegetation is sparse. This high runoff can create sudden local floods of high discharge and short duration called **flash floods.** Flash floods are more common in arid regions than in humid regions. They can turn normally dry streambeds into raging torrents for a short time after a thunderstorm (figure 13.5). Because soil particles are not held in place by plant roots, these occasional floods can effectively erode the land surface in a desert region. As a result, desert streams normally are very heavily laden with sediment. Flash floods can easily erode enough sediment to become *mudflows* (see chapter 9).

Desert stream channels are distinctive in appearance because of the great erosive power of flash floods and the intermittent nature of streamflow. Most stream channels are normally dry and covered with sand and gravel that is moved only during occasional flash floods. Rapid downcutting by sediment-laden floodwaters tends to produce narrow canyons with vertical walls and flat, gravel-strewn floors (see figure 10.35). Such channels are often called *arroyos* or *dry washes.*

Newcomers to deserts sometimes get into serious trouble in desert canyons in rainy weather. Imagine for a moment that you have camped on the canyon floor to get out of the strong desert winds. Later that night, a towering thunderhead cloud forms, and heavy rain falls on the mountains several miles upstream from you. Although no rain has fallen on you, you are awakened several minutes later by a distant roar. The roar grows louder until a 3-meter "wall" of water rounds a bend in the canyon, heading straight for you at the speed of a galloping horse. Boulders, brush, and tree trunks are being swept along in this raging flash flood. The walls of the canyon are too steep to climb. Several hikers died during the summer of 1997 when such a wall of water roared down a side canyon in Grand Canyon National Park in Arizona. In August 2004, two people died in Death Valley National Park in California when their vehicle was carried away by a flash flood that washed out most of the roads in the park (figure 13.5). Stay out of desert canyons if there is any sign of rain; sleeping in such canyons is particularly dangerous!

The resistance of some rocks to weathering and erosion is partly controlled by climate. In a humid (wet) climate, limestone dissolves easily, forming low places on Earth's surface. In a desert climate, the lack of water makes limestone resistant, so it stands up as ridges and cliffs in the desert just as sandstone and conglomerate do. Lava flows and most igneous and metamorphic rock are also resistant. Shale is the least resistant rock in a desert, so it usually erodes more deeply than other rock types and forms gentler slopes or badland topography (figure 13.6).

Although intersecting joints form angular blocks of rock in all climates, desert topography characteristically looks more angular than the gently rounded hills and valleys of a humid region. This may be due indirectly to the low rainfall in deserts. Shortage of water slows chemical weathering processes to the point where few minerals break down to form fine-grained clay minerals. Soils are coarse and rocky with few chemically weathered products. Plants, which help bind soil into a cohesive layer in humid climates, are rare in deserts, and so desert soils are easily eroded by wind and rainstorms. Downhill creep of thick, fine-grained soil is partly responsible for softening the appearance of jointed topography in humid climates. With thin, rocky soil and slow rates of creep, desert topography remains steep and angular.

Climate is only one of many things that determine the shape and appearance of the land. Rock structure is another. As an example, in the next section we will look closely at two different structural regions within the desert of the southwestern United States.

DESERT FEATURES IN THE SOUTHWESTERN UNITED STATES

Much of the southwestern United States has an arid (or semi-arid) climate, partly because it is close to 30° North latitude and partly because of the rain shadow effect of the Sierra Nevada and other mountain ranges. Within this region of low rainfall are two areas of markedly different geologic structure. One area is the Colorado Plateau and the other is the Basin and Range province, a mountainous region centered on the state of Nevada. The boundaries of these two areas are shown in figure 13.7.

The *Colorado Plateau* centers roughly on the spot known as the Four Corners, where the states of Utah, Colorado, Arizona, and New Mexico meet at a common point. The rocks near the surface of the Colorado Plateau are mostly flat-lying beds of sedimentary rock over 1,500 meters above sea level. These rocks are well exposed at the Grand Canyon in Arizona.

Because the rock layers are well above sea level, they are vulnerable to erosion by the little rain that does fall in the region. Flat-lying layers of resistant rock, such as sandstone, limestone, and lava flows, form **plateaus**—broad, flat-topped areas elevated above the surrounding land and bounded, at least in part, by cliffs. As erosion removes the rock at its base, the cliff is gradually eroded back into the plateau (figure 13.8). Remnants of the resistant rock layer may be left behind, forming flat-topped mesas or narrow buttes (figure 13.8*A*). A **mesa** is a broad, flat-topped hill bounded by cliffs and capped with a resistant rock layer. A **butte** is a narrow hill of resistant rock with a flat top and very steep sides. Most buttes form by continued erosion of mesas. (The term *butte* is also used in other parts of the country for any isolated hill.)

The Colorado Plateau is also marked by peculiar, steplike folds (bends in rock layers) called *monoclines*. Erosion of monoclines (and other folds) leaves resistant rock layers protruding above the surface as ridges (figure 13.9). A steeply tilted resistant layer erodes to form a *hogback*, a sharp ridge that has steep slopes. A gently tilted resistant layer forms a *cuesta*, with one steep side and one gently sloping side.

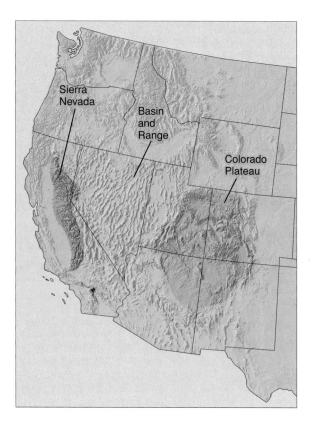

FIGURE 13.7

The Colorado Plateau and the Basin and Range province in the southwestern United States. *After Thelin and Pike, U.S. Geological Survey*

Note that plateaus, monoclines, hogbacks, and cuestas are not unique to deserts. They are surface features found in all climates but are particularly well exposed in deserts because of thin soil and sparse vegetation.

The *Basin and Range province* is characterized by rugged mountain ranges separated by flat valley floors (figure 13.10). The blocks of rock that form the mountain ranges and the valley floors are bounded by **faults,** fractures in the ground along which some movement has taken place. (Chapter 15 discusses faults in more detail.) In the Basin and Range province, movement on the faults has dropped the valleys down relative to the adjacent mountain ranges to accommodate crustal thinning and extension (figure 13.11). Fault-controlled topography is found throughout the Basin and Range province, which covers almost all of Nevada and portions of bordering states as well as New Mexico and a small portion of Texas (figure 13.7). The numerous ranges in this province create multiple rain shadow zones and therefore a very dry climate.

Heavy rainfall from occasional thunderstorms in the mountain ranges causes rapid erosion of the steep mountain fronts and resulting deposition on the valley floors (figure 13.11). Rock debris from the mountains, picked up by flash floods and mudflows, is deposited at the base of the mountain ranges in the form of **alluvial fans.** Alluvial fans (described in chapter 10) build up where stream channels abruptly widen as they flow

A

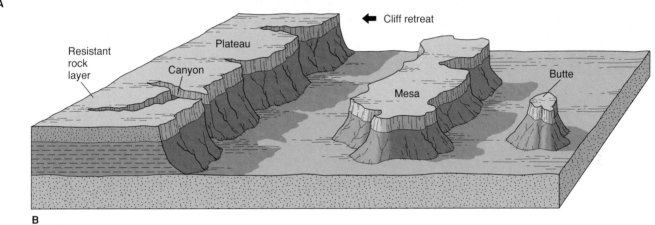

B

FIGURE 13.8

Characteristic landforms of the Colorado Plateau (*A*) Mesas and buttes in Monument Valley, Arizona, an area of eroded, horizontal, sedimentary rocks. (*B*) Erosional retreat of a cliff at the edge of a plateau can leave behind mesas and buttes as erosional remnants of the plateau. *Photo by David McGeary*

FIGURE 13.9

(*A*) Steplike monocline folds often erode so that resistant rock layers form hogbacks and cuestas (these features are not unique to deserts). (*B*) Monocline near Mexican Hat, Utah. Rocks at the top and bottom are horizontal and rocks in the center are steeply tilted and step down toward the right. *Photo © Marli Miller/Visuals Unlimited*

A

B

FIGURE 13.10

Basin and Range topography in Death Valley, California. In the distance, the fault-bounded Panamint Mountains rise more than 3 kilometers (11,000 feet) above Death Valley. Giant alluvial fans at the base of the mountains show a braided stream pattern. Fine-grained sediments and salt deposits underlie the playa in the foreground. The alluvial fans coalesce to form a bajada. *Photo by David McGeary*

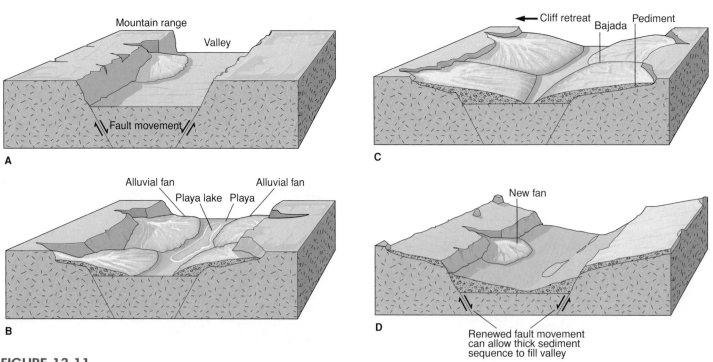

FIGURE 13.11

Development of landforms associated with Basin and Range topography.

346

out of narrow canyons onto the open valley floors, causing a decrease in velocity and rapid deposition of sediment.

Although most of the sediment carried by runoff is deposited in alluvial fans, some fine clay may be carried in suspension onto the flat valley floor. If no outlet drains the valley, runoff water may collect and form a **playa lake** on the valley floor. Playa lakes are usually very shallow and temporary, lasting for only a few days after a rainstorm. After the lake evaporates, a thin layer of fine mud may be left on the valley floor. The mud dries in the sun, forming a **playa,** a very flat surface underlain by hard, mud-cracked clay (figures 13.12 and 13.13). If the runoff contained a large amount of dissolved salt or if seeping ground water brings salt to the surface, the flat playa surface may be underlain by a bright white layer of dried salt instead of mud, as on the Bonneville Salt Flats in Utah (see figure 6.24).

Continued deposition near the base of the mountains may create a **bajada,** a broad, gently sloping depositional surface formed by the coalescing of individual alluvial fans (figures 13.10 and 13.11). A bajada is much more extensive than a single alluvial fan and may have a gently rolling surface resulting from the merging of the cone-shaped fans.

Erosion of the mountain can eventually form a **pediment,** which is a gently sloping surface, commonly covered with a veneer of gravel, cut into the solid rock of the mountain (figure 13.11). A pediment develops uphill from a bajada as the mountain front retreats. It can be difficult to distinguish a pediment from the surface of the bajada downhill, because both have the same slope and gravel cover. The pediment, however, is an erosional surface, usually underlain by solid rock, while the bajada surface is depositional and may be underlain by hundreds of meters of sediment.

An abrupt change in slope marks the upper limit of the pediment, where it meets the steep mountain front. Many geologists who have studied desert erosion believe that as this steep mountain front erodes, it retreats uphill, maintaining a relatively constant angle of slope.

Notice that rock structure, not climate, largely controls the fact that plateaus and cliffs are found in the Colorado Plateau, while mountain ranges, broad valleys, alluvial fans, and pediments are found in the Basin and Range province. Features such as plateaus, mesas, and alluvial fans can also be found in humid climates wherever the rock structure is favorable to their development; they are not controlled by climate. Features such as steep canyons, playa lakes, thin soil, and sparse vegetation, however, *are* typically controlled by climate.

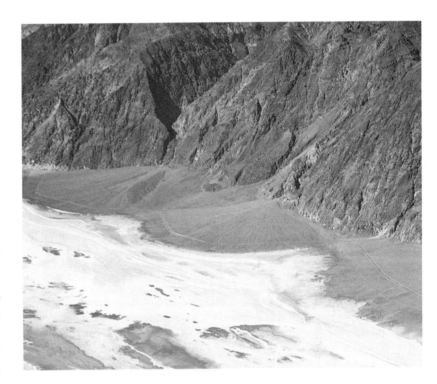

FIGURE 13.12

Alluvial fans at the base of mountain canyons, Death Valley, California. The white salt flat in the foreground is part of a playa. *Photo by Frank M. Hanna*

FIGURE 13.13

Mud-cracked playa surface. *Photo © Bill Ross/Westlight/Corbis*

WIND ACTION

Wind can be an important agent of erosion and deposition in any climate, as long as sediment particles are loose and dry. Wind differs from running water in two important ways. Because air is less dense than water, wind can remove only fine sediment—sand, silt, and clay. But wind is not confined to channels as running water is, so wind can have a widespread effect over vast areas.

In general, the faster the wind blows, the more sediment it can move. Wind velocity is determined by differences in air pressure caused by differences in air temperature. As air warms and cools, it changes density, and these density changes create air pressure differences that cause wind. Wet climates and cloud cover help buffer changes in air temperature, but in dry climates, daily temperature changes can be extreme. In a desert, the temperature may range from 10°C (50°F) at night to more than 40°C (100°F) in the daytime. Because of these temperature fluctuations, wind is generally stronger in deserts than in humid regions, commonly exceeding 100 kilometers per hour (60 miles per hour). The scarcity of vegetation in deserts to slow wind velocity by friction increases the effectiveness of desert winds.

Although strong winds are also associated with rainstorms and hurricanes, these winds seldom erode sediment because rain wets the surface sediment. Wet sediment is heavy and cohesive and will not be blown away. Strong winds in the desert, however, blow over loose, dry sediment, so wind is an effective erosional agent in dry climates. (As we said earlier in the chapter, running water in the form of flash floods is a far more important erosive agent than wind, even though the wind can be very strong.)

Wind Erosion and Transportation

Thick, choking *dust storms* are one example of wind action (figure 13.14). "Dust Bowl" conditions in the 1930s in the agricultural prairie states lasted for several years due to droughts and poor soil-conservation practices. Loose silt and clay are easily picked up from barren dry soil, such as in a cultivated field. Wind erosion is even greater if the soil is disturbed by animals or vehicles. Silt and clay can remain suspended in turbulent air for a long time, so a strong wind may carry a dust cloud hundreds of meters upward and hundreds of kilometers horizontally. Dust storms of the 1930s frequently blacked out the midday sun; fertile soil was lost over vast regions, ruining many farms; and streets and rivers downwind were filled with thick dust deposits.

Wind-blown sediment is sometimes picked up on land and carried out to sea (figure 13.15). Particles from the Sahara Desert in North Africa have been collected from the air over the islands of the Caribbean after having been carried across the entire Atlantic Ocean. A substantial amount of the fine-grained sediment that settles to the sea floor is land-derived sediment that the wind has deposited on the sea surface. Ships 800 kilometers offshore have reported dustfalls a few millimeters thick covering their decks.

Volcanic ash can be carried by wind for very great distances. An explosive volcanic eruption can blast ash more than 15 kilometers upward into the air. Such ash may be caught in

FIGURE 13.15

Satellite image of a dust storm from the Sahara Desert blowing off the coast of Africa northwestward out into the Atlantic Ocean on February 26, 2000. The thick plume of dust is about the size of Spain, and dust particles from this storm were blown all the way to the west side of the Atlantic. Such storms are fairly common in the Sahara Desert and are the world's greatest supplier of dust. *Photo by NASA/Goddard Space Flight Center, The SeaWiFS Project and ORBIMAGE, Scientific Visualization Studio*

FIGURE 13.14

A wall of dust approaches a town in Kansas in October, 1935. Because of the intensity and duration of the storms in the 1930s, parts of the Great Plains became known as the "Dust Bowl." *Photo by National Oceanic and Atmospheric Administration/ Department of Commerce* the high-altitude *jet streams,* narrow belts of strong winds with velocities sometimes greater than 300 kilometers per hour. Following the 1980 eruption of Mount St. Helens in western Washington, a visible ash layer blanketed parts of Washington, Idaho, and Montana to the east. At high altitudes, St. Helens ash could be detected blowing over New York and out over the Atlantic Ocean, 5,000 kilometers from the volcano. But the St. Helens eruption was a relatively small one. Ash from the 1883 Krakatoa eruption in Indonesia circled the globe for two years, causing spectacular sunsets and a slight, but measurable, drop in temperature as the ash reflected sunlight back into space. The 1815 eruption of Tambora, also in Indonesia, put so much ash into the air that there were summer blizzards, crop failures, and famine in New England and northern Europe in 1816, "the year without a summer." Lower temperatures (1°F) and brilliant sunsets also marked the 1991 eruption of Pinatubo in the Philippines.

Because sand grains are heavier than silt and clay, sand moves close to the ground in the leaping pattern called *saltation* (as does some sediment in streams). High-speed winds can cause *sandstorms,* clouds of sand moving rapidly near the land surface. The high-speed sand in such a storm can sandblast smooth surfaces on hard rock and scour the windshields and paint of automobiles. Because of the weight of the sand grains, however, sand rarely rises more than 1 meter above a flat land surface, even under extremely strong winds. Therefore, most of the sandblasting action of wind occurs close to the ground (figure 13.16). Telephone poles in regions of wind-driven sand often are severely abraded near the ground. To prevent such abrasion, desert residents pile stones or wrap sheet metal around the base of the poles.

Wind seldom moves particles larger than sand grains, but wind-blown sand may sculpture isolated pebbles, cobbles, or boulders into **ventifacts**—rocks with flat, wind-abraded surfaces (figure 13.17). If the wind direction shifts or the stone is turned, more than one flat face may develop on the ventifact.

Deflation

The removal of clay, silt, and sand particles from the land surface by wind is called **deflation.** If the sediment at the land

A

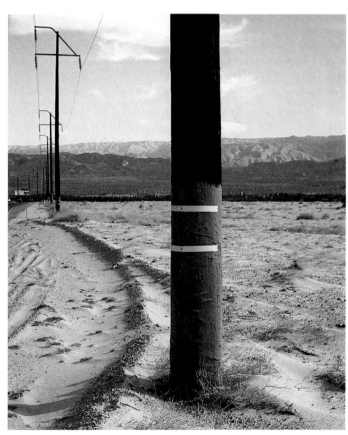

B

FIGURE 13.16

(A) Wind erosion near the ground has sandblasted the lower 1 meter of this chemically weathered basalt outcrop, Death Valley, California. Hammer for scale. (B) Power pole with its base wrapped in an abrasion-resistant material to minimize wind erosion. *Photo A by David McGeary; Photo B courtesy of Paul Bauer*

FIGURE 13.17

Ventifact eroded by sandblasting action of high winds in Death Valley, California. Predominant winds from the south and north (left and right) have sculpted the grooved (fluted) faces. *Photo by Diane Carlson*

surface is made up only of fine particles, the erosion of these particles by the wind can lower the land surface substantially. A **blowout** is a depression on the land surface caused by wind erosion (figure 13.18*A*). A *pillar*, or erosional remnant of the former land, may be left at the center of a blowout.

Blowouts are common in the Great Plains states (figure 13.18*B*). One in Wyoming measures 5 by 15 kilometers and is 45 meters deep. The enormous Qattara Depression

in northwestern Egypt, more than 250 kilometers long and more than 100 meters *below* sea level, has been attributed to wind deflation. Deflation can continue to deepen a blowout in fine-grained sediment until it reaches wet, cohesive sediment at the water table.

Wind Deposition

Loess

Loess is a deposit of wind-blown silt and clay composed of unweathered, angular grains of quartz, feldspar, and other minerals weakly cemented by calcite. Loess has a high porosity, typically near 60%. Deposits of loess may blanket hills and valleys downwind of a source of fine sediment, such as a desert or a region of glacial outwash.

China has extensive loess deposits, more than 100 meters thick in places. Wind from the Gobi Desert carried the silt and clay that formed these deposits. Loess is easy to dig into and has the peculiar ability to stand as a vertical cliff without slumping (figure 13.19), perhaps because of its cement or perhaps because the fine, angular, sediment grains interlock with one another. For centuries, the Chinese have dug cavelike homes in loess cliffs. When a large earthquake shook China in 1920, however, many of these cliffs collapsed, burying alive about 100,000 people.

During the glacial ages of the Pleistocene Epoch, the rivers that drained what is now the midwestern United States transported and deposited vast amounts of glacial outwash. Later, winds eroded silt and clay (originally glacial rock flour) from the flood plains of these rivers and blanketed large areas of the Midwest with a cover of loess (figure 13.20). Soils that have developed from the loess are usually fertile and productive.

A

B

FIGURE 13.18

(*A*) Deflation by wind erosion can form a blowout in loose, dry sediment. Deflation stops at the water table. A pillar, or erosional remnant, may be found in the center of a blowout. (*B*) Large blowout near Harrison, Nebraska. Pillar top is the original level of land before wind erosion lowered the land surface by more than 3 meters. The pillar is the erosional remnant at the center of the blowout. *Photo by N. H. Darton, U.S. Geological Survey*

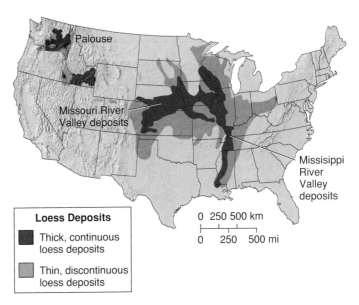

FIGURE 13.20

Major loess-covered areas in the United States. *Source: Data from map from U.S. Department of Agriculture; I. J. Smalley, Loess Lithology and Genesis, fig. 1, p. 768, Halsted Press, 1975*

FIGURE 13.19

Home built into a steep cliff of loess in central China. *Photo © Stephen C. Porter*

The grain fields of much of the Midwest and in the Palouse area in eastern Washington are planted on these rich soils. Wind erosion of cultivated, loess-covered hills in the Palouse region is a serious problem that has locally removed fertile soil from the hilltops.

Sand Dunes

Sand dunes are mounds of loose sand grains piled up by the wind. Dunes are most likely to develop in areas with strong winds that generally blow in the same direction. Patches of dunes are scattered throughout the southwestern United States desert. More extensive dune fields occur on some of the other deserts of the world, such as the Sahara Desert of Africa, which contains vast *sand seas*. Dunes are also commonly found just landward of beaches, where sand is blown inland. Beach dunes are common along the shores of the Great Lakes and along both coasts of the United States. Braided rivers (see chapter 10) can also be sources of sand for dune fields.

The mineral composition of the sand grains in sand dunes depends on both the character of the original sand source and the intensity of chemical weathering in the region. Many dunes, particularly those near beaches in humid regions, are composed largely of quartz grains because quartz is so resistant to chemical weathering. Inland dunes, such as the Great Sand Dunes National Monument in Colorado, often contain unstable feldspar and rock fragments in addition to quartz. Some dunes are formed mostly of carbonate grains, particularly those near tropical beaches. At White Sands, New Mexico, dunes are made of gypsum grains, eroded by wind from playa lake beds.

Sand grains found in dunes are commonly well-sorted and well-rounded because wind is very selective as it moves sediment. Fine-grained silt and clay are carried much farther than sand, and grains coarser than sand are left behind when sand moves. The result is a dune made solely of sand grains, commonly all very nearly the same size. The prevalence of well-rounded grains in many dunes also may be due to selective sorting by the wind. Rounded grains roll more easily than angular grains, and so the wind may remove only the rounded grains from a source to form dunes. Wind will often selectively roll oolitic grains from a carbonate beach of mixed oolitic and skeletal grains.

Most sand dunes are asymmetric in cross section, with a gentle slope facing the wind and a steeper slope on the downwind side. The steep downwind slope of a dune is called the

EARTH SYSTEMS 13.2

Desert Pavement and Desert Varnish

The interaction of the atmosphere and biosphere may result in two intriguing features that can be seen in many deserts, particularly on the surface of old alluvial fans no longer receiving new sediment.

Desert pavement is a thin, surface layer of closely packed pebbles (box figure 1). The pebbles were once thought to be lag deposits, left behind as strong winds blew away all the fine grains of a rocky soil. The pebbles are now thought to be brought to the surface by cycles of wetting and drying, which cause the soil to swell and shrink as water is absorbed and lost by soil particles. Swelling soil lifts pebbles slightly; drying soil cracks, and fine grains fall into the cracks. In this way pebbles move up, while fine grains move down. The surface layer of pebbles protects the land from wind erosion and deflation. When the desert pavement is disturbed (as in the 1991 and 2003 Gulf Wars), dust storms and new sand dunes may result.

Many rocks on the surface of deserts are darkened by a chemical coating known as *desert varnish*. Although the interior of the rocks may be light colored, a hard, often shiny, coating of dark iron and manganese oxides and clay minerals can build up on the rock surface over long periods of time (box figure 2). These paper-thin coatings can be used to obtain a numerical age of the exposed desert surface by measuring cosmogenic helium-3 isotopes preserved in the desert varnish.

Although no one is quite certain how this coating develops, it seems to be added to the rocks from the outside, for even white quartzite pebbles with no internal source of iron, manganese, or clay minerals can develop desert varnish. One hypothesis is that the clay is windblown, perhaps sticking to rocks dampened by dew. A film of clay on a rock may draw iron and manganese-containing solutions upward from the soil by capillary action, and the presence of the clay minerals may help deposit the dark manganese oxide that cements the clay to the rock. Another more recent hypothesis is that the oxide is deposited biologically by manganese-oxidizing bacteria. Regardless of how the varnish forms, the longer a rock is exposed on a desert land surface, the darker it becomes.

BOX 13.2 ■ FIGURE 1

Desert pavement on an old alluvial fan surface in Death Valley, California. The surface pebbles are closely packed; fine sand underlies the pebbles. *Photo by Diane Carlson*

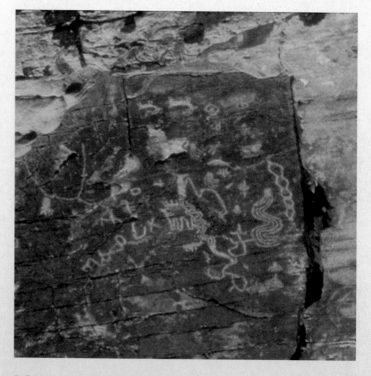

BOX 13.2 ■ FIGURE 2

Petroglyphs carved on this rock cut through the dark desert varnish to show the lighter color of the interior of the rock, Valley of Fire, Nevada. *Photo by J. Freeberg, U.S. Geological Survey*

B

FIGURE 13.21

(A) A sand dune forms with a gentle upwind slope and a steeper slip face on the downward side. Sand eroded from the upwind side of the dune is deposited on the slip face, forming cross-beds. Movement of sand causes the dune to move slowly downwind. (B) Strong desert winds (60 miles per hour) blowing to the right remove sand from the gentle sloping upwind side of this dune. The sand settles onto the steep slip face on the right. *Photo by David McGeary*

slip face (figure 13.21). It forms from loose, cascading sand that generally keeps the slope at the *angle of repose,* which is about 34° for loose, dry sand. Sand is blown up the gentle slope and over the top of the dune. Sand grains fall like snow onto the slip face when they encounter the calm air on the downwind side of the dune. Loose sand settling on the top of the slip face may become oversteepened and slide as a small avalanche down the slip face. These processes form high-angle cross-bedding within the dune. When found in sandstone, such cross-bedding strongly suggests deposition as a dune (see figure 6.27).

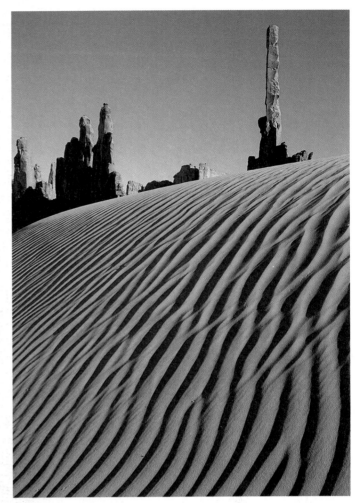

FIGURE 13.22

Wind ripples on sand surface, Monument Valley, Utah. *Photo © Doug Sherman/ Geofile*

In passing over a dune, the wind erodes sand from the gentle upwind slope and deposits it downwind on the slip face. As a result, the entire dune moves slowly in a downwind direction. The rate of dune motion is much slower than the speed of the wind, of course, because only a thin layer of sand on the surface of the dune moves at any one time. The dune may move only 10 to 15 meters per year. Over many years, however, the movement of dunes can be significant, a fact not always appreciated by people who build homes close to moving sand dunes.

If a dune becomes overgrown with grass or other vegetation, movement stops. The Sand Hills of north-central Nebraska are large dunes, formed during the Pleistocene or Holocene Epochs, that have become stabilized by vegetation. The migration of many beach dunes toward beach homes and roads has been stopped by planting a cover of beach grass over the dunes. Dune-buggy tires can uproot and kill the grass, however, and start the dunes moving again.

Sand moving over a dune surface typically forms *wind ripples*—small, low ridges of sand produced by saltation of the grains (figure 13.22). The ripples are similar to those

A Barchans

B Transverse dunes

C Parabolic dunes

D Longitudinal dunes (seifs)

FIGURE 13.23

Types of sand dunes.

formed in sediment by a water current (see chapter 6). Because sand moves perpendicularly to the long dimension of the ripples, a rippled sand surface indicates the direction of sand movement.

Types of Dunes

As figure 13.23 shows, dunes tend to develop certain characteristic shapes, depending on (1) the wind's velocity and direction (that is, whether constant or shifting); (2) the sand supply available; and (3) how the vegetation cover, if any, is distributed.

Where the sand supply is limited, a type of dune called a **barchan** generally develops. The barchan is crescent-shaped with a steep slip face on the inward or concave side. The horns on a barchan dune point in the downwind direction (figure 13.23A). Barchan dunes are usually separated from one another and move across a barren surface (figure 13.24). If more sand is available, the wind may develop a **transverse dune,** a relatively straight, elongate dune oriented perpendicular to the wind direction (figures 13.23B and 13.25).

A **parabolic dune** is somewhat similar in shape to a barchan dune, except that it is deeply curved and is convex in the downwind direction. The horns point upwind and are commonly anchored by vegetation (figure 13.23C). The parabolic dune requires abundant sand and commonly forms around a blowout. Because they require abundant sand and strong winds, parabolic dunes are typically found inland

FIGURE 13.24

Barchan dune formed in the Skeleton Coast National Park in Namibia. Prevailing wind blows from left to right and carries a limited supply of sand. *Photo © Gerry Ellis/Minden Pictures* from an ocean beach (figure 13.26). All three of these dune shapes develop in areas having steady wind direction, and all three have steep slip faces on the downwind side.

One of the largest types of dunes is the **longitudinal dune** or *seif* (figure 13.23*D*), which is a symmetrical ridge of sand that forms parallel to the prevailing wind direction. Longitudinal dunes occur in long, parallel ridges that are exceptionally straight and regularly spaced. They are typically separated by barren ground or desert pavement. Longitudinal dunes in the Sahara Desert (figure 13.27) are as high as 200 meters and more than 120 kilometers in length. Numerous hypotheses have been proposed to explain the development of longitudinal dunes, but none can adequately explain their spectacular size and regular spacing. It appears that crosswinds are important in piling up sand, which adds to the height of longitudinal dunes, whereas the more constant prevailing wind direction redistributes the sand down the length of the dunes. Smoke bomb experiments to analyze airflow have shown that the wind spirals down the intervening troughs between longitudinal dunes and may control the regularity of their spacing.

Not all dunes can be classified by an easily recognizable shape. Many of them are quite irregular.

FIGURE 13.25

Transverse dunes in the Great Sand Dunes National Monument, Colorado. Wind blows from right to left. *Photo © John Shaw/Bruce Coleman*

FIGURE 13.26

Parabolic dunes near Pismo Beach, central California. Wind blows from left to right. The ocean and a sand beach are just to the left of the photo. *Photo by Frank M. Hanna*

FIGURE 13.27

Longitudinal dunes in the Sahara Desert, Algeria. Photo from Gemini spacecraft at an altitude of about 100 kilometers. *Photo by NASA*

PLANETARY GEOLOGY 13.3

Wind Action on Mars

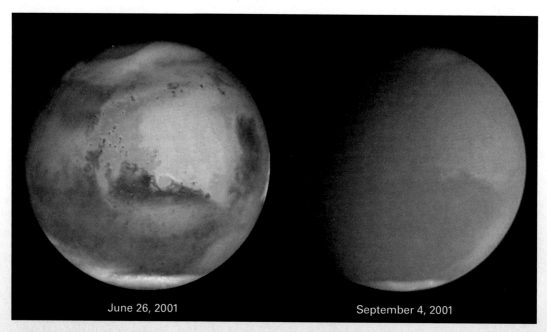

June 26, 2001   September 4, 2001

BOX 13.3 ■ FIGURE 1

Storm watch images of Mars from the Hubble Space Telescope show isolated dust storms on June 26, 2001, in the Hellas basin (lower right edge of Mars) and on the northern polar cap. By the end of July, the entire planet was clouded by dust that obscured its surface for several months. *Photo courtesy of NASA, James Bell (Cornell Univ.), Michael Wolff (Space Science Inst.), and the Hubble Heritage Team (STScI/AURA)*

Mars has an atmosphere only 1/200th as dense as Earth's but with very strong winds that have been recorded at more than 200 kilometers (120 miles) per hour. The sides of Olympus Mons (see box 4.3, figure 1) have been obscured by dust to a height of 15 kilometers (10 miles), a height made possible by the low gravity on Mars. Although dust storms occur throughout the year on Mars, the greatest number and the largest global dust events (those that cover the entire planet) occur during the southern spring and summer, when the southern polar cap of frozen carbon dioxide begins to sublimate. The difference in air temperature between the polar cap and the warmer surrounding landscape creates large pressure differences that produce high winds and trigger isolated dust storms. In June 2001, the Mars Global Surveyor spacecraft and Hubble Space Telescope recorded a sequence of dust storms that began along the retreating margin of the southern polar cap and near the Hellas impact crater (box figure 1). The individual storms intensified and moved north of the equator in only five days. This was the beginning of one of

BOX 13.3 ■ FIGURE 2

False-color image of a dune field in the Endurance crater taken by the Mars Exploration Rover Opportunity. The "blue" tint is caused by the presence of hematite-containing spherules ("blueberries") that accumulate on the flat surface between the dunes. Dunes in the foreground are about 1 meter in height. *Photo by NASA/JPL/Cornell*

BOX 13.3 ■ FIGURE 3

Close-up of the rock named "Moe" from the Pathfinder landing site that shows a smooth but pitted surface similar to wind-abraded rocks, or ventifacts, on Earth; rock is 1 meter in diameter. *Photo by JPL/NASA*

the largest global dust events observed on Mars in decades, and, for the first time, scientists were able to see how the development and progression of regional dust storms resulted in the entire planet being obscured by dust.

Images from the recent Mars Rover missions show that the windswept surface of the planet contains features similar to those found on Earth. Barchan, star, transverse, and longitudinal sand dunes are prevalent, particularly on the floors of impact craters. The 2004 Mars Rover, Opportunity, recorded images of a dune field in Endurance crater and also discovered the presence of hematite-containing spherules ("blueberries") that accumulate on the flat surfaces (box figure 2). What appear to be *yardangs*, wind-eroded round and elliptical knobs, have been observed in the Medussae Fossae region of Mars. The robotic "geologist" Sojourner, launched from the Pathfinder mission, recorded detailed images of rocks with smooth yet pitted surfaces that are similar to wind-scoured ventifacts on Earth (box figure 3).

Additional Resources

For more information and additional images of dust storms and wind features on Mars, visit the NASA Mars Exploration and Planetary Photojournal websites:

- http://mars.jpl.nasa.gov/gallery/duststorms/index.html
- http://mars.jpl.nasa.gov/gallery/sanddunes/index.html
- http://photojournal.jpl.nasa.gov/

SUMMARY

Deserts are located in regions where less than 25 centimeters of rain falls in a year. Such regions are found primarily in belts of descending air at 30° North and South latitude. Arid regions also may be due to the *rain shadow* of a mountain range, great distance from the sea, and proximity to a cold ocean current. Descending air forms cold deserts at the poles.

Desert landscapes differ from those of humid regions in lacking through-flowing streams and having internal drainage and many local, rising base levels. *Flash floods* caused by desert thunderstorms are effective agents of erosion despite the low rainfall. Limestone is resistant in deserts. Thin soil and slow rates of creep may give desert topography an angular look.

Parts of the southwestern United States are desert, the topography determined primarily by rock structure. Flat-lying sedimentary rocks of the Colorado Plateau are sculptured into cliffs, *plateaus, mesas,* and *buttes.* The fault-controlled topography of the Basin and Range province is marked by *alluvial fans, bajadas, playas,* and *pediments.*

Although wind erosion can be intense in regions of low moisture, streams are usually more effective than wind in sculpturing landscapes, regardless of climate.

Fine-grained sediment can be carried long distances by wind, even across entire continents and oceans.

Sand moves by *saltation* close to the ground, occasionally carving *ventifacts.*

Wind can *deflate* a region, creating a *blowout* in fine sediment.

Sand dunes move slowly downwind as sand is removed from the gentle upwind slope and deposited on the steeper *slip face* downwind.

Dunes are classified as *barchans, transverse dunes, parabolic dunes,* and *longitudinal dunes,* but many dunes do not resemble these types. Dune type depends on wind strength and direction, sand supply, and vegetation.

Terms to Remember

alluvial fan 344
bajada 347
barchan 354
blowout 350
butte 344
deflation 349
desert 339
fault 344

flash flood 343
loess 350
longitudinal dune 355
mesa 344
parabolic dune 354
pediment 347
plateau 344
playa 347

playa lake 347
rain shadow 340
sand dune 351
slip face 353
transverse dune 354
ventifact 349

Testing Your Knowledge

Use the following questions to prepare for exams based on this chapter.

1. What are two reasons why parts of the southwestern United States have an arid climate?

2. Sketch a cross section of an idealized dune, labeling the slip face and indicating the wind direction. Why does the dune move?

3. Describe the geologic structure and sketch the major landforms of
    a. the Colorado Plateau   b. the Basin and Range province

4. How does a flash flood in a dry region differ from most floods in a humid region?

5. Give two reasons why wind is a more effective agent of erosion in a desert than in a humid region.

6. Name four types of sand dunes and describe the conditions under which each forms.

7. The defining characteristic of a desert is
    a. shifting sand dunes   b. high temperatures
    c. low rainfall   d. all of the preceding
    e. none of the preceding

8. Which is characteristic of deserts?
    a. internal drainage   b. limited rainfall
    c. flash floods   d. slow chemical weathering
    e. all of the preceding

9. The major difference between a mesa and a butte is one of
    a. shape   b. elevation
    c. rock type   d. size

10. The Basin and Range province covers almost all of
    a. Utah   b. Nevada
    c. Texas   d. Colorado

11. A very flat surface underlain by a dry lake bed of hard, mud-cracked clay is called a
    a. ventifact   b. plateau
    c. playa   d. none of the preceding

12. Rocks with flat, wind-abraded surfaces are called
    a. ventifacts   b. pediments
    c. bajadas   d. none of the preceding

13. The removal of clay, silt, and sand particles from the land surface by wind is called
    a. deflation   b. depletion
    c. deposition   d. abrasion

14. Which is not a type of dune?
    a. barchan   b. transverse
    c. parabolic   d. longitudinal
    e. all of the preceding are dunes

15. Much of the southwestern United States is desert because (choose as many as apply)
    a. it is near 30° North
    b. the western mountains create a rain shadow
    c. cold ocean currents in the Pacific cause high evaporation rates in the land
    d. it is a great distance from the ocean

16. A broad ramp of sediment formed at the base of mountains when alluvial fans merge is
    a. a playa   b. a bajada
    c. a pediment   d. an arroyo

17. A surface layer of closely packed pebbles is called
    a. desert varnish   b. deflation
    c. a blowout   d. desert pavement

Expanding Your Knowledge

1. Study the photos of sand dunes in this chapter. Which way does the prevailing wind blow in each case?

2. Can deserts be converted into productive agricultural regions? How? Are there any environmental effects from such conversion?

3. At what relative depth is ground water likely to be found in a desert? Why? Is the water likely to be drinkable? Why?

Exploring Web Resources

www.mhhe.com/plummer11e

This is the dedicated website for this book. You can go to it for new and updated information. The universal resource locators (URL) listed in this book are also given as links on the website, making it easy to go to those websites without typing in the URL. When you visit our Online Learning Center, go to the Student Center and then to the chapter of interest. Here you will find additional readings and media resources, as well as answers to the Testing Your Knowledge section, more quizzes, animation and video clips, and direct links to the sites listed in this book. Links to additional websites can also be found. We have added questions for some of the sites to allow you to get the most of your exploration of the web. Using the web is an enjoyable way of enhancing your knowledge of geology.

http://pubs.usgs.gov/gip/deserts/contents/

Online version of *Deserts: Geology and Resources* by A. S. Walker provides a good overview of deserts, processes, and mineral resources.

Waves, Beaches, and Coasts

Chapters 9 through 13 have dealt with the sculpturing of the land by mass wasting, streams, ground water, glaciers, and wind. Water waves are another agent of erosion, transportation, and deposition of sediment. Earth's shorelines are continuously changing due to interaction between the hydrosphere and the geosphere. Along the shores of oceans and lakes, waves break against the land, building it up in some places and tearing it down in others.

Ordinary ocean waves (as opposed to tsunamis) are created by wind—interaction between the atmosphere and the hydrosphere. Waves moving across an ocean transfer the energy derived from the wind to shorelines. This energy is used to a large extent in eroding and

Beaches and rocky headlands formed along the Big Sur coastline at Julia Pfeiffer Burns State Park in central California. *Photo © George Ranalli/Photo Researchers* transporting sediment along the shoreline. Understanding how waves travel and move sediment can help you see how easily the balance of supply, transportation, and deposition of beach sediment can be disturbed. Such disturbances can be natural or humanmade, and the changes that result often destroy beachfront homes and block harbors with sand.

Beaches have been called "rivers of sand" because breaking waves, as they sort and transport sediment, tend to move sand parallel to the shoreline. In this chapter, we look at how beaches are formed and examine the influence of wave action on such coastal features as sea cliffs, barrier islands, and terraces.

INTRODUCTION

If you spend a week at the shore during the summer, you may not notice any great change in the appearance of the beach while you are there. Even if you spend the whole summer at the seaside, nothing much seems to happen to the beach during those months. Tides rise and fall every day and waves strike the shore, but the sand that you walk on one day looks very much like the sand that you walked on the previous day. The shape of the beach does not appear to change, nor does the sand seem to move very much.

On most beaches, however, the sand *is* moving, in some cases quite rapidly. The beach looks the same from day to day only because new sand is being supplied at about the same rate that old sand is being removed.

Where is the sand going? Some sand is carried out to deep water. Some is piled up and stored high on the beach. But on most shores, much more of the sand moves along parallel to the beach in relatively shallow water. In this way, loose sand grains travel hundreds of feet per day along some coasts, especially those subject to strong waves.

On some beaches, sand is being removed faster than it is being replenished. When this happens, beaches become narrower and less attractive for swimming. Where erosion is severe, buildings close to the beach can be undermined and destroyed by waves as the beach disappears (figure 14.1). The sand moved from the beach may be redeposited in inconvenient places, such as across the mouth of a harbor, where it must be dredged out periodically. Because moving sand can create many problems for people in coastal towns and cities, it is important to understand something of how and why the sand moves.

WATER WAVES

The energy that moves sand along a beach comes from the wind-driven water waves that break upon the shore. As wind blows over the surface of an ocean or a lake, some of the wind's energy is transferred to the water surface, forming the waves that move through the water. The height of waves (and their length and speed) are controlled by the wind speed, the length of time that the wind blows, and the distance that the wind blows over the water *(fetch)*. The largest waves form where high winds blow over a long expanse of open water for an extended period of time.

Wave shapes can vary. Short, choppy *seas* in and near a storm create a confused sea surface, often with considerable white foam as strong winds blow the tops off of waves. Long, rolling *swells* form a regular series of similar-sized waves on

FIGURE 14.1
Beach houses were built too close to the ocean and destroyed by 2004 Hurricane Ivan on Cape San Blas along the coast of Pensacola, Florida. *Photo © Associated Press, St. Petersburg Times/Wide World Photos*

shores that may be thousands of kilometers from the storms that generated the waves. (Summer surfing waves in southern California can be generated by large storms north of Australia in the Southern Hemisphere winter.) When waves break against the shore as *surf,* a large portion of their energy is spent moving sand along the beach.

The height of waves is the key factor in determining wave energy. **Wave height** is the vertical distance between the **crest,** which is the high point of a wave, and the **trough,** which is the low point (figure 14.2). In the open ocean, normal waves have heights of about 0.3 to 5 meters, although during violent storms, including hurricanes, waves can be more than 15 meters high. The highest wind wave ever measured was 34 meters by the

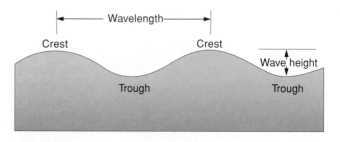

FIGURE 14.2
Wave height is the vertical distance between the wave crest and the wave trough. Wavelength is the horizontal distance between two crests.

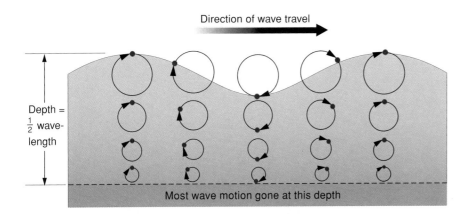

FIGURE 14.3

Orbital motion of water in waves dies out with depth. At the surface, the diameter of the orbits is equal to the wave height.

anxious crew of a ship in the north Pacific in 1933. (The highest tsunami ever measured, caused by a submarine earthquake rather than wind, was 85 meters, in the Ryukyu island chain south of Japan in 1971; see chapter 16.)

Wavelength is the horizontal distance between two wave crests (or two troughs). Most ocean wind waves are between 40 to 400 meters in length and move at speeds of 25 to 90 kilometers per hour (15 to 55 miles per hour) in deep water.

The movement of water in a wave is like the movement of wheat in a field when wind blows across it. You can see the ripple caused by wind blowing across a wheat field, but the wheat does not pile up at the end of the field. Each stalk of wheat bends over when the wind strikes it and then returns to its original position. A particle of water moves in an *orbit,* a nearly circular path, as the wave passes (figure 14.3); the particle returns to its original position after the wave has passed. In deep water, when a wave moves across the water surface, energy moves with the wave; but the water, like the wheat, does not advance with the wave.

At the surface, the diameter of the orbital path of a water particle is equal to the height of the wave (figure 14.3). Below

the surface, the orbits decrease in size until the motion is essentially gone at a depth equal to half the wavelength. This is why a submarine can cruise in deep, calm water beneath surface ships that are being tossed by the orbital motion of large waves.

Surf

As waves move from deep water to shallow water near shore, they begin to be affected by the ocean bottom. A wave first begins to "feel bottom" at the level of lowest orbital motion—that is, when the depth to the bottom equals half the wavelength. For example, a wave 150 meters long will begin to be influenced by the bottom at a water depth of 75 meters.

In shallow water, the presence of the bottom interferes with the circular orbits, which flatten into ovals (figure 14.4). The waves slow down and their wavelength decreases. Meanwhile, the sloping bottom wedges the moving water upward, increasing the wave height. Because the height is increasing while the length is decreasing, the waves become steeper and steeper until they break. A **breaker** is a wave that has become so steep that the crest

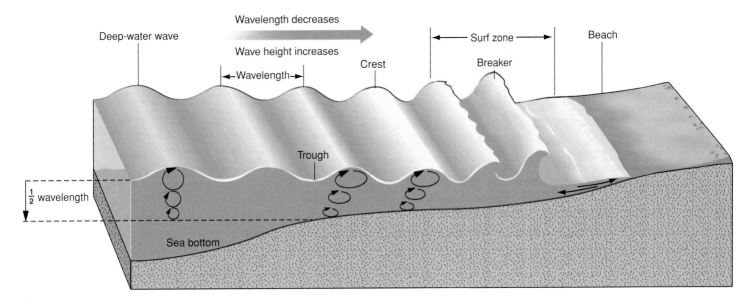

FIGURE 14.4

As a deep-water wave approaches shore, it begins to "feel" the sea bottom and slow down. Circular water orbits flatten and the wave peaks and breaks. In the foamy surf zone, water moves back and forth rather than in orbits.

of the wave topples forward, moving faster than the main body of the wave. The breaker then advances as a turbulent, often foamy, mass. Energy from the wind is transmitted by the wave and finally spent by breakers on the beach. Breakers collectively are called **surf.** Water in the surf zone has lost its orbital motion and moves back and forth, alternating between onshore and offshore flow.

NEAR-SHORE CIRCULATION

Wave Refraction

Most waves do not come straight into shore. A wave crest usually arrives at an angle to the shoreline (figure 14.5A). One end of the wave breaks first, and then the rest breaks progressively along the shore.

This angled approach of a wave toward shore can change the direction of wave travel. One end of the wave reaches shallow water first. This end of the wave "feels bottom" and slows down while the rest of the wave continues at its deep-water speed (figure 14.5B). As more and more of the wave comes into contact with the bottom, more of the wave slows down. As the wave slows progressively along its length, the wave crest changes direction and becomes more nearly parallel to the shoreline. This bending of waves is called **wave refraction** (figure 14.5B).

Longshore Currents

Although most wave crests become nearly parallel to shore as they are refracted, waves do not generally strike *exactly* parallel to shore. Even after refraction, a small angle remains between the wave crest and the shoreline. As a result, the water in the wave is pushed both *up* the beach toward land and *along* the beach parallel to shore.

Each wave that arrives at an angle to the shore pushes more water parallel to the shoreline. Eventually, a moving mass of water called a **longshore current** develops parallel to the shoreline (figure 14.6). The width of the longshore current is about equal to the width of the surf zone. The seaward edge of the current is the outer edge of the surf zone, where waves are just beginning to break; the landward edge is the shoreline. A longshore current can be very strong, particularly when the waves are large. Such a current can carry swimmers hundreds of yards parallel to shore before they are aware that they are being swept along. It is these longshore currents that transport most of the beach sand parallel to shore.

Rip Currents

Rip currents are narrow currents that flow straight out to sea in the surf zone, returning water seaward that breaking waves have pushed ashore (figure 14.6). Rip currents travel at the water surface and die out with depth. They pulsate in strength, flowing most rapidly just after a set of large waves has carried a large amount of water onto shore. Rip currents can be important transporters of sediment, as they carry fine-grained sediment out of the surf zone into deep water.

A

This part of wave stays in deep water and does not slow down

Wave

Refracted wave

Wave touches bottom here and slows down

Longshore   current

Beach

B

FIGURE 14.5

(A) These waves are arriving at an angle to the shoreline. They break progressively along the shore, from the upper right of the photo to the lower left. (B) Wave refraction changes the wave direction, bending the wave so it becomes more parallel to shore. The angled approach of waves to shore sets up a longshore current parallel to the shoreline. *Photo by David McGeary*

As a single wave comes toward shore, its height varies from place to place. Rip currents tend to develop locally where wave height is low. Rip currents that are fixed in position are apt to be found over channels on the sea floor, because depressions on the bottom reduce wave height. Complex wave interactions can also lower wave height, and rip currents that form because of wave interactions tend to shift position along the shore. Such shifting rip currents are usually spaced at regular intervals along the beach.

Rip currents are fed by water within the surf zone. They flow rapidly out through the surf zone and then die out quickly. Where waves are nearly parallel to a shoreline, longshore feeder currents of equal strength develop in the surf zone on either side of a rip current (figure 14.6A). Where waves strike

FIGURE 14.6

Rip currents and their feeder currents can develop regardless of the angle of approach of waves. (A) Waves approach parallel to shore; feeder currents on both sides of rip currents. (B) Waves approach at an angle to the shore; feeder current on only one side of rip current. (C) Rip currents carry dirty water and foam seaward; they can cause incoming waves to break early. *Photo © Sanford Berry/Visuals Unlimited*

the shore at an angle and set up a strong, unidirectional, longshore current, a rip current is fed from one side by the longshore current, which increases in strength as it nears the rip current (figure 14.6*B*). Rip currents are also found alongside points of land and engineered structures such as jetties and piers, which can deflect longshore currents seaward.

You can easily learn to spot rip currents at a beach. Look for discoloration in the water where sediment is being picked up in the surf zone and moved seaward (figure 14.6*C*). Another sign is incoming waves breaking early within a rip current as they meet the opposing flow. The diffuse heads of rip currents

outside the surf zone may be marked at the edge with foam lines. Even on very calm days, rips can often be identified by subtle changes in the water surface, such as a different pattern of water ripples or light reflection off the water.

Getting caught in a rip current and being carried out to sea can panic an inexperienced swimmer—even though the trip will stop some distance beyond the surf zone as the rip dies out. A swimmer frightened by being carried away from land and into breaking waves can grow exhausted fighting the current to get back to shore. The thing to remember is that rip currents are narrow. Therefore, you can get out of a rip easily by

A **B**

FIGURE 14.7

(A) Parts of a beach. (B) The beach face (on the left) and berm (on the right) on a northern California beach. *Photo by Diane Carlson*

swimming *parallel* to the beach instead of struggling against the current.

Surfers, on the other hand, often look for rip currents and paddle intentionally into them to get a quick ride out into the high breakers.

BEACHES

A **beach** is a strip of sediment (usually sand or gravel) that extends from the low-water line inland to a cliff or a zone of permanent vegetation. Waves break on beaches, and rising and falling tides may regularly change the amount of beach sediment that is exposed above water (figure 14.7A).

The steepest part of a beach is the **beach face,** which is the section exposed to wave action, particularly at high tide. Offshore from the beach face there is usually a **marine terrace,** a broad, gently sloping platform that may be exposed at low tide if the shore has significant tidal action. Marine terraces may be *wave-built* terraces constructed of sediment carried away from the shore by waves, or they may be *wave-cut* rock benches or platforms, perhaps thinly covered with a layer of sediment.

The upper part of the beach, landward of the usual high-water line is the **berm,** a wave-deposited sediment platform that is flat or slopes slightly landward (figure 14.7B). It is usually dry, being covered by waves only during severe storms.

Beach sediment is usually sand, typically quartz-rich because of quartz's resistance to chemical weathering. Heavy metallic minerals ("black sands") can also be concentrated on some beaches as less dense minerals such as quartz and feldspar are carried away by waves or wind (titanium-bearing sands are mined on some beaches in Florida and Australia). Tropical beaches may be made of bioclastic carbonate grains from offshore corals, algae, and shells. Some Hawaiian beaches are made of sand-sized fragments of basalt. Gravel beaches are found on coasts attacked by the high energy of large waves (*shingle* is a regional name for disk-shaped gravel). Gravel beaches have a steeper face slope than sand beaches.

In seasonal climates, beaches often go through a summer-winter cycle (figure 14.8). This is due to the greater frequency of storms with strong winds during the winter months, which tend to produce high waves with short wavelengths. These high-energy waves tend to crash onshore and erode sand from the beach face and narrow the berm. Offshore, in less turbulent water, the sand settles to the bottom and builds an underwater sandbar (parallel to the beach) that serves as a "storage facility" for the next summer's sand supply. The following summer, or during calmer weather, low-energy waves with long wavelengths break over the sandbar and gradually push the sand back onto the beach face to widen the berm. Each season the beach changes in shape until it comes into equilibrium with the prevailing wave type.

Many winter beaches can be dangerous because of high waves and narrowed beaches. Several beaches along the Pacific

A Winter beach

B Summer beach

FIGURE 14.8

Seasonal cycle of a beach caused by differing wave types. (A) Narrow winter beach. Waves may break once on the winter sandbar, then reform and break again on the beach face. (B) Wide summer beach.

coast of the United States and Canada are nearly free of accidents in the summer, when they are heavily used, but are regularly marked by drownings in the winter as beach walkers are swept off narrower beaches and out to sea by large storm waves.

LONGSHORE DRIFT OF SEDIMENT

Longshore drift is the movement of sediment parallel to shore when waves strike the shoreline at an angle. Figure 14.9 shows the two ways in which this movement of sediment (usually sand) occurs. Some longshore drift takes place directly on the beach face when waves wash up on land. A wave washing up on the beach at an angle tends to wash sand along at the same angle. After the wave has washed up as far as it can go, the water returns to the sea by running down the beach face by the shortest possible route, that is, straight downhill to the shoreline, not back along the oblique route it came up. (Wave run-up is known as *swash,* the return as *backwash.*) The net effect of this motion is to move the sand in a series of arcs along the beach face.

Much more sand is moved by longshore transport in the surf zone, where waves are breaking into foam. The turbulence of the breakers erodes sand from the sea bottom and keeps it suspended. Even a weak longshore current can move the suspended sand parallel to the shoreline. The sand in the longshore current moves in the same direction as the sand drift on the beach face (figure 14.9).

Vast amounts of sand can be moved by longshore transport. The U.S. Army Corps of Engineers estimates that 436,000 cubic meters of sand per year are moved northward by waves at Sandy Hook, New Jersey, and 1 million cubic meters of sand per year are moved southward at Santa Monica, California.

Eventually, the sand that has moved along the shore by these processes is deposited. Sediment may build up off a point of land to form a **spit,** a fingerlike ridge of sediment that extends out into open water (figures 14.10*A* and *B*). A **baymouth bar,** a

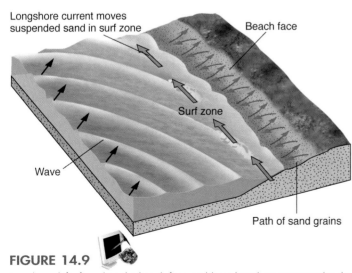

FIGURE 14.9

Longshore drift of sand on the beach face and by a longshore current within the surf zone.

FIGURE 14.10

(A) Longshore drift of sand can form spits and baymouth bars. (B) Curved spit near Victoria, British Columbia. (C) A baymouth bar has sealed off this bay from the ocean as sand migrated across the mouth of the Russian River in northern California. *Photo B by D. A. Rahm, courtesy of Rahm Memorial Collection, Western Washington University; Photo C by Diane Carlson*

FIGURE 14.11

(A) A tombolo has connected this rock, once an island, to the shore. Note the waves bending around the two sides of the rock, near Santa Cruz, California. (B) Formation of a tombolo. Wave refraction around an island interrupts the longshore current and creates a sandbar that connects the island with the mainland. *Photo by David McGeary*

ridge of sediment that cuts a bay off from the ocean, is formed by sediment migrating across what was earlier an open bay (figure 14.10A and C). Off the western coast of the United States, a considerable amount of drifting sand is carried into the heads of underwater canyons, where the sediments slide down into deep, quiet water.

A striking but rare feature formed by longshore drift is a **tombolo,** a bar of sediment connecting a former island to the mainland. As shown in figure 14.11, waves are refracted around an island in such a way that they tend to converge behind the island. The waves sweep sand along the mainland (and from the island) and deposit it at this zone of convergence, forming a bar that grows outward from the mainland and eventually connects to the island.

Human Interference with Sand Drift

Several engineered features can interrupt the flow of sand along a beach (figure 14.12). *Jetties,* for example, are rock walls designed to protect the entrance of a harbor from sediment deposition and storm waves. Usually built in pairs, they protrude above the surface of the water. Figure 14.12A shows how sand piles up against one jetty while the beach next to the other, deprived of a sand supply, erodes back into the shore.

Groins are sometimes built in an attempt to protect beaches that are losing sand from longshore drifting. These short walls

are built perpendicular to shore to trap moving sand and widen a beach (figure 14.12B). However, once a groin is built to capture the sand, beaches down current will erode as the longshore current attempts to replenish its sediment load. The disappearance of neighboring beaches often results in lawsuits and in successive groins being constructed in an attempt to trap the remaining sand.

Sand deposition also occurs when a stretch of shore is protected from wave action by a *breakwater,* an offshore structure built to absorb the force of large, breaking waves and provide quiet water near shore. When the city of Santa Monica in California built a rock breakwater parallel to the shore to create a protected small-boat anchorage, the lessening of wave action on the shore behind the breakwater allowed sand to build up there (figure 14.12C), threatening eventually to fill in the anchorage. The city had to buy a dredge to remove the sand from the protected area and redeposit it farther along the shore where the waves could resume moving sediment.

A beach attempts to come into equilibrium with the waves that strike it. The type and amount of sediment, the position of the sediment, and especially the movement of the sediment, adjust to the incoming wave energy. Whenever human activity interferes with sand drift or wave action, the beach responds by changing its configuration, usually through erosion or deposition in a nearby part of the beach.

A

B

C

FIGURE 14.12

Sand piles up against obstructions and in areas deprived of wave energy. (*A*) Jetties at Manasquan Inlet, New Jersey. Sand drift to the right has piled sand against the left jetty and removed sand near the right jetty. (*B*) Groins at Ocean City, New Jersey. Sand drift is to the right. (*C*) Breakwater at Santa Monica, California, has caused deposition of sand in the wave-protected zone. *Photos A and B by S. Jeffress Williams; Photo C © John S. Shelton*

Sources of Sand on Beaches

Some beach sand comes from the erosion of local rock, such as points of land or cliffs nearby. On a few beaches, replenishment comes from sand stored outside the surf zone in the deeper water offshore. Bioclastic carbonate beaches are formed from the remains of marine organisms offshore. But the greater part of the sand on most beaches comes from river sediment brought down to the ocean. Waves pick up this sediment and move it along the beach by longshore drift.

What happens to a beach if all the rivers contributing sand to it are dammed? Although damming a river may be desirable for many reasons (flood control, power generation, water supply, recreation), when a river is dammed, its sediment load no longer reaches the sea (see box 10.1). The sand that supplied the beach in the past now comes to rest in the quiet waters of the reservoir behind the dam. Longshore drift, however, continues to remove sand from beaches, even though little new sand is being supplied, and the result is a net loss of sand from beaches. Beaches without a sand supply eventually disappear. To prevent this, some coastal communities have set up expensive programs of building pipelines or draining reservoirs and trucking the trapped sand down to the beaches.

COASTS AND COASTAL FEATURES

A beach is just a small part of the **coast,** which is all the land near the sea, including the beach and a strip of land inland from it. Coasts can be rocky, mountainous, and cliffed, as in northern New England and on the Pacific shore of North America, or they can be broad, gently sloping plains, as along much of the southeastern United States. Wave erosion and deposition can greatly modify coasts from their original shapes. Many coasts have been drowned during the past 15,000 years by the rise in sea level caused by the melting of the Pleistocene glaciers (see chapter 12). Other coasts have been lifted up by tectonic forces at a rate greater than the rise in sea level so that sea-floor features are now exposed on dry land.

Erosional Coasts

A great many steep, rocky coasts have been visibly changed by wave erosion. Soluble rocks such as limestone dissolve as waves wash against them, and more durable rocks such as granite are fractured by the enormous pressures caused by

FIGURE 14.13

(A) Wave refraction on an irregular coast. Arrows show transport direction of wave energy, concentrated on headlands and spread out in bays. (B) Wave refraction around a point of land. Rincon Point, California. Note that waves at right center have been bent 90° from their original direction (parallel to the bottom of the photo). *Photo by Frank M. Hanna*

FIGURE 14.14

Coastal straightening of an irregular coastline by wave erosion of headlands and wave deposition of sediment in bays. Continued erosion produces a straight, retreating cliff.

waves slamming into rock (wave impact pressures have been measured as high as 60 metric tons per square meter).

An irregular coast with bays separated by rocky **headlands** (points of land) can be gradually straightened by wave action. Because wave refraction bends waves approaching such a coast until they are nearly parallel to shore, most of the waves' energy is concentrated on the headlands, while the bays receive smaller, diverging waves (figure 14.13). Rocky cliffs form from wave erosion on the headlands. The eroded material is deposited in the quieter water of nearby bays, forming broad beaches. **Coastal straightening** of an irregular shore gradually takes place through wave erosion of headlands and wave deposition in bays (figure 14.14).

Wave erosion of headlands produces **sea cliffs,** steep slopes that retreat inland by mass wasting as wave erosion undercuts them (figure 14.15). At the base of sea cliffs are sometimes found *sea caves,* cavities eroded by wave action along zones of weakness in the cliff rock. As headlands on irregular coasts are eroded landward, sea cliffs enlarge until the entire coast is marked by a retreating cliff (figure 14.14). On

FIGURE 14.15

Retreating wave-cut cliff, north of Bodega Bay, Sonoma County, California. A concrete seawall has been built at the base of the cliff to slow wave erosion and help protect the cliff-edge homes. Note fragments of wave-destroyed structures near the seawall. Seawalls usually increase the erosion of sand beaches. *Photo by David McGeary* some exposed coasts, the rate of cliff retreat can be quite rapid, particularly if the rock is weakly consolidated. Some sea cliffs north of San Diego, California, and at Cape Cod National Seashore in Massachusetts are retreating at an average rate of 1 meter per year. Because sea-cliff erosion in weak rock is often in the form of large, infrequent slumps (see chapter 9), some portions of these coasts may retreat 10 to 30 meters in a single storm. Some of these cliffs have "ocean-view" homes and hotels at their very edges. Sea cliffs in hard, durable rock such as granite and schist retreat much more slowly.

Seawalls may be constructed along the base of retreating cliffs to prevent wave erosion (figure 14.15). Seawalls of giant pieces of broken stone (riprap) or concrete tetrahedrons are designed to absorb wave energy rather than allow it to erode cliff rock. Vertical or concave seawalls of concrete are designed to reflect wave energy seaward rather than allow it to impact the shore. Some of the reflected energy, however, is focused at the base of the seawall, which eventually under-

mines it and causes the seawall to collapse. Reflection of waves from a seawall also increases the amount of wave energy just offshore, often increasing the amount of sand erosion offshore. Thus, a seawall designed to protect a sea cliff (and the buildings at its edge) may in some cases destroy a sand beach at the base of the cliff. Seawalls are difficult and expensive to build and maintain, and they may destroy beaches, but political pressure to build more of them will increase as the sea level rises in the future.

Wave erosion produces other distinctive features in association with sea cliffs. A **wave-cut platform** (or *terrace*) is a horizontal bench of rock formed beneath the surf zone as a coast retreats by wave erosion (figure 14.16). The platform widens as the sea cliffs retreat. The depth of water above a wave-cut platform is generally 6 meters or less, coinciding with the depth at which turbulent breakers actively erode the sea bottom. **Stacks** are erosional remnants of headlands left behind as the coast retreats inland (figure 14.17). They form small, rocky islands off retreating coasts, often directly off headlands (figure 14.14). **Arches** (or *sea arches*) are bridges of rock left above openings eroded in headlands or stacks by waves. The openings are eroded in spots where the rock is weaker than normal, perhaps because of closely spaced fractures.

Depositional Coasts

Many coasts are gently sloping plains and show few effects of wave erosion. Such coasts are found along most of the Atlantic Ocean and Gulf of Mexico shores of the United States. These coasts are primarily shaped by sediment deposition, particularly by longshore drift of sand.

Coasts such as these are often marked by **barrier islands**—ridges of sand that parallel the shoreline and extend above sea level (figure 14.18). These barrier islands may have formed from sand eroded by waves from deeper water offshore, or they may be greatly elongated sand spits formed by longshore drift. The slowly rising sea level associated with the melting of the Pleistocene glaciers may have been a factor in their development. A protected lagoon separates barrier islands from the mainland. Because the lagoon is protected from

A

B

FIGURE 14.16

(*A*) A wave-cut platform (the wide, horizontal bench of dark rock at the base of the cliffs) is exposed at low tide, La Jolla, California. (*B*) A wave-cut platform widens as a sea of cliff retreats. *Photo by David McGeary*

FIGURE 14.17

Sea stacks and an arch mark old headland positions on this retreating, wave-eroded coast in northern California. *Photo by Diane Carlson*

Some barrier islands along the Atlantic and Gulf coasts are densely populated. Atlantic City (New Jersey), Ocean City (Maryland), Miami Beach (Florida), and Galveston (Texas) are examples of cities built largely on barrier islands. In some of these cities, houses, luxury hotels, and condominiums are clustered near the edge of the sea; many are built upon the loose sand of the island (figure 14.19). These developed areas are vulnerable to late-summer hurricanes that sooner or later bring huge storm waves onto these coasts, eroding the sand and undermining the building foundations at the water's edge.

Nonmarine deposition may also shape a coast. Rapid sedimentation in *deltas* by rivers can build a coast seaward (see figure 10.28). *Glacial deposition* can form shoreline features. Several islands off the New England coast were glacially deposited; Long Island, New York, formed from the deposition of a recessional and end moraine.

Drowned Coasts

Drowned (or *submergent*) coasts are common because sea level has been rising worldwide for the past 15,000 years. During the glacial ages of the Pleistocene, sea level was 130 meters below its present level. The shallow sea floor near the continents was then dry land, and rivers flowed across it, cutting valleys. As the great ice sheets melted, sea level began to

waves, it provides a quiet waterway for boats. A series of such lagoons stretches almost continuously from New York to Florida, and many also exist along the Gulf Coast, forming an important route for barge traffic. As tides rise and fall, strong tidal currents may wash in and out of gaps between barrier islands, distributing sand in submerged *tidal deltas* both landward and seaward of the gaps.

FIGURE 14.18

(*A*) A barrier island on a gently sloping coast. A lagoon separates the barrier island from the mainland, and tidal currents flowing in and out of gaps in the barrier island deposit sediment as submerged tidal deltas. (*B*) A barrier island near Pensacola, Florida: open ocean to right, lagoon to left, and mainland Florida on far left. Light-colored lobes of sand within lagoon were eroded from the barrier island by hurricane waves, which washed entirely across the island and into the lagoon. *Photo by Frank M. Hanna*

FIGURE 14.19

Hotels built upon the loose sand of a barrier island, Miami Beach, Florida. *Photo © Werner Bertsch/ Bruce Coleman*

FIGURE 14.20

Landsat satellite photo of estuaries, Albemarle and Pamlico Sounds, North Carolina. Barrier islands are visible in upper right. Infrared image shows vegetation as red. *Photo by NASA*

rise, drowning the river valleys. These drowned river mouths, called **estuaries,** mark many coasts today (figure 14.20). They extend inland as long arms of the sea. Fresh water from rivers mixes with the seawater to make most estuaries brackish. The quiet, protected environment of estuaries makes them very rich in marine life, particularly the larval forms of numerous species. Unfortunately, cities and factories built on many estuaries to take advantage of quiet harbors are severely polluting the water and the sediment of the estuaries. The poor circulation that characterizes most estuaries hinders the flushing away of this pollution, and estuary shellfish are sometimes not safe to eat as a result.

Drowned coasts may be marked by **fiords,** glacially cut valleys flooded by rising sea level (see figure 12.36). They form in the same way as estuaries, except they were cut by glacial ice rather than rivers during low sea-level stands.

Uplifted Coasts

Uplifted (or *emergent*) coasts have been elevated by deep-seated tectonic forces. The land has risen faster than sea level, so parts of the old sea floor are now dry land.

Marine terraces form just offshore from the beach face, as described in the Beaches section of this chapter. These terraces can be wave-cut platforms caused by erosion of rock associated with cliff retreat, or they can be wave-built terraces caused by deposition of sediment. If the shore is elevated by tectonic uplift, these flat surfaces

FIGURE 14.21

Uplifted marine terrace, northern California. The flat land surface at the top of the sea cliff was eroded by wave action, then raised above sea level by tectonic uplift. The rock knob on the terrace was once a stack. *Photo by David McGeary*

will become visible as *uplifted marine terraces* (figure 14.21). They formed below the ocean surface but are visible now because of uplift. The tectonically unstable Pacific coast of the United States and Canada has many areas marked by uplifted terraces, along with the erosional coast features described earlier.

The Biosphere and Coasts

The growth of coral and algal *reefs* offshore can shape the character of a coast. The reefs act as a barrier to strong waves,

protecting the shoreline from most wave erosion (see figure 18.22). Carbonate sediments blanket the sea floor on both sides of a reef and usually form a carbonate sand beach on land (see figure 6.18). Southernmost Florida has a coast of this type.

Branching *mangrove roots* dominate many parts of the southeastern United States coast. The roots dampen wave and current action, creating a quiet environment that provides a haven for the larval forms of many marine organisms, and may trap fine-grained sediment. Mangroves also deposit layers of organic peat on low-lying coasts.

ENVIRONMENTAL GEOLOGY 14.1

The Effects of Rising Sea Level

Long Term

Sea level has risen about 130 meters in the past 15,000 years as the Pleistocene glaciers melted, adding water to the oceans. During this period, sea level initially rose at the rapid rate of 1.3 meters per century, but the rate of rise gradually slowed down, so that for the past 3,000 years, the rise has only been about 4 centimeters per century. Since about 1930, however, the rate of sea-level rise has increased six-fold, to 24 centimeters per century along the Atlantic and Gulf coasts (the reasons for this increase are not clear). When sea level rose this rapidly a few thousand years ago, the coasts were rapidly eroded. If all the glacial ice on Earth were to melt, sea level would rise 60 meters, drowning many coastal areas.

On low-lying coasts, a rise in sea level causes ocean waves (particularly storm waves) to extend much farther inland than before, flooding the land (box figure 1A). Barrier islands migrate landward. In the United States from New Jersey south to Florida and along the Gulf of Mexico to southern Texas, the coastal land is very flat. A small rise in sea level can send seawater many kilometers inland.

On steep coasts, a rise in sea level can accelerate the erosion of coastal cliffs, destroying oceanfront property (box figure 1B).

Some scientists predict that the rise in sea level may accelerate in the next century as a result of global warming by the "green-house effect." Burning of coal and oil has increased the amount of carbon dioxide in the atmosphere by more than 10% in this century. The carbon dioxide and other gases may trap more of the Sun's energy on Earth, warming the air and the sea. Warm air would accelerate the melting of glaciers on Greenland and Antarctica. Warm seawater expands. Together, these two effects could raise sea level.

Predictions on the amount of sea-level rise by the year 2100 vary widely. Early predictions of a rise of 1 to 2 meters have recently been revised downward to .3 to .6 meter as computer models were refined (by considering cloud cover, for example). A rise of even 1 meter on a low-lying coast could destroy thousands of beachfront houses and hotels.

Global warming is an emotional and controversial subject. The recent rise in atmospheric carbon dioxide is indisputable, but temperature data, while convincing some scientists that Earth is warming, indicate to other scientists that Earth is not changing in temperature or is even cooling. The possibility of future warming is widely accepted, but evidence of warming to date is debatable. The debate over global warming does not change the fact that glaciers are now melting (as they have for 15,000 years) and sea level is now rising. The scientific argument is about whether the present rise of 24 centimeters per century will continue or will accelerate.

Short Term

Hurricanes, which are common along the Atlantic and Gulf coasts, tend to cause short-term rises in sea level called *storm surges*. Hurricanes consist of strong winds rotating counter-clockwise (in the Northern Hemisphere) around a region of very low air pressure. They may be 500 kilometers in diameter and contain winds

BOX 14.1 ■ FIGURE 1

Rising sea level can cause erosion of both gentle (A) and steep (B) coasts, leading to destruction of buildings.

up to 300 kilometers (180 miles) per hour. The low air pressure in the center of a hurricane allows the sea surface to rise forming a broad dome, and this rise in sea level is accentuated by strong onshore winds, which pile water against shore (box figure 2). Storm surges can easily raise sea level 5 meters and are most devastating at high tide. A storm surge 8 meters high struck Galveston, Texas, in 1900, completely covering the barrier island on which Galveston is built. High storm waves on top of the high sea level destroyed countless buildings, and 6,000 people died, many by drowning. Hurricane tracking programs cut the death tolls of storm surges today by providing advance warning to coastal communities. In September 1989, hurricane Hugo hit South Carolina with 220-kilometer (135-mile) per hour winds and a 5-meter storm surge north of Charleston, causing $10 billion in damage. Because more than 500,000 people were evacuated along the low-lying coast, however, the death toll was only 29. Early warning and evacuation of residents in Florida prevented major loss of life in August and September 2004 when hurricanes Charley, Frances, Ivan, and Jeanne made landfall. Low-lying areas along the Atlantic and Gulf coasts were repeatedly hit by high winds and flooding from excessive rainfall and storm surges.

Probably the largest natural disaster in the history of the United States occurred the week of August 29, 2005 when Hurricane Katrina grew to a Category 5 storm in the Gulf of Mexico (box figure 3A) and made a direct hit on the Louisiana and the Mississippi coastlines. Katrina weakened slightly before making landfall, but caused a 9-meter high storm surge and sustained winds up to 233-kilometers (145-miles) per hour that completely destroyed low-lying coastal communities in Mississippi (box figure 3B) and Louisiana, and caused floodwaters up to 7-meters high in New Orleans when the storm surge from Lake Pontchartrain broke levees protecting the city. The world watched in horror as many poor and elderly, who could not evacuate the city before Katrina hit, tried to survive for five days before food and water arrived from the government.

Careful planning for the use of coastal land in your lifetime will be necessary to lessen destruction caused by storm surges and long-term sea-level rise. Developed land at the ocean's edge may face a grim choice in the future—abandonment of existing buildings and restoration of coastal wetlands that act as a 'sponge' to storm waves, or expensive "armoring" of the coast with structures such as seawalls in hopes of protecting the buildings from wave damage.

Additional Resource

S. J. Williams, K. Dodd, and K. K. Gohn. 1990. *Coasts in crisis.* U.S. Geological Survey Circular 1075. Online version can be found at http://pubs.usgs.gov/circular/c1075/.

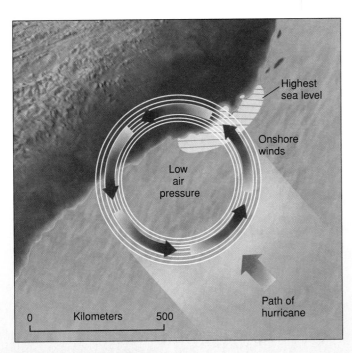

BOX 14.1 ■ FIGURE 2

Strong onshore winds in a hurricane pile water against the shore, forming a storm surge (high sea level) that may cause severe flooding on a low-lying coast. This is particularly a problem where irregular-shaped bays and estuaries (i.e. Lake Pontchartrain) funnel the water onshore. Damage is worse at high tide.

A

B

BOX 14.1 ■ FIGURE 3

(A) Satellite image of Hurricane Katrina taken on August 28, 2005 as it grew to a monster Category 5 hurricane in the Gulf of Mexico with sustained winds up to 233-kilometers (145-miles) per hour. (B) Damage caused by Hurricane Katrina in Gulfport, Mississippi where splintered wood from houses completely destroyed by the high winds and storm surge washed inland along with shipping containers, boats, and recreational vehicles *Photo A by NOAA; Photo B © Paul J. Richards/AFP/Getty Images*

SUMMARY

Wind blowing over the sea surface forms waves, which transfer some of the wind's energy to shorelines. Orbital water motion extends to a depth equal to half the wavelength.

As a wave moves into shallow water, the ocean bottom flattens the orbital motion and causes the wave to slow and peak up, eventually forming a *breaker* whose crest topples forward. The turbulence of *surf* is an important agent of sediment erosion and transportation.

Wave refraction bends wave crests and makes them more parallel to shore. Few waves actually become parallel to the shore, and so *longshore currents* develop in the surf zone. *Rip currents* carry water seaward from the surf zone.

A beach consists of a *berm, beach face,* and *marine terrace.* Summer beaches have a wide berm and a smooth offshore profile. Winter beaches are narrow, with offshore bars.

Longshore drift of sand is caused by the waves hitting the beach face at an angle and by longshore currents.

Deposition of sand that is drifting along the shore can form *spits* and *baymouth bars.* Drifting sand may also be deposited against jetties or groins or inside breakwaters.

Rivers supply most sand to beaches, although local erosion may also contribute sediment. If the river supply of sand is cut off by dams, the beaches gradually disappear.

Coasts may be erosional or depositional, drowned or uplifted, or shaped by organisms such as corals and mangroves.

Coastal straightening by waves is caused by headland erosion and by deposition within bays.

A coast retreating under wave erosion can be marked by *sea cliffs,* a *wave-cut platform, stacks,* and *arches.*

Waves can form *barrier islands* off gently sloping coasts. River and glacial deposition can also shape coasts.

Drowned coasts are marked by *estuaries* and *fiords.* *Uplifted marine terraces* characterize coasts that have risen faster than the recent rise in sea level.

Terms to Remember

arch (sea arch) 372
barrier island 372
baymouth bar 367
beach 366
beach face 366
berm 366
breaker 363
coast 370
coastal straightening 371

crest (of wave) 362
estuary 374
fiord 374
headland 371
longshore current 364
longshore drift 367
marine terrace 366
rip current 364
sea cliff 371

spit 367
stack 372
surf 364
tombolo 368
trough (of wave) 362
wave-cut platform 372
wave height 362
wavelength 363
wave refraction 364

Testing Your Knowledge

Use the following questions to prepare for exams based on this chapter.

1. Show in a sketch how longshore drift of sand can form a baymouth bar.

2. In a sketch, show how and why sand moves along a beach face when waves approach a beach at an angle.

3. How do summer beaches differ from winter beaches? Discuss the reasons for these differences.

4. What would happen to the beaches of most coasts if all the rivers flowing to the sea were dammed? Why?

5. What does the presence of an estuary imply about the recent geologic history of a region?

6. Describe how waves can straighten an irregular coastline.

7. Describe the transition of deep-water waves into surf.

8. Show in a sketch the refraction of waves approaching a straight coast at an angle. Explain why refraction occurs.

9. What is a longshore current? Why does it occur?

10. What is a rip current? Why does it occur? How do you get out of a rip current?

11. The path a water particle makes as a wave passes in deep water is best described as

    a. elliptical   b. orbital

    c. spherical   d. linear

12. The easiest method of escaping a rip current is to

    a. swim toward shore

    b. swim parallel to the shore

    c. swim away from the shore

13. Why is beach sediment typically quartz-rich sand?

    a. other minerals are not deposited on beaches

    b. quartz is the only mineral that can be sand-sized

    c. quartz is resistant to chemical weathering

    d. none of the preceding

14. Longshore drift is
   a. the movement of sediment parallel to shore when waves strike the shoreline at an angle
   b. a type of rip current
   c. a type of tide
   d. the movement of waves

15. Which structure would interfere with longshore drift?
   a. jetties   b. groins
   c. breakwaters   d. all of the preceding

16. What is the most common source of sand on beaches?
   a. sand from river sediment brought down to the ocean
   b. land next to the beach
   c. offshore sediments

17. Which would characterize an erosional coast?
   a. headlands   b. sea cliffs
   c. stacks   d. arches
   e. all of the preceding

18. Which would characterize a depositional coast?
   a. headlands   b. sea cliffs
   c. stacks   d. arches
   e. barrier islands

19. A glacial valley drowned by rising sea level is
   a. a fiord   b. an estuary
   c. a tombolo   d. a headland

20. The surf zone is
   a. the region in which waves break
   b. water less than one-half wavelength in depth
   c. where the longshore current flows
   d. all of the preceding

21. The storm surge of a hurricane is
   a. the highest winds
   b. the tallest waves
   c. the dome of high water in the center of the hurricane
   d. the area of high pressure within the storm

Expanding Your Knowledge

1. Sea level would rise by about 60 meters if *all* the glacial ice on Earth melted. How many U.S. cities would this affect?

2. What happens to a coast if its offshore reef dies?

3. Is a beach a good place to mine sand for construction? Explain your answer carefully.

4. The seaward tip of a headland may be the most rapidly eroding locality on a coast yet also the most expensive building site on a coast. Why is this so?

Exploring Web Resources

www.mhhe.com/plummer11e

This is the dedicated website for this book. You can go to it for new and updated information. The universal resource locators (URL) listed in this book are also given as links on the website, making it easy to go to those websites without typing in the URL. When you visit our Online Learning Center, go to the Student Center and then to the chapter of interest. Here you will find additional readings and media resources, as well as answers to the Testing Your Knowledge section, more quizzes, animation and video clips, and direct links to the sites listed in this book. Links to additional websites can also be found. We have added questions for some of the sites to allow you to get the most of your exploration of the web. Using the web is an enjoyable way of enhancing your knowledge of geology.

http://marine.usgs.gov/

Web page for the *Coastal and Marine Geology Program* of the U.S. Geological Survey contains information about numerous geologic studies of U.S. coastal areas.

http://pubs.usgs.gov/circular/c1075/

Online version of *Coasts in Crisis*, U.S. Geological Survey Circular 1075.

http://woodshole.er.usgs.gov/

Web page for the U.S. Geological Survey *Wood's Hole Field Center* for coastal and marine research contains information and data from ongoing scientific projects.

www-ccs.ucsd.edu/

Web page for the *Center for Coastal Studies at the Scripps Institute of Oceanography* provides information about its research and access to data collected from various coastal studies projects.

www.esdim.noaa.gov/ocean_page.html

Oceans web page from *National Oceanic and Atmospheric Administration* provides numerous links to oceanography research projects and data.

 # Animations

This chapter includes the following animations available on our Online Learning Center at **www.mhhe.com/plummer11e**.

14.8 Seasonal beach cycle
14.9 Wave refraction and longshore movement of sand and water

Geologic Structures

In previous chapters, we have discussed how rock at the surface of Earth is affected by the atmosphere, hydrosphere, and biosphere. We now shift our focus to processes in the solid Earth system, or geosphere. In this chapter, we explain how rocks respond to tectonic forces caused by the movement of lithospheric plates and how geologists study the resulting geologic structures. Studying structural geology is very much like looking at the architecture of the crust and trying to relate how rocks that were once deposited under water in horizontal layers are now found bent (folded) and broken (faulted) many kilometers above sea level.

Subsequent chapters will require an understanding and knowledge of structural geology as presented in this chapter. To understand earthquakes, for instance, one must know about faults. Appreciating how major mountain belts and the continents have evolved (chapter 20) calls for a comprehension of faulting and folding. Understanding plate-tectonic theory as a whole (chapter 19)

Rocks that were once horizontal have been contorted into folds during mountain building, Damaraland, Namibia, Africa. *Photo © Michael Fogden/ Animals, Animals/Earth Scenes* also requires a knowledge of structural geology. (Plate-tectonic theory developed primarily to explain certain structural features.) In areas of active tectonics, the location of geologic structures is important in the selection of safe sites for schools, hospitals, dams, bridges, and nuclear power facilities.

Also, understanding structural geology can help us more fully appreciate the problem of finding more of Earth's dwindling natural resources. Chapter 21 discusses the association of certain geologic structures with petroleum deposits and other valuable resources.

TECTONIC FORCES AT WORK

Stress and Strain in the Earth's Lithosphere

Tectonic forces deform parts of the lithosphere, particularly along plate margins. Deformation may cause a change in orientation, location, and shape of a rock body. In figure 15.1, originally horizontal rock layers have been deformed into wavelike folds that are broken by faults. The layers have been deformed, probably by tectonic forces that pushed or compressed the layers together until they were shortened by buckling and breaking.

When studying deformed rocks, structural geologists typically refer to **stress,** a force per unit area. Where stress can be measured, it is expressed as the force per unit area at a particular point; however, it is difficult to measure stress in rocks that are buried. We can observe the effects of past stress (caused by

tectonic forces and confining pressure from burial) when rock bodies are exposed after uplift and erosion. From our observations, we may be able to infer the principal directions of stress that prevailed. We also can observe in exposed rocks the effect of forces on a rock that was stressed. **Strain** is the change in shape or size (volume), or both, in response to stress.

The relationship between stress and strain can be illustrated by deforming a piece of Silly Putty® (figure 15.2) or any other soft material such as pizza dough. If the Silly Putty® is pushed together or squeezed from opposite directions, we say the stress is **compressive.** Compressive stress results in rocks being *shortened* or *flattened*. In figure 15.2A, an elongate piece of Silly Putty® may shorten by bending, or folding, whereas a ball of Silly Putty® will flatten by shortening in the direction parallel to the compressive stress and elongating or stretching in the direction perpendicular to it. Rocks that have been shortened or flattened are typically found along convergent plate boundaries where rocks have been pushed or shoved together.

FIGURE 15.1

Deformed sedimentary beds exposed in a road cut near Palmdale, California. Squeezing due to movement along the San Andreas fault caused the sedimentary layers to be contorted into folds and broken by smaller faults. *Photo by C. C. Plummer*

A

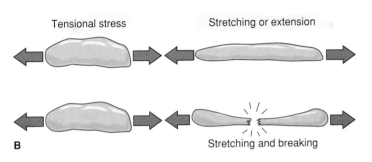

B

FIGURE 15.2

The effects of compressional and tensional stresses on Silly Putty®. (A) Compressing Silly Putty® results in shortening either by folding or flattening. (B) Pulling (tensional stress) Silly Putty® causes stretching or extension; if pulled (strained) too fast, or chilled, the Silly Putty® will break after first stretching.

FIGURE 15.3

Shear strain can be modeled by shearing a deck of cards.

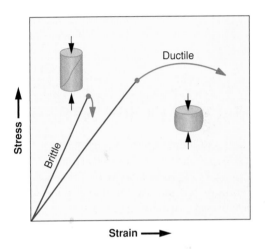

FIGURE 15.4

Graph shows the behavior of rocks with increasing stress and strain. Elastic behavior occurs along the straight line portions (shown in blue) of the graph. At stresses greater than the elastic limit (red points), the rock will either deform as a ductile material or break, as shown in the deformed rock cylinders.

A **tensional stress** is caused by forces pulling away from one another in opposite directions (figure 15.2*B*). Tensional stress results in a *stretching or extension* of material. If we apply a tensional stress on a ball of Silly Putty®, it will elongate or stretch parallel to the applied stress. If the tensional stress is applied rapidly, the Silly Putty® will first stretch and then break apart (figure 15.2*B*). At divergent plate boundaries, the lithosphere is undergoing extension as the plates move away from one another. Because rocks are very weak when pulled apart, fractures and faults are common structures.

When stresses act parallel to a plane, **shear stress** is produced. It is much like putting a deck of cards in your hands and shearing the deck by moving your hands in opposite directions (figure 15.3). A shear stress results in a *shear strain* parallel to the direction of the stresses. Shear stresses occur along actively moving faults.

How Do Rocks Behave When Stressed?

Rocks behave as elastic, ductile, or brittle materials, depending on the amount and rate of stress applied, the type of rock, and the temperature and pressure under which the rock is strained.

If a deformed material recovers its original shape after the stress is reduced or removed, the behavior is **elastic.** For example, if a tensional stress is applied to a rubber band, it will stretch as long as the stress is applied, but once the stress is

removed, the rubber band returns (or recovers) to its original shape and its behavior is elastic. Silly Putty® will behave elastically if molded into a ball and bounced. Most rocks can behave in an elastic way at very low stresses (a few kilobars). However, once the stress applied exceeds the **elastic limit** (figure 15.4), the rock will deform in a permanent way, just as the rubber band will break if stretched too far.

A rock that behaves in a **ductile** or plastic manner will bend while under stress and does not return to its original shape after the stress is removed. Silly Putty® behaves as a ductile material unless the rate of strain is rapid. As discussed in chapter 7, rocks exposed to elevated pressure and temperature during regional metamorphism also behave in a ductile manner and develop a planar texture, or *foliation,* due to the alignment of minerals. As shown in figure 15.4, material behaving in a ductile manner does not require much of an increase in stress to continue to strain (relatively flat curve). Ductile behavior results in rocks that are permanently deformed mainly by folding or bending of rock layers (figure 15.1).

A rock exhibiting **brittle** behavior will fracture at stresses higher than its elastic limit, or once the stresses are greater than the strength of the rock. Rocks typically exhibit brittle behavior at or near Earth's surface, where temperatures and pressure are low. Under these conditions, rocks favor breaking rather than bending. Faults and joints are examples of structures that form by brittle behavior of the crust.

A sedimentary rock exposed at Earth's surface is brittle; it will fracture if you hit it with a hammer. How then do sedimentary rocks, such as those shown in figure 15.1, become bent (or deformed in a ductile way)? The answer is that either stress increased very slowly or that the rock was deformed under considerable confining pressure (buried under more rock) and higher temperatures.

Note, however, that there are some fractures (faults) disrupting the bent layers in figure 15.1. This tells us that although the rock was ductile initially, the amount of stress increased or the rate of strain increased and the rock fractured.

STRUCTURES AS A RECORD OF THE GEOLOGIC PAST

Some geologic structures that give us clues to the past have been described in earlier chapters. Batholiths, stocks, dikes, and sills, for example, are keys to past igneous activity (see chapter 3 on intrusive activity). In this chapter, we are mainly concerned with types of structures that can provide a record of crustal deformation that is no longer active. Very old structures that are now visible at Earth's surface were once buried and are exposed through erosion.

The study of geologic structures is of more than academic interest. The petroleum and mining industries, for example, employ geologists to look for and map geologic structures associated with oil and metallic ore deposits. Understanding and mapping geologic structures is also important for evaluat-

ing problems related to engineering decisions and seismic risk, such as determining the most appropriate sites for building dams, large bridges, or nuclear reactors, and even houses, schools, and hospitals.

Geologic Maps and Field Methods

In an ideal situation, a geologist studying structures would be able to fly over an area and see the local and regional patterns of bedrock from above. Sometimes this is possible, but very often soil and vegetation conceal the bedrock. Therefore, geologists ordinarily use observations from a number of individual *outcrops* (exposures of bedrock at the surface) in determining the patterns of geologic structures (figure 15.5). The characteristics of rock at each outcrop in an area are plotted on a map by means of appropriate symbols. With the data that can be collected, a geologist can make inferences about those parts of the

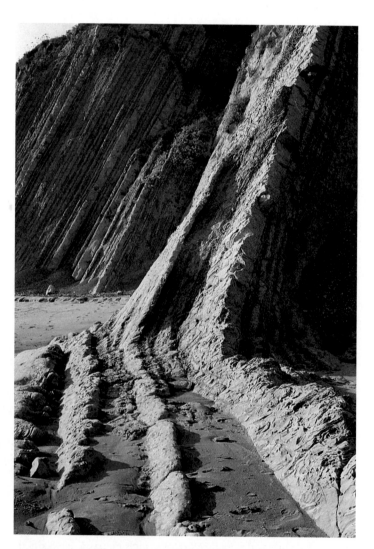

FIGURE 15.6

Tilted sedimentary beds along the coast of northern California near Point Arena. Here, the strike is the line formed by the intersection of the tilted sedimentary beds and the horizontal layer of sand in the foreground. The direction of dip is toward the left. *Photo by Diane Carlson*

FIGURE 15.5

Geology students mapping tilted beds of rocks from a ridge-top vantage point, Mojave Desert, California. *Photo by Diane Carlson* area he or she cannot observe. A **geologic map,** which uses standardized symbols and patterns to represent rock types and geologic structures, is typically produced from the field map for a given area (for example, see the geologic map of North America inside the front cover). On such a map are plotted the type and distribution of rock units, the occurrence of structural features (folds, faults, joints, etc.), ore deposits, and so forth. Sometimes surficial features, such as deposits by former glaciers, are included, but these may be shown separately on a different type of geologic map.

Anyone trained in the use of geologic maps can gain considerable information about local geologic structures because standard symbols and terms are used on the maps and the accompanying reports. For example, the symbol ⊕ on a geologic map denotes horizontal bedding in an outcrop. Different colors and patterns on a geologic map represent distinct rock units.

Strike and Dip

According to the principle of *original horizontality,* sedimentary rocks and some lava flows and ashfalls are deposited as horizontal beds or strata. Where these originally horizontal rocks are found tilted, it indicates that tilting must have occurred after deposition and lithification (figure 15.6). Someone studying a geologic map of the area would want to know the extent and direction of tilting. By convention, this is determined by plotting the relationship between a surface of an inclined bed and an imaginary horizontal plane. You can understand the relationship by looking carefully at figure 15.7, which represents sedimentary beds exposed alongside a lake (the lake surface provides a convenient horizontal plane for this discussion).

Strike is the compass direction of a line formed by the intersection of an inclined plane with a horizontal plane. In this example, the inclined plane is a bedding plane. You can see from figure 15.7 that the beds are striking from north to south. Customarily, only the northerly direction (of the strike line) is given, so we simply say that beds strike north a certain number of degrees east or west (such as N50°E).

Observe that the **angle of dip** is measured downward from the horizontal plane to the bedding plane (an inclined plane). Note that the angle of dip (30° in the figure) is measured within a vertical plane that is perpendicular to both the bedding and the horizontal planes.

The **direction of dip** is the compass direction in which the angle of dip is measured. If you could roll a ball down a bedding surface, the compass direction in which the ball rolled would be the direction of dip.

The dip angle is always measured at a right angle to the strike—that is, perpendicular to the strike line as shown in figure 15.7. Because the beds could dip away from the strike line in either of two possible directions, the general direction of dip is also specified—in this example, west.

A specially designed instrument called a Brunton pocket transit (after the inventor) is used by geologists for measuring the strike and dip (figure 15.8). The pocket transit contains a com-

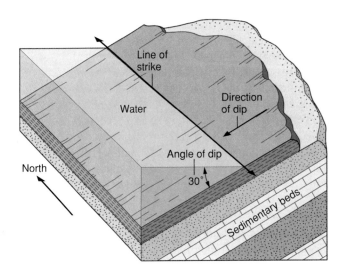

FIGURE 15.7

Strike, direction of dip, and angle of dip. The line of strike is found where an inclined bed intersects a horizontal plane (as shown here by the water). The dip direction is always perpendicular to the strike and in the direction the bed slopes (or a ball would roll down). The dip angle is the vertical angle of the inclined bed as measured from the horizontal.

pass, a level, and a device for measuring angles of inclination. Besides recording strike and dip measurements in a field notebook, a geologist who is mapping an area draws strike and dip symbols on the field map, such as ⅄ or ⅄ for each outcrop with dipping or tilted beds. On the map, the intersection of the two lines at the center of each strike and dip symbol represents the location of the outcrop where the strike and dip of the bedrock were measured. The long line of the symbol is aligned with the compass direction of the strike. The small tick, which is always drawn perpendicular to the strike line, is put on one side or the other, depending on which of the two directions the beds actually

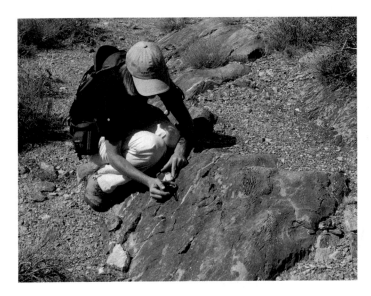

FIGURE 15.8

Geologist determining the strike and dip of an inclined limestone bed in the White Mountains of southern California. *Photo by Diane Carlson* dip. The angle of dip is given as a number next to the appropriate symbol on the map. Thus, ^{25} ⅄ indicates that the bed is dipping 25° from the horizontal toward the northwest and the strike is northeast (assuming that the top of the page is north). The orientation of the bed would be written N45°E, 25°NW. Figure 15.9 is a geologic map with cross section that shows all the sedimentary layers striking north and dipping 30° to the west (N0°, 30°W).

Beds with vertical dip require a unique symbol because they dip neither to the left nor the right of the direction of strike. The symbol used is ⤬, which indicates that the beds are striking northeast and that they are vertical (N30°E, 90°).

Geologic Cross Sections

A **geologic cross section** represents a vertical slice through a portion of Earth. It is much like a road cut (see figure 15.1) or the wall of a quarry in that it shows the orientation of rock units and structures in the vertical dimension. Geologic cross sections are constructed from geologic maps by projecting the dip of rock units into the subsurface (figure 15.9), and are quite useful in helping visualize geology in three dimensions. They are used extensively throughout this book as well as in professional publications.

FIGURE 15.9

A geologic map and cross section of an area with three sedimentary formations. (Each formation may contain many individual sedimentary layers, as explained in chapter 6). Beds strike north and dip 30° to the west. The geologic cross section (vertical cut) is constructed between points A and B on the map.

FIGURE 15.10

Folded sedimentary rock layers exposed at Lulworth Cove, Dorset, England. *Photo © Tom Bean*

FOLDS

Folds are bends or wavelike features in layered rock. Folded rock can be compared to several layers of rugs or blankets that have been pushed into a series of arches and troughs. Folds in rock often can be seen in road cuts or other exposures (figure 15.10). When the arches and troughs of folds are concealed (or when they exist on a grand scale), geologists can still determine the presence of folds by noticing repeated reversals in the direction of dip taken on outcrops in the field or shown on a geologic map.

The fact that the rock is folded or bent shows that it behaved as a ductile material. Yet the rock exposed in outcrops is generally brittle and shatters when struck with a hammer. The rock is not metamorphosed (most metamorphic rock is intensely folded because it is ductile under the high pressure and temperature environment of deep burial and tectonic stresses). Perhaps folding took place when the rock was buried at a moderate depth where higher temperature and confining pressure favor ductile behavior. Alternatively, folding could have taken place close to the surface under a very low rate of strain.

Geometry of Folds

Determining the geometry or shape of folds may have important economic implications because many oil and gas deposits (see box 15.1) and some metallic mineral deposits are localized in folded rocks. The geometry of folds is also important in unraveling how a rock was strained and how it might be related to the movement of tectonic plates. Folds are usually associated with shortening of rock layers along convergent plate boundaries but are also commonly formed where rock has been sheared along a fault.

Because folds are wavelike forms, two basic fold geometries are common—anticlines and synclines (figure 15.11).

An **anticline** is an upward arching fold. Usually the rock layers dip away from the **hinge line** (or *axis*) of the fold. The downward-arching counterpart of an anticline is a **syncline**, a troughlike fold. The layered rock usually dips toward the syncline's hinge line. In the series of folds shown in figure 15.11, two anticlines are separated by a syncline. Each anticline and adjacent syncline share a **limb.** Note the hinge lines on the crests of the two anticlines and bottom of the syncline. Similar hinge lines could be located in the hinge areas at the contacts between any two adjacent folded layers. For each anticline and the syncline, the hinge lines are contained within the shaded vertical planes. Each of these planes is an **axial plane,** an imaginary plane containing all of the hinge lines of a fold. The axial plane divides the fold into its two limbs.

It is important to remember that anticlines are not necessarily related to ridges nor synclines to valleys, because valleys and ridges are nearly always erosional features. In an area that has been eroded to a plain, the presence of underlying anticlines and synclines is determined by the direction of dipping beds in exposed bedrock, as shown in figure 15.12. (In the field, of course, the cross sections are not exposed to view as they are in the diagram.)

Figure 15.12 also illustrates how determining the relative ages of the rock layers, or beds, can tell us whether a structure is an anticline or a syncline. Observe that the oldest exposed rocks are along the hinge line of the anticline. This is because lower layers in the originally flat-lying sedimentary or volcanic rock were moved upward and are now in the core of the anticline. The youngest rocks, on the other hand, which were originally in the upper layers, were folded downward and are now exposed along the synclinal hinge line.

Plunging Fold

The examples shown so far have been of folds with horizontal hinge lines. These are the easiest to visualize. In nature, however, anticlines and synclines are apt to be **plunging folds**—that is, folds in which the hinge lines are not horizontal. On a surface leveled by erosion, the patterns of exposed strata (beds) resemble Vs or horseshoes (figures 15.13 and 15.14) rather than the parallel, striped patterns of layers in nonplunging folds. However, plunging anticlines and synclines are distinguished from one another in the same way as are nonplunging folds—by directions of dip or by relative ages of beds.

A plunging syncline contains the youngest rocks in its center or core, and the V or horseshoe points in the direction opposite of the plunge. Conversely, a plunging anticline contains the oldest rocks in its core, and the V points in the same direction as the plunge of the fold.

FIGURE 15.12

By measuring the strike and dip of exposed sedimentary beds in the field and plotting them on a geologic map (top surface) geologists can interpret the geometry of the geologic structure below the ground surface.

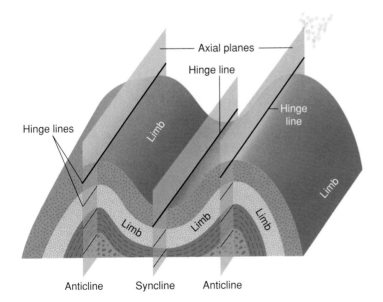

FIGURE 15.11

Diagrammatic sketch of two anticlines and a syncline illustrating the axial planes, hinge lines, and fold limbs.

Hinge line

A

Hinge line of plunging anticline

Hinge line of plunging syncline

Axial plane

Youngest formation

Oldest formation

Axial planes

B

FIGURE 15.13

(*A*) Plunging fold that is cut by a horizontal plane has a V-shaped pattern. (*B*) Plunging anticline on left and right and plunging syncline in center. The hinge lines plunge toward front of block diagram and lie within the axial planes of the folds.

FIGURE 15.14

Rock layers dip away from center of a plunging anticline exposed in Utah. Anticline plunges in the direction of the upper part of the photo. *Photo by Frank M. Hanna*

Geologist's View

388

Structural Domes and Structural Basins

A **structural dome** is a structure in which the beds dip away from a central point. In cross section, a dome resembles an anticline and is sometimes called a doubly plunging anticline. In a **structural basin,** the beds dip toward a central point (figure 15.15); in cross section, it is comparable to a syncline (doubly plunging syncline). A structural basin is like a set of nested bowls. If the set of bowls is turned upside down, it is analogous to a structural dome.

Domes and basins tend to be features on a grand scale (some are more than a hundred kilometers across), formed by uplift somewhat greater (for domes) or less (for basins) than that of the rest of a region. Michigan's lower peninsula and parts of adjoining states and Ontario are on a large structural basin (see map on the inside front cover). Domes of similar size are found in other parts of the Middle West. Smaller domes are found in the Rocky Mountains (figure 15.16).

Domes and anticlines (as well as some other structures) are important to the world's petroleum resources, as described in chapter 21.

FIGURE 15.16

Dome near Casper, Wyoming. The ridges are sedimentary layers that are resistant to erosion. Beds dip away from the center of the dome where the oldest layers are exposed. *Photo by D. A. Rahm, courtesy of Rahm Memorial Collection, Western Washington University*

Oldest layer exposed at the surface

Youngest layer

A

Youngest layer

Oldest layer exposed at the surface

B

FIGURE 15.15

(A) Structural dome. (B) Structural basin.

Further Description of Folds

Folds occur in many varieties and sizes. Some are studied under the microscope, while others can have adjacent hinge lines tens of kilometers apart. Some folds are a kilometer or more in height. Figure 15.17 shows several of the more common types of folds. **Open folds** (figure 15.17*A*) have limbs that dip gently. All other factors being equal, the more open the fold, the less it has been strained by shortening. By contrast, an **isoclinal fold,** one in which limbs are nearly parallel to one another, implies larger shortening strain or shear strain (figure 15.17*B*).

Folds that have vertical axial planes are referred to as upright folds. However, where the axial plane of a fold is not vertical but is inclined or tipped over, the fold may be classified as *asymmetric*. If the axial plane is inclined to such a degree that the fold limbs dip in the same direction, the fold is classified as an **overturned fold** (figure 15.17*C*). Looking at an outcrop where only the overturned limb of a fold is exposed, you would probably conclude that the youngest bed is at the top. The principles of *superposition* (see chapter 8), however, cannot be applied to determine top and bottom for overturned beds. You must either see the rest of the fold or find primary sedimentary structures within the beds such as mud cracks that indicate the original top or upward direction.

Recumbent folds (figure 15.17*D*) are overturned to such an extent that the limbs are essentially horizontal. Recumbent folds are found in the cores of mountain ranges such as the Canadian Rockies, Alps, and Himalayas and record extreme shortening and shearing of the crust typically associated with plate convergence.

A Open folds

B Isoclinal ("hairpin") folds

C Overturned folds

D Recumbent folds

FIGURE 15.17

Various types of folds. The length of the arrows in *A* through *E* is proportional to the amount and direction of shortening and shearing that caused folding. (*A*) Open folds in Spain (they are plunging away from the people). (*B*) Tight to isoclinal folds from the Bighorn Mountains, Wyoming. (*C*) Overturned anticline from northern California. (*D*) Recumbent folds in the Alps. *Photo A by C. C. Plummer; Photos B and C by Diane Carlson; Photo D courtesy of Professor John Ramsay, from J. G. Ramsay & M. J. Huber, The Techniques of Modern Structural Geology, vol. 2., © 1987 Academic Press*

FIGURE 15.18

Vertical joints in sedimentary rock of the Colorado Plateau formed in response to tectonic uplift of the region. *Photo by Frank M. Hanna*

FRACTURES IN ROCK

If a rock is brittle, it will fracture. Commonly, there is some movement or displacement. If essentially no shear displacement occurs, a fracture or crack in bedrock is called a **joint.** If the rock on either side of a fracture moves parallel to the fracture plane, the fracture is a *fault.* Most rock at or near the surface is brittle, so nearly all exposed bedrock is jointed to some extent.

Joints

In discussing volcanoes, we described *columnar jointing,* in which hexagonal columns form as the result of contraction of a cooling, solidified lava flow. *Sheet jointing,* a type of jointing due to expansion (discussed along with weathering in chapter 5), is caused by the pressure release due to removal of overlying rock and has the effect of creating tensional stress perpendicular to the land surface.

Columnar and sheet joints are examples of fractures that form from nontectonic stresses and are therefore referred to as primary joints. In this chapter, we are concerned with joints that form not from cooling or unloading but from tectonic stresses.

Joints are one of the most commonly observed structures in rocks (figure 15.18*A*). A joint is a fracture or crack in a rock body along which essentially no displacement has occurred. Joints form at shallow depths in the crust where rock breaks in a brittle way and is pulled apart slightly by tensional stresses caused by bending or regional uplift. Where joints are oriented approximately parallel to one another, a **joint set** can be defined.

Geologists sometimes find valuable ore deposits by studying the orientation of joints. For example, hydrothermal solutions may migrate upward through a set of joints and deposit quartz and economically important minerals such as gold, silver, copper, and zinc in the cracks (figure 15.19). Accurate information about joints also is important in the planning and construction of large engineering projects, particularly dams and reservoirs. If the bedrock at a proposed location is intensely jointed, the possibility of dam failure or reservoir leakage may make that site too hazardous. The movement of contaminated ground water from unlined landfills and abandoned mines may also be controlled by joints, which results in difficult and costly clean ups.

FIGURE 15.19

Fractures in altered granitic rock are filled with quartz, copper, and iron sulfides (chalcopyrite and pyrite). Silver Bell mine, Arizona. *Photo by Diane Carlson*

Is There Oil Beneath My Property? First Check the Geologic

Structure

An "oil pool" can exist only under certain conditions. Crude oil does not fill caves underground, as the term *pool* may suggest; rather, it simply occupies the pore spaces of certain sedimentary rocks, such as poorly cemented sandstone, in which void space exists between grains. Natural gas (less dense) often occupies the pore spaces above the crude oil, while water (more dense) is generally found saturating the rock below the oil pool (box figure 1).

A **source rock,** which is always a sedimentary rock, must be present for oil to form. The sediment of the source rock has to include remains of organisms buried during sedimentation. This organic matter partially decomposes into petroleum and natural gas. Once formed, the droplets of petroleum tend to migrate, following fractures and interconnecting pore spaces. Being less dense than the rock, the petroleum usually migrates upward, although horizontal migration does occur.

If it is not blocked by impermeable rock, the oil may migrate all the way to the surface, where it is dissipated and permanently lost. Natural oil seeps, where leakage of petroleum is taking place, exist both on land and offshore. Where impermeable rock blocks the oil droplets' path of migration, an oil pool may accumulate below the rock, much like helium-filled balloons might collect under a domed ceiling. For any significant amount of oil to collect, the rock below the impermeable rock must be porous as well as permeable. Such a rock, when it contains oil, is called a **reservoir rock.**

Another necessary condition is that the geologic structure must be one that favors the accumulation and retention of petroleum. An "anticlinal trap" is one of the best structures for holding oil. As oil became a major energy source and the demand for it increased, most of the newly discovered wells penetrated anticlinal traps. Geologists discovered these by looking for indication of anticlines exposed at the surface. As time went by, other types of structures were also found to be oil traps. Many of these were difficult to find because of the lack of telltale surface patterns indicating favorable underground structures. Box figure 2 illustrates some structures other than anticlinal traps that might have a potential for oil production.

At present, oil companies rely on detailed and sophisticated geologic studies of an area they hope may have the potential for an "oil strike." The petroleum industry also depends heavily on geophysical techniques (see chapter 17) for determining, by indirect means, the subsurface structural geology.

Even when everything indicates that conditions are excellent for oil to be present underground, there is no

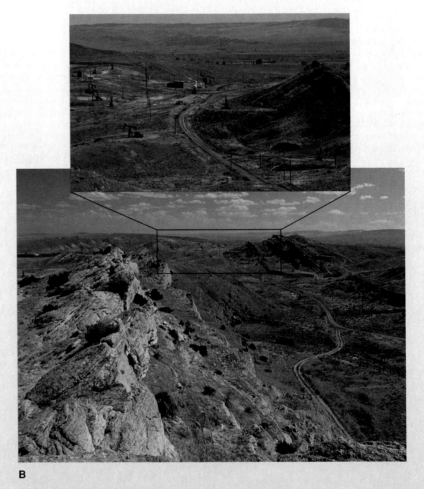

B

BOX 15.1 ■ FIGURE 1

(A) Oil and gas are concentrated or trapped in hinge of anticline. Gas and oil float on water in porous and permeable reservoir rock (sandstone). (B) Eroded anticline forms trap in Lander oil field, Wyoming. *Photos by Diane Carlson* guarantee that oil will be found. Eventually, an oil company must commit a million dollars or more to drill a deep test well, or "wildcat" well. Statistics indicate that the chance of a test well yielding commercial quantities of oil is much less than 1 in 10. As more and more of the world's supply of petroleum is used up, what is left becomes increasingly harder—and costlier—to find.

A Fault

B Unconformity

Permeable rock
C Sedimentary facies change

BOX 15.1 ■ FIGURE 2
Structures other than anticlines that trap oil.

FIGURE 15.20

Fault in Big Horn Mountains, Wyoming, is marked by a 2-meter wide zone of broken, red-stained rocks that offset rock layers. *Photo by Diane Carlson*

Geologist's View

Faults

Faults are fractures in bedrock along which sliding has taken place. The displacement may be only several centimeters or may involve hundreds of kilometers. For many geologists, an active fault is regarded as one along which movement has taken place during the last 11,000 years. Most faults, however, are no longer active.

The nature of past movement ordinarily can be determined where a fault is exposed in an outcrop. The geologist looks for dislocated beds or other features of the rock that might show how much displacement has occurred and the relative direction of movement. In some faults, the contact between the two displaced sides is very narrow. In others, the rock has been broken or ground to a fractured or pulverized mass sandwiched between the displaced sides (figure 15.20).

Geologists describe fault movement in terms of direction of slippage: dip-slip, strike-slip, or oblique-slip (figure 15.21). In a **dip-slip fault,** movement is parallel to the dip of the fault surface. A **strike-slip fault** indicates *horizontal* motion parallel to the strike of the fault surface. An **oblique-slip fault** has both strike-slip and dip-slip components.

Reverse fault

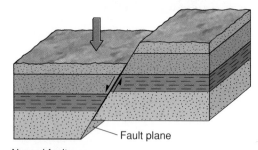

Normal fault

A Dip-slip faults

B Strike-slip fault

C Oblique-slip fault

FIGURE 15.21

Three types of faults illustrated by displaced blocks. Although both blocks probably move when the fault slips, the heavier arrows show only the direction of movement on the left. (A) Dip-slip movement. (B) Strike-slip movement. (C) Oblique-slip movement. Black arrows show dip-slip and strike-slip components of movement.

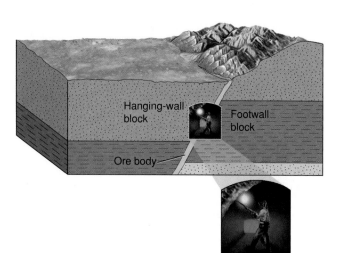

FIGURE 15.22

Relationship between the hanging-wall block and footwall block of a fault. The upper surface where a miner can hang a lantern is the hanging wall. The lower surface below the fault is the footwall.

Dip-slip Faults

In a dip-slip fault, the movement is up or down parallel to the dip of the inclined fault surface. The side of the fault above the inclined fault surface is called the **hanging wall,** whereas the side below the fault is called the **footwall** (figure 15.22). These terms came from miners who tunneled along the fault looking for veins of mineralized rock (ore). As they tunneled, their feet were on the lower *footwall block* and they could hang their lanterns on the upper surface, or *hanging-wall block.*

Normal and reverse faults, the most common types of dip-slip faults, are distinguished from each other on the basis of the relative movement of the footwall block and the hanging-wall block. In a **normal fault** (figures 15.23 and 15.24), the hanging-wall block has moved down relative to the footwall block. The relative movement is represented on a geological cross section by a pair of arrows, because geodetic measurement of normal faults indicate that both blocks move during slip. As shown in figure 15.23, a normal fault results in extension or lengthening of the crust. When there is extension of the crust, the hanging-wall block moves downward along the fault to compensate for the pulling apart of the rocks. Sometimes a block bounded by normal faults will drop down, creating a *graben,* as shown in figure 15.23C. (*Graben* is the German word for "ditch.") *Rifts* are grabens associated with diverging plate boundaries, either along mid-oceanic ridges or on continents (see chapters 18 and 19). The Rhine Valley in Germany and Red Sea are examples of grabens.

If a block bounded by normal faults is uplifted sufficiently, it becomes a fault-block mountain range. (This is also called *a horst,* the opposite of a graben.) The Teton mountains and Sierra Nevada mountains are spectacular examples of fault-block mountain ranges. The Basin and Range province of Nevada and portions of adjoining states are also characterized by numerous mountain ranges (horsts) separated from adjoining valleys by normal faults (see chapter 20). Normal

A Normal fault

B Eroded normal fault

Shaded part eroded
away after faulting

Hanging-
wall block

Footwall
block

Hanging-
wall block

Footwall
block

Graben

C Graben

Horst

D Horst

FIGURE 15.23

Normal faults. (*A*) Diagram shows the fault before erosion and the geometric relationships of the fault. (*B*) The same area after erosion. (*C*) A graben. (*D*) A horst. Arrows in *C* and *D* indicate horizontal extension of the crust.

FIGURE 15.24

Normal faults with prominent horst block offset volcanic ash layers in southern Oregon. *Photo by Diane Carlson*

Horst

Normal faults

Geologist's View

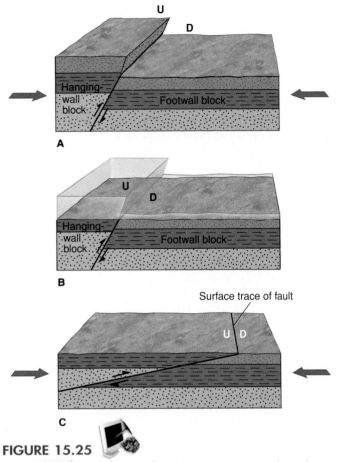

FIGURE 15.25

(A) A reverse fault. The fault is unaffected by erosion. Arrows indicate shortening direction. (B) Diagram shows area after erosion. (C) Thrust fault has a lower angle of dip and accommodates more shortening by stacking rock layers on top of one another.

fault planes typically dip at steep angles (60°) at shallow depths but may become curved or even horizontal at depth (see figure 20.9).

In a **reverse fault,** the hanging-wall block has moved up relative to the footwall block (figures 15.25 and 15.26). As shown in figure 15.25, horizontal compressive stresses cause reverse faults. Reverse faults tend to shorten the crust.

A **thrust fault** is a reverse fault in which the dip of the fault plane is at a low angle (< 30°) or even horizontal (figures 15.25C and 15.27). In some mountain regions, it is not uncommon for the upper plate (or hanging-wall block) of a thrust fault to have overridden the lower plate (footwall block) for several tens of kilometers. Thrust faults typically move or thrust older rocks on top of younger rocks (figure 15.27) and result in an extreme shortening of the crust. Thrust faults commonly form at convergent plate boundaries to accommodate the pushing together and shortening during convergence.

Strike-slip Faults

A fault where the movement (or *slip*) is predominantly horizontal and therefore parallel to the strike of the fault is called a **strike-slip fault.** The displacement along a strike-slip fault is either left-lateral or right-lateral and can be determined by looking across the fault. For instance, if a recent fault displaced a stream (figure 15.28A), a person walking along the stream would stop where it is truncated by the fault. If the person looks across the fault and sees the stream displaced to the right, it is a **right-lateral fault.** In a **left-lateral fault,** a stream or other

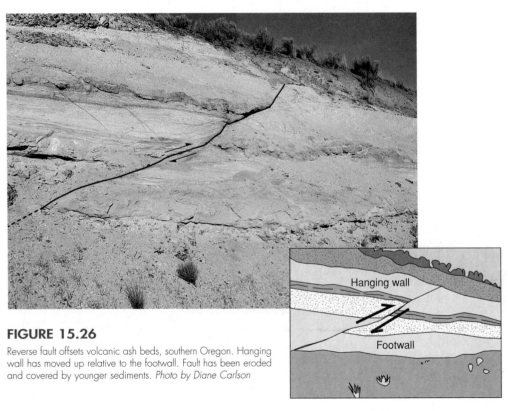

FIGURE 15.26

Reverse fault offsets volcanic ash beds, southern Oregon. Hanging wall has moved up relative to the footwall. Fault has been eroded and covered by younger sediments. *Photo by Diane Carlson*

Geologist's View

A

B

FIGURE 15.27

(A) Chief Mountain in Glacier National Park, Montana, is an erosional remnant of a major thrust fault. *(B)* Cross section of the area. Older (Precambrian) rocks have been thrust over younger (Cretaceous) rocks. Dashed lines show where the Lewis thrust fault has been eroded away. *Photo by Frank M. Hanna*

FIGURE 15.28

(A) Right-lateral strike-slip fault offsets a stream channel. Looking across the fault, you would need to walk to the right to find the continuation of the stream. *(B)* Strike-slip movement along curved faults produces gaps or basins at releasing bends where the lithosphere is pulled apart or shortening and hills where it is pushed together at restraining bends.

IN GREATER DEPTH 15.2

California's Greatest Fault—The San Andreas

The San Andreas fault in California is the best-known geologic structure in the United States, but the geologists and seismologists who study it admit that our knowledge of its activity and its history is far from complete. Actually, the San Andreas is the longest of several, subparallel faults that transect western California (box figure 1). Collectively, these right-lateral faults are known as the San Andreas fault system. The system is in a belt approximately 100 kilometers wide and 1,300 kilometers long that extends into Mexico, ending at the Gulf of California.

Los Angeles is slowly moving toward the San Francisco Bay area because of San Andreas fault motion. At an average rate of movement of about 2 centimeters per year, Los Angeles could be a western suburb of San Francisco (or San Francisco an eastern suburb of Los Angeles) in some 25 million years. Earthquakes are produced by sudden movement within the fault system, as explained in chapter 16. Bedrock along the San Andreas fault shifted as much as 4.5 meters in association with the 1906 quake that destroyed much of San Francisco.

The San Andreas fault is not a simple crack but a belt of broken and ground-up rock, usually a hundred meters or more wide. Its presence is easy to determine throughout most of its length. Along the fault trace are long, straight valleys (formed by erosion and subsidence) that show quite different rocks on either side. Stream channels follow much of the fault zone because the weak, ground-up material along the fault is easily eroded. Locally, elongate lakes (called sag ponds) are found where the ground-up material has settled more than the surface of adjacent parts of the fault zone. The fault was named after one of these ponds, San Andreas Lake, just south of San Francisco (box figure 2).

One can visually follow the fault northward from San Andreas Lake into the southwestern suburbs of San Francisco. There the fault zone is hidden by recently built housing tracts. Apparently the builders and residents have chosen to ignore the hazards of living on the nation's most famous fault.

Geologists have been unable to agree on the total displacement of the fault or on how long it has been active. Some believe movement began in the Mesozoic Era (over 65 million years ago); most think that it began later. The difficulty in establishing an age for the inception of the faulting lies in finding clear evidence of displaced bedrock. What geologists would like to find, if it exists, is a rock unit that can be isotopically dated and that was formed across the fault zone just before the faulting began. Currently displaced rock on both sides of the fault zone would have to be clearly identifiable as having been the same unit.

Geologically young features that cross the fault, such as displaced stream channels (box figure 3), are common. Similarly, ancient rocks that undoubtedly were there before faulting began are recognized as having been displaced. Many California geologists think that the belt of granitic rock just west of the fault was once the southern continuation of the granitic batholiths of the Sierra Nevada (box figure 4A), which are more than 80 million years old. But these extremes tell us only that the age of the San

BOX 15.2 ■ FIGURE 1

California has its faults. Red lines indicate faults that have been active within the last 200 years, and blue lines indicate faults that have been active over the last 2 million years. *From California Division of Mines and Geology*

Andreas fault is somewhere between approximately 80 million years and a few thousand years, when the stream channel in box figure 3 carved its course across the fault.

The strongest evidence for long-term faulting comes from almost identical volcanic sequences now 315 kilometers apart. The volcanic activity took place 23.5 million years ago. Using these figures, we can calculate the average rate of motion as 1.3 centimeters per year for the San Andreas fault. (But movement along other faults means the rate of motion is higher for the fault system.) Older rocks that appear to have been offset 560 kilometers have been correlated with less certainty, suggesting that the total offset for the San Andreas fault is at least 560 kilometers.

How long ago faulting began remains controversial. According to plate-tectonic theory, the San Andreas fault is a transform

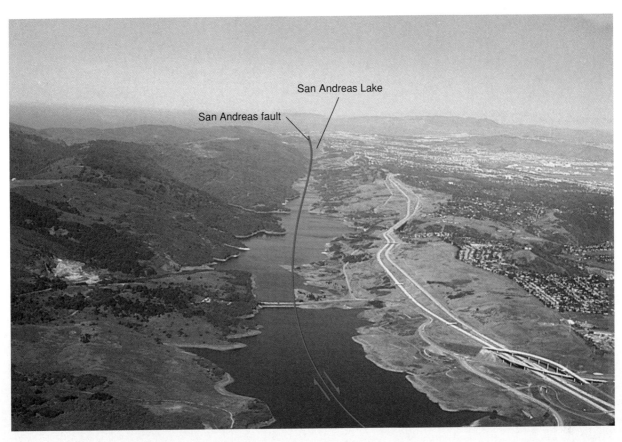

San Andreas Lake

San Andreas fault

BOX 15.2 ■ FIGURE 2
Part of the San Andreas fault. View northward toward San Francisco. Lakes occupy the fault zone. Hills to the left of the fault are moving northward. *Photo by B. Amundson*

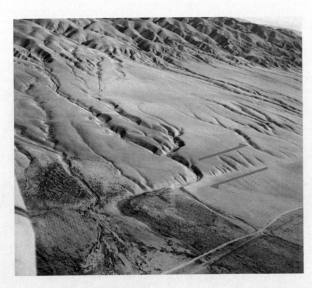

BOX 15.2 ■ FIGURE 3
Stream channel (Wallace Creek) displaced by the San Andreas fault. The arrows on either side of the fault trace indicate relative motion. *Photo by C. C. Plummer*

boundary that separates the North American plate from the Pacific plate. One hypothesis places the beginning of strike-slip movement at about 30 million years ago. According to this hypothesis, the Baja California peninsula split from mainland Mexico as seafloor spreading began (box figure 4*B*). As the Gulf of California widens, the block of crust west of the San Andreas is pushed northward.

Additional Resources

R. E. Wallace, ed. 1990. *The San Andreas fault system, California.* U.S. Geological Survey Professional Paper 1515.
Internet version of the San Andreas fault system.

- http://pubs.usgs.gov/gip/earthq3/

"What have we learned about the San Andreas fault since 1906" web page shows research sites along the San Andreas fault.

- http://quake.usgs.gov/info/1906/index.html

(continued)

(continued)

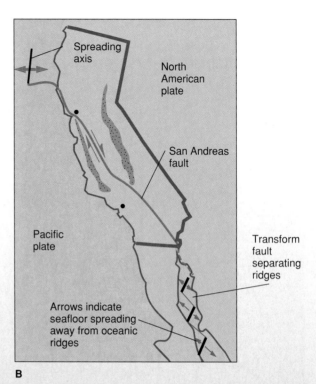

A

B

BOX 15.2 ■ FIGURE 4

(A) Reconstruction of California and Mexico as they may have been before faulting. (B) Continuous opening of the Gulf of California creates motion along the San Andreas fault. *After Tanya Atwater, 1970. Geological Society of American Bulletin. For animations of the tectonic development of California and western North America, visit* www.geol.ucsb.edu/faculty/atwater/

displaced feature would appear to the left across the fault. Again, we cannot tell which side actually moved, so pairs of arrows are used to indicate relative movement.

Large strike-slip faults, such as the San Andreas fault in California, typically define a zone of faulting that may be several kilometers wide and hundreds of kilometers long (see box 15.2). The surface trace of an active strike-slip fault is usually defined by a prominent linear valley that has been more easily eroded where the rock has been ground up along the fault during movement. The linear valley may contain lakes or sag ponds where the impermeable fault rock causes ground water to pond at the surface. The trace of the fault may also be marked by offset surface features such as streams, fences, and roads or by distinctive rock units.

Strike-slip faults that have experienced a large amount of offset typically do not remain straight for long distances.

They may either bend or step over to another fault that is parallel. Depending on the direction of the bend or stepover, the lithosphere is either pulled apart (*releasing bend*) or pushed together (*restraining bend*) (figure 15.28*B*). Normal faults and grabens form in response to the pulling apart at the releasing bends and folds and thrust faults form at the restraining bends to accommodate the pushing or pinching together of the lithosphere.

Strike-slip faults accommodate shearing strain in the brittle, uppermost lithosphere, and may also represent transform plate boundaries where plates slide past one another. One of the most famous examples of a transform fault is the San Andreas fault. The San Andreas fault is a right-lateral strike-slip fault that forms part of the boundary between the North American and Pacific plates (see box 15.2, figure 4).

SUMMARY

Tectonic forces result in deformation of the Earth's crust. *Stress* (force per unit area) is a measure of the tectonic force and confining pressure acting on bedrock. Stress can be *compressive, tensional,* or *shearing. Strained* (changed in shape or size) rock records past stresses, usually as joints, faults, or folds.

A geologic map shows the structural characteristics of a region. *Strike* and *dip* symbols on geologic maps indicate the orientations of inclined surfaces such as bedding planes. The strike and dip of a bedding surface indicate the relationship between the inclined plane and a horizontal plane.

If rock layers bend (ductile behavior) rather than break, they become folded. Rock layers are folded into *anticlines* and *synclines* and recumbent folds. If the hinge line of a fold is not horizontal, the fold is *plunging.* Older beds exposed in the core of a fold indicate an anticline, whereas younger beds in the center of the structure indicate a syncline. In places where folded rock has been eroded to a plain, an anticline can usually be distinguished from a syncline by whether the beds dip toward the center (syncline) or away from the center (anticline). Also, the oldest rocks are found in the center of an eroded anticline whereas the youngest rocks are found in the center or core of a syncline.

Fractures in rock are either *joints* or *faults.* A joint indicates that movement has not occurred on either side of the fracture; displaced rock along a fracture indicates a fault. *Dip-slip* faults are either *normal* or *reverse,* depending on the motion of the hanging-wall block relative to the footwall block. The relative motion of the hanging wall is upward in a reverse fault and downward in a normal fault. A reverse fault with a low angle of dip for the fault plane is a *thrust fault.* Reverse faults accommodate horizontal shortening of the crust, whereas normal faults accommodate horizontal stretching or extension.

In a *strike-slip* fault, which can be either left-lateral or right-lateral, horizontal movement parallel to the strike has occurred.

Terms to Remember

angle of dip 385	geologic map 385	reservoir rock 392
anticline 387	hanging wall 394	reverse fault 396
axial plane 387	hinge line 387	right-lateral fault 396
brittle 383	isoclinal fold 389	shear stress 383
compressive stress 382	joint 391	source rock 392
dip-slip fault 393	joint set 391	strain 382
direction of dip 385	left-lateral fault 396	stress 382
ductile 383	limb 387	strike 385
elastic 383	normal fault 394	strike-slip fault 393
elastic limit 383	oblique-slip fault 393	structural basin 389
fault 393	open fold 389	structural dome 389
fold 386	overturned fold 389	syncline 387
footwall 394	plunging fold 387	tensional stress 383
geologic cross section 386	recumbent fold 389	thrust fault 396

Testing Your Knowledge

Use the following questions to prepare for exams based on this chapter.

1. Most anticlines have both limbs dipping away from their hinge lines. For which kind of fold is this not the case?

2. What is the difference between a joint and a fault?

3. On a geologic map, if no cross sections were available, how could you distinguish an anticline from a syncline?

4. If you locate a dip-slip fault while doing field work, what kind of evidence would you look for to determine whether the fault is normal or reverse?

5. What factors control whether a rock behaves as a brittle material or a ductile material?

6. What is the difference between strike, direction of dip, and angle of dip?

7. Draw a simple geologic map using strike and dip symbols for a syncline plunging to the west.

8. How does a structural dome differ from a plunging anticline?

9. Which of the statements is true?

    a. when forces are applied to an object, the object is under stress

    b. strain is the change in shape or size (volume), or both, while an object is undergoing stress

    c. stresses can be compressive, tensional, or shear

    d. all of the preceding

10. The compass direction of a line formed by the intersection of an inclined plane with a horizontal plane is called

    a. strike

    b. direction of dip

    c. angle of dip

    d. axis

11. Folds in a rock show that the rock behaved in a _____ way.

    a. ductile

    b. elastic

    c. brittle

    d. all of the preceding

12. An anticline is

    a. an upward-arched fold with the youngest rocks exposed along the hinge line

    b. a downward-arched fold with the oldest rocks exposed along the hinge line

    c. an upward-arched fold with the oldest rocks exposed along the hinge line

    d. a downward-arched fold with the youngest rocks exposed along the hinge line

13. A syncline is

    a. an upward-arched fold with the youngest rocks exposed along the hinge line

    b. a downward-arched fold with the oldest rocks exposed along the hinge line

    c. an upward-arched fold with the oldest rocks exposed along the hinge line

    d. a downward-arched fold with the youngest rocks exposed along the hinge line

14. A structure in which the beds dip away from a central point is called a

    a. basin

    b. anticline

    c. structural dome

    d. syncline

15. Which is not a type of fold?

    a. open

    b. isoclinal

    c. overturned

    d. recumbent

    e. thrust

16. Fractures in bedrock along which movement has taken place are called

    a. joints

    b. faults

    c. cracks

    d. folds

17. In a normal fault, the hanging-wall block has moved _____ relative to the footwall block.

    a. upward

    b. downward

    c. sideways

18. Normal faults occur where

    a. there is horizontal shortening

    b. there is horizontal extension

    c. the hanging wall moves up

    d. where the footwall moves down

19. Faults that typically move older rock on top of younger rock are

    a. normal faults

    b. thrust faults

    c. strike-slip faults

Expanding Your Knowledge

1. Why do some rocks fold while others are faulted?

2. In what parts of North America would you expect to find the most intensely folded rock?

3. A subduction zone can be regarded as a very large example of what type of fault?

4. Looking at the San Andreas fault, shown in box 15.2, figure 1, where might restraining bends form? What kind of structures might form there?

5. What features in sedimentary or volcanic rock layers would you look for to tell you that the rock was part of the overturned limb of a fold?

6. Can you identify and name the various geologic structures shown in the figures in chapter 8?

Exploring Web Resources

www.mhhe.com/plummer11e

This is the dedicated website for this book. You can go to it for new and updated information. The universal resource locators (URL) listed in this book are also given as links on the website, making it easy to go to those websites without typing in the URL. When you visit our Online Learning Center, go to the Student Center and then to the chapter of interest. Here you will find additional readings and media resources, as well as answers to the Testing Your Knowledge section, more quizzes, animation and video clips, and direct links to the sites listed in this book. Links to additional websites can also be found. We have added questions for some of the sites to allow you to get the most of your exploration of the web. Using the web is an enjoyable way of enhancing your knowledge of geology.

http://earth.leeds.ac.uk/faultzone/index.html

School of Earth Sciences, University of Leeds website is a dynamic collection of in-depth information and photos and diagrams of geologic structures and virtual field trips throughout the world.

http://www.science.smith.edu/departments/Geology/Structure_Resources/

The *structural geology* page developed by Steven Schimmrich contains many links to online courses, computer software, bibliographies, and many other structural geology resources.

http://www.geology.sdsu.edu/visualgeology/geology101/geo100/strucDIR.htm

Structural Geology website by Gary Girty at San Diego State University contains animations of folds, faults, and interactive exercises dealing with the interpretation and description of geologic structures.

www.geo.cornell.edu/geology/classes/geol326/326.html

Website for structural geology course taught by the Department of Geological Sciences at Cornell University contains images showing structural features and models of thrust-fault movement.

http://craton.geol.brocku.ca/ctg.html

Canadian Tectonics Group website contains structural geology images, computer software, and a newsletter outlining research projects.

 # Animations

This chapter includes the following animations available on our Online Learning Center at **www.mhhe.com/plummer11e**.

15.17 Styles of folding
15.21 Styles of faulting
15.23 Normal faulting
15.25 Reverse and thrust faults

Earthquakes

T his chapter will help you understand the nature and origin of earthquakes. We discuss the seismic waves created by earthquakes and how the quakes are measured and located by studying these waves. We also describe some effects of earthquakes, such as ground motion and displacement, damage to buildings, and quake-caused fires, landslides, and seismic sea waves (tsunami).

Earthquakes commonly affect other parts of Earth systems. Intense shaking associated with an earthquake not only can cause tremendous damage and loss of life but can also trigger landslides that may disperse pathogenic microbes into the atmosphere and cause additional human health concerns. Such was the case after the 1994

Cars are flattened underneath a collapsed ("pancaked") apartment building in Islamabad, Pakistan after a magnitude 7.6 earthquake struck on October 8, 2005 and was centered 110 kilometers to the northeast of the capital city. More than 49,000 people were killed and 74,000 injured when intense shaking collapsed buildings, caused mudslides, and destroyed entire villages. The 2005 Pakistan earthquake was the strongest to hit the region in a century and resulted in the worst devastation in Pakistan history. *Photo © Warrick Page/ Getty Images*

Northridge, California earthquake. Another effect on the biosphere may be the unusual behavior of animals just before an earthquake, as reported by Chinese scientists. Ground breakage associated with earthquakes may affect the hydrosphere by creating new lakes (sag ponds), increasing groundwater flow from springs, and displacing stream channels. Tsunami generated by submarine earthquakes may cause tremendous damage to the coastal environment.

Earthquakes are largely confined to a few narrow belts on Earth. This distribution was once puzzling to geologists, but here we show how the concept of plate tectonics neatly explains it.

As geologists learn more about earthquake behavior, the possibility exists that we will be able to forecast earthquakes. We conclude the chapter with a look at this developing branch of study.

INTRODUCTION

On April 18, 1906, at 5:12 in the morning, part of California slid abruptly past the rest of the state during a great earthquake. A visible scar 450 kilometers (279 miles) long was left where the Earth was torn along coastal northern California. Rock was displaced horizontally as much as 4.5 meters (15 feet); soil above the rock was displaced up to 6.5 meters. The quake, located on a segment of the San Andreas fault near San Francisco, shook the ground for one full minute. Buildings toppled in San Francisco, and broken gas mains fed fires that raged for three days (figure 16.1A). Broken water mains hampered fire fighting. The fires were finally extinguished when buildings were dynamited to create a firebreak. Terrified and homeless people moved to refuge camps set up in city parks. Looters were shot on sight. As the city gradually recovered from the shock of the devastation, it was found that at least 3,000 people had died and $400 million (in 1906 dollars) of damage had been done. Perhaps 90% of the destruction was caused by the fires.

At 5:04 P.M. on October 17, 1989, San Francisco was again severely shaken for fifteen seconds by the Loma Prieta earthquake located to the south on the San Andreas fault near Santa Cruz. Although the quake did not tear the ground surface, it collapsed some buildings and freeway overpasses built upon the soft "bay fill" sediment in San Francisco and Oakland (figure 16.1B). A section of the Bay Bridge collapsed. Just as in 1906, raging fires were fed by broken gas mains in the Marina district of San Francisco and were difficult to fight because of broken water mains; fireboats helped extinguish them. Very severe damage occurred in small towns near the center of the quake. The death toll was sixty-three, and damage was $6 billion.

At 5:30 P.M. on March 27, 1964, southern Alaska was rocked by an earthquake that lasted for three minutes. Although the force of this earthquake was twice that of the 1906 San Francisco earthquake, loss of life and property was relatively low because of Alaska's small population—fifteen people died as a direct result of the shaking, and damage amounted to slightly over $300 million (in 1964 dollars). The tremor was felt over an area of more than 1 million square kilometers (350,000 square miles). A section of the Earth's surface 50 by 200 kilometers was raised as much as 13 meters, and a similar block of land sank 1 to 2 meters. Horizontal movement was slight. In Anchorage, 150 kilometers (93 miles) from the center of the earthquake, landslides wrecked parts of the city. The

greatest loss of life was caused by large sea waves (tsunami) generated by land movement associated with the earthquake—almost 100 people drowned in Alaska (figure 16.1C), and a few people drowned as far away as Oregon and northern California as the waves spread over the Pacific Ocean.

On January 17, 1994, at 4:31 A.M., the Northridge earthquake rocked the San Fernando Valley just north of Los Angeles, California, for forty seconds. The quake, about 3 kilometers (1.8 miles) from California State University, Northridge, damaged or destroyed all fifty-three CSUN buildings, and seriously damaged 300 other schools. Numerous freeway overpasses collapsed (including some that had previously collapsed in a nearby 1971 quake), closing four interstates and seven other highways for months. The two upper stories of the Northridge Meadows Apartments collapsed onto the lower story, killing sixteen people (figure 16.1D). A three-story concrete parking garage at a shopping center pancaked down, trapping a worker for hours under 3 meters of concrete. Gas and water mains were broken, triggering several hard-to-fight fires; about 100 homes burned at a Sylmar mobile-home park. Fifteen thousand newly homeless people had to live in tents, and tens of thousands of people had no water, electricity, or gas for several days. The death toll of seventy-two was very low because the quake occurred early in the morning on the Martin Luther King holiday, so very few commuters were on the collapsed freeways. Damage exceeded $25 billion.

At 10:54 A.M. on February 28, 2001, Seattle was jolted by the second-largest earthquake to strike the state of Washington in recent history. The Nisqually earthquake was centered 65 kilometers (40 miles) southwest of Seattle near the state capital in Olympia. Because the quake occurred deep within the Earth (49 kilometers) on the down-going Juan de Fuca plate, shaking was felt over a broad area. Although classified as a strong earthquake, the depth of the quake minimized the intensity of shaking, and damage was restricted to older buildings constructed of unreinforced brick and concrete (figure 16.1E) and bridges that had not been seismically retrofitted. No one died as a direct result of the earthquake and injuries numbered around 250. This was not the "big one." The probability still exists that a great earthquake with much larger ground motions could strike the Pacific Northwest.

On November 3, 2002, the largest earthquake ever recorded in the interior of Alaska ruptured the ground surface for more than 300 kilometers (200 miles). The rupture (figure 16.1F), mainly along the right-lateral Denali fault, propagated eastward

A

D

B

E

C

F

FIGURE 16.1

Damage from earthquakes in the United States. (A) Damaged buildings and fires in San Francisco after 1906 earthquake. (B) Collapsed double-deck Cypress freeway in Oakland after the 1989 Loma Prieta earthquake. (C) Tsunami damage from the 1964 Alaska earthquake carried a fishing boat inland in Resurrection Bay at Seward. (D) The collapse of the lower story of the Northridge Meadows Apartments killed sixteen people in the 1994 earthquake in San Fernando Valley, southern California. (E) Damage from falling bricks in downtown Seattle after the February 28, 2001, Nisqually earthquake. (F) Scarp formed near the epicenter of the magnitude 7.9 Denali fault earthquake, the largest earthquake to strike the interior of Alaska. This quake resulted in a rupture that broke the surface for over 320 kilometers (200 miles). *Photo A © Arnold Genthe/Corbis; Photo B © Lloyd Cluff/Corbis; Photo C by National Geophysical Data Center; Photo D © Roger Ressmeyer/Corbis Images; Photo E © AP/Wide World Photos; Photo F by Peter Hausler, U.S. Geological Survey* at more than 7,000 miles per hour and offset streams and glaciers and triggered thousands of landslides. The Denali fault earthquake is similar in size and type to the 1906 San Francisco earthquake, but it caused no deaths and minimal damage because it occurred in a remote area of south-central Alaska. Of concern, however, was the Trans-Alaska Oil Pipeline, which crossed the western part of the Denali fault, where 5 meters (16 feet) of offset were recorded. The pipeline suffered only minor damage and did not break, thus preventing a major environmental and economic disaster. This was due in large part to the pipeline's creative engineering design based on input from geologists regarding the location and maximum offset expected along the fault during a major earthquake.

As mentioned in chapter 1, the most deadly tsunami (seismic sea wave) in recorded history occurred on December 26, 2004 in the India Ocean. The tsunami was generated by a 9.3-magnitude earthquake, the second largest recorded on Earth, as the Indian plate shifted under the Eurasian plate off the coast of Sumatra, Indonesia. The earthquake caused major damage to nearby towns in Sumatra but the deadliest consequence was the large tsunami wave that struck within minutes after the quake. The tsunami continued to travel across the Indian Ocean at the speed of a jetliner and caused massive loss of life and destruction in low-lying coastal communities in Indonesia, Thailand, Sri Lanka, India, and east Africa. More than 220,000 people were killed as a series of waves with the force of a locomotive struck the coasts and caused massive flooding and destruction. Trains and cars were overturned, boats washed far inland, and buildings were destroyed by the incredible force of the waves. There were horrific accounts of the tsunami crashing into beaches in Thailand and Sri Lanka that were filled with tourists from around the world over the Christmas holiday.

CAUSES OF EARTHQUAKES

What causes earthquakes? An **earthquake** is a trembling or shaking of the ground caused by the sudden release of energy stored in the rocks beneath Earth's surface. As described in chapter 15, tectonic forces acting deep in the Earth may put a *stress* on the rock, which may bend or change in shape (*strain*). If you bend a stick of wood, your hands put a stress (the force per unit area) on the stick; its bending (a change in shape) is the strain.

Like a bending stick, rock can deform only so far before it breaks. When a rock breaks, waves of energy are released and sent out through the Earth. These are **seismic waves,** the waves of energy produced by an earthquake. It is the seismic

FIGURE 16.2

The elastic rebound theory of the cause of earthquakes. (*A*) Rock with stress acting on it. (*B*) Stress has caused strain in the rock. Strain builds up over a long period of time. (*C*) Rock breaks suddenly, releasing energy, with rock movement along a fault. Horizontal motion is shown; rocks can also move vertically or diagonally. (*D*) Fence offset nearly 3 meters after 1906 San Francisco earthquake. *Photo by G. K. Gilbert, U.S. Geological Survey* waves that cause the ground to tremble and shake during an earthquake.

The sudden release of energy when rock breaks may cause one huge mass of rock to slide past another mass of rock into a different relative position. As you know from chapter 15, the break between the two rock masses is a *fault*. The classic explanation of why earthquakes take place is called the **elastic rebound theory** (figure 16.2). It involves the sudden release of progressively stored strain in rocks, causing movement along a fault. Deep-seated internal forces (*tectonic forces*) act on a mass of rock over many decades. Initially, the rock bends but does not break. More and more energy is stored in the rock as the bending becomes more severe. Eventually, the energy stored in the rock exceeds the breaking strength of the rock, and the rock breaks suddenly, causing an earthquake. Two masses of rock move past one another along a fault. The movement may be vertical, horizontal, or both (figure 16.3). The strain on the rock is released; the energy is expended by moving the rock into new positions and by creating seismic waves.

Recently, some modifications have been suggested for the sequence of events shown in figure 16.2. The classic model implies that existing faults are strong; a very large stress must act to break rocks along a fault. The new idea is that faults are weak and only need a small stress to cause rupture and an earthquake. The evidence for the new idea is suggestive but not yet conclusive, so we currently have two models for fault behavior. The weak-fault model poses serious problems for earthquake prediction, as you will see later in the chapter.

The brittle behavior of breaking rock is characteristic only of rocks near Earth's surface. Rocks at depth are subject to increased temperature and pressure, which tend to reduce brittleness. Deep rocks behave as ductile materials instead of breaking (*brittle* behavior); hence, there is a limit to the depth at which faults can occur.

Most earthquakes are associated with movement on faults, but in some quakes, the connection with faulting may be difficult to establish. Four recent California quakes, including the 1994 Northridge quake, occurred on buried thrust faults, some of which were unknown and none of which involved surface displacement. Most earthquakes in the eastern United States are also not associated with surface displacement. Earthquakes also occur during explosive volcanic eruptions and as magma forcibly fills underground magma chambers prior to many eruptions; these quakes may not be associated with fault movement at all.

Another cause has been recently postulated for deep earthquakes (100 to 670 kilometers below the surface), essentially all of which are found on cold, subducting plates sliding down into the mantle. Although the down-going plates are colder than the surrounding rock, the high temperature and pressure at depth suggest to some geologists that the rock in the plates should behave in a ductile way rather than breaking in the brittle manner of near-surface rocks. The suggested cause of deep quakes is mineral transformations within the down-going rock, as pressure collapses one mineral into a denser form. Lab experiments have shown bodies of the new, denser minerals

FIGURE 16.3

Horizontal offset of trees in an orchard, 1979, El Centro, California. *Photo © John S. Shelton*

along fractures. Whether the process occurs on a large scale to produce large quakes is unknown. Similar suggestions for the cause of deep quakes include the dehydration of water-containing serpentine and the conversion of serpentine into glass. Both of these processes occur suddenly on small fractures in lab experiments.

SEISMIC WAVES

The point within the Earth where seismic waves first originate is called the **focus** (or *hypocenter*) of the earthquakes (figure 16.4). This is the center of the earthquake, the point of initial breakage and movement on a fault. Rupture begins at the focus and then spreads rapidly along the fault plane. The point on the Earth's surface directly above the focus is the **epicenter.**

Two types of seismic waves are generated during earthquakes. **Body waves** are seismic waves that travel through the Earth's interior, spreading outward from the focus in all directions. **Surface waves** are seismic waves that travel on Earth's

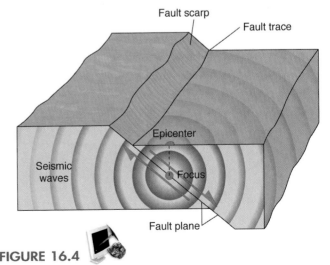

FIGURE 16.4

The focus of an earthquake is the point where rocks first break along a fault; seismic waves radiate from the focus. The epicenter is the point on the Earth's surface directly above the focus.

surface away from the epicenter, like water waves spreading out from a pebble thrown into a pond. Rock movement associated with seismic surface waves dies out with depth into the Earth, just as water movement in ocean waves dies out with depth.

Body Waves

There are two types of body waves, both shown in figure 16.5. A **P wave** is a compressional (or longitudinal) wave in which rock vibrates back and forth *parallel* to the direction of wave propagation. Because it is a very fast wave, traveling through near-surface rocks at speeds of 4 to 7 kilometers per second (9,000 to more than 15,000 miles per hour), a P wave is the first

(or *primary*) wave to arrive at a recording station following an earthquake.

The second type of body wave is called an **S wave** (*secondary*) and is a slower, transverse wave that travels through near-surface rocks at 2 to 5 kilometers per second. An S wave is propagated by a shearing motion much like that in a stretched, shaken rope. The rock vibrates *perpendicular* to the direction of wave propagation, that is, crosswise to the direction the waves are moving.

Both P waves and S waves pass easily through solid rock. A P wave can also pass through a fluid (gas or liquid), but an S wave cannot. We discuss the importance of this fact in chapter 17.

A Primary wave

B Secondary wave

C Love wave

D Rayleigh wave

FIGURE 16.5

Particle motion in seismic waves. (*A*) A P wave is illustrated by a sudden push on the end of a stretched spring or Slinky. The particles vibrate *parallel* to the direction of wave propagation. (*B*) An S wave is illustrated by shaking a loop along a stretched rope. The particles vibrate *perpendicular* to the direction of wave propagation. (*C*) Love waves behave like S waves in that the particle motion is perpendicular to the direction of wave travel along Earth's surface. (*D*) Rayleigh waves are like ocean waves and cause a rolling motion on Earth's surface. The particle motion is elliptical and opposite (counterclockwise) to the direction of wave propagation.

Surface Waves

Surface waves are the slowest waves set off by earthquakes. In general, surface waves cause more property damage than body waves because surface waves produce more ground movement and travel more slowly, so they take longer to pass. The two most important types of surface waves are Love waves and Rayleigh waves, named after the geophysicists who discovered them.

Love waves are most like S waves that have no vertical displacement. The ground moves side to side in a horizontal plane that is perpendicular to the direction the wave is traveling or propagating (figure 16.5C). Like S waves, Love waves do not travel through liquids and would not be felt on a body of water. Because of the horizontal movement, Love waves tend to knock buildings off their foundations and destroy highway bridge supports.

Rayleigh waves behave like rolling ocean waves. Unlike ocean waves, Rayleigh waves cause the ground to move in an elliptical path opposite to the direction the wave passes (figure 16.5D). Rayleigh waves tend to be incredibly destructive to buildings because they produce more ground movement and take longer to pass.

LOCATING AND MEASURING EARTHQUAKES

The invention of instruments that could accurately record seismic waves was an important scientific advance. These instruments measure the amount of ground motion and can be used to find the location, depth, and size of an earthquake.

The instrument used to measure seismic waves is a *seismometer*. The principle of the seismometer is to keep a heavy suspended mass as motionless as possible—suspending it by springs or hanging it as a pendulum from the frame of the instrument (figure 16.6). When the ground moves, the frame of the instrument moves with it; however, the inertia of the heavy mass suspended inside keeps the mass motionless to act as a point of reference in determining the amount of ground motion. Seismometers are usually placed in clusters of three to record the motion along the *x*, *y*, and *z* axes of three-dimensional space.

A seismometer by itself cannot record the motion that it measures. A **seismograph** is a recording device that produces a permanent record of Earth motion detected by a seismometer, usually in the form of a wiggly line drawn on a moving strip of paper (figure 16.7). The paper record of Earth vibration is called a **seismogram.** The seismogram can be used to measure the strength of the earthquake.

A network of seismograph stations is maintained all over the world to record and study earthquakes (and nuclear bomb explosions). Within minutes after an earthquake occurs, distant seismographs begin to pick up seismic waves. A large earthquake can be detected by seismographs all over the world.

Because the different types of seismic waves travel at different speeds, they arrive at seismograph stations in a definite order: first the P waves, then the S waves, and finally the surface waves. These three different waves can be distinguished on the seismograms. By analyzing these seismograms, geologists can learn a great deal about an earthquake, including its location and size.

Determining the Location of an Earthquake

P and S waves start out from the focus of an earthquake at essentially the same time. As they travel away from the quake, the two kinds of body waves gradually separate because they are traveling at different speeds. On a seismogram from a station close to the earthquake, the first arrival of the P wave is separated from the first arrival of the S wave by

A At rest   **B** Ground moves up   **C** Ground moves down

Spring

Heavy weight

Weight stays at same level while the ground moves

FIGURE 16.6

A simple seismograph for detecting vertical rock motion. The pen records the ground motion on the seismogram as the spring stretches and compresses with its up and down movement. Frame and recording drum move with the ground. Inertia of the weight keeps it and the needle relatively motionless.

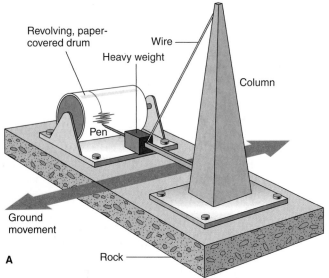

Revolving, paper-covered drum
Heavy weight
Wire
Column
Pen
Ground movement
Rock

A

FIGURE 16.7

(A) A seismograph for horizontal motion. Modern seismographs record Earth motion on moving strips of paper. The mass is suspended by a wire from the column and swings like a pendulum when the ground moves horizontally. A pen attached to the mass records the motion on a moving strip of paper. (B) A seismogram of a magnitude 6.2 earthquake in Taiwan, recorded in Berkeley, California, 6,300 miles away. First arrivals of P, S, and surface waves are shown. *Courtesy of University of California, Berkeley*

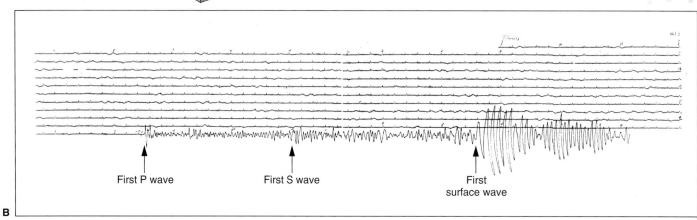

First P wave   First S wave   First surface wave

B

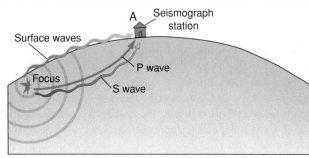

Surface waves
Seismograph station
A
Focus
P wave
S wave

A Station near focus

Time →
Surface
P
S
Time lag
Seismogram from station A

Time of earthquake

Surface waves
Seismograph station
B
Focus
P wave
S wave

B Station far from focus

Time →
P
S
Surface
Time lag
Seismogram from station B

FIGURE 16.8

Because of the difference in travel times, intervals between P waves, S waves, and surface waves increase with distance from the focus.

a short distance on the paper record (figure 16.8). At a recording station far from the earthquake, however, the first arrivals of these waves will be recorded much farther apart on the seismogram. The farther the seismic waves travel, the longer the time intervals between the arrivals of P and S waves and the more they are separated on the seismograms.

Because the time interval between the first arrivals of P and S waves increases with distance from the focus of an earthquake, this interval can be used to determine the distance from the seismograph station to a quake. The increase in the P-S interval is regular with increasing distance for several thousand kilometers and so can be graphed in a **travel-time curve,** which plots seismic-wave arrival time against distance (figure 16.9).

In practice, a station records the P and S waves from a quake, then a seismologist matches the interval between the waves to a standard travel-time curve. By reading directly from the graph, one can determine, for example, that an earthquake has occurred 5,300 kilometers (3,300 miles) away. This determination can often be made very rapidly, even while the ground is still trembling from the quake.

A single station can determine only the distance to a quake, not the direction. A circle is drawn on a globe with the center

of the circle being the station and its radius the distance to the quake (figure 16.10). The scientists at the station know that the quake occurred somewhere on that circle, but from the information recorded, they are not able to tell where. With information from other stations, however, they can pinpoint the location of the quake. If three or more stations have determined the distance to a single quake, a circle is drawn for each station. If this is done on a map, the intersection of the circles locates the epicenter.

Analyses of seismograms can also indicate at what depth beneath the surface the quake occurred. Most earthquakes occur relatively close to Earth's surface, although a few occur much deeper. The maximum **depth of focus**—the distance between focus and epicenter—for earthquakes is about 670 kilometers (416 miles). Quakes are classified into three groups according to their depth of focus:

Shallow focus	0–70 kilometers deep
Intermediate focus	70–350 kilometers deep
Deep focus	350–670 kilometers deep

Shallow-focus earthquakes are most common; they account for 85% of total quake energy released. Intermediate-(12%) and deep-(3%) focus quakes are rarer because most deep rocks

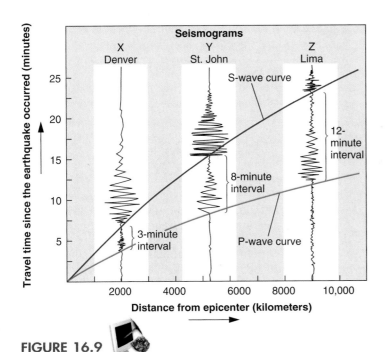

FIGURE 16.9

A travel-time curve is used to determine the distance to an earthquake. Note that the time interval between the first arrival of P and S waves increases with distance from the epicenter. Seismogram X has a 3-minute interval between P and S waves corresponding to a distance of 2,000 km from the epicenter, Y has an interval of 8 minutes, so the earthquake occurred 5,300 km away, and Z an interval of 12 minutes, and is a distance of 9,000 km from the epicenter.

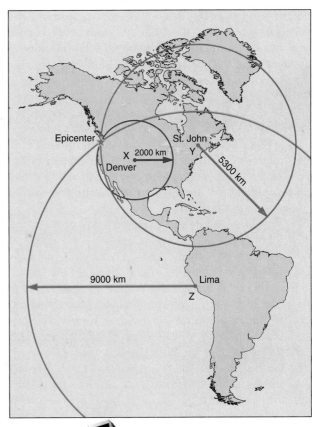

FIGURE 16.10

Locating an earthquake. The distance from each of three stations (Denver, St. John, and Lima) is determined from seismograms and the travel-time curves shown in figure 16.9. Each distance is used for the radius of a circle about the station. The location of the earthquake is just offshore of Vancouver, British Columbia, where the three circles intersect.

flow in a ductile manner when stressed or deformed; they are unable to store and suddenly release energy as brittle surface rocks do.

Measuring the Size of an Earthquake

The size of earthquakes is measured in two ways. One method is to find out how much and what kind of damage the quake has caused. This determines the **intensity,** which is a measure of an earthquake's effect on people and buildings. Intensities are expressed as Roman numerals ranging from I to XII on the **modified Mercalli scale** (table 16.1); higher numbers indicate greater damage.

Although intensities are widely reported at earthquake locations throughout the world, using intensity as a measure of earthquake strength has a number of drawbacks. Because damage generally lessens with distance from a quake's epicenter, different locations report different intensities for the same earthquake (figure 16.11). Moreover, damage to buildings and other structures depends greatly on the type of geologic material on which a structure was built as well as the type of construction. Houses built on solid rock normally are damaged far less than houses built upon loose sediment, such as delta mud or bay fill. Brick and stone houses usually suffer much greater damage than wooden houses, which are somewhat flexible. Damage estimates are also subjective: people may exaggerate damage reports consciously or unconsciously. Intensity maps can be drawn for a single earthquake to show the approximate damage over a wide region (figure 16.11). Intensity maps are useful for assessing how different areas respond to seismic waves and provide valuable information for earthquake planning. But such maps cannot be drawn for uninhabited areas (the open ocean, for instance), so not all quakes can be assigned intensities. The one big advantage of intensity ratings is that no instruments are required, which allows seismologists to estimate the size of earthquakes that occurred before seismographs were available.

The second method of measuring the size of a quake is to calculate the amount of energy released by the quake. This method is usually done by measuring the height (amplitude) of one of the wiggles on a seismogram. The larger the quake, the more the ground vibrates and the larger the wiggle. After measuring a specific wave on a seismogram and correcting for the type of seismograph and for the distance from the quake, scientists can assign a number called the **magnitude.** It is a measure of the energy released during the earthquake.

For the past several decades, magnitude has been reported on the **Richter scale,** a numerical scale of magnitudes. The Richter scale is open-ended, meaning there are no earthquakes too large or too small to fit on the scale. The higher numbers indicate larger earthquakes. Very small earthquakes can have negative magnitudes, but these are seldom reported. The largest Richter magnitude measured so far is 8.6. Smaller earthquakes are much more common than large ones (table 16.2).

There are several methods of measuring magnitude, however. The original Richter scale applied only to shallow earth-

TABLE 16.1	Modified Mercalli Intensity Scale of 1931 (Abridged)

I. Not felt except by a very few under especially favorable circumstances.

II. Felt only by a few persons at rest, especially on upper floors of buildings. Delicately suspended objects may swing.

III. Felt quite noticeably indoors, especially on upper floors of buildings, but many people do not recognize it as an earthquake. Standing motor cars may rock slightly. Vibration like passing of truck. Duration estimated.

IV. During the day felt indoors by many, outdoors by few. At night some awakened. Dishes, windows, doors disturbed; walls made cracking sound. Sensation like heavy truck striking building. Standing motor cars rocked noticeably.

V. Felt by nearly everyone; many awakened. Some dishes, windows, etc., broken; a few instances of cracked plaster; unstable objects overturned. Disturbance of trees, poles, and other tall objects sometimes noticed. Pendulum clocks may stop.

VI. Felt by all; many frightened and run outdoors. Some heavy furniture moved; a few instances of fallen plaster or damaged chimneys. Damage slight.

VII. Everybody runs outdoors. Damage *negligible* in buildings of good design and construction; *slight* to moderate in well-built ordinary structures; *considerable* in poorly built or badly designed structures; some chimneys broken. Noticed by persons driving motor cars.

VIII. Damage *slight* in specially designed structures; *considerable* in ordinary substantial buildings with partial collapse; *great* in poorly built structures. Panel walls thrown out of frame structures. Fall of chimneys, factory stacks, columns, monuments, walls. Heavy furniture overturned. Sand and mud ejected in small amounts. Changes in well water. Persons driving motor cars disturbed.

IX. Damage *considerable* in specially designed structures; well-designed frame structures thrown out of plumb; *great* in substantial buildings, with partial collapse. Buildings shifted off foundations. Ground cracked conspicuously. Underground pipes broken.

X. Some well-built wooden structures destroyed; most masonry and frame structures destroyed with foundations; ground badly cracked. Rails bent. Considerable landslides from river banks and steep slopes. Shifted sand and mud. Water splashed (slopped) over banks.

XI. Few, if any (masonry), structures remain standing. Bridges destroyed. Broad fissures in ground. Underground pipelines completely out of service. Earth slumps and land slips in soft ground. Rails bent greatly.

XII. Damage total. Waves seen on ground surface. Lines of sight and level distorted. Objects thrown upward into the air.

From Wood and Neumann, 1931, *Bulletin of the Seismological Society of America*

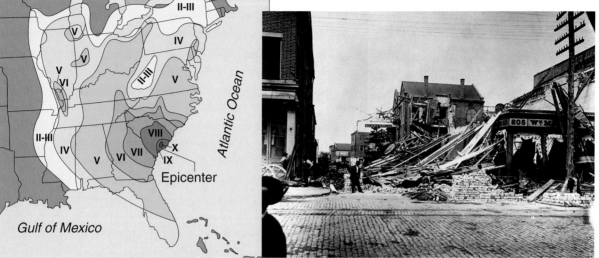

FIGURE 16.11

Zones of different intensity from the 1886 Charleston, South Carolina, earthquake. The map illustrates the general decrease in intensity with increasing distance from the epicenter, as well as the effect of different types of Earth materials. The photo shows damage in Charleston, South Carolina from 1886 earthquake. *Photo by J. K. Hillers, U.S. Geological Survey*

quakes in southern California. Different seismic waves (body or surface) can be measured to make the scale more useful over larger areas, so several different magnitudes are sometimes reported for a single quake. A further complication is that magnitudes calculated from seismograms tend to be inaccurate (usually too low) above magnitude 7.

A better method of calculating magnitude involves the use of the *seismic moment* of a quake, which is determined from the strength of the rock, surface area of the rupture, and the amount of rock displacement along the fault. The **moment magnitude** is the most objective way of measuring the energy released by a large earthquake. The 1964 Alaska quake is estimated to have a moment magnitude of 9.2, the 2004 Sumatra, Indonesia quake a moment magnitude of 9.3, and the 1960 Chile quake 9.5. Unfortunately, the media rarely indicate which type of magnitude they are reporting, and sci-

entists typically revise magnitudes for several weeks after a quake as they receive more information, so trying to find out the "real" magnitude of a recent quake can be confusing.

Because the Richter scale is logarithmic, the difference between two consecutive whole numbers on the scale means an increase of ten times in the amplitude of Earth's vibrations, particularly below magnitude 5. This means that if the measured amplitude of vibration for certain rocks is 1 centimeter during a magnitude-4 quake, these rocks will move 10 centimeters during a magnitude-5 quake occurring at the same location.

It has been estimated that a tenfold increase in the size of Earth vibrations is caused by an increase of roughly 32 times in terms of energy released. A quake of magnitude 5, for example, releases approximately 32 times more energy than one of magnitude 4. A magnitude-6 quake is about 1,000 times (32 × 32) more powerful in terms of energy released than a

TABLE 16.2	Comparison of Earthquake Magnitude, Description, Intensity, and Expected Annual World Occurrence		
Richter Magnitude	Description	Maximum Expected Mercalli Intensity at Epicenter	Annual Expected Number
2.0	Very minor	I Usually detected only by instruments	600,000
2.0–2.9	Very minor	I–II Felt by some indoors, especially on upper floors	300,000
3.0–3.9	Minor	III Felt indoors	49,000
4.0–4.9	Light	IV–V Felt by most; slight damage	6,200
5.0–5.9	Moderate	VI–VII Felt by all; damage minor to moderate	800
6.0–6.9	Strong	VII–VIII Everyone runs outdoors; moderate to major damage	266
7.0–7.9	Major	IX–X Major damage	18
8.0 or higher	Great	X–XII Major and total damage	1 or 2

Source: U.S. Geological Survey

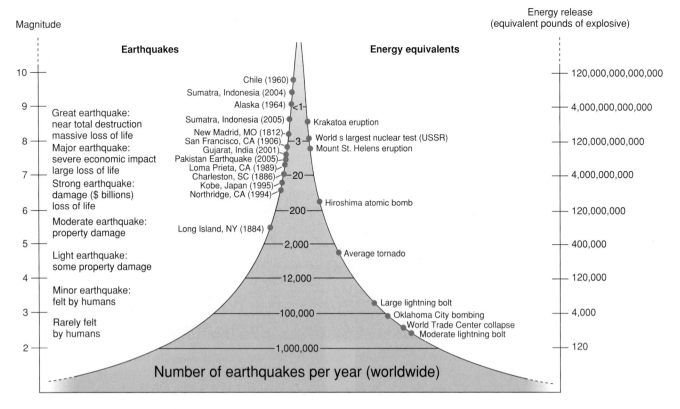

FIGURE 16.12

Diagram shows the relationship between the moment magnitude of an earthquake, the number of earthquakes per year throughout the world, and the energy released during an earthquake. *After IRIS Consortium (www.iris.edu)*

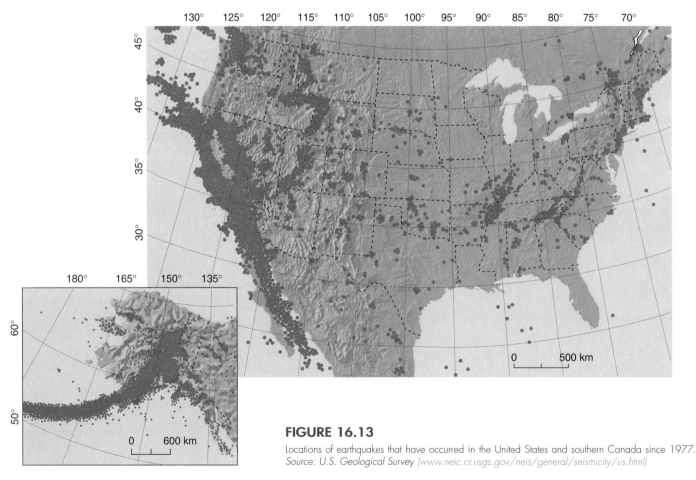

FIGURE 16.13

Locations of earthquakes that have occurred in the United States and southern Canada since 1977. *Source: U.S. Geological Survey (www.neic.cr.usgs.gov/neis/general/seismicity/us.html)*

magnitude-4 quake. The actual energy released in earthquakes of varying magnitudes is shown in figure 16.12.

Although a seismograph is usually required to measure magnitude, this measure has many advantages over intensity as an indicator of earthquake strength. A worldwide network of standard seismograph stations now makes determining magnitude a routine matter; and the media report magnitudes for all earthquakes of interest to the United States. Eventually, a single magnitude number can be assigned to a single earthquake, whereas intensity varies for a single earthquake, depending on the amount and kind of local damage. Magnitudes can be reported for all quakes, even those in distant uninhabited areas where there is no property to affect.

Location and Size of Earthquakes in the United States

Figure 16.13 shows the locations of all damaging earthquakes that have occurred in the United States since 1977. Note that only a few localities are relatively free of earthquakes.

Most of the large earthquakes occur in the western states. Quakes in California, Nevada, Utah, Idaho, Montana, Washington, and other western states are related to known faults and usually (but not always) involve surface rupture of the ground. Earthquakes in Alaska occur mainly below the Aleutian Islands, where the Pacific plate is converging with and being subducted beneath the North American plate.

Earthquakes east of the Rocky Mountains are rarer and generally smaller and deeper than earthquakes in the western United States. They usually are not associated with surface rupture. The quakes may be occurring on the deeply buried, relatively inactive faults of old *divergent plate boundaries* and *failed rifts (aulacogens)*, both of which are described in chapter 19.

Although large quakes are extremely rare in the central and eastern United States, when they do occur, they can be very destructive and widely felt, because Earth's crust is older, cooler, and more brittle in the east than in the west and seismic waves travel more efficiently. The Saint Lawrence River Valley along the Canadian border has had several intensity IX and X earthquakes, most recently in 1944. Plymouth, Massachusetts, had an intensity IX quake in 1638, and a quake of intensity VIII occurred in 1775 near Cambridge, Massachusetts. In 1929, in Attica, New York, an earthquake of intensity IX knocked over 250 chimneys. A series of quakes (intensity XI) that occurred near New Madrid, Missouri, in the winter of 1811–1812 were the most widely felt earthquakes to occur in North America in recorded history. The quakes knocked over chimneys as far away as Richmond, Virginia, and rang church bells in Boston, 700 kilometers (434 miles) away.

The 1886 quake in Charleston, South Carolina (intensity X) was felt throughout almost half the United States (figure 16.11) and killed sixty people; it was sharply felt in New York City. Moderate quakes hit Arkansas and New Hampshire in 1982, and in 1983, a quake of 5.1 magnitude rocked New York's Adirondack Mountains. A 5.0-magnitude quake near Lawrenceville, Illinois, was felt from Kansas to South Carolina to Ontario in 1987. In 1988, a 6.0-magnitude quake north of Quebec City was felt as far away as Indiana and Washington, D.C.

Geologists have mapped regions of seismic risk in the United States (figure 16.14) and elsewhere throughout the world, primarily on the assumption that large earthquakes will occur in the future in places where they have occurred in the past.

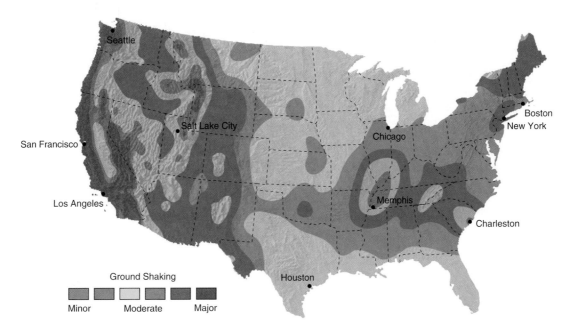

FIGURE 16.14

Map of seismic hazard in the United States based on the expected amount of ground shaking and damage. *USGS-National Seismic Hazard Mapping Project*

A

B

C

D

FIGURE 16.15

Earthquake damage to structures from recent major earthquakes throughout the world. (A) Elevated highway knocked over by a strong horizontal jolt during the 1995 Kobe, Japan, earthquake. Damage exceeded $400 billion and destroyed or severely damaged more than 88,000 buildings. (B) Poorly constructed buildings crumbled during the 1999 Izmit, Turkey, earthquake, while structures built to seismic code and old mosques were left standing. (C) Many high-rise buildings collapsed during the 1999 Taiwan earthquake. The M-7.6 quake was the largest to hit central Taiwan in the past 400 years, and damage exceeded $14 billion. (D) One of the many buildings damaged during the January 2001 Gujarat, India, earthquake that caused over $1.3 billion in damage. *Photo A © Reuter/Sankei/Shimbun; Photo B © AP/Wide World Photo; Photo C © Smith Glenn/SYGMA Corbis; Photo D © Jaswant Arelekar/IITK, Kanpur, India*

EFFECTS OF EARTHQUAKES

Ground motion is the trembling and shaking of the land that can cause buildings to vibrate. During small quakes, windows and walls may crack from such vibration. In a very large quake, the ground motion may be visible. It can be strong enough to topple large structures such as bridges and office and apartment buildings (figure 16.15). Most people injured or killed in an earthquake are hit by falling debris from buildings. Because proper building construction can greatly reduce the dangers, building codes need to be both strict and strictly enforced in earthquake-prone areas. Much of the damage and loss of life in the recent Turkey, El Salvador, and India earthquakes were due to poorly constructed buildings that did not meet building codes. As we have seen, the location of buildings also needs to be controlled; buildings built on soft sediment are damaged more than buildings on hard rock.

Fire is a particularly serious problem just after an earthquake because of broken gas and water mains and fallen elec-trical wires (figure 16.16). Although fire was the cause of most of the damage to San Francisco in 1906, changes in building construction and improved fire-fighting methods have reduced (but not eliminated) the fire danger to modern cities. The stubborn Marina district fires in San Francisco in 1989 attest to modern dangers of broken gas and water mains.

Landslides can be triggered by the shaking of the ground (figure 16.17A). The 1959 Madison Canyon landslide in Montana was triggered by a nearby quake of magnitude 7.7. Landslides and subsidence caused extensive damage in downtown and suburban Anchorage during the 1964 Alaskan quake (magnitude 8.6). The 7.9-magnitude Denali fault earthquake that shook south-central Alaska triggered thousands of landslides. The 1970 Peruvian earthquake (magnitude 7.75) set off thousands of landslides in the steep Andes Mountains, burying more than 17,000 people (see box 9.1). In 1920 in China, over 100,000 people living in hollowed-out caves in cliffs of loess (described in chapter 13) were killed when a quake collapsed the cliffs. The 2001 El Salvador quake resulted in nearly 500

A

B

FIGURE 16.16

(A) Almost 100 homes burned at a Sylmar mobile-home park following the Northridge earthquake, southern California, 1994. (B) People on Sacramento Street watch the smoke rise from fires caused by the 1906 San Francisco earthquake; most of the damage from the earthquake was caused by fires that burned for days. *Photo A © Ken Lubas/Los Angeles Times; Photo B © Arnold Genthe/AP/Wide World Photos*

IN GREATER DEPTH 16.1

Earthquake Engineering

Damage and loss of life can be substantially reduced by siting structures on solid bedrock or dense soils and by building structures that adhere to strict seismic building codes. In the 7.2-magnitude earthquake that struck Armenia in 1988, 50,000 people lost their lives when poorly constructed buildings crumbled. More recently, on January 26, 2001, the 7.7-magnitude Gujarat, India, quake, killed over 18,000, left 600,000 homeless, and destroyed 332,000 houses. In contrast, earthquake resistant structures and enforcement of seismic building codes in the San Francisco Bay area resulted in only sixty-three people dying in the 7.2-magnitude Loma Prieta earthquake.

Buildings that are constructed of strong, flexible, and light materials such as steel, wood, and reinforced concrete (strengthened by steel rebar) are the most resistant to damage by seismic shaking. Houses built with unreinforced concrete block or brick, which are only as strong as the mortar holding the blocks and bricks together, tend to lack flexibility and crumble in large earthquakes. During moderate-sized earthquakes, many houses lose their chimneys or brick facades. Buildings with heavy roofs made of tile or slate also tend to collapse. During the Gujarat, India, earthquake, many reinforced concrete buildings failed because the walls, floors, ceilings, and elevator shafts were not well connected to allow the entire building to flex as one (box figure 1A).

The 1985 Mexico City earthquake, which killed five thousand people and caused over $5 billion in damage, is a classic example of the effect soft soils have on the amplification of earthquake waves. The ground shaking was relatively mild in most parts of Mexico, but it was amplified in some buildings and by the lake-bed sediments beneath Mexico City. Most buildings are not rigid but slightly flexible; they sway gently like large pendulums when struck by wind or seismic waves. The time necessary for a single back-and-forth oscillation is called a period, and it varies with a building's height and mass. Natural bodies of rock and sediment vibrate the same way, with periods that vary with the body's size and density.

When earthquake waves struck Mexico City, many were vibrating with a two-second period. The body of lake sediment beneath the city has a natural period of two seconds also, so the wave motion was amplified by the sediments. This type of amplification, or resonance, occurs when you push a child on a swing—if you push gently in time with the swing's natural period, the swing goes higher and higher. The water-saturated sediment began to move like a sloshing waterbed. The ground moved back and forth by 40 centimeters (15.7 inches) every two seconds, and it did this fifteen to twenty times.

This shaking was devastating to buildings with a natural two-second period (generally those five to twenty stories high), and several hundred buildings in this size range collapsed as they further resonated with the shaking (box figure 1B). Many structures weakened by the main shock collapsed during the large after-

A

B

BOX 16.1 ■ FIGURE 1

(A) Insufficient connection between the reinforced concrete elevator shaft and the rest of the building led to separation and partial collapse during the 2001 Gujarat, India, earthquake. (B) This fifteen-story building collapsed completely, during the 1985 Mexico City earthquake, crushing all its occupants as its reinforced-concrete floors "pancaked" together. *Photo A © C. V. R. Murty/IITK, Kanpur, India; Photo B by M. Celebi, U.S. Geological Survey*

shock. Most shorter and taller buildings had other periods of vibration and rode out the shaking with little damage. Although 800 buildings collapsed or were seriously damaged, most of the city's 600,000 structures survived with little or no damage. The Mexico City quake vividly illustrates the fact that proper building design can greatly lessen earthquake damage.

landslides, the largest of which occurred in Santa Tecla where 1,200 people were missing after tons of soil and rock fell on a neighborhood.

A special type of ground failure caused by earthquakes is *liquefaction.* This occurs when a water-saturated soil or sediment turns from a solid to a liquid as a result of earthquake shaking. Liquefaction may occur several minutes after an earthquake, causing buildings to sink and underground tanks to float as once-solid sediment flows like water (figure 16.17*B*). Liquefaction was responsible for much of the damage in the 1989 Loma Prieta quake and contributed to the damage in the 1906 San Francisco, 1964 and 2002 Alaska, 1985 Mexico City, 1995 Kobe (Japan) and 2001 Puget Sound (Washington) and Gujarat (India) quakes.

Permanent displacement of the land surface may be the result of movement along a fault. Rocks can move vertically, those on one side of a fault rising while those on the other side drop. Rocks can also move horizontally—those on one side of a fault sliding past those on the other side. Diagonal movement with both vertical and horizontal components can also occur during a single quake. Such movement can affect huge areas, although the displacement in a single earthquake seldom exceeds 8 meters. The trace of a fault on Earth's surface may appear as a low cliff, called a *scarp,* or as a closed tear in the ground (figure 16.18). In rare instances, small cracks open during a quake (but not to the extent that Hollywood films often portray). Ground displacement during quakes can tear apart buildings, roads, and pipelines that cross faults. Sudden subsidence of land near the sea can cause flooding and drownings.

Aftershocks are small earthquakes that follow the main shock. Although aftershocks are smaller than the main quake, they can cause considerable damage, particularly to structures weakened by the powerful main shock. A long period of aftershocks can be extremely unsettling to people who have lived through the main shock. *Foreshocks* are small quakes that precede a main shock. They are usually less common and less damaging than aftershocks but can sometimes be used to help predict large quakes (although not all large quakes have foreshocks).

A

B

FIGURE 16.17

(A) Landslide in Pacific Palisades triggered by the Northridge earthquake, 1994. (B) Liquefaction of soil by a 1964 quake in Niigata, Japan, caused earthquake-resistant apartment buildings to topple over intact. *Photo A © Al Seib/Los Angeles Times; Photo B by National Geophysical Data Center*

A

B

C

D

FIGURE 16.18

Varieties of ground displacement caused by earthquakes. (*A*) Sixteen-foot scarp (cliff) formed by vertical ground motion, Alaska, 1964. (*B*) Tearing of the ground near Olema, California, 1906. (*C*) Fence compressed by ground movement, Gallatin County, Montana, 1959. (*D*) Compression of concrete freeway, San Fernando Valley, California, 1971. *Photo A by U.S. Geological Survey; Photo B by G. K. Gilbert, U.S. Geological Survey; Photo C by I. J. Witkind, U.S. Geological Survey; Photo D by David McGeary*

Tsunami

The sudden movement of the sea floor upward or downward during a submarine earthquake can generate very large sea waves, popularly called "tidal waves." Because the ocean tides have nothing to do with generating these huge waves, the Japanese term *tsunami* is preferred by geologists. **Tsunami** are also called **seismic sea waves.** They usually are caused by great earthquakes (magnitude 8+) that disturb the sea floor, but they also result from submarine landslides or volcanic explosions. When a large section of sea floor suddenly rises or falls during a quake, all the water over the moving area is lifted or dropped for an instant. As the water returns to sea level, it sets up long, low waves that spread very rapidly over the ocean (figure 16.19). Because vertical motion of the sea floor is most conducive to the formation of tsunami, most are associated with subduction zone earthquakes, which tend to be some of the strongest quakes.

A tsunami is unlike an ordinary water wave on the sea surface. A large wind-generated wave may have a wavelength of 400 meters and be moving in deep water at a speed of 90 kilometers per hour (55 miles per hour). The wave height when it breaks on shore may be only 0.6 to 3 meters, although in the middle of hurricanes, the waves can be more than 15 meters (49 feet) high. A tsunami, however, may have a wavelength of 160 kilometers and may be moving more than 800 kilometers per hour (500 miles per hour). In deep water, the wave height may be only 0.6 to 2 meters, but near shore, the tsunami may peak at heights of 15 to 30 meters. This great increase in wave height near shore is caused by bottom topography; only a few localities have the combination of gently sloping offshore shelf and funnel-shaped bay that force

1. Before the earthquake: In this case, an oceanic plate subducts under a continental plate. The continental plate bends as stresses between the two plates build over time.

2. Earthquake: Releasing its built-up stress, the continental plate lurches forward over the oceanic crust, lifting the ocean. The displaced water appears as a huge bulge on the sea surface.

3. After the earthquake, gravity collapses the bulge to start a *succession* of waves, or tsunami. The tsunami waves move away in both directions as the mass of water "bobs" up and down over the source of the earthquake.

4. Each wave quickly advances over the land as a sediment-filled wall of water. It stops briefly before retreating, carrying sediments and debris back to the sea. Over time, the intensity of the tsunami subsides.

FIGURE 16.19

Tsunami waves are generated by a submarine earthquake that displaces the sea floor and water column above. Long, low waves are formed above the displaced sea floor to compensate for the momentary rise in sea level and spread very rapidly (at the speed of a jetliner) in the deep ocean. In shallower water, the tsunami slows to highway speeds and builds in height until it breaks and crashes onto the shore with incredible force, causing destructive flooding along low-lying coastal areas.

tsunamis to awesome heights (the record height was 85 meters in 1971 in the Ryukyu islands south of Japan). Along most coastlines tsunami height is very small.

Although the speed of the wave slows drastically as it moves through shallow water, a tsunami can still hit some shores as a very large, very fast wave. Because of its extremely long wavelength, a tsunami does not withdraw quickly as normal waves do. The water keeps on rising for five to ten minutes, causing great flooding before the wave withdraws. The long duration and great height of a tsunami can bring widespread destruction to the entire shore zone.

A tsunami formed by the 1960 Chilean quake crossed the Pacific Ocean (figure 16.20) and did extensive damage in Japan. One of the most destructive tsunamis of modern times was generated April 1, 1946, by a 7.3-magnitude earthquake offshore from Alaska. It devastated the city of Hilo, Hawaii, causing 159 deaths. The tsunami following the 1964 Alaskan quake drowned twelve people in Crescent City, California, a small coastal town near the Oregon border (figure 16.20). Wave damage near an epicenter can be awesome. The 1946 Alaska tsunami destroyed the Scotch Cap lighthouse on nearby Unimak Island, sweeping it off its concrete base, which was 10 meters above sea level, and killing its five occupants. The wave also swept away a radio tower, whose base was 31 meters above sea level.

Another devastating tsunami occurred July 17, 1998, in Papua, New Guinea after a 7.1-magnitude earthquake struck 20 kilometers (12.4 miles) offshore. Three waves were generated twenty minutes after the earthquake, with the highest wave measuring 15 meters, which completely destroyed three coastal villages and killed more than 2,200 people. The tremendous loss of life was caused by the lack of warning as the waves hit at dusk on the three villages located on a low-lying sandbar enclosing a bay. The wave was higher than expected for the size of the earthquake, possibly due to a submarine landslide just offshore that was triggered by the quake. Some of the largest tsunami waves are generated by large landslides caused by unstable slopes or volcanic eruptions that displace a nearby body of water.

After the 1946 Hilo, Hawaii, tsunami, the U.S. Coast and Geodetic Survey established a Tsunami Early Warning System in an attempt to minimize loss of life in Pacific coastal communities. A network of seismic stations reports

A

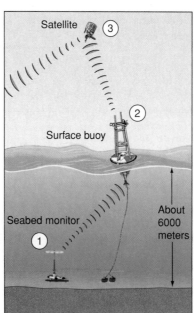

B   NOAA's DART mooring system
(components not to scale)

① Recorder on ocean bottom monitors sea pressure and activity every 15 minutes. An unusual event triggers readings every 15 seconds. Data is transmitted to surface buoy.

② Surface buoy monitors upper level conditions and relays this as well as data from the seabed monitors to satellite in real-time.

③ Data are then relayed via satellite to ground stations for immediate dissemination to NOAA's Tsunami Warning Centers and Pacific Marine Environmental Laboratory.

FIGURE 16.20

(A) Tsunami travel times from the magnitude 9.2 Alaska earthquake and the magnitude 9.5 Chile earthquake to locations within the Pacific. Yellow triangles show the location of tsunami-detection buoys. (B) Details of the Deep-ocean Assessment and Reporting of Tsunamis (DART) system used by NOAA. *From NOAA*

A

B

FIGURE 16.21

(A) Tsunami travel time and wave height (feet) map from December 26, 2004 Sumatra earthquake. The arrival time (hours) of the first wave of the tsunami are shown by the white contour lines. Bandeh Aceh was hit by the tsunami 15 minutes after the earthquake and it took 7 hours to travel across the Indian Ocean to reach the coast of Africa. (B) Many people are surprised while others run for safety as the tsunami caused by the December 26, 2004 9.3 magnitude Sumantra earthquake comes crashing onshore in Koh Raya, Thailand. *Photo © John Russell/AFP/Getty Images*

large earthquakes that are capable of generating tsunamis to the Tsunami Warning Center in Honolulu. An array of tidal gauges and deep-ocean tsunami detectors (figure 16.20) are then read to determine if a tsunami has been generated. Even though tsunamis travel at high speeds, there is usually sufficient time to warn low-lying coastal communities of the impending wave.

The most devastating tsunami in recorded history occurred on December 26, 2004 in the Indian Ocean. The tsunami was triggered by a 9.3-magnitude earthquake that struck off the northern coast of Sumatra, Indonesia as the Indian plate shifted under the Eurasian plate, causing a subduction zone (*megathrust*) earthquake. After the 1960 9.5-magnitude quake in Chile, this is the second-largest earthquake ever recorded in the world. The shift in the sea floor caused a tsunami that hit the towns of Bandeh Aceh and Meulaboh, Sumatra minutes later. The tsunami traveled across the Indian Ocean at the speed of a jetliner and caused massive deaths in low-lying coastal communities (figure 16.21A) that were filled with tourists during the Christmas holiday. The Indian Ocean tsunami caused more casualties than any other tsunami in recorded history. In total, more than 220,272 people were killed and over 20,000 were listed as missing. Horrific accounts of the tsunami coming on shore were told by countless people caught in the low-lying coastal areas (figure 16.21B). As in many large tsunamis, the sea first retreated, luring many for a closer look at exposed coral reefs and stranded fish on the beach. Minutes later, the first wave struck with tremendous force, causing massive loss of life and damage. The first wave retreated with the same force it had when it came onshore and many desperately hung onto trees so they were not swept out to sea. Then, successive waves struck with the same incredible force and caused additional damage and loss of life.

Massive international relief efforts were mounted to help the victims of the tsunami. No tsunami-warning system, or other means of warning communities of the possibility of a tsunami, was in place in the Indian Ocean as it is in the Pacific Ocean. This tragedy has resulted in funding for such a system to minimize loss of life from future tsunamis generated in the Indian Ocean.

WORLD DISTRIBUTION OF EARTHQUAKES

Most earthquakes are concentrated in narrow geographic belts (figure 16.22A), although *some* earthquakes have occurred in most regions on Earth. The boundaries of plates in the plate-tectonic theory are defined by these earthquake belts (figure 16.22B). The most important concentration of earthquakes by far is in the **circum-Pacific belt,** which encircles the rim of the Pacific Ocean. Within this belt occur approximately 80% of the world's shallow-focus quakes, 90% of the intermediate-focus quakes, and nearly 100% of the deep-focus quakes.

Another major concentration of earthquakes is in the **Mediterranean-Himalayan belt,** which runs through the Mediterranean Sea, crosses the Mideast and the Himalayas, and passes through the East Indies to meet the circum-Pacific belt north of Australia.

A number of shallow-focus earthquakes occur in two other significant locations on Earth. One is along the summit or crest of the *mid-oceanic ridge,* a huge underwater mountain range that runs through all the world's oceans (see figure 1.8 and chapter 18). A few earthquakes have also been recorded in isolated spots usually associated with basaltic volcanoes, such as those of Hawaii.

In most parts of the circum-Pacific belt, earthquakes, andesitic volcanoes, and *oceanic trenches* (see chapter 18) appear to be closely associated. Careful determination of the locations and depths of focus of earthquakes has revealed the existence of distinct *earthquake zones* that begin at oceanic trenches and slope landward and downward into Earth at an angle of about 30° to 60° (figure 16.23). Such zones of inclined seismic activity are called **Benioff zones** after the man who first recognized them.

Benioff zones slope under a continent or a curved line of islands called an **island arc.** Andesitic volcanoes may form the islands of the island arc, or they may be found near the edge of a continent that overlies a Benioff zone.

Most of the circum-Pacific belt is made up of Benioff zones associated in this manner with oceanic trenches and andesitic volcanoes. Parts of the Mediterranean-Himalayan belt represent Benioff zones, too, notably in the eastern Mediterranean Sea and the East Indies. Essentially all the world's intermediate- and deep-focus earthquakes occur in Benioff zones.

FIRST-MOTION STUDIES OF EARTHQUAKES

By studying seismograms of an earthquake on a distant fault, geologists can tell which way rocks moved along that fault. First-motion studies played an important role in determining the overall sense of movement along plates boundaries. Rock motion is determined by examining seismograms from many locations surrounding a quake. Each seismograph station can tell whether the first rock motion recorded there was a push or a pull (figure 16.24). If the rock moved toward the station (a push), then the pen drawing the seismogram is deflected up. If the first motion is away from the station (a pull), then the pen is deflected down.

If an earthquake occurs on a fault as shown in figure 16.25A, large areas around the fault will receive a push as first motion, while different areas will receive a pull. Any station within the black area marked A will receive a push, for the rock is moving from the epicenter toward those stations, as shown by the arrows in the figure. All stations in area C will also receive a push, but areas B and D will record a pull as the first motion.

In figure 16.25B, you can see the same pattern of pushes and pulls, but in this case, the pattern is caused by a fault with a different orientation. Either fault can cause the same pattern, if the rock moves in the direction shown by the arrows. In other words, there are two possible solutions to any pattern of first motions.

If the orientation of the fault trace on Earth's surface is known, as it is for most faults on land (and for a few on the ocean floor), the correct choice of the two solutions can be made. But if the orientation of the fault is not known—as is the usual case for earthquakes at sea or at great depth—the choice between the two possible solutions can be difficult. One solution may be more *likely,* based on the study of topography, other faults, or aftershocks. Patterns of aftershocks after the main quake often delineate which of the two solutions is correct.

EARTHQUAKES AND PLATE TECTONICS

One of the great attractions of the concept of plate tectonics is its ability to explain the distribution of earthquakes and the rock motion associated with them.

As described briefly in chapter 1, the concept of plate tectonics is that Earth's surface is divided into a few giant *plates.* Plates are rigid slabs of rock, thousands of kilometers wide and 70 to 125 (or more) kilometers thick, that move across Earth's surface. Because the plates include continents and sea floors on their upper surfaces, the plate-tectonics concept means that the continents and sea floors are moving. The plates change not only position but size and shape (as we discuss in more detail in chapter 19).

A

B

FIGURE 16.22

(A) Distribution of shallow-, intermediate-, and deep-focus earthquakes. (B) The major plates of the world in the theory of plate tectonics. Compare the locations of plate boundaries with earthquake locations shown in figure 16.22A. Double lines show diverging plate boundaries; single lines show transform boundaries. Heavy lines with triangles show converging boundaries; triangles point down subduction zone. *After W. Hamilton, U.S. Geological Survey*

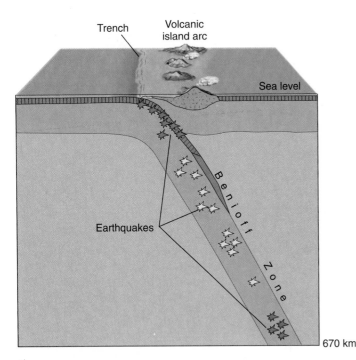

Shallow-focus earthquakes
Intermediate-focus earthquakes
Deep-focus earthquakes

FIGURE 16.23

A Benioff zone of earthquakes begins at an oceanic trench and dips under a continent (such as South America) or a volcanic island arc. Upper part of Benioff zone may extend to a depth of 670 kilometers.

FIGURE 16.24

Seismograms showing how first horizontal motions of rocks along a fault are determined. (A) If the first motion is a push (from the epicenter to the seismograph station), the seismogram trace is deflected upward. (B) If it is a pull (away from the station), the deflection is downward.

Earthquakes occur commonly at the edges of plates but only occasionally in the middle of a plate. The close correspondence between plate edges and earthquake belts can be seen by comparing the map of earthquake distribution in figure 16.22*A* with the plate map in figure 16.22*B*.

This correspondence is hardly surprising—plate boundaries are identified and *defined* by earthquakes. According to plate tectonics, earthquakes are caused by the interactions of plates along plate boundaries. Therefore, narrow bands of earthquakes are used to outline plates on plate maps. This can be clearly seen in the east Pacific Ocean off South America,

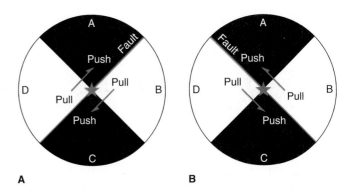

FIGURE 16.25

Map view of two possible solutions for the same pattern of first motion. Each solution has a different fault orientation. If the fault orientation is known, the correct solution can be chosen. The star marks the epicenter, and rock motion is shown by the arrows.

where the Nazca plate (figure 16.22*B*) is almost completely outlined by earthquake epicenters (figure 16.22*A*).

The earthquakes on the western border of the Nazca plate are shallow-focus quakes, and they occur in a narrow belt along the crest of the mid-oceanic ridge here, locally called the East Pacific Rise. The quakes along the eastern boundary occupy a broader belt that lies mostly within South America. This belt includes shallow-, intermediate-, and deep-focus earthquakes in a Benioff zone that begins at the Peru-Chile Trench just offshore and slopes steeply down under South America to the east. The Nazca plate moves eastward, away from the crest of the mid-oceanic ridge and toward the subduction zone at the trench, where the plate plunges down into the mantle. The plate's western boundary is located at the crest of the East Pacific Rise, and its eastern boundary is at the bottom of the Peru-Chile Trench.

Earthquakes at Plate Boundaries

As you have learned, there are three types of plate boundaries, *divergent boundaries* where plates move away from each other, *transform boundaries* where plates move horizontally past each other, and *convergent boundaries* where plates move toward each other. Each type of boundary has a characteristic pattern of earthquake distribution and rock motion.

Divergent Boundaries

At a divergent boundary, where plates move away from each other, earthquakes are shallow, restricted to a narrow band, and much lower magnitude than those that occur at convergent or transform boundaries. A divergent boundary on the sea floor is marked by the crest of the mid-oceanic ridge and the *rift valley* that is often (but not always) found on the ridge crest (figure 16.26). The earthquakes are located along the sides of the rift valley and beneath its floor. The rock motion that is deduced from first-motion studies shows that the faults here are normal faults, parallel to the rift valley. The ridge crest is under tension, which is tearing the sea floor apart, creating the rift valley and causing the earthquakes.

FIGURE 16.26

Distribution of earthquakes at plate boundaries. Shallow-focus earthquakes occur at divergent boundaries where the lithosphere is being pulled apart and also along transform boundaries where slip in the lithosphere accommodates the spreading between oceanic ridges. Shallow- to deep-focus earthquakes occur where a lithosphere subducts during collision of two plates.

A divergent boundary within a *continent* is usually also marked by a rift valley, shallow-focus quakes, and normal faults. The African Rift Valleys in eastern Africa (figure 16.22*B*) seem to be such a boundary. Tensional forces are tearing eastern Africa slowly apart, creating the rift valleys, some of which contain lakes (see figure 19.21). Other areas where the continental crust is being pulled apart, such as the Basin and Range province in the western United States, are also marked by normal faults and shallow earthquakes.

Transform Boundaries

Where two plates move past each other along a transform boundary, the earthquakes are shallow. First-motion studies indicate strike-slip motion on faults parallel to the boundary. The earthquakes are aligned in a narrow band along the transform fault. Although most transform faults occur on the ocean floor and offset ridge segments, some are found in the continental crust. The San Andreas fault in California is the most famous example of a right-lateral transform fault (see box 16.3). The Alpine fault in New Zealand is another example of a right-lateral transform fault.

Convergent Boundaries

Convergent boundaries are of two general types, one marked by the *collision* of two continents, the other marked by *sub-*

duction of the ocean floor under a continent (figure 16.26*B*) or another piece of sea floor. Each type has a characteristic pattern of earthquakes.

Collision boundaries are characterized by broad zones of shallow earthquakes on a complex system of faults (figure 16.26). Some of the faults are parallel to the dip of the suture zone that marks the line of collision; some are not. One continent usually overrides the other slightly (continents are not dense enough to be subducted), creating thick crust and a mountain range. The Himalayas represent such a boundary (figure 16.22*B*). The seismic zone is so broad and complex at such boundaries that other criteria, such as detailed geologic maps, must be used to identify the position of the suture zone at the plate boundary.

During *subduction,* earthquakes occur for several different reasons. As a dense oceanic plate bends to go down at a trench, it stretches slightly at the top of the bend, and normal faults occur as the rocks are subjected to *tension.* This gives a block-faulted character to the outer (seaward) wall of a trench. For some distance below the trench, the subducting plate is in contact with the overlying plate. First-motion studies of earthquakes at these shallow depths show that the quakes are caused by shallow-angle thrust-faulting. This is the motion expected as one plate slides beneath another, a process commonly called *underthrusting.*

ENVIRONMENTAL GEOLOGY 16.2

What to Do Before, During, and After an Earthquake*

Being prepared for an earthquake can reduce the damage to your property and chance of serious injury or loss of life. There is a saying, "earthquakes do not kill people, buildings do." If you live in or visit an earthquake-prone area, you should do the following:

Before an Earthquake

1. Make sure your house is firmly attached to the foundation with anchor bolts; repair any deep cracks in foundations or ceilings. Brick chimneys should be braced and anchored to the roof joists.

2. Check for hazards inside the home. Tall bookshelves should be bolted to the wall, with heavy objects placed on the bottom shelves; flammable items and household chemicals should also be on the bottom shelves in locked cabinets; glass and china should be in lower cabinets secured with strong latches; heavy pictures and mirrors should not be hung where people sit or sleep; the water heater should be strapped to wall studs and bolted to the floor.

3. Learn how to turn off all the utilities at your house; flexible gas lines should be used to avoid breaking. Keep an adjustable wrench near the gas main to shut off the gas immediately after an earthquake to avoid fires.

4. Have disaster supplies on hand (keep in a safe place in large, lockable plastic trash container): flashlight and extra batteries, portable radio, first-aid kit and manual, essential medicines, emergency food and water (one gallon per person per day), nonelectric can opener, sleeping bags and tent, fire extinguisher, matches, portable stove and propane, sturdy shoes, cash and credit cards. *(Check the condition of batteries, water, and food every six months.)*

5. Have an emergency communication plan to reunite family members who may be separated from one another during the earthquake. Because it is often easier to call long distance after an earthquake, establish an out-of-state relative or friend to act as the contact person.

During an Earthquake

1. If you are indoors, DROP, COVER, and HOLD under a heavy piece of furniture positioned against an inside wall or crouch in a room corner, interior hall, or doorway. Stay away from windows or anything that could fall on you. In a high-rise building, do not run to exits or stairways that may be damaged or jammed with people; never use the elevator.

2. If you are in an unreinforced building or otherwise unsafe building, it may be better to leave the building. Because most injuries result from people leaving buildings and being hit by falling debris or downed utility lines, be alert to possible dangers.

3. If you are outdoors, move to an open area away from buildings, street lights, and utility lines until the shaking stops.

4. If you are in a moving vehicle, slow down and drive away from buildings, trees, bridges, ramps, overpasses, and utility lines. Stay in the car until the shaking stops.

After an Earthquake

1. Be prepared for aftershocks.

2. Help anyone who is injured or trapped; do not move seriously injured persons unless they are in immediate danger of further injury.

3. Check for damage to utilities. If you smell gas, turn off gas valves, open the windows, and leave immediately. If electricity is shorting out, turn off the main power switch at the meter box. If water pipes are broken, turn off the supply at the main valve. In an emergency, water from hot water tanks, toilet bowls, and melted ice cubes can be used. Do not flush the toilet until sewage lines are checked.

4. Carefully inspect your chimney for damage to prevent fire and carbon monoxide poisoning.

5. Listen to the radio for the latest emergency information; use your telephone only for emergency calls.

6. Do not travel unnecessarily; avoid low-lying coastal areas (until the threat of a tsunami has passed), landslide areas, and severely damaged structures.

Additional Resources

For additional safety information, visit the American Red Cross website:

- www.redcross.org/services/disaster/keepsafe/readyearth.html

or the U.S. Federal Emergency Management Agency site:

- www.fema.gov/library/quakef.htm

An online version of the Southern California Earthquake Center's handbook *Putting Down Roots in Earthquake Country* gives detailed information on preparing for earthquakes and specific information for southern California seismicity:

- www.scecdc.scec.org/eqcountry.html

*From U.S. Federal Emergency Management Agency (FEMA) and the Red Cross

At greater depths, where the descending plate is not in direct contact with the overlying plate, earthquakes are common, but the reasons for them are not obvious. The quakes are confined to a thin zone, only 20 to 30 kilometers thick, within the lithosphere of the descending plate, which is about 100 kilometers thick. This zone is thought to be near the top of the lithosphere, where the rock is colder and more brittle.

Subduction Angle

The horizontal and vertical distribution of earthquakes can be used to determine the angle of subduction of a down-going plate. Subduction angles vary considerably from trench to trench. Many plates start subducting at a gentle angle, which becomes much steeper with depth. At a few trenches in the open Pacific, subduction begins (and continues) at almost a vertical angle.

Subduction angle is probably controlled by plate density and the rate of plate convergence. Older oceanic lithosphere, such as that in the southeast Pacific, tends to be colder and more dense and therefore subducts at a steeper angle; younger oceanic plates in close proximity to the oceanic ridge are warmer and more buoyant and subduct at a shallower angle. A faster rate of convergence may also result in a shallower angle of subduction.

In summary, earthquakes are very closely related to plate tectonics. Most plate boundaries are defined by the distribution of earthquakes, and plate motion can be deduced by the first motions of the quakes. Analysis of first motions can also help determine the type and orientation of stresses that act on plates, such as tension and compression. Quake distribution with depth indicates the angle of subduction and has shown that some plates change subduction angle and even break up as they descend.

A few quakes, such as those that occur in the center of plates, cannot easily be related to plate motion. These *intraplate* earthquakes probably occurred along older faults that are no longer plate boundaries but remain zones of crustal weakness. Some of the most destructive earthquakes in the United States, such as the 1811–1812 New Madrid, Missouri; 1886 Charleston, South Carolina; and 1755 Boston, Massachusetts, quakes, occurred as intraplate earthquakes.

EARTHQUAKE PREDICTION AND SEISMIC RISK

People who live in earthquake-prone regions are plagued by unscientific predictions of impending earthquakes by popular writers and self-proclaimed prophets. Several techniques are being explored for *scientifically* forecasting a coming earthquake. One group of methods involves monitoring slight changes, or *precursors*, that occur in rock next to a fault before the rock breaks and moves; these methods that assume large amounts of strain are stored in rock before it breaks (figure 16.2).

Just as a bent stick may crackle and pop before it breaks with a loud snap, a rock may give warning signals that it is about to break. Before a large quake, small cracks may open within the rock, causing small tremors, or *microseisms,* to increase. The *properties of the rock* next to the fault may be changed by the opening of such cracks. Changes in the rock's magnetism, electrical resistivity, or seismic velocity may give some warning of an impending quake.

The opening of tiny cracks changes the rock's porosity, so *water levels in wells* often rise or fall before quakes. The cracks provide pathways for the release of radioactive radon gas from rocks (radon is a product of radioactive decay of uranium and other elements). An *increase in radon emission from wells* may be a prelude to an earthquake. The interval between eruptions of geysers may change before and after an earthquake, probably due to porosity changes within the surrounding rock.

In some areas, the *surface of Earth tilts and changes elevation* slightly before an earthquake. Scientists use highly sensitive instruments to measure this increasing strain in hopes of predicting quakes.

Chinese scientists claim successful, short-range predictions by watching *animal behavior*—horses become skittish and snakes leave their holes shortly before a quake. U.S. scientists conducted a few pilot programs along these lines, but remain skeptical because it is difficult to correlate a specific animal behavior to an impending earthquake. It is interesting that very few animals were killed by the Indian Ocean tsunami. Apparently before the tsunami hit, elephants were seen running to higher ground, flamingos left low-lying breeding areas, and dogs refused to go outdoors.

Japanese and Russian geologists were the first to predict earthquakes successfully, and Chinese geologists have made some very accurate predictions. In 1975, a 7.3-magnitude earthquake near Haicheng in northeastern China was predicted five hours before it happened. Alerted by a series of *foreshocks,* authorities evacuated about a million people from their homes; many watched outdoor movies in the open town square. Half the buildings in Haicheng were destroyed, along with many entire villages, but only a few hundred lives were lost. In grim contrast, however, the Chinese program failed to predict the 1976 Tangshan earthquake (magnitude 7.6), which struck with no warning and killed an estimated 250,000 people.

Most of these methods were once considered very promising but have since proved to be of little real help in predicting quakes. A typical quake predictor, such as tilt of the land surface, may precede one quake and then be absent for the next ten quakes. In addition, each precursor can be caused by forces unrelated to earthquakes (land tilt is also caused by mountain building, magmatic intrusion, mass wasting, and wetting and drying of the land).

A fundamentally different method of determining the probability of an earthquake occurring relies on the history of earthquakes along a fault and the amount of tectonic stress building in the rock. Geologists look at the geologic record for evidence of past earthquakes using the techniques of *paleoseismology.* One technique involves digging a trench across the fault zone

FIGURE 16.27

To determine the likelihood of a large earthquake occurring again along an active fault, geologists need to know how often quakes have occurred in the past and how large the last one was. By using the techniques of paleoseismology, geologists dig a trench (A) across or alongside a fault and very carefully map disturbed layers of sediment and soil exposed in the upper few meters of the trench (B). (A) Trench being dug across the southern Hayward fault near Fremont, California, by the U.S. Geological Survey to reevaluate the seismic risk for the San Francisco Bay area.

(B) Photomosaic of the wall of a trench dug across the Coachella Valley section of the San Andreas fault near Thousand Palms Oasis, California reveals evidence for three of the five past earthquakes that struck this area since 825 A.D. (Events TP 1, 2, and 5). Evidence for the three separate earthquakes is shown in the trench either by the fault displacing different channel deposits against one another (*TP 1 offsets channel IV against VI and TP5 offsets channel I against II*) or the fault being buried or terminated by younger channel deposits (*TP2 cuts channel IV but not the overlying channel V sediments*). Based on these relations and on radiocarbon dates obtained from the disrupted layers (shown by small yellow boxes), it has been determined that the average time between earthquakes for this section of the San Andreas fault is 215 ± 25 years. Because the last earthquake (TP 1) occurred sometime after 1520–1680 A.D., more than 233 years have elapsed since the most recent earthquake, and geologists are concerned that the southernmost San Andreas fault zone is overdue for a large earthquake. *Photo A by Jennifer Adleman, U.S. Geological Survey; Photo B from Bulletin Seismological Society of America, 2002, v. 92, no. 7, p. 2851, courtesy of T. E. Fumal, U.S. Geological Survey*

NORTHEAST TRENCH 7 SOUTHEAST WALL SOUTHWEST

B

METERS

to examine sedimentary layers that have been offset and disrupted during past earthquakes (figures 16.27*A* and *B*). If the offset layers contain material such as volcanic ash, pollen, or organic material such as tree roots that can give a numerical age, then the average length of time between earthquakes (*recurrence interval*) can be determined. If the length of time since the last recorded earthquake far exceeds the recurrence interval, the fault is given a high probability of generating an earthquake.

Along some long-active faults are short, inactive segments called *seismic gaps* where earthquakes have not occurred for a long time. These gaps form as part of the seismic cycle and result in a zone of lowered stress, or stress shadow zone, where earthquake activity sharply decreases after a major seismic event. Such was the case after the 1906 San Francisco earthquake and after the 1857 break along the southern section of the San Andreas fault (see box 16.3).

The recurrence interval and likelihood of future earthquakes are also determined by measuring the slip rate along plate boundaries. Exciting new satellite-based techniques such as InSAR (interferometric synthetic aperture radar), in addition to GPS, have allowed seismologists to measure the vertical and horizontal movement along active faults and to determine how long it would take for sufficient stress to build up along the plate boundary to generate rupturing and slip along a fault. For example, if the slip rate along the boundary is determined to be 5 centimeters per year and the last earthquake resulted in 5 meters of slip, then you would expect the next large earthquake to occur in 100 years. Just as a rubber band will break if stretched too far, rock will also break or rupture if a critical level of stress is exceeded. In other cases, the accumulating stress is released aseismically by so-called silent earthquakes where a fault slips very slowly or creeps to gradually relieve the stress. Slip rates and recurrence intervals are used to determine the statistical probability of an earthquake occurring over a given amount of time.

By studying the seismic history of faults, geologists in the United States are sometimes able to forecast earthquakes along some segments of some faults. In 1988, the U.S. Geological Survey estimated a 50% chance of a magnitude-7 quake along the segment of the San Andreas fault near Santa Cruz. In 1989, the magnitude-7 Loma Prieta quake occurred on this very section. Since the techniques are new and in some cases only partly understood, some errors will undoubtedly be made. Many faults are not monitored or studied historically because of lack of money and personnel, so we will never have a warning of impending quakes in some regions. For large urban areas near active faults such as the San Andreas, however, earthquake risk analysis may reduce damage and loss of life.

Another more recent approach to minimize loss of life and reduce damage in a major earthquake is to closely monitor the amount and location of strong shaking by using a dense network of broadband seismometers that digitally relay information via satellites to a central location. At this location, maps showing where the greatest amount of shaking occurred can be generated within minutes to guide emergency personnel to the areas of

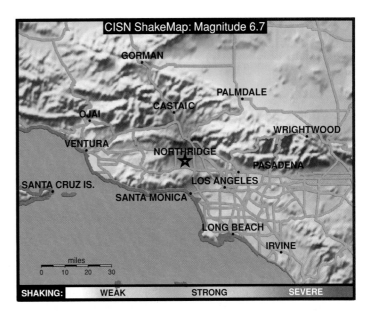

FIGURE 16.28

Map shows the amount of shaking that occurred after the 1994 Northridge earthquake. The ability to create maps within minutes after an earthquake that show the location and severity of maximum ground shaking (ShakeMap) was developed in 1995 by the U.S. Geological Survey. Had this ShakeMap been available minutes after the 1994 Northridge earthquake, emergency personnel could have been immediately directed to the most damaged areas. *Image courtesy David Wald, U.S. Geological Survey*

most damage (figure 16.28). Such a system has been developed in southern California, and there are plans for integrating other regional seismic networks into an Advanced National Seismic System (ANSS) to monitor earthquakes throughout the United States if adequate funding can be obtained.

A major goal of the ANSS program is to locate strong-motion seismometers in buildings, bridges, canals, and pipelines to provide valuable information on how a structure moves during an earthquake to help engineers build more earthquake-resistant structures. One key to reducing damage and loss of life is to create stronger structures that resist catastrophic damage during a major earthquake.

A future goal of the program is to minimize risk by developing an early warning system. With a wide enough distribution of real-time seismometers, it is technically possible for an urban area to get an early warning of an impending earthquake if the earthquake's epicenter is far enough away from the city. For example, if an earthquake occurred 100 kilometers from downtown Los Angeles and its waves are moving at 4 kilometers per second, the system would have 25 seconds to process and analyze the data and broadcast it as an early warning. Even seconds of warning could be enough to shut off main gas pipelines, shut down subway trains, and give schoolchildren time to get under their desks. Japan has successfully used such a system for detecting offshore earthquakes that will shut down the Bullet Train; it is also trying to pursue other ways to use the system to give early warnings to save lives in a major earthquake.

Waiting for the Big One in California

The San Andreas fault, running north–south for 1,300 kilometers (807 miles) through California, is a right-lateral fault capable of generating great earthquakes of magnitude 8 or more. The 1906 earthquake near San Francisco caused a 450-kilometer scar in northern California (box figure 1). The portion of the fault nearest Los Angeles last broke in 1857 in a quake that was probably of comparable size. The ground has not broken in either of these regions since these quakes. Each old break is now a seismic gap, where rock strain is being stored prior to the next giant quake.

Recent California quakes were considerably smaller than the "Big One" long predicted by geologists to be in the magnitude-8 range. The 1906 quake in the north had an estimated Richter magnitude of 8.25, and the southern break in 1857 near Fort Tejon was estimated to have a moment magnitude (M) of 7.8. In contrast, the 1989 Loma Prieta quake on the San Andreas fault near San Francisco was a M7.2 and the 1994 Northridge quake (not on the San Andreas fault) was M6.7. So, recent California quakes have been about magnitude 7 or less, and the Big One should be 8. A magnitude-8 quake has 10 times the ground shaking and 32 times the energy of a magnitude-7 quake. In other words, it would take about 32 Loma Prieta quakes to equal the Big One. Comparing moment magnitudes for 1994 and 1857 in southern California, it would take nearly 64 Northridge quakes to equal the Fort Tejon quake.

A great earthquake of magnitude 8 could strike either the northern section or the southern section of the San Andreas fault tomorrow. Which section will break first? Because the southern section has been inactive longer, it may be the likelier candidate. A magnitude-8 quake here could cause hundreds of billions of dollars in damage and kill thousands of people if it struck during weekday business hours when Los Angeles-area buildings and streets are crowded with people. The M6.7 Northridge quake caused more than $20 billion in damage and was the most costly earthquake in U.S. history. It is daunting to think of an earthquake 64 times more powerful.

Detailed paleoseismology studies suggest that great earthquakes have a recurrence interval of about 105 years on the southern portion of the San Andreas fault near San Bernadino. Historic records in California do not go very far back in time, and much of the evidence involves isotopic dating of broken beds of carbon-rich sediments. A 1,500-year record of earthquake activity is well-preserved in the sedimentary layers along the southern part of the 1857 break and reveals evidence for fourteen separate earthquakes. The 105-year recurrence interval is only an average; two or three quakes spaced about sixty years apart occur in clusters, with the interval between clusters being 200 to 300 years. Prior to 1857, this portion last broke in about 1745. Adding this 112-year difference to 1857 gives 1969, and adding the average of 105 years to 1857 gives 1962, so the present danger to southern California is clear. Because the time elapsed since the most recent 1857 earthquake is much longer than the 105-year average between quakes, geologists are concerned

BOX 16.3 ■ FIGURE 1

The two major breaks on the San Andreas fault in California. Each break occurred during a giant earthquake (break from the 1857 earthquake is shown in green and the 1906 earthquake is shown in red). Each old break is now a seismic gap where the fault is locked and may be the future site for another major earthquake. A creeping segment (blue) separates the two locked portions. *From U.S. Geological Survey*

that the southern part of the fault may rupture again in a M7.6–7.8 earthquake within the next few decades putting the urban San Bernardino–Riverside area at great risk.

But the northern portion of the San Andreas fault is dangerous, too. Prior to 1906, this section of the fault broke in another giant quake in 1838. These quakes were only 68 years apart, and 1906 plus 68 equals 1974, so the northern section may actually be overdue for a big quake.

Another way of estimating the recurrence interval is by rock displacement. In 1906, rocks were displaced about 4.5 meters at the epicenter, and we know that plate motion across the San Andreas fault is about 5 centimeters per year. It should therefore take about 90 years to store enough strain to move rocks 5 meters, so the next quake should have occurred in 1996.

The probability of a repeat of the 1906 quake (8+) on the locked northern section of the San Andreas fault may be very low, less than 21% for the next thirty years. The latest probability studies estimate the chance of a 6.7 or greater quake in the San Francisco Bay area to be 62% from 2003–2032. A likely candidate for the quake is not the San Andreas but the Hayward fault across the bay from San Francisco. Such a quake near or under Bay-area cities such as Oakland and Berkeley would cause far greater death and destruction than the 1989 quake.

The U.S. Geological Survey has estimated that the southern portion of the fault has about a 60% chance of an earthquake of magnitude 7.5 to 8.3 within the next thirty years. The 1992 Landers (M7.3), 1994 Northridge (M6.7) and 1999 Hector Mine (M7.1) quakes have occurred since that prediction, and geologists disagree as to whether those quakes increase or decrease the likelihood of the Big One in southern California. Even more likely (85% chance) is a 7+-magnitude quake on any one of several faults that parallel the San Andreas fault and lie closer to (or even under) Los Angeles.

Because of the destructiveness of the Northridge earthquake and the earlier 1987 Whittier Narrows quake (M5.9), geologists

were concerned that another blind thrust fault (fault that cannot be seen at the surface) might rupture closer to downtown Los Angeles. To determine the underground configuration of the blind thrust faults and to investigate how deep sedimentary basins are that will amplify shaking in the region, the Los Angeles Regional Seismic Experiment (LARSE) was undertaken to predict where the strongest shaking will occur during future earthquakes. The LARSE project involved setting off underground explosive charges across the Los Angeles Basin northward toward the San Andreas fault (box figure 2A) to generate sound waves that could be analyzed by powerful computers to produce images of the subsurface. The experiment revealed a main blind thrust fault 20 kilometers (12 miles) beneath the surface that extends from near the San Andreas fault and transfers stress and strain upward and southward under the San Gabriel Valley and the Los Angeles Basin (box figure 2B). The images also show that the sedimentary basin under the San Gabriel Valley is nearly 5 kilometers (3 miles) deep—much deeper than originally thought—which will increase the potential for strong shaking during the next earthquake in this highly populated area.

A

B

BOX 16.3 ■ FIGURE 2

Los Angeles Regional Seismic Experiment (LARSE). (A) Map showing location of Line 1 where underground explosives (red dots) were set off to create sound waves that were analyzed by powerful computers to generate an image of the surface beneath the San Andreas fault and the Los Angeles Basin. Line 2 has also been studied. Red stars show the epicenters of earthquakes greater than magnitude 5.8 that occurred since 1932. (B) Diagram drawn from the subsurface image generated along Line 1 shows an interpretation of the subsurface structures under the San Andreas fault zone westward under the San Gabriel and Los Angeles Basins. *Image A courtesy LARSE, U.S. Geological Survey*

(continued)

(continued)

Although the chance of a magnitude-8+ Big One along the San Andreas fault may be higher in southern California, the chance of a 7+-quake killing thousands of people and causing extensive damage is about equal in the north and the south.

Additional Resources

U.S. Geological Survey. 1990. *The San Andreas fault system.* Professional Paper 1515.

Special issue on the paleoseismology of the San Andreas fault. 2002. *Bulletin of the Seismological Society of America*, vol. 92, no. 7.

For more information about the San Andreas fault and the likelihood of it creating a large earthquake, visit U.S. Geological Survey websites:

- http://pubs.usgs.gov/gip/earthq3/safaultgip.html and
- http://geopubs.wr.usgs.gov/fact-sheet/fs152-99/

For more details on the Los Angeles Regional Seismic Experiment, visit:

- http://geopubs.wr.usgs.gov/fact-sheet/fs110-99/

SUMMARY

Earthquakes usually occur when rocks break and move along a fault to release strain that has gradually built up in the rock. Volcanic activity can also cause earthquakes. Deep quakes may be caused by mineral transformations.

Seismic waves move out from the earthquake's *focus. Body waves* (P waves and S waves) move through Earth's interior, and *surface waves (Love* and *Rayleigh waves)* move on Earth's surface.

Seismographs record seismic waves on *seismograms,* which can be used to determine an earthquake's strength, location, and depth of focus. Most earthquakes are shallow-focus quakes, but some occur as deep as 670 kilometers below Earth's surface.

The time interval between first arrivals of P and S waves is used to determine the distance between the seismograph and the *epicenter.* Three or more stations are needed to determine the location of earthquakes.

Earthquake *intensity* is determined by assessing damage and is measured on the *modified Mercalli scale.*

Earthquake *magnitude,* determined by the amplitude of seismic waves on a seismogram, is measured on the *Richter scale. Moment magnitudes,* determined by field work, are widely used today and often are larger than Richter magnitudes.

The most noticeable effects of earthquakes are ground motion and displacement (which destroy buildings and thereby injure or kill people), fire, landslides, and *tsunamis. After-shocks* can continue to cause damage months after the main shock.

Earthquakes are generally distributed in belts. The *circum-Pacific belt* contains most of the world's earthquakes. Earthquakes also occur on the Mediterranean-Himalayan belt, the crest of the mid-oceanic ridge, and in association with basaltic volcanoes.

Benioff zones of shallow-, intermediate-, and deep-focus earthquakes are associated with andesitic volcanoes, oceanic trenches, and the edges of continents or island arcs.

The concept of plate tectonics explains most earthquakes as being caused by interactions between two plates at their boundaries. Plate boundaries are generally defined by bands of earthquakes.

Divergent plate boundaries are marked by a narrow zone of shallow earthquakes along normal faults, usually in a rift valley. Transform boundaries are marked by shallow quakes caused by strike-slip motion along one or more faults.

Convergent boundaries where continents collide are marked by a very broad zone of shallow quakes. Convergent boundaries involving deep subduction are marked by Benioff zones of quakes caused by tension, underthrusting, and compression.

The distribution of quakes indicates subduction angles of a down-going plate. The subduction angle is probably controlled by plate density and rate of plate convergence.

Determining the probability of an earthquake occurring uses the measurement of rock properties near faults, slip rate studies, and paleoseismology investigations to determine the recurrence interval of quakes along individual faults.

Terms to Remember

aftershock 421
Benioff zone 426
body wave 409
circum-Pacific belt 426
depth of focus 413
earthquake 408
elastic rebound theory 409
epicenter 409
focus 409

intensity 414
island arc 426
Love wave 411
magnitude 414
Mediterranean-Himalayan belt 426
modified Mercalli scale 414
moment magnitude 415
P wave 410
Rayleigh wave 411

Richter scale 414
seismic sea wave 423
seismic wave 408
seismogram 411
seismograph 411
surface wave 409
S wave 410
travel-time curve 413
tsunami (seismic sea wave) 423

Testing Your Knowledge

Use the following questions to prepare for exams based on this chapter.

1. Describe in detail how earthquake epicenters are located by seismograph stations.

2. What causes earthquakes?

3. Compare and contrast the concepts of intensity and magnitude of earthquakes.

4. Name and describe the various types of seismic waves.

5. Discuss the distribution of earthquakes with regard to location and depth of focus.

6. Show with a sketch how the concept of plate tectonics can explain the distribution of earthquakes in a Benioff zone and on the crest of the mid-oceanic ridge.

7. Describe several techniques that may help scientists predict earthquakes.

8. How may the timing of earthquakes someday be controlled?

9. Describe several ways that earthquakes cause damage.

10. How do earthquakes cause tsunami?

11. What are aftershocks?

12. The elastic rebound theory
    a. explains folding of rocks
    b. explains the behavior of seismic waves
    c. involves the sudden release of progressively stored strain in rocks, causing movement along a fault
    d. none of the preceding

13. The point within Earth where seismic waves originate is called the
    a. focus   b. epicenter
    c. fault scarp   d. fold

14. P waves are
    a. compressional   b. transverse
    c. tensional

15. What is the minimum number of seismic stations needed to determine the location of the epicenter of an earthquake?
    a. 1   b. 2
    c. 3   d. 5
    e. 10

16. The Richter scale measures
    a. intensity
    b. magnitude
    c. damage and destruction caused by the earthquake
    d. the number of people killed by the earthquake

17. Benioff zones are found near
    a. mid-ocean ridges
    b. ancient mountain chains
    c. interiors of continents
    d. oceanic trenches

18. Most earthquakes at divergent plate boundaries are
    a. shallow focus
    b. intermediate focus
    c. deep focus
    d. all of the preceding

19. Most earthquakes at convergent plate boundaries are
    a. shallow focus
    b. intermediate focus
    c. deep focus
    d. all of the preceding

20. A zone of shallow earthquakes along normal faults is typical of
    a. divergent boundaries b. transform boundaries
    c. subduction zones   d. collisional boundaries

21. A seismic gap is
    a. the time between large earthquakes
    b. a segment of an active fault where earthquakes have not occurred for a long time
    c. the center of a plate where earthquakes rarely happen

22. Which of the following is not true of tsunami?
    a. very long wavelength
    b. high wave height in deep water
    c. very fast moving
    d. continued flooding after wave crest hits shore

Expanding Your Knowledge

1. What are some arguments in favor of and against predicting earthquakes? What would happen in your community if a prediction were made today that within a month, a large earthquake would occur nearby?

2. Most earthquakes occur at plate boundaries where plates interact with each other. How might earthquakes be caused in the interior of a rigid plate?

3. How can you prepare for an earthquake in your own home?

4. Suppose you want to check for earthquake danger before buying a new home. How can you check the regional geology for earthquake dangers? The actual building site? The home itself?

Exploring Web Resources

www.mhhe.com/plummer11e

This is the dedicated website for this book. You can go to it for new and updated information. The universal resource locators (URL) listed in this book are also given as links on the website, making it easy to go to those websites without typing in the URL. When you visit our Online Learning Center, go to the Student Center and then to the chapter of interest. Here you will find additional readings and media resources, as well as answers to the Testing Your Knowledge section, more quizzes, animation and video clips, and direct links to the sites listed in this book. Links to additional websites can also be found. We have added questions for some of the sites to allow you to get the most of your exploration of the web. Using the web is an enjoyable way of enhancing your knowledge of geology.

http://quake.wr.usgs.gov/hazprep/BayAreaInsert/

U.S. Geological Survey, 1990. *The next big earthquake.* (This magazine-like pamphlet also is available free from Earthquakes, USGS, 345 Middlefield Road, Menlo Park, CA 94025.)

http://pubs.usgs.gov/gip/earthq3/safaultgip.html

U.S. Geological Survey, 1990. *The San Andreas fault system.* Professional Paper 1515.

www.geophys.washington.edu/seismosurfing.html

Exhaustive list of worldwide Internet sites for information about earthquakes.

http://quake.wr.usgs.gov/

U.S. Geological Survey Earthquake Information. Gives information on reducing earthquake hazards, earthquake preparedness, latest quake information, historical earthquakes, and how earthquakes are studied. Also a good starting place for links to other earthquake sites.

http://quake.wr.usgs.gov/recenteqs/faq.html

Frequently asked questions about recent earthquakes, maintained by the U.S. Geological Survey.

www.seismo.unr.edu/

University of Nevada, Reno Seismological Laboratory site contains information about recent earthquakes, earthquake preparedness, and links to other earthquake sites.

www.seismo.berkeley.edu/seismo/Homepage.html

Seismographic information page maintained by University of California–Berkeley that has many links to other earthquake sites (particularly in California), three-dimensional earthquake movie, Northridge earthquake rupture movies, and information on earthquake preparedness.

http://vquake.calstatela.edu/

California State University, Los Angeles *Virtual Earthquake.* Create and analyze an earthquake.

http://pubs.usgs.gov/gip/earthq4/severitygip.html

General information about the size of an earthquake. Discussion of Richter and Mercalli scales.

http://pubs.usgs.gov/publications/text/dynamic.html

General information about plate tectonics.

http://geopubs.wr.usgs.gov/circular/c1187/

U.S. Geological Survey online version of Tsunami Circular.

http://walrus.wr.usgs.gov/tsunami/PNGhome.html

U.S. Geological Survey web page gives information about the devastating July 17, 1998, tsunami at Papua, New Guinea, and links to other sites.

 # Animations

This chapter includes the following animations available on our Online Learning Center at **www.mhhe.com/plummer11e**.

16.4 Earthquake focus
16.5 Earthquake waves
16.6 Seismometer
16.7 Seismometer
16.8, 16.9, 16.10 Locating earthquake epicenter

CHAPTER

17

Earth's Interior and Geophysical Properties

The only rocks that geologists can study directly in place are those of the crust, and Earth's crust is but a thin skin of rock, making up less than 1% of Earth's total volume. Mantle rocks brought to Earth's surface in basalt flows and in diamond-bearing kimberlite pipes, as well as the tectonic attachment of lower parts of the oceanic lithosphere to the continental crust, give geologists a glimpse of what the underlying mantle might look like. Meteorites also give clues about the possible composition of the core of Earth. But to learn more about the deep interior of Earth, geologists must study it *indirectly,* largely by using the tools of geophysics—that is, seismic waves and the measurement of gravity, heat flow, and Earth magnetism.

The evidence from geophysics suggests that Earth is divided into three major compositional layers—the crust on Earth's surface, the rocky mantle beneath the crust,

Diamonds form in the mantle and are brought to the surface in kimberlite pipes, giving geologists a glimpse of Earth's interior. *Photo © Reuters New-Media Inc./Corbis* and the metallic core at the center of Earth. The study of plate tectonics has shown that the crust and uppermost mantle can be mechanically divided into the brittle lithosphere and the ductile or plastic asthenosphere.

You will learn in this chapter how gravity measurements can indicate where certain regions of the crust and upper man-

tle are being held up or held down out of their natural position of equilibrium. We will also discuss Earth's magnetic field and its history of reversals. We will show how magnetic anomalies can indicate hidden ore and geologic structures. The chapter closes with a discussion of the distribution and loss of Earth's heat.

INTRODUCTION

What *do* geologists know about Earth's interior? How do they obtain information about the parts of Earth beneath the surface? Geologists, in fact, are not able to sample rocks very far below Earth's surface. Some deep mines penetrate 3 kilometers into Earth, and a deep oil well may go as far as 8 kilometers beneath the surface; the deepest scientific well has reached 12 kilometers in Russia (see box 17.1). Rock samples can be brought up from a mine or a well for geologists to study.

A direct look at rocks from deeper levels can be achieved where mantle rocks have been brought up to the surface by basalt flows, by the intrusion and erosion of diamond-bearing kimberlite pipes (see chapter 5), or where the lower part of the oceanic lithosphere (see chapter 18) has been tectonically attached to the continental crust at a convergent plate boundary. However, Earth has a radius of about 6,370 kilometers, so it is obvious that geologists can only scratch the surface when they try to study *directly* the rocks beneath their feet.

Deep parts of Earth are studied *indirectly,* however, largely through the branch of geology called **geophysics,** which is the application of physical laws and principles to a study of Earth. Geophysics includes the study of seismic waves and Earth's magnetic field, gravity, and heat. All of these things tell us something about the nature of the deeper parts of Earth. Together, they create a convincing picture of what makes up Earth's interior.

EVIDENCE FROM SEISMIC WAVES

Seismic waves from a large earthquake may pass through the entire Earth. A nuclear bomb explosion also generates seismic waves. Geologists obtain new information about Earth's interior after every large earthquake and bomb test.

One important way of learning about Earth's interior is the study of **seismic reflection,** the return of some of the energy of seismic waves to Earth's surface after the waves bounce off a rock boundary. If two rock layers of differing densities are separated by a fairly sharp boundary, seismic waves reflect off that boundary just as light reflects off a mirror (figure 17.1). These reflected waves are recorded on a seismogram, which shows the amount of time the waves took to travel down to the boundary, reflect off it, and return to the surface. From the amount of time necessary for the round trip, geologists calculate the depth of the boundary.

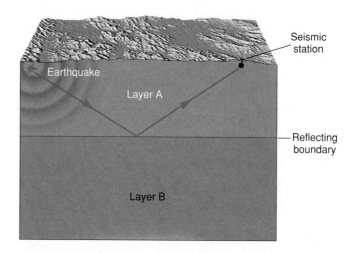

FIGURE 17.1

Seismic reflection. Seismic waves reflect from a rock boundary deep within the Earth and return to a seismograph station on the surface.

Another method used to locate rock boundaries is the study of **seismic refraction,** the bending of seismic waves as they pass from one material to another, which is similar to the way that light waves bend when they pass through the lenses of eyeglasses. As a seismic wave strikes a rock boundary, much of the energy of the wave passes across the boundary. As the wave crosses from one rock layer to another, it changes direction (figure 17.2). This change of direction, or refraction, occurs only if the velocity of seismic waves is different in each layer (which is generally true if the rock layers differ in density or strength).

The boundaries between such rock layers are usually distinct enough to be located by seismic refraction techniques, as shown in figure 17.3. Seismograph station 1 is receiving seismic waves that pass directly through the upper layer (A). Stations farther from the epicenter, such as station 2, receive seismic waves from two pathways: (1) a direct path straight through layer (A) and (2) a refracted path through layer (A) to a higher-velocity layer (B) and back to layer (A). Station 2, therefore, receives the same wave twice.

Seismograph stations close to station 1 receive only the direct wave or possibly two waves, the direct (upper) wave arriving before the refracted (lower) wave. Stations near station 2 receive both the direct and the refracted waves. At some point between station 1 and station 2, there is a transformation from receiving the direct wave first to receiving the *refracted* wave

IN GREATER DEPTH 17.1

Deep Drilling on Continents

The structure and composition of most of the continental crust is unknown. Surface mapping and seismic reflection and refraction suggest that the continents are largely igneous and metamorphic rock, such as granite and gneiss, overlain by a veneer of sedimentary rocks. This sedimentary cover is generally thin, like icing on a cake, but it may thicken to 10 kilometers (6.2 miles) or more in giant sedimentary basins where the underlying "basement rock" has subsided. Although oil companies have drilled as deep as 8 kilometers on land, they drill in the sedimentary basins. The igneous and metamorphic basement, which averages 40 kilometers thick and makes up most of the continental crust, has rarely been sampled deeper than 2 or 3 kilometers (although uplift and erosion have exposed some rocks widely thought to have been formed much deeper in the crust).

Russia has drilled the world's deepest hole on the Kola Peninsula near Murmansk north of the Arctic Circle. The 12-kilometer-deep hole took fifteen years to drill and penetrated ancient Precambrian basement rocks. The second deepest well drilled is the KTB hole in southeastern Germany, which reached a depth of 10 kilometers and cost more than a billion dollars (box figure 1). Deep drilling is as technically complex as space exploration. High pressures and 300°C temperatures require special equipment and techniques. The Russians used a turbodrill that rotated under the pressure of circulating drilling mud. Unlike normal drilling operations, the lightweight aluminum drill pipe does not turn. Because the Kola drilling operation resulted in a crooked hole, the Germans advanced deep-drilling technology by developing a system to keep the hole straight while being drilled.

The drilling at Kola shows that seismic models for this area are wrong. The Russians expected 4.7 kilometers of metamorphosed sedimentary and volcanic rock, then a granitic layer to a depth of 7 kilometers, and a "basaltic" layer below that. The granite, however, appeared at 6.8 kilometers and extends to more than 12 kilometers; the "basalt" has not yet been found. These results, and data from the other deep holes, show that seismic surveys of continental crust are being systematically misinterpreted.

The Russians and Germans unexpectedly found open fractures and circulating fluids throughout the borehole. The fluids include hydrogen, helium, and methane (natural gas), as well as mineralized waters forming ore bodies. Copper-nickel ore was found deeper than theory predicted, and gold mineralization was present from 9.5 to 11 kilometers down. These results will change geologists' models of ore formation and fluid circulation under ground.

BOX 17.1 ■ FIGURE 1

The KTB drilling operation in southeastern Germany reached a depth of 10 kilometers and has advanced the technology of deep drilling. *Photo courtesy of ICDP, GeoForschungsZentrum Potsdam*

Additional Resources

R. A. Kerr. 1993. Looking—deeply—into the Earth's crust in Europe. *Science* 261:295.

Y. A. Kozlovsky. 1987. *The Superdeep Well of the Kola Peninsula*, Springer-Verlag, 558 p.

Scientific Information System for the world's deepest borehole, Kola SDB-3.

IGCP408: Rocks and minerals at great depth and on the surface.
http://icdp.gfz-potsdam.de/html/kola/IGCP408.html

A

Test explosion

Path of seismic wave

New direction of seismic wave

Layer in which seismic waves travel more slowly (low-velocity layer)

Layer in which seismic waves travel more rapidly (high-velocity layer)

B

Test explosion

High-velocity layer

Low-velocity layer

FIGURE 17.2

Seismic refraction occurs when seismic waves bend as they cross rock boundaries. At an interface, seismic (or sound or light) waves will bend toward the lower-velocity material. (*A*) Low-velocity layer above high-velocity layer. (*B*) High-velocity layer above low-velocity layer. Some of the seismic waves will also return to the surface by reflecting off the rock boundary.

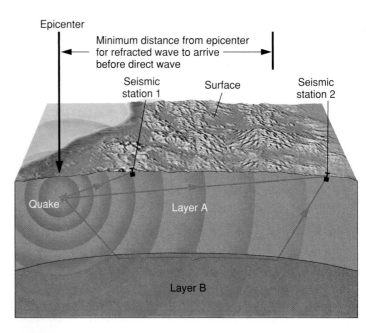

Epicenter

Minimum distance from epicenter for refracted wave to arrive before direct wave

Seismic station 1

Surface

Seismic station 2

Quake

Layer A

Layer B

FIGURE 17.3

Seismic refraction can be used to detect boundaries between rock layers. See text for explanation.

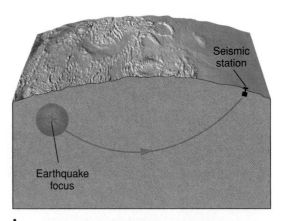

Seismic station

Earthquake focus

A

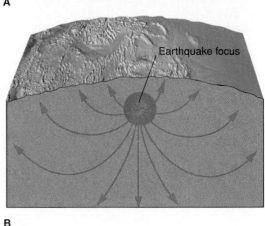

Earthquake focus

B

FIGURE 17.4

Curved paths of seismic waves caused by uniform rock with increasing seismic velocity with depth. (*A*) Path between earthquake and recording station. (*B*) Waves spreading out in all directions from earthquake focus.

first. Even though the refracted wave travels farther, it can arrive at a station first because most of its path is in the high-velocity layer *(B)*.

The distance between this point of transformation and the epicenter of the earthquake is a function of the depth to the rock boundary between layers *(A)* and *(B)*. A series of portable seismographs can be set up in a line away from an explosion (a *seismic shot*) to find this distance, and the depth to the boundary can then be calculated. The velocities of seismic waves within the layers can also be found.

Figure 17.2 shows how waves bend as they travel downward into higher-velocity layers. But why do waves return to the surface, as shown in figure 17.3? The answer is that advancing waves give off energy in all directions. Much of this energy continues to travel horizontally within layer *(B)* (figure 17.3). This energy passes beneath station 2 and out of the figure toward the right. A small part of the energy "leaks" upward into layer *(A)*, and it is this pathway that is shown in the figure. There are many other pathways for this wave's energy that are not shown here.

A sharp rock boundary is not necessary for the refraction of seismic waves. Even in a thick layer of uniform rock, the increasing pressure with depth tends to increase the velocity of the waves. The waves follow curved paths through such a layer, as shown in figure 17.4. To understand the reason for the curving path, visualize the thick rock layer as a stack of very thin layers, each with a slightly higher velocity than the one above. The curved path results from many small changes in direction as the wave passes through the many layers.

EARTH'S INTERNAL STRUCTURE

It was the study of seismic refraction and seismic reflection that enabled scientists to plot the three main zones of Earth's interior (figure 17.5). The **crust** is the outer layer of rock,

which forms a thin skin on Earth's surface. Below the crust lies the **mantle,** a thick shell of rock that separates the crust above from the core below. The **core** is the central zone of Earth. It is probably metallic and the source of Earth's magnetic field.

The Crust

Studies of seismic waves have shown (1) that the crust is thinner beneath the oceans than beneath the continents (figure 17.6) and (2) that seismic waves travel faster in oceanic crust than in continental crust. Because of this velocity difference, it is assumed that the two types of crust are made up of different kinds of rock.

Seismic P waves travel through oceanic crust at about 7 kilometers per second, which is also the speed at which they travel through basalt and gabbro (the coarse-grained equivalent of basalt). Samples of rocks taken from the sea floor by oceanographic ships verify that the upper part of the oceanic crust is basalt and suggest that the lower part is gabbro. The oceanic crust averages 7 kilometers (4.3 miles) in thickness, varying from 5 to 8 kilometers (table 17.1).

FIGURE 17.5

Earth's interior. Seismic waves show the three main divisions of Earth: the crust, the mantle, and the core. *Photo by NASA*

TABLE 17.1	**Characteristics of Oceanic Crust and Continental Crust**	
	Oceanic Crust	Continental Crust
Average thickness	7 km	20 to 70 km (thickest under mountains)
Seismic P-wave velocity	7 km/second	6 km/second (higher in lower crust)
Density	3.0 gm/cm^3	2.7 gm/cm^3
Probable composition	Basalt underlain by gabbro	Granite, other plutonic rocks, schist, gneiss (with sedimentary rock cover)

Seismic P waves travel more slowly through continental crust—about 6 kilometers per second, the same speed at which they travel through granite and gneiss. Continental crust is often called "granitic," but the term should be put in quotation marks because most of the rocks exposed on land are not granite. The continental crust is highly variable and complex, consisting of a crystalline basement composed of granite, other plutonic rocks, gneisses, and schists, all capped by a layer of sedimentary rocks, like icing on a cake. Since a single rock term cannot accurately describe crust that varies so greatly in composition, some geologists use the term *felsic* (rocks high in *feldspar* and *silicon*) for continental crust and *mafic* (rocks high in magnesium and iron) for oceanic crust.

Continental crust is much thicker than oceanic crust, averaging 30 to 50 kilometers (18.6 to 31 miles) in thickness, though it varies from 20 to 70 kilometers. Seismic waves show that the crust is thickest under geologically young mountain ranges, such as the Andes and Himalayas, bulging downward as a *mountain root* into the mantle (figure 17.6). The continental crust is also less dense than oceanic crust, a fact that is important in plate tectonics (table 17.1).

The boundary that separates the crust from the mantle beneath it is called the **Mohorovičić discontinuity** (**Moho** for

short). Note from figure 17.6 that the mantle lies closer to Earth's surface beneath the ocean than it does beneath continents. The idea behind an ambitious program called Project Mohole (begun during the early 1960s) was to use specially equipped ships to drill through the oceanic crust and obtain samples from the mantle. Although the project was abandoned because of high costs, ocean-floor drilling has become routine since then, but not to the great depth necessary to sample the mantle. Perhaps in the future, the original concept of drilling to the mantle through oceanic crust will be revived. (Ocean drilling is discussed in more detail in chapters 18 and 19.)

The Mantle

Because of the way seismic waves pass through the mantle, geologists believe that it, like the crust, is made of solid rock. Localized magma chambers of melted rock may occur as isolated pockets of liquid in both the crust and the upper mantle, but most of the mantle seems to be solid. Because P waves travel at about 8 kilometers per second in the upper mantle, it appears that the mantle is a different type of rock from either oceanic crust or continental crust. The best hypothesis that geologists can make about the composition of the upper mantle is that it consists of ultramafic rock such as peridotite. *Ultramafic rock* is dense igneous rock made up chiefly of ferromagnesian minerals such as olivine and pyroxene. Some ultramafic rocks contain garnet, and feldspar is extremely rare in the mantle.

The crust and uppermost mantle together form the **lithosphere,** the outer shell of Earth that is relatively strong and brittle. The lithosphere makes up the plates of plate-tectonic theory. The lithosphere averages about 70 kilometers (43.4 miles) thick beneath oceans and may be 125 to 250 kilometers thick beneath continents. Its lower boundary is marked by a curious mantle layer in which seismic waves slow down (figure 17.6).

Generally, seismic waves increase in velocity with depth as increasing pressure alters the properties of the rock. Beginning at a depth of 70 to 125 kilometers, however, seismic

FIGURE 17.6

Thin oceanic crust has a P-wave velocity of 7 kilometers per second, whereas thick continental crust has a lower velocity. Mantle velocities are about 8 kilometers per second. The oceanic and continental crust, along with the upper rigid part of the upper mantle, form the lithosphere. The asthenosphere underlies the lithosphere and is defined by a decrease in P-wave velocities.

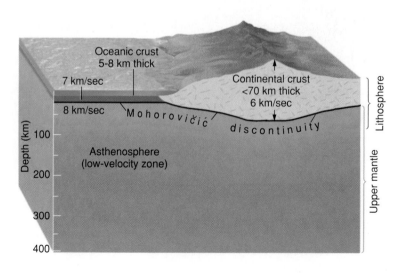

waves travel more slowly than they do in shallower layers, and so this zone has been called the *low-velocity zone* (figure 17.6). This zone, extending to a depth of perhaps 200 kilometers (124 miles), is also called, in plate-tectonic theory, the **asthenosphere.** The rocks in this zone may be closer to their melting point than the rocks above or below the zone. (The rocks are probably not *hotter* than the rocks below—melting points are controlled by pressure as well as temperature.) Some geologists think that these rocks may actually be partially melted, forming a crystal-and-liquid slush; a very small percentage of liquid in the asthenosphere could help explain some of its physical properties.

If the rocks of the asthenosphere are close to their melting point, this zone may be important for two reasons: (1) it may represent a zone where magma is likely to be generated; and (2) the rocks here may have relatively little strength and therefore are likely to flow. If mantle rocks in the asthenosphere are weaker than they are in the overlying lithosphere, then the asthenosphere can deform easily by ductile flow. Plates of brittle lithosphere probably move easily over the asthenosphere, which may act as a lubricating layer below.

There is widespread agreement on the existence and depth of the asthenosphere under oceanic crust but considerable disagreement about asthenosphere under continental crust. Figure 17.6 shows asthenosphere at a depth of 125 kilometers (77.5 miles) below the continents. Some geologists think that the lithosphere is much thicker beneath continents than shown in the figure and that the asthenosphere begins at a depth of 250 kilometers (or even more). A few geologists say that there is *no* asthenosphere beneath continents at all. The reasons for this disagreement are the results of the rapidly developing field of seismic tomography, which is described in box 17.2.

Data from seismic reflection and refraction indicate several concentric layers in the mantle (figure 17.7), with prominent boundaries at 400 and 670 kilometers (248 and 416 miles) (670 kilometers is also the depth of the deepest earthquakes). It is doubtful that the layering is due to the presence of several different kinds of rock. Most geologists think that the chemical composition of the mantle rock is about the same throughout the mantle. Because pressure increases with depth into Earth, the boundaries between mantle layers possibly represent depths at which pressure collapses the internal structure of certain minerals into denser minerals. For example, at a pressure equivalent to a depth of about 670 kilometers, the mineral *olivine* should collapse into the denser structure of the mineral *perovskite*. If the boundaries between mantle layers represent pressure-caused transformations of minerals, the entire mantle may have the same *chemical* composition throughout, although not the same *mineral* composition. However, some geologists think that the 670-kilometer boundary represents a chemical change as well as a physical change and separates the *upper mantle* from the chemically different *lower mantle* below.

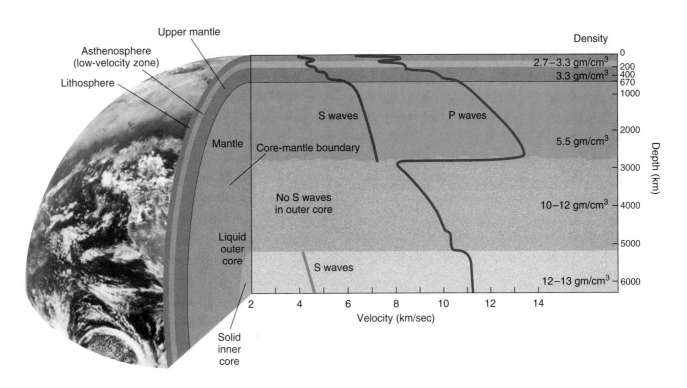

FIGURE 17.7

The concentric shell structure of Earth as defined by variation in S- and P-wave velocities and estimates of density. The velocity of seismic P and S waves generally increases with depth except in the low-velocity zone. The plastic asthenosphere slows down seismic waves. Velocity increases at 400 and 670 kilometers may be caused by mineral collapse. S waves do not pass through the outer core but are thought to travel through the solid inner core.

IN GREATER DEPTH 17.2

A CAT Scan of the Mantle

A new technique for looking at the mantle is similar to the medical technique of CAT scanning (CAT stands for computed axial tomography), which builds up a three-dimensional picture of soft body tissues such as the brain by taking a series of X-ray pictures along successive planes in the body.

Seismic tomography uses earthquake waves and powerful computers to study planar cross sections of the mantle following large earthquakes. Slight variations from expected arrival times at distant seismograph stations can be used to find temperature variations in the mantle. Hot rock slows down seismic waves, so

a late arrival of a seismic wave shows that the wave went through hot rock. Cold rock is dense and strong, so it speeds up seismic waves, resulting in early arrivals. Sophisticated computer analysis of hundreds of sections through the mantle allows maps of seismic-wave velocity (and therefore mantle rock temperature) to be drawn for various depths.

Box figure 1 (top) shows mantle velocities at a depth of 100 kilometers. Red areas show low velocities (probably caused by hot rock) in generally expected positions—along the crest of the mid-oceanic ridge and beneath hot spots. Blue areas show high-velocity (probably cold) rock under continents and old sea

BOX 17.2 ■ FIGURE 1

Map views of seismic-wave velocities in the mantle at depths of 100 and 300 kilometers, as determined by seismic tomography. Blue indicates high velocity (cold rock); red indicates low velocity (hot rock). White lines outline plates; white circles are major hot spots. *From Dziewonski and Anderson, American Scientist, 1984, 72:483–94*

BOX 17.2 ■ FIGURE 2

Vertical cross sections of seismic-wave velocities to a depth of 670 kilometers in the mantle. The orange lines show the locations of the cross sections. *From Dziewonski and Anderson, American Scientist, 1984, 72:483–94*

The Core

Seismic-wave data provide the primary evidence for the existence of the core of Earth. (See chapter 16 for a discussion of seismic P and S waves.) Seismic waves do not reach certain areas on the opposite side of Earth from a large earthquake. Figure 17.8 shows how seismic P waves spread out from a quake until, at 103° of arc (11,500 kilometers) from the epi-

center, they suddenly disappear from seismograms. At more than 142° (15,500 kilometers) from the epicenter, P waves reappear on seismograms. The region between 103° and 142°, which lacks P waves, is called the **P-wave shadow zone.**

The P-wave shadow zone can be explained by the refraction of P waves when they encounter the core boundary deep within Earth's interior. Because the paths of P waves can be accurately calculated, the size and shape of the core can be lat.: 6.0 lon.: 125.0   lat.: −2.5 lon.: 35.0

BOX 17.2 ■ FIGURE 3

Cross section of seismic-wave velocities from the Earth's surface (upper curve) to core. Blue indicates fast seismic velocities (cold rock), and red indicates low velocities (hot rock). There is a presumed cold slab of rock, shown on the left side of the cross section, that is sinking into the lower mantle into other slabs that rest on the core-mantle boundary. Hot rocks, believed to represent mantle plumes, also emanate from the core-mantle boundary, on the right side of the cross section. *Photo courtesy of Stephen Grand, University of Texas at Austin*

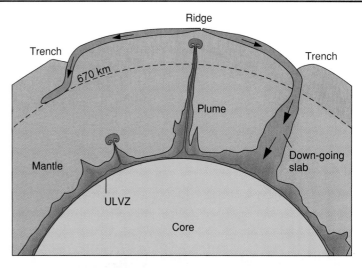

BOX 17.2 ■ FIGURE 4

Seismic data suggest some plates sink to the base of the mantle, whereas other plates are impeded by the increase in density of the mantle at 670 kilometers. Deep mantle plumes emanating from the core-mantle boundary are underlain by an ultralow-velocity zone (ULVZ).

floor such as the western Pacific. Box figure 1 (bottom) shows that these patterns are dramatically different at a depth of 300 kilometers. High-velocity rock extends to this depth below most continents, implying that continents have very deep roots. Some areas that appear hot at 100 kilometers are cold at 300 kilometers, such as the ridge crest just south of Australia. Areas such as the central Pacific and the Red Sea region appear cold at 100 kilometers and hot at 300 kilometers.

In box figure 2, vertical cross sections of seismic velocity are shown to a depth of 670 kilometers for two regions. Note that high-velocity (cold) roots beneath North America, Asia, and Antarctica extend 400 to 600 kilometers downward. This finding casts doubt on our simple lithosphere-asthenosphere model of plate behavior—continental plates here seem to be hundreds of kilometers thick. Notice, too, how some low-velocity hot spots near Greenland (box figure 2, top) and in the south Atlantic and south Pacific (box figure 2, bottom) are underlain by apparently cold rock. This pattern suggests to some geologists that mantle plumes may be quite shallow and may not extend vertically throughout the mantle. On the other hand, plume tails may be too narrow to be detected by this technique.

More recent, deeper CAT scans of the mantle (box figure 3) indicate that some mantle plumes emanate from the core-mantle boundary and are fed by heat loss from the core. The plume under the Hawaiian hot spot was recently found to contain material from the crust, mantle, and core. It is likely that the hot plumes originate from various depths in the mantle.

The new tomographic images also reveal high-velocity areas, which are interpreted as cold sinking slabs of subducted plates, that extend all the way to the core-mantle boundary (box figures 3 and 4).

Other plates stop descending at the 670-kilometer boundary within the mantle. Perhaps the depth of sinking is controlled by plate density. The older the subducting rock is, the colder and denser it is. Old, dense plates may sink to the base of the mantle, while younger plates, being less dense, stop at a depth of 670 kilometers (box figure 4).

It is becoming increasingly apparent that the core-mantle boundary may play an important role in the overall mechanism of plate movement.

Additional Resources

R. A. Kerr. 1991. Do plumes stir Earth's entire mantle? *Science* 252:1068–1069.

_____. 1997. Deep-sinking slabs stir the mantle. *Science* 275: 613–615.

S. P. Grand, R. D. Van der Hilst, and S. Widiyantoro. 1997. Global seismic tomography: A snapshot of convection in the Earth. *GSA Today* 7:1–7.

determined also. In figure 17.8, notice that Earth's core deflects the P waves and, in effect, "casts a shadow" where their energy does not reach the surface. In other words, P waves are missing within the shadow zone because they have been bent (refracted) by the core.

Chapter 16 on earthquakes explains that while P waves can travel through solids and fluids, S waves can travel only through solids. As figure 17.9 shows, an **S-wave shadow zone**

also exists and is larger than the P-wave shadow zone. Direct S waves are not recorded in the entire region more than 103° away from the epicenter. The S-wave shadow zone seems to indicate that S waves do not travel through the core at all. If this is true, it implies that the core of Earth is a liquid, or at least acts like a liquid.

The way in which P waves are refracted within Earth's core (as shown by careful analysis of seismograms) suggests

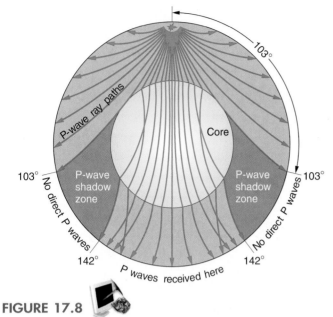

FIGURE 17.8

The P-wave shadow zone, caused by refraction of P waves within the Earth's core.

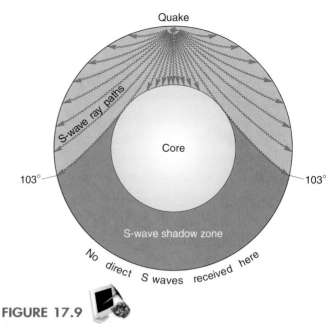

FIGURE 17.9

The S-wave shadow zone. Because no S waves pass through the core, the core is apparently a liquid (or acts like a liquid).

that the core has two parts, a *liquid outer core* and a *solid inner core* (figure 17.7).

Composition of the Core

When evidence from astronomy and seismic-wave studies is combined with what we know about the properties of materials, it appears that Earth's core is made of metal—not silicate rock—and that this metal is probably iron (along with a minor amount of oxygen, silicon, sulfur, or nickel). How did geologists arrive at this conclusion?

The overall density of the earth is 5.5 grams per cubic centimeter, based on calculations from Newton's law of gravitational attraction. The crustal rocks are relatively low density, from 2.7 grams per cubic centimeter for granite to 3.0 grams per cubic centimeter for basalt. The ultramafic rock thought to make up the mantle probably has a density of 3.3 grams per cubic centimeter in the upper mantle, although rock pressure should raise this value to about 5.5 grams per cubic centimeter at the base of the mantle (figure 17.7).

If the crust and the mantle, which have approximately 85% of Earth's volume, are at or below the average density of Earth, then the core must be very heavy to bring the average up to 5.5 grams per cubic centimeter.

Calculations show that the core has to have a density of about 10 grams per cubic centimeter at the core-mantle boundary, increasing to 12 or 13 grams per cubic centimeter at the center of Earth (figure 17.7). This great density would be enough to give Earth an average density of 5.5 grams per cubic centimeter.

Under the great pressures existing in the core, iron would have a density slightly greater than that required in the core. Iron mixed with a small amount of a lighter element, such as oxygen, sulfur or silicon, would have the required density.

Therefore, many geologists think that such a mixture makes up the core.

But a study of density by itself is hardly convincing evidence that the core is mostly iron, for many other heavy substances could be there instead. The choice of iron as the major component of the core comes from looking at meteorites (see box 17.3). Meteorites are thought by some scientists to be remnants of the basic material that created our own solar system. An estimated 10% of meteorites are composed of iron mixed with small amounts of nickel. Material similar to these meteorites may have helped create Earth, perhaps settling to the center of Earth because of metal's high density. The composition of these meteorites, then, may tell us what is in the core. Nickel is denser than iron, however, so a mixture of just iron and nickel would have a density greater than that required in the core. (The other 90% of meteorites is mostly ultramafic rock and perhaps represents material that formed the mantle.)

Seismic and density data, together with assumptions based on meteorite composition, point to a core that is largely iron, with at least the outer part being liquid. The existence of Earth's magnetic field, which is discussed in this chapter, also suggests a metallic core. Of course, no geologist has seen the core, nor is anyone likely to in the foreseeable future. But since so many lines of indirect evidence point to a liquid metal outer core, most scientists accept this theory as the best conclusion that can be made about the core's composition.

The Core-Mantle Boundary

The boundary between the core and mantle is marked by great changes in seismic velocity, density (figure 17.7), and temperature, as we will see in the section on "Heat Within the Earth" in this chapter. Here, there is a transition zone up to 200 kilo-

PLANETARY GEOLOGY 17.3

Meteorites

Small, solid particles of rock, metal, and/or ice orbiting the Sun are called *meteoroids*. When these particles enter Earth's atmosphere, they are heated to incandescence by friction; these glowing particles are called *meteors* (or "shooting stars" or "falling stars"). Most meteors are small and burn up while still in the atmosphere, but about 150 per year are large enough to strike Earth's surface. Those that do are called *meteorites* (box figure 1). The largest fragment of a meteorite found (in Africa) weighs 60 metric tons; much larger meteorites have hit Earth in the past.

Three basic types of meteorite are iron, stony-iron, and stony meteorites. Stony meteorites are by far the most common, but they look like Earth rocks, so they are hard to find. Iron meteorites are rare but look so unique that they are commonly found; most museum meteorites are of the iron type.

Iron meteorites are mostly iron alloyed (mixed) with a small percentage of nickel. Small amounts of other metals or minerals may be present. Iron-nickel meteorites give an important source of information regarding the composition of Earth's core.

Stony-iron meteorites are made of iron-nickel alloy and silicate minerals in about equal parts.

Stony meteorites are made of silicate minerals such as plagioclase, olivine, and pyroxene; they may contain a small amount of iron-nickel alloy. About 90% of stony meteorites contain round silicate grains called *chondrules* and are called *chondrites*. The other 10% are *achondrites*, which lack chondrules.

Chondrules consist mostly of olivine and pyroxene, and range from distinct spheres to large bodies with fuzzy outlines. The composition of chondrite meteorites resembles the ultramafic rock peridotite, but peridotite lacks the chondritic texture and iron-nickel content of the meteorites.

One kind of chondrite is composed mostly of serpentine or pyroxene and contains up to 5% organic materials, including carbon, hydrocarbon compounds, and amino acids. These meteorites are called *carbonaceous chondrites*. All available evidence indicates that the organic compounds were in fact produced by inorganic processes. Carbonaceous chondrites are of particular interest to scientists because they are believed to have the same

BOX 17.3 ■ FIGURE 1

A large meteorite fragment from Meteor Crater, Arizona. Pocketknife for scale. *Photo by Frank M. Hanna*

composition as the original material from which the solar system was formed.

Achondrites are generally similar to terrestrial rocks in composition and texture. In composition, they are most similar to basalts. Some have textures like ordinary igneous rocks, and others are breccias with fragments of different compositions and textures.

Many meteorites have a coarse-grained texture, probably formed by slow cooling within a larger body, such as a planet. The similarity in iron-nickel composition among iron meteorites also suggests that they are fragments from a single, large body. The larger body may have differentiated into a heavy, iron-rich core and a lighter, rocky mantle before it fragmented into meteoroids. Isotopic dating shows that most meteorites have the same age: 4.6 billion years old. No terrestrial rocks have ages greater than 4.03 billion years; therefore, meteorites provide the best clue as to the age of the solar system and the formation of the planets.

---

meters thick, known as the *D" layer*, at the base of the mantle where P-wave velocities decrease dramatically. The *ultralow-velocity zone* (ULVZ) (figure 17.10) that forms the undulating border at the core-mantle boundary may be due to hot core partially melting overlying mantle rock or to part of the liquid outer core reacting chemically with the adjacent mantle. The latest seismic and geodetic studies have hinted that lighter iron alloys from the liquid outer core may react with silicates in the lower mantle to form iron silicates. The less-dense iron silicate "sediment," along with liquid iron in pore spaces, rises and collects in uneven layers along the core-mantle boundary. The pressure of the accumulating "sediment" along the boundary

causes some of the liquid iron to be squeezed out of the pore spaces to form an electrically conductive layer that connects the core and mantle and explains the decrease in seismic velocities at the ULVZ. It may be difficult to prove whether the lowermost mantle is being partially melted by the core or whether the core is instead chemically reacting with the mantle.

Both the mantle and the core are undergoing **convection,** a circulation pattern in which low-density material rises and high-density material sinks. Based on seismic tomography studies, heavy portions of the mantle (including subducted plates) sink to its base but are unable to penetrate the denser core. Light portions of the core may rise to its top and may be

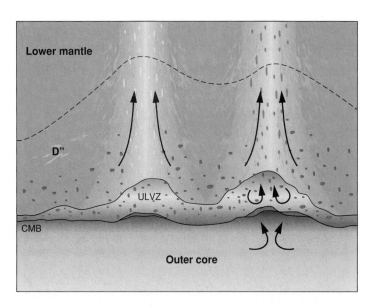

FIGURE 17.10

Recent seismic and geodetic studies are redefining the boundary between the lower 200 kilometers of the mantle (D″ layer) and the outer core. Iron silicate "sediments" (shown in brown) may rise from the underlying liquid core and fill pockets or inverted basins at the core-mantle boundary (CMB). Alternatively, the outer core material (shown in red) may be melting the lowermost mantle (shown in yellow) to form the ultralow-velocity zone (ULVZ). *Modified from Garnero and Jeanloz, Science, 2000*

FIGURE 17.11

Isostatic balance. (A) Wood blocks float in water with most of their bulk submerged. (B) Lithosphere "floats" on asthenospheric mantle in approximately the same way. The thicker the block, the deeper it extends into the asthenosphere.

incorporated into the mantle above. This is suggested by recent isotopic studies of the mantle plume that feeds the Hawaiian hot spot. The resulting Hawaiian volcanic rocks (basalts) contain a light-isotope signature that is characteristic of the core. Continent-sized blobs of liquid and liquid-crystal slush may accumulate at the core-mantle boundary, perhaps interfering with or helping cause heat loss from the core to help drive mantle convection and transfer of heat to the surface, and causing changes in Earth's magnetic field. This boundary is an exciting frontier for geologic study, but data, of course, are sparse and hard to obtain.

ISOSTASY

Isostasy is a balance or *equilibrium* of adjacent blocks of brittle lithosphere "floating" on the asthenosphere. Since lithosphere weighs less than mantle rocks, the crust can be thought of as floating on the denser mantle much as wood floats on water (figure 17.11).

Blocks of wood floating on water rise or sink until they displace an amount of water equal to their own weight. The weight of the displaced water buoys up the wood blocks, allowing them to float. The higher a wood block appears above the water surface, the deeper the block extends under water. Thus, a tall block has a deep "root"—much like an iceberg floating on the water.

In a greatly simplified way, the lithosphere can be thought of as tending to rise or sink gradually until it is balanced by the weight of displaced asthenosphere. This concept of vertical

movement to reach equilibrium is called **isostatic adjustment.** Just as with the blocks of wood, once lithospheric "blocks" have come into isostatic balance, a tall block (a mountain range) extends deep into the mantle (a *mountain root,* as shown in figure 17.11).

Figure 17.11 shows both the blocks of wood and the blocks of lithosphere in isostatic balance. The weight of the wood is equal to the weight of the displaced water. Similarly, the weight of the lithosphere is equal to the weight of the displaced asthenosphere. As a result, the rocks (and overlying seawater) in figure 17.11 can be thought of as separated into vertical columns, each with the same pressure at its base. At some *depth of equal pressure,* each column is in balance with the other columns, for each column has the same weight. A column of thick continental lithosphere (a mountain and its root) has the same weight as a column containing thin continental lithosphere and some of the upper mantle. A column containing seawater, thin oceanic crust, and a thick section of heavy mantle weighs the same as the other two columns.

Figure 17.11 shows the lithosphere as isolated blocks free to move past each other along vertical faults, but this is not really a good picture of the structure of the lithosphere. It is more accurate to think of the lithosphere as bending in broad uplifts and downwarps without vertical faults, as shown in figures 17.12 and 17.13.

Let us look at some examples of isostatic balance (equilibrium) in continental lithosphere rocks. Suppose that two sections of crust of unequal thickness are next to each other, as in figure 17.12. Sediment from the higher part, which is subject to more rapid erosion, is deposited on the lower part. The decrease in weight from the high part causes it to rise, while the increase in weight on the low part causes it to sink. These ver-

A

B

C

FIGURE 17.12

Isostatic adjustment due to erosion and deposition of sediment. Rock within the mantle must flow to accommodate vertical motion of crustal blocks. Mantle flow occurs in the asthenosphere, deeper than shown in C.

A Glacier forms, adding weight to crust

B Subsidence due to weight of ice

C Ice melts, removing weight from crust

D Crustal rebound as crust rises toward original position

FIGURE 17.13

The weight of glaciers depresses the crust, and the crust rebounds when the ice melts.

tical movements (isostatic adjustment) do, in fact, take place whenever large volumes of material are eroded from or deposited on parts of the crust.

Rising or sinking of the lithosphere, of course, requires ductile flow of the asthenosphere to accommodate the motion. By measuring the rate of rising or sinking, the viscosity of the asthenosphere can be calculated.

Another example of isostatic adjustment, caused by ductile mantle flow, is the upward movement of large areas of the crustal lithosphere since the glacial ages. The weight of the thick continental ice sheets during the Pleistocene Epoch depressed the lithosphere underneath the ice (figure 17.13). After the melting of the ice, the crust rose back upward, a process that is still going on in some areas (figure 17.14). This rise of the crustal lithosphere after the removal of the ice is known as **crustal rebound.** The process

FIGURE 17.14

Uplift of land surface in Canada and the northern United States caused by crustal rebound after glaciers melted. Colored lines show the minimum amount of uplift in meters since the ice disappeared. *From Phillip B. King, "Tectonics of Quaternary Time in Middle North America," in* The Quaternary of the United States, *H. D. Wright, Jr. and David G. Frey, eds., fig. 4A, p. 836. Reprinted by permission of Princeton University Press*

of crustal rebound can be easily demonstrated by sitting on a soft couch or bed. The indentation made from the weight of your body gradually disappears or rebounds after you stand up.

Recent geophysical studies have shown that some mountains, such as the Rockies and southern Sierra Nevada, do not have thick roots and are instead buoyed by warm, less-dense mantle. It appears that the upper mantle beneath some continents is not homogeneous but has zones that are quite buoyant due to higher temperatures and less-dense mineral phases.

GRAVITY MEASUREMENTS

According to Newton's law of gravitation, the force of gravity between two objects varies with the masses of the objects and the distance between them (figure 17.15):

$$\text{Force of gravity between } A \text{ and } B = \text{constant} \left(\frac{\text{mass}_A \times \text{mass}_B}{\text{distance}^2} \right)$$

The force increases with an increase in either mass. The gravitational attraction between Earth and the Moon, for example, is vastly greater than the extremely small attraction that exists between two bowling balls. The equation also shows that force decreases with the square of the distance between the two objects. The farther two objects are apart, the less gravitational attraction there is between them.

A useful tool for studying the crust and upper mantle is the **gravity meter,** which measures the gravitational attraction between Earth and a mass within the instrument. One use of the gravity meter is to explore for local variations in rock density

(mass = density × volume). Dense rock such as metal ores and ultramafic rock pulls strongly on the mass inside the meter (figure 17.16). The strong pull stretches a spring, and the amount of stretching can be very precisely determined. So a gravity meter can be used to explore for metallic ore deposits. A cavity or a body of low-density material such as sediment causes a much weaker pull on the meter's mass (figure 17.16).

Another important use of a gravity meter is to discover whether regions are in isostatic equilibrium. If a region is in isostatic balance, as in figure 17.17A, each column of rock has the same mass. If a gravity meter were carried across the rock columns, it would register the same amount of gravitational attraction for each column (after correcting for differences in elevation—gravitational attraction is less on a mountaintop

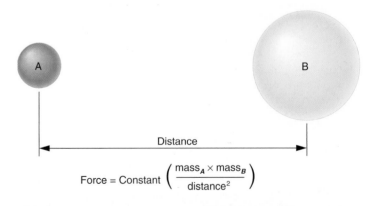

$$\text{Force} = \text{Constant} \left(\frac{\text{mass}_A \times \text{mass}_B}{\text{distance}^2} \right)$$

FIGURE 17.15

The force of gravitational attraction between two objects is a function of the masses of the objects and the distance between the centers of the objects.

than at sea level because the mountaintop is farther from the center of Earth).

Some regions, however, are held up out of isostatic equilibrium by deep tectonic forces. Figure 17.17*B* shows a region with uniformly thick crust. Tectonic forces are holding the center of the region up. This uplift creates a mountain range without a mountain root. There is a thicker section of heavy mantle rock under the mountain range than there is on either side of the mountain range. Therefore, the central "column" has more mass than the neighboring columns, and a gravity meter shows that the gravitational attraction is correspondingly greater over the central than over the side columns.

A gravity reading higher than the normal regional gravity is called a **positive gravity anomaly** (figure 17.17*B*). It can indicate that tectonic forces are holding a region up out of isostatic equilibrium, as shown in figure 17.17*B*. When the forces stop acting, the land surface sinks until it reestablishes isostatic balance. The gravity anomaly then disappears. For the region shown in figure 17.17*B*, equilibrium will be established when the land surface becomes level.

Positive gravity anomalies, particularly small ones, are also caused by local concentrations of dense rock such as metal ore. The gravity meter in figure 17.16 is registering a positive gravity anomaly over ore (the spring inside the meter is stretched). Since there can be more than one cause of a positive gravity anomaly, geologists may disagree about the interpretation of anomalies. Drilling into a region with a gravity anomaly usually discloses the reason for the anomaly.

A region can also be held down out of isostatic equilibrium, as shown in figure 17.17*C*. The mass deficiency in such a region produces a **negative gravity anomaly**—a gravity reading lower than the normal regional gravity. Negative gravity anomalies indicate either that a region is being held down (figure 17.17*C*) or that local mass deficiencies exist for other reasons (figure 17.16).

The greatest negative gravity anomalies in the world are found over oceanic trenches (see chapters 18 and 19). These negative anomalies are interpreted to mean that trenches are actively being held down and are out of isostatic balance.

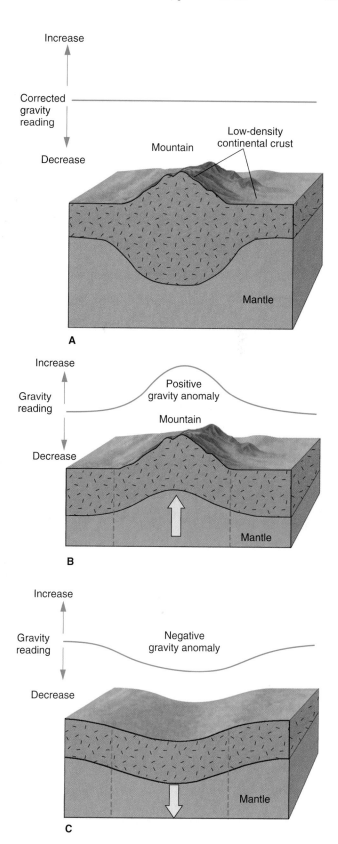

A

B

C

FIGURE 17.17

(*A*) A region in isostatic balance gives a uniform gravity reading (no gravity anomalies), after correcting for differences in elevation. (*B*) A region being held up out of isostatic equilibrium gives a positive gravity anomaly. (*C*) A region being held down out of isostatic equilibrium gives a negative gravity anomaly.

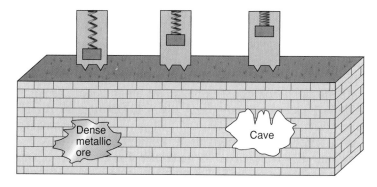

FIGURE 17.16

A gravity meter reading is affected by the density of the rocks beneath it. Dense rock pulls strongly on the mass within the meter, stretching a spring; a cavity exerts a weak pull on the mass. A gravity meter can be used to explore for hidden ore bodies, caves, and other features that have density contrasts with the surrounding rock.

Areas still experiencing isostatic rebound also show negative gravity anomalies. The rebound will end when isostatic equilibrium is once again reached and the gravity anomaly is removed.

EARTH'S MAGNETIC FIELD

A region of magnetic force—a **magnetic field**—surrounds Earth. The invisible lines of magnetic force surrounding Earth deflect magnetized objects, such as compass needles, that are free to move. The field has north and south **magnetic poles,** one near the geographic North Pole, the other near the geographic South Pole. (Because it has two poles, Earth's field is called *dipolar.*) The strength of the magnetic field is greatest at the magnetic poles, where magnetic lines of force appear to leave and enter Earth vertically (figure 17.18).

Because the compass is important in navigation, Earth's magnetism has been observed for centuries. It has long been known that the magnetic poles are displaced about 11½° from the geographic poles (about which Earth rotates). Furthermore, changes in the position of the magnetic poles have been well documented, especially since the time of the great explorations of the globe. Because Earth's field is not 100% dipolar, the magnetic poles appear to be moving slowly around the geographic poles. The separation between the two types of poles has probably never been much greater than it is today.

More recently, geophysical studies have been directed toward the *source* of Earth's magnetism. The rate of the poles' changes in position, together with the strength of the magnetic field, strongly suggest that the magnetic field is generated within the liquid metal of the outer core rather than within the solid rock of the crust or the mantle.

How is Earth's magnetic field generated? A number of hypotheses have been put forth. One widely accepted hypothesis suggests that the magnetic field is created by electric currents within the liquid outer core. The outer core is extremely hot and flows at a rate of several kilometers per year in large convection currents, about 1 million times faster than mantle convection above it. Convecting metal creates electric currents, which in turn create a magnetic field. This hypothesis requires the core to be an electrical conductor. Metals are good conductors of electricity, whereas silicate rock is generally a poor electrical conductor. Indirectly, this is evidence that the core is metallic.

Magnetic Reversals

In the 1950s, evidence began to accumulate that Earth's magnetic field has periodically reversed its polarity in the past. Such a change in the polarity of the magnetic field is a **magnetic reversal.** During a time of *normal polarity,* magnetic lines of force leave Earth near the geographic South Pole and reenter near the geographic North Pole (figure 17.18). This orientation is called "normal" polarity because it is the same as the present polarity. During a time of *reversed polarity,* the

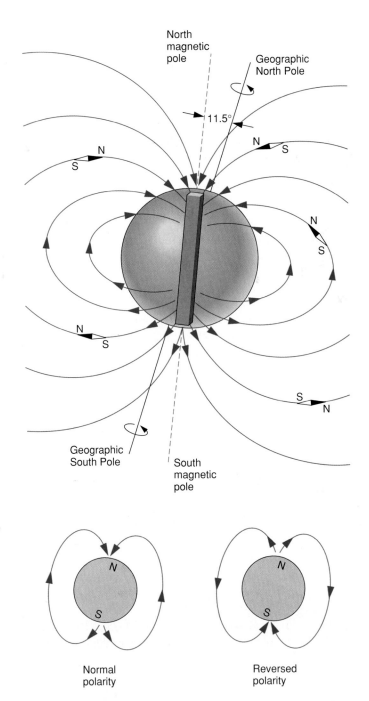

FIGURE 17.18

The Earth's magnetic field. The depiction of the internal field as a large bar magnet is a simplification of the real field, which is more complex. N and S in the two small figures indicate the *geographic* poles.

magnetic lines of force run the other way, leaving Earth near the North Pole and entering near the South Pole (figure 17.18). In other words, during a magnetic reversal, the north magnetic pole and the south magnetic pole exchange positions.

Many rocks contain a record of the strength and direction of the magnetic field *at the time the rocks formed.* When the mineral magnetite, for example, is crystallizing in a cooling lava flow, the iron atoms within the crystals respond to Earth's

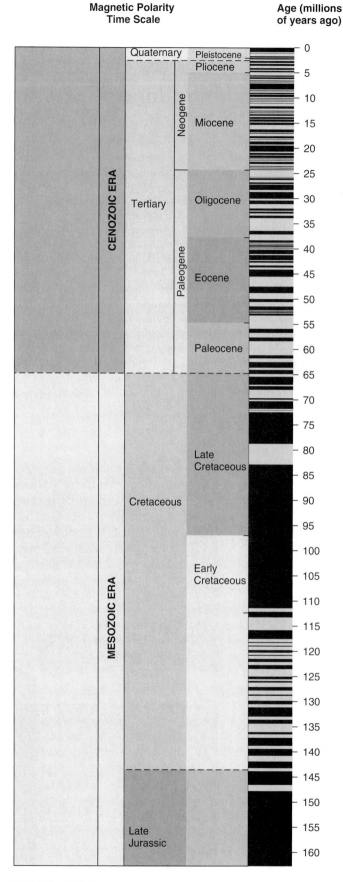

Direction of Earth's
magnetic field

Cooling
lava

Magnetic
alignment
preserved in
magnetite
records orien-
tation of
Earth's field

A

Present orientation of Earth's
magnetic field

Lava flows
showing
reverse
magnetism

B

FIGURE 17.19

(A) Some rocks preserve a record of Earth's magnetic field. (B) Cross section of stacked lava flows showing evidence of magnetic reversals.

magnetic field and form magnetic alignments that "point" toward the north magnetic pole. As the lava cools slowly below the **Curie point** (580°C for magnetite), this magnetic record is permanently trapped in the rock (figure 17.19A). Unless the rock is heated again above the Curie point temperature, this magnetic record is retained and when studied reveals the direction of Earth's magnetic field at the time the lava cooled. Other rock types, including sedimentary rocks stained red by iron compounds, also record former magnetic field directions. The study of ancient magnetic fields is called **paleomagnetism.**

Most of the evidence for magnetic reversals comes from lava flows on the continents. Paleomagnetic studies of a series of stacked lava flows often show that some of the lava flows have a magnetic orientation directly opposite of Earth's present orientation (figure 17.19B). That is, at the time these lava flows cooled, the magnetic poles had exchanged positions. During this time of magnetic reversal, a compass needle would have pointed south rather than north. Many periods of normal and reverse magnetization are recorded in continental lava flows. They are worldwide events. Since lava flows can be dated isotopically, the time of these reversals in Earth's past can be determined. Although reversals appear to occur randomly (figure 17.20), records for tens of millions of years

**Magnetic Polarity
Time Scale**

**Age (millions
of years ago)**

FIGURE 17.20

Worldwide magnetic polarity time scale for the Cenozoic and Mesozoic Eras. Black indicates positive anomalies (and therefore normal polarity). Tan indicates negative anomalies (reverse polarity). *Modified from R. L. Larson and W. C. Pitman, III, 1972, Geological Society of America Bulletin*

IN GREATER DEPTH 17.4

Earth's Spinning Inner Core

Recent studies have led to a new understanding of the dynamics of Earth's inner core and generation of Earth's magnetic field and periodic magnetic reversals. Gary A. Glatzmaier of Los Alamos National Laboratory in New Mexico and Paul H. Roberts of the University of California–Los Angeles developed a very sophisticated computer model of convection in the outer core that has been successful in simulating a magnetic field very similar to that measured on Earth. The model utilizes circulating metallic fluids in the outer core, caused by cooling and heat loss, as the driving force of Earth's magnetic field. The circulation of metallic fluids in the outer core has been theorized for many years, and the computer model was successful in simulating and maintaining a magnetic field similar to that measured on Earth. The model also predicted that Earth's solid inner core spins faster than the rest of the planet, gaining a full lap on the rest of the planet every 150 years. Because the magnetic lines of force penetrate and connect both the inner and outer core, a faster rate of rotation of the inner core would play an important role in the generation of Earth's magnetic field and may influence periodic

magnetic reversals. Interestingly, Glatzmaier and Robert's model produced a magnetic reversal on its own without any additional input from the experimenters after about 35,000 years of simulated time (box figure 1).

The results of this computer model inspired seismologists Xiao Dong Song and Paul Richards from Columbia University's Lamont-Doherty Earth Observatory to look for evidence that the inner core actually spins at a more rapid rate than the rest of the planet. The seismologists knew that previous studies suggested seismic waves pass through the inner core faster along a nearly north-south route. This faster route, or high-velocity pathway, is similar to the grain in a piece of wood. This pathway is not aligned directly with the inner core's spin axis but is tilted about 10° from it (box figure 2). Seismic waves tend to travel slower along other paths, such as in an east-west direction parallel to the inner core's equator.

The seismologists studied seismic wave records from thirty-eight separate, closely spaced earthquakes from 1967 to 1995 near the Sandwich Islands, off Argentina, to determine how long

A Reversed Magnetic Field

C Normal Magnetic Field

B Transitional Magnetic Field

BOX 17.4 ■ FIGURE 1

Computer simulation of Earth's magnetic field and magnetic reversal. (A) Reversed magnetic polarity with magnetic field lines leaving the north magnetic pole (orange) and reentering at the south pole. (B) Transitional magnetic field. (C) Normal magnetic field. *Photos from the Geodynamo Computer Simulation, courtesy of G. A. Glatzmaier, Los Alamos National Laboratory, and P. H. Roberts, University of California, Los Angeles* suggest that Earth's field reverses on average about once every 500,000 years. The present normal orientation has lasted for the past 700,000 years. It takes time for one magnetic orientation to die out and the reverse orientation to build up. Most geologists think that it takes 10,000 years for a reversal to develop, although new evidence suggests that a reversal can occur much faster than that.

What causes magnetic reversals? The question is difficult to answer because no one knows how the magnetic field is generated in the first place. Recent computer modeling and seismological research support the theory that the magnetic field is generated by convection currents in the liquid outer core (see box 17.4). *If* the field is caused by convection currents within the liquid outer core, perhaps a reversal is caused when the currents change direction or by a temporary current building up and then dying out. Some geologists think that reversals may be triggered by the impact of an asteroid or comet with Earth; other geologists disagree.

A magnetic reversal can have some profound effects on Earth. The strength of Earth's magnetic field probably declines to near zero before the orientation reverses; then the field strength increases to its usual values but in the opposite orientation. This collapse of the magnetic field means that deadly cosmic radiation from the Sun would be much more intense at the surface. When the magnetic field is at its usual strength, it shields Earth from these rays, but when the field collapses, this shielding is lost. Cosmic radiation affects organisms; the extinction of some species and the appearance of new species by mutation have been correlated with some magnetic reversals. However, far more reversals than mass extinctions have occurred.

Magnetic Anomalies

A **magnetometer** is an instrument used to measure the strength of Earth's magnetic field. A magnetometer can be carried over the land surface or flown over land or sea. At sea, magnetometers can also be towed behind ships. They are also used as metal detectors in airports.

The strength of Earth's magnetic field varies from place to place. As with gravity, a deviation from average readings is called an *anomaly*. Very broad, regional magnetic anomalies may be due to *circulation patterns in the liquid outer core* or to other deep-seated causes. Smaller anomalies generally reflect *variations in rock type,* for the magnetism of near-surface rocks adds to the main magnetic field generated in the core. Rocks differ in their magnetism, depending upon their content of iron-containing minerals, particularly magnetite.

A **positive magnetic anomaly** is a reading of magnetic field strength that is higher than the regional average. Figure 17.21 shows three geologic situations that can cause positive magnetic anomalies. In figure 17.21*A*, a body of magnetite ore (a highly magnetic ore of the metal iron) has been emplaced in a bed of limestone by hot solutions rising along a fracture. The magnetism of the iron ore adds to the magnetic field of Earth, giving a stronger magnetic field measurement at the surface (a positive anomaly). In figure 17.21*B*, a large dike of gabbro has intruded

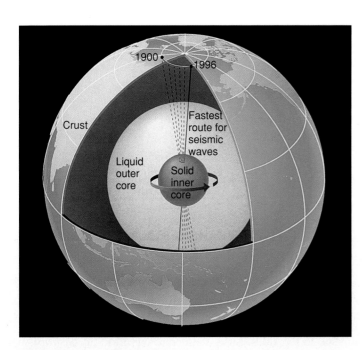

BOX 17.4 ■ FIGURE 2

Seismic waves indicate that Earth's core rotates faster than the rest of the planet by about a degree per year. The solid line indicates the 1996 position of a point in the core relative to the surface of Earth, and the dashed line indicates where the point was in 1900. *Courtesy of Lamont-Doherty Earth Observatory, Columbia University. Data from Michael Carlowicz, Earth Magazine, p. 21, 1996*

it took them to reach a monitoring station in College, Alaska. The waves all took about the same amount of time to reach Alaska; however, the seismic waves in the 1990s arrived in Alaska about 0.3 seconds faster than the seismic waves in the 1960s. Since the seismic waves would have traveled through the inner core, the seismologists have explained the difference in travel time as indicating that the inner core had changed its position relative to the monitoring station in Alaska. That is, the inner core and the high-velocity pathway had rotated slightly with respect to the rest of the planet.

Seismologists at Harvard University looked at additional earthquake records and calculated that the inner core is rotating at approximately the same rate as Glatzmaier and Robert's model predicted. Future studies to examine earthquake records over a longer period of time are needed to confirm whether the inner core has been spinning faster than the rest of the Earth. This is an exciting time for Earth scientists since we may now have a better idea about the inner motion of the core and the generation of Earth's magnetic field.

Additional Resources

M. Carlowicz. 1996. Spin control. *Earth* 12(21):62–63.

Core convection and the Geodynamo website discusses the recent model for reversals of Earth's magnetic field:

• http://ees5-www.lanl.gov/IGPP/Geodynamo.html

FIGURE 17.21

Positive magnetic anomalies can indicate hidden ore and geologic structures.

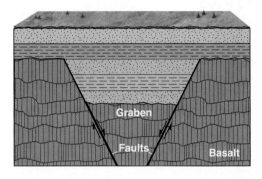

FIGURE 17.22

A graben filled with sediment can give a negative magnetic anomaly if the sediment contains fewer magnetic minerals than the rock beneath it.

iments to the sides. Such a structure can form an *oil trap* (see chapter 21). The granite in the hill contains more iron in its ferromagnesian minerals than the surrounding sedimentary rocks, so a small positive magnetic anomaly occurs where the granite is closer to the surface. Note how each example shows horizontal sedimentary rocks at the surface, with no surface hint of the subsurface geology. The magnetometer helps find hidden ores and geologic structures.

A **negative magnetic anomaly** is a reading of magnetic field strength that is lower than the regional average. Figure 17.22 shows how a negative anomaly can be produced by a down-dropped fault block (a *graben*) in basalt. The thick sedimentary fill above the graben is less magnetic than is the basalt, so a weaker field (a negative magnetic anomaly) develops over the thick sediment.

Not all local magnetic anomalies are caused by variations in rock type. The linear magnetic anomalies found at sea are apparently caused by a *variation in the direction of magnetism,* as you will see in chapter 19.

HEAT WITHIN THE EARTH

Geothermal Gradient

The temperature increase with depth into Earth is called the **geothermal gradient.** The geothermal gradient can be measured on land in abandoned wells or on the sea floor by dropping specially designed probes into the mud. The average temperature increase is 25°C per kilometer (about 75°F per mile) of depth. Some regions have a much higher gradient,

into granitic basement rock. Because gabbro contains more ferromagnesian minerals than granite, gabbro is more magnetic and causes a positive magnetic anomaly. Figure 17.21*C* shows a granitic basement high (perhaps originally a hill) that has influenced later sediment deposits, causing a draping of the layers as the sediments on the hilltop compacted less than the thicker sedindicating concentrations of heat at shallow depths. Such regions have a potential for generating *geothermal energy* (discussed in chapter 11).

The temperature increase with depth creates a problem in deep mines, such as in a 3-kilometer-deep gold mine in South Africa, where the temperature is close to the boiling point of water. Deep mines must be cooled by air-conditioning for the miners to survive. High temperatures at depth also complicate the drilling of deep oil wells. A well drilled to a depth of 7 or 8 kilometers must pass through rock with a temperature of 200°C. At such high temperatures, a tough, steel drilling pipe will become soft and flexible unless it is cooled with a special mud solution pumped down the hole.

Geologists believe that the geothermal gradient must taper off sharply a short distance into Earth. The high values of 25°C per kilometer recorded near Earth's surface could not continue very far into Earth. If they did, the temperature would be 2,500°C at the shallow depth of 100 kilometers. This temperature is above the melting point of all rocks at that depth—even though the increased pressure with depth into Earth raises the melting point of rocks. Seismic evidence seems to indicate a solid, not molten, mantle, so the geothermal gradient must drop to values as low as 1°C per kilometer within the mantle (figure 17.23A).

At the boundary between the inner core and the outer core, there would be some constraints on possible temperatures if the core is molten metal above the boundary and solid metal below. The weight of the thick rock layer of the mantle and the liquid metal of the outer core raises the pressure at this boundary (figure 17.23B) to about 3 million atmospheres. (An *atmosphere* of pressure is the force per unit area caused by the weight of the air in the atmosphere. It is about 1 kilogram per square centimeter, or 14.7 pounds per square inch.)

Using geophysical and geochemical data, in addition to computer modeling and high-pressure experiments, the internal temperature of Earth can be estimated. Recent laboratory experiments with pressure anvils and giant guns have created (for a millionth of a second) the enormous pressures found at the center of Earth. The measured temperature at this pressure was far higher than expected. New estimates of Earth's internal temperatures have resulted: 3,800°C at the core-mantle boundary, 6,300°C ± 800°C at the inner-core/outer-core boundary, and 6,400°C ± 600°C at Earth's center (hotter than the surface of the Sun!).

Heat Flow

A small but measurable amount of heat from Earth's interior is being lost gradually through the surface. This gradual loss of heat through Earth's surface is called **heat flow.** What is the origin of the heat? It could be "original" heat from the time that

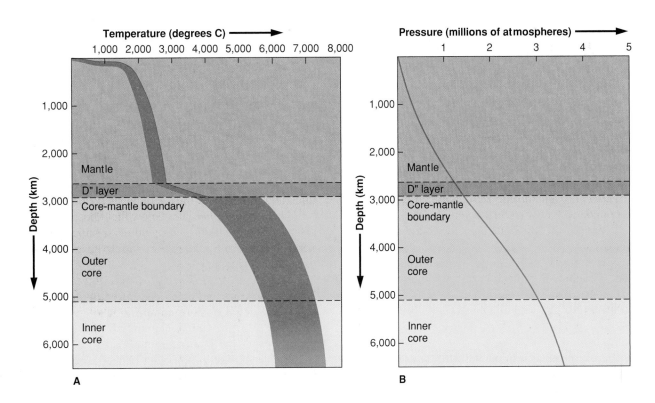

FIGURE 17.23

Estimated (A) temperature and (B) pressure with depth into Earth. The width of the red zone in graph (A) indicates the range of uncertainty of the estimate.

FIGURE 17.24

Some regions have higher heat flow than others; the amount of heat flow is indicated by the length of the arrow. Regions of high heat flow may be underlain by cooling magma or uranium-rich igneous rock.

FIGURE 17.25

The average heat flow from oceans and continents is the same, but the origin of the heat differs from the ocean to continents.

Earth formed, that is, *if* Earth formed as a mass of planetesimals that coalesced and compressed the inner material. Or the heat could be a by-product of the decay of radioactive isotopes inside Earth. Radioactive decay *may* actually be warming up the planet. Geologists are not sure whether Earth formed as a hot or cold mass, or whether the planet is now cooling off or warming up. Changes in Earth's internal temperature are extremely slow (on the order of 100 million years), and trying to work out its thermal history is a slow, often frustrating, job.

Some regions on Earth have a high heat flow. More heat is being lost through the surface in these regions than is normal. High heat flow is usually caused by the presence of a magma body or still-cooling pluton near the surface (figure 17.24). An old body of igneous rock that is rich in uranium and other radioactive isotopes can cause a high heat flow, too, because radioactive decay produces heat as it occurs. High heat flow over an extensive area may be due to the rise of warm mantle rock beneath abnormally thin crust.

The average heat flow from continents is the same as the average heat flow from the sea floor, a surprising fact if you consider the greater concentration of radioactive material in continental rock (figure 17.25). The unexpectedly high average heat flow under the ocean may be due to hot mantle rock rising slowly by convection under parts of the ocean (see chapter 1). Regional patterns of high heat flow and low heat flow on the sea floor (heat flow decreases away from the crest of the mid-oceanic ridge) may also be explained by convection of mantle rock, as we discuss in chapter 19.

SUMMARY

The interior of Earth is studied indirectly by *geophysics*—a study of seismic waves, gravity, Earth magnetism, and Earth heat.

Seismic reflection and *seismic refraction* can indicate the presence of boundaries between rock layers.

Earth is divided into three major zones—the *crust*, the *mantle*, and the *core*.

The crust beneath oceans is 7 kilometers thick and made of basalt on top of gabbro. Continental crust is 30 to 50 kilometers thick and consists of a crystalline basement of granite and gneiss (and other rocks) capped by sedimentary rocks.

The *Mohorovičić discontinuity* separates the crust from the mantle.

The mantle is a layer of solid rock 2,900 kilometers thick and is probably composed of an ultramafic rock such as peridotite. Seismic waves show the mantle has a structure of concentric shells, perhaps caused by pressure transformations of minerals.

The *lithosphere,* which forms plates, is made up of brittle crust and upper mantle. It is 70 to 125 (or more) kilometers thick and moves over the ductile asthenosphere.

The *asthenosphere* lies below the lithosphere and may represent rock close to its melting point (seismic waves slow down here). It is probably the region of most magma generation and isostatic adjustment.

Seismic-wave shadow zones show the core has a radius of 3,450 kilometers and is divided into a liquid outer core and a solid inner core. A core composition of mostly iron is suggested by Earth's density, the composition of meteorites, and the existence of Earth's magnetic field.

Isostasy is the equilibrium of crustal columns "floating" on a ductile mantle. *Isostatic adjustment* occurs when weight is added to or subtracted from a column of rock. *Crustal rebound* is isostatic adjustment that occurs after the melting of glacial ice.

A *gravity meter* can be used to study variations in rock density or to find regions that are out of isostatic equilibrium.

A *positive gravity anomaly* forms over dense rock or over regions being held up out of isostatic balance. A *negative gravity anomaly* indicates low-density rock or a region being held down.

Earth's *magnetic field* has two *magnetic poles,* probably generated by convection circulation and electric currents in the outer core.

Some rocks record Earth's magnetism at the time they form. *Paleomagnetism* is the study of ancient magnetic fields.

Magnetic reversals of polarity occurred in the past, with the north magnetic pole and south magnetic pole exchanging positions. Isotopic dating of rocks shows the ages of the reversals.

A *magnetometer* measures the strength of the magnetic field.

A *positive magnetic anomaly* develops over rock that is more magnetic than neighboring rock. A *negative magnetic anomaly* indicates rock with low magnetism.

Magnetic anomalies can also be caused by circulation patterns in Earth's core and variations in the direction of rock magnetism.

The *geothermal gradient* is about 25°C per kilometer near the surface but decreases rapidly at depth. The temperature at the center of Earth may be 6,400°C ± 600°C. *Heat flow* measurements show that heat loss per unit area from continents and oceans is about the same, perhaps because of convection of hot mantle rock beneath the oceans.

Terms to Remember

asthenosphere 447
convection 451
core 445
crust 445
crustal rebound 453
Curie point 457
geophysics 442
geothermal gradient 460
gravity meter 454
heat flow 461

isostasy 452
isostatic adjustment 452
lithosphere 446
magnetic field 456
magnetic pole 456
magnetic reversal 456
magnetometer 459
mantle 445
Mohorovičić discontinuity (Moho) 446
negative gravity anomaly 455

negative magnetic anomaly 460
paleomagnetism 457
positive gravity anomaly 455
positive magnetic anomaly 459
P-wave shadow zone 448
seismic reflection 442
seismic refraction 442
S-wave shadow zone 449

Testing Your Knowledge

Use the following questions to prepare for exams based on this chapter.

1. Describe how seismic reflection and seismic refraction show the presence of layers within Earth.

2. Sketch a cross section of the entire Earth showing the main subdivisions of Earth's interior and giving the name, thickness, and probable composition of each.

3. What facts make it probable that Earth's core is composed of mostly iron?

4. Describe the differences between continental crust and oceanic crust.

5. What is a gravity anomaly, and what does it generally indicate about the rocks in the region where it is found?

6. Discuss seismic-wave shadow zones and what they indicate about Earth's interior.

7. Describe Earth's magnetic field. Where is it generated?

8. What is the temperature distribution with depth into Earth?

9. Heat flow has been found to be about equal through continents and the sea floor. Why was this unexpected? What might cause this equality?

10. What is the Mohorovičić discontinuity?

11. What is the asthenosphere? Why is it important?

12. How does the lithosphere differ from the asthenosphere?

13. What is a magnetic reversal? What is the evidence for magnetic reversals?

14. What is a magnetic anomaly? How are magnetic anomalies measured at sea?

15. *Felsic* and *mafic* are terms used by some geologists to describe
    a. composition of continental and oceanic crust
    b. behavior of earthquake waves
    c. regions in the mantle

16. The boundary that separates the crust from the mantle is called the
    a. lithosphere
    b. asthenosphere
    c. Mohorovičić discontinuity
    d. none of the preceding

17. The core is probably composed mainly of
    a. silicon   b. sulfur
    c. oxygen   d. iron

18. The principle of continents being in a buoyant equilibrium is called
    a. subsidence   b. isostasy
    c. convection   d. rebound
19. A positive gravity anomaly indicates that
    a. tectonic forces are holding a region up out of isostatic equilibrium
    b. the land is sinking
    c. local mass deficiencies exist in the crust
    d. all of the preceding
20. A positive magnetic anomaly could indicate
    a. a body of magnetic ore
    b. the magnetic field strength is higher than the regional average
    c. an intrusion of gabbro
    d. the presence of a granitic basement high
    e. all of the preceding

21. Which of the following is not an example of the effects of isostasy?
    a. deep mountain roots
    b. magnetic reversals
    c. the postglacial rise of northeastern North America
    d. mountain ranges at subduction zones
22. The S-wave shadow zone is evidence that
    a. the core is made of iron and nickel
    b. the inner core is solid
    c. the outer core is fluid
    d. the mantle behaves as ductile material

Expanding Your Knowledge

1. Why does the heat flow from the continents equal that from the oceans?
2. Subsidence of Earth's surface sometimes occurs as reservoirs fill behind newly built dams. Why?
3. What geologic processes might cause the forces that can hold a region out of isostatic equilibrium?
4. Does the correlation of a species extinction with a magnetic reversal prove that the reversal caused the extinction?
5. If the upper mantle is chemically different from the lower mantle, is mantle convection possible?
6. How could you use geophysical techniques to plan the location of a subdivision in an area containing limestone caverns that could collapse if built on?

Exploring Web Resources

www.mhhe.com/plummer11e
This is the dedicated website for this book. You can go to it for new and updated information. The universal resource locators (URL) listed in this book are also given as links on the website, making it easy to go to those websites without typing in the URL. When you visit our Online Learning Center, go to the Student Center and then to the chapter of interest. Here you will find additional readings and media resources, as well as answers to the Testing Your Knowledge section, more quizzes, animation and video clips, and direct links to the sites listed in this book. Links to additional websites can also be found. We have added questions for some of the sites to allow you to get the most of your exploration of the web. Using the web is an enjoyable way of enhancing your knowledge of geology.

http://rses.anu.edu.au/gfd/Gfd_other_pages/Convection_demo/Demo_page_1.html
The Geophysical Fluid Dynamics Group web page contains images of mantle convection models.

http://ees5-www.lanl.gov/IGPP/Geodynamo.html
Core Convection and Geodynamo web page discusses the recent model for reversals of Earth's magnetic field.

 # Animations

This chapter includes the following animations available on our Online Learning Center at www.mhhe.com/plummer11e.

17.8, 17.9 P and S wave shadow zones
17.11 Isostacy-basic principle
17.12 How isostacy, orogeny, and metamorphism are interrelated
17.13 Isostatic rebound after deglaciation

The Sea Floor

S pace travelers seeing our beautiful, blue planet would likely call it the water planet. The hydrosphere, notably the oceans, dominates the surface of Earth with over 70% of it covered by oceans. Clearly, the part of the geosphere covered by the oceans is important to our understanding of Earth systems, even though most of the sea floor is not readily accessible to direct observation.

The hydrosphere is, of course, vital to the biosphere. Primitive life began in the sea and evolved over billions of years into the rich and diverse plant and animal life that we see today. Life on land is a relative newcomer. It

A variety of life-forms surround the Great Barrier Reef in the Coral Sea. *Photo © Australian Picture Library/Corbis* has been only a few hundred million years since the first creatures ventured out of the sea and land-dwelling life evolved and flourished.

Most of what we know about the sea floor has been discovered during the second half of the twentieth century. Because of the difficulty in accessing the deep-ocean floor, our maps of its surface are not as complete as those of some of our neighboring planets. We do know that the rocks and topography of the sea floor are different from those on land. To understand the evidence for plate tectonics in chapter 19, you need to understand the nature of major seafloor features such as mid-oceanic ridges, oceanic trenches, and fracture zones, as well as the surprisingly young age of the seafloor rocks.

The material discussed in this chapter and chapter 19 is an excellent example of how the scientific method works. This chapter is concerned with the physical *description* of most seafloor features—the data-gathering part of the scientific method. Chapter 19 shows how the theory of plate tectonics explains the *origin* of many of these features. Geologists generally agree on the descriptions of features but often disagree on their interpretations. As you read, keep a clear distinction in your mind between *data* and the *hypotheses* used to explain the data.

ORIGIN OF THE OCEAN

As mentioned in chapter 17, geologists are unsure about the thermal history of Earth. The following scenario for the origin of Earth and its oceans is highly speculative. Many geologists would disagree with some of the statements; they are offered merely as an example of what could have happened in Earth's past.

According to a widely accepted theory, about 4.5 billion years ago, Earth began to form by the accretion of small, cold chunks of rock and metal that surrounded the Sun. As Earth grew, it began to heat up because of the heat of collisional impact, gravitational compaction, and radioactive decay of elements such as uranium. The temperature of Earth rose until the accretions that made up Earth melted and the iron "fell" to the center to form its core. Violent volcanic activity occurred at this time, releasing great quantities of water vapor and other gases from Earth's interior and perhaps even covering the surface with a thick, red-hot sea of lava. Earth began to cool as its growth and internal reorganization slowed down and as the amount of radioactive material was reduced by decay. Eventually, Earth's surface became solid rock, cool enough to permit the condensation of billowing clouds of volcanic water vapor to form liquid water, which fell as rain. Thus, the modern oceans were born, perhaps 4 billion years ago. Evidence for this comes from some of the oldest rocks found on Earth, including pillow basalts, which suggest eruption under water and rocks that were originally sedimentary and formed in a shallow sea. The oceans grew in size as volcanic *degassing* of Earth continued and became salty as the water picked up chlorine from other volcanic gases and sodium (and calcium and magnesium) from the chemical weathering of minerals on Earth's surface.

Some debate exists as to the role comet impacts may have played in the formation of the oceans. These extraterrestrial sources of ice bombarding Earth early in its history could have contributed a substantial amount of water.

Geologists generally agree that the oceans formed early in Earth's history, and most would also agree that the oceans resulted mainly from degassing of Earth's interior. However, geologists were surprised to find that the present oceanic crust is geologically very young.

METHODS OF STUDYING THE SEA FLOOR

Oceans cover more than 70% of Earth's surface. Even though the rocks of the sea floor are widespread, they are difficult to study. Geologists have to rely on small samples of rock taken from the sea floor and brought to the surface, or they must study the rocks indirectly by means of instruments on board ships. Although the sea floor is difficult to study, its overall

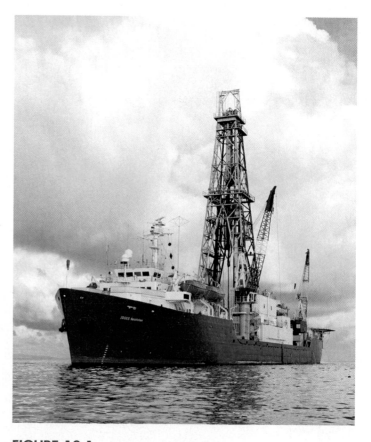

FIGURE 18.1

The JOIDES *Resolution* is a ship built for sampling both sediment and rock from the deep-ocean floor. *Photo by John W. Beck, Ocean Drilling Program*

FIGURE 18.2

The small research submersible *ALVIN* of Woods Hole Oceanographic Institution in Massachusetts; it is capable of taking three oceanographers to a depth of about 4,000 meters. *Photo by Woods Hole Oceanographic Institution*

FIGURE 18.3

Remotely operated submersible *ROPOS* photographs an inactive black smoker along the Explorer Ridge in the northeast Pacific Ocean as it also prepares to take a sample. Yellow color is due to the alteration of iron-rich minerals, and the gray coating is mainly manganese. *Photo by National Oceanic and Atmospheric Administration*

structure is relatively simple (as we discuss later in the "Features of the Sea Floor" section of this chapter), so the small number of samples is not nearly as much a problem as it would be in studying continental regions, where the structure is usually much more complex. The study of seafloor rocks, sediment, and topography provided most of the information that led to the concept of plate tectonics.

Samples of rock and sediments can be taken from the sea floor in several ways. Rocks can be broken from the sea floor by a *rock dredge,* which is an open steel container dragged over the ocean bottom at the end of a cable. Sediments can be sampled with a *corer,* a weighted steel pipe dropped vertically into the mud and sand of the ocean floor.

Both rocks and sediments can be sampled by means of *seafloor drilling.* Offshore oil platforms drill holes in the relatively shallow sea floor near shore. A ship with a drilling derrick on its deck can drill a hole in the deep-sea floor far from land (figure 18.1). The drill cuts long, rodlike rock cores from the ocean floor. Thousands of such holes have been drilled in the sea floor, and the rock and sediment cores recovered from these holes have revolutionized the field of marine geology. In the 1950s, more was known about the Moon's surface than about the floor of the sea. Seafloor drilling has been instru-

mental in expanding our knowledge of seafloor features and history. Small research submarines, more correctly called *submersibles,* can take geologists to many parts of the sea floor to observe, photograph, and sample rock and sediment (figure 18.2). Remotely operated submersibles and specially designed, deep-sea observation and sampling systems allow oceanographers to collect high-quality photographs and samples of the sea floor (figure 18.3).

A basic tool for indirectly studying the sea floor is the single-beam *echo sounder,* which measures water depth and draws profiles of submarine topography (figure 18.4). A sound sent downward from a ship bounces off the sea floor and returns to the ship. The water depth is determined from the time it takes the sound to make the round trip. *Multibeam sonar,* which sends out and records a variety of sound sources, is able to map the sea floor in even more detail than the single-source echo sounder. *Sidescan sonar* measures the intensity of sound (back scatter) reflected back to the tow vehicle from the sea floor to provide detailed images of the sea floor and information about sediments covering the bottom of the sea (figure 18.4).

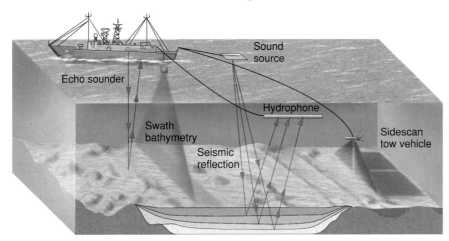

FIGURE 18.4

Diagram showing how echo sounding, seismic reflection, and sidescan sonar are used to study the sea floor. *Modified from U.S. Geological Survey Fact Sheet 039-02*

EARTH SYSTEMS 18.1

Does the Earth Breathe?

Does the Earth breathe to the rhythm of the tides? Maya Tolstoy and her colleagues think it does. Seismic data collected from a mid-oceanic ridge volcano suggest that microseismic events coincide with Earth's low tides. The link between earthquakes and Earth's tidal cycles has been discussed for nearly a century. Some studies have shown that earthquake activity along normal faults can be influenced by tides. The perpetual rise and fall of the sea are the result of the gravitational attraction of the Moon and the Sun on Earth. Because water is not firmly attached to Earth, these gravitational forces cause a disturbance of the ocean's water and generate very long waves, or tides. It is thought that the shifting of water by tides can change the overlying water load or pressure on the ocean crust, which in turn could trigger movement in the crust. For many years, this idea has been difficult to test for lack of a seismometer sensitive enough to record slight crustal movements. In addition, there is a complicated interplay of various stresses that can affect seismic activity, including water depth and crustal properties such as temperature and composition.

Maya Tolstoy and her colleagues set up an array of sensitive, ocean-bottom seismometers to monitor seismic activity continuously over a two-month period at the Axial volcano on the Juan de Fuca Ridge (box figure 1). She reasoned that the crust at mid-oceanic ridges would be vulnerable to tidal energy, because the crust is weakened by networks of fractures filled with circulating water. They recorded 402 microearthquakes (earthquakes of less than 2.5 magnitude) over the two-month period. The ocean tides at the mid-oceanic ridge were recorded as changes in water pressure; low water pressure indicates low tides and high pressure, high tides. Analysis of the occurrence of earthquakes and the

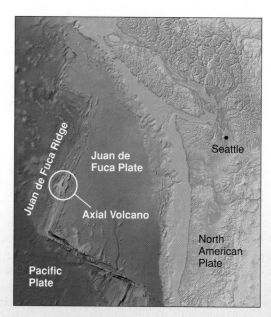

BOX 18.1 ■ FIGURE 1

Map of Juan de Fuca Ridge showing location of Axial volcano where study was conducted.

water pressure data indicate that earthquake activity intensifies near or at low tides. Tolstoy thinks it is likely that the pressure of the overlying water causes normal faults to lock (box figure 2). Once the overlying water pressure drops in response to the ebbing (falling) tide, the reduction in pressure allows slippage along the

A *seismic reflection profiler* works on essentially the same principles as echo sounders but uses a louder noise at lower frequency. This sound penetrates the bottom of the sea and reflects from layers within the rock and sediment. The seismic profiler gives more information than do echo sounders. It records water depth and reveals the internal structure of the rocks and sediments of the sea floor, such as bedding planes, folds and faults, and unconformities.

Magnetic, gravity, and *seismic refraction* surveys (see chapter 17) also can be made at sea. Magnetometers pulled behind navy ships during WW II to look for enemy submarines provided some of the first details of the sea floor that later led to the theory of plate tectonics.

FEATURES OF THE SEA FLOOR

Figure 18.5, a simplified profile of the sea floor, shows that continents have two types of margins. A *passive continental margin,* as found on the east coast of North America, includes a continental shelf, continental slope, and continental rise. An

abyssal plain usually forms a remarkably flat ocean floor beyond the continental rise. An *active continental margin,* found mainly around the Pacific rim, is associated with earthquakes and volcanoes and has a continental shelf and slope, but the slope extends much deeper to form one wall of an *oceanic trench.* Abyssal plains are seldom found off active margins. The deep-ocean floor seaward of trenches is hilly and irregular, lacking the extreme flatness of abyssal plains. Encircling the globe is a *mid-oceanic ridge,* usually (but not always) near the center of an ocean. Conical *seamounts* stick up from the sea floor in some regions. Some important submarine features do not show in this profile view—in particular, fracture zones, submarine canyons, and aseismic ridges.

CONTINENTAL SHELVES AND CONTINENTAL SLOPES

Almost all continental edges are marked by a gently sloping continental shelf near shore and a steeper continental slope that leads down to the deep ocean floor (figure 18.6). Figures such

BOX 18.1 ■ FIGURE 2

Diagram of the Axial volcano area illustrating the affect of tides on earthquakes. (A) Because of their geometry, normal faults remain locked during high tides (red arrows) and unlocked during low tides (green arrows) when less water pressure is pushing down on the faults. (B) The pore pressure in cracks at hydrothermal vents increases at low tide and (C) allows trapped bubbles of gas to expand much like a sealed bottle with air expands on an airplane. The expansion of gas in the cracks increases the pore pressure in the fractured rocks and may help cause slippage (earthquakes) along the normal faults at low tide.

normal faults and fractures in the Axial volcano region. Tolstoy and her colleagues also suggest that as the overlying water pressure drops during low tide, gases in the cracks will expand, increasing the pore pressure of the fractured rocks. This increased pressure may help cause slippage along the faults. The subtle changes in water pressure associated with tidal cycles appear to be strong enough to trigger microearthquakes in the susceptible midocean crust. Tolstoy thinks it is possible that tidal influences on the crust may occur in other ocean ridge systems far from the Axial volcano. This hypothesis needs to be tested with long-term observations of seismicity and water pressure at similar ocean-ridge systems.

Earthquake activity may not be the only deep-sea process that is tidally dependant. The crustal response to tidal forces may also influence the abundant life and flow of nutrients at hydrothermal systems of mid-oceanic ridges. Tolstoy and her colleagues speculate that biological cycles—life and nutrient flux—of the ridges may also be linked to the ebb and flow of the tides. Tolstoy may tell you that the Earth may, indeed, "breathe" to the rhythms of the planets.

Additional Resources

Maya Tolstoy, F. L. Vernon, J. A. Orcutt, and F. K. Wyatt. 2003. Breathing of the sea floor: Tidal correlation of seismicity at Axial volcano. *Geology* vol. 30, no. 6: 503–506. For online version visit:

- http://www.ldeo.columbia.edu/_tolstoy/axial_tides.pdf

To learn more about life as an oceanographer and this study in particular, visit:

- http://www.womenoceanographers.org/doc/MTolstoy/Mayaintro.htm

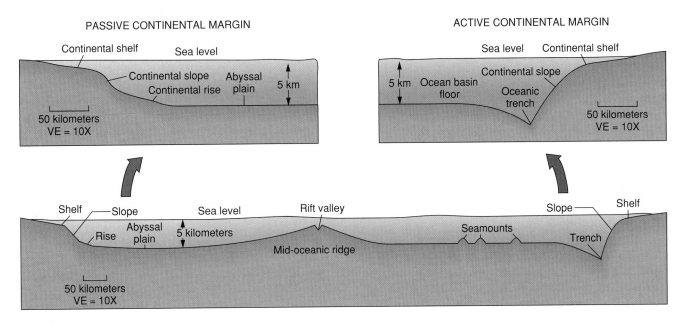

FIGURE 18.5

Profiles of seafloor topography. The vertical scales differ from the horizontal scales, causing vertical exaggeration, which makes slopes appear steeper than they really are. The bars for the horizontal scale are 50 kilometers long, while the same distance vertically represents only 5 kilometers, so the drawings have a vertical exaggeration of 10 times.

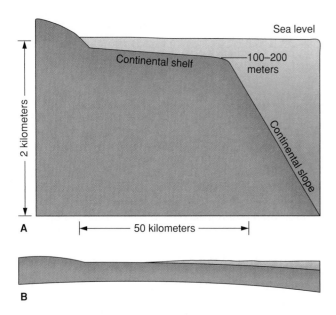

FIGURE 18.6

Continental shelf and continental slope. (A) Vertical exaggeration 25 times. The continental slope has an actual slope of only 4 or 5 degrees, but the great vertical exaggeration of this drawing makes it appear to be sloping about 60°. (B) Same profile with no vertical exaggeration.

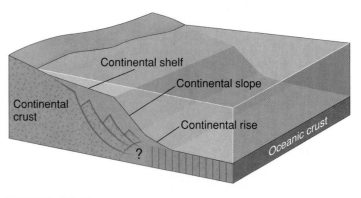

FIGURE 18.7

The continental shelf lies upon continental crust, and the continental rise lies upon oceanic crust. The complex transition from continental crust to oceanic crust lies under the continental slope.

as 18.5 and 18.6 are usually drawn with great *vertical exaggeration,* which makes submarine slopes appear much steeper than they really are.

A **continental shelf,** a shallow submarine platform at the edge of a continent, inclines very gently seaward, generally at an angle of 0.1°. Continental shelves vary in width. On the Pacific coast of North America, the shelf is only a few kilometers wide, but off Newfoundland in the Atlantic Ocean, it is about 500 kilometers (310 miles) wide. Portions of the shelves in the Arctic Ocean off Siberia and northern Europe are even wider. Water depth over a continental shelf tends to increase regularly away from land, with the outer edge of the shelf being about 100 to 200 meters (328 to 656 feet) below sea level.

Continental shelves are *topographic features,* defined by their depth, flatness, and gentle seaward tilt. Their *geologic origin* varies from place to place and is related to plate tectonics, so it will be discussed in chapter 19. Some generalities about shelves, however, are worth noting here.

The continental shelves of the world are usually covered with relatively young sediment, in most cases derived from land. The sediment is usually sand near shore, where the bottom is shallow and influenced by wave action. Fine-grained mud is commonly deposited farther offshore in deeper, quieter water.

The outer part of a wide shelf is often covered with coarse sediment that was deposited near shore during a time of lower sea level. The advance and retreat of continental glaciers during the Pleistocene Epoch caused sea level to rise and fall many times by 100 to 200 meters. This resulted in a complex history of sedimentation for continental shelves as they were alternately covered with seawater and exposed as dry land.

Marine seismic surveys and drilling at sea have shown that the young sediments on many continental shelves are underlain by thick sequences of sandstone, shale, and less commonly limestone. These sedimentary rocks are of late Mesozoic and Cenozoic age, and appear to have been deposited in much the same way as the modern shelf sediments. Beneath these rocks is continental crust (figure 18.7); the continental shelves are truly part of the continents, even though today the shelves are covered by seawater.

A **continental slope** is a relatively steep slope that extends from a depth of 100 to 200 meters at the edge of the continental shelf down to oceanic depths. The average angle of slope for a continental slope is 4° to 5°, although locally, some parts are much steeper.

Because the continental slopes are more difficult to study than the continental shelves, less is known about them. The greater depth of water and the locally steep inclines on the continental slopes hinder rock dredging and drilling and make the results of seismic refraction and reflection harder to interpret. This is unfortunate, for the rocks that underlie the slopes are of particular interest to marine geologists, who believe that in this area, the thick continental crust (beneath the land and the continental shelves) grades into thin oceanic crust (underneath the deep ocean floor), as shown in figure 18.7.

Although relatively little is known about continental slopes, it is clear that their character and origin vary greatly from place to place. Because these variations are associated with the differences between divergent plate boundaries and convergent plate boundaries, we will discuss them in chapter 19.

SUBMARINE CANYONS

Submarine canyons are V-shaped valleys that run across continental shelves and down continental slopes (figure 18.8). On narrow continental shelves, such as those off the Pacific coast of the United States, the heads of submarine canyons may be so close to shore that they lie within the surf zone. On wide shelves, such as those off the Atlantic coast of the United

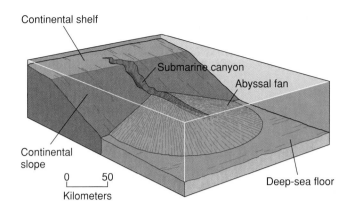

FIGURE 18.8

Submarine canyon and abyssal fan.

States, canyon heads usually begin near the outer edge of the continental shelf tens of kilometers from shore. Great fan-shaped deposits of sediment called **abyssal fans** are found at the base of many submarine canyons (figure 18.8). Abyssal fans are made up of land-derived sediment that has moved down the submarine canyons. Along continental margins that are cut by submarine canyons, many coalescing abyssal fans may build up at the base of the continental slope (figure 18.9).

Submarine canyons are erosional features, but how rock and sediment are removed from the steep-walled canyons is controversial. Erosional agents probably vary in relative importance from canyon to canyon. Divers have filmed *down-canyon movement of sand* in slow, glacierlike flow and in more rapid sand falls (figure 18.10). This sand movement, which has been observed to cause erosion of rock, is particularly common in Pacific coast canyons, which collect great quantities of sand from longshore drift. *Bottom currents* have been measured moving up and down the canyons in a pattern of regularly alternating flow, in some cases apparently caused by ocean tides. The origin of these currents is not well understood, but they often move fast enough to erode and transport sediment. *River erosion* may have helped to cut the upper part of canyons when the drop in sea level during the Pleistocene glaciations left canyon heads above the water, such as the extension of the Hudson River into the Atlantic. Many (but not all) submarine canyons are found off land canyons or rivers, which tends to support the view that river erosion helped shape them. It is unlikely, however, that the deeper parts of submarine canyons were ever exposed as dry land.

Turbidity Currents

In addition to the canyon-cutting processes just described, turbidity currents probably play the major role in canyon erosion. **Turbidity currents** are great masses of sediment-laden water that are pulled downhill by gravity. The sediment-laden water is heavier than clear water, so the turbidity current flows down the continental slope until it comes to rest on the flat abyssal plain at the base of the slope (see figure 6.31). Turbidity currents are thought to be generated by underwater earthquakes and landslides, strong surface storms, and floods of sediment-laden rivers discharging directly into the sea on coasts with a narrow shelf. Although large turbidity currents have not been directly observed in the sea, small turbidity currents can be made and studied in the laboratory.

Indirect evidence also indicates that turbidity currents occur in the sea. The best evidence comes from the breaking of submarine cables that carry telephone and telegraph

FIGURE 18.9

Continental shelf and slope off the coast of southern California are cut by submarine canyons that channel sediments to the deep ocean. Vertical exaggeration is approximately 6X. To view other high-resolution multibeam images of the coastline, visit http://walrus.wr.usgs.gov/pacmaps/la-pers1.html. *Image by Jim Gardner, U.S. Geological Survey*

FIGURE 18.10

A 10-meter-high sand fall in a submarine canyon near the southern tip of Baja California, Mexico. The sand is beach sand, fed into the nearshore canyon head by longshore currents. *From Scripps Institution of Oceanography, University of California, San Diego*

Map

A

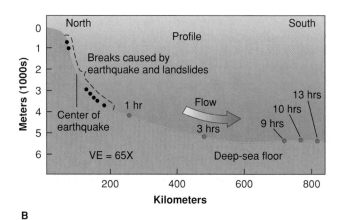

B

FIGURE 18.11

Submarine cable breaks following the Grand Banks earthquake of 1929. (A) Map view of the cable breaks. Black dots near the epicenter show locations of cable breaks that were simultaneous with the earthquake (cables not shown for these breaks). Red dots show cable breaks that followed the earthquake, with the time of each break shown (on the 24-hour clock). Segments of cable more than 100 kilometers long were broken simultaneously at both ends and then carried away. Dashes show seafloor channels that probably concentrated the flow of a turbidity current, increasing its velocity. (B) Profile showing the time elapsed between the quake and each cable break. (A) From H. W. Menard, 1964, Marine Geology of the Pacific, copyright McGraw-Hill, Inc.; (B) from B. C. Heezen and M. Ewing, 1952, American Journal of Science

messages across the ocean floor. Figure 18.11 shows a downhill sequence of cable breaks that followed a 1929 earthquake in the Grand Banks region of the northwest Atlantic. This sequence of cable breaks has been interpreted to be the result of an earthquake-caused turbidity current flowing rapidly down the continental slope.

If cable breaks are caused by turbidity currents, they give good evidence of the currents' dramatic size, speed, and energy. Breaks in Grand Banks cables continued for more than thirteen hours after the 1929 earthquake, the last of the series occurring more than 700 kilometers (430 miles) from the epicenter. The velocity of the flow that caused the breaks has been calculated to be from 15 to 60 kilometers per hour. Sections of cable more than 100 kilometers (160 miles) long were broken off and carried away, both ends of a missing section being broken simultaneously. Attempts to find broken cable sections were fruitless, and it is assumed that they were buried by sediment.

The Grand Banks 1929 cable breaks are not unique. Cables crossing submarine canyons are broken frequently, particularly after river floods and earthquakes. In the submarine canyons off the Congo River of Africa and off the Magdalena River in Colombia, for example, cables break every few years.

Additional indirect evidence for the existence of turbidity currents comes from the graded bedding and shallow-water fossils in the sediments that make up the continental rise and the abyssal plains, as described in the next section.

PASSIVE CONTINENTAL MARGINS

A **passive continental margin** (figure 18.5) includes a continental shelf, continental slope, and continental rise, and generally extends down to an abyssal plain at a depth of about 5 kilometers (3.1 miles). It is called a passive margin because it usually develops on geologically quiet coasts that lack earthquakes, volcanoes, and young mountain belts.

Passive margins are found on the edges of most landmasses bordering the Atlantic Ocean. They also border most parts of the Arctic and Indian Oceans and a few parts of the Pacific Ocean.

The Continental Rise

Along the base of many parts of the continental slope lies the **continental rise,** a wedge of sediment that extends from the lower part of the continental slope to the deep-sea floor. The continental rise, which slopes at about 0.5°, more gently than the continental slope, typically ends in a flat abyssal plain at a depth of about 5 kilometers. The rise rests upon oceanic crust (figure 18.7).

Types of Deposition

Sediments appear to be deposited on the continental rise in two ways—by turbidity currents flowing *down* the continental slope and by *contour currents* flowing *along* the continental slope.

Cores of sediment recovered from most parts of the continental rise show layers of fine sand or coarse silt interbedded with layers of fine-grained mud. The mineral grains and fossils of the coarser layers indicate that the sand and silt came from the shallow continental shelf. Some transporting agent must have carried these sediments from shallow water to deep water. The coarse layers also exhibit graded bedding, which indicates that they settled out of suspension according to size and weight; therefore, the transporting agent for these sediments was most likely turbidity currents. The continental rise in these locations probably formed as turbidity currents deposited abyssal fans at the base of a continental slope.

Sediments in some other parts of the continental rise, however, are uniformly fine-grained and show no graded bedding. This sediment appears to have been deposited by the regular ocean currents that flow along the sea bottom rather than by the intermittent turbidity currents that occasionally flow downslope.

A **contour current** is a bottom current that flows parallel to the slopes of the continental margin—*along* the contour rather than *down* the slope (figure 18.12). Such a current runs south along the continental margin of North America in the Atlantic Ocean. Flowing at the relatively slow speed of a few centimeters per second, this current carries a small amount of fine sediment from north to south. The current is thickest along the continental slope and gets progressively thinner seaward. The thick, landward part of the current carries and deposits the most sediment. The thinner, seaward edge of the current deposits less sediment. As a result, the deposit of sediment beneath the current is wedge shaped, becoming thinner away from land. Similar contour currents apparently shape parts of the continental rise off other continents as well.

Abyssal Plains

Abyssal plains are very flat regions usually found at the base of the continental rise. Seismic profiling has shown that abyssal plains are formed of horizontal layers of sediment. The

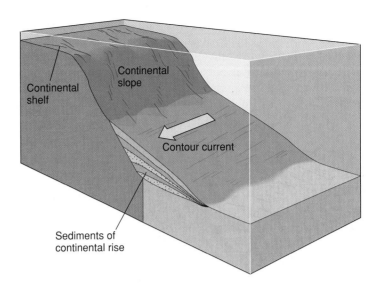

FIGURE 18.12

A contour current flowing along the continental margin shapes the continental rise by depositing fine sediment. Coarse layers within the continental rise were deposited by turbidity currents flowing down the continental slope.

gradual deposition of sediment buried an older, more rugged topography that can be seen on seismic profiler records as a rock basement beneath the sediment layers (figure 18.13). Samples of abyssal plain sediment show that it is derived from land. Graded bedding within sediment layers suggests deposition by turbidity currents.

Abyssal plains are the flattest features on Earth. They generally have slopes less than 1:1,000 (less than 1 meter of vertical drop for every 1,000 meters of horizontal distance) and some have slopes of only 1:10,000. Most abyssal plains are 5 kilometers (3.1 miles) deep.

Not all parts of the deep-ocean basin floor consist of abyssal plains. The deep floor is normally very rugged, broken

FIGURE 18.13

Seismic profiler record of an abyssal plain, showing sediment layers that have buried an irregular rock surface in the Atlantic Ocean. *From Vogt et al. in Hart,* The Earth's Crust and Upper Mantle, *p. 574, American Geophysical Union* by faults into hills and depressions and dotted with volcanic seamounts. Abyssal plains form only where turbidity currents can carry in enough sediment to bury and obscure this rugged relief. If the sediment is not available or if the bottom-hugging turbidity currents are stopped by a barrier such as an oceanic trench, then abyssal plains cannot develop.

ACTIVE CONTINENTAL MARGINS

An **active continental margin,** typically characterized by earthquakes and a young mountain belt and volcanoes on land, consists of a continental shelf, a continental slope, and an oceanic trench (figure 18.5). An active margin usually lacks a continental rise and an abyssal plain and is associated with convergent plate boundaries.

Active margins are found on the edges of most of the landmasses bordering the Pacific Ocean and a few other places in the Atlantic and Indian Oceans (figure 18.14). A notable exception in the Pacific Ocean is most of the coast of North America. The active margin's distinctive combination of oceanic trench and land volcanoes is found in three places along the Pacific coast of North America: southern Central America; Washington, Oregon, and northernmost California (the trench here is sediment-filled); and south-central Alaska. But passive margins with abyssal fans and abyssal plains occur along most of the coasts of Mexico, California, British Columbia, and the Alaskan panhandle. These coasts *do* have earthquakes, mostly along strike-slip faults, and probably were full-fledged active margins in the geologic past but are considered to be passive margins today.

Oceanic Trenches

An **oceanic trench** is a narrow, deep trough parallel to the edge of a continent or an island arc (a curved line of islands like the Aleutians or Japan), as shown in figure 18.14. The continental slope on an active margin forms the landward wall of the trench, its steepness often increasing with depth. The slope is typically 4° to 5° on the upper part, steepening to 10° to 15° or even more near the bottom of the trench. The elongate oceanic trenches, often 8 to 10 kilometers (4.9 to 6.2 miles) deep, far exceed the average depth of abyssal plains on passive margins. The deepest spots on Earth, more than 11 kilometers (6.8 miles) below sea level, are in oceanic trenches in the southwest Pacific Ocean.

Associated with oceanic trenches are the *earthquakes of the Benioff seismic zones* (see chapter 16), which begin at a trench and dip landward under continents or island arcs (figure 18.15). *Volcanoes* are found above the upper part of the Benioff zones and typically are arranged in long belts parallel to oceanic trenches. These belts of volcanoes form island arcs or erupt within young mountain belts on the edges of continents. The rock produced by these volcanoes is usually andesite, a type of extrusive rock intermediate in composition between basaltic oceanic crust and "granitic" continental crust.

Oceanic trenches are marked by abnormally *low heat flow* compared to normal ocean crust. This implies that the crust in trenches may be colder than normal crust.

As you learned in chapter 17, oceanic trenches are also characterized by very large *negative gravity anomalies,* the largest in the world. This implies that trenches are being held down, out of isostatic equilibrium. As you will learn in chapter 19, this is because the oceanic plate is being pulled down or subducted into the mantle.

FIGURE 18.14

The distribution of oceanic trenches. Trenches next to continents mark active continental margins and convergent plate boundaries.

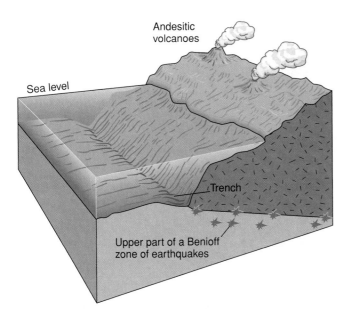

FIGURE 18.15

Active continental margin with an oceanic trench, a Benioff zone of earthquakes (only the upper part is shown), and a chain of andesitic volcanoes on land.

THE MID-OCEANIC RIDGE

The **mid-oceanic ridge** is a giant undersea mountain range that extends around the world like the seams on a baseball (figures 18.16 and 18.17). The ridge, which is made up mostly of basalt, is more than 80,000 kilometers (49,700 miles) long and 1,500 to 2,500 kilometers wide (931 to 1,552 miles). It rises 2 to 3 kilometers (1.2 to 1.8 miles) above the ocean floor.

A **rift valley,** where the crust is undergoing extension, runs down the crest of the ridge (figures 18.16 and 18.17). The rift valley is 1 to 2 kilometers deep and several kilometers wide— about the dimensions of the Grand Canyon in Arizona. The rift valley on the crest of the mid-oceanic ridge is a unique feature—no mountain range on land has such a valley running along its crest. The rift valley is present in the Atlantic and Indian Oceans but is generally absent in the Pacific Ocean. (This absence is related to the faster rate of plate motion in the Pacific, as we will discuss in chapter 19.)

Geologic Activity on the Ridge

Associated with the rift valley at the crest of the mid-oceanic ridge and with the riftless crest of the ridge in the Pacific Ocean, are *shallow-focus earthquakes* from 0 to 20 kilometers (0 to 12.4 miles) below the sea floor.

Careful measurements of the heat loss from Earth's interior through the crust have shown a very *high heat flow* on the crest of the mid-oceanic ridge. The heat loss at the ridge crest is many times the normal value found elsewhere in the ocean; it decreases away from the ridge crest.

Basalt eruptions occur in and near the rift valley on the ridge crest. Sometimes these eruptions build up volcanoes that protrude above sea level as oceanic islands. The large island of Iceland (figure 18.16), which is mostly basaltic, appears to be a section of the mid-oceanic ridge elevated above sea level. Many geologists have studied the active volcanoes, high heat flow, and central rift valley of Iceland to learn about the mid-oceanic ridge. Because Iceland sits on top of a hot spot, it is above sea level and may not represent a typical portion of the ridge.

In the summer of 1974, geologists were able to get a first-hand view of part of the submerged ridge and rift valley. A series of more than forty dives by submersibles, including

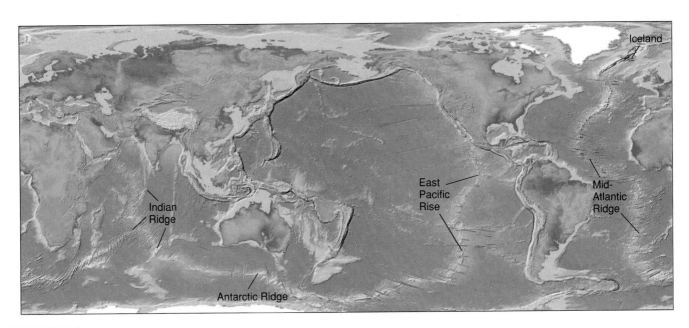

FIGURE 18.16

World map showing oceanic ridges cut by transform faults and fracture zones.

FIGURE 18.17

The floor of the ocean. Note continental margins, mid-oceanic ridge, fracture zones, and oceanic trenches (compare with figures 18.14 and 18.16). Conical mountains are volcanic seamounts; aligned seamounts are aseismic ridges. *From World Ocean Floor by Bruce C. Heezen and Marie Tharp, 1977. Copyright © Marie Tharp 1977. Reproduced by permission of Marie Tharp, 1 Washington Ave., South Nyack, NY 10960*

ALVIN, carried French and American marine geologists directly into the rift valley in the North Atlantic Ocean. The project (called FAMOUS for French-American Mid-Ocean Undersea Study) allowed the ridge rock to be seen, photographed, and sampled directly, rather than indirectly from surface ships (figure 18.18).

The geologists on the FAMOUS project saw clear evidence of extensional faults within the rift valley. These run parallel to the axis of the rift valley and range in width from hairline cracks to gaping fissures that ALVIN dived into. Fresh pillow basalts occur in a narrow band along the bottom of the rift valley, suggesting very recent volcanic activity there, although no active eruptions were observed. It appeared to the geologists that the rift was continuous and that sporadic volcanic activity occurred as a result of the rifting.

Mid-oceanic ridges are often marked by lines of hot springs that carry and precipitate metals. Geologists in submersibles have observed hot springs in several localities along the rift valleys of the mid-oceanic ridges. The hot springs, caused by the high heat flow and shallow basaltic magma beneath the rift valley, range in temperature from about 20°C up to an estimated 350°C (660°F).

As the hot water rises in the rift valley, cold water is drawn in from the sides to take its place. This creates a circulation pattern in which cold seawater is actually drawn *downward* through the cracks in the basaltic crust of the ridge flanks and then moves horizontally toward the rift valley, where it

reemerges on the sea floor after being heated (figure 18.19A). As the seawater circulates, it dissolves metals and sulfur from the crustal rocks. When the hot, metal-rich solutions contact the cold seawater, metal sulfide particles are discharged into the cold seawater at *black smokers*, which precipitate a chimney-like mound around the hot spring (figure 18.19B).

Biologic Activity on the Ridge

The occurrence of black smokers was a surprise to geologists exploring the mid-oceanic ridges, but an even bigger surprise was the presence of exotic, bottom-dwelling organisms surrounding the hot springs. The exotic organisms, including mussels, crabs, starfish, giant white clams, and giant tube worms, are able to survive toxic chemicals, high temperatures, high pressures, and total darkness at depths of more than 2 kilometers (1.2 miles) (figure 18.19C). The organisms live on bacteria that thrive by oxidizing hydrogen sulfide from the hot springs. It is believed that the heat-loving, or *thermophilic bacteria* normally reside beneath the sea floor but are blown out of the hot spring when it erupts. Such sulfur-digesting bacteria have also been found in acidic water in mines containing sulfide minerals. Current research in the new field of *geomicrobiology* is examining the role such bacteria may have had in the precipitation of minerals and the evolution of early life-forms on Earth and possibly other places in the solar system.

FRACTURE ZONES

Fracture zones are major lines of weakness in Earth's crust that cross the mid-oceanic ridge at approximately right angles (figures 18.16 and 18.17). The rift valley of the mid-oceanic ridge is offset in many places across fracture zones (figure 18.20), and the sea floor on one side of a fracture zone is often at a different elevation than the sea floor on the other side, producing steep cliffs. Shallow-focus earthquakes occur on fracture zones but are confined to those portions of the fracture zones between segments of the rift valley. (The portion of the fracture zone that has earthquakes is known as a *transform fault;* the origin of these faults is discussed in chapter 19.)

Fracture zones extend for thousands of kilometers across the ocean floor, generally heading straight for continental margins. Although fracture zones are difficult to trace where they are buried by the sediments of the abyssal plain and the continental rise, some geologists think that they can trace the extensions of fracture zones onto continents. Some major structural trends on continents appear to lie along the hypothetical extension of fracture zones onto the continents.

SEAMOUNTS, GUYOTS, AND ASEISMIC RIDGES

Conical undersea mountains that rise 1,000 meters (3,280 feet) or more above the sea floor are called **seamounts** (figures 18.17

FIGURE 18.18

Underwater photograph of fresh pillow basalt on the floor of the rift valley of the mid-oceanic ridge in the North Atlantic Ocean. A white, tubular sponge grows in the foreground. *Photo by Woods Hole Oceanographic Institution*

A

FIGURE 18.19

(A) Hydrothermal circulation of seawater at ridge crest creates hot springs and metallic deposits in rift valley. Cold seawater is drawn into fractured crust on ridge flanks. (Size of ore deposit is exaggerated.) (B) "Black smoker" or submarine hot spring on the crest of the mid-oceanic ridge in the Pacific Ocean near 21° North latitude. The "smoke" is a hot plume of metallic sulfide minerals being discharged into cold seawater from a chimney 0.5 meters high. The large mounds around the chimney are metallic deposits. The instruments in the foreground are attached to the small submarine from which the picture was taken. (C) Giant worms, crabs, and clams from Galápagos vent. *Photo B by U.S. Geological Survey; Photo C © WHOI/D. Foster/Visuals Unlimited*

B

C

and 18.21). Some rise above sea level to form islands. They are scattered on the flanks of the mid-oceanic ridge and on other parts of the sea floor, including abyssal plains. One area of the sea floor with a particularly high concentration of seamounts—one estimate is 10,000—is the western Pacific. Rocks dredged from seamounts are nearly always basalt, so it is thought that most seamounts are extinct volcanoes. Of the thousands of seamounts on the sea floor, only a few are active

volcanoes. Most of these are on the crests of the mid-oceanic ridges. A few others, such as the two active volcanoes on the island of Hawaii, are found at locations not associated with a ridge.

Guyots are flattopped seamounts (figure 18.21*B*) found mostly in the western Pacific Ocean. Most geologists think that the flat summits of guyots were cut by wave action. These flat tops are now many hundreds of meters below sea level, well

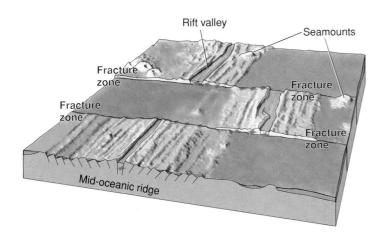

FIGURE 18.20

Fracture zones, which run perpendicular to the ridge crest and connect offset segments of the ridge, are often marked by steep cliffs up to 3 or 4 kilometers high. Shallow-focus earthquakes (shown by stars) occur below the rift valley and on the portion of the fracture zone between the mid-oceanic ridge (transform faults).

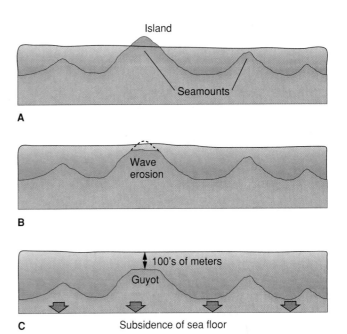

FIGURE 18.21

(A) Seamounts are conical mountains on the sea floor, occasionally rising above sea level to form islands. (B) The flat summit of a guyot was probably eroded by waves when the top of a seamount was above sea level. (C) The present depth of a guyot is due to subsidence.

below the level of wave erosion. If the guyot tops were cut by waves, the guyots must have subsided after erosion took place. Evidence of such subsidence comes from the dredging of dead reef corals from guyot tops. Since such corals grow only in shallow water, they must have been carried to their present depths as the guyots sank.

Many of the guyots and seamounts on the sea floor are aligned in chains. Such volcanic chains, together with some other ridges on the sea floor, are given the name **aseismic ridges** (figure 18.17) that is, they are submarine ridges that are not associated with earthquakes. The name *aseismic* is used to distinguish these features from the much larger mid-oceanic ridge, where earthquakes occur along the rift valley.

REEFS

Reefs are wave-resistant ridges of coral, algae, and other calcareous organisms. They form in warm, shallow, sunlit water that is low in suspended sediment. Reefs stand above the surrounding sea floor, which is often covered with sediment derived from the reef (see figure 6.17). Three important types are *fringing reefs, barrier reefs,* and *atolls* (figure 18.22).

Fringing reefs are flat, tablelike reefs attached directly to shore. The seaward edge is marked by a steep slope leading down into deeper water. Many of the reefs bordering the Hawaiian Islands are of this type.

Barrier reefs parallel the shore but are separated from it by wide, deep lagoons. This type of reef is shown in figures 6.17 and 18.22D. The lagoon has relatively quiet water because the reef shelters it by absorbing the energy of large, breaking waves. A barrier reef lies about 8 kilometers (4.9 miles) offshore of the Florida Keys, a string of islands south of Miami. On a much

grander scale is the Great Barrier Reef off northeastern Australia. It extends for about 2,000 kilometers (1,242 miles) along the coast, and its seaward edge lies up to 250 kilometers (155 miles) from shore. Another long barrier reef lies along the eastern coast of the Yucatan Peninsula in Central America, and others surround many islands in the South Pacific.

Atolls are circular reefs that rim lagoons. They are surrounded by deep water. Small islands of calcareous sand may be built by waves at places along the reef ring. The diameter of atolls varies from 1 to more than 100 kilometers. Numerous atolls dot the South Pacific. Bikini and Eniwetok atolls were used through 1958 for the testing of nuclear weapons by the United States.

Following the four-year cruise of the HMS *Beagle* in the 1830s, Charles Darwin proposed that these three types of reefs are related to one another by subsidence of a central volcanic island, as shown in figure 18.22. A fringing reef initially becomes established near the island's shore. As the volcano slowly subsides because of tectonic lowering of the sea floor, the reef becomes a barrier reef, because the corals and algae grow rapidly upward, maintaining the reef's position near sea level. Less and less of the island remains above sea level, but the reef grows upward into shallow, sunlit water, maintaining its original size and shape. Finally, the volcano disappears completely below sea level, and the reef becomes a circular atoll. Drilling through atolls in the 1950s showed that these reefs were built on deeply buried volcanic cores, thus confirming Darwin's hypothesis of 120 years before.

FIGURE 18.22

Types of coral-algal reefs. (A) Fringing reefs are attached directly to the island. (B) Barrier reefs are separated from the island by a lagoon. (C) Atolls are circular reefs with central lagoons. Charles Darwin proposed that the sequence of fringing, barrier, and atoll reefs forms by the progressive subsidence of a central volcano, accompanied by the rapid upward growth of corals and algae. (D) Reefs on the island of Mooréa in the Society Islands, south-central Pacific (Tahiti in the background). Living corals appear brown. Fringing reef is next to shoreline, barrier reef at breaker line; light blue lagoon between reefs is covered with carbonate sand. The island is an extinct volcano, heavily eroded by stream action. *Photo © David Hiser/Stone/Getty Images*

482

SEDIMENTS OF THE SEA FLOOR

The basaltic crust of the sea floor is covered in many places with layers of sediment. This sediment is either *terrigenous,* derived from land, or *pelagic,* settling slowly through seawater.

Terrigenous sediment is land-derived sediment that has found its way to the sea floor. The sediment that makes up the continental rise and the abyssal plains is mostly terrigenous and apparently has been deposited by turbidity currents or similar processes. Once terrigenous sediment has found its way down the continental slope, contour currents may distribute it along the continental rise. On active continental margins, oceanic trenches may act as traps for terrigenous sediment and prevent it from spreading out onto the deep-sea floor beyond the trenches.

Pelagic sediment is sediment that settles slowly through the ocean water. It is made up of fine-grained clay and the skeletons of microscopic organisms (figure 18.23). Fine-grained pelagic clay is found almost everywhere on the sea floor, although in some places, it is masked by other types of sediments that accumulate rapidly. The clay is mostly derived from land; part of it may be volcanic ash. This sediment is carried out to sea primarily by wind, although rivers and ocean currents also help to distribute it.

Microscopic shells and skeletons of plants and animals also settle slowly to the sea floor when marine organisms of the surface waters die. In some parts of the sea, such as the polar and equatorial regions, great concentrations of these shells have built unusually thick pelagic deposits.

The constant slow rain of pelagic clay and shells occurs in all parts of the sea. Although the rate of accumulation varies

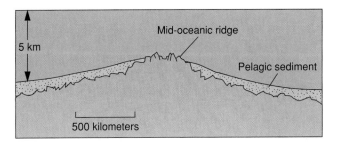

FIGURE 18.24

Pelagic sediment is thin or absent on the crest of the mid-oceanic ridge and becomes progressively thicker away from the ridge crest (distribution of sediment is highly simplified).

from place to place, pelagic sediment should be expected on all parts of the sea floor.

Surprisingly, however, pelagic sediment is almost completely absent on the crest of the mid-oceanic ridge. Pelagic sediment is found on the flanks of the mid-oceanic ridge, often thickening away from the ridge crest (figure 18.24). But its absence on the ridge crest was an unexpected discovery about seafloor sediment distribution.

OCEANIC CRUST AND OPHIOLITES

As you have seen in chapter 17, oceanic crust differs significantly from continental crust; it is both thinner and of a different composition than continental crust. Seismic reflection and seismic refraction surveys at sea have shown the oceanic crust to be about 7 kilometers (4.3 miles) thick and divided into three layers (figure 18.25).

The top layer (Layer 1), of variable thickness and character, is marine sediment. In an abyssal fan or on the continental rise, Layer 1 may consist of several kilometers of terrigenous sediment. On the upper flanks of the mid-oceanic ridge, there may be less than 100 meters (328 feet) of pelagic sediment. An average thickness for Layer 1 might be 0.5 kilometer (1,640 feet).

Beneath the sediment is Layer 2, which is about 1.5 kilometers (about 1 mile) thick. It has been extensively sampled. This layer consists of pillow basalt overlying dikes of basalt. The basalt pillows, rounded masses that form when hot lava erupts into cold water (figure 18.25), are highly fractured at the top of Layer 2. The dikes in the lower part of Layer 2 are closely spaced, parallel, vertical dikes ("sheeted dikes"). Widely sampled by drilling and dredging, Layer 2 has also been observed directly by geologists in submersibles diving into the rift valley of the mid-oceanic ridge during the FAMOUS expedition.

The lowest layer in the crust is Layer 3. It is about 5 kilometers (3.1 miles) thick and thought to consist of sill-like gabbro bodies. The evidence for this interpretation is suggestive but not conclusive. Geologists have drilled through 2 kilometers (1.2 miles) of pillow basalt and basaltic dikes in oceanic

FIGURE 18.23

Photograph (taken through a scanning electron microscope) of pelagic sediment from the floor of the Pacific Ocean. The sediment is made up of microscopic skeletons of single-celled marine organisms (large objects are foraminifera; smaller, sievelike ones are radiolaria about 0.05 mm in diameter). *Photo © Dr. Richard Kessel & Dr. Gene Shih/Visuals Unlimited*

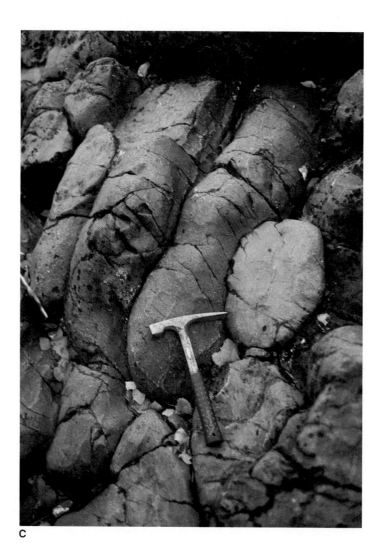

FIGURE 18.25

A comparison of oceanic crust and an ophiolite sequence. (*A*) Structure of oceanic crust, determined from seismic studies and drilling. Gabbro and ultramafic rock drilled and dredged from sea-floor fault blocks may be from Layer 3 and the mantle. (*B*) Typical ophiolite sequence found in mountain ranges on land. Thickness is approximate—sequence is usually highly faulted. (*C*) Pillow basalt from the upper part of an ophiolite, northern California. These rocks formed as part of the sea floor, when hot lava cooled quickly in cold seawater. *Photo C by David McGeary*

crust but have not yet been able to reach gabbro by drilling through basalt. Gabbro (and other rocks) are exposed on some fault blocks in rift valleys and on some steep submarine cliffs in fracture zones; 0.5 kilometer of gabbro has been drilled on one such cliff in the southern Indian Ocean. Some geologists presume that the gabbro represents Layer 3 exposed by faulting, but other geologists are not so sure. Some evidence exists that rocks exposed in fracture zones may not be representative of the entire sea floor. Even if the drilled gabbro *does* represent the upper one-tenth of Layer 3, no one has yet sampled the lower nine-tenths of the layer.

Seismic velocities of 7 kilometers (4.3 miles) per second are consistent with the choice of gabbro for Layer 3, but deep drilling on land (see box 17.1) has shown that seismic reflection and refraction records on land are routinely misinterpreted regarding rock type and depths to rock boundaries. These errors extend to oceanic crust as well. A drillhole in the eastern Pacific Ocean has been reoccupied four times in a twelve-year span and has now reached a total depth of 2,000 meters (6,560 feet) below the sea floor. Seismic evidence suggested that the Layer 2–Layer 3 boundary would be found at a depth of about

1,700 meters (5,576 feet), but the drill went well past that depth without finding the contact between the dikes of Layer 2 and the expected gabbro of Layer 3. Either the seismic interpretation or the model of Layer 3's composition must be wrong.

Geologists' ideas about the composition of oceanic crust are greatly influenced by a study of **ophiolites,** distinctive rock sequences found in many mountain chains on land (figure 18.25). The thin, top layer of an ophiolite consists of marine sedimentary rock, often including thin-bedded chert. Below the sedimentary rock lies a zone of pillow basalt, which in turn is underlain by a sheeted-dike complex that probably served as feeder dikes for the pillowed lava flows above. The similarities between the upper part of an ophiolite and Layers 1 and 2 of oceanic crust are obvious.

Below the sheeted dikes of an ophiolite is a thick layer of podlike gabbro intrusions, perhaps thick sills. Beneath the gabbro lies ultramafic rock such as peridotite, the top part of which has been converted to serpentine by metamorphism. Geologists have long thought that ophiolites represent slivers of oceanic crust somehow emplaced on land. If this is true, then the gabbro of ophiolites may represent oceanic Layer 3,

ENVIRONMENTAL GEOLOGY 18.2

Geologic Riches in the Sea

Many resources are being extracted from the sea floor and seawater, and in some instances, there is a great potential for increased extraction. Realizing this tremendous resource, the United States established the Economic Exclusion Zone, which gives sovereign rights for resources on or below the sea floor in an area that extends 370 kilometers (200 miles) from the coastline.

Offshore oil and gas are the most valuable resources now being taken from the sea. More than one-sixth of U.S. oil production (and more than one-quarter of world production) comes from drilling platforms set up on the continental shelf (box figure 1). Oil and gas have been found within deeper parts of the sea floor, such as the continental slope and continental rise. Producing oil from these deeper regions is much more costly than present production; oil spills from wells in deep water would be especially hard to control. Gas hydrates, which are icelike solids of natural gas and water found trapped in deep marine sediments, are another potential source of energy that is being explored.

Other important resources are dredged from the sea floor. Phosphorite can be recovered from shallow shelves and banks and used for fertilizers. Gold, diamonds, and heavy black sands (which are black because they contain metal-bearing minerals) are being separated from the surface sands and gravels of some continental shelves by specially designed ships.

Manganese nodules (box figure 2) cover many parts of the deep-sea floor, notably in the central Pacific. These black, potato-sized lumps contain approximately 25% manganese, 15% iron, and up to 2% nickel and 2% copper, along with smaller amounts of cobalt. Although there are international legal problems concerning who owns them, larger industrial countries such as the United States may mine them, particularly for copper, nickel, and cobalt. The United States is also interested in manganese; it imports 95% of its manganese, which is critical to producing some types of steel.

Metallic brines and sediments, first discovered in the Red Sea, are deposited by submarine hot springs active at the rift valley of the mid-oceanic ridge crest. The Red Sea sediments contain more than 1% copper and more than 3% zinc, together with impressive amounts of silver, gold, and lead (worth $25 billion, according to a 1983 estimate). Because of their great value, the sediments will probably be mined even though they are at great depth. Deposits similar to those of the Red Sea—although not of such great economic potential—have been found on several other parts of the ridge.

A few substances can be extracted from the salts dissolved in seawater. Approximately two-thirds of the world's production of magnesium is obtained from seawater, and in many regions, sodium chloride (table salt) is obtained by solar evaporation of seawater.

BOX 18.2 ■ FIGURE 1

Offshore oil drilling platform Hibernia is located 320 kilometers (200 miles) off the coast of Newfoundland and began producing oil in 1997. As many as 50 different wells can be drilled from a single platform. Photo courtesy Chevron Canada Resources

BOX 18.2 ■ FIGURE 2

Dense concentrations of manganese nodules on the floor of the abyssal plain (depth 5,350 meters) in the northeast Atlantic Ocean. Photo © Institute of Oceanographic Sciences/ NERC/Photo Researchers, Inc.

and the peridotite may represent mantle rock below oceanic crust. The contact between the peridotite and the overlying gabbro would be the Mohorovičić discontinuity.

Recent work has shown that many, if not most, ophiolites are *not* typical sea floor but a special type of sea floor formed in marginal ocean basins next to continents by the process of backarc spreading (see chapter 19). If ophiolites do not represent typical sea floor, then more extensive and deeper seafloor drilling is needed before a clear picture of oceanic crust can be formed. This is an important goal, for oceanic crust is the most common surface rock, covering 60% of Earth's surface.

THE AGE OF THE SEA FLOOR

As marine geologists began to determine the age of seafloor rocks (by isotopic dating) and sediments (by fossils), an astonishing fact was discovered. All the rocks and sediments of the sea floor proved to be younger than 200 million years. This was true only for rocks and sediments from the *deep*-sea floor, not those from the continental margins. The rocks and sediments presently found on the deep-sea floor formed during the Mesozoic and Cenozoic eras but not earlier.

By contrast, as discussed in chapter 8, Earth is estimated to be 4.5 billion years old. Every continent contains some rocks formed during the Paleozoic Era and the Precambrian. Some of the Precambrian rocks on continents are more than 3 billion years old, and a few are almost 4 billion years old. Continents, therefore, preserve rocks from most of Earth's history. In sharp contrast to the continents is the deep-sea floor, which covers more than half of Earth's surface but preserves less than one-twentieth of its history in its rocks and sediment.

THE SEA FLOOR AND PLATE TECTONICS

As we mentioned at the beginning, this chapter is concerned with the *description* of the sea floor. The *origin* of most seafloor features is related to plate tectonics. Chapter 19 shows how the theory of plate tectonics explains the existence and character of continental shelves and slopes, trenches, the mid-oceanic ridge, and fracture zones, as well as the very young age of the sea floor itself. In the workings of the scientific method, this chapter largely concerns *data*. Chapter 19 shows how *hypotheses* and *theories* account for these data.

SUMMARY

The *continental shelf* and the steeper *continental slope* lie under water along the edges of continents. They are separated by a change in slope angle at a depth of 100 to 200 meters.

Submarine canyons are cut into the continental slope and outer continental shelf by a combination of *turbidity currents,* sand flow and fall, bottom currents, and river erosion during times of lower sea level. Graded bedding and cable breaks suggest the existence of turbidity currents in the ocean.

Abyssal fans form as sediment collects at the base of submarine canyons.

A *passive continental margin* occurs off geologically quiet coasts and is marked by a continental rise and abyssal plains at the base of the continental slope.

The *continental rise* and *abyssal plains* may form from sediment deposited by turbidity currents.

The continental rise may also form from sediment deposited by *contour currents* at the base of the continental slope.

An *active continental margin* is marked by an *oceanic trench* at the base of the continental slope; the continental rise and abyssal plains are absent.

Oceanic trenches are twice as deep as abyssal plains, which generally lie at a depth of 5 kilometers. Associated with trenches are *Benioff zones* of earthquakes and *andesitic volcanism,* forming either an island arc or a chain of volcanoes in a young mountain belt near the edge of a continent. Trenches have low heat flow and negative gravity anomalies.

The *mid-oceanic ridge* is a globe-circling mountain range of basalt, located mainly in the middle of ocean basins. The crest of the ridge is marked by a *rift valley,* shallow-focus earthquakes, high heat flow, active *basaltic volcanism,* hydrothermal activity, and exotic organisms.

Fracture zones are lines of weakness that offset the mid-oceanic ridge. Shallow-focus quakes occur on the portion of the fracture zone between the offset ridge segments.

Seamounts are conical, submarine volcanoes, now mostly extinct. *Guyots* are flattopped seamounts, probably leveled by wave erosion before subsiding.

Chains of seamounts and guyots form *aseismic ridges.*

Corals and algae living in warm, shallow water construct *fringing reefs, barrier reefs,* and *atolls.*

Terrigenous sediment is composed of land-derived sediment deposited near land by turbidity currents and other processes. *Pelagic sediment* is made up of wind-blown dust and microscopic skeletons that settle slowly to the sea floor.

The crest of the mid-oceanic ridge lacks pelagic sediment.

Oceanic crust consists of marine sediments overlying basalt pillows and dikes, probably overlying gabbro.

Ophiolites in continental mountain ranges probably represent slivers of somewhat atypical oceanic crust somehow emplaced on land.

The oldest rocks on the deep-sea floor are 200 million years old. The continents, in contrast, contain some rock that is 3 to 4 billion years old.

Terms to Remember

abyssal fan 473
abyssal plain 475
active continental margin 476
aseismic ridge 481
atoll 481
barrier reef 481
continental rise 475
continental shelf 472

continental slope 472
contour current 475
fracture zone 479
fringing reef 481
guyot 480
mid-oceanic ridge 477
oceanic trench 476
ophiolite 484

passive continental margin 474
pelagic sediment 483
reef 481
rift valley 477
seamount 479
submarine canyon 472
terrigenous sediment 483
turbidity current 473

Testing Your Knowledge

Use the following questions to prepare for exams based on this chapter.

1. What is a submarine canyon? How do submarine canyons form?

2. Discuss the appearance, structure, and origin of abyssal plains.

3. Sketch a cross profile of the mid-oceanic ridge, showing the rift valley. Label your horizontal and vertical scales.

4. Sketch an active continental margin and a passive continental margin, labeling all their parts. Show approximate depths.

5. What is a fracture zone? Sketch the relation between fracture zones and the mid-oceanic ridge.

6. In a sketch, show the association between an oceanic trench, a Benioff zone of earthquakes, and volcanoes on the edge of a continent.

7. Describe two different origins for the continental rise.

8. What is a turbidity current? What is the evidence that turbidity currents occur on the sea floor?

9. Describe the appearance and origin of seamounts and guyots.

10. Describe the two main types of seafloor sediment.

11. How does the age of seafloor rocks compare with the age of continental rocks? Be specific.

12. Sketch a cross section of a fringing reef, a barrier reef, and an atoll.

13. What is the thickness and composition of oceanic crust?

14. Most ocean water probably came from
    a. degassing of Earth's interior
    b. outer space

15. Which is true of the continental shelf?
    a. it is a shallow submarine platform at the edge of a continent
    b. it inclines very gently seaward
    c. it can vary in width
    d. all of the preceding

16. The average angle of slope for a continental slope is
    a. 1°–2°   b. 3°–4°
    c. 4°–5°   d. greater than 10°

17. Great masses of sediment-laden water that are pulled downhill by gravity are called
    a. contour currents   b. bottom currents
    c. turbidity currents   d. traction currents

18. Oceanic trenches
    a. are narrow, deep troughs
    b. run parallel to the edge of a continent or an island arc
    c. are often 8 to 10 kilometers deep
    d. all of the preceding

19. Which is characteristic of mid-oceanic ridges?
    a. shallow-focus earthquakes   b. high heat flow
    c. basalt eruptions   d. all of the preceding

20. Reefs parallel to the shore but separated from it by wide, deep lagoons are called
    a. fringing reefs   b. barrier reefs
    c. atolls   d. lagoonal reefs

21. Pelagic sediment could be composed of
    a. fine-grained clay
    b. skeletons of microscopic organisms
    c. volcanic ash
    d. all of the preceding

22. What part of the continental margin marks the true edge of the continent?
    a. continental shelf   b. continental slope
    c. continental rise   d. abyssal plain

23. Distinctive rock sequences of basalt and marine sedimentary rock that may be slices of the ocean floor are
    a. guyots   b. ophiolites
    c. seamounts   d. fracture zones

Expanding Your Knowledge

1. How many possible origins can you think of for the rift valley on the mid-oceanic ridge?
2. What factors could cause sea level to rise? To fall?
3. Why is the rock of the deep-sea floor (60% of Earth's surface) basalt? Where did the basalt come from?
4. How could fracture zones have formed?
5. How many hypotheses can you think of to explain the relatively young age of the sea floor?

Exploring Web Resources

www.mhhe.com/plummer11e

This is the dedicated website for this book. You can go to it for new and updated information. The universal resource locators (URL) listed in this book are also given as links on the website, making it easy to go to those websites without typing in the URL. When you visit our Online Learning Center, go to the Student Center and then to the chapter of interest. Here you will find additional readings and media resources, as well as answers to the Testing Your Knowledge section, more quizzes, animation and video clips, and direct links to the sites listed in this book. Links to additional websites can also be found. We have added questions for some of the sites to allow you to get the most of your exploration of the web. Using the web is an enjoyable way of enhancing your knowledge of geology.

http://seawifs.gsfc.nasa.gov:80/OCEAN_PLANET/HTML/ocean_planet_ocean_science.html

Smithsonian Ocean Planet Exhibit containing many links to seafloor topics, such as hydrothermal vents and microbiology of deep-sea vents.

www.esdim.noaa.gov/ocean_page.html

National Oceanographic and Atmospheric Administration (NOAA) has many links to numerous sites relating to the oceans.

www.ngdc.noaa.gov/mgg/mggd.html

NOAA Marine Geology and Geophysics Division website contains world seafloor maps and information on ocean-drilling data and samples.

www-odp.tamu.edu/

Ocean Drilling Program website contains information on the nature and history of the ocean crust and provides data from research projects.

www.whoi.edu/VideoGallery/

Woods Hole Oceanographic Institution website contains video clips of black smokers, exotic organisms at mid-oceanic ridges, and oceanographic research using vessels and submersibles.

www.pmel.noaa.gov/vents/

NOAA Vents Program website provides photos, video clips, data, and research program activities about the investigation of submarine volcanoes and hydrothermal venting around the world.

Plate Tectonics

Satellite image of the Red Sea and Arabia. Plate motion has torn the Arabian
Peninsula (right center) away from Africa (left) to form the Red Sea (center).
*Photo provided by the SeaWiFS Project, NASA/Goddard Space Flight
Center, and ORBIMAGE*

s you studied volcanoes; igneous, metamorphic, and sedimentary rocks; and earthquakes, you learned how these topics are related to plate tectonics. In this chapter, we take a closer look at plates and plate motion. We will pay particular attention to plate boundaries and the possible driving mechanisms for plate motion.

The history of the concept of plate tectonics is a good example of how scientists think and work and how a hypothesis can be proposed, discarded, modified, and then reborn. In the first part of this chapter, we trace the evolution of an idea—how the earlier hypotheses of moving continents (continental drift) and a moving sea floor (seafloor spreading) were combined to form the theory of plate tectonics.

Tectonics is the study of the origin and arrangement of the broad structural features of Earth's surface, including not only folds and faults but also mountain belts, continents, and earthquake belts. Tectonic models such as an expanding Earth or a contracting Earth have been used in the past to explain *some* of the surface features of Earth. Plate tectonics has come to dominate geologic thought today because it can explain so *many*

features. The basic idea of **plate tectonics** is that Earth's surface is divided into a few large, thick plates that move slowly and change in size. Intense geologic activity occurs at *plate boundaries* where plates move away from one another, past one another, or toward one another. The eight large lithospheric plates shown in figure 19.1, plus a few dozen smaller plates, make up the outer shell of Earth (the crust and upper part of the mantle).

The concept of plate tectonics was born in the late 1960s by combining two preexisting ideas—continental drift and seafloor spreading. **Continental drift** is the idea that continents move freely over Earth's surface, changing their positions relative to one another. **Seafloor spreading** is a hypothesis that the sea floor forms at the crest of the mid-oceanic ridge, then moves horizontally away from the ridge crest toward an oceanic trench. The two sides of the ridge are moving in opposite directions like slow conveyor belts.

Before we take a close look at plates, we will examine the earlier ideas of moving continents and a moving sea floor because these two ideas embody the theory of plate tectonics.

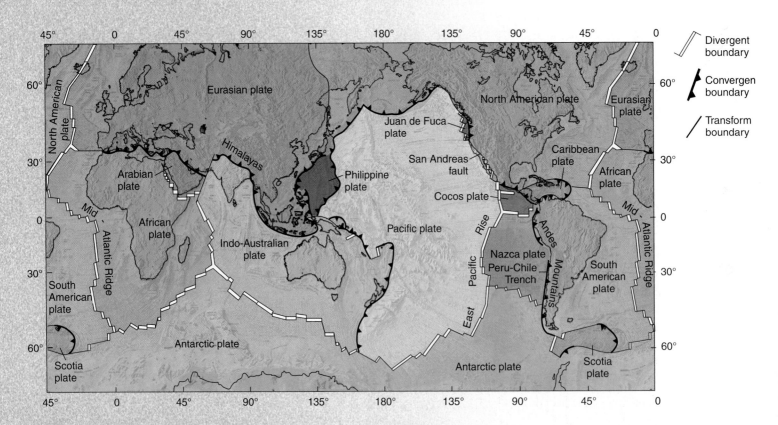

FIGURE 19.1

The major plates of the world. The western edge of the map repeats the eastern edge so that all plates can be shown unbroken. Double lines indicate spreading axes on divergent plate boundaries. Single lines show transform boundaries. Heavy lines with triangles show convergent boundaries, with triangles pointing down subduction zones. *Modified from W. Hamilton, U.S. Geological Survey*

THE EARLY CASE FOR CONTINENTAL DRIFT

Continents can be made to fit together like pieces of a picture puzzle. The similarity of the Atlantic coastlines of Africa and South America has long been recognized. The idea that continents were once joined together and have split and moved apart from one another has been around for more than 130 years (figure 19.2).

In the early 1900s, Alfred Wegener, a German meteorologist, made a strong case for continental drift. He noted that South America, Africa, India, Antarctica, and Australia had almost identical late Paleozoic rocks and fossils.

The plant *Glossopteris* is found in Pennsylvanian and Permian-age rock on all five continents, and fossil remains of *Mesosaurus,* a freshwater reptile, are found in Permian-age rocks only in Brazil and South Africa (figure 19.3). In addition, fossil remains of land-dwelling reptiles *Lystrosaurus* and *Cynognathus* are found in Triassic-age rocks on all five continents.

Wegener reassembled the continents to form a giant supercontinent, *Pangaea* (also spelled *Pangea* today). Wegener thought that the similar rocks and fossils were easier to explain if the continents were joined together, rather than in their present, widely scattered positions.

Pangaea initially separated into two parts. *Laurasia* was the northern supercontinent, containing what is now North America and Eurasia (excluding India). *Gondwanaland* was

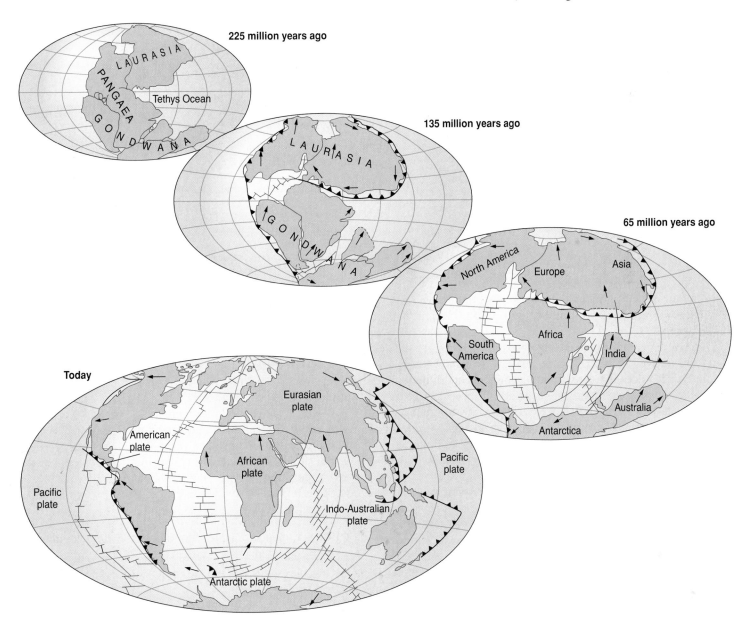

FIGURE 19.2

Pangaea breakup and continental drift. *After C. R. Scotese (www.scotese.com)*

FIGURE 19.3

Distribution of plant and animal fossils that are found on the continents of South America, Africa, Antarctica, India, and Australia give evidence for the southern supercontinent of Gondwana. *Glossopteris* and other fernlike plants are found in Permian- and Pennsylvanian-age rocks on all five continents. *Cynognathus* and *Lystrosaurus* were sheep-sized land reptiles that lived during the Early Triassic Period. Fossils of the freshwater reptile *Mesosaurus* are found in Permian-age rocks on the southern tip of Africa and South America.

the southern supercontinent, composed of all the present-day Southern Hemisphere continents and India (which has drifted north).

The distribution of Late Paleozoic glaciation strongly supports the idea of Pangaea (figure 19.4). The Gondwanaland continents (the Southern Hemisphere continents and India) all have glacial deposits of Late Paleozoic age. If these continents were spread over Earth in Paleozoic time as they are today, a climate cold enough to produce extensive glaciation would have had to prevail over almost the whole world. Yet, no evidence has been found of widespread Paleozoic glaciation in the Northern Hemisphere. In fact, the late Paleozoic coal beds of North America and Europe were being laid down at that time in swampy, probably warm environments. If the continents are arranged according to Wegener's Pangaea reconstruction, then glaciation in the Southern Hemisphere is confined to a much smaller area (figure 19.4), and the absence of widespread glaciation in the Northern Hemisphere becomes easier to explain. Also, the present arrangement of the continents would require that late Paleozoic ice sheets flowed from the oceans toward the continents, which is impossible.

Wegener also reconstructed old climate zones (the study of ancient climates is called *paleoclimatology*). Glacial till

and striations indicate a cold climate near the North or South Pole. Coral reefs indicate warm water near the equator. Cross-bedded sandstones can indicate ancient deserts near 30° North and 30° South latitude. If ancient climates had the same distribution on Earth that modern climates have, then sedimentary rocks can show where the ancient poles and equator were located.

Wegener determined the positions of the North and South Poles for each geologic period. He found that ancient poles were in different positions than the present poles (figure 19.5A). He called this apparent movement of the poles **polar wandering.** Polar wandering, however, is a deceptive term. The evidence can actually be explained in the following ways:

1. The continents remained motionless and the poles actually *did* move—polar wandering (figure 19.5A).

2. The poles stood still and the continents moved—continental drift (figure 19.5B).

3. Both occurred.

Wegener plotted curves of apparent polar wandering (figure 19.6). Since one interpretation of polar wandering data

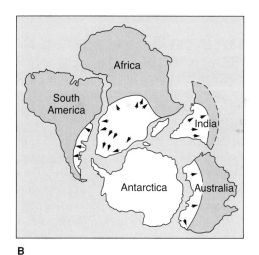

FIGURE 19.4

Distribution of late Paleozoic glaciation; arrows show direction of ice flow. (A) Continents in present positions show wide distribution of glaciation (white land areas with flow arrows). (B) Continents reassembled into Pangaea. Glaciated region becomes much smaller. *From Arthur Holmes, 1965, Principles of Physical Geology, 2d ed., Ronald Press*

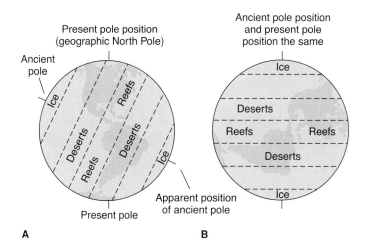

FIGURE 19.5

Two ways of interpreting the distribution of ancient climate belts. (A) Continents are fixed, poles wander. (B) Poles are fixed, continents drift. For simplicity, the continents in (B) are shown as having moved as a unit, without changing positions relative to one another. If continents move, they should change relative positions, complicating the pattern shown.

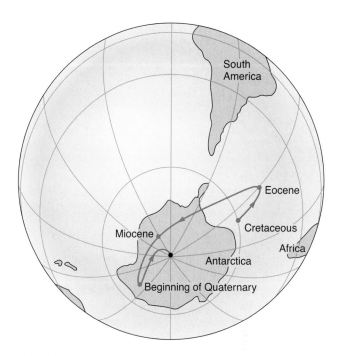

FIGURE 19.6

Apparent wandering of the South Pole since the Cretaceous Period as determined by Wegener from paleoclimate evidence. Wegener, of course, believed that *continents*, rather than the poles, moved. *From A. Wegener, 1928, The Origins of Continents and Oceans, reprinted and copyrighted. 1968, Dover Publications*

was that the continents moved, Wegener believed that this supported his concept of continental drift. (Notice that in only one interpretation of polar wandering do the poles actually move. You should keep in mind that when geologists use the term *polar wandering*, they are referring to an *apparent* motion of the poles, which may or may not have actually occurred.)

Skepticism about Continental Drift

Although Wegener presented the best case possible in the early 1900s for continental drift, much of his evidence was not clear-cut. The presence of land-dwelling reptiles throughout the scattered continents was explained by land bridges, which were postulated to somehow rise up from the sea floor and then subside again. The existence or nonexistence of land bridges was difficult to prove without data on the topography of the sea floor. Also, fossil plants could have been spread from one continent to another by winds or ocean currents. Their distribution over more than one continent does not *require* that the continents were all joined in the supercontinent, Pangaea. In addition, polar wandering might have been caused by moving poles rather than by moving continents. Because his evidence was not conclusive, Wegener's ideas were not widely accepted. This was particularly true in the

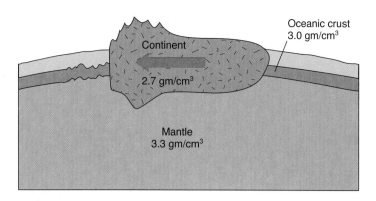

FIGURE 19.7

Wegener's concept of continental drift implied that the less-dense continents drifted *through* oceanic crust, crumpling up mountain ranges on their leading edges as they pushed against oceanic crust.

United States, largely because of the mechanism Wegener proposed for continental drift.

Wegener proposed that continents plowed through the oceanic crust (figure 19.7), perhaps crumpling up mountain ranges on the leading edges of the continents where they pushed against the sea floor. Most geologists in the United States thought that this idea violated what was known about the strength of rocks at the time. The driving mechanism proposed by Wegener for continental drift was a combination of centrifugal force from Earth's rotation and the gravitational forces that cause tides. Careful calculations of these forces showed them to be too small to move continents. Because of these objections, Wegener's ideas received little support in the United States or much of the Northern Hemisphere (where the great majority of geologists live) in the first half of the twentieth century. The few geologists in the Southern Hemisphere, however, where Wegener's matches of fossils and rocks between continents were more evident, were more impressed with the concept of continental drift.

PALEOMAGNETISM AND THE REVIVAL OF CONTINENTAL DRIFT

Much work in the 1940s and 1950s set the stage for the revival of the idea of continental drift and its later incorporation, along with seafloor spreading, into the new concept of plate tectonics. The new investigations were in two areas: (1) study of the sea floor and (2) geophysical research, especially in relation to rock magnetism.

Convincing new evidence about polar wandering came from the study of rock magnetism. Wegener's work dealt with the wandering of Earth's *geographic* poles of rotation. The *magnetic* poles are located close to the geographic poles, as you saw in chapter 17 on Earth's interior. Historical measurements show that the position of the magnetic poles moves from year to year but that the magnetic poles stay close to the geographic poles as they move. As we discuss magnetic evidence for polar wandering, we are referring to an apparent motion of the magnetic poles. Because the magnetic and geographic poles are close together, our discussion will refer to apparent motion of the geographic poles as well.

A Rock near north magnetic pole **B** Rock near equator in Northern Hemisphere **C** Rock in Southern Hemisphere

D

FIGURE 19.8

Magnetic dip (inclination) increases toward the north magnetic pole. Rocks in bottom part of figure are small samples viewed horizontally at locations A, B, and C on the globe. The magnetic dip can therefore be used to determine the distance from a rock to the north magnetic pole. (D) Compass needle showing steep inclination near south magnetic pole in Antarctica. *Photo by C. C. Plummer*

As we discussed in chapter 17, many rocks record the strength and direction of Earth's magnetic field at the time the rocks formed. Magnetite in a cooling basaltic lava flow acts like a tiny compass needle, preserving a record of Earth's magnetic field when the lava cools below the *Curie point.* Iron-stained sedimentary rocks such as red shale can also record Earth's magnetism. The magnetism of old rocks can be measured to determine the direction and strength of the magnetic field in the past. The study of ancient magnetic fields is called *paleomagnetism.*

Because magnetic lines of force dip more steeply as the north magnetic pole is approached, the inclination (dip) of the magnetic alignment preserved in the magnetite minerals in the lava flows can be used to determine the distance from a flow to the pole at the time that the flow formed (figure 19.8).

Old pole positions can be determined from the magnetism of old rocks. The magnetic alignment preserved in magnetite minerals points to the pole, and the dip of the alignment tells how far away the pole was. Figure 19.9 shows how Permian lava flows in North America indicate a Permian pole position in eastern Asia.

For each geologic period, North American rocks reveal a different magnetic pole position; this path of the *apparent* motion of the north magnetic pole through time is shown in figure 19.10. Paleomagnetic evidence thus verifies Wegener's idea of polar wandering (which he based on paleoclimatic evidence).

Like Wegener's paleoclimatic evidence, the paleomagnetic evidence from a *single* continent can be interpreted in two ways: either the continent stood still and the magnetic pole moved, or the pole stood still and the continent moved. At first glance, paleomagnetic evidence does not seem to be a significant advancement over paleoclimatic evidence. But when paleomagnetic evidence from *different* continents was compared, an important discovery was made.

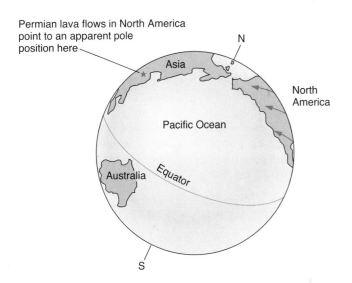

FIGURE 19.9

Paleomagnetic studies of Permian lava flows on North America indicate an apparent position for the north magnetic pole in eastern Asia.

FIGURE 19.10

Apparent polar wandering of the north magnetic pole for the past 520 million years as determined from measurements of rocks from North America (red) and Europe (green).

Although Permian rocks in North America point to a pole position in eastern Asia, Permian rocks in *Europe* point to a different position (closer to Japan), as shown in figure 19.10. Does this mean there were *two* north magnetic poles in the Permian Period? In fact, every continent shows a different position for the Permian pole. A different magnetic pole for each continent seems highly unlikely. A better explanation is that a single pole stood still while continents split apart and rotated as they diverged.

Note the polar wandering paths for North America and Europe in figure 19.10. The paths are of similar shape, but the path for European poles is to the east of the North American path. If we mentally push North America back toward Europe, closing the Atlantic Ocean, and then consider the paths of polar wandering, we find that the path for North America lies exactly on the path for Europe. This strongly suggests that there was one north magnetic pole and that the continents were joined together. There appear to be two north magnetic poles because the rocks of North America moved west; their magnetic minerals now point to a different polar position than they did when the minerals first formed.

Recent Evidence for Continental Drift

As paleomagnetic evidence revived interest in continental drift, new work was done on fitting continents together. By defining the edge of a continent as the middle of the continental slope, rather than the present (constantly changing) shoreline, a much more precise fit has been found between continents (figure 19.11).

FIGURE 19.11

Jigsaw puzzle fit and matching rock types between South America and Africa. Light-blue areas around continents are continental shelves (part of continents). Colored areas within continents are broad belts of rock that correlate in type and age from one continent to another. Arrows show direction of glacier movement as determined from striations.

The most convincing evidence for continental drift came from greatly refined rock matches between now-separated continents. If continents are fitted together like pieces of a jigsaw puzzle, the "picture" should match from piece to piece.

The matches between South America and Africa are particularly striking. Some distinctive rock contacts extend out to sea along the shore of Africa. If the two continents are fitted together, the identical contacts are found in precisely the right position on the shore of South America (figure 19.11). Isotopic ages of rocks also match between these continents.

Glacial striations show that during the late Paleozoic Era, continental glaciers moved from Africa toward the present Atlantic Ocean, while similar glaciers seemingly moved *from* the Atlantic Ocean *onto* South America (figure 19.11). Continental glaciers, however, cannot move from sea onto land. If the two continents had been joined together, the ice that moved off Africa could have been the ice that moved onto South America. This hypothesis has now been confirmed; from their lithology, many of the boulders in South American tills have been traced to a source that is now in Africa.

Some of the most detailed matches have been made between rocks in Brazil and rocks in the African country of Gabon. These rocks are similar in type, structure, sequence, fossils, ages, and degree of metamorphism. Such detailed matches are convincing evidence that continental drift did, in fact, take place.

There is also an abundance of satellite geodetic data from the Global Positioning Satellite (GPS) system, so we can now

watch the continents move—about as eventful as watching your fingernails grow!

History of Continental Positions

Rock matches show when continents were together; once the continents split, the new rocks formed are dissimilar. Paleomagnetic evidence indicates the direction and rate of drift, allowing maps of old continental positions, such as figure 19.2, to be drawn.

Although Pangaea split up 200 million years ago to form our present continents, the continents were moving much earlier. Pangaea was formed by the collision of many small continents long before it split up. Recent work shows that continents have been in motion for at least the past 2 billion years (some geologists say 4 billion years), well back into Precambrian time. For more than half of Earth's history, the continents appear to have collided, welded together, then split and drifted apart, only to collide again, over and over, in an endless, slow dance.

SEAFLOOR SPREADING

At the same time that many geologists were becoming interested again in the idea of moving *continents,* Harry Hess, a geologist at Princeton University, proposed that the *sea floor* might be moving, too. This proposal contrasted sharply with the earlier ideas of Wegener, who thought that the ocean floor remained stationary as the continents plowed through it (figure 19.7). Hess's 1962 proposal was quickly named seafloor spreading, for it suggests that the sea floor moves away from the mid-oceanic ridge as a result of mantle convection (figure 19.12).

According to the initial concept of seafloor spreading, the sea floor is moving like a conveyor belt away from the crest of the mid-oceanic ridge, down the flanks of the ridge, and across the deep-ocean basin, to disappear finally by plunging beneath a continent or island arc (figure 19.12). The ridge crest, with sea floor moving away from it on either side, has been called a *spreading axis* (or *spreading center*). The sliding of the sea floor beneath a continent or island arc is termed **subduction.** The sea floor moves at a rate of 1 to 24 centimeters per year (your fingernail grows at about 1 centimeter per year). Although this may seem to be quite slow, it is rapid compared to most geologic processes.

Hess's Driving Force

Why does the sea floor move? Hess's original hypothesis was that seafloor spreading is driven by deep mantle convection. **Convection** is a circulation pattern driven by the rising of hot material and/or the sinking of cold material. Hot material has a lower density, so it rises; cold material has a higher density and sinks. The circulation of water heating in a pan on a stove is an

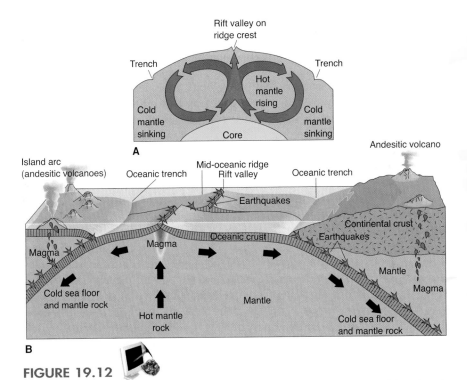

FIGURE 19.12

Seafloor spreading hypothesis of Harry Hess. (A) Hess proposed that convection extended throughout the mantle. (Scale of ridge and trenches is exaggerated.) (B) Hot mantle rock rising beneath the mid-oceanic ridge (a spreading axis) causes basaltic volcanism and high heat flow. Divergence of sea floor splits open the rift valley and causes shallow-focus earthquakes (stars on ridge). Sinking of cold rock causes subduction of older sea floor at trenches, producing Benioff zones of earthquakes and andesitic magma.

example of convection. Convection in the mantle was a controversial idea in 1962; for although convection can be easily demonstrated in a pan of water, it was hard to visualize the solid rock of the mantle behaving as a liquid. Over very long periods of time, however, it is possible for the hot mantle rock to flow in a ductile manner. A slow, convective circulation is set up by temperature differences in the rock, and convection can explain many seafloor features as well as the young age of the seafloor rocks. (The heat that flows outward through Earth to drive convection is both original heat from the planet's formation and heat from the decay of radioactive isotopes, as discussed in chapter 17.)

Explanations

The Mid-Oceanic Ridge

If convection drives seafloor spreading, then hot mantle rock must be rising under the mid-oceanic ridge. Hess showed how the *existence of the ridge* and its *high heat flow* are caused by the rise of this hot mantle rock. The *basalt eruptions* on the ridge crest are also related to this rising rock, for here the mantle rock is hotter than normal and begins to undergo partial melting.

As hot rock continues to rise beneath the ridge crest, the circulation pattern splits and diverges near the surface. Mantle rock moves horizontally away from the ridge crest on each side of the ridge. This movement accompanies tension at the

ridge crest, cracking open the oceanic crust to form the *rift valley* and its associated *shallow-focus earthquakes.*

Oceanic Trenches

As the mantle rock moves horizontally away from the ridge crest, it carries the sea floor (the basaltic oceanic crust) piggyback along with it. As the hot rock moves sideways, it cools and becomes denser, sinking deeper beneath the ocean surface. Hess thought it would become cold and dense enough to sink back into the mantle. This downward plunge of cold rock accounts for the *existence of the oceanic trenches* as well as their *low heat flow* values. It also explains the large *negative gravity anomalies* associated with trenches, for the sinking of the cold rock provides a force that holds trenches out of isostatic equilibrium (see chapter 17).

As the sea floor moves downward into the mantle along a subduction zone, it interacts with the rock above it. This interaction between the moving seafloor rock and the overlying crustal and mantle rock can cause the *Benioff zones of earthquakes* associated with trenches. It can also produce *andesitic volcanism,* which forms volcanoes either on the edge of a continent or in an island arc (figure 19.12).

Hess's ideas have stood up remarkably well over more than thirty years. We now think of lithospheric plates moving instead of sea floor riding piggyback on convecting mantle, and we think that several mechanisms cause plate motion, but Hess's explanation of seafloor topography, earthquakes, and age remains valid today.

Age of the Sea Floor

The *young age of seafloor rocks* (see chapter 18) is neatly explained by Hess's seafloor spreading. New, young sea floor is continually being formed by basalt eruptions at the ridge crest. This basalt is then carried sideways by convection and is subducted into the mantle at an oceanic trench. Thus, old sea floor is continually being destroyed at trenches, while new sea floor is being formed at the ridge crest. (This is also the reason for the puzzling lack of pelagic sediment at the ridge crest. Young sea floor at the ridge crest has little sediment because the basalt is newly formed. Older sea floor farther from the ridge crest has been moving under a constant rain of pelagic sediment, building up a progressively thicker layer as it goes.)

Note that seafloor spreading implies that the youngest sea floor should be at the ridge crest, with the age of the sea floor becoming progressively older toward a trench. This increase in age away from the ridge crest was not known to exist at the time of Hess's proposal but was an important prediction of his hypothesis. This prediction has been successfully tested, as you shall see in the section on "Marine Magnetic Anomalies" in this chapter.

PLATES AND PLATE MOTION

By the mid-1960s, the twin ideas of moving continents and a moving sea floor were causing great excitement and emotional debate among geologists. By the late 1960s, these ideas had been combined into a single theory that revolutionized geology by providing a unifying framework for Earth science—the theory of plate tectonics.

As described earlier, a **plate** is a large, mobile slab of rock that is part of Earth's surface (figure 19.1). The surface of a plate may be made up entirely of sea floor (as is the Nazca plate), or it may be made up of both continental and oceanic rock (as is the North American plate). Some of the smaller plates are entirely continental, but all the large plates contain some sea floor.

Plate tectonics has added some new terms, based on rock behavior, to the zones of Earth's interior, as we have discussed in some previous chapters. The plates are composed of the relatively rigid outer shell of Earth called the **lithosphere.** The lithosphere includes the rocks of the crust and uppermost mantle (figure 19.13).

The lithosphere beneath oceans increases in both age and thickness with distance from the crest of the mid-oceanic ridge. Young lithosphere near the ridge crest may be only 10 kilometers thick, while very old lithosphere far from the ridge crest may be as much as 100 kilometers thick. An average thickness for oceanic lithosphere might be 70 kilometers, as shown in figure 19.13.

Continental lithosphere is thicker, varying from perhaps 125 kilometers thick to as much as 200 to 250 kilometers thick beneath the oldest, coldest, and most inactive parts of the continents.

Below the rigid lithosphere is the **asthenosphere,** a zone of low seismic-wave velocity that behaves in a ductile manner because of increased temperature and pressure. Some geologists think that the asthenosphere is partially molten; the melt-ing of just a few percent of the asthenosphere's volume could account for its properties and behavior. The ductile asthenosphere acts as a lubricating layer under the lithosphere, allowing the plates to move. The asthenosphere, made up of upper mantle rock, is the low-velocity zone described in chapter 17. It may extend from a depth of 70 to 200 kilometers beneath oceans; its thickness, depth, and even existence under continents are vigorously debated. Below the asthenosphere is more rigid mantle rock.

The idea that plates move is widely accepted by geologists, although the reasons for this movement are debated. Plates move away from the mid-oceanic ridge crest or other spreading axes. Some plates move toward oceanic trenches. If the plate is made up mostly of sea floor (as are the Nazca and Pacific plates), the plate can be subducted down into the mantle, forming an oceanic trench and its associated features. If the leading edge of the plate is made up of continental rock (as is the South American plate), that plate will not subduct. Continental rock, being less dense (specific gravity 2.7) than oceanic rock (specific gravity 3.0), is too light to be subducted.

To a first approximation, a plate may be viewed as a rigid slab of rock that moves as a unit. As a result, the interior of a plate is relatively inactive tectonically (but see box 19.2). Plate interiors generally lack earthquakes, volcanoes, young mountain belts, and other signs of geologic activity. According to plate-tectonic theory, these features are caused by plate interactions at plate boundaries.

Plate boundaries are of three general types, based on whether the plates move away from each other, move toward each other, or move past each other. A **divergent plate boundary** is a boundary between plates that are moving apart. A **convergent plate boundary** lies between plates that are moving toward each other. A **transform plate boundary** is one at which two plates move horizontally past each other.

HOW DO WE KNOW THAT PLATES MOVE?

The proposal that Earth's surface is divided into moving plates was an exciting, revolutionary hypothesis, but it required testing to win acceptance among geologists. You have seen how the study of paleomagnetism supports the idea of moving continents. In the 1960s, two critical tests were made of the idea of a moving sea floor. These tests involved marine magnetic anomalies and the seismicity of fracture zones. These two, successful tests convinced most geologists that plates do indeed move.

Marine Magnetic Anomalies

In the mid-1960s, magnetometer surveys at sea disclosed some intriguing characteristics of marine magnetic anomalies. Most magnetic anomalies at sea are arranged in bands that lie parallel to the rift valley of the mid-oceanic ridge. Alternating positive and negative anomalies (chapter 17) form a stripelike pattern parallel to the ridge crest (figure 19.14).

FIGURE 19.13

The rigid lithosphere includes the crust and uppermost mantle; it forms the plates. The ductile asthenosphere acts as a lubricating layer beneath the lithosphere. Oceanic lithosphere averages 70 kilometers thick; continental lithosphere varies from 125 to 250 kilometers thick. Asthenosphere may not be present under continents.

A

B

FIGURE 19.14

Marine magnetic anomalies. (A) The red line shows positive and negative magnetic anomalies as recorded by a magnetometer towed behind a ship. Positive anomalies are shown in black and negative anomalies are shown in tan. Notice how magnetic anomalies are parallel to the rift valley and symmetric about the ridge crest. (B) Symmetric magnetic anomalies ("stripes") from the mid-Atlantic ridge south of Iceland.

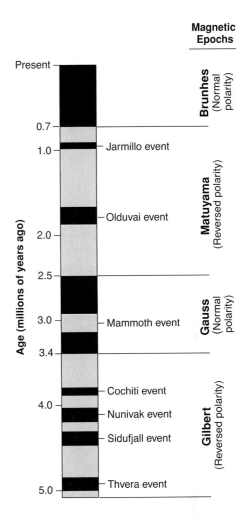

FIGURE 19.15

Magnetic reversals during the past 5 million years determined from lava flows that have been radiometrically dated. Black represents normal magnetism; tan represents reverse magnetism. *After Mankinen, E. A. and Dalrymple, G. B., 1979. Revised geomagnetic polarity time scale for the interval 0–5 m.y. B.P. Journal of Geophysical Research, v. 84, p. 615–626.*

The Vine-Matthews Hypothesis

Two British geologists, Fred Vine and Drummond Matthews, made several important observations about these anomalies. They recognized that the pattern of magnetic anomalies was symmetrical about the ridge crest. That is, the pattern of magnetic anomalies on one side of the mid-oceanic ridge was a mirror image of the pattern on the other side (figure 19.14).

Vine and Matthews also noticed that the same pattern of magnetic anomalies exists over different parts of the mid-oceanic ridge. The pattern of anomalies over the ridge in the northern Atlantic Ocean is the same as the pattern over the ridge in the southern Pacific Ocean.

The most important observation that Vine and Matthews made was that the pattern of magnetic *anomalies* at sea matches the pattern of magnetic *reversals* already known from studies of lava flows on the continents (figure 19.15 and chapter 17). This correlation can be seen by comparing the pattern of colored bands in figure 19.15 (reversals) with the pattern in figure 19.14 (anomalies).

Putting these observations together with Hess's concept of seafloor spreading, which had just been published, Vine and Matthews proposed an explanation for magnetic anomalies. They suggested that there is continual opening of tensional cracks within the rift valley on the mid-oceanic ridge crest. These cracks on the ridge crest are filled by basaltic magma from below, which cools to form dikes. Cooling magma in the dikes records Earth's magnetism at the time the magnetic minerals crystallize. The process is shown in figure 19.16.

When Earth's magnetic field has a *normal polarity* (the present orientation), cooling dikes are normally magnetized. Dikes that cool when the field is reversed (figure 19.16) are reversely magnetized. So each dike preserves a record of the polarity that prevailed during the time the magma cooled. Extension produced by the moving sea floor then cracks a dike in two, and the two halves are carried away in opposite directions down the flanks of the ridge. New magma eventually intrudes the newly opened fracture. It cools, is magnetized, and forms a new dike, which in turn is split by continued extension. In this way, a system of reversely magnetized and normally magnetized dikes forms parallel to the rift valley. These dikes, in the Vine-Matthews hypothesis, are the cause of the anomalies.

The magnetism of normally magnetized dikes adds to Earth's magnetism, and so a magnetometer carried over such dikes registers a stronger magnetism than average—a *positive*

magnetic anomaly. Dikes that are reversely magnetized subtract from the present magnetic field, and so a magnetometer towed over such dikes measures a weaker magnetic field—a *negative* magnetic anomaly. Since seafloor motion separates these dikes into halves, the patterns on either side of the ridge are mirror images.

Measuring the Rate of Plate Motion

There are two important points about the Vine-Matthews hypothesis of magnetic anomaly origin. The first is that it allows us to measure the *rate of seafloor motion* (which is the same as plate motion, since continents and the sea floor move together as plates).

Because magnetic reversals have already been dated from lava flows on land (figure 19.15), the anomalies caused by these reversals are also dated and can be used to discover how fast the sea floor has moved (figure 19.16D). For instance, a piece of the sea floor representing the reversal that occurred 4.5 million years ago may be found 45 kilometers away from the rift valley of the ridge crest. The piece of sea floor, then, has traveled 45 kilometers since it formed 4.5 million years ago. Dividing the distance the sea floor has moved by its age gives 10 kilometers per million years, or 1 centimeter per year, for the rate of seafloor motion here. In other words, on each side of the ridge, the sea floor is moving away from the ridge crest at a rate of 1 centimeter per year. Such measured rates generally range from 1 to 24 centimeters per year.

Predicting Seafloor Age

The other important point of the Vine-Matthews hypothesis is that it *predicts the age of the sea floor* (figure 19.16D). Magnetic reversals are now known to have occurred back into Precambrian time. Sea floor of *all* ages is therefore characterized by parallel bands of magnetic anomalies. Figure 17.20 shows the pattern of marine magnetic anomalies (and the reversals that caused them) during the past 160 million years. The distinctive pattern of these anomalies through time allows them to be identified by age, a process similar to dating by tree rings.

Now, even before they sample the sea floor, marine geologists can predict the age of the igneous rock of the sea floor by measuring the magnetic anomalies at the sea surface. Most sections of the sea floor have magnetic anomalies. By matching the measured anomaly pattern with the known pattern that is shown in figure 17.20, the age of the sea floor in the region can be predicted, as shown on the map in figure 19.17.

This is a very powerful test of the hypothesis that the sea floor moves. Suppose, for example, that the sea floor in a particular spot is predicted to be 70 million years old from a study of its magnetic anomalies. If the hypothesis of seafloor motion and the Vine-Matthews hypothesis of magnetic anomaly origin are correct, a sample of igneous rock from that spot *must* be 70 million years old. If the rock proves to be 10 million years old or 200 million years old or 1.2 billion years old, or any other age except 70 million years, then both of these hypotheses are wrong. But if the rock proves to be 70 million years old, as predicted, then both hypotheses have been successfully tested.

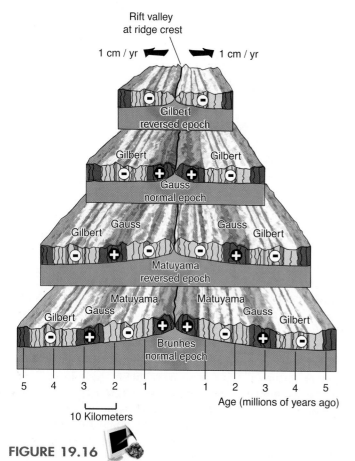

FIGURE 19.16

The origin of magnetic anomalies. During a time of reversed magnetism (Gilbert reversed epoch), a series of basaltic dikes intrudes the ridge crest, becoming reversely magnetized. The dike zone is torn in half and moved sideways, as a new group of normally magnetized dikes forms at the ridge crest. A new series of reversely magnetized dikes forms at the ridge crest. The dike pattern becomes symmetric about the ridge crest. Correlating the magnetic anomalies with magnetic reversals allows anomalies to be dated. Magnetic anomalies can therefore be used to predict the age of the sea floor and to measure the rate of seafloor spreading (plate motion).

	Miocene	Oligocene	Eocene	Paleo-cene	Late Cretaceous	Middle Cretaceous	Early Cretaceous	Late Jurassic
Ocean Geology								

0 m.y.   50   100   150

FIGURE 19.17

The age of the sea floor as determined from magnetic anomalies. *After* The Bedrock Geology of the World *by R. L. Larson, W. C. Pitman, III, et al., W. H. Freeman*

Hundreds of rock and sediment cores recovered from holes drilled in the sea floor were used to test these hypotheses. Close correspondence has generally been found between the predicted age and the measured age of the sea floor. (The seafloor age is usually measured by fossil dating of sediment in the cores rather than by isotopic dating of igneous rock.) This evidence from deep-sea drilling has been widely accepted by geologists as verification of the hypotheses of plate motion and magnetic anomaly origin. Most geologists now think that these concepts are no longer hypotheses but can now be called theories. (A *theory*, as discussed in box 1.4 in connection with the scientific method, is a hypothesis that has been tested and found to explain observations.)

Another Test: Fracture Zones and Transform Faults

Cores from deep-sea drilling tested plate motion by allowing us to compare the actual age of the sea floor with the age predicted from magnetic anomalies. Another rigorous test of plate motion has been made by studying the seismicity of fracture zones.

The mid-oceanic ridge is offset along fracture zones (see figure 19.1). Conceivably, the mid-oceanic ridge was once continuous across a fracture zone but has been offset by strikeslip motion along the fracture zone (figure 19.18A). If such motion is occurring along a fracture zone, we would expect to find two things: (1) earthquakes should be distributed along the entire length of the fracture zone, and (2) the motion of the rocks on either side of the fracture zone should be in the direction shown by the arrows in figure 19.18A.

In fact, these things are not true about fracture zones. Earthquakes do occur along fracture zones, but only in those segments between offset sections of ridge crest. In addition, first-motion studies of earthquakes (see chapter 16) along fracture zones show that the motion of the rocks on either side of the fracture zone during an earthquake is exactly opposite to the motion shown in figure 19.18A. The actual motion of the rocks as determined from first-motion studies is shown in figure 19.18B. The portion of a fracture zone between two offset portions of ridge crest is called a **transform fault.**

The motion of rocks on either side of a transform fault was predicted by the hypothesis of a moving sea floor. Note that sea floor moves away from the two segments of ridge crest (figure 19.18B). Looking along the length of the fracture

FIGURE 19.18

Two possible explanations for the relationship between fracture zones and the mid-oceanic ridge. (A) The expected rock motions and earthquake distribution assuming that the ridge was once continuous across the fracture zone. (B) The expected rock motions and earthquake distribution assuming that the two ridge segments were never joined together and that the sea floor moves away from the rift valley segments. Only explanation (B) fits the data. The portion of the fracture zone between the ridge segments is a transform fault.

zone, you can see that blocks of rock move in opposite directions only on that section of the fracture zone between the two segments of ridge crest. Earthquakes, therefore, occur only on this section of the fracture zone, the transform fault. The direction of motion of rock on either side of the transform fault is exactly predicted by the assumption that rock is moving away from the ridge crests. Verification by first-motion studies of this predicted motion along fracture zones was another successful test of plate motion.

Measuring Plate Motion Directly

In recent years, the motion of plates has been directly measured using satellites, radar, lasers, and the Global Positioning System (GPS). These techniques can measure the distance between two widely separated points to within 1 centimeter. GPS is now routinely used to measure the relative motion between plates because of its accuracy and because the receivers are relatively inexpensive and fairly portable (figure 19.19A). Plate motions are now recorded on a yearly basis throughout the world (figure 19.19B).

If two plates move toward each other at individual rates of 2 centimeters per year and 6 centimeters per year, the combined rate of convergence is 8 centimeters per year. The measurement techniques are sensitive enough to easily measure such a rate if measurements are repeated each year. Such measured rates match closely the predicted rates from magnetic anomalies.

DIVERGENT PLATE BOUNDARIES

Divergent plate boundaries, where plates move away from each other, can occur in the middle of the ocean or in the middle of a continent. The result of divergent plate boundaries is to create, or open, new ocean basins. This dynamic process has occurred throughout the geologic past.

When a supercontinent such as Pangaea breaks up, a divergent boundary can be found in the middle of a continent. The divergent boundary is marked by rifting, basaltic volcanism, and uplift. During rifting, the continental crust is stretched and thinned. This extension produces shallow-focus earthquakes on normal faults, and a rift valley forms as a central *graben* (a downdropped fault block). The faults act as pathways for basaltic magma, which rises from the mantle to erupt on the surface as cinder cones and basalt flows. Uplift at a divergent boundary is usually caused by the upwelling of hot mantle beneath the crust; the surface is elevated by the thermal expansion of the hot, rising rock and of the surface rock as it is warmed from below.

Figure 19.20 shows how a continent might rift to form an ocean. The figure shows rifting before uplift, because recent work indicates that this was the sequence for the opening of the Red Sea. The crust is initially stretched and thinned. Numerous normal faults break the crust, and the surface subsides into a central graben (figure 19.20A). Shallow earthquakes and basalt eruptions occur in this rift valley, which also has high heat flow. An example of a boundary at this stage is the African Rift Valleys in eastern Africa (figure 19.21). The valleys are grabens that may mark the site of the future breakup of Africa.

As divergence continues, the continental crust on the upper part of the plate clearly separates, and seawater floods into the linear basin between the two divergent continents (figure 19.20B). A series of fault blocks have rotated along curved fault planes at the edges of the continents, thinning the continental crust. The rise of hot mantle rock beneath the thinned crust causes continued basalt eruptions that create true oceanic crust between the two continents. The center of the narrow ocean is marked by a rift valley

FIGURE 19.19

(A) Global Positioning System (GPS) station being installed in Iceland that will collect signals from orbitting GPS satellites to determine plate motions. (B) Yearly plate motions from stations around the world as measured by GPS. *Photo (A) © J-sef H-lm j-rn, Icelandic Meteoric Office; (B) from NASA (http://slideshow.jpl.nasa.gov/mbh/series.html)*

A Continent undergoes extension. The crust is thinned and a rift valley forms.

B Continent tears in two. Continent edges are faulted and uplifted. Basalt eruptions form oceanic crust.

C Continental sediments blanket the subsiding margins to form continental shelves. The ocean widens and a mid-oceanic ridge develops, as in the Atlantic Ocean.

FIGURE 19.20

A divergent plate boundary forming in the middle of a continent will eventually create a new ocean.

with its typical high heat flow and shallow earthquakes. The Red Sea is an example of a divergent margin at this stage (figure 19.21).

After modest widening of the new ocean, uplift of the continental edges may occur. As continental crust thins by stretching and faulting, the surface initially subsides. At the same time, hot mantle rock wells up beneath the stretched crust (figure 19.20B). The rising diapir of hot mantle rock would cause uplift by thermal expansion.

The new ocean is narrow, and the tilt of the adjacent land is away from the new sea, so rivers flow away from the sea

(figure 19.20B). At this stage, the seawater that has flooded into the rift may evaporate, leaving behind a thick layer of rock salt overlying the continental sediments. The likelihood of salt precipitation increases if the continent is in one of the desert belts or if one or both ends of the new ocean should become temporarily blocked, perhaps by volcanism. Not all divergent boundaries contain rock salt, however.

The plates continue to diverge, widening the sea. Thermal uplift creates a mid-oceanic ridge in the center of the sea (figure 19.20C). The flanks of the ridge subside as the seafloor rock cools as it moves.

FIGURE 19.21

(A) The East African Rift Valleys and the Red Sea. (B) Satellite photo of Red Sea. Gulf of Suez is on the upper left and Gulf of Aqaba on upper right. Note the similarities in the shorelines of the Arabian Penninsula (right) and Africa (left) suggesting that the Red Sea was formed by splitting of the continent. *Photo by Jeff Schmaltz, MODIS Rapid Response Team, NASA/GSFC (http://visibleearth.nasa.gov/)*

The trailing edges of the continents also subside as they are lowered by erosion and as the hot rock beneath them cools. Subsidence continues until the edges of the continents are under water. A thick sequence of marine sediment blankets the thinned continental rock, forming a *passive continental margin* (figures 19.20C and 19.22; see also chapter 18). The sediment forms a shallow continental shelf, which may contain a deeply buried salt layer. The deep continental rise is formed as sediment is carried down the continental slope by turbidity currents and other mechanisms. The Atlantic Ocean is currently at this stage of divergence (see figure 18.17).

A divergent boundary on the sea floor is located on the crest of the mid-oceanic ridge. If the spreading rate is slow, as it is in the Atlantic Ocean (1 centimeter per year), the crest has a rift valley. Fast spreading, as along the East Pacific Rise (18 centimeters per year) and along other ridges in the Pacific Ocean, prevents a rift from forming. A divergent boundary at sea is marked by the same features as a divergent boundary on land—tensional cracks, normal faults, shallow earthquakes, high heat flow, and basaltic eruptions. The basalt forms dikes within the cracks and pillow lavas on the sea floor, creating new oceanic crust on the trailing edge of plates.

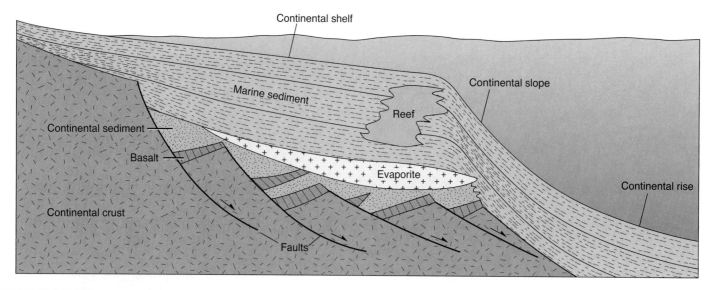

FIGURE 19.22

A passive continental margin formed by continental breakup and divergence. Downfaulted continental crust forms basins, which fill with basalt and sediment. A layer of rock salt may form if a narrow ocean evaporates. A thick sequence of marine sediments covers these rocks and forms the continental shelf, slope, and rise. A reef may form at the shelf edge if the water is warm; buried reefs occur on many parts of the Atlantic shelf of North America.

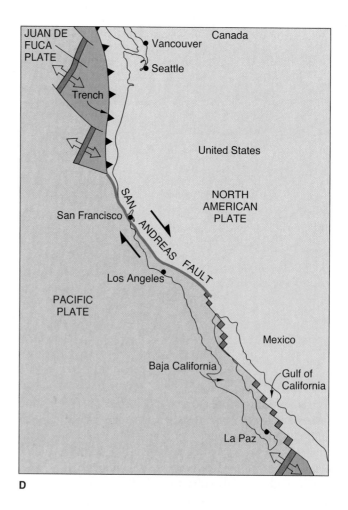

FIGURE 19.23

Transform boundaries (A) between two ridges; (B) between a ridge and a trench; and (C) between two trenches. Triangles on trenches point down subduction zones. Trench-trench transform boundaries are common in the southeast Pacific. Color tones show two plates in each case. (D) The San Andreas fault is a ridge-ridge transform plate boundary between the North American plate and the Pacific plate. The south end of the San Andreas fault is a ridge segment (shown in red) near the U.S.-Mexico border. The north end of the fault is a "triple junction" where three plates meet at a point. The relative motion along the San Andreas fault is shown by the large black arrows, as the Pacific plate slides horizontally past the North American plate. (D) *Modified from U.S. Geological Survey*

TRANSFORM BOUNDARIES

At transform boundaries, where one plate slides horizontally past another plate, the plate motion can occur on a single fault or on a group of parallel faults. Transform boundaries are marked by shallow-focus earthquakes in a narrow zone for a single fault or in a broad zone for a group of parallel faults (see figure 16.27). First-motion studies of the quakes indicate strikeslip movement parallel to the faults.

The name *transform fault* comes from the fact that the displacement along the fault abruptly ends or transforms into another kind of displacement. The most common type of transform fault occurs along fracture zones and connects two divergent plate boundaries at the crest of the mid-oceanic ridge (figures 19.23 and 19.18B). The spreading motion at one ridge segment is transformed into the spreading motion at the other ridge segment by strike-slip movement along the transform fault.

Not all transform faults connect two ridge segments. As you can see in figure 19.23, a transform fault can connect a ridge to a trench (a divergent boundary to a convergent boundary), or it can connect two trenches (two convergent boundaries). The San Andreas fault in California is a transform fault with a complex history (figure 19.23D).

What is the origin of the offset in a ridge-ridge transform fault? The offsets appear to be the result of irregularly shaped divergent boundaries (figure 19.24). When two oceanic plates begin to diverge, the boundary may be curved on a sphere. Mechanical constraints prevent divergence along a curved boundary, so the original curves readjust into a series of right-angle bends. The ridge crests align perpendicular to the spreading direction, and the transform faults align parallel to the spreading direction. An old line of weakness in a continent may cause the initial divergent boundary to be oblique to the spreading direction when the continent splits. The boundary will then readjust into a series of transform faults parallel to the spreading direction.

A

B

FIGURE 19.24

Divergent boundaries form ridge crests perpendicular to the spreading direction and transform faults parallel to the spreading direction. (A) Oceanic plates. (B) Continental plates.

Ocean-Ocean Convergence

Where two plates capped by sea floor converge, one plate subducts under the other (the Pacific plate sliding under the western Aleutian Islands is an example). The subducting plate bends downward, forming the outer wall of an oceanic trench, which usually forms a broad curve convex to the subducting plate (figures 19.25 and 19.26).

CONVERGENT PLATE BOUNDARIES

At convergent plate boundaries, two plates move toward each other (often obliquely). The character of the boundary depends partly on the type of plates that converge. A plate capped by oceanic crust can move toward another plate capped by oceanic crust, in which case one plate dives (subducts) under the other. If an oceanic plate converges with a plate capped by a continent, the dense oceanic plate subducts under the continental plate. If the two approaching plates are both carrying continents, the continents collide and crumple, but neither is subducted.

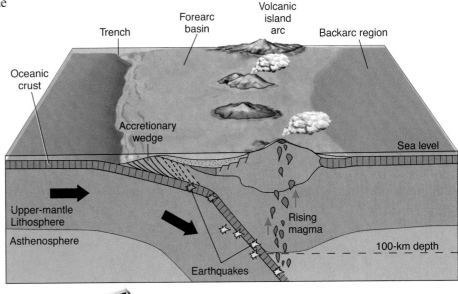

FIGURE 19.25

Ocean-ocean convergence forms a trench, a volcanic island arc, and a Benioff zone of earthquakes.

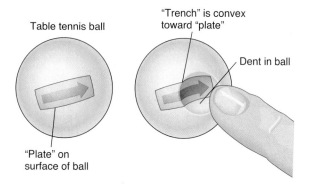

FIGURE 19.26

A dented table tennis ball can show why trenches are curved on a sphere.

As one plate subducts under another, a Benioff zone of shallow-, intermediate-, and deep-focus earthquakes is created within the upper portion of the down-going lithosphere (see figure 16.23). The reasons for these quakes are discussed in chapter 16. The existence of deep-focus earthquakes to a depth of 670 kilometers tells us that brittle plates continue to (at least) that depth. The pattern of quakes shows that the angle of subduction changes with depth, usually becoming steeper (figure 19.25). Some plates crumple or break into segments as they descend.

As the descending plate reaches depths of at least 100 kilometers, magma is generated in the overlying asthenosphere (figure 19.25). The magma probably forms by partial melting of the asthenosphere, perhaps triggered by dewatering of the downgoing oceanic crust as it is subducted, as described in chapter 3. Differentiation and assimilation may also play an important role in the generation of the magma, which is typically andesitic to basaltic in composition.

The magma works its way upward to erupt as an **island arc,** a curved line of volcanoes that form a string of islands parallel to the oceanic trench (figure 19.25). Beneath the volcanoes are large plutons in the thickened arc crust.

The distance between the island arc and the trench can vary, depending upon where the subducting plate reaches the 100-kilometer depth. If the subduction angle is steep, the plate reaches this magma-generating depth at a location close to the trench, so the horizontal distance between the arc and trench is short. If the subduction angle is gentle, the arc-trench distance is greater. A thick, buoyant plate (such as a subducting aseismic ridge) may subduct at such a gentle angle that it merely slides horizontally along under another plate. Because the top of the subducting plate never reaches the 100-kilometer depth, such very shallow subduction zones lack volcanism.

When a plate subducts far from a mid-oceanic ridge, the plate is cold, with a low heat flow. Oceanic plates form at ridge crests, then cool and sink as they spread toward trenches. Eventually, they become cold and dense enough to sink back into the mantle. Oceanic trenches are marked by strong negative gravity anomalies. These show that trenches are not currently in isostatic equilibrium but are being actively pulled down. Hess thought that this pulling was caused by a down-turning convection current in the mantle. Today, most geologists think that the pulling is caused by the sinking of cold, dense lithosphere.

The inner wall of a trench (toward the arc) consists of an *accretionary wedge* (or *subduction complex*) of thrust-faulted and folded marine sediment and pieces of oceanic crust (figure 19.25). The sediment is "snowplowed" off the subducting plate by the overlying plate. New slices of sediment are continually added to the bottom of the accretionary wedge, pushing it upward to form a ridge on the sea floor. A relatively undeformed *forearc basin* lies between the accretionary wedge

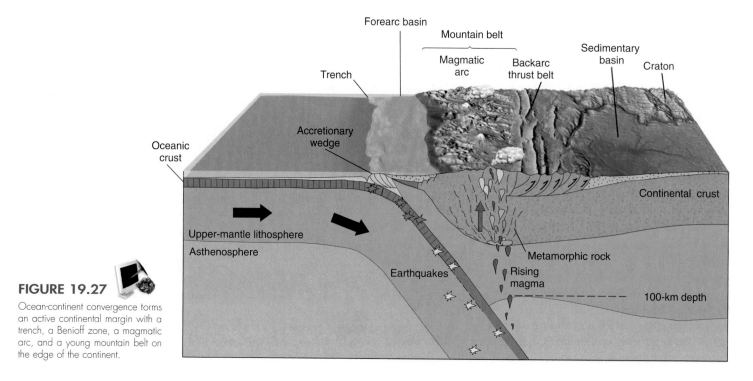

FIGURE 19.27

Ocean-continent convergence forms an active continental margin with a trench, a Benioff zone, a magmatic arc, and a young mountain belt on the edge of the continent.

Plate Tectonics and Sea Level

Geologists have long known that at certain times in the geologic past, the sea covered vast areas of the continents that are now dry land. Much of the interior of the United States, for example, is underlain by marine limestones deposited during parts of the Paleozoic Era. Were the continents lower at these times, or was sea level higher?

As you have seen, the subsidence of the craton during subduction can allow vast regions of the continental interior to be flooded with seawater. Some marine deposits on the craton, however, are so extensive that they probably were caused by a rise in sea level.

Although several mechanisms, such as glaciation, can change sea level, the development of plate tectonics has led to a hypothesis that may explain some of the ancient sea-level fluctuations.

During an episode of rapid plate motion, an active spreading axis will be marked by a mid-oceanic ridge caused by the thermal expansion of rock on the rising limb of a convection current. When plate motion stops, convection also stops, and the rock at an old spreading axis cools off and contracts. This means that the mid-oceanic ridge subsides and eventually becomes level sea floor. That is, when plates move, a ridge is present; and when plates stop, the ridge is absent.

When a ridge is present, it displaces seawater, raising sea level and causing the sea to flood land areas. When the ridge is absent, the water returns to the ocean basin and the continents are dry once again.

The plates need not stop completely. A rapid spreading rate would cause a large ridge and a sea-level rise, and a slower spreading rate would cause a smaller ridge and a lower sea level. There is good evidence that some sea-level fluctuations can be correlated to changes in the rate of the sea-floor motion. Not all changes in sea level can be explained by this mechanism. Glaciation and other factors clearly affect sea level, too.

and the volcanic arc. (The trench side of an arc is the forearc; the other side of the arc is the backarc.)

Trench positions change with time. As one plate subducts, the overlying plate may be moving toward it. The motion of the leading edge of the overlying plate will force the trench to migrate horizontally over the subducting plate. The Peru-Chile Trench is moving over the Nazca plate in this manner as South America moves westward (figure 19.1). There is another reason that trenches move. It is now widely believed that a subducting plate does not sink in a direction parallel to the length of the plate but falls through the mantle at an angle that is *steeper* than the dip of the down-going plate. This steep sinking pulls the subducting plate progressively away from the overlying plate and causes the hinge line of bending and the oceanic trench to migrate seaward onto the subducting plate. The location at which the subducting plate contacts the 100-kilometer depth where magmas are generated in the asthenosphere also migrates seaward toward the subducting plate and may cause the position of the island arc to migrate toward the subducting plate as well.

Ocean-Continent Convergence

When a plate capped by oceanic crust is subducted under the *continental* lithosphere, an accretionary wedge and forearc basin form an *active continental margin* between the trench and the continent (figure 19.27). A Benioff zone of earthquakes dips under the edge of the continent, which is marked by andesitic volcanism and a young mountain belt. Examples of this type of boundary are the subduction of the Nazca plate under western South America and the Juan de Fuca plate under North America.

The magma that is created by ocean-continent convergence forms a **magmatic arc,** a broad term used both for island arcs at sea and for belts of igneous activity on the edges of continents. The surface expression of a magmatic arc is either a line of andesitic islands (such as the Aleutian Islands) or a line of andesitic continental volcanoes (such as the Cascade volcanoes of the Pacific Northwest). Beneath the volcanoes are large plutons in thickened crust. We see these plutons as batholiths on land when they are exposed by deep erosion. The igneous processes that form the granitic and intermediate magmas of batholiths are described in chapter 3.

The hot magma rising from the subduction zone thickens the continental crust and makes it weaker and more mobile than cold crust. Regional metamorphism takes place within this hot, mobile zone. Crustal thickening causes uplift, so a young mountain belt forms here as the thickened crust rises isostatically.

Another reason for the growth of the mountain belt is the stacking up of thrust sheets on the continental (backarc) side of the magmatic arc (figure 19.27). The thrust faults, associated with folds, move slivers of mountain-belt rocks landward over the continental interior (the *craton*). Underthrusting of the rigid craton beneath the hot, mobile core of the mountain belt may help form the fold-thrust belt.

Inland of the backarc fold-thrust belt, the craton subsides to form a sedimentary basin (sometimes called a *foreland basin*). The weight of the stacked thrust sheets depresses the craton isostatically. The basin receives sediment, some of which may be marine if the craton is forced below sea level. This basin extends the effect of subduction far inland. Subduction of the sea floor off California during the Mesozoic Era produced basin sedimentation as far east as the central Great Plains.

FIGURE 19.28

The collision of two continents forms a young mountain belt in the interior of a new, larger continent. The most famous example of continent-continent collision is the collision of India with Asia. (A) India is moving toward Asia due to ocean-continent convergence. (B) India collides with Asia to form the Himalayas, the highest mountain range on Earth. (C) Map view of the northward movement of India through time.

Continent-Continent Convergence

Two continents may approach each other and collide. They must be separated by an ocean floor that is being subducted under one continent and that lacks a spreading axis to create new oceanic crust (figure 19.28). The edge of one continent will initially have a magmatic arc and all the other features of ocean-continent convergence.

As the sea floor is subducted, the ocean becomes narrower and narrower until the continents eventually collide and destroy or close the ocean basin. Oceanic lithosphere is heavy and can sink into the mantle, but continental lithosphere is less dense and cannot sink. One continent may slide a short distance under another, but it will not go down a subduction zone. After collision, the heavy oceanic lithosphere breaks off the continental lithosphere and continues to sink, leaving the continent behind.

The two continents are welded together along a dipping *suture zone* that marks the old site of subduction (figure 19.28B).

Thrust belts and subsiding basins occur on both sides of the original magmatic arc, which is now inactive. The presence of the original arc thickens the crust in the region of impact. The crust is thickened further by the shallow underthrusting of one continent beneath the other and by the stacking of thrust sheets in the two thrust belts. The result is a mountain belt in the interior of a continent (a new, large continent formed by the collision of the two, smaller continents). The entire region of impact is marked by a broad belt of shallow-focus earthquakes along the numerous faults, as shown in figure 16.26. A few deeper quakes may occur within the sinking oceanic lithosphere beneath the mountain range.

The Himalayas in central Asia are thought to have formed in this way, as India collided with and underthrust Asia to produce exceptionally thick crust and high elevations. Paleomagnetic studies show that India was once in the Southern Hemisphere and drifted north to its present position. The collision with Asia occurred after an intervening ocean was destroyed by subduction (figure 19.2).

IN GREATER DEPTH 19.2

Backarc Spreading

Regional extension occurs within or behind many arcs. This extension can tear an arc in two, moving the two halves in opposite directions (box figure 1). If it occurs behind an arc, it can move the arc away from a continent. It can split the edge of a continent, moving a narrow strip of the continent seaward (this is apparently how Japan formed). In each case, the spreading creates new oceanic crust that is similar, but not identical, to the oceanic crust formed at the crest of mid-oceanic ridges. This backarc oceanic crust is apparently the type of crust found in most ophiolites (chapter 18).

The reason for backarc extension is energetically debated. One suggestion is that extension is caused by a rising and spreading mantle diapir of hot rock or magma somehow generated by the down-going plate (box figure 2). The spreading diapir tears open the backarc basin, and the rising magma forms new oceanic crust. Another suggestion is that the sub-ducting plate drags on the overlying asthenosphere, causing it to move in secondary convection cells that stretch and fracture the overlying oceanic crust. A third suggestion, which seems the best explanation for the most rapidly spreading backarc basins in the Pacific, is that the overlying plate is retreating away from the subducting plate. If the arc on the overlying plate stays fixed near the subducting plate, the retreat of the overlying plate will tear open the backarc basin.

A

B

C

BOX 19.2 ■ FIGURE 1

Backarc spreading. Regional extension in the overlying plate of a subduction zone can split an arc, move an arc offshore, or split a continent.

BOX 19.2 ■ FIGURE 2

Causes of backarc spreading. Extension may be caused by a rising mantle diapir, secondary convection, or relative plate motions.

IN GREATER DEPTH 19.3

Indentation Tectonics and "Mushy" Plate Boundaries

While it is easiest to conceive of Earth's skin as being made up of rigid tectonic plates that interact narrowly along their edges, the reality is more complicated—and interesting. The forces that cause plates to be geologically active along their boundaries may extend far into their interiors as well. Consider the following two examples:

The Collision of India with Asia

Between 40 and 60 million years ago, India began colliding with Asia to form the Himalaya Mountains, the biggest mountain system in the world. The sea that once separated India from Asia drained away as the former ocean floor rose into ridges and peaks as much as 5 miles high. The stresses of the continent-continent convergence extend far to the north of the Himalaya plate boundary, however—perhaps as far as 5,000 kilometers into Central Asia. Huge strike-slip fault systems with roughly east-west orientation break China apart (box figure 1). These have formed as Central Asia shifts out of the way of India, with the lithosphere moving primarily eastward to override the Pacific and Philippine plates in a series of very active subduction zones. India, in other words, has greatly "indented" Asia by colliding with it. The world's most destructive earthquakes have occurred in

China, far from any plate boundaries. The Shaanxi earthquake in 1556 alone killed over 800,000 people.

Near the northern edge of the zone of collisional stress, huge grabens have opened up, including the Baikal Rift, which contains the deepest lake in the world. A hot spot lies near the southern end of the Baikal Rift system, creating the Hangay Mountain Range in western Mongolia, which has been volcanically active within the past few thousand years—smack in the middle of Eurasia.

In Tibet, the northern prow of the Indian landmass has slipped beneath Central Asia along a series of Himalayan thrust faults, causing a doubling up of the lithosphere and uplift of the largest high-elevation plateau in the world. Lhasa, the capital of Tibet, lies in a fertile valley at an elevation of 3,700 meters (12,000 feet) above sea level.

The San Andreas Transform Boundary

The San Andreas fault is a 1,100-kilometer-long rupture marking the border between the Pacific and North American plates in California. But only about *one-third* of the approximately 2,000 kilometers of total slip between the two places, the biggest plates on Earth, has taken place along the fault during its 25 to 30-million-year history. How then do the plates actually move

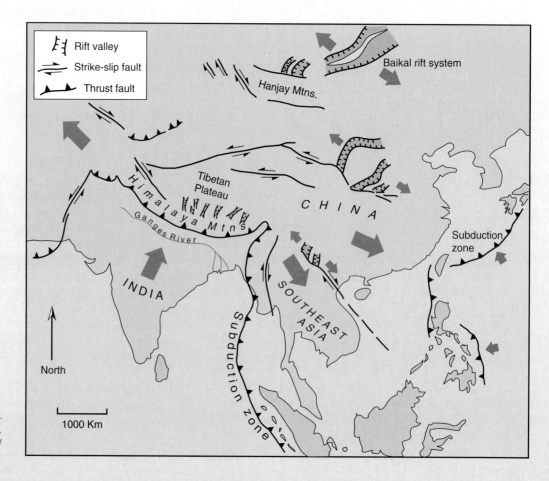

BOX 19.3 ■ FIGURE 1

Central Asia has adjusted to the broadside collision of India through uplift of the Tibetan plateau, and by stretching and slipping along major "intraplate" faults.

past one another in this region? The answer is that the San Andreas belongs to a much larger system of related parallel faults. Other ruptures, such as the Death Valley fault in eastern California and the Brothers fault zone in Oregon, also take up components of plate motion, so that the plate boundary is actually a *zone* of slippage about 600 kilometers wide rather than the single line you see on a map in an introductory geology textbook (box figure 2). The western side of North America is sliced up like a giant stack of dominoes—in other words, each domino slides past another in a right-lateral sense.

The San Andreas itself bends in places, so that it is not always parallel to the vectors of plate movement. In southern California, local plate convergence along the fault has shoved up the mountains bordering Los Angeles. Here, the San Andreas is a dynamic, evolving structure that will almost certainly wane as new faults inland more efficiently ease the plates past one another in the not-so-distant geological future.

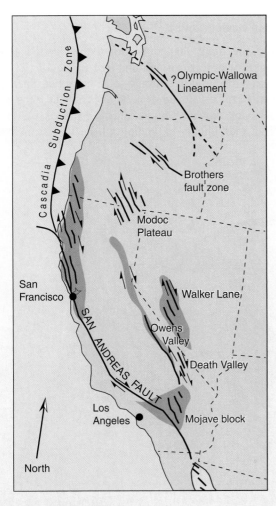

BOX 19.3 ■ FIGURE 2

Many faults participate in easing North America past the Pacific plate.

THE MOTION OF PLATE BOUNDARIES

Almost nothing is fixed in plate tectonics. Not only do plates move, but plate boundaries move as well. Plates may move away from each other at a divergent boundary on a ridge crest for tens of millions of years, but the ridge crest can be migrating across Earth's surface as this occurs. Ridge crests can also jump to new positions. The original ridge crest may suddenly become inactive; the divergence will jump quickly to a new position and create a new ridge crest (the evidence lies in the seafloor magnetic anomaly pattern).

Convergent boundaries migrate, also. As they do, trenches and magmatic arcs migrate along with the boundaries. Convergent boundaries can also jump; subduction can stop in one place and begin suddenly in a new place.

Transform boundaries change position, also. California's San Andreas fault has been in its present position about 5 million years. Prior to that, the plate motion was taken up on seafloor faults parallel to the San Andreas. In the future, the San Andreas may shift eastward again. The 1992 Landers earthquake, on a new fault in the Mojave Desert, and its pattern of aftershocks extending an astonishing 500 miles northward, suggest that the San Andreas may be trying to jump inland again. Geodetic studies have shown that more than 25% of the plate motion between the Pacific and North American plates is accommodated along faults in eastern California and western Nevada (see box 19.3, figure 2). If more motion is taken up along this zone, most of California will be newly attached to the Pacific plate instead of the North American plate, and California will slide northwestward relative to the rest of North America.

PLATE SIZE

Plates can change in size. For example, new sea floor is being added on the trailing edge of the North American plate at the spreading axis in the central Atlantic Ocean. Most of the North American plate is not being subducted along its leading edge because this edge is made up of lightweight continental rock. Thus, the North American plate is growing in size as it moves slowly westward.

The Nazca plate is getting smaller. The spreading axis is adding new rock along the trailing edge of the Nazca plate, but the leading edge is being subducted down the Peru-Chile Trench. If South America were stationary, the Nazca plate might remain the same size, because the rate of subduction and the rate of spreading are equal. But South America is slowly moving westward because of spreading on the Atlantic Ridge, pushing the Peru-Chile Trench in front of it. This means that the site of subduction of the Nazca plate is gradually coming closer to its spreading axis to the west, and so the Nazca plate is getting smaller. The same thing is probably happening to the Pacific plate as the Eurasian plate moves eastward into the Pacific Ocean.

THE ATTRACTIVENESS OF PLATE TECTONICS

The theory of plate tectonics is attractive to geologists because it can explain in a general way the distribution and origin of many Earth features. These features are discussed throughout this book, and we summarize them here.

The distribution and composition of the world's *volcanoes* can be explained by plate tectonics. *Basaltic* volcanoes and lava flows form at divergent plate boundaries when hot mantle rock rises at a spreading axis. *Andesitic* volcanoes, particularly those in the circum-Pacific belt, result from subduction of an oceanic plate beneath either a continental plate or another oceanic plate. Although most of the world's volcanoes occur at plate margins, some do not (Hawaii being an example). We will discuss some of these isolated volcanoes in the "Mantle Plumes and Hot Spots" section of this chapter.

Earthquake distribution and first motion can largely be explained by plate tectonics. Shallow-focus earthquakes along normal faults are caused by extension at divergent plate boundaries. Shallow-focus earthquakes also occur on transform faults when plates slide past one another. Broad zones of shallow-focus earthquakes are located where two continents collide. Dipping Benioff zones of shallow-, intermediate-, and deep-focus quakes are found along the giant thrust faults formed when an oceanic plate is subducted beneath another plate. Most of the world's earthquakes (like most volcanoes) occur along plate boundaries, although a few take place within plates and are difficult to explain in terms of plate tectonics.

Young mountain belts—with their associated igneous intrusions, metamorphism, and fold-thrust belts—form at convergent boundaries. "Subduction mountains" form at the edges of continents where sea floor is sliding under continents. Examples include the Andes and Cascade Mountains. "Continental-collision" mountains such as the Himalayas form in continental interiors when two continents collide to form a larger continent. Old mountain belts such as the Urals in Russia mark the position of old, now inactive, plate boundaries.

The major features of the sea floor can also be explained by plate tectonics. The *mid-oceanic ridge* with its rift valley forms at divergent boundaries. *Oceanic trenches* are found where oceanic plates are subducted at convergent boundaries. *Fracture zones* are created at transform boundaries.

WHAT CAUSES PLATE MOTIONS?

A great deal of speculation currently exists about why plates move. There may be several reasons for plate motion. Any mechanism for plate motion has to explain why:

1. mid-oceanic ridge crests are hot and elevated, while trenches are cold and deep;

2. ridge crests have tensional cracks; and

3. the leading edges of some plates are subducting sea floor, while the leading edges of other plates are continents (which cannot subduct).

There is no doubt that convection in the mantle is linked in some crucial way to plate motions (see figure 19.12). Mantle convection—the slow overturning of Earth's hot, ductile interior as heated rock wells up from below, cools near the surface, and sinks back down again—could take place as a series of giant cells, individually extending all the way from the heat

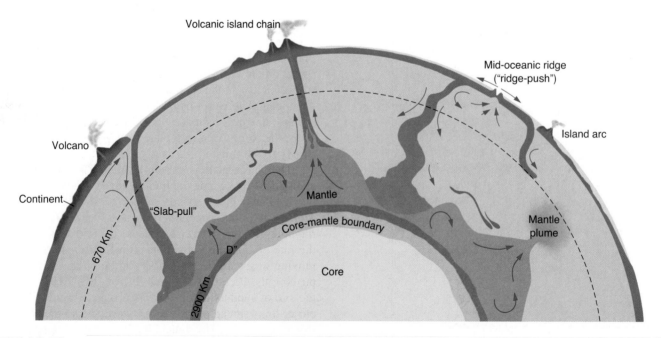

FIGURE 19.29

A possible model of mantle convection. *Modified from L. H. Kellogg, B. H. Hager, and R. D. van der Hilst, 1999, Science, 283:1881–84* source at the core-mantle boundary to the base of the litho-sphere itself. Recent studies using seismic tomography and computer modeling indicate that this idea of "whole mantle convection" is too simplistic, however (figure 19.29). Change in density with depth in the Earth, the property of large continents to trap mantle heat, and the "stirring" of the mantle from the sinking of subducted oceanic lithosphere all contribute to a more complex pattern of convective heat loss. Cold lithospheric plates may subduct down to the core-mantle boundary, whereas other, less-dense (younger) plates may only reach the 670-kilometer boundary. One of the most recent models suggests that the lowermost part of the mantle does not mix with the upper and middle mantle but acts as a "lava lamp" turned on low, fueled by internal heating and heat flow across the core-mantle boundary. Variation in the thickness of this dense layer may control where mantle plumes rise and subducted plates ultimately rest.

Some geologists think that mantle convection is a *result* of plate motion rather than a cause of it. The sinking of a cold, subducting plate can create mantle convection (convection can be driven by either hot, rising material or by cold, sinking material). Hot mantle rock rises at divergent boundaries to take the place of the diverging plates; however, such plate-caused convection would be shallow rather than mantle-deep.

The basic question in plate motion is, why do plates diverge and sink? Two or three different mechanisms may be at work here.

One proposal is called *"ridge-push."* As a plate moves away from a divergent boundary, it cools and thickens. Cooling sea floor subsides as it moves, and this subsidence forms the broad side slopes of the mid-oceanic ridge. An even more important slope forms on the base of the lithosphere mantle. The mantle thickens as cooling converts asthenospheric mantle to lithospheric mantle. Therefore, the boundary between them is a slope down which the lithosphere slides

(figure 19.30). The oceanic plate is thought to slide down this slope at the base of the lithosphere, which may have a relief of 80 to 100 kilometers.

Another mechanism is called *"slab-pull"* (figure 19.30). Cold lithosphere sinking at a steep angle through hot mantle should pull the surface part of the plate away from the ridge crest and then down into mantle as it cools. A subducting plate sinks because it is denser than the surrounding mantle. This density contrast is partly due to the fact that the sinking lithosphere is cold. The subducting plate may also increase its density while it sinks, as low-density materials such as water are lost and as plate minerals collapse into denser forms during subduction. Slab-pull is thought to be at least twice as important as ridge-push in moving an oceanic plate away from a ridge crest. Slab-pull causes rapid plate motion.

If subducting plates fall into the mantle at angles steeper than their dip (figure 19.30), then trenches and the overlying plates are pulled horizontally seaward toward the subducting plates. This mechanism has been termed *"trench-suction."* It is probably a minor force but it may be important in moving continents apart. Divergent continents at the leading edges of plates cannot be moved by slab-pull, because they are not on subducting plates. They might be moved by ridge-push from the rear, or trench-suction from the front, or both (figure 19.30). They move much more slowly than subducting plates.

All three of these mechanisms (ridge-push, slab-pull, and trench-suction), particularly in combination, are compatible with high, hot ridges; cold, deep trenches; and tensional cracks at the ridge crest. They can account for the motion of both oceanic and continental plates. In this scheme, plate motions are controlled by variations in lithosphere density and thickness, which, in turn, are controlled largely by cooling. In other words, the reasons for plate motions are the properties of the plates themselves and the pull of gravity. This idea is in sharp contrast to most convection models, which assume that plates are dragged along by the movement of mantle rock beneath the plates.

FIGURE 19.30

Other possible mechanisms for plate motion. Plates are pushed apart at the ridge *(ridge-push)* by sliding downhill on the sloping boundary between the lithosphere and asthenosphere. Plates may also be pulled *(slab-pull)* as the dense leading edge of a subducting plate sinks down into the asthenosphere. If the subducting plate falls into the asthenosphere at angles steeper than its dip *(slab rollback)* then the trench and overlying plate are pulled horizontally seaward toward the subducting plate by *trench suction.*

Mantle Plumes and Hot Spots

A modification of the convection process was suggested by W. Jason Morgan of Princeton University. Morgan proposed that convection occurs in the form of **mantle plumes,** narrow columns of hot mantle rock that rise through the mantle, much like smoke rising from a chimney (figure 19.31). Mantle plumes are now thought to have large, spherical or mushroom-shaped heads above a narrow, rising tail. They are essentially stationary with respect to moving plates and to each other.

Plumes may form "hot spots" of active volcanism at Earth's surface. Note in figure 19.32 that many hot spots are located in volcanic regions such as Iceland, Yellowstone, and Hawaii. Recent seismic tomography images of the mantle suggest that not all hot spots are fed by mantle plumes. Of the forty-five hot spots identified on Earth, only twelve show evidence of a deep, continuous plume in the underlying mantle.

According to one hypothesis, when the head of a large plume ("super plume") nears the surface, it may cause uplift and the eruption of vast fields of flood basalts. As the head widens beneath the crust, the flood-basalt area widens and the crust is stretched. The tail that follows the head produces a narrow spot of volcanic activity, much smaller than the head.

The outward, radial flow of the expanding head may be strong enough to break the lithosphere and start plates moving. In Morgan's view, a few plumes, such as those underlying some of the hot spots on the mid-oceanic ridge in the Atlantic Ocean in figure 19.32, are enough to drive plates apart (in this case, to push the American plates westward). Lithospheric tension set up by trench suction or slab-pull could combine with mantle plume action to break a large plate (such as the former Pangaea) into smaller, diverging fragments.

A mantle plume rising beneath a continent should heat the land and bulge it upward to form a dome marked by volcanic eruptions. As the dome forms, the stretched crust typically fractures in a three-pronged pattern (figure 19.33). Continued radial flow outward from the rising plume eventually separates the crust along two of the three fractures but leaves the third fracture inactive. In this model of continental breakup, the two active fractures become continental edges as new sea floor forms between the divergent continents. The third fracture is a *failed rift* (or *aulacogen*), an inactive rift that becomes filled with sediment.

An example of this type of fracturing may exist in the vicinity of the Red Sea (figure 19.33*D*). The Red Sea and the Gulf of Aden are active diverging boundaries along which the Arabian Peninsula is being separated from northeastern Africa. The third, less-active rift is the northernmost African Rift Valley, lying at an angle about 120° to each of the narrow seaways.

Figure 19.34 shows how two plumes might split a continent and begin plate divergence. Local uplift *causes* rifting over each plume. The rifts lengthen with time until the land is torn in two. The two halves begin to diverge from being dragged along from below by the outward radial flow of the plume. Along the long rift segments between plumes, rifting occurs *before* uplift.

Some plumes rise beneath the centers of oceanic plates. A plume under Hawaii rises in the center of the Pacific plate. As the plate moves over the plume, a line of volcanoes forms, creating an aseismic ridge (figure 19.35). The volcanoes are gradually carried away from the eruptive center, isostatically sinking as they go because of cooling. The result is a line of extinct volcanoes (seamounts and guyots) increasing in age away from an active volcano directly above the plume.

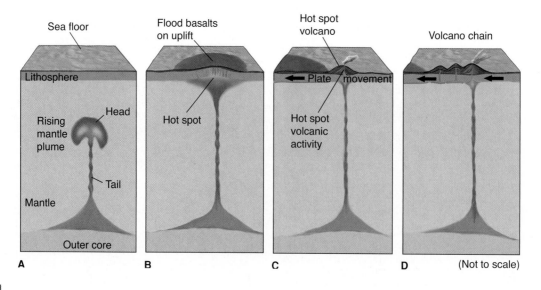

FIGURE 19.31

Model of mantle plume rising upward through the mantle to form a hot spot and associated flood basalts and volcanic chain. (*A*) Rising mantle plume contains a hot, mushroom-shaped plume head and a narrow tail. (*B*) Plume head forms a broad hot spot when it reaches the top of the mantle and causes uplift and stretching of the crust and eruption of flood basalts. (*C*) When the tail rises to the surface, a narrower hot spot forms a volcano. (*D*) Continued plate motion over the hot spot creates a trail or chain of volcanoes.

FIGURE 19.32

Distribution of hot spots, identified by volcanic activity and structural uplift within the past few million years. The hot spots near the poles are not shown.

FIGURE 19.33

Continental breakup caused by a mantle plume. (*A*) A dome forms over a mantle plume rising beneath a continent. (*B*) Three radial rifts develop due to outward radial flow from the top of the mantle plume. (*C*) Continent separates into two pieces along two of the three rifts, with new ocean floor forming between the diverging continents. The third rift becomes an inactive "failed rift" (or aulacogen) filled with continental sediment. (*D*) An example of radial rifts. The Red Sea and the Gulf of Aden are the active rifts, as the Arabian peninsula drifts away from Africa. The Gulf of Aden contains a mid-oceanic ridge and central rift valley. The less active, failed rift (aulacogen) is the rift valley shown in Africa.

A

B

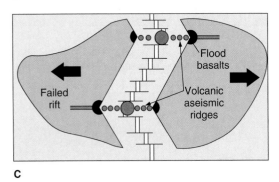
C

In the Hawaiian island group, the only two active volcanoes are in the extreme southeastern corner (figure 19.36). The isotopic ages of the Hawaiian basalts increase regularly to the northwest, and a long line of submerged volcanoes forms an aseismic ridge to the northwest of Kauai (see figure 18.17). Most aseismic ridges on the sea floor appear to have active volcanoes at one end, with ages increasing away from the eruptive centers. Deep-sea drilling has shown, however, that not all aseismic ridges increase in age along their lengths. This evidence has led to alternate hypotheses for the origin of aseismic ridges. It may pose difficulties for the plume hypothesis itself.

FIGURE 19.34

(A) Two mantle plumes beneath a continent. (B) The rifts lengthen and flood basalts erupt over the plumes. (C) The continent splits and failed rifts form. The new ocean is marked by ridge crests, fracture zones, and aseismic ridges (chains of volcanoes).

FIGURE 19.35

Fringing reef forms around volcanic island as it moves off hot spot. Waves erode and flatten top of volcanic islands to form guyots that progressively sink and become submerged away from hot spot.

FIGURE 19.36

Ages of volcanic rock of the Hawaiian island group. Ages increase to northwest. Two active volcanoes on Hawaii are shown by red dots. The plume is currently offshore under the Loihi seamount (red dot), where recent underwater eruptions have been documented.

EARTH SYSTEMS 19.4

The Relationship between Plate Tectonics and Ore Deposits

The plate-tectonic theory provides an overall model for the origin of metallic ore deposits that has been used to explain the occurrence of known deposits and in exploration for new deposits. Because many ore deposits are associated with igneous activity, a close relationship exists between plate boundaries and metallic ore deposits.

As discussed in chapter 18, *divergent plate boundaries* are often marked by lines of active hot springs in rift valleys that carry and precipitate metallic minerals in mounds around the hot springs. The metals in rift-valley hot springs are predominantly iron, copper, and zinc, with smaller amounts of manganese, gold, and silver. Although the mounds are nearly solid metal sulfide, they are small and widely scattered on the sea floor, so commercial mining of them may not be practical. Occasionally, the ore minerals may be concentrated in richer deposits. On the floor of the Red Sea, metallic sediments have precipitated in basins filled with hot-spring solutions. Although the solutions are hot (up to 60°C or 140°F), they are very dense because of their high salt content (they are seven times saltier than seawater), so they collect in seafloor depressions instead of mixing with the overlying seawater. Although not currently mined, the metallic sediments have an estimated value of $25 billion.

Hot metallic solutions are also found along some divergent continental boundaries. Near the Salton Sea in southern California, which lies along the extension of the mid-oceanic ridge inland, hot water very similar to the Red Sea brines has been discovered underground. The hot water is currently being used to run a geothermal power plant. The high salt and metal content is corrosive to equipment, but metals such as copper and silver may one day be recovered as valuable by-products.

Seafloor spreading carries the metallic ores away from the ridge crest (box figure 1), perhaps to be subducted beneath island arcs or continents at *convergent plate boundaries*. Slivers of *ophiolite* on land may contain these rich ore minerals in relatively intact form. A notable example of such ores occurs on the island of Cyprus in the Mediterranean Sea (box figure 2). Banded chromite ores may also be contained in the serpentinized ultramafic rock at the bottom of ophiolites.

Volcanism at *island arcs* can also produce hot-spring deposits on the flanks of the andesitic volcanoes. Pods of very rich ore collect above local bodies of magma, and the ore is sometimes distributed as sedimentary layers in shallow basins (box figure 3).

It is tempting to think that *mantle plumes* might cause ore deposition, for plumes provide a source of both magma and hydrothermal solutions. The locations of supposed plumes, however, such as Yellowstone and Hawaii, are notable for their *lack* of ore deposits.

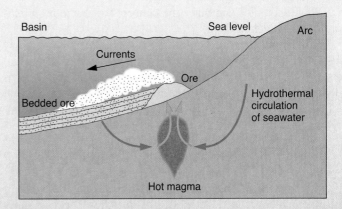

BOX 19.4 ■ FIGURE 2

Second largest copper mine in Cyprus, Greece mines copper from the Troodos ophiolite. Copper deposits were initially formed on the sea floor where active hot springs precipitated metallic minerals in rift valleys where the oceanic lithosphere was being pulled apart. *Photo © Jonathan Blair/Corbis*

BOX 19.4 ■ FIGURE 1

Divergent oceanic plates carry metallic ores away from rift valley. (Size of ore deposits is exaggerated.)

BOX 19.4 ■ FIGURE 3

On island arcs, metallic ores can form over hot springs and be redistributed into layers by currents in shallow basins.

A FINAL NOTE

Geologists, like other people, are susceptible to fads. Most geologists believe plate tectonics is an exciting theory and accept it as a working model of Earth. There will undoubtedly continue to be modifications and refinements to the theory. But widespread belief in a theory does not make it true. Two hundred years ago, geologists "knew" that basalt crystallized out of seawater. In the 1800s, glacial deposits were thought to be deposited by Noah's flood. Both of these incorrect ideas were finally disproved by decades of exacting field work and often bitter debate.

Forty years ago, continental drift rated only a footnote in most introductory textbooks. Now, good evidence exists of continental motion, and textbooks use it as a framework for the entire field of physical geology. Although the idea of continental stability provided the framework for many past textbooks, today the idea that continents are fixed in position rates only a footnote as an outmoded concept.

Objections were raised to the concept of plate tectonics after it was proposed in the late 1960s. Some seafloor features did not seem compatible with a moving sea floor. The geology of many continental regions did not seem to fit into the theory of plate tectonics—in some cases, not even slightly (see box 19.3). But a revolutionary idea in science is always controversial. As it progresses from an "outrageous hypothesis" to a more widely accepted theory, after much discussion and testing, a new idea evolves and changes. The newness of the idea wears off, and successful tests and predictions convert skeptics to supporters (sometimes grudgingly). Perhaps equally important, dissenters die off.

As refinements were made to plate tectonics and as more was learned about the puzzling seafloor features and continental regions, they began to seem more compatible with plate tectonics. Objections died out, and plate tectonics became widely accepted.

It is wise to remember that at the time of Wegener, most geologists vehemently disagreed with continental drift. Because Wegener proposed that continents plow through seafloor rock and because his proposed forces for moving continents proved inadequate, most geologists thought that continental drift was wrong. Although these geologists had sound reasons for their dissent, we now know, due to overwhelming evidence, that lithospheric plates move and the early geologists were wrong.

The evidence for plate tectonics is very convincing. The theory has been rightly called a revolution in Earth science, comparable to the development of the theory of evolution in the biological sciences. It has been an exciting time to be a geologist. Our whole concept of Earth dynamics has changed in the last forty years.

SUMMARY

Plate tectonics is the idea that Earth's surface is divided into several large plates that change position and size. Intense geologic activity occurs at plate boundaries.

Plate tectonics combines the concepts of *seafloor spreading* and *continental drift.*

Alfred Wegener proposed continental drift in the early 1900s. His evidence included coastline fit, similar fossils and rocks in now-separated continents, and paleoclimatic evidence for *apparent polar wandering.* Wegener proposed that all continents were once joined together in the supercontinent *Pangaea.*

Wegener's ideas were not widely accepted until the 1950s, when work in paleomagnetism revived interest in polar wandering.

Evidence for continental drift includes careful fits of continental edges and detailed rock matches between now-separated continents. The positions of continents during the past 200 million years have been mapped.

Hess's hypothesis of *seafloor spreading* suggests that the sea floor moves away from the ridge crest and toward trenches as a result of mantle convection.

According to the concept of seafloor spreading, the high heat flow and volcanism of the ridge crest are caused by hot mantle rock rising beneath the ridge. Divergent *convection* currents in the mantle cause the rift valley and earthquakes on the ridge crest, which is a *spreading axis* (or *center*). New sea floor near the rift valley has not yet accumulated pelagic sediment.

Seafloor spreading explains trenches as sites of seafloor *subduction,* which causes low heat flow and negative gravity anomalies. Benioff zones and andesitic volcanism are caused by interaction between the subducting sea floor and the rocks above.

Seafloor spreading also explains the young age of the rock of the sea floor as caused by the loss of old sea floor through subduction into the mantle.

Plates are composed of blocks of *lithosphere* riding on a ductile *asthenosphere.* Plates move away from spreading axes, which add new sea floor to the trailing edges of the plates.

An apparent confirmation of plate motion came in the 1960s with the correlation of marine *magnetic anomalies* to *magnetic reversals* by Vine and Matthews. The origin of magnetic anomalies at sea apparently is due to the recording of normal and reverse magnetization by dikes that intrude the crest of the mid-oceanic ridge, then split and move sideways to give anomaly patterns a mirror symmetry.

The Vine-Matthews hypothesis gives the rate of plate motion and can predict the age of the sea floor before it is sampled.

Deep-sea drilling has apparently verified plate motions and the age predictions made from magnetic anomalies.

Earthquake distribution and first-motion studies on *transform faults* on fracture zones also verify plate motions.

Divergent plate boundaries are marked by rift valleys, shallow-focus earthquakes, high heat flow, and basaltic volcanism.

Transform boundaries between plates sliding past one another are marked by strike-slip (transform) faults and shallow-focus earthquakes.

Convergent plate boundaries can cause *subduction* or *continental collision*. Subducting plate boundaries are marked by trenches, low heat flow, Benioff zones, andesitic volcanism, and young mountain belts or island arcs. Continental-collision boundaries have shallow-focus earthquakes and form young mountain belts in continental interiors.

The distribution and origin of most volcanoes, earthquakes, young mountain belts, and major seafloor features can be explained by plate tectonics.

Plate motion was once thought to be caused by *mantle convection* but is now attributed to the cold, dense, leading edge of a subducting plate pulling the rest of the plate along with it *(slabpull)*. Plates near mid-oceanic ridges also slide down the sloping lithosphere-asthenosphere boundary at the ridge *(ridgepush)*. *Trench-suction* may help continents diverge.

Mantle plumes are narrow columns of hot, rising mantle rock. They cause flood basalts and may split continents, causing plate divergence.

An aseismic ridge may form as an oceanic plate moves over a mantle plume acting as an eruptive center (hot spot).

Terms to Remember

asthenosphere 500
continental drift 492
convection 498
convergent plate boundary 500
divergent plate boundary 500
island arc 510

lithosphere 500
magmatic arc 511
mantle plume 518
plate 500
plate tectonics 492
polar wandering 494

seafloor spreading 492
subduction 498
transform fault 503
transform plate boundary 500

Testing Your Knowledge

Use the following questions to prepare for exams based on this chapter.

1. What was Wegener's evidence for continental drift?

2. What is polar wandering? What is the paleoclimatic evidence for polar wandering? What is the magnetic evidence for polar wandering? Does polar wandering require the poles to move?

3. What is the evidence that South America and Africa were once joined?

4. In a series of sketches, show how the South Atlantic Ocean might have formed by the movement of South America and Africa.

5. What is Pangaea?

6. In a single cross-sectional sketch, show the concept of seafloor spreading and how it relates to the mid-oceanic ridge and oceanic trenches.

7. How does seafloor spreading account for the age of the sea floor?

8. What is a plate in the concept of plate tectonics?

9. Define *lithosphere* and *asthenosphere*.

10. What is the origin of marine magnetic anomalies according to Vine and Matthews?

11. Why does the pattern of magnetic anomalies at sea match the pattern of magnetic reversals (recorded in lava flows on land)?

12. How has deep-sea drilling tested the concept of plate motion?

13. How has the study of fracture zones tested the concept of plate motion?

14. Explain how plate tectonics can account for the existence of the mid-oceanic ridge and its associated rift valley, earthquakes, high heat flow, and basaltic volcanism.

15. Explain how plate tectonics can account for the existence of oceanic trenches as well as their low heat flow, their negative gravity anomalies, the associated Benioff zones of earthquakes, and andesitic volcanism.

16. What is a transform fault?

17. Discuss possible driving mechanisms for plate tectonics.

18. Describe the various types of plate boundaries and the geologic features associated with them.

19. What is a mantle plume? What is the geologic significance of mantle plumes?

20. The southern supercontinent is called
    a. Gondwanaland
    c. Laurasia
    b. Pangaea
    d. Glossopteris

21. The sliding of the sea floor beneath a continent or island arc is called
    a. rotation
    c. subduction
    b. tension
    d. polar wandering

22. In cross section, the plates are part of a rigid outer shell of the Earth called the
    a. lithosphere
    c. crust
    b. asthenosphere
    d. mantle

23. The Vine-Matthews hypothesis explains the origin of
    a. polar wandering
    b. seafloor magnetic anomalies
    c. continental drift
    d. mid-ocean ridges

24. The San Andreas fault in California is a
    a. normal fault
    c. transform fault
    b. reverse fault
    d. thrust fault

25. What would you most expect to find at ocean-ocean convergence?
    a. suture zone
    c. mid-ocean ridge
    b. island arc

26. What would you most expect to find at ocean-continent convergence?
    a. magmatic arc
    c. island arc
    b. suture zone
    d. mid-ocean ridge

27. What would you most expect to find at continent-continent convergence?
    a. magmatic arc
    c. island arc
    b. suture zone
    d. mid-ocean ridge

28. Passive continental margins are created at
    a. divergent plate boundaries
    b. transform faults
    c. convergent plate boundaries

29. The Hawaiian Islands are thought to be the result of
    a. subduction
    b. mid-ocean ridge volcanics
    c. mantle plumes
    d. ocean-ocean convergence

30. Metallic ores are created at diverging plate boundaries
    a. through hydrothermal processes
    b. in lava flows
    c. in sedimentary deposits
    d. through metamorphism

Expanding Your Knowledge

1. Plate tectonics helps cool Earth as hot mantle rock rises near the surface at ridge crests and mantle plumes. What can we assume about the internal temperature of other planets that do not seem to have plate tectonics? What would happen to Earth's internal temperature if the plates stopped moving?

2. Are ridge offsets along fracture zones easier to explain with mantle-deep convection *causing* plate motion or with shallow convection occurring as a *result* of plate motion?

3. Why are mantle plumes narrow? What conditions at the core-mantle boundary could cause the formation and rise of a mushroom-shaped plume?

4. The slab-pull and ridge-push mechanisms of plate motion may operate only after a plate starts to move. What starts plate motion?

5. If subducting plates can penetrate the 670-kilometer mantle boundary and sink all the way to the base of the mantle, why are there no earthquakes deeper than 670 kilometers?

Exploring Web Resources

This is the dedicated website for this book. You can go to it for new and updated information. The universal resource locators (URL) listed in this book are also given as links on the website, making it easy to go to those websites without typing in the URL. When you visit our Online Learning Center, go to the Student Center and then to the chapter of interest. Here you will find additional readings and media resources, as well as answers to the Testing Your Knowledge section, more quizzes, animation and video clips, and direct links to the sites listed in this book. Links to additional websites can also be found. We have added questions for some of the sites to allow you to get the most of your exploration of the web. Using the web is an enjoyable way of enhancing your knowledge of geology.

http://pubs.usgs.gov/publications/text/dynamic.html

This Dynamic Earth: The Story of Plate Tectonics. U.S. Geological Survey online book by W. J. Kious and R. Tilling provides general information about plate tectonics.

http://cddisa.gsfc.nasa.gov/926/slrtecto.html

Tectonic Plate Motion (explains how plate motion is calculated).

http://vishnu.glg.nau.edu/rcb/globaltext.html

View images of plate-tectonic reconstructions by R. Blakely at Northern Arizona University.

www.scotese.com

View images and animations by C. R. Scotese of the assembly and breakup of Pangaea through time.

Animations

This chapter includes the following animations available on our Online Learning Center at **www.mhhe.com/plummer11e**.

- 19.12 Seafloor spreading
- 19.14 Magnetic reversals at MO Ridge
- 19.16 How seafloor spreading creates magnetic polarity stripes
- 19.17 Age of ocean floor
- 19.18 Transform faults
- 19.20 Continental rifting and early drift
- 19.25 Convergence of plates—Ocean-Ocean
- 19.27 Convergence of plates—Ocean-Continent
- 19.28 Convergence of plates—Continent-Continent
- 19.35 Formation of Hawaiian Island chain by hotspot volcanism

Mountain Belts and the Continental Crust

Mountain belts can evolve from marine-deposited rocks to towering peaks during periods of tens of millions of years. Ultimately, the peaks are eroded to plains and become part of the stable interior of a continent. To appreciate the long and complex process of mountain building, you need to know much of the material covered in previous chapters. For instance, you must understand structural geology to appreciate what a particular pattern of folds and faults can tell us about the history of mountain building in a particular region. To understand how the rocks formed during the various stages of a mountain belt's history, you must know about volcanism, plutonism, sedimentation, and metamorphism. Your earlier study of weathering and erosion will help you understand how mountains are worn away.

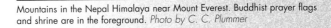

Mountains in the Nepal Himalaya near Mount Everest. Buddhist prayer flags and shrine are in the foreground. *Photo by C. C. Plummer*

527

Plate-tectonic theory has been strikingly effective in helping geologists make sense of often complex aspects of mountain belts and the continental crust. For this reason, you may need to go back to some of the material in chapter 19 to appreciate how continents evolve.

In this chapter, we first point out what geologists have observed of mountain belts. Next, we describe how these observations are interpreted, particularly in light of plate-tectonic theory. Finally, we discuss current perceptions of how continents change and grow.

INTRODUCTION

A mountain, as you know, is a large terrain feature that rises more or less abruptly from surrounding levels. Volcanoes are mountains, so are erosional remnants of plateaus (mesas). In this chapter, we will not focus on individual mountains; rather, we are concerned here with Earth's **major mountain belts,** chains thousands of kilometers long composed of numerous mountain ranges. A **mountain range** is a group of closely spaced mountains or parallel ridges (figure 20.1). A mountain range is likely to be composed of tectonically deformed sedimentary, volcanic, or metamorphic rocks. It may also show a history of intrusive igneous activity.

The map in figure 20.2 shows that most of the world's mountains are in long chains that extend for thousands of kilometers. The Himalaya, the Andes, the Alps, and the Appalachians are examples of major mountain belts, each comprising numerous mountain ranges.

Geologists find working in mountain ranges to be physically challenging and intellectually stimulating. High mountains have steep faces and broad exposures of bedrock. This is good because they allow a geologist to decipher complex interrelationships between rock units. But the geologist may have to become a proficient mountain climber to access the good exposures. (Conversely, mountain climbers who develop an interest in the rocks they climb sometimes become geologists.) On the other hand, exposures of bedrock critical to interpreting the local geology may be buried beneath glaciers or talus from rockfall. Furthermore, even in the highest and best exposed mountains, we never see bedrock representative of all of a mountain range. Significant amounts of rock (usually thousands of meters) once overlying the rocks we now see have been eroded away. Moreover, the present exposures are like the proverbial tips of icebergs—there is much more rock below the exposed mountain range that we cannot observe. For instance, the Himalaya, Earth's highest mountain range, rise to 8,000 meters above sea level; yet their roots (Earth's crust beneath the mountains) extend downward 65,000 meters. In other words, at best, we have exposed to us less than one-eighth of the thickness of a mountain range.

Our models of how major mountain belts evolve use data from over a century of studying the geologic structures and rocks exposed in the world's many mountain ranges. Often, a particular study aims to piece together the geologic history of a single mountain range or part of a range. In other field studies, a geologist focuses on a particular type of rock exposed in a mountain range. For instance, a geologist may study the variations in metamorphic rocks in a mountainous area to determine the temperature, depth of burial, and nature of deformation during metamorphism. Geologists working on the "big picture," developing hypotheses of how major mountain belts evolve, might employ the published results of hundreds or thousands of local studies, using them as pieces of a puzzle. (Science works largely because scientists build on the work of others.) Models that currently are widely accepted regarding the evolution of mountain belts are cast within the broader framework of plate-tectonic theory and will be described later in this chapter.

Mountain belts differ from one another because each has undergone a unique combination of events that contributed to its present characteristics. The major controlling factors that interact with one another during a mountain belt's long history are:

- *Intense deformation.* This is mainly compression and results in intense folding and faulting of sedimentary and volcanic rocks. At depth, deformation results in foliation accompanying metamorphism. Such an episode (usually lasting millions of years) of intense deformation is known as an **orogeny.** We now attribute orogenies mainly to plate convergence.

- *Isostasy.* Vertical movement of mountain belts, both during and after an orogeny, is accounted for by isostasy (described in chapters 1 and 17). Isostatic adjustment means that thicker continental crust tends to "float" higher on the mantle than does thinner crust.

- *Weathering and erosion.* The rate and nature of weathering and erosion are affected by many factors, such as the climate, type of rock, and height of a landmass above sea level. An Earth systems approach is used to understand the interaction of the geosphere, atmosphere, and hydrosphere in forming mountains (see box 20.1).

FIGURE 20.1

View of glaciated peaks in one of the mountain ranges in the Andes mountain belt. A parallel but much lower range is visible at the extreme right skyline of the picture. *Photo by C. C. Plummer*

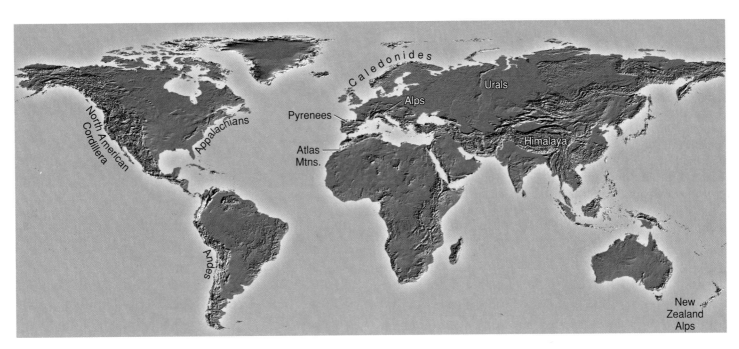

FIGURE 20.2

Map of the world showing major mountain belts.

A System Approach to Understanding Mountains

During recent years, geologists have used a system approach to gain insight into the growth and wearing away of mountains. This approach regards mountains as products of three closely interdependent components: (1) tectonics (plate tectonics and isostasy), (2) climate, and (3) erosion.

The tendency in the past has been to concentrate on tectonics to explain the growth of a mountain belt and to relegate climate and erosion to relatively minor roles. Through mountain system analyses, we gain insight into the extent to which each of the three components interacts with and changes the other two components. Climate influences erosion in obvious ways. For instance, if there is a wet climate, there will be erosion due to abundant running water at lower elevation and heavy glaciation at higher elevation. If the climate is arid, erosion will be much slower. Tectonics affects climate because if a region is uplifted to a high elevation, the climate there will be cold and glaciers can develop. With less uplift and lower mountains, erosion will be mainly due to running water. A moist climate can also result in heavy vegetation at lower elevations, which would tend to retard erosion.

Erosion and climate can influence tectonics as well. For example, the extent and type of erosion can help determine whether a highland grows higher or lower with time. If a high plateau, dissected by only a few valleys, undergoes erosion, the plateau is eroded downward uniformly (box figure 1). Following erosion, isostasy results in the plateau floating upward but not up to its original level. Its average surface, which essentially is its actual surface, is at a lower elevation than before erosion took place. If erosion carves many deep valleys and leaves relatively few mountains between the valleys, the entire regional block will have less mass and will float isostatically upward. As in the case of the plateau, its average surface would rise to a level lower than before erosion; however, its average surface is somewhere between the peaks and bottoms of valleys. Although the average height of the block rises to a level below its previous average height, the mountains rise to heights greater than before.

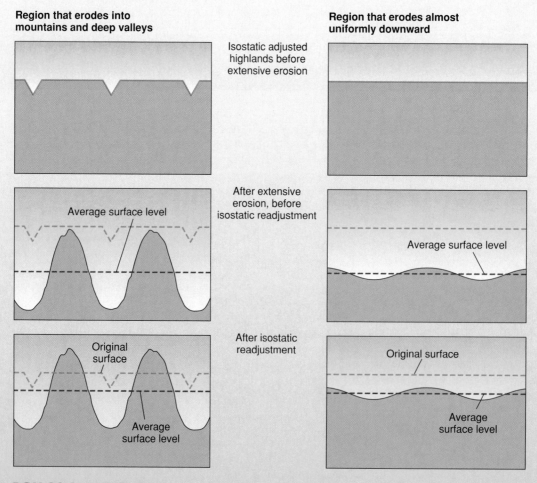

Region that erodes into mountains and deep valleys

Region that erodes almost uniformly downward

Isostatic adjusted highlands before extensive erosion

After extensive erosion, before isostatic readjustment

Average surface level

Average surface level

After isostatic readjustment

Original surface

Average surface level

Original surface

Average surface level

BOX 20.1 ■ FIGURE 1

Comparison between two regions before extensive erosion, after extensive erosion, and after isostatic readjustment. Region on the left erodes into mountains and deep valleys. Region on the right remains a plateau after approximately uniform erosion. Steepness of mountains is exaggerated.

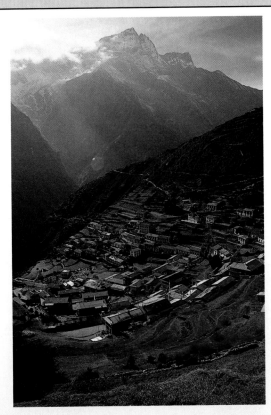

BOX 20.1 ■ FIGURE 2

Namche Bazaar in the Nepalese Himalaya. Mountain peaks rise thousands of meters above the town. Streams have carved deep valleys below the town. *Photo by C. C. Plummer*

Climate enters the picture because, interacting with tectonics, it helps control the type and extent of weathering that takes place. For instance, heavy precipitation takes place in the Himalaya because of the flow of very humid air from the south during the summer monsoon. At the higher elevations, the precipitation in the form of heavy snowfall contributes to extensive and very active glaciation. As described in chapter 12 on glaciation, glaciers are extremely effective agents of erosion. Sharp peaks are separated by glacially carved valleys. At a lower level, streams fed by meltwater (and rainfall) deepen stream-carved valleys (box figure 2). The mountains will grow higher during isostatic adjustment at the same time the region as a whole is lowered by erosion. The Tibetan Plateau is north of the Himalaya. It is the highest, largest plateau in the world, with an average elevation of around 5 kilometers—higher than any mountain in the United States except for Alaska. The plateau has not been carved into mountains because the climate is quite different from that of the Himalaya. The moist air from the south is blocked by the Himalaya and the Tibetan Plateau is in its rain shadow (see chapter 13). Without water, there are no glaciers or large rivers to carve the plateau into mountains and valleys. So this region is slowly being eroded downward, getting progressively lower as erosion and isostasy balance each other out.

Additional Resource

N. Pinter and M. T. Brandon. How erosion builds mountains. *Scientific American* (April 1997): pp. 74–79.

CHARACTERISTICS OF MAJOR MOUNTAIN BELTS

Size and Alignment

Major mountain belts are very long compared to their width. Figure 20.3 shows the two major mountain belts of North America, the *Appalachian Mountains* along the East Coast and the *North American Cordillera* in the West. Some of the better known ranges in the North American Cordillera, such as the Sierra Nevada and the Rocky Mountains, are labeled. Note that the mountain belts in North America tend to be parallel to the coast lines. However, some mountain belts elsewhere in the world (most notably the Himalaya) are not parallel to a coastline.

Ages of Mountain Belts and Continents

Major mountain belts with higher mountain ranges tend to be geologically younger than those where mountains are lower. Mountain building (orogeny) for the Himalaya, Earth's highest mountain belt, began only about 45 million years ago and is still taking place. Mountain building, other than isostatic adjustment, in the much lower Appalachians ceased around 250 million years ago. Individual ranges within a mountain belt, however, may vary considerably in height even though they are about the same age.

Mountain regions commonly show evidence that they were once high above sea level during an orogeny, were eroded to hills or low plains, and then rose again in a later episode of isostatic uplift. Such episodes of uplift and erosion may occur a number of times during the long history of a mountain range. Ultimately, mountain ranges stabilize and are eroded to plains.

On the North American continent, the Appalachian Mountains extend from eastern Canada southward through the eastern United States into Alabama (figure 20.3). In the Appalachians, fossils and isotopic ages of rocks indicate that these mountains began to evolve earlier than the mountain belt along the western coast of North America. The interior plains between the Appalachians and the Cordillera are considered to have evolved from mountain belts in the very distant geologic past (during the Precambrian). The once deep-seated roots of the former Precambrian mountain belts are the *basement* rock for the now stable, central part of the continent. Layers of Paleozoic and younger sedimentary rock cover most of that basement. The great age of the orogenic episodes that preceded the Paleozoic sedimentation is confirmed by isotopically determined dates of over 1 billion years obtained from plutonic and metamorphic rocks in the few scattered locations where the basement is exposed. (The most noteworthy are the Grand Canyon in Arizona, the Ozark dome in Missouri, the Black Hills of South Dakota, some ranges in the Rocky Mountains, and the

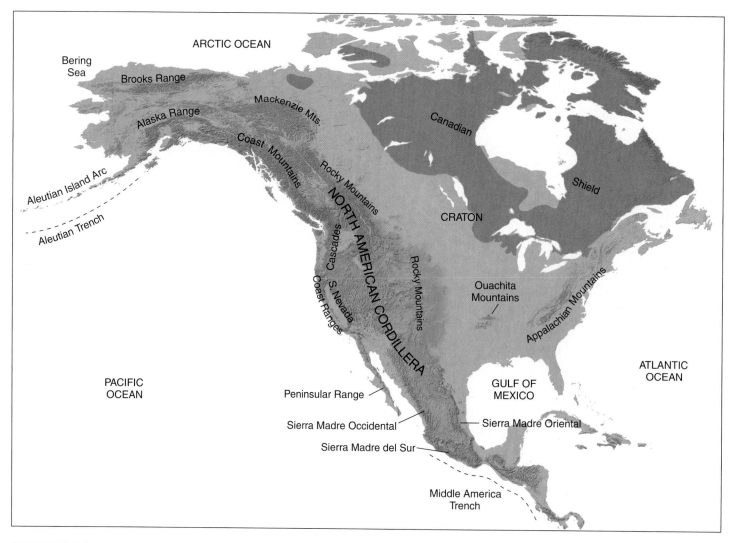

FIGURE 20.3

The mountain belts and craton (including Canadian Shield) of North America. Major ranges in the Cordillera are labeled.

Adirondacks in New York.) The region of a continent that has been structurally stable for a prolonged period of time is called a **craton** (figures 20.3, 20.4, and 20.5). The central part of the United States and Canada is all part of a craton. Other continents similarly have a craton at their core.

Most of the craton in the central United States has a very thin blanket—only 1,000 to 2,000 meters—of sedimentary rock layers overlying its basement. Sediment was mostly deposited in shallow inland seas during Paleozoic time; however, for the craton in much of eastern and northern Canada, as well as Greenland, no sedimentary rocks cover the eroded remnants of old mountain ranges. This region is a **Precambrian shield**—that is, a complex of Precambrian metamorphic and plutonic rocks exposed over a large area. Such shields and basement complexes of cratons represent the roots of mountain ranges that completed the deformation process more than a billion years ago.

Thickness and Characteristics of Rock Layers

The relatively thin cover of sedimentary rock overlying the basement in the craton contrasts sharply with the thick sedimentary sequence typical of a mountain belt (figure 20.4). In mountain belts, layered sedimentary rock commonly is more than 10 kilometers thick. We now know that the thick sequence of mostly marine-deposited sedimentary rock was originally deposited on a continental margin (continental shelf and continental slope—see chapter 18). If the sedimentary rock is mainly shale, sandstone, and limestone, we can infer that marine deposition was at a passive continental margin. If the sediment has a significant component of volcanic material, the depositional environment was an active continental margin.

The sedimentary rock in cratons may show no deformation, or it may have been gently warped into basins and

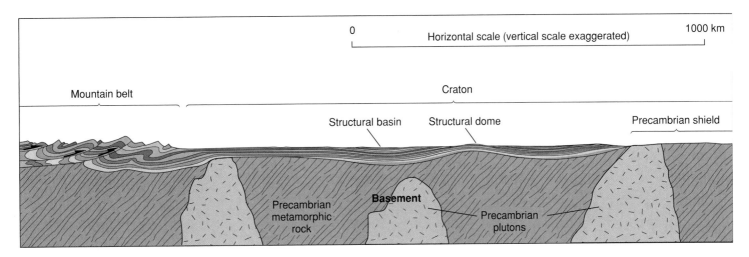

FIGURE 20.4
Schematic cross section through part of a mountain belt (left) and part of a continental interior (craton). Vertical scale is exaggerated.

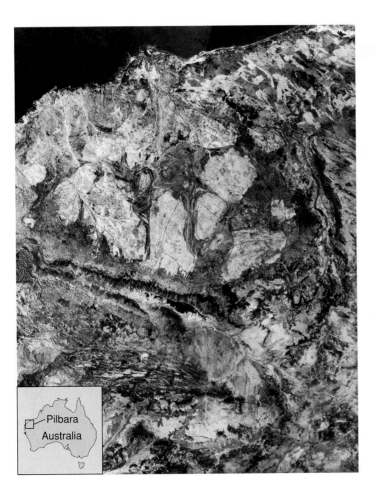

FIGURE 20.5
Satellite image of part of a craton in Western Australia. Metamorphic rock (dark gray) that is 3.5 billion to 3 billion years old surrounds oval-shaped domes of granite and gneiss (white) that are 2.8 billion to 3.3 billion years old. Gently dipping sedimentary and volcanic rocks (tan and reddish) unconformably overlie the granite-metamorphic basement complex. The area is 400 kilometers across. *Landsat mosaic produced by the Remote Sensing Applications Centre, Department of Land Administration, Western Australia*

domes above the basement. By contrast, mountain belts are characterized by a variety of folds and faults that indicate moderate to very intense orogenic deformation.

Patterns of Folding and Faulting

Reconstructing the original position and determining the thickness of layers of sedimentary and volcanic rock in mountain belts are complicated because, in most instances, the layered rocks have been folded and faulted after they were deposited. (Refer to figure 20.6 as you read through the following paragraphs.) Folds will be open in those parts of a mountain belt where deformation was not very intense. Tighter folds (figure 20.7) indicate greater deformation. Large overturned and recumbent folds (figure 20.8) may be exposed in more intensely deformed portions of mountain belts. Reverse faults are common, particularly in the intensely folded regions. Especially noteworthy are the **fold and thrust belts** found in many mountainous regions. These are characterized by large thrust faults (reverse faults at a low angle to horizontal), stacked one upon another; the intervening rock usually was folded while it was being transported during faulting.

Overall, the folds and thrust faults in a mountain belt suggest tremendous squeezing or *crustal shortening* and *crustal thickening*. The sedimentary rocks of the Alps, for instance, are estimated to have covered an area of ocean floor about 500 kilometers wide when they were deposited. During the Alpine orogeny, they were compressed into the present width of the Alps, which is less than 200 kilometers (see figure 20.13).

Metamorphism and Plutonism

A complex of regional metamorphic and plutonic rock is generally found in the mountain ranges of the most intensely deformed portions of major mountain belts. Most of the metamorphic rocks were originally sedimentary and volcanic rocks

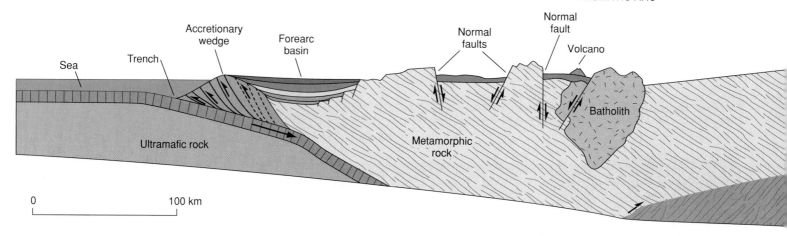

FIGURE 20.6

Cross section of an "Andean type" mountain belt; that is, one whose orogeny is due to oceanic-continental convergence. For simplicity, only a few of the many layers of sedimentary rock are shown. The size of some features is exaggerated for illustrative purposes.

that had been deeply buried and subjected to intense stress and high temperature during an orogeny. *Migmatites* (interlayered granitic and metamorphic rocks) may represent those parts of the mountain belts that were once at even deeper levels in the crust, where higher temperatures caused partial melting of the rocks (as described in chapters 3 and 7). These must have been transported into much higher levels of the crust during and after an orogeny. Batholiths, largely granitic, also have their origin in the lower crust (or uppermost mantle). Rather than remaining behind and forming migmatites, the magma generated from partial melting collects in large blobs (diapirs) that work their way upward into an upper level of Earth's crust.

Normal Faulting

Older portions of some major mountain belts have undergone normal faulting (figure 20.9). Crosscutting relationships show that the normal faulting occurs after the orogeny that resulted in tight folding, thrust faulting, and metamorphism, and after most batholiths had formed. This late stage of normal faulting (described in chapter 15 on geologic structures) is a result of *vertical uplift* or *horizontal extension*. Either of these contrasts with the overall shortening that prevailed during the orogeny.

Normal faulting may also take place in the high, central part of a mountain belt during an orogeny while folding and

FIGURE 20.7

False-color satellite image of part of the Valley and Ridge province of the Appalachian mountain belt, near Harrisburg, Pennsylvania. The ridges are sedimentary beds resistant to erosion. The pattern indicates tight and open folding occurred prior to erosion. *Photos by Jacques Descloitres, MODIS Land Rapid Response Team, NASA/GSFC*

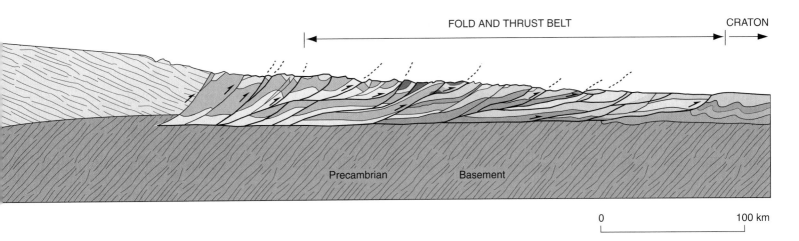

FOLD AND THRUST BELT   CRATON

Precambrian   Basement

0   100 km

thrust faulting are taking place at the outer parts of the belt (see figure 20.10). As the mountain belt is being compressed and shortening takes place, the central portion is pushed upward. Extension, along with normal faulting, takes place as rock at high levels flows outward over the rock being compressed at the lower level.

Thickness and Density of Rocks

Geophysical investigations yield additional information about mountains and the continental crust. As discussed in chapter 17 about Earth's interior, gravity measurements indicate that the rocks of the continental crust (including mountain belts) are

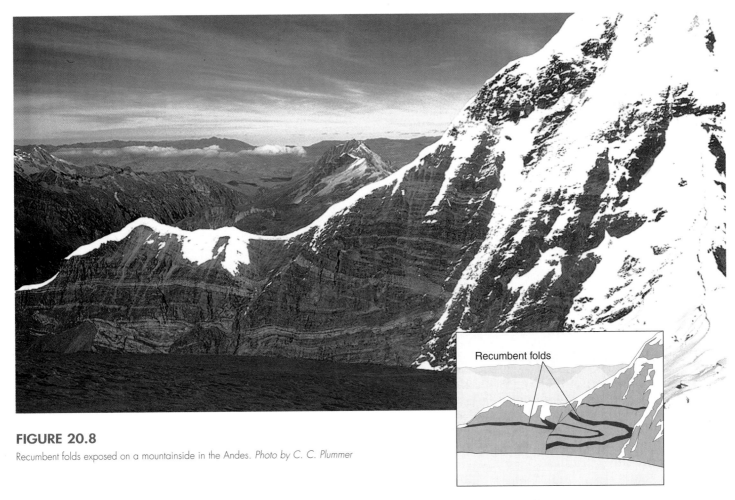

FIGURE 20.8

Recumbent folds exposed on a mountainside in the Andes. *Photo by C. C. Plummer*

Recumbent folds

Geologist's View

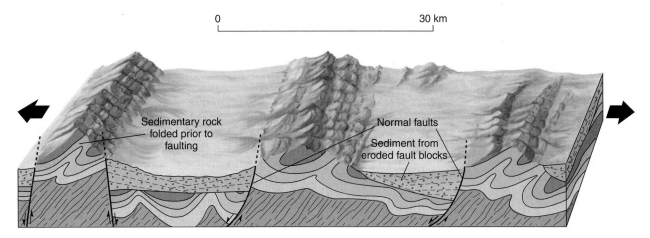

FIGURE 20.9

Fault-block mountains with movement along normal faults.

lighter (less dense) than those of the oceanic crust. Seismic velocities indicate a composition approximating that of granite for continental crust. Furthermore, evidence from seismic studies supports the view that this lighter crust is much thicker beneath mountain belts than under the craton and that the crust is thicker under younger mountain belts than under older ones.

Features of Active Mountain Ranges

Frequent earthquakes are characteristic of portions of mountain belts that are geologically young and considered still active. Also, deep-ocean trenches are found parallel to many young mountain belts (the Andes, for example). Trenches lie off the coasts of island arcs, which can be regarded as very young mountain ranges. Isolated active volcanoes perched on top of older rock in a mountain range suggest that melting is still taking place at depth.

EVOLUTION OF MOUNTAIN BELTS

Orogenies and Plate Convergence

As described earlier, each orogeny is an episode of intense deformation of the rocks in a region; the deformation is usually

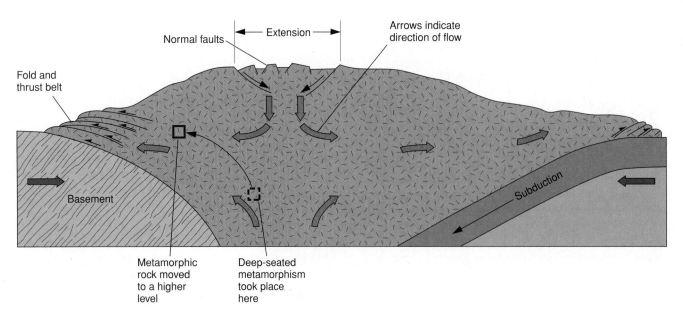

FIGURE 20.10

Schematic cross section of a mountain belt in which gravitational collapse and spreading are taking place during plate convergence. Red arrows indicate flowage of rock. Faulting occurs in brittle rock near the surface. Rock that was metamorphosed at depth flows to a higher level in the mountain belt. Not drawn to scale.

IN GREATER DEPTH 20.2

Ultramafic Rocks in Mountain Belts— From the Mantle to Talcum Powder

Ultramafic rocks (described in chapter 3 on intrusive rocks) occur commonly in the portions of mountain belts occupied by metamorphic and plutonic rocks. Ultramafic rocks tend to crop out in long, narrow zones that parallel the trend of a mountain belt. Geologists regard most bodies of ultramafic rocks as being former mantle material that was faulted into the crust during the mountain-building process. Some of the ultramafic bodies are found associated with marine-deposited volcanic and sedimentary rocks in an *ophiolite sequence* (described in chapter 18 about the sea floor). An ophiolite is regarded as a segment of a former oceanic crust together with its underlying mantle.

Ultramafic rocks in mountain belts commonly show the effect of the metamorphism that has altered adjacent rock units. Two of the foliated metamorphic products of ultramafic rocks are of special interest. One is *serpentinite,* a rock composed of the mineral serpentine. Another is a rock composed mostly of the mineral talc, commonly known as *soapstone.* Serpentine and talc are both hydrated magnesium silicates. They are products of metamorphic recrystallization of ultramafic rock when water is present. Metamorphism takes place in the crust under cooler and lower pressure conditions than those under which the original ultramafic rock formed in the mantle.

Serpentinite is a shiny, mottled, dark green and black rock that looks rather like a snake's skin. It splits apart easily along irregular, slippery, foliation surfaces. Hillsides or slopes with serpentinite as bedrock are sparsely vegetated because constant sliding prevents soil and vegetation from building up. Houses built on serpentinite hillsides (by people without a knowledge of geology) also slide downslope. Serpentinite is the official state rock of California—a state in which a large number of homes have been destroyed because they were built on sliding hillsides. (Serpentinite, however, is seldom to blame.)

Soapstone, which is less common than serpentinite, is valuable mainly because of talc's softness (number 1 on Mohs scale). Many sculptures (most notably, Inuit carvings; box figure 1) are made from soapstone because of the ease with which it can be cut. The best-known product of talc, however, is talcum powder.

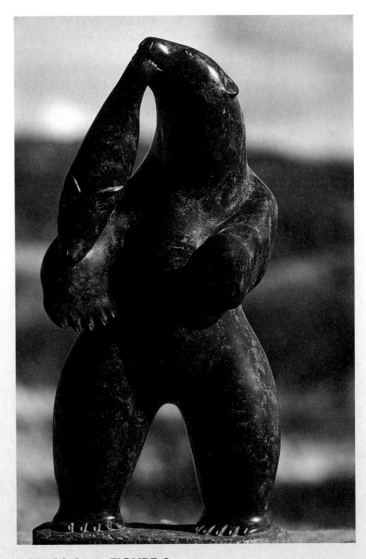

BOX 20.2 ■ FIGURE 1

Soapstone (talc) sculpture, polar bear and killed seal, by Inuit artist Nalinek Temela, Baffin Island, Canada. *Photo © Fred Bruemmer/Peter Arnold*

accompanied by metamorphism and igneous activity. Layered rocks are compressed into folds. Reverse faulting (especially thrust faulting) is widespread during an orogeny. Normal faulting may also occur but is not as widespread.

The more deeply buried rocks, subjected to regional metamorphism, are converted to schists and gneisses (see chapter 7). Magma generated in the deep crust or the upper mantle works its way upward to erupt in volcanoes or form batholiths (see chapter 3).

One important aspect of an orogeny is that the continental crust becomes thicker. This is achieved by the intense compression that results in tight folds and reverse faults. The addition of batholiths in the crust also helps thicken the crust and make it more buoyant. The thicker crust will isostatically "float" higher on the underlying mantle, resulting in higher mountains.

Each mountain belt has its own characteristics and history. However, by understanding which one of three kinds of plate convergence took place (described in chapter 19), we can better understand the mountain-building processes that a mountain belt underwent during an orogeny. The three types of convergence are discussed next.

Orogenies and Ocean-Continent Convergence

Figure 20.6 shows a hypothetical mountain belt that has ongoing ocean-continent convergence. The Andes, in which the South America plate is overriding the Nazca plate, is an example.

Figures 3.26 and 7.17 show igneous and metamorphism processes during oceanic-continental convergence. Plate convergence also accounts for the folded and reverse-faulted layered rocks found in mountain belts. An *accretionary wedge* develops where newly formed layers of marine sediment are folded and faulted as they are snowplowed off the subducting oceanic plate (see figure 19.27 and explanation in chapter 19).

Rock caught in and pulled down the subduction zone is subjected to intense shearing. If rock is carried further down the subduction zone, it becomes metamorphosed (as described in chapter 7).

Fold and thrust belts may develop on the craton (backarc) side of the mountain belt (figures 20.6 and 20.10). Thrusting is away from the magmatic arc toward the craton. The magmatic arc is at a high elevation because the crust is thicker and composed largely of hot igneous and metamorphic rocks. The large thrust sheets move toward and sometimes over the craton. (In the Rocky Mountains, thrust faulting of the craton itself has taken place.) The thrusting probably is largely due to the crustal shortening caused by convergence. There is, however, some controversy among geologists over additional processes that may take place. Some geologists regard gravity flow (from the high and mobile magmatic arc outward over the low and rigid craton) to contribute significantly to the process. Others think that the expanding magmatic arc pushes the sedimentary (and sometimes metamorphic and igneous) rocks outward to become the fold and thrust belt. (The magmatic arc is likened to a bulldozer pushing a wedge of loose material outward.)

In the late 1980s and early 1990s, geologists developed a model that explains (1) fold and thrust belts, (2) simultaneous normal faulting, and (3) how once deep-seated metamorphic rocks rise to an upper level in a mountain belt. What is believed to occur is that the thick and high part of the mountain belt becomes too high and gravitationally unstable, resulting in **gravitational collapse and spreading.** The mobile portion becomes increasingly elevated during plate convergence. This is due to compression of sedimentary and metamorphic rocks, as well as to volcanic eruptions and emplacement of plutons. After some time, the welt in the mountain belt becomes too high to be supported by the underlying rocks and collapse begins. (Geologist John Dewey, then at Oxford University, suggested that collapse begins when the welt exceeds 3 kilometers above sea level.) As shown in figure 20.10, the gravitational collapse forces rock outward as well as downward. At deeper levels in the mountain belt, the rock is *ductile* (or *plastic*) and flows; nearer the surface, rock fractures, so movement is through faulting. The rock is pushed outward and helps create, along with crustal shortening, the fold and thrust belt.

In the high part, the outward flowing rock results in extension (figure 20.10). Therefore, the brittle, near-surface rocks fracture and normal faulting takes place.

The flowage pattern (as shown in figure 20.10) can also explain how once deep-seated metamorphic rocks (migmatites, for example) are found in upper levels of a mountain belt. Lower crustal rocks are squeezed, forcing them to flow upward and outward, bringing them closer to the surface.

Arc-Continent Convergence

Sometimes an island arc collides with a continent (figure 20.11). As the arc and continent converge, the intervening ocean floor is destroyed by subduction. When collision occurs, the arc, like a continent, is too buoyant to be subducted. Continued convergence of the two plates may cause the remaining sea floor to break away from the arc and create a new site of subduction and a new trench seaward of the arc (figure 20.11C). Note that the direction of the new subduction is opposite the direction of the original subduction (this is sometimes called a *flipping subduction zone*), but it still may supply the arc with magma. The arc

FIGURE 20.11

Arc-continent convergence can weld an island arc onto a continent. The direction of subduction can change after impact.

North America   Africa

Precambrian supercontinent breaks up.
Seafloor spreading begins.

A Late Precambrian

North
American
craton   Piedmont Arc   Carolina Arc   Ancestral Atlantic   Africa

B Cambrian

North
American
craton   Accreted
Piedmont
terrane   Carolina Arc   Ancestral Atlantic   Africa

C Ordovician – Silurian (Taconic Orogeny)

FIGURE 20.12

The geologic evolution of the southern Appalachians. *Modified from "Tectonic syn-thesis of the U.S. Appalachians," figure 9, volume F-2, from* Geology of North America: An Overview, *by R. D. Hatcher, Jr. © 1989 by Geological Society of America. Reproduced with permission of the Geological Society of America. See also U.S. Geological Survey pamphlet* Birth of the Mountains. The Geologic Story of the South-ern Appalachian Mountains. *pdf version available through* http://pubs.usgs.gov/gip/birth/birth.pdf

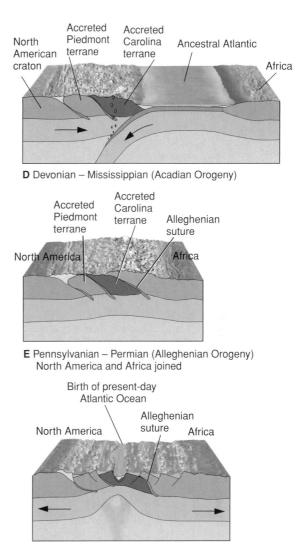

North
American
craton   Accreted
Piedmont
terrane   Accreted
Carolina
terrane   Ancestral Atlantic   Africa

D Devonian – Mississippian (Acadian Orogeny)

Accreted
Piedmont
terrane   Accreted
Carolina
terrane   Alleghenian
suture

North America   Africa

E Pennsylvanian – Permian (Alleghenian Orogeny)
North America and Africa joined

Birth of present-day
Atlantic Ocean
Alleghenian
suture
North America   Africa

F Triassic – Rifting begins. Breakup of Pangaea starts

has now become welded to the continent, increasing the size of the continent.

This type of collision apparently occurred during recent geologic time in northern New Guinea (north of Australia). A similar collision may have added an island arc to the Sierra Nevada complex in California during Mesozoic time, when a subduction zone may have existed in what now is central California. Many geologists think that much of westernmost North America has formed from a series of arcs colliding with North America (discussed later in this chapter in the "Displaced Ter-ranes" section). During the Paleozoic, arc-continent collision played a significant role in the building of the Appalachians (figure 20.12).

Orogenies and Continent-Continent Convergence

As described in chapter 19 (see figure 19.28), some mountain belts form when an ocean basin closes and continents collide along a suture zone. Mountain belts that we find within conti-nents (with cratons on either side) are believed to be products of continent-continent convergence. The Ural Mountains resulted from the collision of Asia and Europe. Convergence of the African and European plates created the Alps (figure 20.13). Our highest mountains are in the Himalayan belt. The Himalayan orogeny started around 45 million years ago as India began colliding with Asia (India was originally in the Southern Hemisphere). The thick sequences of sedimentary rocks that had built up on both continental margins were intensely faulted and folded. Fold and thrust belts developed and were carved by erosion into the mountain ranges that make up the Himalaya. The mountains are still rising, and frequent earthquakes indicate continuing tectonic activity. North of the Himalaya, Tibet rose to become what is now the highest plateau in the world. Normal faults in the Tibetan Plateau tell us that gravitational collapse is taking place.

The collision of India and Asia has affected parts of Asia well beyond the Himalaya and Tibet. Figure 20.14 shows

FIGURE 20.13

Cross section through part of the Alps. Thicker lines are thrust faults. Lesser folds are not shown. Movement is from the right to the left of the diagram (southeast to northwest). Only a few arrows are shown to indicate movement of the overriding thrust block. *From S. E. Boyer and D. Elliot, 1982. AAPG Bulletin. Reprinted by permission of American Association of Petroleum Geologists*

present-day crustal motion in and around the Himalaya and Tibet. As India continues to push northward, some of the motion is deflected from Tibet eastward into China. Box 19.3, figure 1 in the plate tectonics chapter shows major fault systems extending through China and neighboring countries and their relationships to the ongoing plate collision.

The Appalachian Mountains are an example of continent-continent convergence but with a more complicated history. Arc-continent convergence and oceanic-continental conver-

FIGURE 20.14

Motion in and around Tibetan plateau as determined through Global Positioning System (GPS) measurements. Blue arrows point in the direction of motion and their lengths indicate velocities (millimeters/year). The scale in the lower left corner indicates the length of an arrow for 20 millimeters/year. Dashed yellow polygons show regions—for instance, 1 is the Himalaya, 6 the Tibet Plateau. *From Zhang, et al. Continuous deformation of the Tibetan Plateau from global positioning system data. Geology 2004 Vol. 32, No. 9, pp. 809–812. Fig 1. on pg. 810.*

gence were also involved, and the mountain belt was later split apart by plate divergence.

A condensed version of orogeny in the Appalachians is as follows (figure 20.12): During late Precambrian and earliest Paleozoic time, the ancestral Atlantic Ocean developed as seafloor spreading forced the passive margins of North America, Europe, and Africa away from one another. During the Paleozoic, plate motion shifted, subduction began, and the ocean basin began closing. Island arcs developed between the continents. These became plastered onto the North American craton as the ancestral Atlantic basin closed. A couple hundred million years after subduction began, the ocean basin closed completely, first with Europe and later with Africa crashing into North America. By the end of the Paleozoic, the three continents were sutured together. The Appalachians and what is now the Caledonide mountain belt of Great Britain and Norway and the Atlas Mountains of North Africa were part of a single mountain belt within the supercontinent *Pangaea*. The mountain belt was comparable to the present-day Himalaya.

Early in the Mesozoic Era, the supercontinent split, roughly parallel to the old suture zone. The present continents moved (and continue to move) farther and farther away from their present divergent boundary, the mid-Atlantic ridge.

What happened to the Appalachians seems implausible. Yet, if one accepts the principles of plate-tectonic theory and examines the rocks and structures in the Appalachians (and their counterparts in Europe and Africa), the argument for this sequence of events is not only plausible but convincing.

The cycle of splitting of a supercontinent, opening of an ocean basin, followed by closing of the basin and collision of continents is known as the *Wilson Cycle.* Canadian geologist J. Tuzo Wilson proposed the cycle in the 1960s for the tectonic history of the Appalachians.

The Wilson Cycle apparently has occurred before. A question raised is why would a continent split apart more or less along a suture zone where one would expect the crust to be thickest? One recently proposed hypothesis is that this zone is weakened and thinned somewhat by outward flow of rock during gravitational collapse and spreading. (Another hypothesis involves *delamination,* detachment and sinking downward of the underlying lithospheric mantle, as described on p. 543.)

Post-Orogenic Uplift and Block-Faulting

After an orogeny ceases to affect part or all of a mountain belt and the prevailing compressive force is relaxed, there is a long period of uplift accompanied by erosion. Isostatic adjustment took place during orogeny, but was overshadowed by the compressive horizontal forces of plate convergence. With horizontal forces becoming insignificant at the end of an orogeny, isostatic adjustment takes over as the dominant process. For many millions of years, large regions in the mountain belt move vertically upward. Erosion may keep pace with uplift and the area remain low. Alternatively, uplift may temporarily outpace erosion, resulting in plateaus or mountain ranges. The present Appalachian Mountains are the result of uplift and erosion that have taken place long after the last orogeny ended more than 250 million years ago. It is likely that the Appalachian Mountains eroded down to a plain after the last, Paleozoic orogeny. The coastal plain east of the Appalachians is made of young sedimentary rock unconformably overlying metamorphic and igneous rocks that were part of the original mountain belt. This region has remained a plain. The present Appalachians represent rejuvenation following relatively recent uplift in late Tertiary time. The uplift may have been due to reactivation of ancient thrust faults caused by compressive stress within the westward-moving North American plate. The coastal plains have not moved upward, probably due to a lack of thrust faults. So the topography of the Appalachians is geologically quite young, while the original structures due to orogenic deformation are quite old. The Adirondack Mountains of northern New York also participated in the uplift and rejuvenation, but the orogeny that they went through is Precambrian—much older than the Appalachian orogenies. Eventually, the entire Appalachian mountain belt will be eroded to a plain and become part of the North American craton.

Isostasy

According to the concept of isostasy, lighter, less-dense continental crust "floats" higher on the mantle than the denser oceanic crust. The craton has achieved an equilibrium and is floating at the proper level for its thickness. Mountains, being

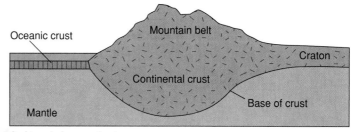

A At end of orogenic stage

B Mountain belt moves upward

C Erosion and renewed uplift continue until crust beneath mountain belt is the same thickness as that of the craton

FIGURE 20.15

Isostasy in a mountain belt. The thickness of the continental crust is exaggerated.

thicker continental crust, "float" higher than the stable continent. As material is removed from mountains by erosion, the range floats upward to regain its isostatic balance (figure 20.15). This process can be thought of as "the pull of erosion." Isostatic adjustment does not take place instantaneously. Usually, there will be a considerable time lag between erosion and isostatic adjustment. As the mountains wear down to a low plain, erosion becomes virtually ineffective and the now thin crust achieves isostatic balance; the former mountain belt becomes part of the craton. The reason the craton consists of plutonic and metamorphic rock is that these were the rocks that formed the deep roots of the former mountain belt.

At most places on continents, the altitude above sea level is related to local crustal thickness. Beneath the 5-kilometer-high Tibetan Plateau, the crust is 75 kilometers thick. Under Kansas, the crust is 44 kilometers thick, and beneath Denver, the "mile-high city," the crust is 50 kilometers thick. (If the United States ever joins the rest of the world and goes metric, Denver will be known as the "1.6 kilometer-high city.") Just west of Denver, the altitude of the Rocky Mountains jumps to 2 kilometers higher than that at Denver. Scientists expected to find a corresponding thickening of the crust beneath these mountains. They were surprised by 1995 seismic studies that

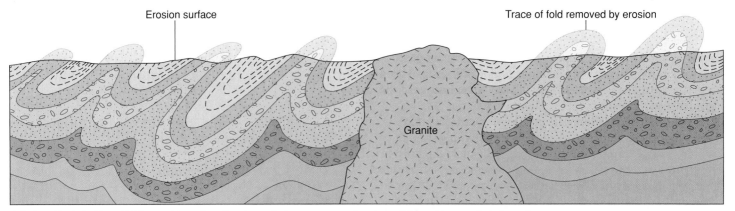

Erosion surface

Trace of fold removed by erosion

Granite

A Before block faulting. Folding and intrusion of a pluton during an orogeny has been followed by a period of erosion.

0   10 kilometers

Mountain peak

Sediment deposited

Portion of uplifted block removed by erosion

B The same area after block-faulting. Tilted fault-block mountain range on left. Range to right is bounded by normal faults.

FIGURE 20.16

Development of fault-block mountain ranges.

FIGURE 20.17

The Teton Range, Wyoming, a tilted fault-block range. The rocks exposed are Precambrian metamorphic and igneous rocks that were faulted upward. Extensive past glaciation is largely responsible for their rugged nature. For more information, go to www.winona.msus.edu/geology/travels/tetons/travel.html. *Photo by C. C. Plummer* indicated that the crust is no thicker beneath the Rockies than at Denver. (Similar discrepancies between crustal thickness and mountain elevations have been reported for the southern Sierra Nevada.) Geologists explain the higher elevations by regarding the mantle as hotter and therefore less dense beneath that part of the Rockies. The crust plus less-dense mantle are floating on deeper, denser mantle. Seismic wave studies verify that the mantle here is hot and appears to be asthenosphere that is at a shallower level in Earth than usual.

Normal Faulting

Normal faults develop after orogenies in several settings. One is when a continent is split and a divergent boundary forms (see chapter 19). An example is the breakup of Pangea in the early Mesozoic, after the final orogeny in the Appalachians (figure 20.12*F*) and counterpart mountains in Africa and Europe. A normal fault will also develop if part of the crust moves upward isostatically more than does adjoining crust.

In parts of some mountain belts, the crust breaks into fault-bounded blocks. If an upthrown block is large enough, it becomes a **fault-block mountain range.** The normal faulting implies a *horizontal extension strain,* the regional pulling apart of the crust. Isostatic vertical adjustment of a fault block probably occurs at the same time.

Fault-block mountain ranges are bounded by normal faults on either side of the range, or, more commonly, are tilted fault blocks in which the uplift has been great along one side of the range, while the other side of the range has pivoted as if hinged (figure 20.16). The Sierra Nevada (California) and Teton (Wyoming) Range are tilted fault-block mountains (figure 20.17).

Isolated volcanic activity may be associated with some faults. Eruptions occur along faults extending deep into the crust or the upper mantle.

Uplift is neither rapid nor continuous. Part of a mountain range may suddenly move upward a few centimeters (or, more rarely, a few meters) and then not move again for hundreds of years. Erosion works relentlessly on newly uplifted mountains, carving the block into peaks during the long, spasmodic rise. Over the long time period, the later episodes of renewed faulting and uplift involve successively less and less vertical movement.

Block-faulting is taking place in much of the western United States—the Basin and Range province (also called the Great Basin) of Nevada and parts of Utah, Arizona, New Mexico, Idaho, and California (figure 20.18). Hundreds of small, block-faulted mountain ranges are in evidence. They are separated by valleys that are filling with debris eroded from the mountains. Extension in the Basin and Range is probably due to hotter mantle beneath the crust as shown in figure 20.19.

Delamination

The hypothesis of lithospheric delamination is used to explain the block-faulting, thin crust, and geologically young volcanic activity of the Basin and Range. **Lithospheric delamination** (or simply **delamination**) is the detachment of part of the mantle portion of the lithosphere beneath a mountain belt (figure 20.20). As you know, the lithosphere consists of the crust and the underlying, rigid mantle. Beneath the lithosphere is the hotter, ductile mantle of the asthenosphere. During an orogeny, the crust as well as the underlying lithosphere mantle thickens. The lithosphere mantle is cooler and denser than the asthenosphere mantle. As indicated in figure 20.20, the thickened portion of the lithosphere mantle is gravitationally unstable, so after it is softened from convecting asthenosphere, it breaks off and sinks through the asthenosphere to a lower level in the mantle. Hot asthenosphere mantle flows in to replace the foundered, colder mantle. Heating of the crust follows, allowing the lower crust to flow. The once-thick crust becomes thinner than that of adjoining regions of the mountain belt. Extension results in block-faulting in the upper part of the crust (as in figure 20.19).

FIGURE 20.18

The Basin and Range and adjoining geological provinces.

FIGURE 20.19

Upwelling, hot, buoyant mantle (asthenosphere) causes extension, thinning, and block-faulting of the overlying crust.

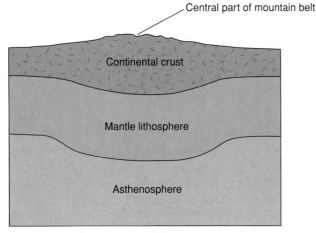

A Thick continental crust of a mountain belt produced during orogeny.

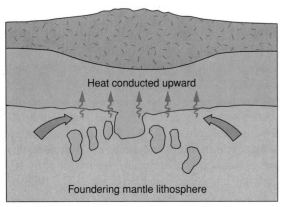

B Delamination of gravitationally unstable mantle lithosphere. Hot asthenosphere flows and replaces foundering lithosphere, heating overlying lithosphere.

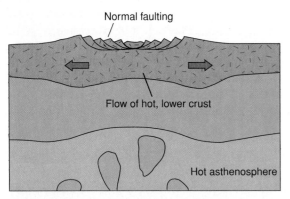

C Extension with hot, lower crust flows outward.

FIGURE 20.20

Delamination and thinning of continental crust following orogeny. (Not drawn to scale.)

Dance of the Continents (with SWEAT)

Go to the Online Learning Center (www.mhhe.com/plummer11e) and read how building followed by breakup of supercontinents has taken place throughout geologic time. The process is compared to a dance. Each dance cycle, which takes about a billion years, is set to a symphony in four movements in which the continental fragments come together and later "dance" away. The creation, in the Paleozoic, and breakup, in the Mesozoic, of Pangaea is only the latest of the several dance cycles.

Related to this is a hypothesis called SWEAT, which is an acronym for southwest United States-East Antarctic connection. According to this hypothesis, in one of the Precambrian supercontinents, the craton in the southwestern United States adjoined what is now Antarctica.

For the full story, go to:
www.mhhe.com/plummer11e

Delamination beneath the Basin and Range helps explain the extensive rhyolitic and basaltic eruptions that occurred tens of million years after the end of the last orogeny. Heating in the lower part of the crust to 700°C would have generated silicic magma that erupted as rhyolite. Basaltic magma would have formed from partial melting of the asthenosphere when it moved upward (replacing the foundered lithosphere mantle) and pressure was reduced (as explained in chapter 3). That the crust was once thicker in the Basin and Range is supported by recent studies of fossil plants indicating that the Basin and Range was 3 kilometers higher than at present.

Delamination is also being invoked to help explain why, when Pangaea broke up, North America split from Europe and Africa more or less along the old suture zone. *Gravitational collapse* could have contributed to the weakening and thinning of this once-thick part of the mountain belt during its last orogeny. The breakup of the supercontinent began around 30 million years after the late Paleozoic orogeny ended. Delamination of the underlying lithosphere mantle would have resulted in heating and thinning of the overlying, remaining lithosphere. Rifting of the supercontinent began with normal faulting (see chapter 19 and figure 19.20) and was accompanied by basaltic eruptions and intrusions. The Appalachians split from the European Caledonides. Europe, Africa, and North America went their separate ways as the Atlantic opened and widened.

Delamination (like gravity collapse) is an example of an hypothesis that builds on plate-tectonic theory. It was proposed because it explains data better than other concepts do. It still needs further testing to become widely accepted as a theory.

THE GROWTH OF CONTINENTS

Continents grow bigger as mountain belts evolve along their margins. Accumulation of sediment and igneous activity add new continental crust beyond former coastlines. In the Paleozoic Era, the Appalachians were added to eastern North America, and during the Mesozoic and Cenozoic Eras, the continent grew westward because of accumulation and orogenic processes in many parts of what is now the Cordillera. Therefore, if we isotopically dated rocks that had been through an orogeny, starting in the Canadian Shield and working toward the east and west coasts, we should find the rocks to be progressively younger. In a very general way, this seems to be the case; however, there are some rather glaring exceptions.

Displaced Terranes

In some regions of mountain belts, the age and characteristics of the bedrock appear unrelated to that of adjacent regions. To help understand the geology of mountain belts, geologists have in recent years begun dividing major mountain belts into **tectonostratigraphic terranes** (or, more simply, **terranes**), regions within which there is geologic continuity. The geology in one terrane is markedly different from a neighboring terrane. Terrane boundaries are usually faults. Typically, a terrane covers thousands of square kilometers, but some terranes are considerably smaller. Alaska and western Canada have been subdivided by some geologists into over fifty terranes (figure 20.21). Terranes are named after major geographic features; for instance, Wrangellia, parts of which are now in Alaska and Canada (with fragments in Washington and Idaho, according to some geologists) was named after the Wrangell Mountains of Alaska.

Many terranes appear to have formed essentially in place as a result of accumulation and orogeny along the continent's margin. Other terranes have rock types and ages that do not seem related to the rest of the geology of the mountain belt and have been called **suspect terranes,** that is, terranes that may not have formed at their present site. If evidence indicates that a terrane did not form at its present site on a continent, it is regarded as an **accreted terrane.** Accreted terranes that can be shown to have traveled great distances are known as *exotic terranes.*

A suspect terrane will have rock types and ages different from adjoining terranes, but to prove that it came from elsewhere in the world (and therefore is an accreted terrane), geologists may compare fossil assemblages or determine the paleomagnetic poles (see chapter 19) of the terrane's rocks. If the terrane is exotic, its fossil assemblage should indicate a very different climatic or environmental setting compared to that of the adjoining terrane. (For example, the Cache Creek terrane in western Canada contains fossils from the Permian period that indicate a marine equatorial environment.) For an exotic terrane, the paleomagnetic poles for the rocks in the terrane will plot at some part of the world very distant from poles

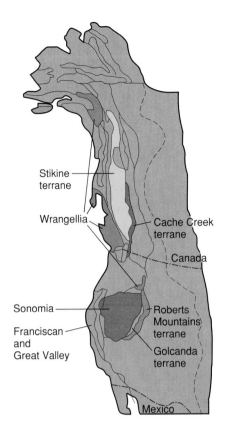

FIGURE 20.21

Some terranes in western North America. Note: Wrangellia is in Alaska, British Columbia, and Idaho. *After U.S. Geological Survey Open File Map 83–716*

of adjoining terranes that formed in place. This indicates that a particular terrane formed in a different part of Earth and drifted into the continent of which it is now a part. Some accreted terranes were island arcs, and some might have been *microcontinents* (such as present-day New Zealand) that moved considerable distances before crashing into other landmasses. Others may have been fragments of distant continents that split off and moved a long distance because of transform faulting. Imagine what might happen if the San Andreas fault remains active for another 100 million years or so. Not only would Los Angeles continue northward toward San Francisco and bypass it in about 25 million years (see box 15.2), but the block of coastal California west of the fault would continue moving out to sea, becoming a large island with continental crust that drifts northward across the Pacific. Ultimately, it would crash into and suture onto Alaska.

Figure 20.22 shows a tentative reconstruction of how parts of Alaska might have migrated in time. This is based on paleomagnetic data that indicate that parts of Alaska originally formed south of the equator and moved many thousands of kilometers to become part of the North American Cordillera. Note from the diagram that the path of migration was not simple. Plates split, plates joined, and the direction of movement changed from time to time.

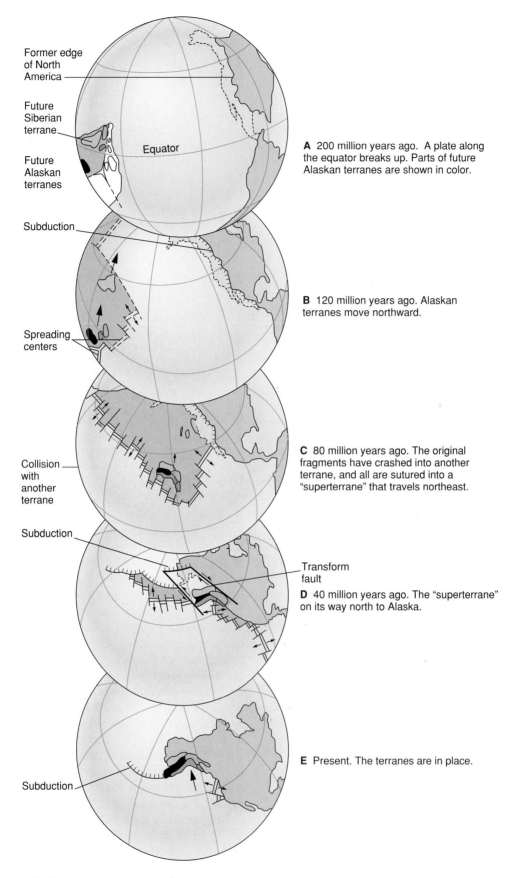

Former edge of North America

Future Siberian terrane

Future Alaskan terranes

Equator

A 200 million years ago. A plate along the equator breaks up. Parts of future Alaskan terranes are shown in color.

Subduction

Spreading centers

B 120 million years ago. Alaskan terranes move northward.

Collision with another terrane

C 80 million years ago. The original fragments have crashed into another terrane, and all are sutured into a "superterrane" that travels northeast.

Subduction

Transform fault

D 40 million years ago. The "superterrane" on its way north to Alaska.

Subduction

E Present. The terranes are in place.

FIGURE 20.22

How fragments of the Southern Pacific Crust may have become part of Alaska. Modified from D. B. Stone, B. C. Panuska, and D. R. Packer, 1982, "Paleolatitudes versus time for southern Alaska," Journal of Geophysical Research, vol. 87 (pp. 3697–3707), copyrighted by American Geophysical Union

The Appalachians as well as mountain belts in other continents have also been divided into terranes. Even the Canadian Shield has been subdivided into terranes. Some geologists think they can determine, despite the great age and complexity of the shield's rocks, the extent to which some terranes traveled before crashing together.

We should caution the reader that geologists do not always agree on the nature and boundaries of terranes. While most would probably agree that some terranes are exotic, many geologists think the subdividing of Alaska and western Canada into fifty terranes is overdoing it and not supported by sufficient evidence. Only time and more painstaking gathering of evidence will allow geologists to determine the history of each alleged terrane.

Concluding Comment

Only a few decades ago, many geologists thought that through the application of plate-tectonic theory, we could easily determine the processes at work in each mountain belt and work back in time to understand the history of each of the continents. Some suggested that there would hardly be major problems for Earth scientists to solve in the future. Plate tectonics was a breakthrough, and a great many problems were solved; but with this great leap forward in the science we have identified new problems. New generations of geologists will have no shortage of challenges and no less excitement from solving newly discovered problems than did their predecessors who saw the dawn of the plate-tectonics breakthrough. Science present builds on science past.

SUMMARY

Major mountain belts are made up of a number of *mountain ranges*. Mountain belts are generally several thousand kilometers long but only a few hundred kilometers wide.

The major factors that control the growth and development of mountain ranges are *intense deformation* (during an *orogeny*), *isostasy,* and *weathering and erosion.* An orogeny involves folding and faulting of sedimentary and volcanic rock, regional metamorphism, and igneous activity. Orogenies are associated with plate convergence. After an orogeny, there is a long period of uplift, often with block-faulting, and erosion. Eventually, the mountain belt is eroded down to a plain and incorporated into the *craton,* or stable interior of the continent.

According to plate-tectonic theory, mountains on the edge of continents are formed by continent-oceanic convergence, and mountains in the interior of continents are formed by continent-continent collisions.

The uplift of a region following termination of an orogeny is generally attributed to isostatic adjustment of continental crust.

Continents grow larger when new mountain belts evolve along continental margins. They may also grow by the addition of terranes that may have traveled great distances before colliding with a continent.

Terms to Remember

accreted terrane 545
craton 532
fault-block mountain range 543
fold and thrust belts 533
gravitational collapse and spreading 538

lithospheric delamination
    (or delamination) 543
major mountain belt 528
mountain range 528
orogeny 528

Precambrian shield 532
suspect terrane 545
terrane (tectonostratigraphic terrane) 545

Testing Your Knowledge

Use the following questions to prepare for exams based on this chapter.

1. What does a fold and thrust belt tell us about what occurred during an orogeny?
2. What is the difference between the forces that could explain fault-block mountains and the forces that could account for an orogenic stage?

3. Explain how erosion and isostasy eventually produce stable, relatively thin, continental crust.
4. How do the sequences of sedimentary rocks in cratons differ from those in mountain belts?

5. What sequence of events accounts for a mountain belt that is bounded on either side by cratons?

6. The mountain belt that forms the western part of North America is called the
   a. Appalachians
   b. North American Cordillera
   c. Himalaya
   d. Andes
   e. Rockies

7. The craton
   a. covers the central part of the United States and Canada
   b. has only 1,000–2,000 meters of sedimentary rock overlying basement rock
   c. has rock beneath any sedimentary rock that is old plutonic and metamorphic rock
   d. all of the preceding

8. The Precambrian shield
   a. contains geologically young rocks
   b. occurs only in mountainous regions
   c. is a complex of Precambrian metamorphic and plutonic rocks exposed over a large area
   d. all of the preceding

9. Folds and reverse faults in a mountain belt suggest
   a. crustal shortening
   b. tensional stress
   c. deep-water deposition of the sediment
   d. all of the preceding

10. Which of the following is *not* one of the major controlling factors that determine the characteristics of a mountain range?
    a. age of rocks
    b. intensity of deformation
    c. weathering and erosion
    d. isostasy

11. To explain fold and thrust belts, simultaneous normal faulting, and how once deep-seated metamorphic rocks rise to an upper level in a mountain belt, geologists use a model called
    a. tectonism
    b. gravitational collapse and spreading
    c. orogeny
    d. faulting

12. The Wilson Cycle describes
    a. the cycle of uplift and erosion of mountains
    b. the movement of asthenosphere
    c. the block-faulting that occurs at mountains
    d. the cycle of splitting of a supercontinent, opening of an ocean basin, followed by closing of the basin and collision of continents

13. The detachment of part of the mantle portion of the lithosphere beneath a mountain belt is called
    a. gravitational collapse
    b. rifting
    c. lithospheric delamination
    d. none of the preceding

14. Which is not a type of terrane?
    a. accumulated
    b. exotic
    c. suspect
    d. tectonostratigraphic
    e. accreted

15. Which is a source for terranes?
    a. microcontinents
    b. fragments of distant continents
    c. island arcs
    d. all of the preceding

16. Block-faulting may be due to (choose all that apply)
    a. isostatic adjustment
    b. subduction
    c. gravitational collapse
    d. lithospheric delamination

17. Which of the following is not characteristic of a mountain belt that formed through ocean-continent convergence?
    a. fold and thrust belts
    b. thick accumulations of marine sediment
    c. prevalence of normal faults over reverse faults
    d. metamorphism

18. The present elevation of the Appalachian Mountains is due mainly to
    a. post-orogenic uplift
    b. break-up of Pangea
    c. an orogeny during early Paleozoic time
    d. the Wilson cycle

Expanding Your Knowledge

1. How are unconformities used to determine when orogenies occurred?

2. How has seismic tomography contributed to our understanding of mountain belts?

3. How do basalt and ultramafic rocks from the oceanic lithosphere become part of mountain belts?

4. Why is a craton locally warped into basins and domes?

5. How could fossils in a terrane's rocks be used to indicate that it is an exotic terrane?

Exploring Web Resources

www.mhhe.com/plummer11e

This is the dedicated website for this book. You can go to it for new and updated information. The universal resource locators (URL) listed in this book are also given as links on the website, making it easy to go to those websites without typing in the URL. When you visit our Online Learning Center, go to the Student Center and then to the chapter of interest. Here you will find additional readings and media resources, as well as answers to the Testing Your Knowledge section, more quizzes, animation and video clips, and direct links to the sites listed in this book. Links to additional websites can also be found. We have added questions for some of the sites to allow you to get the most of your exploration of the web. Using the web is an enjoyable way of enhancing your knowledge of geology.

http://pubs.usgs.gov/gip/birth/birth.pdf

Birth of the Mountains. The Geologic Story of the Southern Appalachian Mountains. This is the pdf version of a publication by the U.S. Geologic Survey.

http://info.hartwick.edu/geology/vft/VFT-so-far/VFT.html

Hartwick College Virtual Field Trip. A field trip through part of the Appalachians in central New York. The site includes a geologic history for this part of the Appalachians.

www.winona.msus.edu/geology/travels/tetons/travel.html

Geology of Grand Teton National Park, Wyoming. A photo, map, and text description of the spectacular Grand Teton Range and its geologic history.

Geologic Resources

Throughout this book, we have mentioned human use of Earth materials, most of which are nonrenewable. Water is a resource that is an important exception. It is a renewable and was discussed in chapters 10 and 11. Our purpose in this chapter is to survey briefly some important geologic resources, other than water, of economic value.

We first look at energy resources to see which ones might help replace our disappearing supplies of oil. Then we discuss metals and their relation to igneous rocks and plate tectonics and conclude with nonmetallic resources such as sand and gravel.

Evaporation ponds at a potassium mine near Moab, Utah. Water is pumped underground to dissolve potassium salts, then pumped to the evaporation ponds. Green dye is added to hasten the evaporation process. *Photo © Peter Turner/Image Bank/Getty Images*

Relationships to Earth Systems

Many geologic resources originate in the hydrosphere. Petroleum and coal come from organisms that live and die in water. Oil is derived from the organic remains of tiny creatures that lived in the seas. Coal originates from dense vegetation that grew in ancient swamps. (Both examples were, of course, part of the biosphere.) The atmosphere interacts with the geosphere (and the hydrosphere) to produce some resources. For instance, aluminum ore is a product of weathering in tropical climates. Aluminum oxides form from chemical reactions involving air, water, and aluminum silicate minerals.

The hydrosphere, the atmosphere, and the geosphere are profoundly altered by one member of the biosphere—the human species. In the processes of extracting, manufacturing, and using Earth's resources, we change the air, land, and water with which we live.

INTRODUCTION

Geologic resources sustain life, and the most fundamental of these resources are soil and water. Industrial civilization, however, draws a much greater variety of resources from the Earth. In many ways, our modern lives have come to depend upon dozens of different kinds of Earth materials as "essential," from the coal that powers our electricity to the europium that provides the red pigment we use for color balance in our television screens and computer monitors. In early times, even ordinary people knew directly what resources they needed to support their lives, and where to find them. Today, few people fully appreciate the multitude of resources and the complex web of supply and demand that enables each of us to live.

Excluding soil and water, which we discuss elsewhere in the book, we group geologic resources into three general categories: (1) sources of energy, (2) sources of metals, and (3) nonmetallic sources of construction materials. *Energy resources,* like petroleum and uranium, provide the power that drives the modern world. *Metallic resources,* such as iron ore, enable us to create metals, which provide strength to modern construction and help many technologies operate—for example, by conducting electricity or sparking motors. The *nonmetallic resources,* including building stones and road gravels, have a longer history in the development of civilization, but nonetheless are vital to the modern world as well. Consider a highway overpass: It consists of a nonmetallic resource exterior (the concrete) with a metallic core (the rebar and girders). The overpass supports a road of asphalt—fossil organic matter—that provides access to cars made of metals (body and engine) and biological matter (the rubber), powered by petroleum.

Every year in the United States, it takes over 15,500 pounds of nonmetallic material, 650 pounds of metals, and over 21,000 pounds of geologic fuels to provide for the standard of living of each American. The energy needed to maintain a typical lifestyle in the United States is equivalent to the power of 300 servants working around the clock. Try to imagine how much mining, blasting, drilling, and pumping all of this requires, and it is easy to believe the fact that human activity now moves more earth—perhaps as much as 3 trillion tons annually—than all of the rivers of the world combined (a *mere* 24 billion tons per year of transported sediment).

Some geologic resources are *renewable,* that is, they are replenished by natural processes fast enough that people can use them continuously. Water is the best example. Under sustainable conditions, the supply of water is never-ending, provided that we extract water no faster than it is replenished naturally by precipitation, runoff, and infiltration. Most geologic resources, however, are **nonrenewable resources.** They form very slowly, often over millions of years under unusual conditions in restricted geographic settings. They are "happy accidents" of nature. Humans extract nonrenewable resources much faster than nature replaces them. The annual rate of extraction of crude oil, for example, is on the order of a million times faster than natural rates of replenishment.

ENERGY RESOURCES

The state of our energy supplies is very much on the minds of many people today. How much oil do we have left? What will power the airplane of the future? Is a new Hydrogen Economy on the way? To answer these questions, it's first important that we give some thought to basic thermodynamics—the branch of science that deals specifically with energy and its transformation through the natural world.

Energy is simply "the ability to do work." Without energy, nothing could exist—in fact, it is basic to everything. There are two basic principles of thermodynamics that sum up what you need to know about energy, anywhere you go: The First, or "Conservation," Law of Thermodynamics states that *energy is neither created nor destroyed; it is merely transformed from one state to another.* In human terms, some of these states are more useful and exploitable than others. The Second Law takes this one step further, by stating that *whenever a transformation of energy takes place, some of it is lost* ("dissipated") *to the surrounding environment,* usually in the form of "waste" heat. It is theoretically possible to collect all of this dissipated energy and re-concentrate it for use, but to do so would require even more inputs of energy and would not make any economic sense.

Car engines illustrate energy dissipation well. A motor grows quite hot when operated for a long time. On cold winter days, you can exploit some of this waste heat to warm your passenger compartment. *Efficiency* is simply a measure of how much energy actually goes into doing useful work versus energy dissipated into the surroundings. Although cars do us great service in getting us around, they are actually quite inefficient energy users. A conventional internal combustion engine is only about 10–15% energy efficient.

Coal

Now let's apply these thermodynamic principles to study the geology of one of our most important energy resources: coal. (**Coal,** as described in chapter 6, is a sedimentary rock that forms from the compaction of plant material that has not completely decayed.) Imagine a swampy, coastal environment in a tropical setting. Sunlight filters through the still trees onto dark, stagnant water below, providing the energy for photosynthesis to take place even in the shadows. Less than 2% of all sunlight striking the many leaves in this sultry jungle actually drives photosynthetic reactions; so you may think of plants as very inefficient energy users, despite all of their biological success. But this modest amount of energy is enough to make the forest thrive and support the complex web of terrestrial life through the food chain. The plant matter stores absorbed energy in molecular bonds, holding together hydrocarbon (hydrogen + carbon) molecules, the building blocks of cellulose and other plant tissues:

The photosynthetic reaction:

$$CO_2 + H_2O \rightarrow CH_2O + O_2$$

(atmospheric carbon dioxide + water from the soil combine in sunlight to make cellulose and other plant tissues, releasing oxygen by plant respiration)

When the plant dies, it will decay readily in the atmosphere, and its stored solar energy will return to the atmosphere. The reaction taking place during decay is essentially the reverse of the photosynthetic reaction, with oxygen reacting with dead tissues to release pungent gases and water. But if the dead plant matter settles into stagnant, oxygen-depleted water and becomes buried by sediment, it takes that stored solar energy with it. In time, the inherited energy may become even more concentrated as the molecules in the dead plant break down into less-complex forms. Under pressure and heat, the fossil plant remains transform into coal.

There is a succession of stages in coal development, from relatively low-energy forms with a small amount of concentrated carbon inside, to higher-energy forms with high relative carbon contents. The more carbon that is present, the more combustible—and economically desirable—the coal. The initial stage of coal development begins as a mat of densely packed, spongy, moist, unconsolidated plant material called *peat* (figure 21.1). When dried out, peat can be burned as a fuel, as in Britain and ancient Rome. With compaction, peat transforms into solid *lignite (brown coal),* which may still contain visible pieces of wood. Lignite is soft and often crumbles as it

FIGURE 21.1

A layer of peat being cut and dried for fuel on the island of Mull, Scotland. Coal often forms from peat. *Photo by David McGeary*

dries in air. It is subject to spontaneous combustion as it oxidizes in air, and this somewhat limits its use as a fuel. *Subbituminous coal* and *bituminous coal (soft coal)* are black and often banded with layers of different plant material (figure 21.2). They are dusty to handle, ignite easily, and burn with a smoky flame. *Anthracite (hard coal),* the highest "grade" or "rank" of coal, has the most concentrated stored solar energy, is hard to ignite, but is dust-free and smokeless. If the coal is squeezed and heated any further, its hydrocarbon molecules break down altogether under essentially metamorphic conditions, and all that remains is pure carbon—graphite, the stuff that we put in pencil leads.

We measure the heat released by burning coal in terms of BTUs, or *British Thermal Units.* One BTU is equivalent to the

FIGURE 21.2

Coal embedded with sandstone. *Photo © Parvinder Sethi*

TABLE 21.1	Varieties (Ranks) of Coal				
	Color	Water Content (%)	Other Volatiles (%)	Fixed Carbon[2] (%)	Approximate Heat Value (BTUs of heat per pound of dry coal)
Peat[1]	Brown	75	10	15	Varies
Lignite	Brown to brownish-black	45	25	30	7,000
Subbituminous coal	Black	25	35	40	10,000
Bituminous coal (soft coal)	Black	5 to 15	20 to 30	45 to 86	10,500 to 15,000
Anthracite (hard coal)	Black	5 to 10	5	86 to 98	14,000 to 15,000

1. Peat is not truly a coal, but may be thought of as "pre-coal."
2. "Fixed carbon" means solid combustible material left after water, volatiles, and ash (noncombustible solids) are removed.

amount of heat energy it takes to raise one pound of water from 62°F to 63°F. This is equivalent to 252 calories of heat. A pound of ordinary bituminous coal, the most common type of coal in the United States, typically contains 45–86% carbon and releases 10,500–15,000 BTUs (table 21.1). This is sufficient to produce electricity and is equivalent to one to two times the food energy consumed by a typical person every day. The highest-grade anthracite coal (86–98% carbon) will produce no more than 15,000 BTUs per pound. In contrast, a pound of high-octane gasoline produces 18,500 BTUs; and a pound of diesel fuel, the most "energy-dense" of all fossil fuels, produces about 19,000 BTUs. It takes an especially strong, heavy engine to combust diesel fuel; hence, its use in trucks, trains, and other bulky machinery.

Coal beds are typically interlayered with ordinary sedimentary rocks, including sandstones and shales. Beds typically range in thickness from a few centimeters to 30 meters or more. Miners dig up beds that lie close to the surface—within a few tens of meters—by **strip mining,** the complete removal of overlying rock and vegetation (figure 21.3). Strip mining is an environmentally harmful activity that destroys topsoil and leaves behind open pits that must be filled back in and replanted to curb further erosion and water pollution. But strip mining is the only way much of the world's coal supply can be safely mined. Shaft and tunnel mining provide access to deeper coal deposits (figure 21.4). This form of "deep-rock" mining is especially dangerous because of the weakness of coal beds and high concentrations of flammable gas and coal dust. In the decade before Congress established the U.S. Bureau of Mines in 1910, 2,000 persons died each year in coal-mining accidents in the United States alone. Collapse of mine walls still accounts for 50% of all accidents in the coal fields. Once the coal is mined, it is shipped as raw rubble by train, truck, barge, or freighter to power plants, foundries, smelters, and other distributors; little additional processing is needed to make it usable right away.

Coal became the major substitute for wood as a source of energy in Europe beginning in the fifteenth century. Efforts to mine coal from greater depths, even below the water table, led

FIGURE 21.3

Coal strip mine near Gillette, Wyoming. The upper half of a 30-meter-thick bed of subbituminous coal is shown here, overlain by 3 to 10 meters of overburden. *Photo by David McGeary*

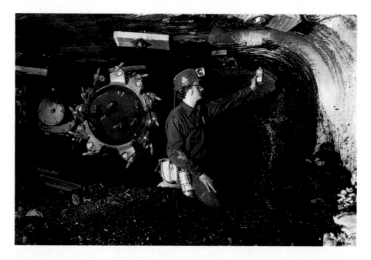

FIGURE 21.4

An underground coal mine. *Photo by Larry Lee/West Light/Corbis*

Englishman Thomas Newcomen to invent a pump in 1712 to drain the deep coal mines—the ancestor of the steam engine that started the Industrial Revolution two centuries ago. Coal was the main fuel of industrial civilization until people discovered that large amounts of petroleum could be pumped from the Earth, and that petroleum provided a less dirty, more transportable fuel with all sorts of new and exciting uses. The Coal Age gave way to the Petroleum Interval almost a century ago. In 1900, more than 90% of American energy needs were satisfied by burning coal. Today, domestic coal use is only about 25%, but this use has grown steadily again since 1975 as heavy demand for petroleum and natural gas are beginning to stretch supplies of those other two fuel sources thin.

About 88% of the present use of coal in the United States is for generating electricity. Consider the energy transformation chain from a thermodynamic perspective—solar energy is converted with great inefficiency into chemically stored energy in plants, then is burned to boil water in a coal-fired power plant—two additional energy transformations. The steam spins turbines to generate electrical energy for industry, yet another transformation of the original solar energy.

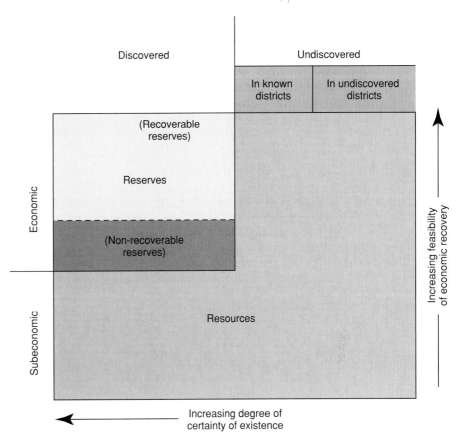

FIGURE 21.5

Important factors in the classification of reserves and resources. (Reserves are subsets of resources.)

Finally, the electricity speeds along wires through transformers and circuits to power businesses, schools, and homes. Only a fraction of 1% of the original solar energy striking our coastal swamp ends up driving a human civilization in a much-altered form millions of years later. By far, the largest percentage of this energy is lost as it is converted into forms that we find useful. While matter may be recyclable, energy is not.

When lightly burned, the most volatile ingredients in coal—particularly noxious sulfur fumes—escape to leave a new form of coal called *coke*. Coke releases more intense heat in a furnace than does ordinary coal. And it is hardly smoky at all. Because of these fortunate properties, it has become one of the most important substances in our industrial civilization, serving as the main fuel for producing steel in foundaries. Without coke, our metals would be too brittle to use in building skyscrapers, bridges, and other infrastructure.

Coal has also been converted into liquid fuels. The South Africans and Chinese, in fact, are looking at their immense coal reserves as a potential source of future automobile fuels. Less-refined, liquefied coal—*slurry*—can be flushed through pipelines stretching up to several hundred miles from mine to factory or power plant.

Resource is the term used to describe the total amount of any given geologic material of *potential economic interest,* whether discovered or not (figure 21.5). A resource can be measured directly through mining or drilling ("measured," "demonstrated," or "proved resources"), or simply inferred based upon reasonable geological guesswork and statistical modeling ("inferred," "unproved," "hypothetical," or "speculative resources"). The size of a nonrenewable resource does not change in time; it is a value that is fixed and theoretically determinable. That portion of a resource that has been discovered and is economically and legally extractable is called a **reserve.** Unlike a resource, the size of a reserve generally changes all of the time, depending upon a variety of factors (box 21.1). For example, mining or drilling of a substance will cause a reserve to shrink, especially if no new discoveries are made. Wage increases and a drop in market price could make it too expensive to continue mining, which would also reduce reserve size. On the other hand, new discoveries and new technologies making it easier to locate and mine a resource make a reserve larger. Changes in laws can also affect reserve sizes. Large areas of government-owned land are off limits to mining and drilling, so any geologic materials under these areas are not legally extractable and cannot be included in reserves. Opening more land to extraction can therefore increase reserves.

Not all of a reserve may be recoverable. For example, in the United States, the total coal resource is on the order of 2 trillion tons. Only about 25% of this (470 billion tons) makes up the U.S. coal reserve, however. But not all of this reserve can actually be extracted. Some has to be left behind during mining for safety reasons to support mine pillars, to prevent landslides, or to avoid water pollution problems, and so on. In fact, only about 60% of the coal in any bed that is mined deeply can be removed.

IN GREATER DEPTH 21.1

Copper and Reserve Growth

As the prices of metals and the energy used to extract them fluctuate, so do the potential profits from minerals. For example, in 1900, copper could be mined at a profit only if its concentration in ore exceeded 5%. By the early 1980s, this profit level dropped to 0.5%, and the world's recoverable copper reserves rose to a half billion tons. Since then, the world has consumed about 150 million tons of copper, but introduction of recycling (which now provides the United States with almost as much copper as direct mining), the introduction of substitutes for copper (such as fiber-optic cable) and the discovery of new reserves actually increased world copper reserves to 650 million tons by 2001. Mineral markets and reserves are volatile and erratic, with reserves generally shrinking as market process declines and swelling as prices rise, most often due to investor speculation and temporary supply shortages and gluts. Nevertheless, there are some persistent trends: The world almost always appears to have large reserves of iron and aluminum, moderate reserves of copper, lead, and zinc, and scanty reserves of gold and silver. These levels reflect the relative abundance of these resources in nature. There is no sign that we are about to "run out" of any of these metals.

Other challenges, however, loom on the horizon because the sizes of mineral reserves are tied critically to the price of energy. It takes very large amounts of energy to mine, refine, process, and transport minerals for use. Mineral extraction, in fact, is the most energy-intensive industry in world. Over most of the past 120 years, overall unit energy costs (adjusted for inflation) have not grown appreciably and have held generally steady, providing a reliable platform for industrial growth. If the long-term cost of energy increases, however, we might expect the sizes of mineral reserves to drop in response, simply because it will become so much more expensive to mine at a profit.

The value for strip and open-pit mines is somewhat greater—80–90%. The *recoverable reserve* for coal in the United States, hence, is only about 13% (270 billion tons) of the total known coal resource in the country. We will never be able to exhaust all of coal in the world, but this is only because we will never be able to mine it all safely *at a profit*.

Considering all of these definitions together, it is easy to see why people become confused by arguments about geologic resources. The fact that many individual companies and governments often refine the classification of resources to suit their own needs makes this topic even more challenging. Nevertheless, these terms are very important in the business world. Whether an investor will put money into developing a mine, drilling a well, or exploring further depends upon the extent to which geologic prospecting is able to unravel the structure and strata of the crust, and the degree to which any identified reserve can actually be recovered.

About 25% of the world's total coal reserves lie in the United States (figure 21.6). Russia has the second-most-abundant reserve base (20–25%), with China and India making up another 25%. At current rates of consumption, U.S. recoverable reserves will be exhausted in about 250 years. A similar level of depletion is occurring worldwide. Of course, the many factors just mentioned may alter the sizes of the world's coal reserves, and logarithmic economic growth will significantly impact the longevity of any resource supply. Given present *trends*, we are likely to have consumed most of the reserves in the U.S. well before 250 years have passed, leaving only expensive residues behind for future mining.

Petroleum and Natural Gas

A current argument rages between economists, who believe that abundant supplies of new petroleum can yet be discovered, and many resource geologists, who caution that most of the regions that contain "new" oil have already been explored and, in fact, are being rapidly depleted. Who is right? We won't

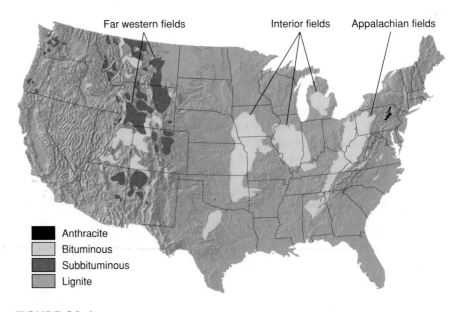

FIGURE 21.6

Coal fields of the United States. Alaska also has coal. *From U.S. Geological Survey* know for sure until we complete global exploration and experience a peak in global petroleum production, but odds are that the geologists know something that many more optimistic business people don't: The geologic factors responsible for creating a rich petroleum reservoir are special indeed, and greatly limit the chances for petroleum to form under natural conditions.

The origin of **petroleum,** or "oil," (the words are synonyms) differs significantly from that of coal. Instead of a coastal swamp, imagine well-lit, coastal seawater, or a sparkling, tropical lagoon, light-green with suspended microscopic life-forms including plankton, foraminifera, diatoms, and other organisms. These life-forms thrive continuously in waters well supplied with nutrients from upwelling marine currents and rivers entering the sea nearby. Plankton-rich coastal seas are the "rain forests" of the ocean. As on land, the food web depends primarily on photosynthesis, the capture of solar energy in the making of living tissues. Dead organic matter shed from this floating rain forest can decay underwater just as readily as it does onshore, but some of it will also settle to the sea bed where sediment accumulates and become sealed in a shallow, oxygen-deprived burial. This is where new oil is born.

Not all shorelines feature the conditions optimal for concentrating dead organic matter in this way. Shallow-bottom currents commonly sweep away organic debris and disperse it across the deeper ocean floor farther offshore. In certain kinds of coastal marine basins, however, nutrients can become highly concentrated (figure 21.7). These basins are separated from the wider ocean by gentle underwater ridges, termed marine *sills* (not to be confused with "igneous sills").

If the coastal basin lies in a hot, arid climate, considerable evaporation will take place, drawing in shallow surface water from the ocean beyond to make up for the evaporative loss within the waters of the basin. This inflow displaces deeper water, forcing it seaward across the sill as a bottom current that carries away all nutrients and fine organic matter with it. The bottom of the basin becomes a *nutrient desert* that will never generate an oil deposit (figure 21.7A).

However, if a large river pours into the basin, especially in a cooler, less-evaporative climate, the opposite pattern of circulation will take place: Brackish river water spreads as a sheet across the basin, driving a shallow current that heads away from the shore. To replace this water, the bottom currents flow in the opposite direction, sweeping across the sill toward the shore. Organic debris may be swept up by this inflowing current, but only temporarily. The organic matter soon rains back down again, trapped within the basin because the current is not strong enough to carry particles in suspension very far. Sediment from the river mouth pours across the basin floor, sealing much of this debris in place as well. A perfect *nutrient trap* is formed—a future oil reservoir (figure 21.7B).

Not many places in the world combine all of the features that make a good nutrient trap; in fact, the only large area today is the Eastern Mediterranean basin. Many smaller areas, notably coastal lagoons with organic-rich muds, provide candidates for future oil as well. In terms of a tectonic framework, nutrient traps may form most easily in young marine seaways

A Marine Nutrient Desert

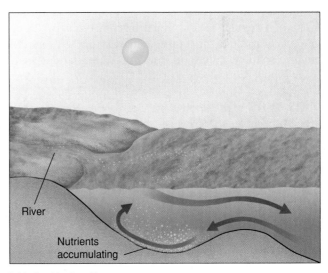

B Marine Nutrient Trap

FIGURE 21.7

Contrasting coastal basins. In "nutrient deserts," the pattern of currents disperses nutrients (A). In "nutrient traps," they become concentrated (B).

and in areas where ocean basins are closing as continental landmasses come into collision, as in the Mediterranean.

A nutrient trap is only the first step in developing petroleum. Next, the buried hydrocarbons must be heated enough to break down, or "crack," into simpler molecules that are useful for human purposes. This takes place as a result of anoxic sedimentary burial combined with isostatic subsidence, such as beneath a river delta or along a passive continental margin. As more and more sediment accumulates, the lithosphere beneath compensates by sinking, accommodating even larger amounts of accumulating sediment.

Initially, the buried hydrocarbons will partly disintegrate thanks to the activity of hungry bacteria. Chemical reactions then set up with clay minerals in the sediment. A gooey, hydrocarbon-rich sediment, termed a *sapropel,* forms. The total

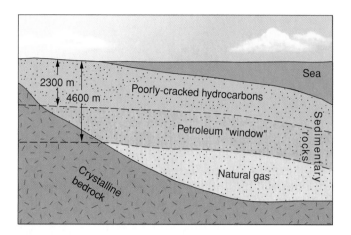

FIGURE 21.8

Typical depths of hydrocarbon cracking. The "petroleum window" lies between 2,300 and 4,600 meters (7,500–15,000 feet) down. Depth will vary somewhat, depending on the geothermal gradient.

hydrocarbon content of a sapropel layer is typically no more than a few percent of the total sedimentary deposit, but this is enough to supply a future oil reservoir. As burial deepens the sapropel, it heats up, at a rate of 8°C (14°F) for every thousand feet beneath the surface. At a depth of 2,300 meters (7,500 feet), it is hot enough—around 82°C (180°F)—for the complex hydrocarbon molecules to crack into petroleum. Ever-simpler molecules form with greater natural cooking, until at around 4,600 meters (15,000 feet) deep, the hydrocarbons have broken down altogether into **natural gas.** The *petroleum window* is the space between 2,300 and 4,600 meters (7,500 and 15,000 feet) down through which a sapropel must pass if it is to transform into oil (figure 21.8). Obviously, not all sapropels become so deeply buried and, in some instances, some sapropels may be too deeply buried to yield up anything but natural gas.

Once petroleum has formed, it next must accumulate in concentrations that can be drilled and pumped. Under deep burial conditions, pressure easily squeezes the fluid up into overlying permeable rocks, and it may continue to migrate all the way to the surface to issue from the Earth as tar and oil seeps *(breas)*. The first uses of oil, in providing mortar for mud bricks in ancient Sumeria, exploited such sites (figure 21.9).

Natural oil seeps do not concentrate enough oil, however, to be of interest in the modern economy. Instead, petroleum geologists look for places where upward-infiltrating oil concentrates in specific locations *underground*. These concentrations develop readily in regions where tectonic activity has deformed older sedimentary rocks—for example, along former passive margins that become active thanks to later onset of plate convergence. The geologists seek to identify three specific features before drilling: (1) a **source rock**—the original sapropel, such as an oil shale, containing organic matter that is converted to petroleum; (2) a **reservoir rock,** usually sandstone or limestone, that is sufficiently permeable and porous to transmit and store the petroleum as it migrates surface-ward; and (3) a **structural (or oil) trap,** a place where impermeable rock (called "trap rock") prevents any further upward percola-

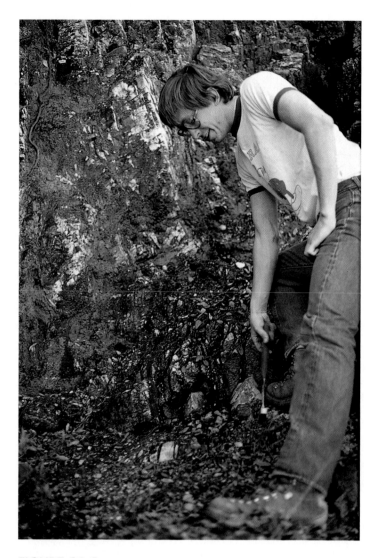

FIGURE 21.9

A brea, or natural oil seep in a hill slope near Santa Paula, California. *Photo by Richard Hazlett*

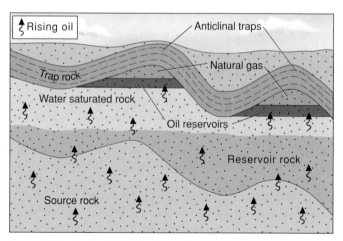

Accumulation of petroleum into reservoirs

FIGURE 21.10

Features related to petroleum reservoirs.

tion of petroleum (figure 21.10). Natural gas requires the same conditions as oil for accumulation, and drillers can never be quite certain how much natural gas they may encounter when they first begin exploiting a potential petroleum deposit. In fact, as you might suppose, some prospects yield up nothing but natural gas.

Figure 21.11 depicts several types of structural traps for oil and gas. Some types of traps are more abundant in particular regions than in others. For example, anticlines and domes (described in chapter 15) create the most common oil traps in the Persian Gulf; anticlines and faults are important trap-formers in southern California's oil fields; and salt domes account for most of the petroleum reserve in the Gulf of Mexico. Where oil and water occur together in folded sandstone beds, the oil droplets, being less dense than water, rise within the permeable sandstones toward the top of the fold. There, the oil may be trapped by impermeable shale overlying the sandstone reservoir rocks. Because natural gas is less dense than oil, the gas collects in a pocket under fairly high pressure, on top of the oil. It is important to bear in mind that this layered pool of fluids does not fill a hollow, underground chamber, like a flooded cave, but is merely *filling all of the pore spaces* in a highly permeable sedimentary rock (figure 21.10). Nevertheless, we refer to this accumulation as an *oil reservoir.* Because of the surface tension of fluids clinging to the walls of the pores within this rock, only a certain fraction of the petroleum may ever be extracted and, in most cases, this is less than half of the oil present.

The gaseous pocket at the top of a reservoir is an especially valuable feature that drillers can put to good use. When a drill hole first penetrates an oil reservoir, the pent-up gases within may drive the petroleum all the way to the surface so that no pumping has to be done whatsoever. This "fluid-pressure effect" saves oil companies a tremendous amount of money and, in fact, may make all the difference between a successful drilling operation and one that needs to be abandoned. The celebrated gushers of many oil photos showing the discovery of new oil reservoirs illustrate the very high fluid pressures that gas may help generate in some deposits.

In time, fluid pressure diminishes and an oil field becomes less economical to operate. Remaining oil may be flushed out of the ground by "flooding" the reservoir with injected ground

A Anticline

B Normal fault

C Thrust fault

D Sandstone lenses

E Sandstone pinchout

F Unconformity

G Reef (a small "patch" reef)

H Salt dome

FIGURE 21.11

Major types of petroleum traps. In all cases, impermeable rock encloses or caps the petroleum.

water. The ground water drives the petroleum ahead of it from the area of injection wells toward oil wells for removal (figure 21.12). Developers have also used steam to drive out the oil. As much as one-third of the original reserve in an oil field may be extracted using these *secondary recovery* methods.

One important consideration in operating an oil field is a factor called *energy return on energy invested (EROEI).* This is the ratio of the amount of energy extracted versus the amount of energy put into the extraction process. If it takes more energy to get petroleum out of the ground than is derived from the sale and consumption of that petroleum, then it is no longer worth operating the oil field. During the heyday of petroleum exploration in the 1940s, EROEI for newly discovered oil fields was typically around 100:1. Since then, less and less oil has been discovered. The peak in global new discoveries occurred in the mid-1960s. At present, for every four barrels of oil that we consume, only one barrel of new replacement oil is

"Horsehead" beam oil pumping unit

Water-injection well

Water-injection well

Impermeable rock

Natural gas

Oil

Water

FIGURE 21.12

Water is often injected to drive additional, hard-to-get oil out of the ground.

discovered. Because of the surface tension effect in petroleum-bearing pores, and because oil drilling is increasingly taking place at greater depths or in more remote places (thanks to the exhaustion of older fields and rising global demand), mean global EROEI has declined to around 8:1. When it reaches 1:1, the industrial world will have to find a new energy source. Long before this happens, one would hope, we will have turned to new energy technologies and a much different kind of civilization. One of the implications of EROEI is that we will never really run out of oil. However, it will become too expensive for us to continue exploiting oil in large quantities, perhaps within your lifetime.

Oil fields are regions underlain by one or more oil reservoirs. Over 60% of the world's oil comes from exploitation of oil fields in the Middle East. Most major oil fields outside of the Middle East are presently in production decline, including the two largest in the United States in West Texas and Alaska (figure 21.13). Hope for large oil finds in Central Asia has not paid off as well as expected, though there may still be prospects for finding large amounts of oil in Siberia and around the Asian Arctic coast. To develop such reservoirs will be quite costly, however. The geopolitical implications of increasing dependency of the industrialized world on politically unstable regions, and of a world scrambling to meet demand for a shrinking resource, are ominous portends of conflict. The greatest conflict of the last century, World War II, was essentially an oil war, and this serves as a reminder, especially to

FIGURE 21.13

Major oil fields in North America. The amount of oil in a field is not necessarily related to its areal extent on a map. It is also governed by the vertical "thickness" of the oil pools in a field. The fields with the most oil are in Alaska and east Texas. *From U.S. Geological Survey and other sources*

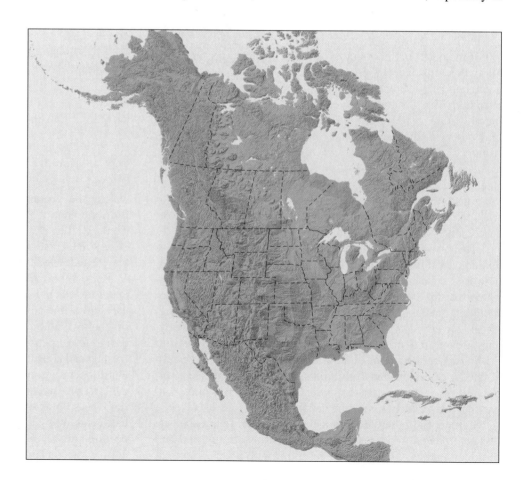

politicians, that geology matters to world affairs in the most basic way imaginable. The peace and organization of human society are basically dictated by the choice of the resources we choose to exploit.

Society demands a wide range of oil and gas products. Table 21.2 lists the various types of hydrocarbons that we artificially crack from oil delivered to refineries. In the booming early days of oil, from the late 1850s until around 1900, the sole product of interest was kerosene for lighting lamps. No more than about 40% of any barrel of oil pumped would go to market as this product. Since there was no use for the heavier compounds of oil, such as asphalt and diesel fuel, this material was often simply burned near the well, creating awful palls of smog. The lighter stuff, including gasoline, was often dumped in rivers and streams. Gases simply vented to the air, worsening the already severe environmental impact. Subsequent demand for oil products arose with invention of the automobile and the conversion of military forces to petroleum-based transport. Asphalt—essentially the dead bodies of countless, tiny marine organisms—ended up paving roads to minimize dust and facilitate high-speed driving. Kerosene became aviation fuel. Of the gasolines, octane (C_8H_{18}) proved best for performance (speed, power) in car engines and produced the least exhaust upon combustion. But because refineries have never been able to produce enough pure octane to meet demand, we have introduced substitutes ("reformulated fuels") to provide the same fuel services. Some of these substitutes (e.g., leaded fuels and MTBE—methyl tributyl alcohol) have proven to be costly environmental hazards. Highly complex hydrocarbons, such as polyethylene, end up in supermarket "plastic" bags. Indeed, from nylon and computer components to food production and pharmaceuticals, it is hard to see where petroleum products *aren't* used in the modern world.

The yield on the original solar energy that produces oil products is amazingly low, thanks to the laws of thermodynamics. Less than .01% of the energy originally contained in the marine plant life that ends up as petroleum provides the power we use to furnish our civilization, especially transportation.

While coal resources and reserves may be estimated or ascertained with reasonable confidence, geologists are less confident about the amount of petroleum in the world because it is more widely dispersed through the crust and difficult to locate. The U.S. Geological Survey (USGS) assessed the world's recoverable oil for the year 2000 and estimated the global reserve to be 870 billion barrels (a barrel contains 42 gallons, or 159 liters, of oil). The total resource, of course, is much larger than the reserve. The USGS estimate is around 2,300 billion barrels (including the reserve); much of this in difficult-to-reach localities under the sea floor or in subarctic settings (figure 21.14). At present, the world consumes 1 billion barrels of oil every ten days, giving us thirty to forty years before the current reserve disappears. Forecasts like this, however, fail to take into account changing reserve sizes and consumption rates. Nevertheless, they are useful in expressing how much ready reserve we have at hand.

TABLE 21.2	Types of Cracked Petroleum-Related Hydrocarbons and Their Uses (Listed in Order of Increasing Complexity)		
Name	Chemical Formula	Type of Hydrocarbon	Use
Methane	CH_4	Natural gas	Fertilizer manufacture; source of hydrogen for fuel cells
Ethane	C_2H_6	Natural gas	Fertilizer manufacture; source of hydrogen for fuel cells
Propane	C_3H_8	Gas condensates*	Cooking stoves, home heating, lanterns
Butane	C_4H_{10}	Gas condensates*	Cooking stoves, lanterns, lighters, home heating, soldering irons
Hexane	C_6H_{14}	Gasoline	
Heptane	C_7H_{16}	Gasoline	
Octane	C_8H_{18}	Gasoline	Isooctane, a form of octane, is the best kind of gasoline for internal combustion engines
Nonane	C_9H_{20}		
Decane	$C_{10}H_{22}$		

Successively Heavier Hydrocarbon Molecules:

1 Kerosenes and heating oils—aviation fuel, home heating

2 Diesel fuels—transportation fuel for trucks, trains, ships

3 Heavy crude oils ($C_{17}H_{36}$-$C_{22}H_{46}$)—lubricating and engine oils

4 Asphalts, waxes, greases, paraffins—paving, machinery lubricating

5 Plastics, polyethylene—computer frames, shopping bags, toys, CDs, etc.

*Also called "natural gas liquids," "drip gases," or "white gold"

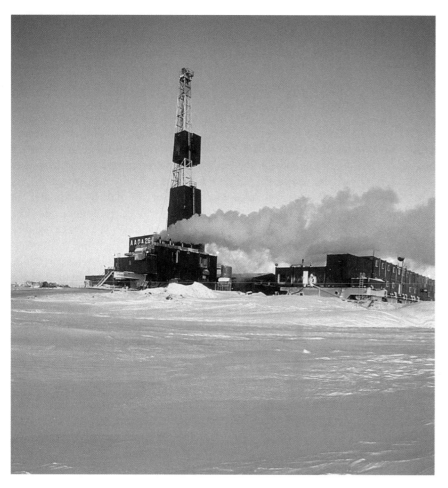

FIGURE 21.14

Drilling rig on Alaska's North Slope. *Photo by B. P. America, Inc.*

World natural gas resources are even more difficult to estimate. The Energy Information Agency estimates that nearly 1,300 trillion cubic feet (tcf) of recoverable natural gas exist in the United States. Current U.S. levels of consumption amount to around 22 tcf per year—around 25% of the world's total—giving a seventy-five-year reserve of natural gas from domestic sources alone. Most natural gas comes from just six states; Louisiana, Texas, Oklahoma, New Mexico, Colorado, and Wyoming. Reserve estimates for natural gas in the western United States would certainly increase—though it's not known by how much—by opening certain public lands to gas extraction, including some national forests and monuments. This has made the further development of gas reserves a controversial issue.

The ultimate global recoverable reserve of natural gas is presently thought to be on the order of 5,200 tcf, 35% of which lies in the Middle East and 42% in Europe and Russia and its neighbors. Unfortunately, it is much harder to transport natural gas than petroleum, requiring pipelines and LNG (liquid-natural gas) tankers for widespread distribution. Nevertheless, natural gas has become vital in the modern world. It is used to heat homes, cook food, make synthetic NPK (nitrogen-phosphorus-potassium) fertilizers for agriculture, produce electricity, and power fuel cells. In fact, the much-discussed "Hydrogen Revolution" may be launched thanks to cheap supplies of natural gas. Another bonus is that it is a less-polluting fuel than coal or petroleum, and has a high EROEI—from 10.3 to 6.3, depending upon whether extraction is done from onshore wells or offshore.

As with petroleum, the mean productivity of gas wells in the United States is on the decline—from 435,000 tcf/well/day in 1971 to only about 150,000 tcf/well/day today. Some geologists warn of a production "cliff"—a precipitous sudden drop in supply—when gas fields near exhaustion. They caution that it is best for the nation to maintain a mix of energy and material resources to dampen the impact of such an event on the U.S. economy.

Coal Bed Methane

Coal beds themselves may prove to be a major source of natural gas in the future. When coal forms, water and natural gas in the form of methane are trapped in the fine pores, pockets, and fractures that speckle and lace the interior of the coal. Pumping the water out lowers pressure and releases gas in huge quantities. Coal can store six to seven times more gas than an equivalent amount of rock in an ordinary natural gas field. A problem arises with respect to the water removed during pumping. Coal-water gets saltier the deeper the deposit, and disposal of salty water into surface watersheds seriously degrades water quality. Ground water supplies may also be contaminated during gas extraction. In any event, there is a considerable amount of **coal bed methane** in the United States. The overall resource may exceed 700 tcf, but only about 100 tcf is likely to be economically recoverable. The U.S. Geological Survey estimates that the total coalbed methane resource worldwide might be as high as 7,500 tcf.

Heavy Crude and Oil Sands

Heavy crude is dense, viscous petroleum. It may flow into a well, but its rate of flow is too slow to be economical. As a result, heavy crude is left out of reserve and resource estimates of less viscous "light oil," or regular oil. Heavy crude can be made to flow faster by injecting steam or solvents down wells, and if it can be recovered, it can be refined into gasoline and many other products just as light oil is. Most California oil is heavy crude.

Oil sands (or **tar sands**) are asphalt-cemented sand or sandstone deposits. The asphalt is solid, so oil sands are often mined rather than drilled into, although the techniques for

ENVIRONMENTAL GEOLOGY 21.2

Flammable Ice: Gas Hydrate Deposits—Solution to Energy Shortage or Major Contributor to Global Warming?

Gas hydrates (also called *methane hydrates*) are an unusual mixture of ice and gas in which methane (one of the gases in natural gas) is trapped in ice crystals (box figure 1). These are found in extreme environments, notably permafrost in polar regions and in the deep ocean floor. If lit, a piece of gas hydrate ice will burn with a red flame. The amount of gas hydrate in the ocean floors is staggering. Although estimates of gas hydrate resources vary, it appears that there is at least twice the amount of potential fuel tied up in gas hydrates as in all petroleum and coal combined.

Commercially exploiting gas hydrate deposits presents formidable challenges. Most of the deposits are in lenses frozen in sediment at deep-ocean floors, beneath a kilometer or more of water. There, gas hydrate is stable because of the cold and high pressure. If pressure is reduced or the substance heated, it becomes unstable and the gas escapes. Mining anything at this depth is difficult, but trying to remove the icy substance from the sediment and get it to the surface without losing the methane is an extreme technological problem.

Gas hydrate could significantly exasperate global warming. Methane, like carbon dioxide, is a greenhouse gas. However, unlike CO_2, methane will only stay in the atmosphere for around ten years. But methane reacts with oxygen in the atmosphere and produces CO_2, which remains in the atmosphere indefinitely. Significant volumes of methane could be released if gas hydrate sediments are disrupted by submarine landslides or other means. They could also be released if the oceans warm even slightly.

Additional Resources

E. Suess, G. Bohrmann, J. Greinert, and E. Lausch. 1999. Flammable Ice. *Scientific American* (November): pp. 76–83.

The November 2004 issue of *Geotimes* features three articles about gas hydrates: G. R. Dickens. Methane hydrate and abrupt climate change: pp. 18–22. T. S. Collett. Gas hydrates as a future energy resource: pp. 24–27. N. Lubick. Detecting marine gas hydrates: pp. 28–30.

These may be accessed on the web at www.geotimes.org Go to "Archives" to find the articles.

BOX 21.2 ■ FIGURE 1

Gas hydrate burning. Methane or other gases are trapped in ice-like cages. The chunks of ice sustain their own combustion. *Photo by J. Pinkston and L. Stern, U.S. Geological Survey*

reducing the viscosity of heavy crude often work on oil sands as well.

The origin of heavy crude and oil sands is uncertain. They may form from regular oil if the lighter components are lost by evaporation or other processes. Oil sands and asphalt seeps at Earth's surface (such as the Rancho La Brea Tar Pits in Los Angeles) probably formed from evaporating oil. But some heavy crudes and oil sands are found as far as 4,000 meters underground. Most have much higher concentrations of sulfur and metals, such as nickel and vanadium, than does regular oil.

Some geologists believe that oil sands represent oil that arose from source rocks but never became trapped and concentrated by structural traps.

The best-known oil sand deposit in the world is the Athabasca Tar Sand in northern Alberta, Canada. Almost 20% of Canada's present oil production comes from these oil sands. Counting this unconventional resource, this gives Canada the second-largest oil reserves in the world, after Saudi Arabia. Venezuela has even more oil sand than Canada. Oil-hungry countries such as the United States view these deposits with keen interest. Unfortunately, the EROEI of extracting petroleum from oil sands is substantially lower than that of conventional oil—close to 1:1 by some estimates. This is mostly due to the need to dilute the viscous, heavy oil ("bitumen") to get it out of the sand. Natural gas, gas condensates, hot water, steam, and naptha (an aromatic solvent) are all used—and each of these flushing compounds is, in and of itself, an energy material or requires energy to produce.

The time-tested method of extracting the bitumen requires mining the ground directly. For every 2 tons of earth processed, only one barrel of oil can be made. Disposing of this material, not to mention dealing with environmental concerns such as natural habitat destruction and water pollution, has raised serious questions about the future of the industry. New technology more recently has allowed miners to extract the oil from deep underground without disturbing the surface. This process involves mixing the oil sands in place with hot water and then pumping the slurry through a pipeline to the processing plant. Questions of pollution and low EROEI remain, but this is a definite improvement, and it appears that the oil sand industry has established a firm future for itself in the world economy.

Oil Shale

Oil shale is a black or brown shale with a high content of solid organic matter from which oil may be extracted by distillation. The best-known oil shale in the United States is the Green River Formation, which covers more than 40,000 square kilometers in Colorado, Wyoming, and Utah, with deposits up to 650 meters thick (figures 21.15 and 21.16). The oil shale, which includes numerous fossils of fish skeletons, formed from mud deposited on the bottom of large, shallow Eocene lakes. The organic matter came from algae and other organisms that lived in the lakes.

The Green River Formation includes more than 400 billion barrels of oil in rich beds that yield over 25 gallons of oil per ton of rock. Another 1,400 billion barrels of oil occur in lower-grade beds yielding 10 to 25 gallons per ton. An estimated 300 to 600 billion barrels may be recoverable.

Relatively low-grade oil shales in Montana contain another 180 billion barrels of recoverable oil in shale that should be economical to mine because of its high content of vanadium, nickel, and zinc. Therefore, oil shale can supply potentially vast amounts of oil in the future as our liquid petroleum runs out.

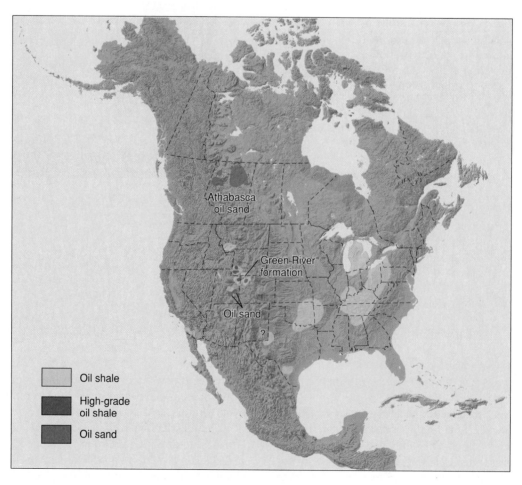

FIGURE 21.15

Distribution of major deposits of oil sand and oil shale in the United States and Canada. *From U.S. Geological Survey and other sources*

FIGURE 21.16

Cliffs of oil shale that have been mined near Rifle, Colorado. *Photo by William W. Atkinson, Jr.*

A few distillation plants extract shale oil, but the current price for oil makes shale oil uneconomical. A large price increase for oil may make large-scale production of shale oil feasible in the future.

The mining of oil shale can create environmental problems. During distillation, the shale expands, creating a space problem. Spent shale could be piled in valleys and compacted, but land reclamation would be troublesome. A great amount of water is required, for both distillation and reclamation, and water supply is always a problem in the arid American West. New processing techniques that extract the oil in place without bringing the shale to the surface may eventually help solve some of the problems and lower the water requirements. It is possible to burn fractured oil shale in large, underground excavations. The heat separates most of the oil from the rock; the oil can be collected as a liquid. (The fires, however, would be hard to control and would affect groundwater levels.) Another proposal involves heating the shale with radio waves or microwaves to separate the liquid oil from the rock, but to use either technique, a substantial amount of shale must first be removed.

Uranium

The metal *uranium,* which powers nuclear reactors, occurs as *pitchblende,* a black uranium oxide found in hydrothermal veins, or, much more commonly in the United States, as yellow *carnotite,* a complex, hydrated oxide found as incrustations in sedimentary rocks. Ground water easily transports oxidized uranium, which is highly soluble. Organic matter reduces uranium, making it relatively insoluble, so uranium precipitates in association with organic matter.

Most of the easily recoverable uranium in the United States is found in sandstone in New Mexico, Utah, Colorado, and Wyoming, some of it in and near-petrified wood. In the 1950s uranium boom, western prospectors looked for petrified

logs and checked them with Geiger counters. Some individual logs contained tens of thousands of dollars worth of uranium. Some petrified logs have so much uranium that it would be dangerous to keep them as souvenirs. Most of the uranium, however, is in sandstone channels that contain plant fragments.

Organic phosphorite deposits of marine origin in Idaho and Florida also contain uranium. The uranium is not very concentrated, but the deposits are so large that overall, they contain a substantial amount of uranium. The black Devonian shales of the eastern United States also contain uranium. These shales are really low-grade oil shales (figure 21.15). Uranium may be recovered from phosphorites or shales as a by-product or another mining operation.

The principal use of uranium at present is to provide power for electricity-generating nuclear reactors, although uranium was also used to make tens of thousands of nuclear warheads during the Cold War. Some naval craft use uranium to power ship-borne nuclear engines, but concerns about accidents and radioactive pollution greatly limit expansion of this form of transportation. At present, nuclear reactors produce 10% of the energy needs of the United States. France is the industrialized nation most dependent upon nuclear power, which generates 75% of its electricity.

To be useful in a nuclear reactor, the uranium minerals must be processed to concentrate the unstable isotope uranium-235 (^{235}U), the main reactor fuel, which as we saw in chapter 8 is much less abundant than the stable form of uranium, ^{238}U. Only 0.7% of the mass of natural uranium consists of ^{235}U. The refining of uranium ore to concentrate ^{235}U requires considerable energy, metals, and other resources. This contributes to the rather moderate EROEI level of nuclear power—around 4.5. In all, about 65,000 tons of uranium ore must be produced every year to satisfy the needs of the world's 430 nuclear reactors, which supply about 7% of the world's total energy needs.

Nuclear bombs require much richer concentrations of ^{235}U than do reactors—about 300 times as much; hence, bombs require more resources to produce. Diversion of resources for nuclear weapons programs has been huge in the past; from 1940 through 1996, about 11% of all federal spending (30% of all military spending) in the United States went into the nuclear weapons program—a whopping $6 trillion. The former Soviet Union may have invested even more heavily in building nuclear weapons.

Once processed, the ^{235}U goes into *fuel rods,* which are inserted into the cores of nuclear reactors (figure 21.17). A reactor core consists of a *containment vessel,* or reservoir, lined with steel and concrete to protect outsiders from harmful radiation. The vessel is filled with a fluid called a *moderator,* usually water. Each fuel rod is about the size of an automobile, but because of the high atomic weight of its constituent materials, weighs around 65,000 pounds. It contains only about 2–3% ^{235}U (most of the uranium is harmless ^{238}U), but this is enough to get a chain reaction going among the many fuel rods immersed in the moderator.

As the nucleus of a ^{235}U atom disintegrates, heat and neutrons escape and the uranium atom transforms into thorium-231,

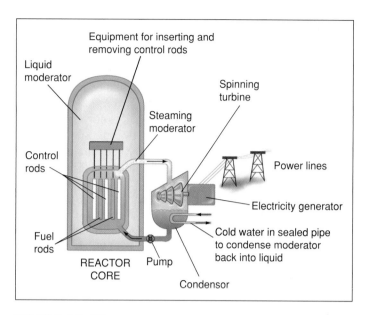

FIGURE 21.17

Basic design of a nuclear reactor. This shows operation of a BWR (boiling water) type reactor, the simplest design.

which in turn disintegrates into radium-227. (Over a period of twenty-two years, radioactive decay eventually changes the original uranium atom into lead, which itself is a toxic, though stable, substance.) The escaping neutrons may be slow-moving or fast. The slower neutrons penetrate the nuclei of neighboring ^{235}U atoms, triggering their decay as well and, in short order, the full, heat-generating chain reaction is underway. The moderator fluid helps to slow the neutrons, thus stimulating the chain reaction even further. The fluid also gets very hot, reaching the boiling point and generating steam to drive turbines for electricity just as in an ordinary coal-fired power plant. (In some nuclear reactors the moderator is passed through a heat-exchanger to boil a secondary supply of water for the turbines. This keeps radioactivity out of the turbines.)

To control the nuclear reaction, *control rods* made of carbon, boron, or cadmium (which is quite poisonous) must be inserted into the moderator. These soak up neutrons and can stop a chain reaction altogether. Careful manipulation of control and fuel rods brings a reactor to just the right level of heat production for energy-generation purposes. Nuclear accidents may take place if this balance in control is lost.

There are tremendous environmental problems associated with the disposal of fuel rods once their ^{235}U has depleted. A typical fuel rod has a service lifetime of only about three years. Even though most of its ^{235}U may be gone, neutrons released during a chain reaction will transform ^{238}U into deadly plutonium-239 (^{239}Pu). Spent fuel rods, hence, are quite dangerous. Temporary storage of used fuel rods takes place in a pool near the reactor core at a plant site. Later, the spent rods may be transported and reprocessed to extract the ^{239}Pu for nuclear weapons development. This is a special concern for persons who monitor nuclear proliferation—"rogue states" that have built nuclear reactors, ostensibly for "peace-

ful purposes," can easily create the fuel to build nuclear weapons. The ^{239}Pu may also be used as a fuel in a different kind of reactor, a *breeder reactor,* which greatly enhances the ability of uranium to produce energy. Full conversion of nuclear reactors to breeder designs could extend the effective uranium reserve lifetime sixty-fold at current consumption rates, providing the world with electricity for several millennia to come. Breeder reactors, unfortunately, have notably higher potential than conventional nuclear power plants for disastrous accidents. The moderator in conventional breeder reactors, liquid sodium, is very sensitive and explosive, and reactor core blasts into the kiloton range are possible.

Spent fuel rods and other nuclear waste, particularly ^{239}Pu, must be stored someplace out of contact from people for a long period of time—as long as 250,000 years. Many proposals have arisen to do this, including "science-fiction" scenarios that require shooting this deadly waste into the sun or placing it in subduction zones. In the United States, the "permanent" waste repository at Yucca Mountain in southern Nevada will house the nation's reactor wastes in welded pyroclastic rocks nearly 300 meters (1,000 feet) underground. The tuff contains zeolites, which are natural sponges that absorb escaping radioactivity. The Yucca Mountain site is not without controversy—the tuff is not impermeable, young faults lace the region and zeolites can move in ground water in colloidal form. Nevertheless, this well-studied repository is certainly far better than the alternative of leaving fuel rods in storage sites next to over 100 nuclear power plants scattered across the country. In Scandinavia, a similar repository stores nuclear wastes in a granite bedrock vault underneath the Baltic Sea. The advantage of this site is that any escaping waste will remain confined in saline ground water beneath the ocean floor, rather than infiltrate water supplies tapped by people on surrounding lands. Many other waste facilities have been established, but it is beyond the scope of this book to consider them all. In all cases, they involve the shallow, underground storage of nuclear materials.

Present world reserves of uranium (at about $130 per 2 kilos of uranium) are around 4.4 million tons. Kazakhstan and Australia have about half of these reserves, but Canada is the largest producer of uranium in the world. Reserves at current consumption rates would last about seventy years. The size of the total uranium resource is unknown given incomplete exploration and the wide dispersion of uranium-bearing minerals, even in ordinary rocks such as granite. In any case, it is likely that the largest, most easily accessible uranium deposits have already been identified and are undergoing exploitation.

Geothermal Power

In discussing geothermal ("Earth-heat") power, it is important to understand the difference between *power* and *energy.* Energy is the "ability to do work." Power is a *concentration of energy,* or the rate at which energy may be consumed. There is an enormous amount of heat *energy* escaping from Earth's interior every day. In the upper 10 kilometers of Earth's surface within the United States alone, there is the heat equivalent of 1,000 trillion tons of coal—enough energy to satisfy the country's needs for 100,000 years! But this energy is too dispersed to be exploitable anywhere but in a few places. It would take an area about the size of a football field to provide the escaping heat energy needed to power a single, 60-watt, high-efficiency (20%) light bulb. That is not enough *power* to be of any practical use.

Happily, there are areas of unusually hot, water-saturated rock around young, cooling plutons and volcanic areas that can be exploited for energy development. The temperature of the crust in these "high-power" areas rises to as high as 350°C or more at depths of 1 to 3 kilometers—enough to turn ground water into steam if it is pumped all the way to the surface. (Water under great pressure will not boil.) The escaping steam, in turn, can be channeled into turbines to generate electricity. If the hot ground water occurs in a confined aquifer, so much the better; the ground water will require little, if any, effort to extract as high fluid pressure drives it surfaceward through boreholes.

The world's largest geothermal power plant is at the Geysers in the Coast Range of northern California. This 1,000-megawatt facility provides the energy needs for a million people, though its level of production has been declining steadily since 1980, underscoring the point that geothermal energy is a somewhat fragile resource. The most important reason for this is that groundwater supplies are withdrawn faster than nature can replenish—and reheat—them. Water is an excellent natural carrier of heat energy. Dry rock acts as an insulator, or a slow conductor of heat. Water can be injected into hot, dry rocks, and artificial fracturing of these rocks can create reservoir space for a considerable volume of water. But it takes energy and diverts water, often from more urgent needs elsewhere. This defeats the purpose of developing geothermal power in this manner.

Even where water is recycled by pumping it directly back into the Earth after passing through turbines, costs of operating a geothermal power plant can be quite high. An ordinary geothermal station may be expected to function at a profit over only a matter of decades—a few centuries at the most. In some active volcanic areas, such as Kilauea volcano on the Island of Hawaii, and Krafla in Iceland, frequent intrusions of shallow magma will restore heat to the crust, and influx of water may be sufficient to guarantee a long-lasting resource. But a volcanic eruption will readily destroy a power plant, and major development of energy resources in such areas is risky at best.

Another problem with developing geothermal power is that hot ground water often carries with it dissolved minerals, brines, and acids that corrode or clog pipes and turbines. Maintenance costs to keep a plant in operation may be prohibitive for this reason alone. Furthermore, released steam may be naturally polluted, requiring scrubbing and purification of escaping vapors to avoid irritating people downwind.

Despite all of these limitations, energy production from geothermal power is on the whole quite competitive with that of fossil fuels. EROEI levels as high as 13 exist, and for lightly populated regions such as Iceland and New Zealand, there is great incentive to develop Earth-heat rather than import oil or build nuclear reactors. The world production of geothermal power, which takes place in about thirty countries, provides 12,000 megawatts of electricity—equivalent to that of around twenty conventional nuclear reactors. In some nations, including parts of the United States, geothermal waters provide heat as well as electricity. The hot water passes through pipes and walls to warm ("space-heat") home interiors, and can even be used directly in showers and taps (Figure 21.18).

It is difficult to know whether to call geothermal energy a "renewable" form of energy or not. Given that this resource is exhausted readily as most plants worldwide, geothermal energy must be regarded as nonrenewable. On the other hand, ground water and heat are both readily replenished on a time scale of centuries or millennia, and in a few places there seems to be a never-ending supply of extractable Earth-heat. Unlike fossil fuels, however, geothermal power is not transportable worldwide. Geothermal energy is likely to have only local or regional applications.

FIGURE 21.18
A geothermal station about 50 kilometers to the east of Reykjavik, Iceland, extracts hot ground water heated by shallow magma intrusions and pumps it to Reykjavik for commercial and residential use. *Photo by Richard Hazlett*

Renewable Energy Sources

Some energy resources are unquestionably renewable and easily tapped. These include solar, wind, wave (tidal), and hydroelectric power, which together may provide the world with as much as 9% of its electricity by 2010. At present, the growth in renewable energy supply is about 2.3% per year, compared to an increase of 1.6% in nonrenewable resource extraction.

Unlike fossil fuels, which require huge industries to mine, transport, and distribute to users, solar and wind power can be generated locally—even in a backyard or atop the roof of one's own home. The development of solar and wind power is stimulating a transformation from highly centralized power production to much more distributed, smaller-scale, "neighborhood-scale" sources of power. At present, the EROEIs of solar and wind power are rather low—about 2 for wind and somewhat less than 1 for solar, in part because of high unit costs of production associated with low demand for these energy resources. Solar collectors and windmills require large amounts of steel, copper, and other mineral resources to make. They are also rather inefficient sources of energy (5–15% for solar, usually around 20% for wind farms). These problems will diminish with improving technology and rising demand. Government tax incentives and practices, such as net-metering, have also given these industries a real boost. The introduction of new, dry-sensitized cell collectors to solar panels could boost EROEI to as high as 10. The Danish government, which has invested heavily in developing wind power for its national energy supply, estimates that EROEI may climb as high as 50 in harvesting the strong sea breezes of that country. Nevertheless, it is highly unlikely that solar and wind power alone will ever substitute completely for the other energy resources we presently use.

Hydroelectric power facilities transform gravitational energy in the form of falling water into electrical energy. Most hydroelectric power stations lie at the foot of dams where water spilling from reservoirs into rivers downstream spins turbines. These stations produce electricity somewhat more cheaply than fossil fuels (EROEI is around 10), especially in regions where oil, coal, or gas has to be imported from afar. Downstream bank erosion, disruption of fish migrations, the flooding of land, and displacement of populations by filling reservoirs are major environmental concerns.

Hydroelectric power is by far the largest of the developed renewable energy resources at present, accounting for 96% of all renewable energy production worldwide—about half of it in Europe and North America. Hydropower provides 3% of the energy and 11% of the electricity consumed in the United States today (figure 21.19). It can be generated locally; a small station placed on a creek or stream can provide the power needs of a home, farm, or ranch. Hydropower does not directly produce any air pollution, and can even be tapped during times of low energy demand to recycle water back upriver by pumping into reservoirs.

Tidal power is a variation of hydropower. A barrier, called a *barrage*, must be constructed across the mouth of an estuary or bay (figure 21.20). Gates in the barrage allow water to pass through as the tide rises, spinning turbines to produce electricity. The gates close when the tide is in, capturing the water inshore from the barrage. The gates reopen after the tide falls on the seaward side, and the water pouring out spins the turbines again in reverse.

A

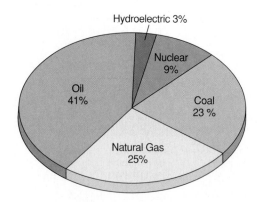

FIGURE 21.19

Major sources of energy for the United States in 2001. Minor sources such as geothermal, wind, and solar are not included. For more recent information, go to www.eia.doe.gov/. *Source: U.S. Energy Information Administration*

B

FIGURE 21.20

Generation of tidal power.

The world's largest tidal-generating station, at Rance in France, generates 320 megawatts, enough to supply several hundred thousand users. High costs; concerns about impacts on fish, bird life, and ecosystem health; and irregular power supply have greatly impeded development of tidal mills elsewhere in the world. In fact, only four countries at present—France, China, Russia, and Canada—have tidal power facilities.

METALLIC RESOURCES

Modern industrial society stands upon two feet—one of fossil fuels and the other of metal. While civilizations have always had access to basic construction materials—and always will—the production of high-quality, high-strength metals and exploitation of fossil fuels make our times stand out in all of human history.

The successful search for metals depends on finding **ores,** which are naturally occurring materials that can be profitably mined (table 21.3). It is important to recognize that the local concentration of a metal must be greater (usually much greater) than its average crustal abundance to be a potential ore body. Metals must be concentrated in a particular place in a large enough amount to be viable ore bodies. Take gold in seawater. You could become fabulously wealthy if you could extract a fraction of the gold in seawater, because there are over 10^{11} troy ounces of gold—around $52 trillion worth in the world's oceans. But the concentration is 4 grams per 1 million tons of water. It would cost you far more to remove that gold than you could sell it for.

Whether or not a mineral (or rock) is considered a metal ore depends on its chemical composition, the percentage of extractable metal, and the market value of the metal. The mineral hematite (Fe_2O_3), for example, is usually a good *iron ore* because it contains 70% iron by weight; this high percentage is profitable to extract at current prices for iron. Limonite ($Fe_2O_3 \cdot nH_2O$) contains less iron than hematite and, hence, is not as extensively mined. Even a mineral containing a high percentage of metal is not described as an *ore* if the metal is too difficult to extract or the site is too far from a market; profit is part of what defines an ore.

Many different kinds of geologic processes can accumulate ores, from weathering and sedimentation to the settling of crystals deep within magma chambers (table 21.4). We'll survey the possibilities, moving from Earth's interior to the surface, in the pages that follow. In all cases, note that people mined ore minerals long before we understood how they form. The field of *economic geology* developed, in large part, to study the origin of ore deposits and to expand our ability to locate and develop new reserves more easily.

Ores Formed by Igneous Processes

Crystal Settling

Crystal settling occurs as early-forming minerals crystallize and settle to the bottom of a cooling body of magma (figure 21.21). This process was described under differentiation in chapter 3. The metal chromium comes from chromite ore bodies near the base of sills and other intrusions. Most of the world's chromium and platinum come from a single intrusion, the huge Bushveldt Complex in South Africa. In Montana, another huge

TABLE 21.3	Common Ore Minerals	
Metal	**Ore Mineral**	**Composition**
Aluminum	Bauxite (a mineral mixture)	AlO(OH) and Al(OH)$_3$
Chromium	Chromite	FeCr$_2$O$_4$
Copper	Native copper	Cu
	Chalcocite	Cu$_2$S
	Chalcopyrite	CuFeS$_2$
Gold	Native gold	Au
Iron	Hematite	Fe$_2$O$_3$
	Magnetite	Fe$_3$O$_4$
Lead	Galena	PbS
Manganese	Pyrolusite	MnO$_2$
Mercury	Cinnabar	HgS
Nickel	Pentlandite	(Fe, Ni)S
Silver	Native silver	Ag
	Argentite	Ag$_2$S
Tin	Cassiterite	SnO$_2$
Uranium	Pitchblende	U$_3$O$_8$
	Carnotite	K(UO$_2$)$_2$(VO$_4$)$_2$ · 3H$_2$O
Zinc	Sphalerite	ZnS

TABLE 21.4	Some Ways Ore Deposits Form
Type of Ore Deposit	**Some Metals Found in This Type of Ore Deposit**
Crystal settling within cooling magma	Chromium, platinum, iron
Hydrothermal deposits (contact metamorphism, hydrothermal veins, disseminated deposits, hot-spring deposits)	Copper, lead, zinc, gold, silver, iron, molybdenum, tungsten, tin, mercury, cobalt
Pegmatites	Lithium, rare metals
Chemical precipitation as sediment	Iron, manganese, copper
Placer deposits	Gold, tin, platinum, titanium
Concentration by weathering and ground water	Aluminum, nickel, copper, silver, uranium, iron, manganese, lead, tin, mercury

FIGURE 21.21

Early-forming minerals such as chromite may settle through magma to collect in layers near the bottom of a cooling sill.

Precambrian sill called the Stillwater Complex contains similar, but lower-grade deposits, of these two metals.

Hydrothermal Fluids

Hydrothermal fluids, discussed in chapter 7, are the most important source of metallic ore deposits other than for iron and aluminum. The hot water and other fluids are part of the magma itself, injected into the surrounding country rock during the last stages of magma crystallization (figure 21.22). Atoms of metals such as copper and gold, which do not fit into the growing crystals of feldspar and other minerals in the cooling pluton, are concentrated residually in the remaining water-rich magma. Eventually, a hot solution, rich in metals and silica (quartz is the lowest-temperature mineral on Bowen's reaction series), moves into the country rock to create ore deposits. Most hydrothermal ores are metallic sulfides, often mixed with milky quartz. The origin of the sulfur is widely debated.

A magma body or hot rock may heat ground water and cause convection circulation. This water may mix with water given off from solidifying magma, or it may leech metals from solid rock and deposit metallic minerals elsewhere as the water cools. However the hydrothermal solutions form, they tend to create four general types of hydrothermal ore deposits: (1) contact metamorphic deposits, (2) hydrothermal veins, (3) disseminated deposits, and (4) hot-spring deposits.

Contact metamorphism can create ores of iron, tungsten, copper, lead, zinc, silver, and other metals in country rock. The country rock may be completely or partially removed and replaced by ore (figures 21.23A). This is particularly true of limestone beds, which react readily with hydrothermal solutions. (The metasomatic addition of ions to country rock is described in chapter 7.) The ore bodies can be quite large and very rich.

Hydrothermal veins are narrow ore bodies formed along joints and faults (figure 21.23B). They can extend great distances from their apparent plutonic sources. Some extend so far that it is questionable whether they are even associated with plutons. The fluids can precipitate ore (and quartz) within cavities along the fractures and may replace the wall rock of the fractures with ore. Hydrothermal veins form most of the world's great deposits of lead, zinc, silver, gold, tungsten, tin, mercury, and, to some extent, copper (figure 21.24).

Hot solutions can also form *disseminated deposits* in which metallic sulfide ore minerals are distributed in very low concentration through large volumes of rock, both above and within a pluton (figure 21.23C and box 7.5). Most of the world's copper comes from disseminated deposits (also called *porphyry copper deposits* because the associated pluton is usually porphyritic). Along with the copper are deposited many

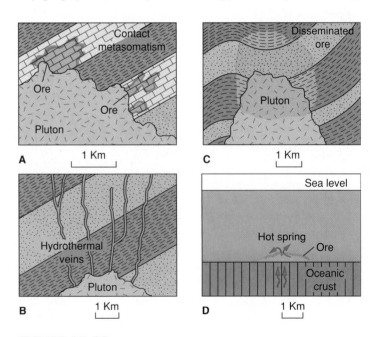

FIGURE 21.23

Hydrothermal ore deposits. (A) Contact metamorphism in which ore replaces limestone. (B) Ore emplaced in hydrothermal veins. (C) Disseminated ore within and above a pluton (porphyry copper deposits, for example). (D) Ore precipitated around a submarine hot spring (size of ore deposit is exaggerated).

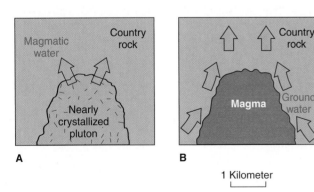

FIGURE 21.22

Two possible origins of hydrothermal fluids. (A) Residually concentrated magmatic water moves into country rock when magma is nearly all crystallized. (B) Ground water becomes heated by magma (or by a cooling solid pluton), and a convective circulation is set up.

FIGURE 21.24

Hydrothermal quartz veins in granite. *Photo by David McGeary*

FIGURE 21.25

2,250 million year old banded iron ore from Michigan. *Photo © Parvinder Sethi*

other metals, such as lead, zinc, molybdenum, silver, and gold (and iron, though not in commercial quantities).

Where hot solutions rise to Earth's surface, *hot springs* form (see chapters 11 and 18). Hot springs on land may contain large amounts of dissolved metals. More impressive are hot springs on the sea floor (figure 21.23*D*), which can precipitate large mounds of metallic sulfides, sometimes in commercial quantities (chapters 7 and 18).

Pegmatites (see box 3.1), very coarse-grained plutonic rocks, are another type of ore deposit associated with igneous activity. They may contain important concentrations of minerals containing lithium, beryllium, and other rare metals, as well as gemstones such as emeralds and sapphires.

Ores Formed by Surface Processes

Chemical precipitation in layers is the most common origin for ores of iron and manganese. A few copper ores form in this way, too. Banded iron ores, usually composed of alternating layers of iron minerals and chert, formed as sedimen-

tary rocks in many parts of the world during the Precambrian, apparently in shallow, water-filled basins (figure 21.25). Later folding, faulting, metamorphism, and solution have destroyed many of the original features of the ore, so the origin of the ore is difficult to interpret. The water may have been fresh or marine, and the iron may have come from volcanic activity or deep weathering of the surrounding continents. The alternating bands may have been created by some rhythmic variation in volcanic activity, river runoff, basin water circulation, growth of organisms, or some other factor. Since banded iron ores are all Precambrian, their origin might be connected to an ancient atmosphere or ocean chemically different from today's.

Placer deposits in which streams have concentrated heavy sediment grains in a river bar are described in chapter 10. Wave action can also form placers at beaches. Placers include gold nuggets and dust, native platinum, diamonds and other gemstones, and worn pebbles or sand grains composed of the heavy oxides of titanium and tin.

Ore deposits due to *concentration by weathering* were described in chapter 5. Aluminum (in bauxite) forms through weathering in tropical climates.

Another type of concentration by weathering is the *supergene enrichment* of disseminated ore deposits. Through supergene enrichment, low-grade ores of 0.3% copper in rock can be enriched to a minable 1% copper. The major ore mineral in a disseminated copper deposit is chalcopyrite, a copper-iron sulfide containing about 35% copper. Near Earth's surface, downward-moving ground water can leach copper and sulfur from the ore, leaving the iron behind (figure 21.26). At or below the water

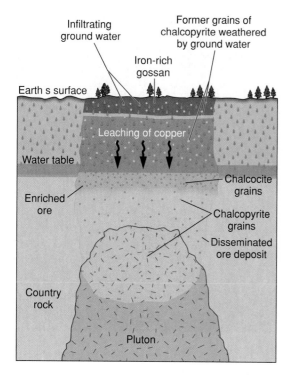

FIGURE 21.26

Supergene enrichment. Ground water leaches copper from upper part of disseminated deposit and precipitates it at or below the water table, forming rich ore.

table, the dissolved copper can react with chalcopyrite in the
lower part of the disseminated deposit, forming a richer ore mineral such as chalcocite, which is about 80% copper:

$$3\,Cu^{++} \quad + \quad CuFeS_2 \quad \rightarrow \quad 2\,Cu_2S \quad + \quad Fe^{++}$$

Copper	Chalcopyrite	Chalcocite	Iron in
dissolved			solution
in ground water			

In this way, copper is removed from the top of the deposit and
added to the lower part (Figure 21.26). The ore below the water
table may be several times richer than the ore in the rest of the
deposit, with silver concentrating as readily as copper. The iron
remains behind, staining the surface as it oxidizes to form a
gossan (defined in the next section).

MINING

As in the case of coal, miners use both surface and underground techniques to extract ore minerals (figure 21.27). Strip
mining—the wholesale removal of large areas of soil and shallow rock cover—has already been mentioned in connection
with coal beds. Aluminum ore (bauxite), which forms in
weathered soil beds under tropical conditions, is often most
easily extracted this way. *Open pit mining* is related to strip
mining, but concentrates on the removal of valuable deposits
from a specific, relatively small area. Open-pit mines often dig

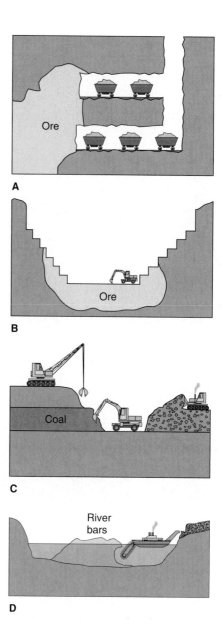

FIGURE 21.27

Types of mines: (A) Underground. (B) Open pit. (C) Strip. (D) Placer (being mined by
a floating dredge).

much deeper than strip mines. *Placer mines* are localized to
ancient or modern river bar or beach deposits.

Underground, or *bedrock mining* must be done to excavate
many valuable mineral deposits. Bedrock mining of ores typically extends to much greater depths than ordinary coal mines,
and this presents its own set of technical challenges. The
world's deepest mines, in South Africa, extend to depths of
1,500–2,500 meters. The walls grow hot to the touch so deep
underground, and pumping of fresh, cool air and water must be
done to make working conditions tolerable. Mines have notori-
ously poor air circulation, and the use of dynamite to blast
openings releases toxic gases that must be removed. Ammo-
nium nitrate (NH_4NO_3) mixed with fuel oil (CH_2) is a typical

blasting agent. The explosive reaction generates poisonous carbon monoxide whenever there is a slight excess of oil in the mixture. Carbon monoxide (CO) is a heavier-than-air gas that sinks deep into the mine:

$$3 \, NH_4NO_3 + CH_2 \rightarrow CO_2 \, (or \, CO \, plus \, oxygen) + 7 \, H_2O + 3 \, N_2 + heat$$

This is one of the reasons why abandoned mines should never be explored.

Where the mines extend beneath the water table, the water that seeps in must be pumped out to avoid flooding. An active mine consumes large amounts of energy as well as material resources.

The design of a mine takes into consideration three vital factors: (1) the geometry of the underground ore body; (2) the need for safety; and (3) the need to maximize profit. It is typically easiest for miners to construct a set of vertical and horizontal passages to access and remove the ore (figure 21.28). The vertical openings, called *shafts* (or *winzes,* if they do not open all the way to the surface) allow elevators to take miners underground and bring ore up to the surface. Shafts are also conduits for electrical cables, water hoses, and air lines. Miners can blast and dig open a shaft at a rate of 30 feet every day. The horizontal tunnels, termed *adits* (or *drifts,* if they do not open to the surface), are the pathways through which ore is directly excavated. In larger mines, multiple drifts radiate off of shafts. *Ramps* are slanted tunnels, many of which have tracks for winching ore carts. In some places, the ore may be so rich that miners excavate a giant underground chamber called a

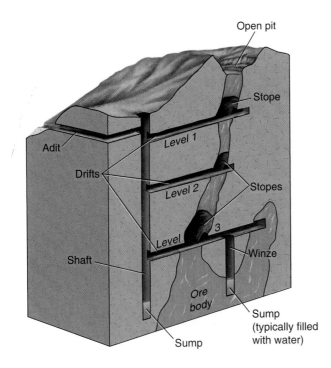

FIGURE 21.28
Design of a typical bedrock mine.

stope. To avoid collapse, the walls of the stope may have to be shored up with timbers or other construction material.

In earlier times, prospectors located potential ore bodies by looking at the eroded rock fragments ("float") in streams and on hillsides, hoping to find telltale minerals such as white quartz with gold or oxidized sulfide minerals, red jasper, or bright-blue crysocholla. Following this debris upslope, the treasure-seekers might discover what they were looking for—a *gossan,* an area of yellow or ruddy orange, oxidized ground marking the intersection of an ore body or vein with the surface. The Spanish termed such areas "colorados," and the name of the state, in fact, derives from this origin. "Gossan" is itself a Cornish mining term, meaning "iron hat," in reference to the fact that gossans cap deeper ore bodies.

Today, more sophisticated—and expensive—prospecting techniques are applied to locating and determining the shape of an underground ore body. Exploration geologists must study the structural geology (stratigraphy and deformation of the surrounding rocks), examine the evidence brought up in preliminary boreholes, and conduct geophysical surveys, including the use of gravimeters, magnetometers, and electrical-resistivity equipment. Some ore bodies are excellent electrical conductors and may be highly magnetic. Exploration geologists have the ultimate say on whether or not a company should proceed with mining.

Given the dangers and economic factors involved, whole technical schools have been established to train mining engineers (e.g., the Colorado School of Mines). Today, these schools must also consider environmental factors because in the past, mines have been terrible sources of watershed pollution, among other problems. Ground water running or being pumped out of a mine causes *acid mine drainage.*

Sulfide ore minerals (table 21.3) and pyrite (FeS_2) are most often the source of the trouble. Ground water transports oxygen to the sulfides, which are then oxidized to iron oxide and sulfuric acid. In some mines, expensive programs of holding and neutralizing drainage water in ponds or artificial wetlands prevent pollution of surface streams and harm to forests and wildlife. The worst problem is with long-abandoned mines that are still draining acid waters. Many of these may never be neutralized.

Some Important Metals

Iron

Iron is one of the three essential ingredients required to make steel. Coal (coke) and limestone, the other two, are needed to melt the iron in furnaces before molding. Iron can be mixed with other metals ("alloyed"), including silicon, chromium, and nickel, to make special kinds of steel. In fact, there are over 2,000 kinds of steel used in the world today, representing various mixtures of many metals. But in all, iron is dominant. Steel is used to make everything from skillets to locomotives.

The major iron ore minerals are hematite and magnetite. Most of the iron in the United States comes from the region

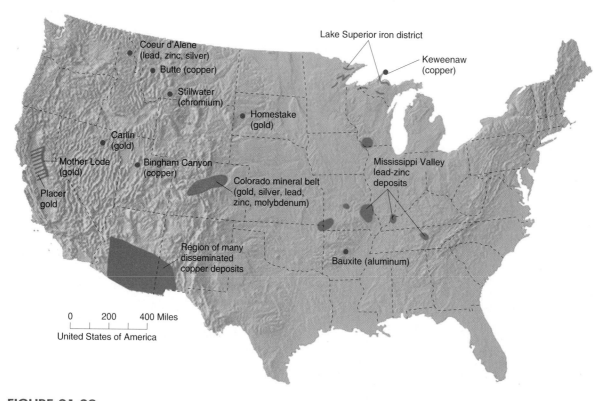

FIGURE 21.29

Some metallic ore localities mentioned in the text.

FIGURE 21.30

Open-pit copper mine in Morenci, Arizona. Heavy equipment in the bottom of the pit gives a sense of scale. *Photo by David McGeary* around Lake Superior (figure 21.29), most notably, Minnesota. The ores are banded iron ores of Precambrian age, typical of most iron ores in the world. Mining is done mostly by open-pit methods.

Copper

Less abundant is *copper,* another important metal for industry. More than half the copper used in the United States goes into electrical wire and equipment and one-third into the manufacture of brass, a copper-zinc alloy.

Most copper ores are sulfides. Chalcopyrite is the most important copper ore mineral. Some vein deposits of copper exist (as at Butte, Montana), but most major deposits are disseminated through large volumes of rock; so most copper mines are open pits (figure 21.30). Arizona, Utah, and a few other western states are the major producers in the United States (figure 21.29).

Aluminum

Aluminum is consumed in the manufacture of beer and soft drink cans, airplanes, electrical cable, and many other products. The use of aluminum is increasing rapidly.

The ore of aluminum is bauxite, which forms under tropical weathering conditions. The United States has very little bauxite, so it imports 90% of its aluminum ore from tropical countries. The largest aluminum mine in the United States is in Arkansas (figure 21.29). Open-pit mining is the usual technique for extracting bauxite. Aluminum was the first metal to be subject to widespread recycling, in part because so much of North America's aluminum ore is imported.

Lead

The main use of *lead* (79%) is in batteries. Substantial amounts are recycled, largely from automobile batteries.

The most important ore of lead is galena. Major deposits occur in Missouri, Idaho, Utah, and Colorado (figure 21.29). The Missouri deposits are mostly found in limestone beds; their precise origin is a matter of some controversy. The ore is mined both underground and from open pits. Deposits in Idaho occur mostly as veins and are usually mined underground.

Zinc

Widely used in industry, *zinc* is necessary for galvanizing and the manufacture of brass and other alloys.

The major zinc ore is sphalerite. As sphalerite usually is found closely associated with galena, most lead mines also extract zinc. Zinc occurs without lead in some areas, however.

Silver

Coins, tableware, jewelry, photographic film, and many other products are made of *silver.*

Silver, found as a native metal and in sulfide ores, is a common by-product of lead and copper mining. The lead-zinc mines of Idaho (the Coeur d'Alene district) are the largest silver producers in the United States (figure 21.29).

Gold

The rare and valuable metal *gold* is used in coins, jewelry, decoration, dentistry, electronics, and the space program. Gold bars are stored to back national currency, although this use is rapidly disappearing.

Gold is found as a native element in the form of nuggets and grains. In some parts of the world, these are concentrated in placer deposits (California's Gold Rush of 1849 was triggered by discoveries of placer gold). Gold nuggets, flakes, and dust can be separated from the other sediments by (1) *panning;* (2) *sluice boxes* (figure 21.31), which catch the heavy gold on the bottom of a box as gravel is washed through it; (3) *hydraulic mining* (figure 21.32), which washes gold-bearing gravel from a hillside into a sluice box;

FIGURE 21.31

Sluice box used to separate gold from gravel in Alaska. *Photo by D. J. Miller, U.S. Geological Survey*

FIGURE 21.32

Hydraulic mining for gold in Alaska. *Photo by T. L. Péwé, U.S. Geological Survey*

FIGURE 21.33

A gold dredge separates gold from gravel. *Photo by David McGeary*

or (4) floating *dredges* (figure 21.33), which separate gold from gravel aboard a large barge, piling the spent gravel behind. When gold is found in hydrothermal veins associated with milky quartz, as it is in parts of Colorado and in California's Mother Lode (figure 21.29), it is mined underground. A large amount of disseminated gold is mined in open pits near Carlin, Nevada, and at several other localities in Nevada and California.

Other Metals

Many other metals are vital to our economy. *Chromium, nickel, cobalt, manganese, molybdenum, tungsten,* and *vanadium* are all important in the steel industry, particularly in the manufacture of specialty products such as stainless steel and armor plate. Most of these metals have other uses as well. *Tin* is used in solder and for plating steel in "tin cans." *Mercury* is used in thermometers, silent electrical switches, medical compounds, and batteries. *Magnesium* is used in aircraft. *Titanium,* as strong as steel but weighing half as much, is used in aircraft. *Platinum* is used in catalytic converters to clean automobile exhaust.

NONMETALLIC RESOURCES

Nonmetallic resources are Earth resources that are not mined to extract a metal or as a source of energy. Most rocks and minerals contain metals, but when nonmetallic resources are mined, it is usually to use the rock (or mineral) as is (for example, gravel and sand for construction projects); whereas, metallic ores are processed to extract metal. With the exception of the gemstones such as diamonds and rubies, nonmetallic resources do not have the glamour of many metals or energy resources. Nonmetallic resources are generally inexpensive and are needed in large quantities (again, except for gemstones); however, their value exceeds that of all mined metals. The large demand and low unit price means that these resources are best taken from local sources. Transportation over long distances would add significantly to the cost.

Construction Materials

Sand and *gravel* are both needed for the manufacture of concrete for building and highway construction. Sand is also used in mortar, which holds bricks and cement blocks together. The demand for sand and gravel in the United States has more than doubled in the last twenty-five years. Sand dunes, river channel and bar deposits, glacial outwash, and beach deposits are common sources for sand and gravel. Cinder cones are mined for "gravel" in some areas. Sand and gravel are ordinarily mined in open pits (figure 21.34).

Stone refers to rock used in blocks to construct buildings or crushed to form roadbed. Most stone in buildings is limestone or granite, and most crushed stone is limestone. Huge quantities of stone are used each year in the United States. Stone is removed from open pits called *quarries* (figure 21.35).

FIGURE 21.34

Sand and gravel pit in a glacial esker near Saranac Lake, New York. *Courtesy Ward's Natural Science Est., Inc., Rochester, NY*

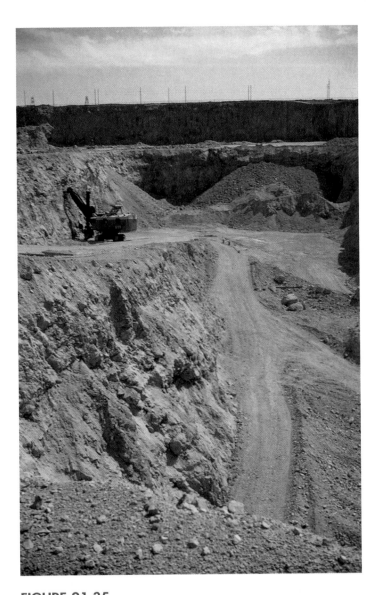

FIGURE 21.35

A limestone quarry in northern Illinois. The horizon marks the original land surface before the rock was removed. *Photo by David McGeary*

Limestone has many uses other than building stone or crushed roadbed. Cement, used in concrete and mortar, is made from limestone and is vital to an industrial economy. Pulverized limestone is in demand as a soil conditioner and is the principal ingredient of many chemical products.

Fertilizers and Evaporites

Fertilizers (phosphate, nitrate, and potassium compounds) are extremely important to agriculture today, so much so that they are one of the few nonmetallic resources transported across the sea. *Phosphate* is produced from phosphorite, a sedimentary rock formed by the accumulation and alteration of the remains of marine organisms. Major phosphate deposits in the United States are in Idaho, Wyoming, and Florida. *Nitrate* can form

directly as an evaporite deposit but today is usually made from atmospheric nitrogen. *Potassium* compounds are often found as evaporites.

Rock salt is coarsely crystalline halite formed as an evaporite. Salt beds are mined underground in Ohio and Michigan; underground salt domes are mined in Texas and Louisiana. (Some salt is also extracted from seawater by evaporation.) Rock salt is used in many ways—deicing roads in winter, preserving food, as table salt, and in manufacturing hydrochloric acid and sodium compounds for baking soda, soap, and other products. Rock salt is heavily used by industry.

Gypsum forms as an evaporite. Beds of gypsum are mined in many states, notably California, Michigan, Iowa, and Texas. Gypsum, the essential ingredient of plaster and wallboard (Sheetrock), is used mainly by the construction industry, although there are other uses.

Sulfur occurs in bright-yellow deposits of elemental sulfur. Most of its commercial production comes from the cap rock of salt domes. Sulfur is widely used in agriculture as a fungicide and fertilizer and by industry to manufacture sulfuric acid, matches, and many other products.

Other Nonmetallics

Gemstones (called *gems* when cut and polished) include precious stones such as diamonds, rubies, emeralds, and sapphires and semiprecious stones such as beryl, garnet, jade, spinel, topaz, turquoise, and zircon. Gems (see box 2.7) are used for jewelry, bearings, and abrasives (most are above 7 on Mohs' scale of hardness). Diamond drills and diamond saws are used to drill and cut rock. Old watches and other instruments often have hard gems at bearing points of friction ("17-jewel watches"). Gemstones are often found in pegmatites or in close association with other igneous intrusives. Some are recovered from placer deposits.

Asbestos is a fibrous variety of serpentine or chain silicate minerals. The fibers can be separated and woven into fireproof fabric used for firefighters' clothes and theater curtains. It is also used in manufacturing ceiling and sound insulation, shingles, and brake linings, although the use of asbestos is being rapidly curtailed because of concern about its connection with lung cancer (see box 2.5). The United States no longer produces asbestos. Large amounts are mined in Canada, chiefly in Quebec. *Talc*, used in talcum powder and other products, is often found associated with asbestos (see box 20.2).

Other nonmetallic resources are important. *Mica* is used in electrical insulators. *Barite* ($BaSO_4$), because of its high specific gravity, is used to make heavy drilling mud to prevent oil gushers. *Borates* are boron-containing evaporites used in fiberglass, cleaning compounds, and ceramics. *Fluorite* (CaF_2) is used in toothpaste, Teflon finishes, and steel-smelting. *Clays* are used in ceramics, manufacturing paper, and as filters and absorbents. *Diatomite* is used in swimming pool filters and to filter out yeast in beer and wine. *Glass sand*, which is over 95%

IN GREATER DEPTH 21.3

Substitutes, Recycling, and Conservation

Substitutes for many geologic resources now exist, and others will be found. Aluminum is replacing the more expensive copper in many electrical uses, particularly in transmission lines. Glass fibers also are replacing copper for telephone lines. Aluminum has largely replaced tin-coated steel for beverage containers. Cotton and wool use could increase, replacing polyester and other petroleum-based synthetic fibers in clothes.

Suitable substitutes, however, seem unlikely for some resources. Nothing yet developed can take the place of steel in bridges or mercury in thermometers, although electronic thermometers are being more widely used. Cobalt is vital for strong, permanent magnets. Although substitutes may help prolong the life of the supplies of some resources, they are not available for many others.

Recycling helps augment the supply of some resources such as aluminum, gold, silver, lead, plastics, and glass. No resource, however, receives even half its supply from recycling. Increased volunteer recycling on the part of the public and waste reclaiming of urban trash could increase these percentages. New ore will always be needed, however, because some uses of products prevent the material from being recyclable. A steel can that rusts away beside a road will never be recycled. The iron oxides are scattered in low percentage in the soil and can never be recov-

ered. Many resources, such as petroleum and coal, are consumed during use and cannot be recycled. Some people argue against recycling when newly derived materials are cheaper. This is a short-sighted view and neglects the fact that there is a finite supply of all nonrenewable resources in Earth's crust. The more we conserve now, the longer our reserves will last. By squandering resources now, we may be depriving future generations of resources.

Conservation of scarce resources is extremely important. The United States' use of oil from 1979 to 1985 declined as a result of conservation efforts such as the 55 mile-per-hour speed limit, more fuel-efficient automobiles, the upgrading of insulation in buildings, and the elimination of unneeded heating and lighting. But in 1986, oil use rose sharply as its price declined. Politically caused shortages and gluts of oil can obscure the fact that the supply of oil in the United States is severely limited. There will be difficult times ahead as U.S. domestic oil becomes scarcer. Therefore, the need for conserving fuel should be repeatedly stressed. The price of gasoline will rise substantially as the United States becomes increasingly dependent on foreign oil. Conservation of metals, particularly those imported in large quantities, will become increasingly important. Smaller automobiles and more durable appliances can help conserve metals.

quartz, is the main component of glass. *Graphite* is used in foundries, lubricants, steelmaking, batteries, and pencil "lead."

THE HUMAN PERSPECTIVE

Humans have a tendency to take a one-sided view of geologic resources and the problems created by the extraction and transportation of those resources. The conflicts are between (1) maintaining or increasing our standard of living and raising the quality of life not only for ourselves but for people in underdeveloped nations; (2) maintaining the environment; and (3) making sure that we do not deprive future generations of the resources that sustain us (box 21.3).

The extreme position for each of the three concerns could be stated as follows: (1) Extreme exploiter: "Let's mine what we can and make ourselves as rich as possible now. Technological breakthroughs will provide for future generations. Damage to the environment due to extraction is insignificant (or it's where it doesn't bother me)." (2) Extreme environmen-

talist: "Any mine or oil well does environmental damage and therefore should not be permitted. We can maintain our lifestyles by recycling or by leading less technologically dependent lifestyles." (3) Extreme conservationist: "Let's not mine anything now because there are many future generations that need to rely on these resources."

You probably agree that none of the three extreme positions is reasonable. The middle ground among the three is where almost everyone thinks we should be—the challenge is deciding where in the middle ground. Should we lean toward more exploitation and away from environmental concerns so that underdeveloped countries can raise their standards of living? Should we minimize mining and energy consumption so as to reduce any impact on air pollution or wildlife? Our hope is that we can at least understand the perspective of those who may disagree to strike a balance and try to deal with each case with enlightenment. Your understanding of geology is an important step in your being able to help resolve moral dilemmas that we face to which there is no ideal solution.

SUMMARY

Most people interact with the Earth primarily through their interaction with *geologic resources*; soil, water, metals, non-metals, and fuels, most of which are *nonrenewable* and form under particular and transient natural conditions.

Coal, petroleum, and *natural gas* are fossil fuels that are the main sources of energy in the modern world. They are also important in nonenergy applications, such as making fertilizer, steel, and many other products. These resources are essentially ancient *solar energy,* unlocked by combustion in power plants after mining or drilling. *Coal beds* occur in areas of ancient swamps and marshes, and derive from the accumulation of dead land-based plant matter. *Oil* (petroleum) and natural gas derive from certain nearshore marine settings where dead, microscopic, floating organisms accumulate. Coal and oil both require heat and burial to develop. Oil and natural gas also need to be sealed into reservoirs as they percolate toward the surface. Anticlines, faults, and other structural traps provide this lid.

Reserves are known deposits that can be legally and economically recovered now—the short-term supply. Resources include reserves as well as other known and undiscovered deposits that might be economically extractable in the future. There is evidence that the world's reserves of petroleum are nearing critical depletion, given the high level of demand. This will force the world into finding energy alternatives, including the burning of more coal and increased development of *nuclear power.*

Uranium-235 is the primary fuel of *nuclear reactors.* Through breeder designs, nuclear energy could provide us with electricity long into the future. But risk of accidents and waste-disposal issues raise questions about the long-term viability of this energy source.

Geothermal power benefits only a few localized areas around the world. This resource can be easily exhausted and it is never likely to become a principal source of world-energy, despite the enormous amount of heat contained inside the Earth.

Renewable energy strategies, such as *solar* and *wind power,* are an appealing, environmentally clean substitute for "conventional" energy sources. *Hydroelectric power* is the most successful and extensively used type of renewable energy, though it is localized to areas with abundant flowing water. Other approaches are difficult to develop for ecological reasons, or because the cost of producing the energy is greater than the return. New technologies and economies of scale may change this in time.

Metallic ores, which can be profitably mined, are often associated with igneous rocks, particularly their *hydrothermal fluids,* which can form in contact metamorphic deposits, hydrothermal veins, disseminated deposits, and submarine hot-spring deposits. Iron occurs in sedimentary layers, and aluminum forms from weathering. Metallic ores form from hot springs at divergent plate boundaries, on the flanks of island arcs, and in belts on the edges of continents above subduction zones.

Ores are mined at the surface in *strip* and *open-pit mining,* and in costly, potentially dangerous, and carefully executed underground mining. *Placer* mining takes advantage of the sedimentary reworking and concentration of ore minerals.

Metals are vital to an industrial economy, particularly iron for steel production and copper for electrical equipment.

Nonmetallic resources such as sand and gravel and limestone for crushed rock and cement are used in huge quantities. Fertilizers, rock salt, gypsum, sulfur, and clays are also widely used.

Terms to Remember

coal 553
coal bed methane 562
heavy crude 562
hydrothermal veins 570
natural gas 558
nonrenewable resources 552

oil field 560
oil (tar) sands 562
oil shale 564
ore 569
petroleum 557
placer deposits 571

reserve 555
reservoir rock 558
resource 555
source rock 558
strip mining 554
structural (or oil) trap 558

Testing Your Knowledge

Use the following questions to prepare for exams based upon this chapter.

1. State the First and Second Laws of Thermodynamics and explain why they are important in understanding our ability to extract and use energy resources.

2. Why is it likely that we will never run out of oil? If we never run out, why is it likely that we stop extracting oil as a major energy resource someday?

3. Contrast the geologic conditions responsible for the formation of coal, oil, and natural gas.

4. How does a nuclear reactor work?

5. What are the plusses and minuses of exploiting geothermal and hydroelectric power?

6. Why is it important to distinguish between *energy* and *power* in discussing energy resources?

7. Describe several ways in which ore deposits related to igneous processes form.

8. How can surface processes create ore deposits?

9. Describe four ways in which resources are mined.

10. Discuss common uses for iron, copper, oil extraction, and coal mining.

11. Under what conditions could ground water (and geothermal energy) be regarded as a renewable resource? Under what circumstances are these resources nonrenewable?

12. What limitations exist on present widespread development of solar, tidal, and wind power?

13. Why are the amounts of some resources better known than others?

14. Which is not a type of coal?
    a. Natural gas   c. bituminous
    b. lignite   d. anthracite

15. Which metal would most likely be found in an ore deposit formed by crystal settling?
    a. copper   c. silver
    b. gold   d. chromium

16. Which metal would not be found in hydrothermal veins?
    a. aluminum   d. silver
    b. lead   e. gold
    c. zinc

17. Coal forms
    a. by crystal settling
    b. through hydrothermal processes
    c. by compaction of plant material
    d. on the ocean floor

18. The main use of lead is in
    a. coins   c. batteries
    b. gasoline   d. pencils

19. What factors can increase reserves of Earth resources (choose all that apply)?
    a. extraction of the resource   c. price controls
    b. new discoveries   d. new mining technology

20. The largest use of sand and gravel is
    a. glassmaking   c. construction
    b. extraction of quartz   d. ceramics

21. Oil accumulates when the following conditions are met (choose all that apply)
    a. source rock where oil forms   c. impermeable oil trap
    b. permeable reservoir rock   d. shallow burial

Expanding Your Knowledge

1. Many underdeveloped countries would like to have the standard of living enjoyed by the United States, which has 6% of the world population and uses 15% to 40% of the world's production of many resources. As these countries become industrialized, what happens to the world demand for geologic resources? Where will these needed resources come from?

2. If driven 12,000 miles per year, how many more gallons of gasoline per year does a sports utility vehicle or pickup truck rated at 12 miles per gallon use than a minicompact car rated at 52 mpg? Over five years, how much more does it cost to buy gasoline at $2 per gallon for the low-mileage car? At $5 per gallon (the price in many European countries)?

Exploring Web Resources

www.mhhe.com/plummer11e

This is the dedicated website for this book. You can go to it for new and updated information. The universal resource locators (URL) listed in this book are also given as links on the website, making it easy to go to those websites without typing in the URL. When you visit our Online Learning Center, go to the Student Center and then to the chapter of interest. Here you will find additional readings and media resources, as well as answers to the Testing Your Knowledge section, more quizzes, animation and video clips, and direct links to the sites listed in this book. Links to additional websites can also be found. We have added questions for some of the sites to allow you to get the most of your exploration of the web. Using the web is an enjoyable way of enhancing your knowledge of geology.

www.klws.com/gold/gold.html

Gold prospecting site. This site is mainly aimed at the amateur gold prospector but contains facts and information on gold.

www.NRCan.gc.ca/

Natural Resources Canada. Use this site to get information on Canada's mineral and energy resources.

http://minerals.usgs.gov/

U.S. Geological Survey Mineral Resources Program. Provides current information on occurrence, quality, quantity, and availability of mineral resources.

www.eia.doe.gov/

U.S. Energy Information Administration. Provides data, analysis, and forecasts of energy and issues related to energy.

www.api.org

American Petroleum Institute. Information on all aspects of petroleum from the industry's perspective.

The Earth's Companions

by Tom Arny

Jupiter's moon Io seen against the cloud tops of the giant planet by the Cassini spacecraft. It is about the same size as our moon—about one quarter the size of earth. *Photo by NASA/JPL/University of Arizona*

THE EARTH IN SPACE

The Earth is not alone in space. It is but one of nine **planets** that orbit the Sun. We cannot yet explore these other planets directly, but with robotic spacecraft and Earth-based telescopes, astronomers can study and compare these other worlds to our home planet. These other planets differ greatly and are fascinating in their own right. Moreover, they give us a better understanding and appreciation of Earth.

Starting from the Sun and working out, the nine planets are Mercury, Venus, Earth, Mars, Jupiter, Saturn, Uranus, Neptune, and Pluto. Figure 22.1 shows pictures of these nine distinctive objects and illustrates their relative size and appearance. Some are far smaller and others vastly larger than Earth, but all are dwarfed by the Sun.

The Sun

The Sun is a **star,** a huge ball of gas over 100 times the diameter of the Earth (figure 22.2) and over 300,000 times more massive. If the Sun were a volleyball, the Earth would be about the size of a pinhead, and Jupiter roughly the size of a nickel. The Sun's great mass is not just of academic interest: it generates the gravity that holds the planets in their orbits, preventing them from flying off into the cold of interstellar space.

The Sun differs from Earth not just in size but also in its composition. Like most stars, the Sun is made mainly of hydrogen (about 75% by mass). Hydrogen plays a critical role in the Sun because it is the fuel that keeps it shining. Within the Sun's core, the temperature and pressure are so high that hydrogen atoms can fuse together to form helium atoms. That process

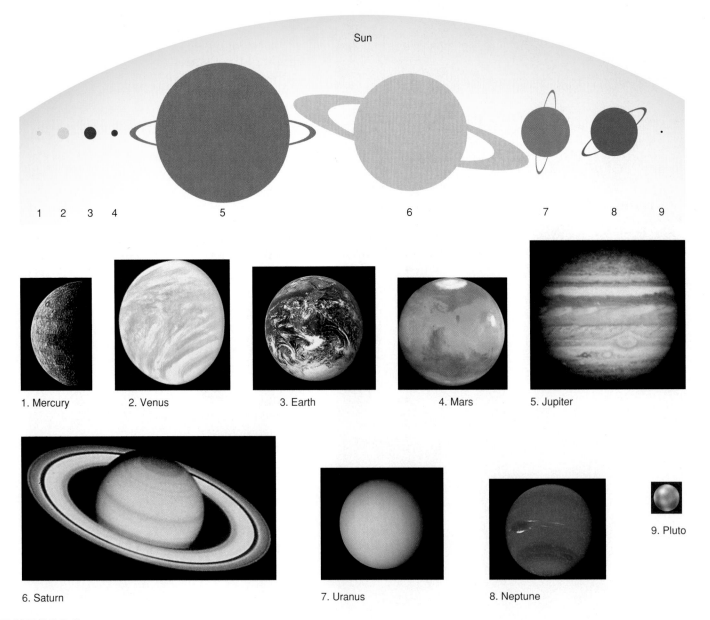

FIGURE 22.1

Images of the nine planets and a sketch at top showing them and Sun to approximately correct relative size. *Photos courtesy of NASA/JPL* mately circular orbits centered on the Sun (figure 22.3), and these orbits all lie approximately in the same plane. Only Mercury's and Pluto's orbits are tilted strongly, and even they are inclined by only 7° and 17°, respectively, to Earth's orbit. Thus, the planets, omitting Pluto, lie in a disk (figure 22.4) whose flatness is comparable to that of a U.S. twenty-five-cent piece.

Apart from their orbital regularities, the planets also show a regularity of composition and size. The four planets nearest the Sun (Mercury, Venus, Earth, and Mars) are basically balls of rock. That is, they are composed mainly of silicates surrounding an iron core.

The next four planets out from the Sun (Jupiter, Saturn, Uranus, and Neptune) are composed mainly of hydrogen gas and its compounds, such as ordinary water (H_2O), methane (CH_4), and ammonia (NH_3). These gases form thick atmospheres that cloak their liquid interior. In fact, they have no solid surface at all. Moreover, they dwarf the inner planets. For example, Jupiter, the largest of the four, has 300 times the mass of Earth and ten times its diameter. The outermost planet, Pluto, is in a class by itself. It is far and away the smallest planet and is composed of a mix of ice and silicates. As such, it is really just the largest of a family of icy bodies that formed at the outer reaches of our solar system.

The many differences between the planets nearest the Sun and those farther out have led astronomers to put them in two categories: the **inner planets** (because they lie close to the Sun) and the **outer planets.** The inner planets are also sometimes

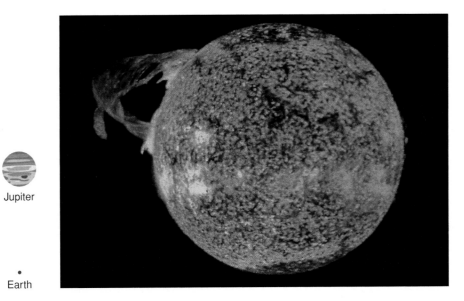

FIGURE 22.2
The Sun as viewed through a special filter that allows its outer gases to be seen. The Earth and Jupiter are shown to scale beside it. *Photo courtesy of Naval Research Laboratory*

releases energy, which flows to the Sun's surface and from there pours into space to illuminate and warm the planets.

The Solar System

The Sun, its nine planets and their moons, together with smaller objects such as asteroids and comets, form the **solar system.** Despite the diversity of its members, the solar system shows many regularities. For example, all the planets move around the Sun in the same direction. They all follow approxi-

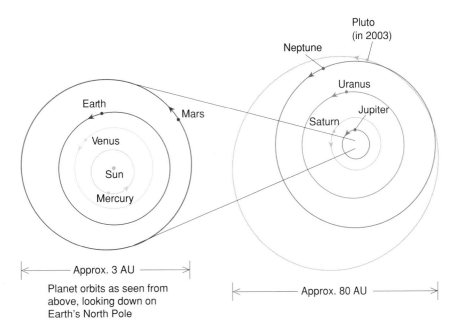

FIGURE 22.3
An artist's view of the solar system from above. The orbits are shown at the correct relative scale in the two drawings. Note: 1 AU is the Earth-Sun distance.

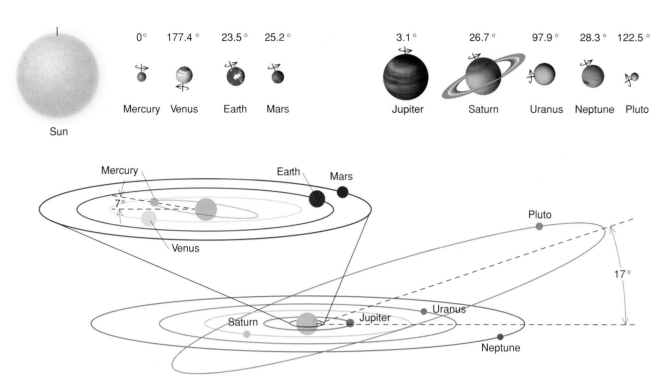

FIGURE 22.4

Planets and their orbits from the side. Sketches also show the orientation of the rotation axes of the planets and Sun. Orbits and bodies are not to the same scale. (Technically, Venus's tilt angle is about 177°, but it can also be described as 3° [180°–177°], with a "backwards," or retrograde spin.)

called terrestrial planets because of their similarity to Earth. Similarly, the four big, outer planets are sometimes called the Jovian planets because of their similarity to Jupiter. They are also often referred to as gas giants. Their gaseous nature means they lack surface features such as we see on the inner planets. Their moons, on the other hand, have many extremely curious surface features, some of which are as yet not well understood. Figure 22.5 shows cutaway views of the nine planets and illustrates their structural differences.

The smaller objects in the solar system—the asteroids and comets—also form two very different families. **Asteroids** are rocky or metallic objects, while **comets** are icy objects. Most asteroids orbit between Mars and Jupiter. Comets come from two sources. Some are from the Kuiper Belt, a region that extends from a little past Neptune's orbit to well beyond Pluto's. Most comets, however, come from a huge spherical region over a thousand times farther from the Sun than Pluto. This remote, frigid zone is called the Oort cloud, in honor of the Dutch astronomer, Jan Oort, who hypothesized its existence (figure 22.6). Comets are far too dim for us to see them while they are in the Oort cloud, and they only become visible if they move into the inner solar system.

You might wonder how astronomers know the composition of these distant objects. One of the most powerful ways to learn that information is from analysis of the sunlight reflected from their surface or atmosphere. When light passes through a gas or reflects from a surface, atoms there absorb some of the colors. The missing colors may then be matched against laboratory samples to give information about what kinds of atoms are pres-

ent. Yet another way to infer a planet's composition is to measure its density. You may recall from chapter 2 that an object's density is its mass divided by its volume. We saw there that density is an important clue to the identity of minerals. Similarly, astronomers can use density to deduce a planet's composition and structure. In the case of Venus, Mars, and Jupiter, we have additional confirmation from space probes that have penetrated their atmospheres or landed on their surface.

Our knowledge of the solar system would be more complete if the planets were not so remote. For example, Pluto's orbit lies about 4 billion miles from the Sun. To measure such a vast distance in miles or kilometers is as meaningless as to measure the distance between New York and Tokyo in inches. Thus, astronomers use a much larger unit based on the Earth's average distance from the Sun. That distance, about 150 million kilometers, we call the **astronomical unit** (AU). The solar system out to Pluto has a diameter of about 80 AU. If we include the Oort cloud at the fringes of the solar system, where the comets generally orbit, the system is approximately 300,000 AU across (figure 22.6).

The Milky Way and the Universe

Just as the Earth is but one of many planets orbiting the Sun, so too is the Sun but one of a vast swarm of stars orbiting within our galaxy, the **Milky Way.** Astronomers estimate that the Milky Way contains some 100 billion stars spread throughout a roughly disk-shaped volume. That volume is so huge it is difficult to imagine, even by analogy. For example, if the solar

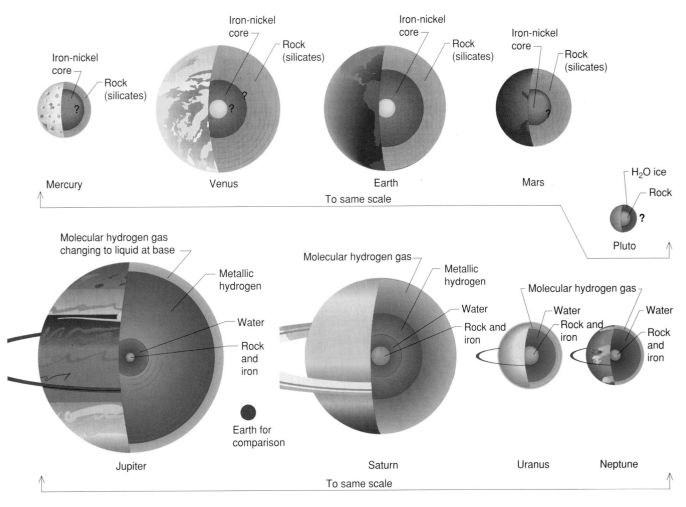

Mercury

Iron-nickel core

Rock (silicates)

Venus

Iron-nickel core

Rock (silicates)

Earth

Iron-nickel core

Rock (silicates)

Mars

Iron-nickel core

Rock (silicates)

To same scale

Pluto

H₂O ice

Rock

Jupiter

Molecular hydrogen gas changing to liquid at base

Metallic hydrogen

Water

Rock and iron

Earth for comparison

Saturn

Molecular hydrogen gas

Metallic hydrogen

Water

Rock and iron

Uranus

Molecular hydrogen gas

Water

Rock and iron

Neptune

Molecular hydrogen gas

Water

Rock and iron

To same scale

FIGURE 22.5

Sketches of the interiors of the planets. Details of sizes and composition of inner regions are uncertain for many of the planets. (Pluto is shown with the terrestrial planets for scale reasons only.)

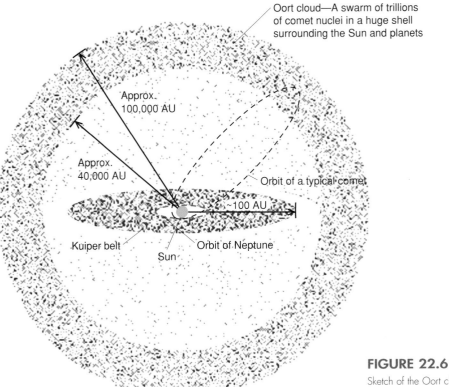

Oort cloud—A swarm of trillions of comet nuclei in a huge shell surrounding the Sun and planets

Approx. 100,000 AU

Approx. 40,000 AU

Orbit of a typical comet

~100 AU

Kuiper belt

Sun

Orbit of Neptune

FIGURE 22.6

Sketch of the Oort cloud and the Kuiper belt. The dimensions shown are known only approximately. Orbits and bodies are not to scale.

system were the size of a cookie, the Milky Way would be the size of the Earth. In fact, even the astronomical unit is too small to be a sensible unit for measuring the size of galaxies. Accordingly, astronomers use a far larger unit of distance, the light-year, to measure such sizes.

One light-year is the distance light travels in a year—about 10 trillion kilometers. In these immense units, the Milky Way is roughly 100,000 light-years in diameter. But the Milky Way is just one among myriads of other galaxies. Together, these galaxies with all their stars and planets form the visible **universe.**

ORIGIN OF THE PLANETS

The Solar Nebula

Speculation about how the planets formed dates back to pre-historic times and the creation myths of virtually all peoples. But among the first scientific theories were those of the eighteenth century by Immanuel Kant, a German philosopher, and Pierre Simon Laplace, a French mathematician. Kant and Laplace independently proposed that the solar system originated from a rotating, flattened disk of gas and dust called the **solar nebula.** In this model, the outer part of the disk became the planets, and the center became the Sun. This simple picture explains nicely the flattened shape of the system and the common direction of motion of the planets around the Sun.

Astronomers today have additional evidence that supports this basic picture. First, we now know that the inner planets are rich in silicates and iron, and the outer planets are rich in hydrogen. Second, to the best of our knowledge, the Earth, Sun, and Moon are all about the same age—about 4.6 billion years old—suggesting they formed in a single event. Third, the surfaces of objects like our Moon, Mercury, and the moons of the outer planets are heavily scarred with craters, suggesting that they were bombarded with infalling objects in the distant past. With this evidence, we can now discuss how astronomers think the planets were born.

The modern form of this **Nebular Hypothesis** proposes that our solar system was born about 4.6 billion years ago from an interstellar cloud, an enormous aggregate of gas and dust like the one shown in figure 22.7A. Such clouds are common between the stars in our galaxy even today, and astronomers now think most, if not all, stars and their planets have formed from them. Although our main concern in this chapter is with the birth of our own solar system, we should bear in mind that our hypothesis applies more broadly and implies that most stars could have planets or at least surrounding disks of dust and gas from which planets might form. In fact, astronomers observe just such disks around many young stars (figure 22.7B), lending the hypothesis additional support.

Although interstellar clouds are found in many shapes and sizes, the one that became the Sun and its planets was probably a few light-years in diameter and was made mostly of hydrogen (71%) and helium (27%) gas, with tiny traces of other chemical elements. In addition to gas, the cloud also contained microscopic dust particles. From analysis of the radiation of such dust particles, astronomers know they are made of a mixture of silicates, iron compounds, carbon compounds, and water frozen into ice.

The cloud that became our solar system began its transformation into the Sun and planets when the gravitational attraction between the particles in the cloud caused it to collapse inward, as shown in figure 22.8A. The collapse may have been triggered by a star exploding nearby or by a collision with another cloud. But because the cloud was rotating, it flattened, as shown in figure 22.8B. Flattening occurred because rotation retarded the collapse perpendicular to the cloud's rotation axis. You can see a similar effect in an old-fashioned pizza parlor where the chef flattens the dough by tossing it into the air with a spin.

It probably took a few million years for the cloud to collapse and form a rotating disk with a bulge in

FIGURE 22.7

(A) Photograph of an interstellar cloud (the dark region at top) that may be similar to the one from which the solar system formed. This dark cloud is known as Barnard 86. The star cluster is NGC 6520. (B) The small blobs in this picture are protostars and their surrounding disk of dust and gas. They are in the Orion Nebula, a huge gas cloud about 1,500 light-years from Earth. *Photo A by David Malin, courtesy of Anglo-Australian Telescope Board; Photo B courtesy of Space Telescope Science Institute*

Interstellar cloud

A

B

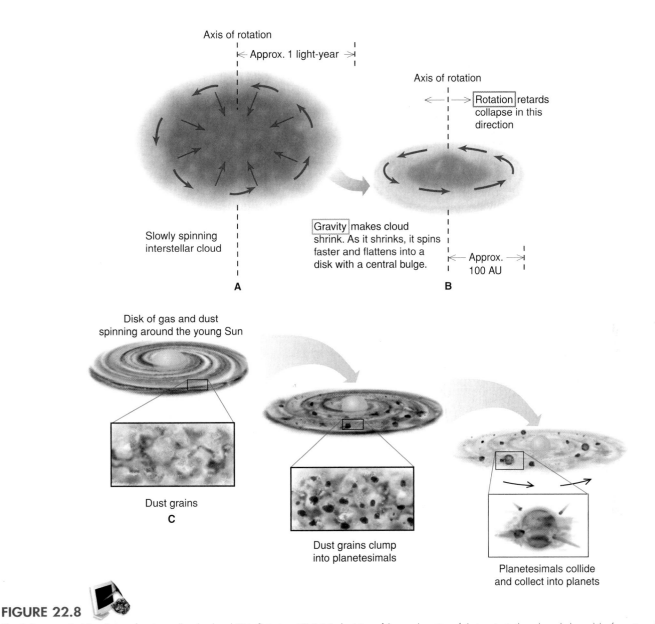

Axis of rotation

Approx. 1 light-year

Slowly spinning
interstellar cloud

A

Axis of rotation

Rotation retards
collapse in this
direction

Gravity makes cloud
shrink. As it shrinks, it spins
faster and flattens into a
disk with a central bulge.

Approx.
100 AU

B

Disk of gas and dust
spinning around the young Sun

Dust grains

C

Dust grains clump
into planetesimals

Planetesimals collide
and collect into planets

FIGURE 22.8

Sketch illustrating the (A) collapse of an interstellar cloud and (B) its flattening. (C) Artist's depiction of the condensation of dust grains in the solar nebula and the formation of planetesimals.

the center that became the Sun. Over a few more million years, dust particles in the disk began to stick together, perhaps helped by static electric forces such as those that make lint cling to clothes in a drier. As the particles slowly grew in size, they were more likely to collide with other particles, hastening their growth. But the composition of these particles depended strongly on where in the disk they formed. In the inner part of the disk (where the particles were near the Sun), it was too warm for water-ice to condense. Solid particles there were thus composed almost entirely of silicate and iron-rich material. Only at about Jupiter's distance from the Sun was the disk cold enough for water-ice to condense on the particles. Thus, particles in those outer regions consisted of silicate and iron-rich material *plus* frozen water. Although water molecules were relatively abundant throughout the cloud, they could only freeze onto particles in the outer disk. As a result, those particles grew

much larger than the particles in the inner disk. Thus, the nebula became divided into two regions: an inner zone of silicate and iron particles, and an outer zone of similar particles onto which ices also condensed.

Within each zone, the growing particles began to collide as they orbited in the disk. If the collisions were not too violent, the particles would stick (much as gently squeezing two snowballs together fuses them). By such processes, smaller particles grew steadily in size until they were kilometers across, as illustrated schematically in figure 22.8C. These larger objects are called **planetesimals** (that is, small, planetlike bodies). Because the planetesimals near the Sun formed from silicate and iron particles, while those farther out were cold enough that they could incorporate ice and frozen gases, two main types of planetesimals developed: rocky-iron ones near the Sun and icy-rocky-iron ones farther out.

Formation of the Planets

As time passed, the planetesimals themselves began to collide. Computer simulations show that some collisions led to the shattering of both bodies, but less-violent collisions led to merging, with the planetary orbits gradually becoming approximately circular.

Merging of the planetesimals increased their mass and thus their gravitational attraction. That, in turn, helped them grow even more massive. At this stage, objects large enough to be called planets were orbiting in the disk. But because there were two types of planetesimals (rocky and icy) according to their location in the inner or outer disk, two types of planets formed.

Planet growth was especially rapid in the outer parts of the solar nebula. Planetesimals there had more material from which to grow because ice was about ten times more abundant than silicate and iron compounds. Additionally, once a planet grew somewhat larger than the diameter and mass of the Earth, it was able to attract and retain gas by its own gravity. Because hydrogen was overwhelmingly the most abundant material in the solar nebula, planets large enough to tap that reservoir could grow vastly larger than those that formed only from solid material. Thus, Jupiter, Saturn, Uranus, and Neptune may have begun as slightly larger than Earth-sized bodies of ice and rock, but their gravitational attraction resulted in their becoming surrounded by the huge envelopes of hydrogen-rich gases that we see today. Alternatively, these outer planets are large enough that they may have formed directly by gravity drawing gas into clumps in the outer solar nebula. In the inner solar system, the smaller and warmer bodies could not capture hydrogen directly and therefore remained small.

As planetesimals struck the growing planets, their impact released gravitational energy that heated both the planetesimal and the planet. Gravitational energy is liberated whenever something falls. For example, if you drop a bowling ball into a box of tennis balls, the impact scatters the tennis balls in all directions, giving them kinetic energy—energy of motion. In much the same manner, planetesimals falling onto a planet's surface give energy to the atoms in the crustal layers, energy that appears as heating. You can easily demonstrate that motion can generate heat by hitting an iron nail a dozen or so times with a hammer and then carefully touching the nail: the metal will feel distinctly hot. Imagine now the vastly greater heating resulting from mountain-sized masses of rock plummeting onto a planet. The heat so liberated, in combination with radioactive heating (as occurs even today within the Earth), partially or completely melted the planets and allowed matter with high density (such as iron) to sink to their cores, while matter with lower density (such as silicate rock) "floated" to their surfaces, a process known as planetary **differentiation.** The Earth's iron core probably formed by this process, and astronomers believe that the other terrestrial planets formed iron cores and rocky crusts and mantles in much the same way. A similar process probably occurred for the outer planets when rock and iron material sank to their cores.

Formation of Moons

Once a planet grew massive enough so that its gravitational force could draw in additional material, it became ringed with debris. This debris could then collect into lumps to form moons. Thus, moon formation was a scaled-down version of planet formation. Although most of the large moons probably formed this way (they revolve and spin in the same plane and direction as the planets they orbit), the smaller moons of the outer planets were probably captured asteroids and small planetesimals (they have orbital planes and directions that are random with respect to the planets they orbit).

Final Stages of Planet Formation

During the last stage of planet formation, leftover planetesimals bombarded the planets and blasted out the huge craters such as those we see on the Moon and on all other bodies with solid surfaces in the solar system. Occasionally, an impacting body was so large that it did more than simply leave a crater. For example, Earth's Moon may have been created when an object a few times the size of Mars struck the Earth, as we will discuss further in the section titled "Portraits of the Planets." Likewise, Mercury may have suffered a massive impact that blasted away its crust. The extreme tilt of the rotation axes of Uranus and Pluto may also have arisen from planetesimal collisions. In short, planets and satellites were brutally battered by the remaining planetesimals early in the history of our solar system.

Formation of Atmospheres

Atmospheres were the last part of the planet-forming process. The inner and outer planets are thought to have formed atmospheres differently, a concept that explains their very different atmospheric compositions. The outer planets probably captured most of their atmospheres directly from the solar nebula, as mentioned in the "Formation of the Planets" section. Because the nebula was rich in hydrogen and helium, so are the atmospheres of the outer planets.

The inner planets were not massive enough and were too hot to capture gas directly from the solar nebula and are therefore deficient in hydrogen and helium. The atmospheres of Venus, Earth, and Mars probably formed by a combination of processes: volcanic eruptions releasing gases from their interiors; vaporization of comets and icy planetesimals that struck them in their youth; and gases freed from within the planetesimals out of which the planets themselves formed. Objects such as Mercury and the Moon, which have only traces of atmosphere, show few, if any, signs of volcanic activity within the last few billion years. Moreover, these smaller bodies have such weak gravity that any atmospheric gases they originally possessed escaped easily from them.

Other Planetary Systems

This theory for the origin of the solar system explains many of its features, but astronomers still have many questions about the Sun and its family of planets and moons. Answers to some of these questions are beginning to emerge from the discovery of planets around dozens of other stars. These planets are far too dim to be seen directly with present telescopes. Rather, they are detected by the slight gravitational tug they exert on their parent stars. That tug can be detected by analyzing the light from such stars. Over 100 of these *extrasolar* planets have been detected so far. Such observations have delighted astronomers by confirming that planets have indeed formed around other stars. However, they have also led to surprises. For example, many of these systems have huge, Jupiter-like planets very close to their parent stars. One hypothesis for this puzzling observation is that the planets formed farther out and "migrated" inward toward the

star. Such shifts in orbits may also have happened in our solar system. For example, many astronomers think that Neptune has migrated outward from an original orbit that was nearer the Sun. Thus, although we know that our solar system is not unique, we have yet to fully understand its properties and origin.

PORTRAITS OF THE PLANETS

What is the nature of these planets whose birth we have discussed? In this section, we will work our way through our solar system, describing each of the planets in turn. Table 22.1 below summarizes the physical properties, such as size and mass, of the nine planets. Table 22.2 summarizes their orbital properties. Before we consider planets far from Earth, however, it will help us to look at an object that is not officially a planet but a satellite—our Moon.

TABLE 22.1 Physical Properties of the Planets and Our Moon

Name	Radius (compared to Earth)	Radius (Equator) (km)	Mass (compared to Earth)	Mass (kg)	Average Density (g/cm^3)
Mercury	0.382	2,439	0.055	3.30×10^{23}	5.43
Venus	0.949	6,051	0.815	4.87×10^{24}	5.25
Earth	1.00	6,378	1.00	5.97×10^{24}	5.52
Our Moon	0.27	1,738	0.012	7.35×10^{22}	3.3
Mars	0.533	3,397	0.107	6.42×10^{23}	3.93
Jupiter	11.19	71,492	317.9	1.90×10^{27}	1.33
Saturn	9.46	60,268	95.18	5.68×10^{26}	0.69
Uranus	3.98	25,559	14.54	8.68×10^{25}	1.32
Neptune	3.81	24,764	17.13	1.02×10^{26}	1.64
Pluto	0.181	1,160	0.00216	1.29×10^{22}	2.05

TABLE 22.2 Orbital Properties of the Planets

Name	Distance from Sun* (AU)	(10^6 km)	Period Years	(Earth Days)	Orbital Inclination† Degrees	Orbital Eccentricity
Mercury	0.387	57.9	0.2409	(87.97)	7.00	0.206
Venus	0.723	108.2	0.6152	(224.7)	3.39	0.007
Earth	1.00	149.6	1.0	(365.26)	0.00	0.017
Mars	1.524	227.9	1.8809	(686.98)	1.85	0.093
Jupiter	5.203	778.3	11.8622	(4,332.59)	1.31	0.048
Saturn	9.539	1,427.0	29.4577	(10,759.22)	2.49	0.056
Uranus	19.19	2,869.6	84.014	(30,685.4)	0.77	0.046
Neptune	30.06	4,496.6	164.793	(60,189)	1.77	0.010
Pluto	39.53	5,900	247.7	(90,465)	17.15	0.248

*These values are half the long diameter of the orbit.

†With respect to the Earth's orbit.

Our Moon

The Moon is our nearest neighbor in space, a natural satellite orbiting the Earth. It is the frontier of direct human exploration and of great interest to astronomers. Unlike Earth, it is essentially a dead world, with neither plate tectonic nor current volcanic activity. Although extensive volcanism occurred early in the Moon's geologic history, it was largely confined to lower elevations of the lunar *nearside* (the side that permanently faces toward the Earth). This localized volcanism, coupled with the Moon's lack of atmosphere, means that many of its surface features are virtually unaltered since its youth. Its surface thus bears a record of events in the early solar system that gives clues not only to the Moon's birth but to that of the solar system as well.

Description of the Moon

General Features

The Moon is about one-fourth the diameter of the Earth and is a barren ball of rock possessing no air, water,[1] or life. In the words of lunar astronaut Buzz Aldrin, the Moon is a place of "magnificent desolation." But you don't have to walk on the Moon to see its desolation. Even to the naked eye, the Moon is a world of grays; yet even gray has its variety, as you can see where the dark, roughly circular areas stand out from the lighter background, as shown in figure 22.9.

Surface Features

Through a small telescope or even a pair of binoculars, you can see that the dark areas of the Moon are smooth while the bright areas are covered with numerous large, circular pits called *craters,* as illustrated in figures 22.10*A* and *B.* Craters usually have a raised rim and range in size from microscopic holes to gaping scars in the Moon's crust as much as 240 kilometers across. The larger craters have mountain peaks or peak rings at their center (figure 22.10*C*). Still larger lunar *impact basins* range up to 2,500 kilometers in diameter, with multiple concentric rings of mountain peaks and fault scarps (figure 22.10*D*).

The large, smooth, dark areas are called *maria* (pronounced *MAR-ee-a*), from the Latin word for "seas." This usage comes from early observers who believed the maria looked like oceans and who gave them poetic names such as Mare (pronounced *MAR-ay*), Serenitatis (Sea of Serenity), or Mare Tranquillitatis (Sea of Tranquility), the site where astro-

FIGURE 22.9

Photograph showing the different appearance of the lunar highlands and maria. The highlands are bright, rough, and heavily cratered. The maria are dark, smooth, and have few craters. *Photo © University of California Regents/Lick Observatory. Unauthorized use prohibited*

nauts first landed on the Moon. These regions, like the rest of the Moon, are essentially devoid of water and were formed by extensive volcanic activity. The maria are essentially low areas (primarily impact craters and basins) flooded by dark-colored basaltic lava flows.

The bright areas that surround the maria are called *highlands.* The highlands and maria differ in brightness because they are composed of different rock types. The maria are composed of basalt, a dark, congealed lava rich in iron, magnesium, and titanium silicates. The highlands, on the other hand, are mainly anorthosite, a rock type rich in calcium and aluminum feldspar silicates. This difference has been verified from rock samples obtained by astronauts. Moreover, the samples also show that the highland material is considerably older than the mare material. The highlands are also more rugged than the maria, being pitted with more and larger craters. In fact, highland craters are so abundant that they often overlap.

Craters are not the only type of lunar surface feature. From many craters, long, light streaks of pulverized rock called rays radiate outward, as shown in figure 22.11*A.* A particularly bright set spreads out from the crater Tycho near the Moon's South Pole and can be seen easily with a pair of binoculars when the Moon is full. A small telescope reveals still other surface features. Lunar canyons known as sinuous rilles, channels

---

[1]Astronomers have found some evidence for extremely small amounts of water frozen in perpetually shadowed craters near the Moon's poles.

FIGURE 22.10

(A) Overlapping craters in the Moon's highlands. (B) Isolated craters in the smooth mare. (C) Tycho—an 85-kilometer crater with a central peak, located on the south-central nearside of the Moon. (D) The 930-kilometer impact basin Orientale, located on the western limb of the lunar nearside. *Photos A and B © University of California Regents/Lick Observatory. Unauthorized use prohibited; Photos C and D by NASA and the Lunar and Planetary Institute*

used by ancient lava flows, wind away from some craters, as shown in figure 22.11*B*. Elsewhere, straight rilles break the surface, probably the result of tectonic activity.

Origin of Lunar Surface Features

Most of the surface features we see on the Moon—craters, rays, and basins (which bound the larger maria)—were made by the impact of solid objects on its surface. When such an object hits a solid surface at high speed, it disintegrates in a cloud of vapor-ized rock and fragments. The resulting explosion blasts a hole whose diameter depends on the mass and velocity of the impacting object. The shape of the hole is circular, however, unless the impact is grazing.

As the vaporized rock expands from the point of impact, it forces surrounding rock outward, piling it into a raised rim. Pulverized rock spatters in all directions, forming rays. Sometimes the impact compresses the rock below the crater sufficiently that it rebounds upward, creating a central peak.

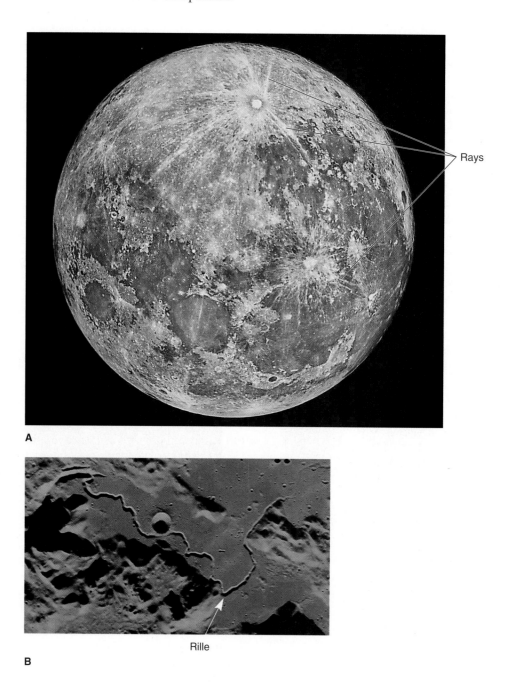

A

B

Rays

Rille

FIGURE 22.11

(A) The long, narrow, white streaks radiating away from the craters at the top are lunar rays. (B) Photograph of Hadley Rille, a large lava channel at the Apollo 15 landing site (just to the right of the arrow). *Photo A © UC Regents/Lick Observatory. Unauthorized use prohibited; Photo B by NASA*

The maria fill large impact features, but to understand their formation, we must briefly describe the early history of the Moon. From the great age of the highland rocks (in some cases as old as 4.5 billion years), astronomers deduce that these rugged uplands formed shortly after the Moon's birth. At that time, the Moon was probably molten throughout, allowing dense, iron-rich material to sink to its interior while less-dense material floated to the lunar surface. On reaching the surface, the less-dense rock cooled and congealed, forming the Moon's crust. A similar process probably formed the ancient cores of the Earth's continents. The Moon's highlands were then heav-

ily bombarded by solid bodies from space, creating the numerous craters we see there.

As the Moon continued to cool, its crust thickened. But before the crust grew too thick, a small number of exceptionally large objects—~100 kilometers across—struck the surface, blasting huge holes to create the largest lunar basins. Over the next billion years or so, molten material from deep within the Moon flooded these vast craters and congealed to form the smooth, dark lava plains that we see now. Because it was molten more recently, the mare material is younger than the highlands. Moreover, by the time the maria formed, most of the small, numerous impacting bodies were gone—swept up and captured by the Earth and Moon in earlier collisions. Thus, few bodies remained to crater the maria, which therefore remain mostly smooth (that is, far less heavily cratered) to this day.

Our home planet Earth furnishes additional evidence that most lunar craters formed early in the Moon's history. Like the Moon, Earth too was presumably battered by impacts in its youth. Although the vast multitude of these craters have been obliterated by erosion and plate tectonics, a few remain in ancient rock whose measured age is typically hundreds of millions or billions of years. From the scarcity of such craters, astronomers can deduce that the main bombardment must have ended almost 4 billion years ago.

Craters and maria so dominate the lunar landscape that we might not notice the absence of folded mountain ranges and the great rarity of volcanic peaks, landforms common on Earth. Why have such features not formed on the Moon?

Structure of the Moon

The Moon's small size relative to the Earth explains the differences between the two. The Moon, like Earth, was born hot and, as we mentioned before, was perhaps completely molten at some point in its youth. However, because the Moon is so small relative to the Earth, heat escapes from it more easily. Thus, the Moon has cooled far more than the Earth. (Think of how a small baked potato taken from a hot oven cools much faster than a big one.) Being cooler, the Moon lacks the internal convection currents that drive plate-tectonic activity on the Earth. Confirmation of this comes from studies of the Moon's interior.

Crust and Interior

The Moon's interior can be studied by seismic waves just as Earth's can. One of the first instruments set up on the Moon by the *Apollo* astronauts was a seismic detector. Measurements from that and other seismic detectors placed by later *Apollo* crews show that the Moon's interior is essentially inactive and has a much simpler structure than Earth's.

The Moon's surface layer is shattered rock from countless impacts that forms a regolith (meaning "blanket of rock") about 10 meters deep. Below this rocky rubble is the Moon's solid crust, which is about 60 to 70 kilometers thick, on the average. The Moon's crust, like Earth's, is composed of silicate rocks, relatively rich in aluminum and poor in iron. Beneath the crust is a thick mantle of solid rock, extending down about 1,000 kilometers. Unlike the Earth's mantle, however, it appears too cold and rigid to be stirred by the Moon's feeble heat.

The Moon's low average density (3.3 grams per cubic centimeter) tells us its interior contains relatively little iron. Moreover, because its small size allowed heat to escape relatively rapidly, its core is cool compared to Earth's. These deductions are also consistent with the Moon's lack of a magnetic field. Thus, a compass would be of no use to an astronaut lost on the Moon.

The Absence of a Lunar Atmosphere

The Moon's surface is never hidden by lunar clouds or haze. In fact, astronomers detect only the barest traces of an atmosphere, less than 100-trillionth (10^{-14}) the density of Earth's atmosphere. Two reasons explain this lack. First, the Moon's interior is too cool to cause ongoing volcanic activity (probably the source of much of the Earth's early atmosphere). Second, and more important, even if volcanoes had created an atmosphere in its youth, the Moon's small mass creates too weak a gravitational force for it to retain the erupted gas.

Origin and History of the Moon

Lunar rocks brought back to Earth by the *Apollo* astronauts have led astronomers to radically revise their ideas of how the Moon formed. Before the *Apollo* program, lunar scientists had three hypotheses of the Moon's origin. In one, the Moon was originally a small planet orbiting the Sun that approached the Earth and was captured by its gravity (capture hypothesis). In another, the Moon and Earth were "twins," forming side by side from a common cloud of dust and gas (twin formation hypothesis). In the third, the Earth initially spun enormously faster than now and formed a bulge that ripped away from it to become the Moon (fission hypothesis).

The failure of evidence based on lunar surface samples to confirm any of the three hypotheses led astronomers to consider alternatives, and a completely different picture of the Moon's origin has emerged. According to the new hypothesis, the Moon formed from debris blasted out of the Earth by the impact of an object about the size of Mars, as shown in figure 22.12*A*. The great age of lunar rocks and the absence of any impact feature on the Earth indicate that this event must have occurred during the Earth's own formation, at least 4.5 billion years ago. The colliding body melted and vaporized millions of cubic kilometers of the Earth's surface rock and hurled it into space in an incandescent plume. Although most of this debris would have rained back down on Earth, some of it remained in orbit and its gravity gradually drew it together into what we now see as the Moon.

After the Moon's birth, stray fragments of the ejected rock pelted its surface, creating the craters that blanket the highlands. A few huge fragments plummeting onto the Moon later in its formation process blasted enormous holes that later flooded with molten interior rock to become the maria. Since that time, the Moon has experienced no major changes. It has been a virtually dead world for all but the earliest times in its history.

Mercury

Mercury is the smallest inner planet and resembles our Moon in both size and appearance. Mercury is covered with circular craters caused by impacts like those on the Moon (figure 22.13*A*), but its surface is not totally moonlike. Enormous scarps—formed where the crust has shifted—run for hundreds of kilometers

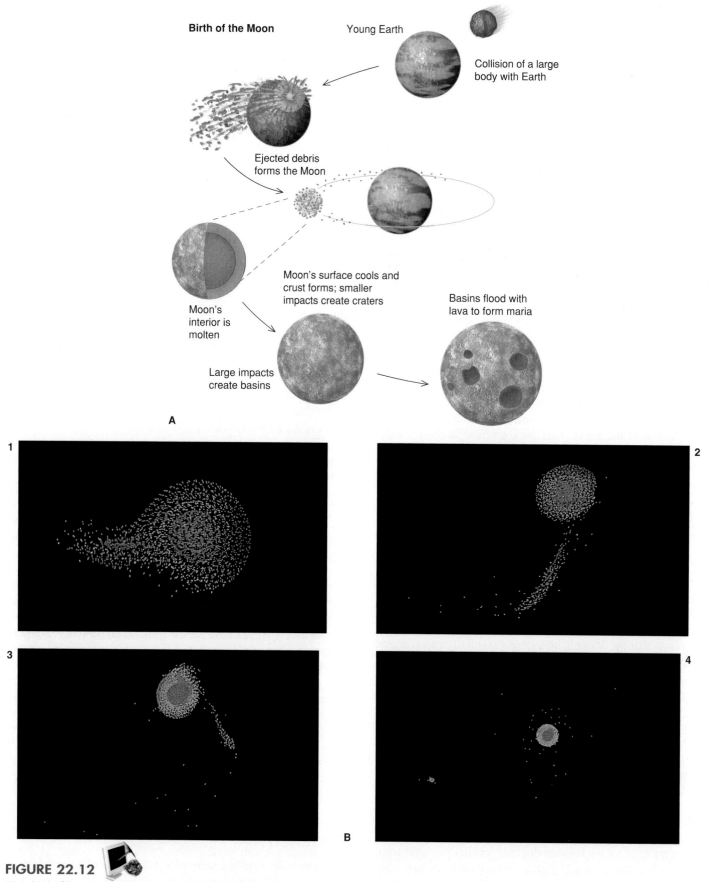

Birth of the Moon

Young Earth

Collision of a large body with Earth

Ejected debris forms the Moon

Moon's interior is molten

Moon's surface cools and crust forms; smaller impacts create craters

Basins flood with lava to form maria

Large impacts create basins

A

1

2

3

4

B

FIGURE 22.12

(A) A sketch of how the Moon may have formed. (B) A computer simulation showing how the collision of a roughly Mars-sized body with the young Earth can splash out debris, which then is drawn together by its own gravity to make Earth's Moon. The scale changes between the frames: (1) is a close-up of the impact; (4) shows the newly assembled Moon (lower left) orbiting Earth, which is still surrounded by debris. *Photos B courtesy of A. G. W. Cameron, Lunar and Planetary Laboratory, University of Arizona*

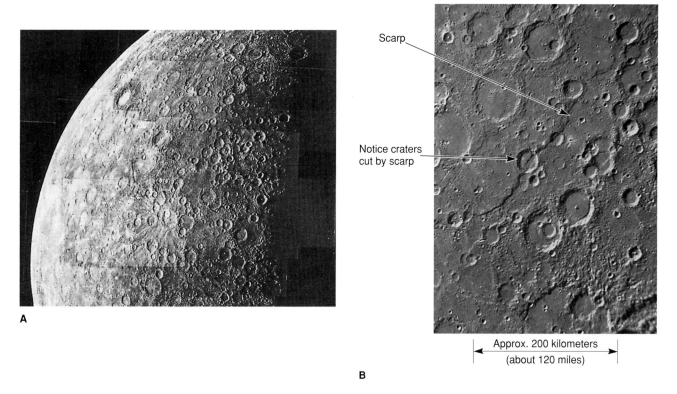

FIGURE 22.13

(A) Mosaic of pictures of Mercury's cratered surface (taken by the Mariner 10 spacecraft). (B) Photograph of a scarp, (a cliff), running across Mercury's surface. *Photo A by JPL/California Institute of Technology/NASA; Photo B by NASA*

across Mercury's surface, as seen in figure 22.13*B*. These scarps may have formed as the planet cooled and shrunk, wrinkling like a dried apple.

Mercury's Temperature and Atmosphere

Mercury's surface is one of the hottest places in the solar system. At its equator, noon temperatures reach approximately 430°C (about 800°F). On the other hand, nighttime temperatures are among the coldest in the solar system, dropping to approximately −173°C (about −280°F). These extremes result from Mercury's closeness to the Sun, its very slow rotation, and its almost complete lack of atmosphere: only traces of gas, perhaps captured from interplanetary space, have been detected. Without any effective atmosphere, nothing moderates the inflow of sunlight during the day or retains heat during the night.

Mercury lacks an atmosphere for the same reason the Moon does: its small mass makes its gravity too weak to retain much gas around it. Moreover, its proximity to the Sun makes keeping an atmosphere especially difficult because the resulting high temperature causes gas molecules to move so fast that they readily escape into space. The intense *solar wind* (rapid outflow of ionized gas from the Sun) at Mercury's proximity to the Sun would also tend to strip away any existing atmosphere quite rapidly.

Mercury's Interior

Mercury probably has an unusually large (~75% of its diameter) iron core beneath its silicate crust, but planetary scientists have little proof because no spacecraft has landed there to deploy seismic detectors. Their conclusion is based on Mercury's high density, which indicates an iron-rich interior with only a thin rock (silicate) mantle. Why Mercury is so relatively rich in iron is unclear. One possibility is that Mercury once had a thicker rocky crust that was blasted off by the impact of an enormous planetesimal.

Mercury's Rotation

Mercury spins very slowly. Its rotation period is 58.646 Earth days, exactly two-thirds its orbital period around the Sun. This means that it spins exactly three times for each two trips it makes around the Sun. This proportion is not just coincidence. The Sun's gravity exerts a force on Mercury, which tends to twist the planet and make it rotate with the same period in which it orbits the Sun. Mercury's orbit, however, is very elliptical, so its orbital speed changes as it moves around the Sun. Because of that changing speed, the Sun cannot lock Mercury into a perfect match between its spin and orbital motion: The closest to a match that it can achieve is to make three spins for each two orbits.

Venus

Of all the planets, Venus is most like the Earth in diameter and mass. We might therefore expect it to be like the Earth in other ways. However, Venus and the Earth have radically different surfaces and atmospheres.

The Venusian Atmosphere

The atmosphere of Venus is mainly (96%) carbon dioxide, as deduced from analysis of reflected sunlight and from measurements with space probes. These methods also show that, in addition to carbon dioxide, Venus's atmosphere contains about 3.5% nitrogen and very small amounts of water vapor and other gases. (For comparison, Earth's atmosphere is about 78% nitrogen and 21% oxygen, with CO_2 being a mere 0.04%).

Dense clouds perpetually cloak Venus and prevent us from ever seeing its surface directly. These clouds are composed of sulfuric acid droplets that formed when sulfur compounds—perhaps ejected from volcanoes—combined with the traces of water in the atmosphere. Robot spacecraft that have landed on Venus show that below the clouds, the Venusian atmosphere is relatively clear, and some sunlight penetrates to the surface. The light is tinged orange, however, because the blue colors are absorbed in the deep cloud layer.

Venus's atmosphere is extremely dense. It exerts a pressure roughly ninety times that of the Earth's. Its lower atmosphere is also extremely hot: the surface temperature is more than 480°C (about 900°F)—hotter than the daylit surface of Mercury! What makes the atmosphere of a planet so similar to the Earth in size and only slightly nearer the Sun so very hot? The answer is the greenhouse effect.

Venus's Surface Temperature and the Greenhouse Effect

A planet's surface temperature is set by a balance between the energy it receives from the Sun and the energy it re-radiates back into space. Certain gases, such as carbon dioxide, freely transmit visible light but strongly absorb infrared (heat) radiation. Thus, if a planet's atmosphere contains carbon dioxide (or other gases such as water or methane), sunlight can reach the surface to warm the planet, but the radiation of that heat back into space is impeded, which makes the planet warmer than it would be if it had no atmosphere. This phenomenon, whereby an atmosphere warms a planet's surface, is called the **greenhouse effect.** Here on Earth, carbon dioxide creates a weak greenhouse effect. Venus, however, has about 300,000 times more carbon dioxide than Earth, and so its greenhouse effect is much stronger.

The Surface of Venus

Because the surface of Venus is hidden by its clouds (see figure 22.1), scientists cannot see it directly. However, just as aircraft pilots use radar to penetrate terrestrial clouds and fog to show a runway, astronomers use radar telescopes to penetrate the Venusian clouds and map the planet's surface. Such maps show that Venus is generally less mountainous and rugged than Earth. Most of its surface is low, gently rolling plains, but two major highland regions, Ishtar and Aphrodite, rise above the lowlands to form continent-size landmasses. Ishtar is largely composed of mountain belts, the highest part of which, Maxwell Montes, rises more than 11 kilometers (roughly the height of Mount Everest here on Earth) above the average level of the planet. The parallel ridges, troughs, and fractures in these mountain belts indicate they were formed by tectonic processes resulting from both compressional and tensional forces.

Venus is so similar in diameter and mass to the Earth that planetary geologists expected to see landforms there similar to those on the Earth. Yet Venus has a surface almost totally unlike our planet's. Although Venus has some folded mountains, volcanic landforms dominate. Surface features include peaks with immense lava flows, "blisters" of uplifted rock, grids of long, narrow faults, and continent-scale uplands bounded by mountain belts (figure 22.14). All of these features indicate a young and active surface. From the small number of impact craters on Venus, planetary geologists deduce that virtually all of Venus's original old surface has been covered over by volcanic activity. (Small numbers of craters indicate a young surface just as few potholes in a road indicate a recent paving job.) The surface we see is probably at most half a billion years old, much younger than much of Earth's continental regions. Whether Venus's volcanoes are still active is unknown.

The numerous volcanic features on Venus suggest that heat flows less uniformly within Venus than within the Earth. Although a couple of dozen locations on Earth (Yellowstone National Park, for example) are heated by "plumes" of rising hot rock, such plumes seem far more common on Venus. We may be viewing on our sister planet what Earth looked like as its lithosphere began to form and before smooth heat flows were established. Why do Venus and Earth have such different surface features and subsurface activity? One set of explanations is based on differences in the amount of water in the rocks of these two worlds.

Water trapped in rock lowers the rock's melting temperature and makes the magma less viscous than magma with little or no water. These effects in turn lead to differences in the thickness of a planet's lithosphere. In particular, on a planet whose rocks are rich with water, the molten rock flows more easily, so it leads to a thinner lithosphere than if the rocks contained little water. Other things being equal, we therefore expect wet Earth to have a thinner lithosphere than dry Venus. This thinness of Earth's outer layers allows a steady loss of heat, so our planet's interior cools steadily. Moreover, a thin lithosphere breaks easily into many small plates. This breaking allows crustal motion and activity to occur more or less continuously and at many places on the surface. For Venus, the thicker lithosphere holds heat in, keeping the interior hot and limiting the crust's break up.

Ultimately, however, the trapped heat must escape, and it is here that the hypotheses differ strongly. According to one hypothesis, at points where the hot, rising material reaches the crust, the surface bulges upward and weakens, and volcanoes

Congealed lava domes at the eastern edge of
the Alpha Regio Highlands

20 kilometers (about 12 miles)

Maat Mons volcano
(vertical scale exaggerated)

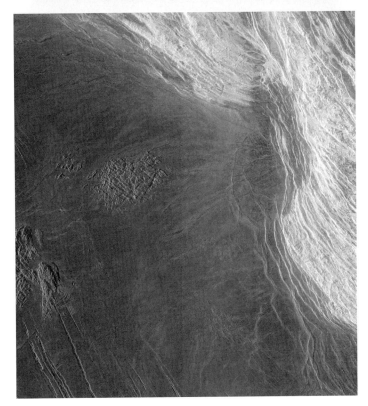

Craters in the Lavinia region

Approx. 50 kilometers
(about 31 miles)

Fractured plains in the Lakshmi region

Approx. 37 kilometers (about 23 miles)

FIGURE 22.14

Venus surface features imaged by *Magellan* radar. The orange color has been added to match the color of the landscape observed by the Russian *Venera* lander. Features are also exaggerated vertically for clarity. *Photos by NASA/JPL*

Complex mountain belts in western Ishtar terra (bright ridges in the northeastern half of the image) produced by compressional and tensional tectonic forces. Image covers an area approximately 200 kilometers (125 miles) wide.

FIGURE 22.15

Images of the surface of Venus returned by the Soviet *Venera* 13 lander in 1982. The upper two images show panoramic views of flat surface rocks, soil, and the sampling arm and ejected camera cover fragments (semicircular objects in each frame). The bottom image is an approximately true color image of an area within the second panorama. *Photo by NSSDC, NASA, and the Soviet Venera 13 Mission.*

A

B

C

FIGURE 22.16

(*A*) The north Martian polar cap and (*B*) the south Martian polar cap. (*C*) Note the layered structure visible in this enlarged view of the north polar cap. *Photo A by A. S. McEwen, U.S. Geological Survey; Photo B by NASA/JPL; Photo C by NASA/JPL/ Malin Space Science Systems*

may form. Where cooler material sinks, the thick crust crumples into a continent-like region. This creates a planet with few and small continents and whose surface is tectonically and volcanically active but only in isolated spots.

According to another hypothesis, the trapped heat gradually melts the bottom of the thick crust, thinning it and allowing it to break up. This may happen over widespread areas, flooding almost the entire planetary surface with lava in a brief time. The heat then rapidly escapes to space, the interior cools, and the crust again thickens. This process could occur cyclically.

Although most of what planetary scientists know of the Venusian surface comes from radar maps, several Russian *Venera* spacecraft have landed there and transmitted pictures to Earth (figure 22.15). These pictures show a barren surface covered with flat, broken rocks and lit by the pale, orange glow of sunlight diffusing through the deep clouds. The robotic spacecraft have also sampled the rocks, showing them to be volcanic.

The Interior of Venus

The deep interior of Venus is probably like the Earth's, an iron core and silicate rock mantle. However, astronomers have no seismic information to confirm this conjecture and, as with Mercury, must rely on deductions from its gravity and density, which are similar to the Earth's. However, Venus's lack of a global magnetic field suggests significant differences between the interior dynamics of these two sister planets.

The next planet out from the Sun after Venus is our own Earth, the subject of the earlier chapters of this book. Let us, therefore, continue our journey to the next planet, Mars.

Mars

Mars is named for the Roman god of war, presumably because of its distinctly reddish color.[2] Although Martian winds sweep dust and patchy clouds of ice crystals through its sky, its atmosphere is generally clear enough for astronomers to view its surface directly. Such views from Earth and from spacecraft orbiting Mars show a world of mostly familiar features.

Perhaps the most obvious of Mars' surface features are its polar caps (figure 22.1). These frigid regions are not just frozen water, as on Earth. The south polar cap is frozen carbon dioxide—dry ice—and the north cap has a surface layer of frozen CO_2 over deep layers of ordinary water-ice. These numerous separate layers (easily seen in figure 22.16*C*) indicate that the Martian climate changes cyclically.

---

[2]Mars' red color comes from its very iron-rich surface, as occurs in deserts on Earth.

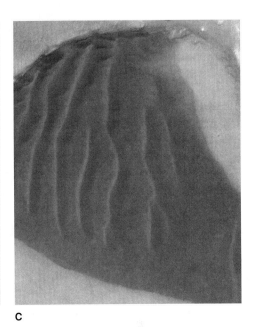

A   **B**   **C**

FIGURE 22.17

(A) Transverse and barchanoid dunes in a crater in the southern hemisphere of Mars. (B) Barchan dunes near the edge of Mars's north polar cap. (C) Small field of transverse dunes in the southern hemisphere of Mars. Note that, unlike on Earth where most sand dunes are light colored, dunes on Mars are typically dark. This is due to different minerals present in the dunes—primarily light quartz on Earth, versus darker ferromagnesian minerals on Mars. *Images by NASA, JPL, Malin Space Science System*

Many locations on Mars, particularly around the northern polar area and in craters in the southern hemisphere, exhibit dunes blown into parallel ridges by the Martian winds and crescentic barchans, as illustrated in figure 22.17. At mid-latitudes, a huge upland called the Tharsis bulge is dotted with volcanic peaks. One of these peaks, Olympus Mons, rises about 22 kilometers above its surroundings, two-and-a-half times the height of Earth's highest peaks (figure 22.18) and is the highest known mountain in the solar system. The base of this immense peak spans a distance slightly greater than the width of Arizona.

Some planetary geologists think the Tharsis bulge may also have created the gigantic Valles Marineris (figure 22.19), which lies to the southeast, near the Martian equator. This enormous canyon is 5,000 kilometers long, 100 kilometers wide, and 10 kilometers deep. It dwarfs Earth's Grand Canyon and would span the continental United States. Some planetary geologists think it formed as the Tharsis region swelled, stretching and cracking the crust. Others believe that this vast chasm is evidence for plate-tectonic activity, like that of Earth, and that the Martian crust began to split, but the motion ceased as the planet aged and cooled. In either case, this immense canyon system is primarily the result of tectonic activity rather than erosional forces.

Perhaps the most surprising features revealed by spacecraft are the huge channels and what look like dry riverbeds, such as those seen in figure 22.20. Planetary geologists infer from these features that liquid water once flowed on Mars, even though no surface liquid is present now. In fact, many planetary scientists now believe that huge lakes or small

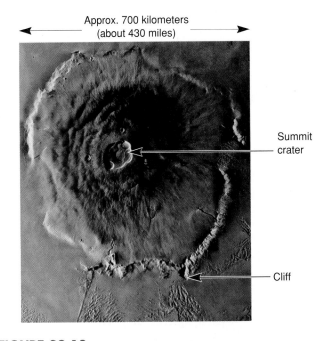

Approx. 700 kilometers
(about 430 miles)

Summit crater

Cliff

FIGURE 22.18

Photograph of Olympus Mons (on Mars), the largest known volcano (probably inactive) in the solar system. *Photo by A. S. McEwen, U.S. Geological Survey*

oceans once existed on Mars. The strongest evidence for ancient bodies of water is layered rocks, some of which have been recently confirmed by surface rovers to have formed in the presence of standing water (figure 22.21), within many craters and basins. Moreover, some of these features are cut by

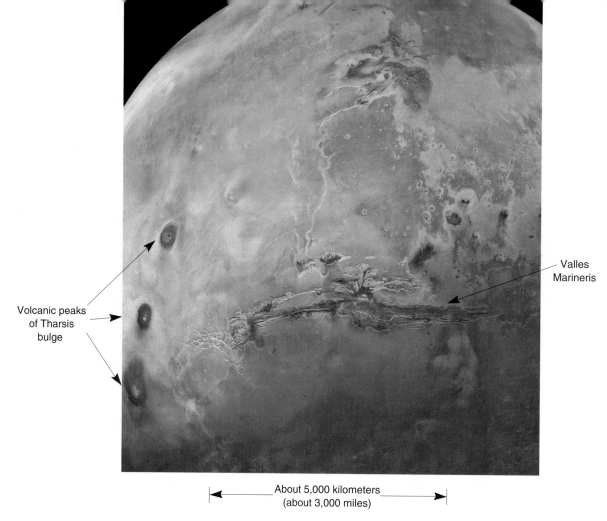

Volcanic peaks
of Tharsis
bulge

Valles
Marineris

About 5,000 kilometers
(about 3,000 miles)

FIGURE 22.19

Photograph of Valles Marineris, the Grand Canyon of Mars. This enormous gash in Mars's crust may be a rift that began to split apart the Martian crust but failed to open farther. The canyon is about 5,000 kilometers long. Were it on Earth, it would stretch from California to Florida. The canyon was named for the *Mariner* spacecraft, which sent the first pictures of it back to Earth. *Photo by A. S. McEwen, U.S. Geological Survey*

Approx. 100 kilometers
(about 60 miles)

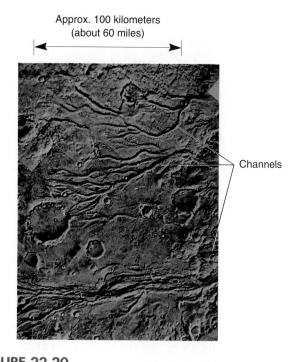

Channels

FIGURE 22.20

Picture of channels probably carved by running water on Mars. *Photo by NASA*

narrow canyons (figure 22.22) that breach their rims and appear to "drain" into lowland areas. But the atmospheric pressure on Mars now is so low that liquid water is not stable at the surface. However, recently acquired images suggest it may even today flow across the surface briefly (perhaps beneath blankets of water frost) before boiling off into the thin atmosphere (figure 22.23). We thus are faced with at least two puzzles: where has the water gone, and what has led to the change of conditions that makes liquid water no longer stable there? A first step toward solving these puzzles lies in examining the Martian atmosphere.

The Martian Atmosphere

Clouds and wind-blown dust (figure 22.24) are visible evidence that Mars has an atmosphere. Analysis of sunlight reflected from the planet shows that this atmosphere is mostly (95%) carbon dioxide with small amounts (3%) of nitrogen and traces of oxygen and water. The atmosphere is very tenuous—only about 1% the density of Earth's—and its composition and low density have been verified by spacecraft that have landed on Mars.

The Martian atmosphere is so tenuous that the carbon dioxide creates only a very weak greenhouse effect, which, combined with Mars's greater distance from the Sun, makes

A

FIGURE 22.21

This approximate true-color image taken by NASA's Mars *Exploration Rover Opportunity* (October 2004) shows an unusual, lumpy rock informally named "Wopmay" on the lower slopes of "Endurance Crater." Scientists think the lumps in Wopmay may be related to cracking and alteration processes, possibly caused by exposure to water. The area between intersecting sets of cracks eroded in a way that created the lumpy appearance. *Image by NASA/JPL*

the planet very cold. Temperatures at noon at the equator may reach a little above the freezing point of water, but at night they plummet to far below zero. Thus, although water exists on Mars, most of it is frozen, locked up either below the surface in the form of permafrost or in the polar caps as solid water-ice.

Clouds of dry ice (frozen CO_2) and water-ice crystals (H_2O) drift through the atmosphere carried by the Martian winds. The Martian winds are generally gentle, but seasonally and near the poles, they become gales, which sometimes pick up large amounts of dust from the surface. The resulting vast dust storms occasionally envelop the planet completely and turn its sky pink.

No rain falls from the Martian sky, despite its clouds, because the atmosphere is too cold and contains too little water. Mars has not always been so dry, as the numerous channels in its highlands show. The existence of such channels carved by liquid water is therefore strong evidence that in its past, Mars was warmer and had a thicker atmosphere. But that milder climate must have ended billions of years ago. From the large number of craters on Mars, planetary scientists deduce that its surface has not been significantly eroded by rain or flowing water for about 3 billion years. Where, then, has its water gone?

Some water may lie buried below the Martian surface as ice. If the Martian climate was once warmer and then cooled drastically, water would condense from its atmosphere and freeze, forming sheets of surface ice. Wind might then bury this ice under protective layers of dust, as happens in polar and high mountain regions of Earth. In fact, recent observations suggest

B

FIGURE 22.22

Martian valley systems cutting through the cratered plains of Mars. (*A*) *Viking* orbiter image showing valleys that drained water toward the northern Martian lowlands. (*B*) Mars *Global Surveyor* image showing close-up view of part of one of these valleys, outlined by the white box in (*A*). *Photos by NASA/JPL/Malin Space Science Systems*

FIGURE 22.23

Mars Global Surveyor image of recently formed gullies in a crater wall on Mars. These gullies appear to have been formed by the transient flow of liquid water on the surface. Their downslope deposits overlap dark (dust-free) sand dune deposits at bottom right, indicating a young age for the gullies. *Photo by NASA/JPL/Malin Space Science System*

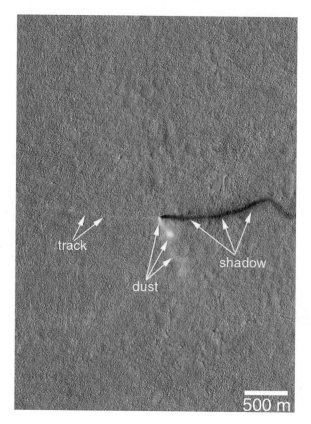

FIGURE 22.24

Mars Global Surveyor image of a dust devil on the surface of Mars. *Photo by NASA/JPL/Malin Space Science System*

widespread deposits of such dust-covered ice, and figure 22.25 shows indirect evidence of it. In this image, you can see an area where planetary geologists think that subsurface ice has melted, perhaps from volcanic activity, and drained away in a catastrophic flood. The removal of the ice has then caused the ground to collapse and leave the lumpy, fractured terrain visible at the broad end of the channel. The released meltwater then flowed downstream, carving the 20-kilometer-wide canyon. But how did Mars's water come to be buried and frozen, and why did the planet cool so?

If Mars had a denser atmosphere in the past, then the greenhouse effect might have made the planet significantly warmer than it is now. The loss of such an atmosphere would then weaken the greenhouse effect and permanently cool the planet. One way such a loss could happen is that Mars may have been struck by a huge asteroid whose impact blasted its atmosphere off into space. Such impacts, although rare, do occur, and our own planet may have been struck about 65 million years ago with results nearly as dire, as we shall learn in the section on "Giant Impacts." Alternatively, Mars may have gradually lost its atmospheric gases by solar wind erosion. Mars's current lack

of a magnetic field allows solar wind particles to directly impact its atmosphere. Over billions of years, much of Mars's atmosphere could have been lost to space in this way.

The Martian Interior

Planetary scientists believe that the interior of Mars, like that of Earth, consists of a crust, mantle, and iron core. Because Mars is smaller than Earth, however, it would be expected to lose heat more rapidly, and thus, its interior is now probably cooler than Earth's. Unfortunately, scientists have no direct seismic confirmation of Mars's interior structure, although recent analysis of the orbital motion of the *Mars Global Surveyor* satellite suggests that Mars's core may be partly molten.

Mars has no folded mountains, so planetary geologists think that the interior of Mars has cooled and its lithosphere thickened to perhaps twice the thickness of the Earth's lithosphere. As a result, its now-weak interior heat sources can no longer drive crustal motions. Mars's current low level of crustal activity is also demonstrated by its many craters, whose numbers imply that Mars has been relatively geologically quiet for billions of years.

The Martian Moons

Mars has two tiny moons, Phobos and Deimos. They are named for the minor gods of fear and panic, respectively, in Greek mythology. First discovered in 1877, these irregular and

FIGURE 22.25

Evidence for buried ice on Mars. (A) A channel cut by water released as subsurface ice melted. Note the chaotic terrain at the broad end of the channel where the surface collapsed as the water drained away. (B) Crater with surrounding flow patterns. Heat released by the impact that formed the crater has melted subsurface ice. The thawed material has oozed out. *Photos by NASA*

cratered bodies are only about 20 kilometers across and are probably captured asteroids.

Life on Mars?

Scientists have long wondered whether life developed on Mars. To search for life there, the United States landed two *Viking* spacecraft on Mars in 1976. These craft carried instruments to look for signs of carbon chemistry in the soil and metabolic activity in soil samples that were put in a nutrient broth carried on the lander. All tests either were negative or ambiguous.

Then, in 1996, a group of American and English scientists reported possible signs of life in meteorites blasted out of Mars. These objects had traveled through interplanetary space and happened to land on Earth. Over the last few years, however, scientists have shown that ordinary chemical weathering can form structures similar to the supposed Martian "fossils." As a result, most scientists today are unconvinced that any meteorite yet studied shows persuasive evidence of Martian life. Future Mars lander missions will no doubt revisit this subject with more sensitive instruments.

Why are the Terrestrial Planets So Different?

We have seen that the four terrestrial planets have little in common apart from being rocky spheres. Astronomers think their great differences arise from their different masses, radii, and distance from the Sun.

Role of Mass and Radius

As discussed earlier, a planet's mass and radius affect its interior temperature and thus its level of tectonic activity, with low-mass, small-radius planets being cooler inside than larger bodies. We see, therefore, a progression of activity from small, relatively inert Mercury, to slightly larger and once-active Mars, to the larger and far more active interiors and surfaces of Venus and Earth. Mercury's surface still bears the craters from its earliest history. In contrast, Earth and Venus retain essentially none of their original crust: Their surfaces have been enormously modified by activity in their interiors over the lifetimes of these planets. Such internal activity in turn affects a planet's atmosphere. Recall that volcanoes probably supplied much of our atmosphere. Thus, small, volcanically inactive[3] Mercury probably never had much atmosphere, and Mars, active once but now quiescent, likewise could create and retain only a thin atmosphere. Moreover, the low mass of these planets implies a weak surface gravity relative to Venus and Earth, which makes retaining an atmosphere difficult. As a result, Mercury (and the Moon) is virtually without an atmosphere today, and Mars has only a vestige of its original atmosphere. On the other hand, highly active Venus and Earth have extensive atmospheres.

---

[3]Some smooth plains on Mercury may have been produced by volcanism very early in the planet's history.

Role of Distance from the Sun

A planet's distance from the Sun, other things being equal, would determine its surface temperature. Thus, we expect that Venus will be warmer than Earth and that Earth will be warmer than Mars. These expected temperature differences are increased, however, by the atmospheres of these bodies, which trap heat to differing extents. Even relatively small differences in temperature can lead to large differences in physical behavior and chemical reactions within an atmosphere. For example, Venus is just enough nearer the Sun than the Earth that, even without a strong greenhouse effect, most of its atmosphere would be so warm that water would have difficulty condensing and turning to rain. Moreover, with the atmosphere being warmer throughout, water vapor can rise to great heights in the Venusian atmosphere, whereas on cooler Earth, water vapor condenses to ice at about 10 kilometers, making our upper atmosphere almost totally devoid of water. The great difference in the water content of the upper atmospheres of Earth and Venus has led to a drastic difference between their atmospheres at lower levels. At high altitudes, ultraviolet light from the Sun breaks apart any water molecules present into their component oxygen and hydrogen atoms. Being very light, the hydrogen atoms so liberated escape into space, while the heavier oxygen atoms remain. Because water can rise to great heights in the Venusian atmosphere, over billions of years it has been almost completely lost from our sister planet. In our atmosphere, water has survived.

Water makes possible chemical reactions that profoundly alter the composition of our atmosphere. For example, CO_2 dissolves in liquid water and creates carbonic acid, which in dilute form we drink as "soda water." In fact, the bubbles in soda water are just carbon dioxide that is coming out of solution. As rain falls through our atmosphere, it picks up CO_2, making it slightly acidic, even in unpolluted air. As the rain falls on the ground, it reacts chemically with silicate rocks to form carbonates (as discussed in chapter 5), locking some CO_2 into the rock.

Role of Biological Processes

Biological processes also remove some CO_2 from the atmosphere. For example, plants use it to make cellulose, the large organic molecule that is their basic structural material. This CO_2 is usually stored for only short periods of time, however, before decay or burning releases it back into the atmosphere. More permanent removal occurs when rain carrying dissolved CO_2 runs off into the oceans, where sea creatures use it to make shells of calcium carbonate. As these creatures die, they sink to the bottom, where their shells form sediment that eventually is changed to rock (see chapter 6). Thus, carbon dioxide is swept from our atmosphere and locked up both chemically and biologically in the crust of our planet. With most of the carbon dioxide removed from our atmosphere, mostly nitrogen is left. In fact, our atmosphere contains roughly the same total amount of nitrogen as the atmosphere of Venus.

Our atmosphere is also rich in oxygen, a gas found in such relative abundance nowhere else in the solar system. Our planet's oxygen is almost certainly the product of green plants breaking down the H_2O molecule during photosynthesis. What evidence supports the idea that liquid water and living things removed most of the early Earth's carbon dioxide? It is the carbonate rock we mentioned earlier. If the carbon dioxide locked chemically in that rock were released back into our atmosphere and the oxygen were removed, our planet's atmosphere would closely resemble Venus's.

If water is so effective at removing carbon dioxide from our atmosphere, why does *any* CO_2 remain? Although human activities add small amounts of CO_2 by burning wood and fossil fuels, the major contribution is from natural processes. In particular, geological activity gradually releases CO_2 from rock back into our atmosphere. At plate boundaries, all rock, including sedimentary material, is carried downward into the mantle, where it is melted. Heating breaks down the carbonate rock into carbon dioxide, which then rises with the magmas to the surface and reenters the atmosphere. A similar process may once have occurred on Mars to remove carbon dioxide from its atmosphere, locking it up in rock there. Mars's present low level of tectonic activity, however, prevents its CO_2 from being recycled. Thus, with so little of its original carbon dioxide left, Mars has grown progressively colder. Our Earth, because it is active, has retained enough CO_2 in its atmosphere to maintain a moderate greenhouse effect, making our planet habitable. Thus, poised between one planet that is too hot and another that is too cold, Earth is fortunate, indeed, to have a relatively stable atmosphere, one factor in the complex web of an environment on which our existence depends.

Jupiter

To the ancient Romans, Jupiter was the king of the gods. Although they did not know how immense this planet is, they nevertheless chose its name appropriately. Jupiter is the largest planet in both radius and mass. In fact, its mass is larger than that of all other planets in the solar system combined. It is about eleven times the Earth's diameter and more than 300 times its mass. Dense, richly colored parallel bands of clouds cloak the planet, as shown in figure 22.26. Analysis of the sunlight reflected from these clouds shows that Jupiter's atmosphere consists mostly of hydrogen, helium, and hydrogen-rich gases such as methane (CH_4), ammonia (NH_3), and water (H_2O). These gases were also directly detected in December 1995, when the *Galileo* spacecraft parachuted a probe into Jupiter's atmosphere. The clouds themselves are harder to analyze, but their bright colors may come from complex organic molecules whose composition is still uncertain.

Jupiter's Interior

Astronomers cannot see through Jupiter's cloud layers to its interior, nor can they probe its interior with seismic detectors. Instead, they rely on theory to tell them what lies inside this

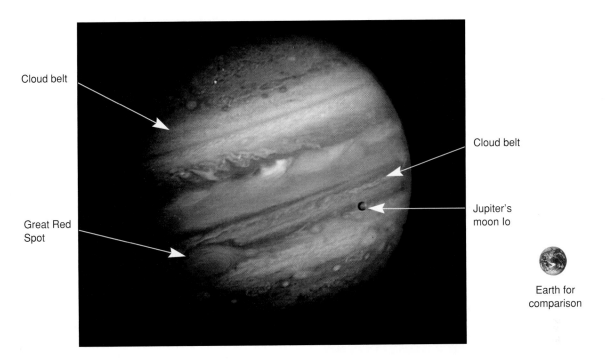

Cloud belt

Cloud belt

Great Red
Spot

Jupiter's
moon Io

Earth for
comparison

FIGURE 22.26

Jupiter's Red Spot and cloud belts. *Photo by NASA*

giant planet. For example, despite its great mass, Jupiter is much less dense, on the average, than the Earth. Its average density is only 1.3 times that of water, showing that the bulk of the planet, not just its atmosphere, must be composed mainly of very light elements, such as hydrogen and helium.

Deep within Jupiter, its gravity compresses molecules so close together that the gas changes to liquid. Thus, about 10,000 kilometers below the cloud tops—about one-sixth of the way into the planet—lies a vast sea of liquid hydrogen. Deeper still, the weight of the liquid hydrogen compresses matter below it into a state known as liquid metallic hydrogen, a form of hydrogen that scientists on Earth have only recently been able to create in tiny high-pressure chambers. Astronomers hypothesize that if the material from which Jupiter formed has the same overall composition as the Sun's, then Jupiter must also contain heavy elements such as silicon and iron. Because of their high density, these elements have probably sunk to Jupiter's center, forming a core of iron and rocky material a few times larger than that of the Earth (see figure 22.5).

Jupiter's interior is extremely hot, perhaps 30,000°C, or about five times hotter than the Earth's core. Some of this heat may be left over from Jupiter's formation. As we saw in discussing the origin of the solar system, planet building is a hot process. Most of that heat comes from the gravitational energy released by the hail of gas and planetesimals onto the forming world. Subsequently, additional heat may be generated by slow but steady shrinkage of the planet as it adjusts to the greater gravity created by the mass added in the last stages of its formation. Whatever its source, the heat generates circulating currents (convection) similar to those in the Earth's interior. These currents stir both Jupiter's deep interior and its atmosphere.

Jupiter's Atmosphere

As warm gas rises to the top of Jupiter's atmosphere, it carries heat, which escapes to space. Having lost heat, the now-cooler gas sinks and is reheated to rise again. These up-and-down motions are then modified by the Coriolis effect, which draws the rising and sinking gas into winds that blow *around* the planet. The Coriolis effect operates whenever material moves over the surface of a spinning planet. On Earth, the effect deflects air currents into our global wind system, such as the jet streams. On Jupiter, the Coriolis effect also creates global winds, which we see as cloud belts, illustrated in figure 22.26. Winds in adjacent regions may blow in opposite directions, as also occurs on Earth, where tropical winds generally blow from east to west, while mid-latitude winds generally blow from west to east. As these winds circle Jupiter, gas between them is spun into huge, whirling atmospheric vortices, much as a pencil held between your palms twirls when you rub your hands together. Some of these spinning regions are brightly colored, such as the Great Red Spot (figure 22.26). For a few months in 1994, Jupiter bore a number of other dark markings created by the debris from Comet Shoemaker-Levy, which struck the giant planet in July of that year.

Matter also circulates in Jupiter's deep interior, where combined with the planet's rapid rotation, it generates a magnetic field much as the Earth does. Jupiter's magnetic field is by far the strongest of any planet in the solar system.

Jupiter's Moons

When the seventeenth-century Italian scientist Galileo first viewed Jupiter with his telescope, he saw four moons orbiting the planet. Today, some sixty are known, although most of these are too small to be readily seen from Earth and are probably captured asteroids.

The four large satellites discovered by Galileo—Io, Europa, Ganymede, and Callisto—are very large: all but Europa are larger than our Moon, and Ganymede (with a diameter bigger than Mercury) is the largest moon in the solar system. The large moons all orbit approximately in Jupiter's equatorial plane, forming a flattened disk, rather like a miniature solar system, suggesting they condensed in

orbit around Jupiter. Many of the smaller moons move along highly inclined orbits, suggesting they were captured later on.

Io (pronounced *eye-oh*) is the nearest to Jupiter of the Galilean moons and therefore is subject to a strong distorting force created by Jupiter's gravity, which heats Io by internal friction. (You can see the effect of such distorting on a smaller scale if you repeatedly bend a metal paperclip until it breaks and then touch the freshly broken end: it will feel hot.) As molten matter oozes to the surface, it erupts, creating volcanic plumes and lava flows, as shown in figure 22.27A. In fact, Io is the most volcanically active object in our solar system at this time. Io's volcanic gases are very rich in sulfur, which colors its surface with red and yellows.

Europa, the smallest of the Galilean moons, looks rather like a cracked egg. Long, thin lines score its surface, as shown in figure 22.27B. The white material is a crust of ice, while the red material is probably mineral-rich water that has oozed to the surface through the cracks and froze. Europa, too, may be heated by Jupiter's gravitational forces, although not as strongly as is Io. That heat, in combination with a small amount from radioactive decay of rocky material in its core, may be sufficient to keep a layer of water melted beneath Europa's crust, forming an ocean. Some scientists have suggested that

← **Cloud from volcanic eruption**

A

FIGURE 22.27

(A) Jupiter's moon Io during an eruption of one of its volcanoes. (B) Jupiter's moon Europa as imaged by the *Voyager* spacecraft; a model of Europa's interior (Note possible liquid water ocean below icy crust and close-up of ice chunks on Europa's surface. All photos are color enhanced to bring out details. *Photos by NASA/JPL*

Ice covering

Liquid ocean under ice

Metallic core

Rocky interior

Water layer

B

this ocean might harbor primitive life-forms, but no evidence exists to support this idea.

Ganymede and Callisto lie still farther from Jupiter and show proportionally less evidence of internal heating and geologic activity (figure 22.28). Ganymede has numerous faults and fractures slicing across its heavily cratered surface, whereas the surface of Callisto shows almost nothing but impact scars.

A

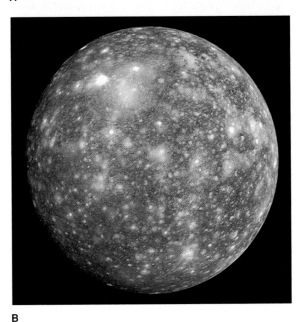

B

FIGURE 22.28

(A) Global color view of Ganymede, Jupiter's (and the solar system's) largest moon. Note the many linear features, which are sets of tectonic fractures, amongst the bright, circular impact craters. (B) Global color view of Callisto, Jupiter's second-largest moon. Note how its ancient surface is dominated by almost innumerable impact craters. *Photos by Calvin J. Hamilton*

Saturn

Saturn is the second-largest planet and lies about 10 AU from the Sun. Surrounded by its lovely rings, Saturn bears the name of an ancient Roman harvest god. In later mythology, Saturn came to be identified with Cronus (also spelled Kronos), whom the ancient Greeks considered the father of the gods.

Saturn's Appearance and Physical Properties

Saturn's diameter is about nine times larger than Earth's. Saturn's mass is about ninety-five times that of the Earth, and its average density is very low—only 0.7 grams per cubic centimeter. Such a low density suggests that Saturn, like Jupiter, is composed mostly of hydrogen and hydrogen-rich compounds. However, Saturn's lower mass produces less compression (and lower density) of the gases in its interior. Analysis of sunlight reflected from its clouds further suggests to astronomers that these two giant planets have very similar compositions and internal structures.

Saturn's Rings

Saturn's rings were first seen by Galileo. Through his small, primitive telescope, however, they looked like "handles" on each side of the planet, and it was not until 1659 that Christiaan Huygens, a Dutch scientist, observed that the rings were detached from Saturn and encircled it.

The rings are very wide but astonishingly thin. The main band is 70,000 kilometers across and extends to a little more than twice the planet's radius. Yet despite the rings' immense breadth, they are probably less than 100 meters thick—thin enough to allow stars to be seen through them.

The rings are not solid structures but a swarm of small, solid chunks of mostly water-ice and bits of rock in individual orbits. These pieces are relatively small, averaging only a few centimeters to a few meters across, as deduced from radar signals bounced off them. What formed the rings? Present thinking is that they are debris from collisions between tiny moons orbiting Saturn or between a moonlet and a comet that strayed too close to the giant planet.

Saturn's Moons

Saturn has several large moons and about two dozen known smaller ones. Its largest moon is Titan, whose diameter is about 5,000 kilometers, making it slightly bigger than Mercury. Because Titan is so far from the Sun and is thus very cold, gas molecules that leak from its interior move relatively slowly and are unable to escape Titan's gravitational attraction. Titan thus possesses its own atmosphere, which analysis shows to be mostly nitrogen. Clouds in Titan's atmosphere perpetually hide its surface. On the basis of chemical calculations, some astronomers believe that Titan's surface may have lakes or oceans of liquid methane or the hydrocarbon ethane (C_2H_6), and "continents" of water-ice, and that ethane "rain" may fall from its clouds. The first images of Titan's surface, recently returned by the *Cassini* spacecraft, show sharp boundaries between light and dark terrain and very few impact craters,

FIGURE 22.29

(A) The icy surface of Titan, as seen by the *Cassini* spacecraft in October 2004. The bright patch toward the bottom of the image is clouds over the southern polar region. Note the sharp boundaries between light and dark (carbon-rich?) areas on the surface, and the lack of obvious impact craters. (B) Close-up of dendritic channel system on Titan, probably carved by liquid methane (mosaic of three images taken by Huygens Probe as it descended to Titan's frozen surface). *Images courtesy (A) NASA/JPL/ Space Science Institute, and (B) European Space Agency*

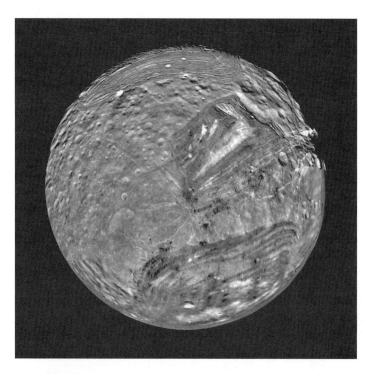

FIGURE 22.30

View of the south polar region of Uranus's moon Miranda. This moon has some of the most bizarre and least understood surface features in our solar system. The patchwork of chevron-shaped terrains includes fault scarps over 10 kilometers high! *Photo by NASA/JPL/Space Science Institute*

suggesting a relatively young (and perhaps geologically active) surface (figure 22.29).

Uranus

Uranus, although small compared with Jupiter and Saturn, is nonetheless much larger than the Earth. Its diameter is about four times that of the Earth, and its mass is about fifteen Earth masses. Lying more than twice Saturn's distance from the Sun, Uranus is difficult to study from Earth. Even pictures of it taken by the *Voyager* spacecraft show few details.

Uranus was unknown to the ancients, even though it is just visible to the naked eye. It was discovered in 1781 by Sir William Herschel. The planet's name comes from ancient Greek mythology, where Uranus was god of the heavens and the father of Cronus.

Uranus's atmosphere is rich in hydrogen and methane. In fact, methane gives the planet its deep blue color. When sunlight falls on Uranus's atmosphere, the methane gas strongly absorbs the red light. The remaining light, now blue, scatters from cloud particles in the Uranian atmosphere and is reflected into space.

Astronomers rely on theoretical models to understand the interior of Uranus. Such models suggest that the planet has a deep, liquid-water ocean below its methane-rich clouds. During its formation, Uranus must have collected heavy elements such as silicon and iron. These denser materials probably sank to its center (as they did in the other giant planets), where they now form a core of rock and iron a little larger than Earth.

One final property of interest is that Uranus's rotation axis is tipped so that its equator is nearly perpendicular to its orbit. Thus, it spins nearly "on its side." Moreover, the orbits of its moons are also similarly tilted. To explain this tilt, some astronomers hypothesize that during its formation, Uranus was struck by an enormous planetesimal whose impact tilted the planet and splashed out material to create its family of moons. In addition to its moons (figure 22.30), Uranus is encircled by a set of narrow rings that, like those of Saturn, are composed of a myriad of small particles.

Neptune

Neptune is the outermost of the large planets and is very similar to Uranus in size, with a diameter about 3.9 times that of the Earth and a mass about seventeen times the Earth's. Like Uranus, it is a lovely blue color, and for the same reason: strong absorption of red light by the methane in its atmosphere. Pictures show that cloud bands encircle it, and it even has briefly shown a Great "Dark" Spot, a huge, dark blue atmospheric vortex similar to Jupiter's Great Red Spot.

Neptune, named for the Roman god of the sea, was discovered in the 1840s from predictions made independently by a young English astronomer, John Couch Adams, and a French astronomer, Urbain Leverrier. Adams and Leverrier both noticed that Uranus was not precisely following its predicted orbit, and they therefore inferred that its motion was being dis-

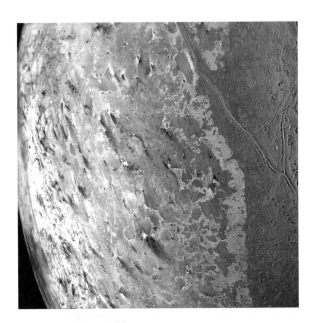

FIGURE 22.31

The strange icy surface of Neptune's largest moon, Triton, shows evidence of icy geysers whose deposits have been blown downwind (dark streaks) by winds in its thin atmosphere. Triton is thus one of only three bodies in the solar system (Earth and Io are the others) where active volcanism has been observed. *Photo by NASA/JPL/ Space Science Institute*

moons follow orbits approximately parallel to their planet's equator). These orbital peculiarities lead many astronomers to think that Triton is a surviving icy planetesimal captured from what was originally a more distant orbit around the Sun.

Triton intrigues astronomers for more than its orbital oddity. It is massive enough that its gravity, in combination with its low temperature, allows it to retain gases around it. Triton, along with Saturn's moon Titan, is thus one of two moons in the solar system with an atmosphere. Moreover, it shows signs of eruptions of gases (mostly nitrogen) from its interior (figure 22.31).

Pluto

Pluto, named for the Greek and Roman god of the underworld, was the last of the nine planets to be discovered. It was found in 1930 by the American astronomer, Clyde Tombaugh. Pluto's great distance from the Sun combined with its small size makes it very dim. Even in the largest telescopes on the ground, Pluto looks like a dim star.

Pluto has a moon, Charon, named for the mythological boatman who ferries dead souls across the river Styx to the underworld. Charon's diameter is approximately one-half Pluto's, making these two objects almost a "double planet." From Charon's motion around Pluto, astronomers can calculate Pluto's mass using the law of gravity. They find that Pluto has only about 0.002 times the Earth's mass, making it by far the least massive planet.

From Pluto's mass and radius, astronomers can deduce its density to be about 2.1 grams per cubic centimeter, a value suggesting that the planet must be a mix of water, ice, and rock. They know little, however, about its surface. Images of Pluto made with the *Hubble* Space Telescope (figure 22.32) show only that Pluto's polar regions are much brighter than its equator, implying the presence of polar caps there, probably composed of frozen methane, carbon monoxide, or nitrogen.

By analyzing the spectrum of sunlight reflected from Pluto, astronomers have detected that it has a very tenuous atmosphere composed mostly of nitrogen and carbon monoxide, with traces of methane.

turbed by the gravitational force of an as yet unknown planet. From the size of these orbital disturbances, Adams and Leverrier predicted where the unseen body must lie, and a search by Johann Galle, a German astronomer, found that planet exactly where predicted.

Neptune's structure is probably similar to Uranus's. That is, the planet is composed mostly of ordinary water (with a small rock and iron core) surrounded by a thin atmosphere rich in hydrogen and hydrogen compounds, such as methane.

Neptune, like the other giant planets, has rings, but they are very narrow, more like those of Uranus than those of Saturn. It also has a number of small moons and one immense moon, Triton. Triton orbits "backwards" (counter to Neptune's rotation) compared with the motion of most other satellites, and its orbit is highly tilted with respect to Neptune's equator (most large

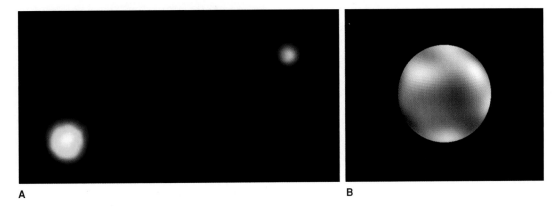

A   B

FIGURE 22.32

Best images from *Hubble* space telescope of (A) Pluto and its smaller moon, Charon, and (B) surface brightness patterns on Pluto. *Images by NASA/Space Telescope Science Institute*

Pluto's orbit is peculiar. It crosses Neptune's, but because it is also more highly tilted than the orbits of the other planets, Pluto passes well above or below Neptune when the crossings occur. Its odd orbit in combination with its small mass once led some astronomers to hypothesize that Pluto was originally a satellite of Neptune that escaped and now orbits the Sun independently. Today, however, astronomers think almost the reverse—that Neptune has "captured" Pluto. The reason for this hypothesis is that Pluto's orbital period of 247.7 years is nearly exactly one-and-one-half times Neptune's. Thus, Pluto makes two orbits around the Sun for every three made by Neptune. This match of orbital periods gives Neptune's gravitational attraction on Pluto a cumulative effect that has probably "tugged" it into its current orbit. In fact, several dozen other much smaller, presumably icy, objects a few hundred kilometers in diameter have been found orbiting at nearly the same distance from the Sun as Pluto.

MINOR OBJECTS OF THE SOLAR SYSTEM

Orbiting the Sun and scattered throughout the solar system are numerous bodies much smaller than the planets—asteroids and comets. Asteroids are generally rocky objects in the inner solar system. Comets are icy bodies and spend most of their time in the outer solar system. These small members of the Sun's family, remnants from the formation of the solar system, are of great interest to planetary scientists and astronomers because they are our best source of information about how long ago and under what conditions the planets formed. In fact, many (if not most) asteroids and comets may be planetesimals—the solid bodies from which the planets were assembled—that have survived nearly unchanged from the birth of the solar system. Apart from their scientific value, asteroids and comets are fascinating objects both for their beauty and potential for danger to life on Earth.

Meteors and Meteorites

If you have spent even an hour looking at the night sky in a dark location away from city lights, you have probably seen a "shooting star," a streak of light that appears in a fraction of a second and as quickly fades. Astronomers call this brief but lovely phenomenon a **meteor.**

A meteor is the glowing trail of hot gas and vaporized debris left by a solid object heated by friction as it moves through the Earth's atmosphere. The solid body itself, while in space and before it reaches the atmosphere, is called a *meteoroid.*

Heating of Meteors

Meteors heat up on entering the atmosphere for the same reason a reentering spacecraft does. When an object plunges from space into the upper layers of our atmosphere, it collides with

atmospheric molecules and atoms. These collisions convert some of the body's energy of motion (kinetic energy) into heat and vaporize matter from its surface. Most of the heating occurs between about 100 and 50 kilometers altitude, in the outer fringes of the atmosphere. The trail of hot evaporated matter and atmospheric gas emits light, making the glow that we see.

Most meteors that we see last only a few seconds and are made by meteoroids the size of a pea or smaller. These tiny objects are heated so strongly that they completely vaporize. Larger pieces, though heated and partially vaporized, are so drastically slowed by air resistance that they may survive the ordeal and reach the ground. We call these fragments found on the Earth **meteorites.**

Meteorites

Astronomers classify meteorites into three broad categories based on their composition: iron, stony (that is, composed mainly of silicate minerals), and stony-iron. Most stony meteorites are composed of smaller, rounded chunks of rocky material stuck together. The grains are called chondrules, and meteorites that have this lumpy structure are called chondritic meteorites (figure 22.33). Chondrules appear to have been rapidly melted and cooled in the solar nebula. The cause of the heating is not known.

Chondrules contain traces of radioactive material, which can be used to measure their age. They are extremely old, 4.5 billion years, and are believed to be the first solid material that condensed within the solar nebula and have been relatively unaltered. Thus, chondritic meteorites offer us valuable information about the early history of the solar system.

Some chondritic meteorites contain a black, carbon-rich, coal-like substance and are called carbonaceous chondrites. This carbonaceous matter contains organic compounds, includ-

FIGURE 22.33

A chondrite meteorite. Note the many circular chondrules of which it is composed. Such meteorites are made primarily of dust to sand-sized grams of silicate minerals. *Photo by NASA/JPL* ing amino acids, the same complex molecules used by living things for the construction of their proteins and genetic material. Thus, the presence of amino acids in meteoritic matter indicates that the raw material of life can form in space and that it might, therefore, have been available right from the start within the solar system. Astronomers think that most meteors are fragments of asteroids and comets.

Asteroids

Asteroids are small, generally rocky bodies that orbit the Sun. Most lie in the asteroid belt, a region between the orbits of Mars and Jupiter. Asteroids range tremendously in diameter, from Ceres—about 1,000 kilometers (about the size of Texas) across—down to bodies only a few meters in diameter. Most are irregularly shaped (figure 22.34). Only Ceres and a few dozen other large asteroids are approximately spherical. Because of their large size, their gravitational force is strong enough to crush their material into a sphere. Small bodies with weak gravities remain irregular and are made more so by collisions blasting away pieces. Collisions leave the parent body pitted and irregularly shaped, and the fragments become smaller asteroids in their own right.

Origin of Asteroids

The properties of asteroids that we have discussed (composition, size, and their location between Mars and Jupiter) give us clues to their origin and support the solar nebula hypothesis for the origin of the solar system. In fact, the asteroids are probably fragments of planetesimals, the bodies from which the

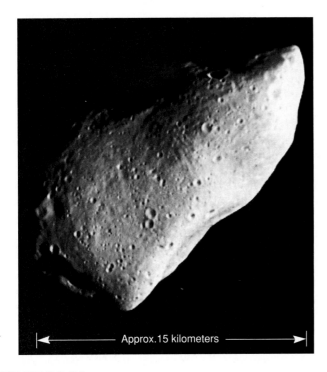

FIGURE 22.34

The asteroid Gaspra, as imaged by the *Galileo* spacecraft. *Photo by NASA/JPL*

Approx.15 kilometers

planets were built. Although science-fiction movies and TV shows like to show the asteroids as a dense swarm, they are actually very widely separated—typically, thousands of kilometers apart. The ones we see today failed to be incorporated into a planet, probably as the result of their nearness to Jupiter, whose immense gravity disturbed their paths sufficiently to keep them from aggregating into a planet.

Farther from the main belt are the Apollo asteroids, whose orbits carry them into the inner solar system across the Earth's orbit. Although there are about 1,000 such Earth-crossing asteroids, larger than one kilometer in diameter, the chance of collision in the near future is slim. Nevertheless, on the average, one such body hits the Earth about every 10,000 years.

Comets

A bright comet is a stunning sight, as you can see in figure 22.35A. But such sights are now sadly rare for most people because light pollution from our cities drowns the view. Comets have long been held in fear and reverence, and their sudden appearance and equally sudden disappearance after a few days or so have added to their mystery.

Structure of Comets

Comets consist of two main parts, as illustrated in figure 22.35B. The largest part is the long tail, a narrow column of dust and gas that may stretch across the inner solar system for as much as 100 million kilometers (nearly an AU!).

The tail emerges from a cloud of gas called the *coma*, which may be some 100,000 kilometers in diameter (ten times or so the size of the Earth). Despite the great volume of the coma and the tail, these parts of the comet contain very little mass. The gas and dust are extremely tenuous, and so a cubic centimeter of the gas contains only a few thousand atoms and molecules. By terrestrial standards, this would be considered a superb vacuum. This extremely rarified gas is matter that the Sun's heat has sublimated off the heart of the comet, its nucleus.

The comet nucleus is a block of ice and gases that have frozen in the extreme cold of interplanetary space into an irregular mass averaging perhaps a bit less than 10 kilometers in diameter. The nucleus of a comet (figure 22.35C) has been described as a giant "iceberg" or "dirty snowball," and it contains most of the comet's mass. Our best information about the nucleus comes from studies of Comets Halley, Borelly, and Wild-2, made by the *Giotto, Deep Space-1,* and *Stardust* spacecraft, respectively. All three comets appear to have densities well under 1 gram per cubic centimeter, a value implying that the icy material of the nucleus is "fluffy," like snow, not hard and compacted like pure ice. Unfortunately, the mass estimates are not very accurate, and so the density we infer from them is uncertain.

Origin of Comets

Astronomers think that most comets come from the Oort cloud, the swarm of trillions of icy bodies believed to lie far beyond the orbit of Pluto, as we discussed in the section titled "The

A

|← ——————— Approx. 5 kilometers ——————— →|

C

B

FIGURE 22.35

(A) Photograph of Comet Hale-Bopp in the evening sky. Note the blue tail (gas) and the whitish tail (sunlight reflected from tiny dust particles). (B) Artist's depiction of the structure of a comet showing the tiny nucleus, surrounding coma, and the long tail. (C) Image of the nucleus of comet Borelly. *Photo (A) courtesy of Mike Skrutskie; Photo (C) by NASA/JPL.*

Solar System." Astronomers think the Oort cloud formed from planetesimals that originally orbited near the giant planets and were tossed into the outer parts of the solar system by the gravitational force of those planets. Some comets also seem to come from a flatter, less-remote region—the so-called Kuiper Belt. The Kuiper Belt begins near the orbit of Neptune and appears to extend to about 55 AU from the Sun. Planetary astronomers are greatly interested in this region because it appears to be the birthplace of Pluto, Charon, and Neptune's moon Triton.

Each comet nucleus moves along its own path, and those in the Oort cloud take millions of years to complete an orbit. With orbits so far from the Sun, these icy bodies receive essentially no heat from the Sun, so their gases and ices remain deeply frozen.

These distant objects are invisible to us on Earth, but sometimes a comet's orbit alters, carrying it closer to us and the Sun. Such orbital changes may arise from the chance passage of a star far beyond the outskirts of the solar system during its orbit around the center of the Milky Way galaxy.

As the comet falls inward toward the inner solar system, the Sun's radiation heats it and begins to sublimate the ices. At a distance of about 5 AU from the Sun (Jupiter's orbit), the heat is enough to vaporize the ices, forming gas that escapes to make a coma around the comet nucleus. The escaping gas carries tiny dust grains that were frozen into the nucleus with it. The comet then appears through a telescope as a dim, fuzzy ball. As the comet falls ever nearer the Sun, its gas escapes even faster, but now the Sun begins to exert additional forces on the cometary gas and dust.

Formation of the Comet's Tail

Sunlight striking dust grains imparts a tiny force to them, a process known as radiation pressure. We don't feel radiation pressure when sunlight falls on us because the force is tiny and the human body is far too massive to be shoved around by sunlight. However, the microscopic dust grains in the comet's coma do respond to radiation pressure and are pushed away from the Sun.

The tail pushed out by radiation pressure is made of dust particles, but figure 22.35A shows that comets often have a second tail. That tail is created by the outflow of the **solar wind.**

The solar wind blows away from the Sun at about 400 kilometers per second. It is very tenuous, containing only a few atoms per cubic centimeter. But the material in the comet's coma is tenuous, too, and the solar wind is dense enough by comparison to blow it into a long plume. Thus, two forces, radiation pressure and the solar wind, act on the comet to drive out a tail. Because those forces are directed away from the Sun, the comet's tail always points more or less away from the Sun.

Although most comets that we see from Earth swing by the Sun on orbits that will bring them back to the inner solar system only after millions of years, a small number (including Halley's) reappear at time intervals of less than 200 years. Astronomers think that these objects come from the Kuiper Belt. They may also be responsible for meteor showers because as they repeatedly orbit the Sun, its heat gradually whittles them away: all the ices and gases eventually evaporate, and only the small amount of solid matter, dust and grit, remains. This fate is like that of a snowball made from snow scooped up alongside the road, where small amounts of gravel have been packed into it. If such a snowball is brought indoors, it melts and evaporates, leaving behind only the grit accidentally incorporated in it. So, too, the evaporated comet leaves behind in its orbit grit that continues to circle the Sun. The material left by the comet produces a delightful benefit: it is a source of meteors.

Meteor Showers

If you go outside on a clear night and have an unobscured view of the sky, you will see, on the average, one meteor every fifteen or so minutes. Most of these meteors are stray fragments of asteroids that arrive at the Earth randomly. At some times of the year, however, meteors are far more frequent. For example, on about August 12 (the exact date varies slightly from year to year), you may see roughly one meteor every minute or so, an event called a *meteor shower.* Moreover, if you watch carefully, you will see that these meteors appear to come from the same general direction in the sky.

During the August shower, meteoroids rain into our atmosphere from a direction that lies toward the constellation Perseus. This shower is therefore called the *Perseids.* The Perseid meteors themselves have no association with Perseus. Rather, they are following an orbit around the Sun that happens to lie roughly in that direction, and the Earth happens to cross

their orbits in mid-August. Thus, at that time, we encounter far more meteoroids than usual.

On rare occasions, the Earth will pass through a particularly dense clump of material left by a comet. If that happens, thousands of meteors per hour may spangle the sky. Such a display happened in November 2001 during the Leonid shower. Spectacles of this kind are a delight. Occasionally, however, far more sinister meteoritic events may occur.

GIANT IMPACTS

Every few thousand years, the Earth is hit by a huge meteoroid, a body tens of meters or more in size. A large meteoroid will produce not only a spectacular glare as it passes through the atmosphere but also an enormous blast on impact. The violence of such events arises because a meteoroid has a very large kinetic energy, that is, energy of motion, which is released when it hits the ground or when it breaks up in the atmosphere.

The energy so released can be huge. A body 20 meters in diameter, the size of a small house, would have on impact the explosive power of a thermonuclear bomb and make a crater about one-half kilometer in diameter. Were such a body to hit a heavily populated area, it would obviously have catastrophic results.

Giant Meteor Impacts

One of the most famous meteor impacts was the event that formed the giant crater in northern Arizona. About 50,000 years ago, a meteoroid estimated to have been some 50 meters in diameter hit the Earth about 40 miles east of what is now Flagstaff. Its impact vaporized tons of rock, which expanded and peeled back the ground, creating a crater about 1.2 kilometers across and 200 meters deep (see box figure 7.1 in chapter 7).

More recently, in 1908, a small comet or asteroid broke up in our atmosphere over a largely uninhabited part of north-central Siberia. This so-called *Tunguska event,* named for the region where it hit, leveled trees radially outward from the blast point to a distance of some 30 kilometers. The blast was preceded by a brilliant fireball in the sky and was followed by clouds of dust that rose to the upper atmosphere. Casualties were few because the area was so remote.

Not all impacts have so few casualties, however. In the distant past, about 65 million years ago, at the end of the Cretaceous period, an asteroid hit the Earth. Its impact and the subsequent disruption of the atmosphere are blamed for exterminating the dinosaurs and many less conspicuous but widespread creatures and plants (see box 8.2)

Other mass extinctions have occurred earlier and later than the Cretaceous event. These may have resulted from similar impact events, but many scientists believe that massive volcanic eruptions or drastic changes in sea level may have played a role as well. Thus, like so many of the most interesting issues in science, this story has no definitive explanation at this time.

SUMMARY

The Earth is just one of nine planets orbiting our Sun. The planets closest to the Sun (the inner planets—Mercury, Venus, Earth, and Mars) are rocky and are similar in size. They differ greatly from the outer planets (Jupiter, Saturn, Uranus, Neptune, and Pluto). Pluto is small and icy, while the other four are huge and rich in hydrogen and helium, as well as other hydrogen compounds. Smaller objects—asteroids and comets—also orbit the Sun, and all these objects formed in our solar system.

The motion of the planets around the Sun within a flat, disk-shaped region and their compositional differences give clues to their origin. Astronomers think that our solar system formed from the collapse of a slowly spinning interstellar cloud of dust and gas. Rotation flattened the cloud into a disk—the *solar nebula*. Dust particles within the disk stuck together and gradually grew in size, eventually becoming solid objects a few kilometers across—*planetesimals*. The planetesimals, aided by the gravity of their large mass, drew together and formed planets. Near the Sun, it was too warm for the planetesimals to capture or incorporate much water. The objects near the Sun,

therefore, are composed mainly of silicates and iron particles. Farther from the Sun, it was cold enough for water-ice to be captured, which allowed larger planetesimals to form. These eventually grew big enough to capture gas directly from the solar nebula.

In the late stages of their growth, the planets were bombarded with surviving planetesimals still orbiting the Sun. The impact of these objects created craters on their surfaces. Some planetesimals (or their fragments) exist even today as *asteroids* and *comets*.

Size strongly affects a planet's history. All planets were probably born hot, but small ones (such as Mercury) cooled quickly and are inactive today, lacking volcanic and tectonic activity. Larger planets (such as Earth and Venus) remain hot and have geologically active surfaces. Their surfaces are altered now mainly by processes driven by their internal heat, but impacts of asteroidal or cometary objects still occur and on rare occasions have devastating results.

Terms to Remember

asteroid 586
astronomical unit 586
comet 586
differentiation 590
greenhouse effect 598
inner planet 585

meteor 612
meteorite 612
Milky Way 586
Nebular Hypothesis 588
outer planet 585
planet 584

planetesimal 589
solar nebula 588
solar system 585
solar wind 615
star 584
universe 588

Testing Your Knowledge

1. What is the approximate shape of the solar system?
2. What evidence supports the hypothesis that the solar system formed from a rotating disk of dust and gas?
3. What is meant by a planetesimal?
4. Why is the surface of Venus hotter than the surface of Mercury, despite Mercury being closer to the Sun?
5. What has created the craters on the Moon?
6. Why is the Earth much less cratered than the Moon?
7. What has led to the formation of two very different types of planets in the solar system?
8. Where do comets come from?
9. What created the rings of Saturn and the other large planets?
10. Which of the following is evidence that Jupiter's composition is rich in hydrogen?
    a. Its large mass
    b. Its large density

c. Its low density

d. Its distance from the Sun

11. Why do Mercury and the Moon lack an atmosphere?

    a. They formed after all the gas had been used up.

    b. They are so cold that all their gases have frozen into deposits below their surface.

    c. They formed before the solar nebula had captured any gas.

    d. They are so small that their gravity is too weak to retain an atmosphere.

12. Why would it be difficult to land a spacecraft on Jupiter?

    a. Jupiter has no solid surface.

    b. Jupiter's immense gravity would squash it.

    c. Jupiter's intense magnetic field would destroy it.

    d. The clouds are so thick it would be hard to navigate to a safe spot.

13. Scientists think that water may once have flowed on Mars. The evidence for this is

a. fossil fish found in several craters

b. channels that resemble dried-up riverbeds

c. frozen lakes along the Martian equator

d. photographs made in the late 1800s that show blue spots that were probably lakes

14. The core of a comet is composed of

a. molten iron

b. frozen gases

c. liquid hydrogen

d. uranium

15. Why are most asteroids not spherical?

a. Their gravity is too weak to pull them into a sphere and they have been fragmented by impacts.

b. The Sun's gravity distorts them.

c. Strong magnetic fields in their molten core makes them lumpy.

d. The statement is false. Nearly all asteroids are spherical.

16. What causes meteor showers such as occur near August 12?

a. The breakup of an asteroid in our upper atmosphere

b. Bursts of particles ejected from the Sun

c. A comet being captured into orbit around the Earth

d. The Earth passing through the trail of debris left by a comet

Expanding Your Knowledge

1. What obvious evidence suggests that the lunar highlands are older than the maria?

2. What has led to the inner planets having iron cores?

3. What is the likely origin of the main asteroid belt and the Kuiper Belt?

4. Why is it not surprising that the Moon lacks folded mountain ranges like we have on Earth?

Exploring Web Resources

www.mhhe.com/plummer11e

This is the dedicated website for this book. You can go to it for new and updated information. The universal resource locators (URL) listed in this book are also given as links on the website, making it easy to go to those websites without typing in the URL. When you visit one Online Learning Center, go to the Student Center and then to the chapter of interest. Here you will find additional readings and media resources, as well as answers to the Testing Your Knowledge section, more quizzes, animation and video clips, and direct links to the sites listed in this book. Links to additional websites can also be found. We have added questions for some of the sites to allow you to get the most of your exploration of the web. Using the web is an enjoyable way of enhancing your knowledge of geology.

photojournal.jpl.nasa.gov
NASA Planetary Photojournal

www.pdsa.jpl.nasa.gov/planets
Welcome to the Planets

www.seds.org
Students for the Exploration and Development of Space

www.hawastsoc.org/solar/homepage.html
Views of the Solar System, by Calvin Hamilton

 # Animations

This chapter includes the following animations available on our Online Learning Center at **www.mhhe.com/plummer11e**.

22.8 Formation of the Solar System

22.12 Impact formation of the Moon

Each mineral is identified by a unique set of physical or chemical properties. Determining some of these properties requires specialized equipment and techniques. Most common minerals, however, can be distinguished from one another by tests involving simple observations. Cleavage is an especially useful property. If cleavage is present, you should determine the number of cleavage directions, estimate the angles between cleavage directions, and note the quality of each direction of cleavage. Other easily performed tests and observations check for hardness (abbreviated H), luster, and color, and determining crystal form (if present). A simple chemical test can be made using dilute hydrochloric acid to see if the mineral effervesces.

The identification tables included here can be used to identify the most common minerals (the rock-forming minerals) and some of the most common ore minerals. For identifying less common minerals, refer to one of the websites on mineralogy listed at the end of chapter 2. Mineral identification takes practice, and you will probably want to verify your mineral identifications with a geology instructor.

Because the common rock-forming minerals are the ones you are most likely to encounter, we have included a simple key for identifying them. The key is based on first determining whether or not the mineral is harder than glass and then checking other properties that should lead to identification of the mineral. You should verify your identification by seeing whether other properties of your sample correspond to those listed for the mineral in table A.1.

Ore minerals are usually distinctive enough that a key is unnecessary. To identify an ore mineral, read through table A.2 and determine which set of properties best fits the unknown mineral.

KEY FOR IDENTIFYING COMMON ROCK-FORMING MINERALS

Determine whether a fresh surface of the mineral is harder or softer than glass. If you can scratch the mineral with a knife blade, the mineral is softer than glass.

I. Harder than glass—knife will not scratch mineral. (If softer than glass, go to II of this key.)

  A. Determine if cleavage is present or absent (this may require careful examination). If cleavage is absent proceed to 1; if cleavage is present, proceed to B.
    1. Vitreous luster
      a. Olive green or brown—*olivine*
      b. Reddish brown or in equidimensional crystals with twelve or more faces—*garnet*
      c. Usually light-colored or clear—*quartz*
    2. Metallic luster
      a. Bright yellow—*pyrite*
    3. Greasy or waxy luster
      a. Mottled green and black—*serpentine*

  B. Cleavage present. Determine the number of directions of cleavage in an individual grain or crystal.
    1. Two directions, good, at or near 90°—*feldspar*
      a. If striations are visible on cleavage surfaces—*plagioclase*
      b. If pink or salmon-colored—*potassium feldspar* (or *orthoclase*)
      c. If white or light gray without striations, it could be either type of feldspar
    2. Two directions, fair, at 90°
      a. Dark green to black—*pyroxene* (usually augite)
    3. Two directions, excellent, not near 90°
      a. Dark green to black—*amphibole* (usually hornblende)

II. Softer than glass—knife scratches mineral.

  A. No cleavage detectable
    1. Earthy luster, in masses too fine to distinguish individual grains—*clay group* (for instance, *kaolinite*)

  B. Cleavage present
    1. One direction
      a. Perfect cleavage in flexible sheets—*mica:*
        Clear or white—*muscovite mica*
        Black or dark brown—*biotite mica*
    2. Three directions
      a. All three perfect and at 90° to each other (cubic cleavage)—*halite*
      b. All three perfect and not near 90° to each other:
        If effervesces in dilute acid—*calcite*
        If effervesces in dilute acid only after being pulverized—*dolomite*

Table A.1	Diagnostic Properties of the Common Rock-Forming Minerals

Name (mineral groups shown in capitals)	Chemical Composition	Chemical Group	Diagnostic Properties	Other Properties
AMPHIBOLE (A mineral group in which *hornblende* is the most common member.)	$XSi_8O_{22}(OH)_2$ (X is a combination of Ca, Na, Fe, Mg, Al)	Chain silicate	2 good cleavage directions at 60° (120°) to each other.	H = 5–6 (barely scratches glass). Hornblende is dark green to black; tends to form in needlelike or elongate crystals; vitreous luster.
Augite (*see* Pyroxene)				
Biotite (*see* Mica)				
Calcite	$CaCO_3$	Carbonate	3 excellent cleavage directions, *not* at right angles (they define a rhombohedron). H = 3. Effervesces vigorously in weak acid.	Usually white, gray, or colorless; vitreous luster. Clear crystals show double refraction.
CLAY MINERALS (*Kaolinite* is a common example of this large mineral group.)	Compositions include $XSi_4O_{10}(OH)_8$ (X is Al, Mg, Fe, Ca, Na, K)	Sheet silicate	Generally microscopic crystals. Masses of clay minerals are softer than fingernail. Earthy luster. Claylike smell when damp.	Seen as a chemical weathering product of feldspars and most other silicate minerals. A constituent of most soils.
Dolomite	$CaMg(CO_3)_2$	Carbonate	Identical to calcite (rhombohedral cleavage, H = 3) except effervesces in weak acid only when pulverized.	Usually white, gray, or colorless. Vitreous luster.
FELDSPAR (Most common group of minerals.)	Framework	Framework silicates	H = 6 (scratch glass). 2 good cleavage directions at about 90° to each other.	Vitreous luster but surface may be weathered to clay, giving an earthy luster. Perfect crystal, shaped like an elongated box.
The group includes:				
Potassium feldspar (orthoclase)	$KAlSi_3O_8$		White, pink, or salmon-colored.	Never has striations on cleavage surfaces.
Plagioclase (sodium and calcium feldspar)	Mixture of: $CaAl_2Si_2O_8$ and $NaAlSi_3O_8$		White, light to dark gray, rarely other colors. *May* have striations on cleavage surfaces.	Calcium-rich varieties generally a darker gray and may show an iridescent play of colors.
GARNET	$XAl_2Si_3O_{12}$ (X is a combination of Ca, Mg, Fe, Al, Mn)	Isolated silicate	No cleavage. Usually reddish brown. Tends to occur in perfect equidimensional crystals, usually 12 sided. H = 7.	Rarely yellow, green, or black. Usually found in metamorphic rocks. Vitreous luster.
Gypsum	$CaSO_4 \cdot 2H_2O$	Sulfate	H = 2. 1 good and 2 perfect cleavage directions. Vitreous or silky luster.	Clear, white, or pastel colors. Flexible cleavage fragments.

Table A.1 | **Diagnostic Properties of the Common Rock-Forming Minerals (continued)**

Name (mineral groups shown in capitals)	Chemical Composition	Chemical Group	Diagnostic Properties	Other Properties
Halite	NaCl	Halide	3 excellent cleavage directions at 90° to each other (cubic). H = 2 1/2. Salty taste. Soluble in water.	Usually clear or white.
Hematite (see Table A.2)				
Hornblende (see Amphibole)				
Kaolinite (see Clay)				
MICA	$K(X)(AlSi_3O_{10})(OH)_2$	Sheet silicate	1 perfect cleavage direction (splits easily into flexible sheets).	H = 2–3. Vitreous luster.
The group includes:				
Biotite	(X is Mg, Fe, and Al)		Black or dark brown.	
Muscovite	(X is Al)		White or transparent.	
Olivine	X_2SiO_4 (X is Fe, Mg)	Isolated silicate	No cleavage. Generally olive green or brown. H = 6–7 (scratches glass). Vitreous luster.	Usually as small grains in mafic or ultramafic igneous rocks.
Orthoclase (see Feldspar)				
Plagioclase (see Feldspar)				
Pyrite ("fool's gold")	FeS_2	Sulfide	H = 6 (scratches glass). Bright, yellow, metallic luster. Black streak.	Commonly occurs as perfect crystals: cubes or crystals with five-sided faces. Weathers to brown.
PYROXENE (A mineral group; Augite is most common member.)	$XSiO_3$ (X is Fe, Mg, Al, Ca)	Chain silicate	2 fair cleavage directions at 90° to each other.	H = 6. Augite is dark green to black. Vitreous luster; usually stubby crystals.
Quartz	SiO_2	Framework silicate	H = 7. No cleavage. Vitreous luster. Does not weather to clay.	Almost any color but commonly white or clear. Good crystals have six-sided "column" with complex "pyramid" on top.
Serpentine	$Mg_6Si_4O_{10}(OH)_8$	Sheet silicate	Hardness variable but softer than glass. Mottled green and black. Greasy luster. Fractures along smooth curved surfaces.	Sometimes fibrous (asbestos).

Table A.2   Diagnostic Properties of the Most Common Ore Minerals

Name	Chemical Composition	Diagnostic Properties	Other Properties
Azurite	$Cu_3(CO_3)_2(OH)_2$	Azure blue; effervesces in weak acid.	H = 3–4.
Bauxite	$Al_2O_3 \cdot nH_2O$	Earthy luster. A variety of clay. Generally pea-sized spheres included in a fine-grained mass.	
Bornite	Cu_3FeS_4	Metallic luster, tarnishes to iridescent purple color.	Gray streak; H = 3 (softer than glass).
Chalcopyrite	$CuFeS_2$	Metallic luster, brass-yellow. Softer than glass.	Black streak.
Cinnabar	HgS	Scarlet red, bright red streak.	Softer than glass. Generally an earthy luster.
Galena	PbS	Metallic luster, gray; 3 directions of cleavage at 90° (cubic). High specific gravity.	Softer than glass; gray streak.
Gold	Au	Metallic luster, yellow. H = 3 (softer than glass, can be pounded into thin sheets, easily deformed).	Yellow streak; high specific gravity.
Halite	$NaCl$	Salty taste; 3 cleavage directions at 90° (cubic)	Clear or white; easily soluble in water.
Hematite	Fe_2O_3	Red-brown streak.	Either in earthy reddish masses or in metallic, silver-colored flakes or crystals.
Limonite	$Fe_2O_3 \cdot nH_2O$	Earthy luster; yellow-brown streak.	Yellow to brown color; softer than glass.
Magnetite	Fe_3O_4	Metallic luster, black; magnetic.	Harder than glass; black streak.
Malachite	$Cu_2(CO_3)(OH)_2$	Bright-green color and streak.	Softer than glass; effervesces in weak acid.
Sphalerite	ZnS	Brown to yellow color; 6 directions of cleavage.	Lusterlike resin; yellow or cream-colored streak; softer than glass.
Talc	$Mg_3Si_4O_{10}(OH)_2$	White, gray, or green; softer than fingernail (H = 1).	Greasy feel.

IGNEOUS ROCKS

Igneous rocks are classified on the basis of texture and composition. For some rocks, texture alone suffices for naming the rock. For most igneous rocks, composition as well as texture must be taken into account. Ideally, the mineral content of the rock should be used to determine composition; but for fine-grained igneous rocks, accurate identification of minerals may require a polarizing microscope or other special equipment. In the absence of such equipment, we rely on the color of fine-grained rocks and assume the color is indicative of the minerals present.

To identify a common igneous rock, use either table 3.1 or follow the key given below.

Key for Identifying Common Igneous Rocks

I. What is the texture of the rock?

A. Is it glassy (a very vitreous luster)? If so, it is *obsidian,* regardless of its chemical composition. Obsidian exhibits a pronounced conchoidal fracture.

B. Does it have a frothy appearance? If so, it is *pumice.* Pumice is light in weight and feels abrasive (it probably will float on water).

C. Does it have angular fragments of rock embedded in a volcanic-derived matrix? If so, it is a *volcanic breccia.* If the precise nature of the rock fragments and matrix can be identified, modifiers may be used; for instance, the rock may be an *andesite* breccia or a *rhyolite* breccia.

D. Is the rock composed of interlocking, very coarse-grained minerals? (The minerals should be more than 1 centimeter across.) If so, the rock is a *pegmatite.* Most pegmatites are mineralogically equivalent to granite, with feldspars and quartz being the predominant minerals.

E. Is the rock entirely coarse-grained? (That is, does it have an interlocking crystalline texture in which nearly all grains are more than 1 millimeter across?) If so, go to part II of this key.

F. Is the rock *entirely* fine-grained? (Are grains less than 1 millimeter across or too fine to distinguish with the naked eye?) If so, go to part III of this key.

G. Is the matrix fine-grained with some coarse-grained minerals visible in the rock? If so, go to part III and add the adjective *porphyritic* to the name of the rock.

II. Igneous rocks composed of interlocking coarse-grained minerals.

A. Is quartz present? If so, the rock is a *granite.* Confirmation: Granite should be composed predominantly of feldspar—generally white, light gray, or pink (indicating high amounts of potassium or sodium in the feldspar). Rarely are there more than 20% ferromagnesian minerals in a granite.

B. Are quartz and feldspar absent? If so, the rock should be composed entirely of ferromagnesian minerals and is *ultramafic.* Confirmation: Identify the minerals as being olivine or pyroxene (or less commonly, amphibole or biotite).

C. Does the rock have less than 50% feldspar and no quartz? If so, the rock should be a *gabbro.* Confirmation: Most of the rock should be ferromagnesian minerals. Plagioclase can be medium or dark gray. There would be no pink feldspars.

D. Is the rock composed of 30% to 60% feldspar (and no quartz)? If so, the rock is a *diorite.* Confirmation: Feldspar (plagioclase) is usually white to medium gray but never pink.

III. Igneous rocks that are fine-grained.

A. Can quartz be identified in the rock? If so, the rock is a *rhyolite.*

B. If the rock is too fine-grained for you to determine whether quartz is present but is white, light gray, pink, or pale green, the rock is most likely a *rhyolite.*

C. Is the rock composed predominantly of ferromagnesian minerals? If so, the rock is *basalt.*

D. If the rock is too fine-grained to identify ferromagnesian minerals but is black or dark gray, the rock is probably a *basalt.*
   1. Does the rock have rounded holes in it? If so, it is a *vesicular basalt* or *scoria.*

E. Is the rock composed of roughly equal amounts of white or gray feldspar and ferromagnesian minerals (but no quartz)? If so, the rock is an *andesite.* Confirmation: Most andesite is porphyritic, with numerous identifiable crystals of white or light gray feldspar and lesser amounts of hornblende crystals within the darker, fine-grained matrix. Andesite is usually medium to dark gray or green.

SEDIMENTARY ROCKS

The following key shows how sedimentary rocks are classified on the basis of texture and composition. The descriptions of the rocks in the main body of the text provide additional information, such as common rock colors. *Equipment* needed for identification of sedimentary rocks includes a bottle of dilute hydrochloric acid, a hand lens or magnifying glass, a millimeter scale, a glass plate for hardness tests, and a pocketknife or rock hammer.

Begin by testing the rock for carbonate minerals by applying a small amount of dilute hydrochloric acid to the surface of the rock.

1. The rock does not effervesce (fizz) in acid, or effervesces weakly, but when powdered by a knife or hammer, the powder effervesces strongly. If so, the rock is *dolomite.*

2. The rock does not effervesce at all, even when powdered, or effervesces only in some places, such as the cement between grains. Go to part I of this key.

3. The rock effervesces strongly. The rock is *limestone.* Go to part II of this key to determine limestone type.

I. With a hand lens or magnifying glass, determine if the rock has a clastic texture (grains cemented together) or a crystalline texture (visible, interlocking crystals).
   A. If clastic:
      1. Most grains are more than 2 millimeters in diameter.
         a. Angular grains *sedimentary breccia.*
         b. Rounded grains—*conglomerate.*
      2. Most grains are between 1/16 and 2 millimeters in diameter. Rock feels gritty to the fingers. *Sandstone.*
         a. More than 90% of the grains are quartz—*quartz sandstone.*
         b. More than 25% of the grains are feldspar—*arkose.*
         c. More than 25% of the grains are fine-grained rock fragments, such as shale, slate, and basalt—*lithic sandstone.*
         d. More than 15% of the rock is fine-grained matrix—*graywacke.*
      3. Rock is fine-grained (grains less than 1/16 millimeter in diameter). Feels smooth to fingers.
         a. Grains visible with a hand lens—*siltstone.*
         b. Grains too small to see, even with a hand lens.
            1. Rock is laminated, fissile—*shale.*
            2. Rock is unlayered, blocky—*mudstone.*
   B. If crystalline:
      1. Crystals fine to coarse, hardness of 2—*rock gypsum.*
      2. Coarse crystals that dissolve in water—*rock salt.*
   C. Hard to determine if clastic or crystalline:
      1. Very fine-grained, smooth to touch, conchoidal fracture, hardness of 6 (scratches glass), nonporous—*chert* (*flint* if dark).
      2. Very fine-grained, smooth to touch, breaks into flat chips—*shale.*
      3. Black or dark brown, readily broken, soils fingers—*coal.*

II. *Limestone* may be clastic or crystalline, fine- or coarse-grained, and may or may not contain visible fossils. Usually gray, tan, buff, or white. Some distinctive varieties are:
   A. *Bioclastic limestone*—clastic texture, grains are whole or broken fossils. Two relatively rare varieties are:
      1. *Coquina*—very coarse, recognizable shells, much open pore space.
      2. *Chalk*—very fine-grained, white or tan, soft and powdery.
   B. *Oolitic limestone*—grains are small spheres (less than 2 millimeters in diameter), all about the same size.
   C. *Travertine*—coarsely crystalline, no pore space, often contains different-colored layers (bands).

METAMORPHIC ROCKS

The characteristics of a metamorphic rock are largely governed by (1) the composition of the parent rock and (2) the particular combination of temperature, confining pressure, and directed pressure. These factors cause different textures in rocks formed under different sets of conditions. For this reason, texture is usually the main basis for naming a metamorphic rock. Determining the composition (e.g., mineral content) is necessary for naming some rocks (e.g., *quartzite*), but for others, the minerals present are used as adjectives to describe the rock completely (e.g., *biotite* schist).

Metamorphic rocks are identified by determining first whether the rock has a *foliated* or *nonfoliated* texture.

Key for Identifying Metamorphic Rocks

I. If the rock is *nonfoliated,* then it is identified on the basis of its mineral content.
   A. Does the rock consist of mostly quartz? If so, the rock is a *quartzite.* A quartzite has a mosaic texture of interlocking grains of quartz and will easily scratch glass.
   B. Is the rock composed of interlocking coarse grains of calcite? If so, it is *marble.* (The individual grains should exhibit rhombohedral cleavage; the rock is softer than glass.)

C. Is the rock a dense, dark mass of grains mostly too fine to identify with the naked eye? If so, it probably is a *hornfels.* A hornfels may have a few larger crystals of uncommon minerals enclosed in the fine-grained mass.

II. If the rock is *foliated,* determine the type of foliation and then, if possible, identify the minerals present.

A. Is the rock very fine-grained and does it split into sheetlike slabs? If so, it is *slate.* Most slate is composed of extremely fine-grained sheet silicate minerals, and the rock has an earthy luster.

B. Does the rock have a silky sheen but otherwise appear similar to slate? If so, it is a *phyllite.*

C. Is the rock composed mostly of visible grains of platy or needlelike minerals that are approximately parallel to one another? If so, the rock is a *schist.* If the rock is composed mainly of mica, it is a *mica schist.* If it also contains garnet, it is called a *garnet mica schist.* If hornblende is the predominant mineral, the rock is a *hornblende schist.* If talc prevails, it is a *talc schist* (sometimes called soapstone).

D. Are dark and light minerals found in separate lenses or layers? If so, the rock is a *gneiss.* The light layers are composed of feldspars and perhaps quartz, whereas the darker layers commonly are formed of biotite, amphibole, or pyroxene. A gneiss may appear similar to granite or diorite but can be distinguished from the igneous rocks by the foliation.

Table C.1

Atomic Number	Name	Symbol	Atomic Weight	Some Usual Charge of Ions	Atomic Number	Name	Symbol	Atomic Weight	Some Usual Charge of Ions
1	Hydrogen	H	1.0	+1	29	Copper	Cu	63.5	+2
2	Helium	He	4.0	0 inert	30	Zinc	Zn	65.4	+2
3	Lithium	Li	6.9	+1	33	Arsenic	As	74.9	+3
4	Beryllium	Be	9.0	+2	35	Bromine	Br	79.9	—
5	Boron	B	10.8	+3	37	Rubidium	Rb	85.5	+1
6	Carbon	C	12.0	+4	38	Strontium	Sr	87.3	+2
7	Nitrogen	N	14.0	+5	40	Zirconium	Zr	91.2	—
8	Oxygen	O	16.0	−2	42	Molybdenum	Mo	95.9	+4
9	Fluorine	F	19.0	−1	47	Silver	Ag	107.9	+1
10	Neon	Ne	20.2	0 inert	48	Cadmium	Cd	112.4	—
11	Sodium	Na	23.0	+1	50	Tin	Sn	118.7	+4
12	Magnesium	Mg	24.3	+2	51	Antimony	Sb	121.8	+3
13	Aluminum	Al	27.0	+3	52	Tellurium	Te	127.6	—
14	Silicon	Si	28.1	+4	55	Cesium	Cs	132.9	—
15	Phosphorus	P	31.0	+5	56	Barium	Ba	137.4	+2
16	Sulfur	S	32.1	−2	60	Neodymium	Nd	144	+3
17	Chlorine	Cl	35.5	−1	62	Samarium	Sm	150	+3
18	Argon	Ar	39.9	0 inert	74	Tungsten	W	183.9	—
19	Potassium	K	39.1	+1	78	Platinum	Pt	195.2	—
20	Calcium	Ca	40.1	+2	79	Gold	Au	197.0	—
22	Titanium	Ti	47.9	+4	80	Mercury	Hg	200.6	+2
23	Vanadium	V	50.9	—	82	Lead	Pb	207.2	+2
24	Chromium	Cr	52.0	—	83	Bismuth	Bi	209.0	—
25	Manganese	Mn	54.9	+4, +3	86	Radon	Rn	222	0 inert
26	Iron	Fe	55.8	+2, +3	88	Radium	Ra	226.1	—
27	Cobalt	Co	58.9	—	90	Thorium	Th	232.1	—
28	Nickel	Ni	58.7	+2	92	Uranium	U	238.1	—
					94	Plutonium	Pu	239.0	—

Periodic Table of Elements

1 Group IA																	18 VIIIA
1 **H** 1.008	2 IIA											13 IIIA	14 IVA	15 VA	16 VIA	17 VIIA	2 **He** 4.00
3 **Li** 6.94	4 **Be** 9.01											5 **B** 10.81	6 **C** 12.01	7 **N** 14.01	8 **O** 16.00	9 **F** 19.00	10 **Ne** 20.18
11 **Na** 22.99	12 **Mg** 24.31	3 IIIB	4 IVB	5 VB	6 VIB	7 VIIB	8	9 VIIIB	10	11 IB	12 IIB	13 **Al** 26.98	14 **Si** 28.09	15 **P** 30.97	16 **S** 32.06	17 **Cl** 35.45	18 **Ar** 39.95
19 **K** 39.10	20 **Ca** 40.08	21 **Sc** 44.96	22 **Ti** 47.90	23 **V** 50.94	24 **Cr** 52.00	25 **Mn** 54.94	26 **Fe** 55.85	27 **Co** 58.93	28 **Ni** 58.69	29 **Cu** 63.54	30 **Zn** 65.37	31 **Ga** 69.72	32 **Ge** 72.61	33 **As** 74.92	34 **Se** 78.96	35 **Br** 79.91	36 **Kr** 83.80
37 **Rb** 85.47	38 **Sr** 87.62	39 **Y** 88.91	40 **Zr** 91.22	41 **Nb** 92.91	42 **Mo** 95.94	43 **Tc** (98)	44 **Ru** 101.07	45 **Rh** 102.90	46 **Pd** 106.42	47 **Ag** 107.87	48 **Cd** 112.41	49 **In** 114.82	50 **Sn** 118.69	51 **Sb** 121.75	52 **Te** 127.60	53 **I** 126.90	54 **Xe** 131.29
55 **Cs** 132.91	56 **Ba** 137.34	57 **La** 138.91	72 **Hf** 178.49	73 **Ta** 180.95	74 **W** 183.85	75 **Re** 186.21	76 **Os** 190.2	77 **Ir** 192.22	78 **Pt** 195.08	79 **Au** 196.97	80 **Hg** 200.59	81 **Tl** 204.37	82 **Pb** 207.19	83 **Bi** 208.98	84 **Po** (209)	85 **At** (210)	86 **Rn** (222)
87 **Fr** (223)	88 **Ra** 226.03	89 **Ac** 227.03	104 **Rf** (261)	105 **Ha** (262)	106 **Sg** (263)	107 **Ns** (262)	108 **Hs** (265)	109 **Mt** (266)	110 **Ds** (281)	111 **Rg** (272)							

Key:
6 **C** 12.01 — Atomic number / Symbol / Atomic weight

Representative elements · Noble gases · Transition metals · Metalloids · Lanthanides · Actinides

Transition metals

58 **Ce** 140.12	59 **Pr** 140.91	60 **Nd** 144.24	61 **Pm** (145)	62 **Sm** 150.36	63 **Eu** 151.96	64 **Gd** 157.25	65 **Tb** 158.92	66 **Dy** 162.50	67 **Ho** 164.93	68 **Er** 167.26	69 **Tm** 168.93	70 **Yb** 173.04	71 **Lu** 174.97
90 **Th** 232.04	91 **Pa** 231.04	92 **U** 238.03	93 **Np** (237)	94 **Pu** (244)	95 **Am** (243)	96 **Cm** (247)	97 **Bk** (247)	98 **Cf** (251)	99 **Es** (252)	100 **Fm** (257)	101 **Md** (258)	102 **No** (259)	103 **Lr** (260)

Go to www.webelements.com for an excellent periodic table on which you can click any element and get a detailed description of its properties.

Elements with an atomic number greater than 92 are not naturally occurring.

APPENDIX E
Selected Conversion Factors

TABLE E.1

	English Unit	Conversion Factor	Metric Unit	Conversion Factor	English Unit
Length and Distance	inch (in)	2.54	centimeters (cm)	0.4	inch (in)
	foot (ft)	0.3048	meter (m)	3.28	feet (ft)
	inch (in)	0.026	meter (m)	39.4	inches (in)
	mile, statute (mi)	1.61	kilometers (km)	0.62	mile (mi)
Area	square inch (in^2)	6.45	square centimeters (cm^2)	0.16	square inch (in^2)
	square foot (ft^2)	0.093	square meter (m^2)	10.8	square feet (ft^2)
	square mile (mi^2)	2.59	square kilometer (km^2)	0.39	square mile (mi^2)
	acre	0.4	hectare	2.47	acres
Volume	cubic inch (in^3)	16.4	cubic centimeters (cm^3)	0.06	cubic inch (in^3)
	cubic yard (yd^3)	0.76	cubic meter (m^3)	1.3	cubic yards (yd^3)
	cubic foot (ft^3)	0.0283	cubic meter (m^3)	35.5	cubic feet (ft^3)
	quart (qt)	0.95	liter	1.06	quarts (qt)
Weight	ounce (oz)	28.3	grams (g)	0.04	ounce (oz)
	pound (lb)	0.45	kilogram (kg)	2.2	pounds (lb)
	ton, short (2,000 lb)	907	kilograms (kg)	0.001	ton, short
	ton, short	0.91	ton, metric	1.1	ton, short
Temp.	degrees Fahrenheit (°F)	$-32° \times 5/9$	degrees Celsius (°C)	$\times 1.8 + 32°$	degrees Fahrenheit (°F)

Go to http://www.speckdesign.com/Tools2a.html for a unit conversion calculator.

APPENDIX F
Rock Symbols

Shown below are the rock symbols used in the text. In general, these symbols are used by all geologists, although they sometimes are modified slightly.

Granite

Metamorphic basement rock

Shale

Conglomerate

Basalt

Limestone

Sandstone

Breccia

Crystalline continental crust

Dolomite

Cross-bedded sandstone

Rock salt

abyss   deep (Greek)

alluvium   deposited by flowing water (Latin)

anti-   opposite (Greek)

archea (archaeo)-   ancient (Greek)

astheno-   weak, lack of strength (Greek)

ceno   recent (Greek)

circum-   about, around, round about (Latin)

clast   broken (Greek)

-cline   tilted, gradient (Greek)

de-   lower, reduce, take away (Latin)

dis-   separation, opposite of (Latin)

ex-   out of, away from (Greek)

feld   field (Swedish, German)

folium   leaf (Latin)

geo-   Earth (Greek)

glomero-   cluster (Latin)

hydro-   water (Greek)

iso-   equal (Greek)

-lith   stone or rock (Greek)

meso-   middle (Greek)

meta-   change (Greek)

-morph   form, shape (Greek)

paleo-   ancient (Greek)

ped-   foot (Latin)

pelagic   pertaining to the ocean (Greek)

petro-   stone or rock (Greek)

phanero-   visible, evident (Greek)

pheno-   large, conspicuous ("to show" in Greek)

pluto-   deep-seated (from Roman god of the underworld or infernal regions)

pre-   before, in front (Latin)

proto-   first, primary, primitive (Greek)

pyro-   fire (Greek)

spar   crystalline material (German)

-sphere   ball (Greek)

strat-   layer or layered (Latin)

stria   small groove, streak, band (Latin)

sub-   under, less than (Latin)

super-   above, more than, in addition to (Latin)

syn-   together, at the same time (Latin, Greek)

tecto-   means building or constructing in Greek and Latin; in geology, it means movement of structures caused by internal forces.

terra, terre   pertaining to Earth (Latin)

thermal   pertaining to heat (Greek)

trans-   over, beyond, through, across (Latin)

xeno-   strange, foreign (Greek)

zoo, zoic-   animal (Greek)

GLOSSARY

A

aa   A lava flow that solidifies with a spiny, rubbly surface.

ablation   The loss of the glacial ice or snow by melting, evaporation, or breaking off into icebergs. (Also called *wastage.*)

abrasion   The grinding away of rock by friction and impact during transportation.

absolute age   Age given in years or some other unit of time. (Also known as *numerical age.*)

abyssal fan   Great fan-shaped deposit of sediment on the deep-sea floor at the base of many submarine canyons.

abyssal plain   Very flat, sediment-covered region of the deep-sea floor, usually at the base of the continental rise.

accreted terrane   Terrane that did not form at its present site on a continent.

accretionary wedge (subduction complex)   A wedge of thrust-faulted and folded sediment scraped off a subducting plate by the overlying plate.

active continental margin   A margin consisting of a continental shelf, a continental slope, and an oceanic trench.

actualism   The principle that the same processes and natural laws that operated in the past are those we can actually observe or infer from observations as operating at present. Under present usage, *uniformitarianism* has the same meaning as actualism for most geologists.

advancing glacier   Glacier with a positive budget, so that accumulation results in the lower edges being pushed outward and downward.

aftershock   Small earthquake that follows a main shock.

A horizon   The top layer of soil, characterized by the downward movement of water. (Also called *zone of leaching.*)

alkali soil   Soil containing such a great quantity of sodium salts precipitated by evaporating ground water that it is generally unfit for plant growth.

alluvial fan   Large, fan-shaped pile of sediment that usually forms where a stream's velocity decreases as it emerges from a narrow canyon onto a flat plain at the foot of a mountain range.

alpine glaciation   Glaciation of a mountainous area.

amphibole group   Mineral group in which all members are double-chain silicates.

amphibolite   Amphibole (hornblende), plagioclase schist.

andesite   Fine-grained igneous rock of intermediate composition. Up to half of the rock is plagioclase feldspar with the rest being ferromagnesian minerals.

angle of dip   A vertical angle measured downward from the horizontal plane to an inclined plane.

angular   Sharp-edged; lacking rounded edges or corners.

angular unconformity   An unconformity in which younger strata overlie an erosion surface on tilted or folded layered rock.

anorthosite   A crystalline rock composed almost entirely of calcium-rich plagioclase feldspar.

antecedent stream   A stream that maintains its original course despite later deformation of the land.

anthracite   Coal that has undergone low-grade metamorphism. Burns dust-free and smokeless.

anticline   An arched fold in which the rock layers usually dip away from the axis of the fold.

aquifer   A body of saturated rock or sediment through which water can move readily.

arch (sea arch)   Bridge of rock left above an opening eroded in a headland by waves.

Archean Eon   The oldest eon of Earth's history.

arête   A sharp ridge that separates adjacent glacially carved valleys.

arid region   An area with less than 25 centimeters of rain per year.

arkose   A sandstone in which more than 25% of the grains are feldspar.

artesian aquifer   *See* confined aquifer.

artesian well   A well in which water rises above the aquifer.

artificial recharge   Groundwater recharge increased by engineering techniques.

aseismic ridge   Submarine ridge with which no earthquakes are associated.

ash (volcanic)   Fine pyroclasts (less than 4 millimeters).

assimilation   The process in which very hot magma melts country rock and assimilates the newly molten material.

asteroid   A small, generally rocky, solid body orbiting the Sun and ranging in diameter from a few meters to hundreds of kilometers.

asthenosphere   A region of Earth's outer shell beneath the lithosphere. The asthenosphere is of indeterminate thickness and behaves plastically.

astronomical unit (AU)   A distance unit based on the average distance of the Earth from the Sun.

atmosphere   Gases that envelop Earth.

atoll   A circular reef surrounding a deeper lagoon.

atom   Smallest possible particle of an element that retains the properties of that element.

atomic mass number   The total number of neutrons and protons in an atom.

atomic number   The total number of protons in an atom.

atomic weight   The sum of the weight of the subatomic particles in an average atom of an element, given in atomic mass units.

augite   Mineral of the pyroxene group found in mafic igneous rocks.

aulacogen   *See* failed rift.

aureole   Zone of contact metamorphism adjacent to a pluton.

axial plane   A plane containing all of the hinge lines of a fold.

axis   *See* hinge line.

B

backarc spreading   A type of seafloor spreading that moves an island arc away from a continent, or tears an island arc in two, or splits the edge of a continent, in each case forming new sea floor.

backshore   Upper part of the beach, landward of the high-water line.

bajada   A broad, gently sloping, depositional surface formed at the base of a mountain range in a dry region by the coalescing of individual alluvial fans.

bar   A ridge of sediment, usually sand or gravel, that has been deposited in the middle or along the banks of a stream by a decrease in stream velocity.

barchan   A crescent-shaped dune with the horns of the crescent pointing downwind.

barrier island   Ridge of sand paralleling the shoreline and extending above sea level.

barrier reef   A reef separated from the shoreline by the deeper water of a lagoon.

basal sliding   Movement in which the entire glacier slides along as a single body on its base over the underlying rock.

basalt   A fine-grained, mafic, igneous rock composed predominantly of ferromagnesian minerals and with lesser amounts of calcium-rich plagioclase feldspar.

base level   A theoretical downward limit for stream erosion of Earth's surface.

batholith   A large discordant pluton with an outcropping area greater than 100 square kilometers.

bauxite   The principal ore of aluminum; $Al_2O_3 \cdot nH_2O$.

baymouth bar   A ridge of sediment that cuts a bay off from the ocean.

beach   Strip of sediment, usually sand but sometimes pebbles, boulders, or mud, that extends from the low-water line inland to a cliff or zone of permanent vegetation.

beach face   The section of the beach exposed to wave action.

bedding   An arrangement of layers or beds of rock.

bedding plane   A nearly flat surface separating two beds of sedimentary rock.

bed load   Heavy or large sediment particles in a stream that travel near or on the stream-bed.

bedrock   Solid rock that underlies soil.

Benioff zone   Distinct earthquake zone that begins at an oceanic trench and slopes landward and downward into Earth at an angle of about 30° to 60°.

bergschrund   The crevasse that develops where a glacier is pulling away from a cirque wall.

berm   Platform of wave-deposited sediment that is flat or slopes slightly landward.

B horizon   A soil layer characterized by the accumulation of material leached downward from the A horizon above; also called *zone of accumulation.*

biochemical   Precipitated by the action of organisms.

bioclastic limestone   A limestone consisting of fragments of shells, corals, and algae.

biosphere   All of the living or once-living material on Earth.

biotite   Iron/magnesium-bearing mica.

block   Large angular pyroclast.

blowout   A depression on the land surface caused by wind erosion.

body wave   Seismic wave that travels through Earth's interior.

bomb   Large spindle- or lens-shaped pyroclast.

bonding   Attachment of an atom to one or more adjacent atoms.

bottomset bed   A delta deposit formed from the finest silt and clay, which are carried far out to sea by river flow or by sediments sliding downhill on the sea floor.

boulder   A sediment particle with a diameter greater than 256 millimeters.

Bowen's reaction series   The sequence in which minerals crystallize from a cooling basaltic magma.

braided stream   A stream that flows in a network of many interconnected rivulets around numerous bars.

breaker   A wave that has become so steep that the crest of the wave topples forward, moving faster than the main body of the wave.

breakwater   An offshore structure built to absorb the force of large breaking waves and provide quiet water near shore.

brittle strain   Cracking or rupturing of a body under stress.

butte   A narrow pinnacle of resistant rock with a flat top and very steep sides.

C

calcareous   Containing calcium carbonate.

calcite   Mineral with the formula $CaCO_3$.

caldera   A volcanic depression much larger than the original crater.

capacity (of stream)   The total load that a stream can carry.

capillary action   The drawing of water upward into small openings as a result of surface tension.

capillary fringe   A thin zone near the water table in which capillary action causes water to rise above the zone of saturation.

carbonaceous chondrite   Stony meteorite containing chondrules and composed mostly of serpentine and large quantities of organic materials.

carbonic acid H_2CO_3, a weak acid common in rain and surface waters.

cave (cavern)   Naturally formed underground chamber.

cement   The solid material that precipitates in the pore space of sediments, binding the grains together to form solid rock.

cementation   The chemical precipitation of material in the spaces between sediment grains, binding the grains together into a hard rock.

Cenozoic Era   The most recent of the eras; followed the Mesozoic Era.

chain silicate structure   Silicate structure in which two of each tetrahedron's oxygen ions are shared with adjacent tetrahedrons, resulting in a chain of tetrahedrons.

chalk   A very fine-grained bioclastic limestone.

channel (*Mars*)   Feature on the surface of the planet Mars that very closely resembles certain types of stream channels on Earth.

chaotic terrain (*Mars*)   Patch of jumbled and broken angular slabs and blocks on the surface of Mars.

chemical sedimentary rock   A rock composed of material precipitated directly from solution.

chemical weathering   The decomposition of rock resulting from exposure to water and atmospheric gases.

chert   A hard, compact, fine-grained sedimentary rock formed almost entirely of silica.

chill zone   In an intrusion, the finer-grained rock adjacent to a contact with country rock.

chondrule   Round silicate grain within some stony meteorites.

C horizon   A soil layer composed of incompletely weathered parent material.

cinder (volcanic)   Pyroclast approximately the size of a sand grain. Sometimes defined as between 4 and 32 millimeters in diameter.

cinder cone   A volcano constructed of loose rock fragments ejected from a central vent.

circum-Pacific belt   Major belt around the edge of the Pacific Ocean on which most composite volcanoes are located and where many earthquakes occur.

cirque   A steep-sided, amphitheaterlike hollow carved into a mountain at the head of a glacial valley.

clastic texture   An arrangement of rock fragments bound into a rigid network by cement.

clay   Sediment composed of particles with diameter less than 1/256 millimeter.

clay mineral   A hydrous aluminum-silicate that occurs as a platy grain of microscopic size with a sheet-silicate structure.

clay mineral group   Collective term for clay minerals.

cleavage   The ability of a mineral to break along preferred planes.

coal   A sedimentary rock formed from the consolidation of plant material. It is rich in carbon, usually black, and burns readily.

coal-bed methane   Gas trapped in coal.

coarse-grained rock   Rock in which most of the grains are larger than 1 millimeter (igneous) or 2 millimeters (sedimentary).

coast   The land near the sea, including the beach and a strip of land inland from the beach.

coastal straightening   The gradual straightening of an irregular shoreline by wave erosion of headlands and wave deposition in bays.

cobble   A sediment particle with a diameter of 64 to 256 millimeters.

column   A dripstone feature formed when a stalactite growing downward and a stalagmite growing upward meet and join.

columnar structure   Volcanic rock in parallel, usually vertical columns, mostly six-sided; also called *columnar jointing.*

comet   Small object in space, no more than a few kilometers in diameter, composed of frozen methane, frozen ammonia, and water-ice, with small solid particles and dust imbedded in the ices.

compaction   A loss in overall volume and pore space of a rock as the particles are packed closer together by the weight of overlying material.

competence   The largest particle that a stream can carry.

composite volcano (stratovolcano)   A volcano constructed of alternating layers of pyroclastics and rock solidified from lava flows.

compressive stress   A stress due to a force pushing together on a body.

conchoidal fracture   Curved fracture surfaces.

concordant   Parallel to layering or earlier developed planar structures.

concretion   Hard, rounded mass that develops when a considerable amount of cementing material precipitates locally in a rock, often around an organic nucleus.

cone of depression   A depression of the water table formed around a well when water is pumped out; it is shaped like an inverted cone.

confined aquifer (artesian aquifer)   An aquifer completely filled with pressurized water and separated from the land surface by a relatively impermeable confining bed, such as shale.

confining pressure   Pressure applied equally on all surfaces of a body; also called *lithostatic pressure.*

conglomerate   A coarse-grained sedimentary rock (grains coarser than 2 millimeters) formed by the cementation of rounded gravel.

consolidation   Any process that forms firm, coherent rock from sediment or from liquid.

contact   Boundary surface between two different rock types or ages of rocks.

contact (thermal) metamorphism   Metamorphism under conditions in which high temperature is the dominant factor.

continental crust   The thick, granitic crust under continents.

continental drift   A concept suggesting that continents move over Earth's surface.

continental glaciation   The covering of a large region of a continent by a sheet of glacial ice.

continental rise   A wedge of sediment that extends from the lower part of the continental slope to the deep-sea floor.

continental shelf   A submarine platform at the edge of a continent, inclined very gently seaward generally at an angle of less than 1°.

continental slope   A relatively steep slope extending from a depth of 100 to 200 meters at the edge of the continental shelf down to oceanic depths.

contour current   A bottom current that flows parallel to the slopes of the continental margin (along the contour rather than down the slope).

contour line   A line on a topographic map connecting points of equal elevation.

convection (convection current)   A very slow circulation of a substance driven by differences in temperature and density within that substance.

convergent plate boundary   A boundary between two plates that are moving toward each other.

coquina   A limestone consisting of coarse shells.

core   The central zone of Earth.

correlation   In geology, correlation usually means determining time equivalency of rock units. Rock units may be correlated within a region, a continent, and even between continents.

country rock   Any rock that was older than and intruded by an igneous body.

covalent bonding   Bonding due to the sharing of electrons by adjacent atoms.

crater (of a volcano)   A basinlike depression over a vent at the summit of a volcanic cone.

craton   Portion of a continent that has been structurally stable for a prolonged period of time.

creep   Very slow, continuous downslope movement of soil or debris.

crest (of wave)   The high point of a wave.

crevasse   Open fissure in a glacier.

cross-bedding   An arrangement of relatively thin layers of rock inclined at an angle to the more nearly horizontal bedding planes of the larger rock unit.

crosscutting relationship   A principle or law stating that a disrupted pattern is older than the cause of disruption.

cross section   *See* geologic cross section.

crude oil   A liquid mixture of naturally occurring hydrocarbons.

crust   The outer layer of rock, forming a thin skin over Earth's surface.

crustal rebound   The rise of Earth's crust after the removal of glacial ice.

crystal   A homogeneous solid with an orderly internal atomic arrangement.

crystal form   Arrangement of various faces on a crystal in a definite geometric relationship to one another.

crystalline   Describing a substance in which the atoms are arranged in a regular, repeating, orderly pattern.

crystalline texture   An arrangement of interlocking crystals.

crystallization   Crystal development and growth.

crystal settling   The process whereby the minerals that crystallize at a high temperature in a cooling magma move downward in the magma chamber because they are denser than the magma.

cuesta   A ridge with a steep slope on one side and a gentle slope on the other side.

Curie point   The temperature below which a material becomes magnetized.

D

data   What scientists regard as facts.

daughter product   The isotope produced by radioactive decay.

debris   Unconsolidated material (soil) in which coarse-grained fragments predominate.

debris avalanche   Very rapid and turbulent mass wasting of debris, air, and water.

debris flow   Mass wasting involving the flow of soil (unconsolidated material) in which coarse material (gravel, boulders) is predominant.

decompression melting   Partial melting of hot mantle rock when it moves upward and the pressure is reduced to the extent that the melting point drops to the temperature of the body.

deflation   The removal of clay, silt, and sand particles from the land surface by wind.

delamination   *See* lithospheric delamination.

delta   A body of sediment deposited at the mouth of a river when the river velocity decreases as it flows into a standing body of water.

dendritic pattern   Drainage pattern of a river and its tributaries that resembles the branches of a tree or veins in a leaf.

density   Weight per given volume of a substance.

deposition   The settling or coming to rest of transported material.

depth of focus   Distance between the focus and the epicenter of an earthquake.

desert   A region with low precipitation (usually defined as less than 25 centimeters per year).

desertification   The expansion of barren deserts into once-populated regions.

desert pavement   A thin layer of closely packed gravel that protects the underlying sediment from deflation; also called *pebble armor.*

detachment fault   Major fault in a mountain belt above which rocks have been intensely folded and faulted.

detrital sedimentary rock   A sedimentary rock composed of fragments of preexisting rock.

diapir   Bodies of rock (e.g., rock salt) or magma that ascend within Earth's interior because they are less dense than the surrounding rock.

differential stress   When pressures on a body are not of equal strength in all directions.

differential weathering   Varying rates of weathering resulting from some rocks in an area being more resistant to weathering than others.

differentiation   Separation of different ingredients from an originally homogeneous mixture.

dike   A tabular, discordant intrusive structure.

diorite   Coarse-grained igneous rock of intermediate composition. Up to half of the rock is plagioclase feldspar and the rest is ferromagnesian minerals.

dip   *See* angle of dip, direction of dip.

dip-slip fault   A fault in which movement is parallel to the dip of the fault surface.

directed pressure   *See* differential stress.

direction of dip   The compass direction in which the angle of dip is measured.

discharge   In a stream, the volume of water that flows past a given point in a unit of time.

disconformity   A surface that represents missing rock strata but beds above and below that surface are parallel to one another.

discordant   Not parallel to any layering or parallel planes.

dissolved load   The portion of the total sediment load in a stream that is carried in solution.

distributary   Small shifting river channel that carries water away from the main river channel and distributes it over a delta's surface.

divergent plate boundary   Boundary separating two plates moving away from each other.

divide   Line dividing one drainage basin from another.

dolomite   A sedimentary rock composed mostly of the mineral dolomite.

dolomitic marble   Marble in which dolomite, rather than calcite, is the prevalent mineral.

dome   *See* structural dome.

double refraction   The splitting of light into two components when it passes through certain crystalline substances.

downcutting   A valley-deepening process caused by erosion of a streambed.

drainage basin   Total area drained by a stream and its tributaries.

drainage pattern   The arrangement in map view of a river and its tributaries.

drawdown   The lowering of the water table near a pumped well.

dripstone   Deposits of calcite (and, rarely, other minerals) built up by dripping water in caves.

drumlin   A long, streamlined hill made of till.

ductile   Capable of being molded and bent under stress.

ductile strain   Strain in which a body is molded or bent under stress and does not return to its original shape after the stress is removed.

dust (volcanic)   Finest-sized pyroclasts.

E

E horizon   Soil horizon that is the zone of leaching, characterized by the downward movement of water and removal of fine-grained soil components.

earth In mass wasting, soil in which fine-grained particles are predominant.

Earth systems Study of Earth by analyzing how its components, or subsystems, interrelate.

earthflow Slow-to-rapid mass wasting in which fine-grained soil moves downslope as a very viscous fluid.

earthquake A trembling or shaking of the ground caused by the sudden release of energy stored in the rocks beneath the surface.

earthy luster A luster giving a substance the appearance of unglazed pottery.

echo sounder An instrument used to measure and record the depth to the sea floor.

elastic limit The maximum amount of stress that can be applied to a body before it deforms in a permanent way by bending or breaking.

elastic rebound theory The sudden release of progressively stored strain in rocks results in movement along a fault.

elastic strain Strain in which a deformed body recovers its original shape after the stress is released.

electron A single, negative electric charge that contributes virtually no mass to an atom.

element A substance that cannot be broken down to other substances by ordinary chemical methods. Each atom of an element possesses the same number of protons.

emergent coast A coast in which land formerly under water has recently been placed above sea level, either by uplift of the land or by a drop in sea level.

end moraine A ridge of till piled up along the front edge of a glacier.

environment of deposition The location in which deposition occurs, usually marked by characteristic physical, chemical, or biological conditions.

eon The largest unit of geological time.

epicenter The point on Earth's surface directly above the focus of an earthquake.

epoch Each period of the standard geologic time scale is divided into epochs (e.g., Pleistocene Epoch of the Quaternary Period).

equilibrium Material is in equilibrium if it is adjusted to the physical and chemical conditions of its environment so that it does not change or alter with time.

equilibrium line An irregular line marking the highest level to which the winter snow cover on a glacier is lost during a melt season; also called *snow line.*

era Major subdivision of the standard geologic time scale (e.g., Mesozoic Era).

erosion The physical removal of rock by an agent such as running water, glacial ice, or wind.

erratic An ice-transported boulder that does not derive from bedrock near its present site.

esker A long, sinuous ridge of sediment deposited by glacial meltwater.

estuary Drowned river mouth.

etch-pitted terrain *(Mars)* A terrain on the surface of Mars characterized by small pits.

evaporite Rock that forms from crystals precipitating during evaporation of water.

exfoliation The stripping of concentric rock slabs from the outer surface of a rock mass.

exfoliation dome A large, rounded landform developed in a massive rock, such as granite, by the process of exfoliation.

exotic terrane Terrane that did not form at its present site on a continent and traveled a great distance to get to its present site.

expansive clay Clay that increases in volume when water is added to it.

extension Strain involving an increase in length. Extension can cause crustal thinning and faulting.

extrusive rock Any igneous rock that forms at Earth's surface, whether it solidifies directly from a lava flow or is pyroclastic.

F

faceted A rock fragment with one or more flat surfaces caused by erosive action.

failed rift (aulacogen) An inactive, sediment-filled rift that forms above a mantle plume. The rift becomes inactive as two other rifts widen to form an ocean.

fall The situation in mass wasting that occurs when material free-falls or bounces down a cliff.

fault A fracture in bedrock along which movement has taken place.

fault-block mountain range A range created by uplift along normal or vertical faults.

faunal succession A principle or law stating that fossil species succeed one another in a definite and recognizable order; in general, fossils in progressively older rock show increasingly greater differences from species living at present.

feldspar Group of most common minerals of Earth's crust. All feldspars contain silicon, aluminum, and oxygen and may contain potassium, calcium, and sodium.

felsic rock Silica-rich igneous rock with a relatively high content of potassium and sodium.

ferromagnesian mineral Iron/magnesium-bearing mineral, such as augite, hornblende, olivine, or biotite.

fine-grained rock A rock in which most of the mineral grains are less than 1 millimeter across (igneous) or less than 1/16 millimeter (sedimentary).

fiord A coastal inlet that is a glacially carved valley, the base of which is submerged.

firn A compacted mass of granular snow, transitional between snow and glacier ice.

firn limit *See* equilibrium line.

fissility The ability of a rock to split into thin layers.

flank eruption An eruption in which lava erupts out of a vent on the side of a volcano.

flash flood Flood of very high discharge and short duration; sudden and local in extent.

flood plain A broad strip of land built up by sedimentation on either side of a stream channel.

flow A type of movement that implies that a descending mass is moving downslope as a viscous fluid.

flowstone Calcite precipitated by flowing water on cave walls and floors.

focus The point within Earth from which seismic waves originate in an earthquake.

fold Bend in layered bedrock.

fold and thrust belt A portion of a major mountain belt characterized by large thrust faults, stacked one upon another. Layered rock between the faults was folded when faulting was taking place.

fold axis *See* hinge line.

foliation Parallel alignment of textural and structural features of a rock.

footwall The underlying surface of an inclined fault plane.

foreland basin A sediment-filled basin on a continent, landward of a magmatic arc, and caused indirectly by ocean-continent convergence.

foreset bed A sediment layer in the main part of a delta, deposited at an angle to the horizontal.

foreshock Small earthquake that precedes a main shock.

foreshore The zone that is regularly covered and uncovered by the rise and fall of tides.

formation A body of rock of considerable thickness that has a recognizable unity or similarity making it distinguishable from adjacent rock units. Usually composed of one bed or several beds of sedimentary rock, although the term is also applied to units of metamorphic and igneous rock. A convenient unit for mapping, describing, or interpreting the geology of a region.

fossil Traces of plants or animals preserved in rock.

fossil assemblage Various different species of fossils in a rock.

fracture The way a substance breaks where not controlled by cleavage.

fracture zone Major line of weakness in Earth's crust that crosses the mid-oceanic ridge at approximately right angles.

fracturing Cracking or rupturing of a body under stress.

framework silicate structure Crystal structure in which all four oxygen ions of a silica tetrahedron are shared by adjacent ions.

fretted terrain *(Mars)* Flat lowland with some scattered high plateaus on the surface of Mars.

fringing reef A reef attached directly to shore. *See* barrier reef.

frost action Mechanical weathering of rock by freezing water.

frost heaving The lifting of rock or soil by the expansion of freezing water.

frost wedging A type of frost action in which the expansion of freezing water pries a rock apart.

G

gabbro A mafic, coarse-grained igneous rock composed predominantly of ferromagnesian minerals and with lesser amounts of calcium-rich plagioclase feldspar.

gaining stream A stream that receives water from the zone of saturation.

geode Partly hollow, globelike body found in limestone or other cavernous rock.

geologic cross section A representation of a portion of Earth in a vertical plane.

geologic map A map representing the geology of a given area.

geologic resources Valuable materials of geologic origin that can be extracted from Earth.

geology The scientific study of the planet Earth.

geophysics The application of physical laws and principles to a study of Earth.

geosphere Solid Earth system. The rock and other inorganic material that make up the bulk of the planet.

geothermal energy Energy produced by harnessing naturally occurring steam and hot water.

geothermal gradient Rate of temperature increase associated with increasing depth beneath the surface of Earth (normally about 25°C per kilometer).

geyser A type of hot spring that periodically erupts hot water and steam.

geyserite A deposit of silica that forms around many geysers and hot springs.

glacier A large, long-lasting mass of ice, formed on land by the compaction and recrystallization of snow, which moves because of its own weight.

glacier ice The mosaic of interlocking ice crystals that form a glacier.

glassy (or vitreous) luster A luster that gives a substance a glazed, porcelainlike appearance.

gneiss A metamorphic rock composed of light and dark layers or lenses.

gneissic The texture of a metamorphic rock in which minerals are separated into light and dark-layers or lenses.

goethite The commonest mineral in the limonite group; $Fe_2O_3 \cdot nH_2O$.

Gondwanaland The southern part of *Pangaea* (*see* definition) that formed South America, Africa, India, Australia, and Antarctica.

graben A downdropped block bounded by normal fault.

graded bed A single bed with coarse grains at the bottom of the bed and progressively finer grains toward the top of the bed.

graded stream A stream that exhibits a delicate balance between its transporting capacity and the sediment load available to it.

granite A felsic, coarse-grained, intrusive igneous rock containing quartz and composed mostly of potassium- and sodium-rich feldspars.

gravel Rounded particles coarser than 2 millimeters in diameter.

gravitational collapse and spreading When part of a mountain belt becomes too high, it moves vertically downward forcing rock at depth to spread out laterally.

gravity The force of attraction that two bodies exert on each other that is proportional to the product of their masses and inversely proportional to the square of the distance from the centers of the two bodies.

gravity anomaly A deviation from the average gravitational attraction between Earth and an object. *See* negative gravity anomaly, positive gravity anomaly.

gravity meter An instrument that measures the gravitational attraction between Earth and a mass within the instrument.

graywacke A sandstone with more than 15% fine-grained matrix between the sand grains.

greenhouse effect The trapping of heat by a planet's atmosphere, making the planet warmer than would otherwise be expected. Generally, the greenhouse effect operates if visible sunlight passes freely through a planet's atmosphere, but the infrared radiation produced by the warm surface cannot escape readily into space.

groin Short wall built perpendicular to shore to trap moving sand and widen a beach.

ground moraine A blanket of till deposited by a glacier or released as glacier ice melted.

ground water The water that lies beneath the ground surface, filling the cracks, crevices, and pore space of rocks.

guyot Flat-topped seamount.

H

Hadean Eon The oldest eon.

half-life The time it takes for a given amount of a radioactive isotope to be reduced by one-half.

hanging valley A smaller valley that terminates abruptly high above a main valley.

hanging wall The overlying surface of an inclined fault plane.

hardness The relative ease or difficulty with which a smooth surface of a mineral can be scratched; commonly measured by Mohs scale.

headland Point of land along a coast.

headward erosion The lengthening of a valley in an uphill direction above its original source by gullying, mass wasting, and sheet erosion.

heat engine A device that converts heat energy into mechanical energy.

heat flow Gradual loss of heat (per unit of surface area) from Earth's interior out into space.

heavy crude Dense, viscous petroleum that flows slowly or not at all.

hematite A type of iron oxide that has a brick-red color when powdered; Fe_2O_3.

highland (*Moon*) A rugged region of the lunar surface representing an early period in lunar history when intense meteorite bombardment formed craters.

hinge line Line about which a fold appears to be hinged. Line of maximum curvature of a folded surface.

hinge plane *See* axial plane.

hogback A sharp-topped ridge formed by the erosion of steeply dipping beds.

Holocene Epoch The youngest epoch which began around 10,000 years ago and is continuing presently.

horn A sharp peak formed where cirques cut back into a mountain on several sides.

hornblende Common amphibole frequently found in igneous and metamorphic rocks.

hornfels A fine-grained, unfoliated metamorphic rock.

horst An up-raised block bounded by normal faults.

hot spot An area of volcanic eruptions and high heat flow above a rising mantle plume.

hot spring Spring with a water temperature warmer than human body temperature.

hydraulic action The ability of water to pick up and move rock and sediment.

hydrologic cycle The movement of water and water vapor from the sea to the atmosphere, to the land, and back to the sea and atmosphere again.

hydrology The study of water's properties, circulation, and distribution.

hydrosphere The water on or near Earth's surface.

hydrothermal rock Rock deposited by precipitation of ions from solution in hot water.

hydrothermal vein Quartz or other minerals that have been deposited in a crack by hot fluids.

hypocenter Synonym for the focus of an earthquake.

hypothesis A tentative theory.

I

iceberg Block of glacier-derived ice floating in water.

ice cap A glacier covering a relatively small area of land but not restricted to a valley.

icefall A chaotic jumble of crevasses that split glacier ice into pinnacles and blocks.

ice sheet A glacier covering a large area (more than 50,000 square kilometers) of land.

igneous rock A rock formed or apparently formed from solidification of magma.

incised meander A meander that retains its sinuous curves as it cuts vertically downward below the level at which it originally formed.

inclusion A fragment of rock that is distinct from the body of igneous rock in which it is enclosed.

inclusion, principle of Fragments included in a host rock are older than the host rock.

index fossil A fossil from a very short-lived species known to have existed during a specific period of geologic time.

inner planet A planet orbiting in the inner part of the Solar System. Sometimes taken to mean Mercury, Venus, Earth, and Mars.

intensity A measure of an earthquake's size by its effect on people and buildings.

intermediate rock Rock with a chemical content between felsic and mafic compositions.

intrusion (intrusive structure) A body of intrusive rock classified on the basis of size, shape, and relationship to surrounding rocks.

intrusive rock Rock that appears to have crystallized from magma emplaced in surrounding rock.

ion An electrically charged atom or group of atoms.

ionic bonding   Bonding due to the attraction between positively charged ions and negatively charged ions.

iron meteorite   A meteorite composed principally of iron-nickel alloy.

island arc   A curved line of islands. 403

isoclinal fold   A fold in which the limbs are parallel to one another.

isolated silicate structure   Silicate minerals that are structured so that none of the oxygen atoms are shared by silica tetrahedrons.

isostasy   The balance or equilibrium between adjacent blocks of crust resting on a plastic mantle.

isostatic adjustment   Concept of vertical movement of sections of Earth's crust to achieve balance or equilibrium.

isotherm   A line along which the temperature of rock (or other material) is the same.

isotopes   Atoms (of the same element) that have different numbers of neutrons but the same number of protons.

isotopic dating   Determining the age of a rock or mineral through its radioactive elements and decay products (previously and somewhat inaccurately called *radiometric* or *radioactive dating*).

J

jetty   Rock wall protruding above sea level, designed to protect the entrance of a harbor from sediment deposition and storm waves; usually built in pairs.

joint   A fracture or crack in bedrock along which essentially no displacement has occurred.

joint set   Joints oriented in one direction approximately parallel to one another.

K

kame   Low mound or irregular ridge formed of outwash deposits on a stagnating glacier.

kame and kettle topography   Irregular, bumpy landscape of hills and depressions associated with many moraines.

karst topography   An area with many sinkholes and a cave system beneath the land surface and usually lacking a surface stream.

kettle   A depression caused by the melting of a stagnant block of ice that was surrounded by sediment.

kimberlite   An ultramafic rock that contains olivine along with mica, garnet, or both. Diamonds are found in some kimberlite bodies.

L

laccolith   A concordant intrusive structure, similar to a sill, with the central portion thicker and domed upward. Laccoliths are not common and are not discussed in this textbook.

laminar flow   Slow, smooth flow, with each drop of water traveling a smooth path parallel to its neighboring drops.

laminated terrain *(Mars)*   Area where series of alternating light and dark layers can be seen on the surface of Mars.

lamination   A thin layer in sedimentary rock (less than 1 centimeter thick).

landform   A characteristic shaped feature of Earth's surface, such as a hill or a valley.

lapilli (plural)   Pyroclasts in the 2–64 millimeter size range (singular, *lapillus*).

landslide   The general term for a slowly to very rapidly descending mass of rock or debris.

lateral continuity   Principle that states that an original sedimentary layer extends laterally until it tapers or thins at its edges.

lateral erosion   Erosion and undercutting of stream banks caused by a stream swinging from side to side across its valley floor.

lateral moraine   A low, ridgelike pile of till along the side of a glacier.

laterite   Highly leached soil that forms in regions of tropical climate with high temperatures and very abundant rainfall.

lava   Magma on Earth's surface.

lava flow   Flow of lava from a crater or fissure.

lava tube   Tunnel-like cave within a lava flow. It forms during the late stages of solidification of a mafic lava flow.

left-lateral fault   A strike-slip fault in which the block seen across the fault appears displaced to the left.

limb   Portion of a fold shared by an anticline and a syncline.

limestone   A sedimentary rock composed mostly of calcite.

limonite   A type of iron oxide that is yellowish-brown when powdered; $Fe_2O_3 \cdot nH_2O$.

liquefaction   A type of ground failure in which water-saturated sediment turns from a solid to a liquid as a result of shaking, often caused by an earthquake.

lithification   The consolidation of sediment into sedimentary rock.

lithosphere   The rigid outer shell of Earth, 70 to 125 or more kilometers thick.

lithospheric delamination   The detachment of part of the mantle portion of the lithosphere beneath a mountain belt.

lithostatic pressure   Confining pressure due to the weight of overlying rock.

loam   Soil containing approximately equal amounts of sand, silt, and clay.

loess   A fine-grained deposit of wind-blown dust.

longitudinal dune (seif)   Large, symmetrical ridge of sand parallel to the wind direction.

longitudinal profile   A line showing a stream's slope, drawn along the length of the stream as if it were viewed from the side.

longshore current   A moving mass of water that develops parallel to a shoreline.

longshore drift   Movement of sediment parallel to shore when waves strike a shoreline at an angle.

losing stream   Stream that loses water to the zone of saturation.

Love waves   A type of surface seismic wave that causes the ground to move side to side in a horizontal plane perpendicular to the direction the wave is traveling.

low-velocity zone   Mantle zone at a depth of about 100 kilometers where seismic waves travel more slowly than in shallower layers of rock.

luster   The quality and intensity of light reflected from the surface of a mineral.

M

mafic rock   Silica-deficient igneous rock with a relatively high content of magnesium, iron, and calcium.

magma   Molten rock, usually mostly silica. The liquid may contain dissolved gases as well as some solid minerals.

magmatic arc   A line of batholiths or volcanoes. Generally the line, as seen from above, is curved.

magmatic underplating   *See* underplating.

magnetic anomaly   A deviation from the average strength of Earth's magnetic field. *See* negative magnetic anomaly, positive magnetic anomaly.

magnetic field   Region of magnetic force that surrounds Earth.

magnetic pole   An area where the strength of the magnetic field is greatest and where the magnetic lines of force appear to leave or enter Earth.

magnetic reversal   A change in Earth's magnetic field between normal polarity and reversed polarity. In normal polarity, the north magnetic pole, where magnetic lines of force enter Earth, lies near the geographic North Pole. In reversed polarity, the south magnetic pole, where lines of force leave Earth, lies near the geographic North Pole (the magnetic poles have exchanged positions).

magnetite   An iron oxide that is attracted to a magnet.

magnetometer   An instrument that measures the strength of Earth's magnetic field.

magnitude   A measure of the energy released during an earthquake.

major mountain belt   A long chain (thousands of kilometers) of mountain ranges.

mantle   A thick shell of rock that separates Earth's crust above from the core below.

mantle diapir   A body of mantle rock, hotter than its surroundings, that ascends because it is less dense than the surrounding rock.

mantle plume   Narrow column of hot mantle rock that rises and spreads radially outward.

marble   A coarse-grained rock composed of interlocking calcite (or dolomite) crystals.

maria *(Moon)*   Lava plains on Moon's surface (singular, *mare*).

marine terrace   A broad, gently sloping platform that may be exposed at low tide.

mass wasting (or mass movement)   Movement, caused by gravity, in which bedrock, rock debris, or soil moves downslope in bulk.

matrix   Fine-grained material found in the pore space between larger sediment grains.

meander   A pronounced sinuous curve along a stream's course.

meander cutoff   A new, shorter channel across the narrow neck of a meander.

meander scar   An abandoned meander filled with sediment and vegetation.

mechanical weathering   The physical disintegration of rock into smaller pieces.

medial moraine   A single long ridge of till on a glacier, formed by adjacent lateral moraines joining and being carried downglacier.

Mediterranean-Himalayan belt (Mediterranean belt)   A major concentration of earthquakes and composite volcanoes that runs through the Mediterranean Sea, crosses the Mideast and the Himalaya, and passes through the East Indies.

melt   Liquid rock resulting from melting in a laboratory.

Mercalli scale   *See* modified Mercalli scale.

mesa   A broad, flat-topped hill bounded by cliffs and capped with a resistant rock layer.

Mesozoic Era   The era that followed the Paleozoic Era and preceded the Cenozoic Era.

metallic bonding   Bonding, as in metals, whereby atoms are closely packed together and electrons move freely among atoms.

metallic luster   Luster giving a substance the appearance of being made of metal.

metamorphic facies   Metamorphic rocks that contain the same set of pressure or temperature sensitive minerals are regarded as belonging to the same facies, implying that they formed under broadly similar pressure and temperature conditions.

metamorphic rock   A rock produced by metamorphism.

metamorphism   The transformation of preexisting rock into texturally or mineralogically distinct new rock as a result of high temperature, high pressure, or both but without the rock melting in the process.

metasomatism   Metamorphism coupled with the introduction of ions from an external source.

meteor   Fragment that passes through Earth's atmosphere, heated to incandescence by friction; sometimes incorrectly called "shooting" or "falling" stars.

meteorite   Meteor that strikes Earth's surface.

meteoroid   Small solid particles of stone and/or metal orbiting the sun.

mica group   Group of minerals with a sheet-silicate structure.

microcline (potassium) feldspar   A feldspar with the formula $KAlSi_3O_8$.

mid-oceanic ridge   A giant mountain range that lies under the ocean and extends around the world.

migmatite   Mixed igneous and metamorphic rock.

Milky Way galaxy   The galaxy to which the Sun belongs. Seen from Earth, the galaxy is a pale, milky band in the night sky.

mineral   A crystalline substance that is naturally occurring and is chemically and physically distinctive.

mineraloid   A substance that is not crystalline but otherwise would be considered a mineral.

model   In science, a model is an image—graphic, mathematical, or verbal—that is consistent with the known data.

modified Mercalli scale   Scale expressing intensities of earthquakes (judged on amount of damage done) in Roman numerals ranging from I to XII.

Mohorovičić discontinuity   The boundary separating the crust from the mantle beneath it (also called **Moho**).

Mohs' hardness scale   Scale on which ten minerals are designated as standards of hardness.

molecule   The smallest possible unit of a substance that has the properties of that substance.

moment magnitude   An earthquake magnitude calculated from the strength of the rock, surface area of the fault rupture, and the amount of rock displacement along the fault.

monocline   A local steeping in a gentle regional dip; a steplike fold in rock.

moraine   A body of till either being carried on a glacier or left behind after a glacier has receded.

mountain range   A group of closely spaced mountains or parallel ridges.

mud   Term loosely used for silt and clay, usually wet.

mud crack   Polygonal crack formed in very fine-grained sediment as it dries.

mudflow   A flowing mixture of debris and water, usually moving down a channel.

mudstone   A fine-grained sedimentary rock that lacks shale's laminations and fissility.

muscovite   Transparent or white mica that lacks iron and magnesium.

N

natural gas   A gaseous mixture of naturally occurring hydrocarbons.

natural levee   Low ridges of flood-deposited sediment formed on either side of a stream channel, which thin away from the channel.

nebula   A large volume of interstellar gas and dust.

Nebular Hypotheses   The hypothesis that the Solar System formed from a rotating cloud of gas and dust, the solar nebula.

negative gravity anomaly   Less than normal gravitational attraction.

negative magnetic anomaly   Less than average strength of Earth's magnetic field.

neutron   A subatomic particle that contributes mass to an atom and is electrically neutral.

nonconformity   An unconformity in which an erosion surface on plutonic or metamorphic rock has been covered by younger sedimentary or volcanic rock.

nonmetallic luster   Luster that gives a substance the appearance of being made of something other than metal (e.g., glassy).

nonrenewable resource   A resource that forms at extremely slow rates compared to its rate of consumption.

normal fault   A fault in which the hanging-wall block moved down relative to the footwall block.

nucleus   Protons and neutrons form the nucleus of an atom. Although the nucleus occupies an extremely tiny fraction of the volume of the entire atom, practically all the mass of the atom is concentrated in the nucleus.

numerical age   Age given in years or some other unit of time.

O

oblique-slip fault   A fault with both strike-slip and dip-slip components.

obsidian   Volcanic glass.

oceanic crust   The thin, basaltic crust under oceans.

oceanic trench   A narrow, deep trough parallel to the edge of a continent or an island arc.

O horizon   Dark-colored soil layer that is rich in organic material and forms just below surface vegetation.

oil   *See* crude oil.

oil field   An area underlain by one or more oil pools.

oil pool   Underground accumulation of oil.

oil sand   Asphalt-cemented sand deposit.

oil shale   Shale with a high content of organic matter from which oil may be extracted by distillation.

oil trap   A set of conditions that hold petroleum in a reservoir rock and prevent its escape by migration.

olivine   A ferromagnesian mineral with the formula $(Fe, Mg)_2SiO_4$.

oolite (ooid)   A small sphere of calcite precipitated from seawater.

oolitic limestone   A limestone formed from oolites.

opal   A mineraloid composed of silica and water.

open fold   A fold with gently dipping limbs.

open-pit mine   Mine in which ore is exposed at the surface in a large excavation.

ophiolite   A distinctive rock sequence found in many mountain ranges on continents.

ore   Naturally occurring material that can be profitably mined.

ore mineral   A mineral of commercial value.

organic sedimentary rock   Rock composed mostly of the remains of plants and animals.

original horizontality   The deposition of most water-laid sediment in horizontal or near-horizontal layers that are essentially parallel to Earth's surface.

orogeny   An episode of intense deformation of the rocks in a region, generally accompanied by metamorphism and plutonic activity.

orthoclase (potassium) feldspar   A feldspar with the formula $KAlSi_3O_8$.

outcrop   A surface exposure of bare rock, not covered by soil or vegetation.

outer planet   A planet whose orbit lies in the outer part of the Solar System. Jupiter, Saturn, Uranus, Neptune, and Pluto are outer planets.

outwash   Material deposited by debris-laden meltwater from a glacier.

overburden   The upper part of a sedimentary deposit. Its weight causes compaction of the lower part.

overturned fold   A fold in which both limbs dip in the same direction.

oxbow lake   A crescent-shaped lake occupying the abandoned channel of a stream meander that is isolated from the present channel by a meander cutoff and sedimentation.

P

pahoehoe   A lava flow characterized by a ropy or billowy surface.

paired terraces   *Stream terraces* (*see* definition) that occur at the same elevation on each side of a river.

paleomagnetism   A study of ancient magnetic fields.

Paleozoic Era   The era that followed the Precambrian and began with the appearance of complex life, as indicated by fossils.

Pangaea   A supercontinent that broke apart 200 million years ago to form the present continents.

parabolic dune   A deeply curved dune in a region of abundant sand. The horns point upwind and are often anchored by vegetation.

parent rock   Original rock before being metamorphosed.

partial melting   Melting of the components of a rock with the lowest melting temperatures.

passive continental margin   A margin that includes a continental shelf, continental slope, and continental rise that generally extends down to an abyssal plain at a depth of about 5 kilometers.

paternoster lakes   A series of rock-basin lakes carved by glacial erosion.

peat   A brown, lightweight, unconsolidated or semi-consolidated deposit of plant remains.

pebble   A sediment particle with a diameter of 2 to 64 millimeters.

pediment   A gently sloping erosional surface cut into the solid rock of a mountain range in a dry region; usually covered with a thin veneer of gravel.

pegmatite   Extremely coarse-grained igneous rock.

pelagic sediment   Sediment made up of fine-grained clay and the skeletons of microscopic organisms that settle slowly down through the ocean water.

perched water table   A water table separated from the main water table beneath it by a zone that is not saturated.

peridotite   Ultramafic rock composed of pyroxene and olivine.

period   Each era of the standard geologic time scale is subdivided into periods (e.g., the Cretaceous Period).

permafrost   Ground that remains permanently frozen for many years.

permeability   The capacity of a rock to transmit a fluid such as water or petroleum.

petrified wood   A material that forms as the organic matter of buried wood is either filled in or replaced by inorganic silica carried in by ground water.

petroleum   Crude oil and natural gas. (Some geologists use petroleum as a synonym for oil.)

Phanerozoic   Eon of geologic time. Includes all time following the Precambrian.

phenocryst   Any of the large crystals in porphyritic igneous rock.

phyllite   A metamorphic rock in which clay minerals have recrystallized into microscopic micas, giving the rock a silky sheen.

physical continuity   Being able to physically follow a rock unit between two places.

physical geology   A large division of geology concerned with Earth materials, changes of the surface and interior of Earth, and the forces that cause those changes.

pillow structure   Rocks, generally basalt, formed in pillow-shaped masses fitting closely together; caused by underwater lava flows.

placer mine   Surface mines in which valuable mineral grains are extracted from stream bar or beach deposits.

plagioclase feldspar   A feldspar containing sodium, calcium, or both, in addition to aluminum, silicon, and oxygen.

planet   A body in orbit around a star.

planetesimal   Small, planet-like body.

plastic   Capable of being molded and bent under stress.

plastic flow   Movement within a glacier in which the ice is not fractured.

plate   A large, mobile slab of rock making up part of Earth's surface.

plateau   Broad, flat-topped area elevated above the surrounding land and bounded, at least in part, by cliffs.

plateau basalts   Layers of basalt flows that have built up to great thicknesses.

plate tectonics   A theory that Earth's surface is divided into a few large, thick plates that are slowly moving and changing in size. Intense geologic activity occurs at the plate boundaries.

playa   A very flat surface underlain by hard, mud-cracked clay.

playa lake   A shallow temporary lake (following a rainstorm) on a flat valley floor in a dry region.

Pleistocene Epoch   An epoch of the Quaternary Period characterized by several glacial ages.

plunging fold   A fold in which the hinge line (or axis) is not horizontal.

pluton   An igneous body that crystallized deep underground.

plutonic rock   Igneous rock formed at great depth.

pluvial lake   A lake formed during an earlier time of abundant rainfall.

point bar   A stream *bar* (*see* definition) deposited on the inside of a curve in the stream, where the water velocity is low.

polarity   *See* magnetic reversal.

polar wandering   An apparent movement of the Earth s poles.

polymorphs   Substances having the same chemical composition but different crystal structures (e.g., diamond and graphite).

pore space   The total amount of space taken up by openings between sediment grains.

porosity   The percentage of a rock's volume that is taken up by openings.

porphyritic rock   An igneous rock in which large crystals are enclosed in a matrix (or ground mass) of much finer-grained minerals or obsidian.

positive gravity anomaly   Greater than normal gravitational attraction.

positive magnetic anomaly   Greater than average strength of the Earth s magnetic field.

potassium feldspar   A feldspar with the formula $KAlSi_3O_8$.

pothole   Depression eroded into the hard rock of a streambed by the abrasive action of the stream's sediment load.

Precambrian   The vast amount of time that preceded the Paleozoic Era.

Precambrian shield   A complex of old Precambrian metamorphic and plutonic rocks exposed over a large area.

pressure release   A significant type of mechanical weathering that causes rocks to crack when overburden is removed.

prograde metamorphism   Metamorphism in which progressively greater pressure and temperature act on a rock type with increasing depth in Earth's crust.

Proterozoic   Eon of Precambrian time.

proton   A subatomic particle that contributes mass and a single positive electrical charge to an atom.

pumice   A frothy volcanic glass.

P wave   A compressional wave (seismic wave) in which rock vibrates parallel to the direction of wave propagation.

P-wave shadow zone   The region on Earth's surface, 103° to 142° away from an earthquake epicenter, in which P waves from the earthquake are absent.

pyroclast   Fragment of rock formed by volcanic explosion.

pyroclastic debris   Rock fragments produced by volcanic explosion.

pyroclastic flow   Turbulent mixture of pyroclastics and gases flowing down the flank of a volcano.

pyroxene group   Mineral group, all members of which are single-chain silicates.

Q

quartz   Mineral with the formula SiO_2.

quartzite   A rock composed of sand-sized grains of quartz that have been welded together during metamorphism.

quartz sandstone   A sandstone in which more than 90% of the grains are quartz.

Quaternary Period   The youngest geologic period; includes the present time.

R

radial pattern   A drainage pattern in which streams diverge outward like spokes of a wheel.

radioactive decay   The spontaneous nuclear disintegration of certain isotopes.

radioactivity   The spontaneous nuclear disintegration of atoms of certain isotopes.

radon   A radioactive gas produced by the radioactive decay of uranium.

rain shadow   A region on the downwind side of mountains that has little or no rain because of the loss of moisture on the upwind side of the mountains.

rampart crater *(Mars)*   Meteorite crater that is surrounded by material that appears to have flowed from the point of impact.

rayed crater *(Moon)*   Crater with bright streaks radiating from it on the Moon's surface.

Rayleigh waves   A type of surface seismic wave that behaves like a rolling ocean wave and causes the ground to move in an elliptical path.

receding glacier   A glacier with a negative budget, which causes the glacier to grow smaller as its edges melt back.

Recent (Holocene) Epoch   The present epoch of the Quaternary Period.

recessional moraine   An end moraine built during the retreat of a glacier.

recharge   The addition of new water to an aquifer or to the zone of saturation.

reclamation   Restoration of the land to usable condition after mining has ceased.

recrystallization   The development of new crystals in a rock, often of the same composition as the original grains.

rectangular pattern   A drainage pattern in which tributaries of a river change direction and join one another at right angles.

recumbent fold   A fold overturned to such an extent that the limbs are essentially horizontal.

reef   A resistant ridge of calcium carbonate formed on the sea floor by corals and coralline algae.

regional (dynamothermal) metamorphism   Metamorphism that takes place at considerable depth underground.

regolith   Loose, unconsolidated rock material resting on bedrock.

relative time   The sequence in which events took place (not measured in time units).

relief   The vertical distance between points on Earth's surface.

reserves   The discovered deposits of a geologic material that are economically and legally feasible to recover under present circumstances.

reservoir rock   A rock that is sufficiently porous and permeable to store and transmit petroleum.

residual clay   Fine-grained particles left behind as insoluble residue when a limestone containing clay dissolves.

residual soil   Soil that develops directly from weathering of the rock below.

resources   The total amount of a geologic material in all its deposits, discovered and undiscovered. *See* reserves.

reverse fault   A fault in which the hanging-wall block moved up relative to the footwall block.

rhyolite   A fine-grained, felsic, igneous rock made up mostly of feldspar and quartz.

Richter scale   A numerical scale of earthquake magnitudes.

ridge push   The concept that oceanic plates diverge as a result of sliding down the sloping lithosphere-asthenosphere boundary.

rift valley   A tensional valley bounded by normal faults. Rift valleys are found at diverging plate boundaries on continents and along the crest of the mid-oceanic ridge.

right-lateral fault   A strike-slip fault in which the block seen across the fault appears displaced to the right.

rigid zone   Upper part of a glacier in which there is no plastic flow.

rille *(Moon)*   Elongate trenched or cracklike valley on the lunar surface.

rip current   Narrow currents that flow straight out to sea in the surf zone, returning water seaward that has been pushed ashore by breaking waves.

ripple mark   Any of the small ridges formed on sediment surfaces exposed to moving wind or water. The ridges form perpendicularly to the motion.

rock   Naturally formed, consolidated material composed of grains of one or more minerals. (There are a few exceptions to this definition.)

rock avalanche   A very rapidly moving, turbulent mass of broken-up bedrock.

rock-basin lake   A lake occupying a depression caused by glacial erosion of bedrock.

rock cycle   A theoretical concept relating tectonism, erosion, and various rock-forming processes to the common rock types.

rockfall   Rock falling freely or bouncing down a cliff.

rock flour   A powder of fine fragments of rock produced by glacial abrasion.

rock gypsum   An evaporite composed of gypsum.

rock salt   An evaporite composed of halite.

rockslide   Rapid sliding of a mass of bedrock along an inclined surface of weakness.

rotational slide   In mass wasting, movement along a curved surface in which the upper part moves vertically downward while the lower part moves outward. Also called a slump.

rounded knobs (glacial)   Bedrock that is more resistant to glacial erosion stands out as rounded knobs, usually elongated parallel to the direction of glacier flow. These are also known as *roche moutonnées* (French for "rock sheep").

rounding   The grinding away of sharp edges and corners of rock fragments during transportation.

rubble   Angular sedimentary particles coarser than 2 millimeters in diameter.

S

saltation   A mode of transport that carries sediment downcurrent in a series of short leaps or bounces.

sand   Sediment composed of particles with a diameter between 1/16 and 2 millimeters.

sand dune   A mound of loose sand grains heaped up by the wind.

sandstone   A medium-grained sedimentary rock (grains between 1/16 and 2 millimeters) formed by the cementation of sand grains.

saturated zone   A subsurface zone in which all rock openings are filled with water.

scale   The relationship between distance on a map and the distance on the terrain being represented by that map.

schist   A metamorphic rock characterized by coarse-grained minerals oriented approximately parallel.

schistose   The texture of a rock in which visible platy or needle-shaped minerals have grown essentially parallel to each other under the influence of directed pressure.

scientific method   A means of gaining knowledge through objective procedures.

scoria   A basalt that is highly vesicular.

sea cave   A cavity eroded by wave action at the base of a sea cliff.

sea cliff   Steep slope that retreats inland by mass wasting as wave erosion undercuts it.

seafloor metamorphism   Metamorphism of rock along a mid-oceanic ridge caused by circulating hot water.

seafloor spreading   The concept that the ocean floor is moving away from the mid-oceanic ridge and across the deep-ocean basin, to disappear beneath continents and island arcs.

seamount   Conical mountain rising 1,000 meters or more above the sea floor.

seawall   A wall constructed along the base of retreating cliffs to prevent wave erosion.

sediment   Loose, solid particles that can originate by (1) weathering and erosion of preexisting rocks, (2) chemical precipitation from solution, usually in water, and (3) secretion by organisms.

sedimentary breccia   A coarse-grained sedimentary rock (grains coarser than 2 millimeters) formed by the cementation of angular rubble.

sedimentary facies   Significantly different rock types occupying laterally distinct parts of the same layered rock unit.

sedimentary rock   Rock that has formed from (1) lithification of any type of sediment, (2) precipitation from solution, or (3) consolidation of the remains of plants or animals.

sedimentary structure   A feature found within sedimentary rocks, usually formed during or shortly after deposition of the sediment and before lithification.

seismic gap   A segment of a fault that has not experienced earthquakes for a long time; such gaps may be the site of large future quakes.

seismic profiler   An instrument that measures and records the subbottom structure of the sea floor.

seismic reflection   The return of part of the energy of seismic waves to Earth's surface after the waves bounce off a rock boundary.

seismic refraction   The bending of seismic waves as they pass from one material to another.

seismic sea wave   *See* tsunami.

seismic wave   A wave of energy produced by an earthquake.

seismogram   Paper record of Earth vibration.

seismograph   A seismometer with a recording device that produces a permanent record of Earth motion.

seismometer   An instrument designed to detect seismic waves or Earth motion.

serpentine   A magnesium silicate mineral. Most asbestos is a variety of serpentine.

shale   A fine-grained sedimentary rock (grains finer than 1/16 millimeter in diameter) formed by the cementation of silt and clay (mud). Shale has thin layers (laminations) and an ability to split (fissility) into small chips.

shear force   In mass wasting, the component of gravitational force that is parallel to an inclined surface.

shearing   Movement in which parts of a body slide relative to one another and parallel to the forces being exerted.

shear strength   In mass wasting, the resistance to movement or deformation of material.

shear stress   Stress due to forces that tend to cause movement or strain parallel to the direction of the forces.

sheet erosion   The removal of a thin layer of surface material, usually topsoil, by a flowing sheet of water.

sheet joints   Cracks that develop parallel to the outer surface of a large mass of expanding rock, as pressure is released during unloading.

sheet-silicate structure   Crystal structure in which each silica tetrahedron shares three oxygen ions.

sheetwash   Water flowing down a slope in a layer.

shield volcano   Broad, gently sloping cone constructed of solidified lava flows.

silica   A term used for oxygen plus silicon.

silicate   A substance that contains silica as part of its chemical formula.

silica tetrahedron   See silicon-oxygen tetrahedron.

silicic rock or magma   Silica-rich igneous rock or magma with a relatively high content of potassium and sodium.

silicon-oxygen tetrahedron   Four-sided, pyramidal object that visually represents the four oxygen atoms surrounding a silicon atom; the basic building block of silicate minerals. Also called a silica tetrahedron or a silicon tetrahedron.

sill   A tabular intrusive structure concordant with the country rock.

silt   Sediment composed of particles with a diameter of 1/256 to 1/16 millimeter.

siltstone   A sedimentary rock consisting mostly of silt grains.

sinkhole   A closed depression found on land surfaces underlain by limestone.

sinter   A deposit of silica that forms around some hot springs and geysers.

slab pull   The concept that subducting plates are pulled along by their dense leading edges.

slate   A fine-grained rock that splits easily along flat, parallel planes.

slaty   Describing a rock that splits easily along nearly flat and parallel planes.

slaty cleavage   The ability of a rock to break along closely spaced parallel planes.

slide   In mass wasting, movement of a relatively coherent descending mass along one or more well-defined surfaces.

slip face   The steep, downwind slope of a dune; formed from loose, cascading sand that generally keeps the slope at the angle of repose (about 34°).

slump   In mass wasting, movement along a curved surface in which the upper part moves vertically downward while the lower part moves outward. Also called a rotational slide.

snow line   See equilibrium line.

soil   A layer of weathered, unconsolidated material on top of bedrock; often also defined as containing organic matter and being capable of supporting plant growth. In mass wasting, soil means unconsolidated material, regardless of particle size or composition (also called engineering soil).

soil horizon   Any of the layers of soil that are distinguishable by characteristic physical or chemical properties.

solar nebula   The rotating disk of gas and dust from which the Sun and planets formed.

solar system   The Sun, planets, their moons, and other bodies that orbit the Sun.

solar wind   The outflow of low-density, hot gas from the Sun's upper atmosphere. It is partially this wind that creates the tail of a comet by blowing dust and gas away from the comet's immediate surroundings.

solid solution   The substitution of atoms of one element for those of another element in a particular mineral.

solifluction   Flow of water-saturated soil over impermeable material.

solution   Usually slow but effective process of weathering and erosion in which rocks are dissolved by water.

sorting   Process of selection and separation of sediment grains according to their grain size (or grain shape or specific gravity).

source area   The locality that eroded to provide sediment to form a sedimentary rock.

source rock   A rock containing organic matter that is converted to petroleum by burial and other postdepositional changes.

spatter cone   A small, steep-sided cone built from lava spattering out of a vent.

specific gravity   The ratio of the mass of a substance to the mass of an equal volume of water, determined at a specified temperature.

speleothem   Dripstone deposit of calcite that precipitates from dripping water in caves.

spheroidally weathered boulder   Boulder that has been rounded by weathering from an initial blocky shape.

spit   A fingerlike ridge of sediment attached to land but extending out into open water.

spreading axis (or **spreading center**)   The crest of the mid-oceanic ridge, where sea floor is moving away in opposite directions on either side.

spring   A place where water flows naturally out of rock onto the land surface.

stable   Describing a mineral that will not react with or convert to a new mineral or substance, given enough time.

stack   A small rock island that is an erosional remnant of a headland left behind as a wave-eroded coast retreats inland.

stalactite   Iciclelike pendant of dripstone formed on cave ceilings.

stalagmite   Cone-shaped mass of dripstone formed on cave floors, generally directly below a stalactite.

standard geologic time scale   A worldwide relative scale of geologic time divisions.

star   A massive, gaseous body held together by gravity and generally emitting light. Normal stars generate energy by nuclear reactions in their interiors.

static pressure   See confining pressure.

stock   A small discordant pluton with an outcropping area of less than 100 square kilometers.

stony-iron meteorite   A meteorite composed of silicate minerals and iron-nickel alloy in approximately equal amounts.

stony meteorite   A meteorite made up mostly of plagioclase and iron-magnesium silicates.

stoping   Upward movement of a body of magma by fracturing of overlying country rock. Magma engulfs the blocks of fractured country rock as it moves upward.

storm surge   High sea level caused by the low pressure and high winds of hurricanes.

strain   Change in size (volume) or shape of a body (or rock unit) in response to stress.

stratigraphy   The field of geology concerning layered rocks and their interrelationships.

stratovolcano   See composite volcano.

streak   Color of a pulverized substance; a useful property for mineral identification.

stream   A moving body of water, confined in a channel and running downhill under the influence of gravity.

stream capture   See stream piracy.

stream channel   A long, narrow depression, shaped and more or less filled by a stream.

stream discharge   Volume of water that flows past a given point in a unit of time.

stream-dominated delta   A delta with fingerlike distributaries formed by the dominance of stream sedimentation; also called a birdfoot delta.

stream gradient   Downhill slope of a stream's bed or the water surface, if the stream is very large.

stream headwaters   The upper part of a stream near the source.

stream mouth   The place where the stream enters the sea, a large lake, or a larger stream.

stream piracy   The natural diversion of the headwaters of one stream into the channel of another.

stream terrace   Steplike landform found above a stream and its flood plain.

stream velocity   The speed at which water in a stream travels.

stress   A force acting on a body, or rock unit, that tends to change the size or shape of that body, or rock unit. Force per unit area within a body.

striations   (1) On minerals, extremely straight, parallel lines; (2) Glacial straight scratches in rock caused by abrasion by a moving glacier.

strike   The compass direction of a line formed by the intersection of an inclined plane (such as a bedding plane) with a horizontal plane.

strike-slip fault   A fault in which movement is parallel to the strike of the fault surface.

strip mine   A mine in which the valuable material is exposed at the surface by removing a strip of overburden.

structural basin   A structure in which the beds dip toward a central point.

structural dome   A structure in which beds dip away from a central point.

structural geology   The branch of geology concerned with the internal structure of bedrock and the shapes, arrangement, and interrelationships of rock units.

structural [or oil] trap   *See* oil trap.

subduction   The sliding of the sea floor beneath a continent or island arc.

subduction complex   *See* accretionary wedge.

subduction zone   Elongate region in which subduction takes place.

submarine canyon   V-shaped valleys that run across the continental shelf and down the continental slope.

submergent coast   A coast in which formerly dry land has been recently drowned, either by land subsidence or a rise in sea level.

subsidence   Sinking or downwarping of a part of the Earth s surface.

superposed stream   A river let down onto a buried geologic structure by erosion of overlying layers.

superposition   A principle or law stating that within a sequence of undisturbed sedimentary rocks, the oldest layers are on the bottom, the youngest on the top.

surf   Breaking waves.

surface wave   A seismic wave that travels on Earth's surface.

suspect terrane   A terrane that may not have formed at its present site.

suspended load   Sediment in a stream that is light enough in weight to remain lifted indefinitely above the bottom by water turbulence.

S wave   A seismic wave propagated by a shearing motion, which causes rock to vibrate perpendicular to the direction of wave propagation.

S-wave shadow zone   The region on Earth's surface (at any distance more than 103° from an earthquake epicenter) in which S waves from the earthquake are absent.

swelling clay   *See* expansive clay.

syncline   A fold in which the layered rock usually dips toward an axis.

T

talus   An accumulation of broken rock at the base of a cliff.

tarn   *See* rock-basin lake.

tectite   Small, rounded bits of glass formed from rock melting and being thrown into the air due to a meteorite impact.

tectonic forces   Forces generated from within Earth that result in uplift, movement, or deformation of part of Earth's crust.

tectonostratigraphic terrane   *See* terrane.

tensional stress   A stress due to a force pulling away on a body.

tephra   *See* pyroclastic debris.

terminal moraine   An end moraine marking the farthest advance of a glacier.

terminus   The lower edge of a glacier.

terrane (tectonostratigraphic terrane)   A region in which the geology is markedly different from that in adjoining regions.

terrigenous sediment   Land-derived sediment that has found its way to the sea floor.

texture   A rock's appearance with respect to the size, shape, and arrangement of its grains or other constituents.

theory   An explanation for observed phenomena that has a high possibility of being true.

theory of glacial ages   At times in the past, colder climates prevailed during which significantly more of the land surface of Earth was glaciated than at present.

thermal metamorphism   *See* contact (thermal) metamorphism.

thrust fault   A reverse fault in which the dip of the fault plane is at a low angle to horizontal.

tidal delta   A submerged body of sediment formed by tidal currents passing through gaps in barrier islands.

"tidal wave"   An incorrect name for a tsunami.

tide-dominated delta   A delta formed by the reworking of sand by strong tides.

till   Unsorted and unlayered rock debris carried by a glacier.

tillite   Lithified till.

time-transgressive rock unit   An apparently continuous rock layer in which different portions formed at different times.

tombolo   A bar of marine sediment connecting a former island or stack to the mainland.

topographic map   A map on which elevations are shown by means of contour lines.

topset bed   In a delta, a nearly horizontal sediment bed of varying grain size formed by distributaries shifting across the delta surface.

trace fossil   Trail, track or burrow resulting from animal movement preserved in sedimentary rock.

traction   Movement by rolling, sliding, or dragging of sediment fragments along a stream bottom.

transform fault   The portion of a fracture zone between two offset segments of a mid-oceanic ridge crest.

transform plate boundary   Boundary between two plates that are sliding past each other.

translational slide   In mass wasting, movement of a descending mass along a plane approximately parallel to the slope of the surface.

transportation   The movement of eroded particles by agents such as rivers, waves, glaciers, or wind.

transported soil   Soil not formed from the local rock but from parent material brought in from some other region and deposited, usually by running water, wind, or glacial ice.

transverse dune   A relatively straight, elongate dune oriented perpendicular to the wind.

travel-time curve   A plot of seismic-wave arrival times against distance.

travertine   A porous deposit of calcite that often forms around hot springs.

trellis pattern   A drainage pattern consisting of parallel main streams with short tributaries meeting them at right angles.

trench   *See* oceanic trench.

trench suction   The concept that overlying plates move horizontally toward oceanic trenches as subducting plates sink at an angle steeper than their dip.

tributary   Small stream flowing into a large stream, adding water to the large stream.

trough (of wave)   The low point of a wave.

truncated spur   Triangular facet where the lower end of a ridge has been eroded by glacial ice.

tsunami   Huge ocean wave produced by displacement of the sea floor; also called seismic sea wave.

tufa   A deposition of calcite that forms around a spring, lake, or percolating ground water.

tuff   A rock formed from fine-grained pyroclastic particles (ash and dust).

turbidity current   A flowing mass of sediment-laden water that is heavier than clear water and therefore flows downslope along the bottom of the sea or a lake.

turbulent flow   Eddying, swirling flow in which water drops travel along erratically curved paths that cross the paths of neighboring drops.

U

ultramafic rock   Rock composed entirely or almost entirely of ferromagnesian minerals.

unconfined aquifer   A partially filled aquifer exposed to the land surface and marked by a rising and falling water table.

unconformity   A surface that represents a break in the geologic record, with the rock unit immediately above it being considerably younger than the rock beneath.

unconsolidated   In referring to sediment grains, loose, separate, or unattached to one another.

underplating   The pooling of magmas at the base of the continental crust.

uniformitarianism   Principle that geologic processes operating at present are the same processes that operated in the past. The principle is stated more succinctly as, "The present is the key to the past." *See* actualism.

Universe   The largest astronomical structure we know of. The Universe contains all matter and radiation and encompasses all space.

unloading   The removal of a great weight of rock.

unpaired terraces   *Stream terraces (see* definition) that do not have the same elevation on opposite sides of a river.

unsaturated zone   A subsurface zone in which rock openings are generally unsaturated and filled partly with air and partly with water; above the saturated zone.

U-shaped valley   Characteristic cross-profile of a valley carved by glacial erosion.

V

valley glacier   A glacier confined to a valley. The ice flows from a higher to a lower elevation.

varve   Two thin layers of sediment, one dark and the other light in color, representing one year's deposition in a lake.

vein   *See* hydrothermal vein.

vent   The opening in Earth's surface through which a volcanic eruption takes place.

ventifact   Boulder, cobble, or pebble with flat surfaces caused by the abrasion of wind-blown sand.

vertical exaggeration   An artificial steepening of slope angles on a topographic profile caused by using a vertical scale that differs from the horizontal scale.

vesicle   A cavity in volcanic rock caused by gas in a lava.

viscosity   Resistance to flow.

vitreous luster   *See* glassy luster.

volcanic breccia   Rock formed from large pieces of volcanic rock (cinders, blocks, bombs).

volcanic dome   A steep-sided, dome- or spine-shaped mass of volcanic rock formed from viscous lava that solidifies in or immediately above a volcanic vent.

volcanic neck   An intrusive structure that apparently represents magma that solidified within the throat of a volcano.

volcanism   Volcanic activity, including the eruption of lava and rock fragments and gas explosions.

volcano   A hill or mountain constructed by the extrusion of lava or rock fragments from a vent.

W

wastage   *See* ablation.

water table   The upper surface of the zone of saturation.

wave crest   *See* crest.

wave-cut platform   A horizontal bench of rock formed beneath the surf zone as a coast retreats because of wave erosion.

wave-dominated delta   A delta formed by the reworking of sand by wave action.

wave height   The vertical distance between the crest (the high point of a wave) and the trough (the low point).

wavelength   The horizontal distance between two wave crests (or two troughs).

wave refraction   Change in direction of waves due to slowing as they enter shallow water.

wave trough   *See* trough.

weathering   The group of processes that change rock at or near Earth's surface.

welded tuff   A rock composed of pyroclasts welded together.

well   A hole, generally cylindrical and usually walled or lined with pipe, that is dug or drilled into the ground to penetrate an aquifer below the zone of saturation.

Wilson cycle   The cycle of splitting of a continent, opening of an ocean basin, followed by closing of the basin and collision of the continents.

wind ripple   Small, low ridge of sand produced by the saltation of wind-blown sand.

wrinkle ridge *(Moon)*   Wrinkle on lunar maria surface.

X

xenolith   Fragment of rock distinct from the igneous rock in which it is enclosed.

Z

zone of ablation   That portion of a glacier in which ice is lost.

zone of accumulation   (1) That portion of a glacier with a perennial snow cover; (2) *See* B horizon (a soil layer).

zone of leaching   *See* A horizon (a soil layer).

zone of saturation   *See* saturated zone.

zoning   Orderly variation in the chemical composition within a single crystal.

INDEX

Note: Figures and tables are indicated by *f* and *t* respectively, and are cited only when they appear outside related text discussions. An A or G preceding a page number refers to a page(s) in the Appendices or Glossary.

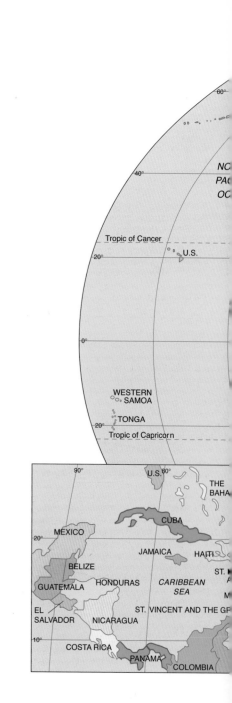

NO___
PA___
OC___

Tropic of Cancer

U.S.

WESTERN
SAMOA

TONGA

Tropic of Capricorn

90°   U.S. 80°   THE
                        BAHA

CUBA

MEXICO

JAMAICA   HAITI

BELIZE

GUATEMALA   HONDURAS   CARIBBEAN
                          SEA

EL
SALVADOR   NICARAGUA

ST. VINCENT AND THE GR

COSTA RICA

PANAMA

COLOMBIA

≋USGS
science for a changing world

U.S. DEPARTMENT OF THE INTERIOR
U.S. GEOLOGICAL SURVEY

Hawaiian Islands

Geologic Time Scale
Escala de tiempo geológico
Échelle de temps géologiques

Millions of Years Ago (non-linear)
Millones de años (escala no lineal)
Millions d'années dans le geand (echelle non linéaire)

Geological Survey of Canada
Commission géologique du Canada

Consejo de Recursos Minerales

NOF
SE

NETHE

FRANC

SWI

MONA

MEDITE

0°

55°

40°

GREENLAND
(DENMARK)

ARCTIC OCEAN

Arctic Circle

80°

160°   140°   120°   100°   80°   60°   40°   20°   0°   20°   40°   60°   80°

NORWAY   SWEDEN   FINLAND

ICELAND

U.S.

CANADA

UNITED KINGDOM

IRELAND

UNITED STATES

FRANCE
ANDORRA

PORTUGAL   SPAIN

NORTH
ATLANTIC
OCEAN

TURKEY   UZE
TURKME

TUNISIA   CYPRUS   SYRIA
LEBANON   IRAQ   IRAN
ISRAEL   KUWA
MOROCCO   JORDAN   QAT
ALGERIA   LIBYA   EGYPT   BAHRAIN
                    SAUDI
                    ARABIA
MAURITANIA   MALI   NIGER   YEMEN   OM
                ERITREA
CAPE   CHAD   SUDAN   DJI
VERDE
        NIGERIA   SOMALIA
                UGANDA   ETHIOPIA

MEXICO

GUYANA
SURINAME
FRENCH
GUIANA
(FR)

CAMEROON
CENTRAL AFRICAN REPUBLIC
SÃO TOMÉ AND PRÍNCIPE
EQUATORIAL GUINEA
GABON
CONGO REPUBLIC

COLOMBIA

VENEZUELA

Equator

ECUADOR

P
E
R
U

B R A Z I L

BOLIVIA

PARAGUAY

RWANDA
DEM. REP.
OF THE   KENYA
CONGO   BURUNDI   SE
        TANZANIA   MALAWI
                COMOROS
ANGOLA   MOZA
        ZAMBIA
                MADAGASC

NAMIBIA   BOTSWANA

ZIMBABWE
SWAZILAND
SOUTH   LESOTHO
AFRICA

CHILE

A
R
G
E
N
T
I
N
A

URUGUAY

SOUTH
ATLANTIC
OCEAN

70°

0   300 Miles

0   300 Kilometers

MINICAN
UBLIC   PUERTO RICO

AND NEVIS
GUA AND BARBUDA
        DOMINICA
NIQUE   ST. LUCIA
DINES
    BARBADOS
    GRENADA

NIDAD AND TOBAGO

VENEZUELA

SOUTH
PACIFIC
OCEAN

Antarctic Circle

Scale: 1 to 125,000,000

0   1000   2000 Miles

0   1000   2000   3000 Kilometers

MAURITANIA

SENEGAL

MALI

GAMBIA

GUINEA-
BISSAU

GUINEA

BURKINA FAS

SIERRA
LEONE

ATLANTIC
OCEAN

LIBERIA

IVORY
COAST

GHANA

10°

10°

0   150   300 Miles

0   150   300 Kilometers

The North America Tapestry of Time and Terrain

Cobertura de Tiempo y Terrenos de Norte América

L'Amérique du Nord : un collage de terrains d'âges différents

Tapestry Compiled by

Kate E. Barton[1], David G. Howell[1], and José F. Vigil[1]

Geology Compiled by

John C. Reed, Jr.[1] and John O. Wheeler[2]

[1]U.S. Geological Survey, [2]Geological Survey of Canada

2003

Sedimentary Rocks
Rocas sedimentarias
Roches sédimentaires

Volcanic Rocks
Rocas volcánicas
Roches volcaniques

Plutonic Rocks
Rocas plutónicas
Roches plutoniques

Metamorphic Rocks
Rocas metamórficas
Roches métamorphiques

Scale · Escala · Échelle 1:8,000,000

Projection Lambert Azimuthal Equal Area
Proyección Lambert Acimutal de Igual Área
Projection azimutale équivalente de Lambert

For sale by U.S. Geological Survey, Map Distribution Center, Box 25286,
Federal Center, Denver, Colorado, 80225, or 1-800-ASK-USGS
Available on the Web at http://geopubs.wr.usgs.gov/i-map/i2781
Manuscript approved for publication on December 4, 2002